Biotechnological Applications of Quorum Sensing Inhibitors

Vipin Chandra Kalia

Editor

Biotechnological Applications of Quorum Sensing Inhibitors

 Springer

Editor
Vipin Chandra Kalia
Microbial Biotechnology and Genomics
CSIR – Institute of Genomics and Integrative Biology (IGIB)
Delhi, India

Molecular Biology Lab 502; Department of Chemical Engineering
Konkuk University
Seoul, Republic of Korea

ISBN 978-981-10-9025-7   ISBN 978-981-10-9026-4   (eBook)
https://doi.org/10.1007/978-981-10-9026-4

Library of Congress Control Number: 2018941847

Printed on acid-free paper

This Springer imprint is published by the registered company Springer Nature Singapore Pte Ltd.
The registered company address is: 152 Beach Road, #21-01/04 Gateway East, Singapore 189721, Singapore

Dedicated to my Lovely wife Amita

Preface

Plants, animals, and humans live in association with microbes. Some of these relationships are viewed as beneficial whereas others seem harmful. The ubiquitous existence of microbes provides unique clues about the diversity of their metabolic activities and abilities to evolve in response to environmental stress. They have also been bestowed with a great reservoir of arsenal to infect other organisms including eukaryotes. This infection leads to drastic changes in physiological conditions of the host metabolic machinery resulting in inefficient performance. This condition is termed as disease. A variety of strategies have been tried to counter microbial attack: (1) the vaccines, (2) the antibiotics, and (3) the antipathogens. Microbial pathogens are a constant cause of worry among the health departments and agriculturists. The big question is: How to handle this ever-evolving battle between eukaryotes and pathogenic microbes? Vaccines are biological responses, which enable living beings to acquire immunity against specific diseases. Vaccines have proved effective in fighting infectious diseases; however, there are quite a few limitations – inadequate immune response, genetic inability to generate antibodies, and potential adverse effects. The discovery of antibiotics brought great relief from pathogenic organisms. However, hope started fading with the emergence of antibiotic resistant strains. This provoked medical practitioners to prescribe higher doses of antibiotics. Bacteria reacted to this unwarranted stress and evolved drug resistance. This has discouraged researchers and pharmaceutical companies also to invest in this area. One of the major ways by which microbes gain resistance to antibiotics is through a unique characteristic. Here, microbes simply multiply when their population density is low. At high cell density, they become virulent by expressing genes responsible for pathogenicity: a phenomenon termed as quorum sensing (QS). Microbes exploit QS to produce antibiotics and toxins, which have the potential to kill other microbes in their vicinity. On the contrary, these rival organisms produce bioactive molecules, which act as QS inhibitors (QSIs). Research efforts have revealed that microbes such as *Pseudomonas, Bacillus, Oceanobacillus,* and *Streptomyces* produce QSIs. In addition, plants including legumes and medicinal plants have also been bestowed with abilities to produce secondary metabolites, which act as QSIs. These QSIs can disrupt QS mediated biofilms thereby exposing bacteria, which thus become susceptible to even low doses of antibiotics. These QSIs can thus treat bacterial infections on human beings and plants. Since these molecules do not cause any unnecessary stress on microbial growth, the risk of their developing resistance to QSIs is quite low. The unique properties of QSIs have been extended to act as antipathogens for treating diseases and reduce economic losses to humans and plants. This book is intended to present the status of these diverse possibilities, our views, and opinions and to finally provide mankind with novel, innovative, and long-lasting strategies, in the book entitled *Biotechnological Applications of Quorum Sensing Inhibitors*. This book has been based on the contributions of scientists who are passionate about science and gaining insights and share their knowledge with young curious minds and provide strategies to bestow economic benefits to the human society. This book is a vivid reflection of the sincerity with which scientific minds are dedicated to the welfare of the community. Scientist are too busy to participate in most societal activities but here they agreed to spare time in order to serve the society and readily agreed to put their creation for the benefit of newcomers. This contribution will take the researchers around the world a step further into a beautiful and healthy future. I am extremely thankful to the contributing authors for paying attention to my request and showing faith in my abilities. I will remain indebted to all of them forever. These words cannot justify the worthiness of their efforts. I have been inspired by my associates as this work stems from their faith in me and the constant support of – Late Mrs. Kanta Kalia and Mr. R.B. Kalia (parents), my sons Daksh and Bhrigu, Sunita Bhardwaj, Ravi Bhardwaj, Sunita Dahiya, Sanjay Dahiya, Kumud bua, Devender ji, Kusum bua, my friends – Rup, Hemant, Yogendra, Naveen, Prince, Virender K. Sikka, Atya, Jyoti, Neeru, Ritusree, Malabika, Chinoo, Renu Paul Smart, and Vijay Lakshmi Joshi, my young friends – Anil, Balvinder, Bela, Bhartendu, Devanshi, Meera, Mukesh, Neelam, Rajan, Rajiv, Sadhana, Sanjeev, Satyender, and Shanty of DeshBandhu76 clan. I must also acknowledge the support of my student friends – Sanjay Patel, Sanjay Yadav, Sadhana, Mamtesh, Subhasree, Shikha, and Jyotsana.

Delhi, India   Vipin Chandra Kalia

Contents

About the Editor

Vipin Chandra Kalia is a Professor at Molecular Biotechnology Lab, Department of Chemical Engineering, Konkuk University, Seoul, Korea. He has been an Emeritus Scientist and ex-Chief Scientist at Microbial Biotechnology and Genomics, CSIR Institute of Genomics and Integrative Biology, Delhi. He received his M.Sc. and Ph.D. degrees in Genetics from the Indian Agricultural Research Institute, New Delhi. He has been elected as a Fellow of the Association of Microbiologists of India (FAMSc), National Academy of Sciences (FNASc), and National Academy of Agricultural Sciences (FNAAS). His main areas of research are microbial biodiversity, genomics and evolution, bioenergy, biopolymers, antimicrobials, quorum sensing, and quorum quenching. He has published over 100 papers in scientific journals such as *Nature Biotechnology*, *Biotechnology Advances*, and *Trends in Biotechnology*.

He has edited 11 books, is currently Editor in Chief of the *Indian Journal of Microbiology*, and serves as an editor for several other journals. He is a life member of, e.g., the Society of Biological Chemists of India, the Society for Plant Biochemistry and Biotechnology, India, and the Biotech Research Society of India (BRSI).

He has been conferred numerous awards, including the Prof. S.R. Vyas Memorial Award, Association of Microbiologists of India (2016), ASM-IUSSF Indo-US Professorship Program, American Society of Microbiologists, USA (2014), and the Dr. J.V Bhat Award, Association of Microbiologists of India (2012, 2015, 2016, 2017). He can be contacted at: vckalia@igib.in or vc_kalia@yahoo.co.in.

Part I
Human Health

Chapter 1
Inhibition of Quorum-Sensing: A New Paradigm in Controlling Bacterial Virulence and Biofilm Formation

Aleksandra Ivanova, Kristina Ivanova, and Tzanko Tzanov

Abstract Bacterial pathogens coordinate the expression of multiple virulence factors and formation of biofilms in cell density dependent manner, through a phenomenon named quorum sensing (QS). Protected in the biofilm community, bacterial cells resist the antibiotic treatment and host immune responses, ultimately resulting in difficult to treat infections. The high incidence of biofilm-related infections is a global concern related with increased morbidity and mortality in healthcare facilities, prolonged time of hospitalization and additional financial cost. This has led to the urgent need for innovative strategies to control bacterial diseases and drug resistance. In this review, we outline the disruption of QS pathways as a novel strategy for attenuation of bacterial virulence and prevention of resistant biofilms formation on medical devices and host tissues. Unlike the traditional antibiotics, inhibiting the QS signaling in bacteria will not kill the pathogen or affect its growth, but will block the targeted genes expression, making the cells less virulent and more vulnerable to host immune response and lower dosage of antimicrobials. We summarize the recent successes and failures in the development of novel anti-QS drugs as well as their application in controlling bacterial infections in healthcare facilities. The inhibitory targeting of the production of QS signals, their transduction and recognition by the other cells in the surrounding are discussed. Special focus is also given to the anti-QS nanomaterials with improved effectiveness and specificity towards the pathogens.

Keywords Quorum sensing inhibition · Bacterial biofilm · Drug resistance · Nanomaterials

Aleksandra Ivanova and Kristina Ivanova have contributed equally with all other contributors.

A. Ivanova · K. Ivanova · T. Tzanov (✉)
Group of Molecular and Industrial Biotechnology, Department of Chemical Engineering,
Universitat Politècnica de Catalunya, Terrassa, Spain
e-mail: Kristina.ivanova@upc.edu; tzanko.tzanov@upc.edu

© Springer Nature Singapore Pte Ltd. 2018
V. C. Kalia (ed.), *Biotechnological Applications of Quorum Sensing Inhibitors*,
https://doi.org/10.1007/978-981-10-9026-4_1

1.1   Introduction

Bacteria have the ability to "talk" in a cell density dependent manner through a process called quorum sensing (QS) (Hawver et al. 2016). In QS, bacteria release signaling molecules – autoinducers (AIs) – into their surrounding that after reaching high concentrations are recognized by the intracellular receptors of Gram-negative or membrane receptors of Gram-positive bacteria. As a result of the signals recognition, bacteria express group related behaviors such as virulence factors production, bioluminescence, biofilm formation and dispersal (Tan et al. 2015). Bacterial biofilms are highly organized clusters of cells, encased in a self-produced extracellular polymeric matrix (EPM) that can grow on any biotic and abiotic surface (Grant and Hung 2013). Biofilm growth on medical devices such as indwelling urinary catheters, contact lenses, endotracheal tubes, intrauterine devices, pacemakers, mechanical heart valves, or living human tissues, frequently cause opportunistic and chronic infections (Taraszkiewicz et al. 2013).

Since their discovery, the antibiotics have been widely used to heal and prevent microbial colonization and infection occurrence. However, the formation of high densely packed EPM makes the cells within the biofilm 100–1000 times more insusceptible to the current antibacterials, than their free-floating counterparts (Taraszkiewicz et al. 2013). The alarming statistics shows that the bacterial communities are at the root of 80% of all infections occurring in the hospitals (Römling and Balsalobre 2012). The Gram-negative *Escherichia coli, Klebsiella pneumoniae, Pseudomonas aeruginosa* and *Proteus mirabilis,* and the Gram-positive *Enterococcus faecalis, Staphylococcus aureus, Staphylococcus epidermidis,* as well as the yeast *Candida albicans,* are among the most common biofilm forming species found in the severe and persistent nosocomial infections (Percival et al. 2015). Despite growing in a single strain biofilms, these bacterial species are also able to communicate and interact, forming even more resistant sessile structures. Therefore, bacterial colonization and biofilm growth on medical devices and living tissues is a serious health concern responsible for the increased mortality and morbidity, as well as longer time of hospitalization and higher medical care costs (Miquel et al. 2016).

Several groups have already reported the significance of QS in the regulation of microbial pathogenesis and the establishment of steady biofilms by bacteria. For examples, QS regulates the genes expression in Gram-negative *P. aeruginosa* encoding the secretion of numerous virulence factors such as pyocyanin, elastase, lectin, exotoxin A, as well as the production of essential tools for the bacterium growth in biofilm form (Tay and Wen Yew 2013).The produced virulence factors are important for the colonization of the host as they help the cells to evade the immune system and establish severe infections which in most cases are resistant to the current antibiotic treatments. The production of fibronectin-binding protein, hemolysins, protein A, lipases and enterotoxins by Gram-positive *S. aureus* is also under the control of QS. The signaling process in this pathogen plays key role for acute virulence and incidence of common skin infections (Tong et al. 2013). In the design of adequate anti-virulent strategy, it is important to take into account that the role of the QS process is not only specific for the *S. aureus* infections, but also for the infecting strain (Khan et al. 2015).

The inhibition of QS has appeared as attractive and potentially broad spectrum active therapy to control bacterial virulence, prevent microbial colonization and infections occurrence (Rampioni et al. 2014). The less virulent, individual cells will be more sensitive to the common antibacterial agents at lower dosage. Herein, we first discuss the difference between the QS systems used by Gram-positive and Gram-negative bacteria and the most promising QS targets for the design of anti-virulent and antibiofilm strategies. Then, the most recent examples of the QS inhibitors and their application for controlling the bacterial pathogenesis and drug resistant biofilm occurrence are highlighted.

1.2   Quorum Sensing Systems in Bacteria

QS is involved in the regulation of a wide array of bacterial phenotypes including pathogenesis, plasmid conjugation, motility and antibiotic resistance that aid bacteria to adapt and survive in harsh environmental conditions. The signals for cells communication are different in Gram-negative and Gram-positive bacteria. Gram-negative bacteria use acyl-homoserine lactones (AHLs) (Lee and Zhang 2014; Bortolotti et al. 2015), while Gram-positive bacteria produce oligopeptides, termed autoinducing peptides (AIPs) (Fig. 1.1) (Monnet et al. 2016). There is also a third QS signal type, AI-2, which is found in both bacteria, and have gained the interest as target to control multispecies infections (Scott and Hasty 2016).

The QS signals of Gram-negative bacteria are produced by LuxI-type AHL synthases, using AHL precursors acyl-acyl carrier protein and S-adenosyl-methionine (SAM). The signals accumulate and saturate in the extracellular environment and when the concentrations are above the threshold level they pass through the cell membrane via diffusion (Montebello et al. 2014; Papenfort and Bassler 2015). Once in the cells the AHLs conjugate to specific QS transcription regulators, frequently from the LuxR family, which in turn promote the target genes expression (Fig. 1.1, left) (Papenfort and Bassler 2015). Unlike the communication system in Gram-negative bacteria, the signals for Gram-positive bacteria are small AIPs that are post-translationally synthesized in the bacterial cells and subsequently exported through specific membrane bound transporters. When their concentration exceeds the threshold level, these peptides bind to a two-component histidine kinase sensor, which autophophorylates and in turns alter the target genes expression and corresponding group related behaviors (Fig. 1.1, right). The QS system in *S. aureus*, for example, is under the control of the accessory gene regulator (*agr*) locus which encodes the AIPs secretion (Yerushalmi et al. 2013; Pollitt et al. 2014). The *agr* system regulates the expression of many toxins and degradative exoenzymes, predominantly controlled by P2 and P3 promoters, producing two divergent transcripts respectively RNAII and RNAIII (Le and Otto 2015).The RNAII transcript is composed of four genes -*agrB, agrD, agrC* and *agrA* – encoding the production of AIP

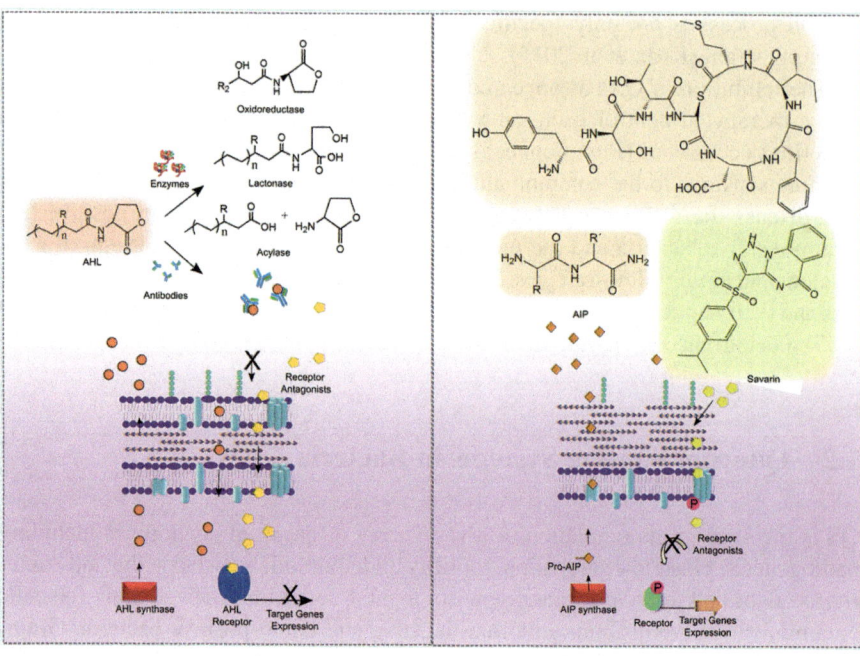

Fig. 1.1 QS signaling and inhibition in Gram-negative (left) and Gram-positive bacteria (right). In Gram-negative bacteria, AHL QS signals (orange circles) are produced by AHL synthases (red rectangle) and secreted in the cells surrounding. When the concentration of AHLs in the surrounding is above the threshold, the signals pass through the cell membrane via diffusion and activate specific AHL receptors (blue motif), promoting the target genes expression. QS inhibition with enzymes and antibodies in the extracellular environment and receptor antagonists (yellow pentagons) is shown. In Gram-positive bacteria, small AIPs (orange pentagons) are post-translationally synthesized in the cells and exported through specific membrane bound transporters (light blue motif). At threshold level, AIPs bind to a two-component histidine kinase sensor, which autophosphorylates and alters the target genes expression. The developed QSIs (e.g. savirin) can block the function of the transcriptional regulator (light green ellipses) and inhibit the expression of QS regulated behaviours

precursors, the signal maturation and its transduction by histidine kinase. The RNAIII is an intracellular effector molecule and is responsible for the upregulation of α-haemolysin and proteases, as well as for the downregulation of surface proteins, such as protein A among others (Tay and Yew 2013; Cheung et al. 2014).

Except the interspecies way of talking, bacteria possess the ability to "understand" the messages sent by other species present in the environment (Xue et al. 2012). In this case, they use another type QS molecules procured from 4, 5-dihydroxy-2, 3-pentanedione (DPD), also termed AI-2 interspecies signal. The synthesis of AI-2 signal, obtained from S-ribosylhomocysteine is catalyzed by the LuxS type synthase involved in the regulation of the AI-2 QS system in multiple bacterial species (Rezzonico et al. 2012). LuxS, for example controls the expression of over 400 genes involved in surface adhesion, motility and toxin production by *E. coli* (Antunes et al. 2010). Therefore, interfering with bacterial communication by blocking the AIs production, breaking down the AIs signal in the cells surrounding, or inhibiting the interaction with the cognate signal receptor could be an attractive anti-virulent/antibiofilm strategy, which is less invasive and unlikely to exert selective pressure for resistance development (Ivanova et al. 2013).

1.3 Quorum Sensing Inhibition for Controlling Bacterial Pathogenesis and Biofilm Formation

In the past years, the misuse or overuse of antimicrobial agents progressively led to the occurrence of drug resistance, thus, imposing the need for new more efficient compounds with different mechanisms of antimicrobial action. The capability of pathogens to colonize the host, produce virulence factors and establish sessile communities as a function of the population density, suggested that interfering in this interrelated process by interruption of bacterial "conversation" will make the pathogens more susceptible to host immune response and antibiotics (Beceiro et al. 2013; Munita et al. 2016). Herein, we will discuss the anti-infective and antibiofilm potential of the current anti-QS agents as well as their biomedical application either as therapeutic formulations or coatings on the surfaces medical devices (Table 1.1).

1.3.1 Inhibitors of Quorum Sensing Signal Synthesis in Pathogens

Inhibition of QS synthesis is an attractive strategy to reduce the QS controlled virulence factors expression and prevent inflammation. LuxI, HdtS and LuxM are identified as signals producers in Gram-negative bacteria (Li and Nair 2012). In particular, LuxI has been found in many bacterial strains and has been deeply investigated as a potential target to control toxins secretion and biofilm formation. It has been reported that compounds such as S-adenosylhomocysteine, butyryl-SAM and sinefungin inhibit the *in vitro* AHL synthesis in pathogenic *P. aeruginosa* (Hentzer and Givskov 2003). Structural analogs of acyl-SAM have been developed and their binding potential to the AHL producing enzymes demonstrated. However, in this study the focus is on the ligand-receptor interactions and the role of the analogs in identification of novel AHL signals producers, than on the applicability in attenuating bacterial virulence (Kai et al. 2014).

In another study, Chang and co-authors have reported that trans-cinnamaldehyde was able to efficiently inhibit the AHL production by Rhl QS system in *P. aeruginosa*. In addition, trans-cinnamaldehyde could reduce the pyocyanin production in *P. aeruginosa* by up to 40% without affecting the normal bacterial growth (Chang et al. 2014). Despite the very promising anti-virulence and antibiofilm properties of the above mentioned inhibitors, their *in vivo* application was not demonstrated.

Table 1.1 Common QS inhibitors and their application in infections control

Compound	Mechanism of action	Main achievements
Adenosyl-homocysteine	Inhibition of AHL synthesis in *P. aeruginosa*	Attenuation of QS regulated toxins secretion and prevention of bacterial infections (Hentzer and Givskov 2003)
Lactonase	AHLs lactone ring hydrolyzation	Inhibition of *Vibrio parahaemolyticus* biofilm growth and intestinal colonization in shrimp (Vinoj et al. 2014)
Acylase	AHLs amide bond hydrolyzation	Coating of medical devices for controlling biofilm related urinary tract infections (Ivanova et al. 2015a, b)
Flavanoids	Block the QS receptors	Suggested as alternatives to the traditional antibiotics for treatment of *P. aeruginosa* associated infectious diseases (Paczkowski et al. 2017)
Structural analogue of AHL signal	Block the QS receptors	Attenuation of *P. aeruginosa* pathogenesis through the decreased production of rhamnolipids, elastase and protease in *P. aeruginosa* (Yang et al. 2012)
RNAIII-inhibiting peptide	Inactivate staphylococcal TRAP/agr system	Grafting on urinary stents to counteract *S. aureus* urinary tract infections (Cirioni et al. 2007)
Antibody XYD-11G2	Catalyze the hydrolysis of Gram-negative 3-oxo-C12-HSL	Reduction of pyocyanin secretion and potential to prevent and treat *P. aeruginosa* of infections (De Lamo Marin et al. 2007)
Curcumin with antibiotics	Synergistic mechanism of action	Development of combined therapy against bacterial pathogens (Bahari et al. 2017)
Solid lipid NPs loaded with QS inhibitors	Site specific delivery of QS inhibitors	Incorporation in nanostructures for better penetration into the mucus and treatment of *P. aeruginosa* pulmonary infections (Nafee et al. 2014)
Gold NPs with lactonase	AHLs lactone ring hydrolyzation	Nanoparticles with improved anti-QS activity; decreased production of the biofilm exopolysaccharides in *Proteus species* and potential to prevent urinary tract infections (Vinoj et al. 2015)

1.3.2   Quorum Sensing Signals Degradation As Alternative for Controlling Infectious Diseases

Altering the QS signals in the extracellular environment is among the most promising anti-QS approaches, avoiding the need to penetrate the cells and reach the specific signal receptors. Several enzymes degrading the AIs have been reported till now. They are produced by bacterial strains such as *Bacillus* strain *COT1*, *Bacillus sp. 240B1*, *Agrobacterium tumefaciens*, *Arthrobacter sp. IBN110*, *Bacillus cereus*, *Bacillus mycoides*, *Rhodococcus*, *Anabaena*, *Ralstonia*, *Bacillus thuringiensis*, and *Streptomyces sp.* (Lade et al. 2014).

Bacillus sp. possess aiiA that encodes the well know AHL-inactivating enzyme lactonase. Lactonase inhibits the QS regulated virulence in various Gram-negative pathogens via the hydrolysis of the AHLs lactone ring (Fig. 1.1, left) (Cheng et al. 2014). Lactonase encoding genes from *Bacillus* spp. or *A. tumefaciens* have been introduced into tobacco and potato to enhance plants defense towards common plant pathogens such as *Pectobacterium carotovorum* (Dong et al. 2001). This QS disrupting enzyme have also been used in controlling bacterial infections in fishes. For instance, the enzyme administration resulted in a significant reduction of the *Aeromonas hydrophila* infection in zebrafish (Cao et al. 2012). Lactonase isolated from *Bacillus licheniformis* DAHB1 was able to interrupt the QS and inhibit the *Vibrio parahaemolyticus* biofilm growth as well as intestinal colonization in shrimp (Vinoj et al. 2014).

AHL lactonase, from metallo-β-lactamase superfamily has been identified recently and characterized by Tang et al. and co-workers. The enzyme has the ability to decompose AHLs with different side chain length, with or without *oxo*-group at the position C-3 (Tang et al. 2015). The authors demonstrated the anti-infective potential of the enzyme through the reduction of pyocyanin and elastase expression by *P. aeruginosa*, and delayed *in vivo* infection occurrence in *Caenorhabditis elegans* model system (Tang et al. 2015). The lactonase did not affect the bacterial cells viability, but has the ability to increase bactericidal activity of the antibiotics and reduce the sessile growth of *P. aeruginosa* and *Acinetobacter baumannii* (Zhang et al. 2017). The QS inhibition by lactonase have been utilized as a tool to boost the activity of the antibiotics gentamicin and ciprofloxacin towards the drug resistant biofilm forms of *P. aeruginosa* (Kiran et al. 2011). Furthermore, the synergistic effect with ciprofloxacin has been demonstrated against *P. aeruginosa* PAO1 in burn infection model on mice. Topical application of gels containing the anti-QS enzyme lactonase inhibited the spread of skin pathogens and reduced to minimum the mice mortality after the addition of lower antibiotic dosage (Gupta et al. 2015).

Enzymes with lactonase activity have been identified in the human serum as well. They are called paraoxonases (PONs) and have been able to hydrolyze the N-3-oxododecanyol homoserine lactone (3-oxo-C12-HSL), leading to inactivation of the molecule and disruption of QS (Bar-Rogovsky et al. 2013). However, these have not been extensively studied as anti-infective antibiotic alternatives.

Some bacterial strains such as *Variovorax paradoxus* and *Ralstonia sp.* produce another AHL-degrading enzyme, named acylase. Acylase hydrolyzes the amide bond of the AHLs and interrupts the QS pathways in plant and human pathogens (Fig. 1.1, left) (Lade et al. 2014). Acylase expression in pathogenic *P. aeruginosa* PAO1 significantly reduced the production of pyocyanin and elastase, declined the bacterium swarming motility and weakened its pathogensis *in vivo* in *C. elegans* infection model (Lin et al. 2003). In another study, the enzyme was used by the soil pseudomonad, PAI-A, to degrade the AHLs with long side chain, and inhibit the QS in pathogenic *P. aeruginosa* (Huang et al. 2003). Recently, Koch et al. reported that the combination of two single mutations in the AHL-degrading enzyme PvdQ from *P. aeruginosa,* led to higher activity toward N-octanoyl-L-homoserine lactone (C8-HSL). These mutations inhibited the QS regulated phenotypes of cystic fibrosis pathogen *Burkholderia cenocepacia* and significantly decreased the amount of C8-HSL signal in the bacterial surrounding (Koch et al. 2014). In our group, silicone urinary catheter has been coated with fungal acylase from *Aspergillus meleus* using layer-by-layer technique. The acylase-coated catheters disrupted the AHLs based QS signaling and inhibited *P. aeruginosa* biofilm development *in vitro* under conditions simulating the real situation during catheterization (Ivanova et al. 2015a, b). In a more recent work, the enzyme was covalently immobilized on biomedical grade polyurethane surface. The presence of acylase led to considerable reduction in the virulence factors expression and 60% decrease of the sessile *P. aeruginosa* ATCC 10145 and PAO1 growth. The developed enzyme coatings were stable in simulated urine, demonstrating the feasibility of these approaches for controlling catheter-related bacterial infections (Ivanova et al. 2015a, b, Grover et al. 2016). The anti-QS enzyme acylase has been also used as an antifouling agent after chemical and enzymatic immobilization on carboxylated polyaniline nanofibers (Lee (J) et al. 2017). Our group developed multi enzymatic coatings on silicone catheters comprised of acylase and another hydrolytic enzyme amylase, which acts on the bacterial exopolysaccharides. The generated enzyme coatings demonstrated better antibiofilm activity depending on the last layer (acylase vs. amylase). The coatings with acylase as uppermost layer inactivated the QS signals and demonstrated higher antibiofilm activity against *P. aeruginosa* and *E. coli* single and dual species cultures. The enzymes multilayers could delay the *in vivo* biofilm formation in catheterizes rabbit model. These enzyme activated surfaces may find potential application for preventing difficult-to-treat biofilm-associated infections (Ivanova et al. 2015a, b). AHL-degrading acylase which is produced by soil isolate *Ochrobactrum sp.* A44 has been identified (Czajkowski et al. 2011). The extract from the bacterium degraded a broad range of AHLs and was used to attenuate the virulence of *P. carotovorum* pathogen in plants (Czajkowski et al. 2011).

Oxidoreductases are another type enzymes interfering with bacterial communication via inactivation of the AHL signals (Fig. 1.1, left) (Reuter et al. 2016). The enzyme modifies the acyl side chain and affects the AHLs specificity to the cognate intracellular receptor, disturbing the expression of QS-related virulent genes. Hong et al. reported the production of oxidoreductases as defense mechanisms in bacteria such as *Burkholderia* sp. GG4 and *Rhodococcus erythropolis* (Uroz et al. 2005; Hong et al. 2012). Bijtenhoorn and co-workers identified novel type NADP-dependent oxidoreductase derived from metagenome. This novel BpiB09 oxidoreductase was able to inactivate the 3-oxo-C12-HSL and its expression in *P. aeruginosa* PAO1 decreased the secretion of pyocyanin, swarming and swimming motility, as well as the formation of resistant biofilms (Bijtenhoorn et al. 2011). Protein with oxidoreductase activity on both AHL and AI-2 signaling molecules have been recently reported. Its chemical immobilization onto glass surface resulted in 75% inhibition of the *Klebsiella oxytoca* and *K. pneumoniae* biofilm growth (Weiland-Bräuer et al. 2016). Cytochrome P450 from *Bacillus megaterium,* named CYP102A1 is widely studied due to its ability to oxidize compounds in AHL QS pathways such as fatty acids and acyl homoserines at positions ω-1, ω-2 and ω-3. The AHLs oxidation by CYP102A1 produced hydroxy-AHL and consequently inhibited the QS activity in *A. tumefaciens* NTL4 (Hong et al. 2012).

Besides of the AHLs inactivating enzymes, there is another type of enzymes that are capable to impair the quinolone signals involved the *P. aeruginosa* QS circuit. The enzyme 3-Hydroxy-2-methyl-4(1H) – quinolone 2, 4-dioxygenase, Hod, for example, degrades the 2-heptyl-3-hydroxy-4(1H)-quinolone (PQS) signal and decrease its accumulation in the surrounding. Although, the 2-heptyl-4(1H)-quinolone precursor of the PQS signal can inhibit Hod, the enzyme decreased the secretion of pyocyanin, rhamnolipids, and lectin A toxins, protecting plants from tissue damage and inflammation (Pustelny et al. 2009). Anti-QS enzymes are very attractive antibiotic alternatives for controlling bacterial infections and surface colonization, imposing minimal risk for selection of resistance. However, the enzymes stability *in vivo* is ever existing concern for their biomedical applications. Therefore, strategies aiming at developing resistant and stable at use enzyme proteins are of great importance (Ivanova et al. 2015a, b).

1.3.3 Prevention of Bacterial Diseases via Inactivation of Signals Receptors

Recently, it has been demonstrated that flavonoids can bind to the QS receptors and significantly reduce their ability to activate the corresponding genes expression in *P. aeruginosa* (Fig. 1, left) (Paczkowski et al. 2017). Yang et al. reported a structural analogue of AHL signal, named N-decanoyl-L-homoserine benzyl ester, that blocks the cognate receptor in *P. aeruginosa* and counteract the production of rhamnolipids, elastase and protease, without killing the bacterial cells. The receptor antagonist also potentiated the antibacterial activity on various antibiotics, lowering their therapeutic concentrations for *P. aeruginosa* infections (Yang et al. 2012). However, the inhibitor application for treatment of pathogenic diseases is limited by the instability of the homoserine lactone moiety in alkaline condition and its degradability by mammalians lactonases. Loughlin et al. synthesized metachloro-thiolactone, and meta-bromo-thiolactone, and demonstrated their anti-virulent potential through the inhibition of pyocyanin production. The last one not only could downregulate the virulence factors expression and biofilm formation, but also could protect *C. elegans* and human lung epithelial cells from the virulent *P. aeruginosa* (Loughlin et al. 2013). Geske and co-authors have developed AHLs analogues that can bind with the LuxR, TraR and LasR receptors in *Vibrio fischeri*, *A. tumefaciens* and *P. aeruginosa*, respectively (Geske et al. 2008).

In another investigation, Kim et al. evaluated the potential of 6-gingerol for controlling bacterial pathogenesis and infections occurrence. The authors confirmed that this anti-QS compound was effective and can repress the QS genes involved in the exoprotease, rhamnolipid and pyocyanin production in *P. aeruginosa* (Kim et al. 2015). These finding are considered significant, because the decrease of the toxins production will increase host defense as well as the efficiency of the antimicrobials, which is a promising strategy for treatment of infectious diseases. Packiavathyet al. assessed the anti-QS potential of curcumin, a major component of turmeric (*Curcuma longa*) rhizomes, towards *E. coli, Serratia marcescens,* and *Proteus mirabilis.* The QS regulated violacein production by the reporter strain *Chromobacterium violaceum* was decreased, confirming the anti-QS activity of this compound. Moreover, curcumin inhibited the biofilm growth and disturbed the swimming motility of the tested pathogens. The treatment of uropathogens with curcumin led to significant reduction of biofilm EPM and therefore could increase the pathogens susceptibility to conventional therapies (Packiavathy et al. 2014).

Nowadays, targeting the *agr* operon in Gram-positive bacteria has received noteworthy attention. Changing the amino acids or removing the side chain of the AIPs have been shown as efficient methods for designing analogues that block the *agr* receptors (LaSarre and Federle 2013). Significant contributions have been made by Sully et al., who in a high throughput screening identified that the small molecule savirin is capable to disrupt the *agr*-depended QS behaviors in *S. aureus* (Fig. 1.1, right). The *in vivo* murine models of skin infections demonstrated the efficiency of savirin against *S. aureus,* without evidences for resistance emergence. Taken together, these findings demonstrated that savirin is an excellent QS inhibitor for controlling infectious diseases (Sully et al. 2014). In another work, Lee et al. reported the anti-virulence and antibiofilm properties of cananga, black pepper, and myrrh oils towards *S. aureus*. The black pepper significantly downregulated the *hla, nuc1, nuc2,* and *sar* genes expressing key proteins for infection establishment, attracting the attention towards them as alternative anti-virulence approaches against *S. aureus* infections (Lee (K) et al. 2014).

RNAIII-inhibiting peptide (RIP) is another promising inhibitor targeting the staphylococcal TRAP/agr system. Animal model of rats treated with RIP demonstrated its role in preventing methicillin-resistant *S. aureus* infections. RIP has also been shown to be very effective in controlling biofilm related staphylococcal infections caused by the use of ureteral stents, central orthopedic implants, and venous catheters. The RIP deposition onto surface of the medical devices resulted in suppression of *S. aureus* biofilm formation (Cirioni et al. 2007). Recently, Broderick and co-workers coated surfaces with macrocyclic peptide-based inhibitor that target *agr* QS system. The peptide was able to regulate the expression of toxic shock syndrome and attenuate the virulence of *S. aureus* and was further applied as a coating of tampons (Broderick et al. 2014).

Despite the very promising results, the utilization of the QSIs as anti-virulent drugs may have adverse effects on *S. aureus*. There are evidences that the *agr* QS system negatively controls the biofilm growth and inhibits the colonizing factors expression. Therefore, instead of preventing, the QSIs may actually enhance the *S. aureus* surface colonization and biofilm formation. In such scenario, the inhibitors of the *agr* receptors would be more suitable for acute *S. aureus* infection (Gray et al. 2013). Even so, *agr*-deficient strains *S. aureus* are frequently involved in acute bacteraemia, bringing up the question whether the QSIs can be used as alternative to cure *S. aureus* infections.

Antagonists of the AI-2 based intracellular signaling have been developed as a broader spectrum anti-virulent strategy. For instance, analogues of DPD were found to exhibit strong antagonistic activity against AI-2 based QS signaling in Gram-negative *E. coli*. However, most of the tested analogues, except one called butyl-DPD, demonstrated agonistic activity in *Salmonella typhimurium* (Gamby et al.

2012; LaSarre and Federle 2013). Halogenated furanones (fimbrolides) are natural QS inhibitors produced by red macroalga *Deliseapulchra*. Fimbrolides have been also reported as inhibitors of the QS regulated swarming motility of pathogenic *Serratia liquefaciens* and light emission by *V. fischeri* and *Vibrio harveyi* (Hentzer and Givskov 2003; Galloway et al. 2011). Covalent modification of these natural inhibitors affected the AI-2 signals production, inactivating the LuxS and consequently disrupting the interspecies bacterial communication in biofilms and then prevented the infection establishment (Zang and Lee 2010).

1.3.4 Antibody Specific Disruption in Counteraction Bacterial Virulence

Despite the numerous synthetic and natural compounds with anti-QS activity, there are several reports on the QS inhibiting potential of mammalians antibodies. It is known that the immune system in the mammalians produces antibodies in the presence of antigens as a response. Although, it is believed that the AI signals should not induce immunity response due to their low molecular weight and non-proteinaceous nature, the AHL and DPD signals evoke apoptosis and modulate the inflammation and host immune system (Kalia et al. 2015). In the past few years, Kaufmann et al. have reported antibody, named RS2-1G9, as alternative therapeutic agent for sequestering the 3-oxo-C12-HSL in the extracellular surrounding of *P. aeruginosa* (Kaufmann et al. 2008). Marin et al. demonstrated that XYD-11G2 antibody was able to catalyze the hydrolysis of the 3-oxo-C12-HSL signal, inhibiting the QS regulated pyocyanin production in Gram-negative bacteria (De Lamo Marin et al. 2007). On the other hand, Park and co-workers have investigated the monoclonal AP4-24H11 antibody, for its ability to disrupt the QS in Gram-positive *S. aureus*. They observed that this antibody efficiently suppressed the QS via the AIP IV sequestration and strongly inhibited the *S. aureus* pathogenesis. *In vivo*, in mouse model of skin and soft tissue infection of *S aureus* RN4850 the AP4-24H11 significantly reduced the severity of necrosis. Despite, the ability of the monoclonal antibodies to neutralize the AIPs and impede the QS pathways in pathogens, their anti-infective potential and application for treatment of infection diseases is still in infancy (Park et al. 2007).

1.3.5 Novel Therapeutic Combinations of Quorum Sensing Inhibitors and Antibiotics

The combination antibiotics QS inhibitors, is thought very promising strategy that may enhance bacterial susceptibility and eliminate multi-drug resistant bacteria (Kuo et al. 2015). Several studies have been performed to evaluate and demonstrate the synergistic effect of QS inhibitors with conventional antibiotics. Ajoene is a compound found to downregulate the QS-regulated genes expression in *P. aeruginosa in vitro* and *in vivo* in a pulmonary infection murine model. When Christensen and his collaborators, treated the peritoneal cavity of mice, pre-colonized with *P. aeruginosa*, with QSIs (e.g. ajoene, furanone C-30 and horseradish extract) and antibiotic tobramycin a synergistic effect, demonstrated by the significant decrease in bacterial load was observed. It is suggested that the QS inhibitors downregulated the expression of virulence factors, making the cell less virulent and more vulnerable to the antibiotic at lower amounts (Christensen et al. 2012). Other investigations have been carried out to demonstrate the synergism of curcumin with antibiotics against *P. aeruginosa*. Drastic decrease of virulent QS-genes expression and minimum inhibitory concentrations of gentamicin and azithromycin was obtained when the actives were applied simultaneously (Bahari et al. 2017). Natural phenolic compounds such as epigallocatechin-3-gallate and caffeic acid, also demonstrated enhanced activity with tetracycline, ciprofloxacin, and gentamicin against *K. pneumoniae* clinical isolates (Dey et al. 2016). Furiga et al. demonstrated enhanced activity against *P. aeruginosa* after mixing the N-(2- pyrimidyl) butanamide QS inhibitor and tobramycin, colistin, and ciprofloxacin (Furiga et al. 2016).

A synergism between the 2, 3-pyrazine dicarboxylic acid derivatives, targeting the LuxO receptors, and chloramphenicol, tetracycline, doxycycline, and erythromycin was also reported. This multiple-acting antimicrobial therapy improved antibiofilm activity and potentiated the bactericidal effect of the antibiotics (Hema et al. 2016). Jabra-Rizk et al. found that the QS inhibitor farnesol is able to attenuate the *S. aureus* virulence and its further combination with gentamicin reduced the bacterium growth with more than 2 logs (Jabra-Rizk et al. 2006). Recently, Brackman and co-authors showed that hamamelitannin (HAM) disrupts the QS in *S. aureus* and enhances the susceptibility of *S. aureus* biofilm to several antibiotics (Brackman et al. 2011). The HAM also increased the antibiotics effect *in vivo* in *C. elegans* and murine infection models. These findings suggested the great potential of the combined anti-virulent/antimicrobial therapy for prevention and control of bacterial diseases (Brackman et al. 2016).

1.3.6 Quorum Sensing Inhibitors in Nanotechnology

Particular attention is directed to the nanotechnology as a potential platform that may solve the failure of the existing antimicrobial treatments. Nanoparticles (NPs) exhibit unique physical and chemical properties at nanometer scale, which frequently differ from their bulk counterparts (Singh (BN) et al. 2017). The utility of nanotechnology as a tool to improve the bactericidal activity of drugs and fight antimicrobial resistance mostly rely on nanocapsules and liposomes as drug carriers for delivering the antibacterial agents to the site of infection (Singh (BN) et al. 2017). We have reported that antibacterial NPs of cationic biopolymers and antibiotics are more active than the free counterpart, eradicating bacteria through the cell membrane disruption (Fernandes et al. 2016, 2017; Francesko et al. 2016).

Recently, the generation of nanomaterials of QS inhibitors has been investigated to improve the efficacy of these anti-infective agents and control the active agents release at the site infection. Nafee et al. developed solid lipid NPs loaded with anti-QS compounds able to penetrate the mucus and exhibited up to seven fold higher anti-virulence activity against *P. aeruginosa* compared to the free agent (Nafee et al. 2014). Kasper et al. encapsulated *S*-phenyl-L-cysteine sulfoxide into zein NPs in order to inhibit the biofilm growth of the oral pathogen *Streptococcus mutants* (Kasper et al. 2016). In another work, Garcia-Lara and co-workers reported the potential of ZnO NPs to decrease the QS dependent pyocyanin secretion and biofilm establishment of *P. aeruginosa* (García-Lara et al. 2015). Further, chitosan NPs carrying the kaempferol showed enhanced anti-QS properties reducing the violacein production by *C. violaceum* CV026 (Ilk and Sa 2017). Miller and co-authors engineered silicone dioxide NPs functionalized with β-cyclodextrin to target the signals in QS. The β-cyclodextrin binds the HSL molecules and its grafting onto NPs surface demonstrated enhanced capability to quench the signals in *V. fisheri* (Miller et al. 2015). Using rhizopusarrhizus BRS-07 biomass extract, Singh et al. have modified silver NPs (AgNPs) to obtain surface protective NPs. These bio-silver NPs demonstrated capability to suppress the biofilm formation and the virulence of *P. aeruginosa PAO1* by disrupting its QS system (Singh (BR) et al. 2015). Vinoj et al. decorated gold NPs with the enzyme lactonase against the biofilm formation of pathogenic *Proteus* species. The enzyme coated NPs exhibited increased anti-biofilm activity and decreased exopolysaccharides production compared to the bulk enzyme form, which is promising strategy for controlling urinary tract infections (Vinoj et al. 2015).

1.4   Conclusions

Targeting the QS systems in bacteria represents next-generation anti-infective strategy to control the virulence factors production and biofilm formation. The development of novel non-antibiotic therapeutics aiming to suppress the virulent genes expression and prevent infection, without affecting bacterial cell viability has attracted the attention due to the decreased risk of drug resistance occurrence. Although, the QS inhibitors have a great potential to control numerous bacterial pathogenic phenotypes most of them are still at preclinical stage and need to be translated to human trails. To date, only three clinical trials conducted with QS inhibitors are described. The practical use of the discovered anti-QS compounds is frequently limited because of their potential toxicity, instability and reduced therapeutic effect compared to the antibiotics. Nano-size transformation of the anti-virulent and antibacterial compounds have gained the attention for engineering new class of antimicrobials with unique properties, improved stability and higher efficacy *in-vitro* and *in-vivo* compared to the bulk solutions. Furthermore, combining the QS inhibitors and conventional antibiotics may have substantial benefits to increase the effect and prolong the life span of the therapeutic drugs.

Acknowledgements This work was supported by the European project PROTECT – "Pre-commercial lines for production of surface nanostructured antimicrobial and anti-biofilm textiles, medical devices and water treatment membranes" (H2020 – 720851). A. I. wish to acknowledge Generalitat de Catalunya for providing her Ph.D. grant (2017FI_B_00524).

References

Antunes LCM, Ferreira RBR, Buckner MMC, Finlay BB (2010) Quorum sensing in bacterial virulence. Microbiology 156:2271–2282. https://doi.org/10.1099/mic.0.038794-0

Bahari S, Zeighami H, Mirshahabi H, Roudashti S, Haghi F (2017) Inhibition of *Pseudomonas aeruginosa* quorum sensing by subinhibitory concentrations of curcumin with gentamicin and azithromycin. J Global Antimicrob Resis 10:21–28. https://doi.org/10.1016/j.jgar.2017.03.006

Bar-Rogovsky H, Hugenmatter A, Tawfik AS (2013) The evolutionary origins of detoxifying enzymes: the mammalian serum paraoxonases (PONs) relate to bacterial homoserine lactonases. J Biol Chem 288:23914–23927. https://doi.org/10.1074/jbc.M112.427922

Beceiro A, Tomás M, Bou G (2013) Antimicrobial resistance and virulence: a successful or deleterious association in the bacterial world? Clin Microbiol Rev 26:185–230. https://doi.org/10.1128/CMR.00059-12

Bijtenhoorn P, Mayerhofer H, Muller-Dieckmann J, Utpatel C, Schipper C, Hornung C, Szesny M, Grond S, Thurmer A, Brzuszkiewicz E, Daniel R, Dierking K, Schulenburg H, Streit WR (2011) A novel metagenomic short-chain dehydrogenase/reductase attenuates *Pseudomonas aeruginosa* biofilm formation and virulence on *Caenorhabditis elegans*. PLoS One 6:e26278. https://doi.org/10.1371/journal.pone.0026278

Bortolotti D, Le Maoult J, Trapella C, Di Luca D, Carosella ED, Rizzo R (2015) *Pseudomonas aeruginosa* quorum sensing molecule N-(3-oxododecanoyl)-L-homoserine-lactone induces HLA-G expression in human immune cells. Infect Immun 83:3918–3925. https://doi.org/10.1128/IAI.00803-15

Brackman G, Cos P, Maes L, Nelis HJ, Coenye T (2011) Quorum sensing inhibitors increase the susceptibility of bacterial biofilms to antibiotics *in vitro* and *in vivo*. Antimicrob Agents Chemother 55:2655–2661. https://doi.org/10.1128/AAC.00045-11

Brackman G, Breyne K, De Rycke R, Vermote A, Van Nieuwerburgh F, Meyer E, Van Calenbergh S, Coenye T (2016) The quorum sensing inhibitor hamamelitannin increases antibiotic susceptibility of *Staphylococcus aureus* biofilms by affecting peptidoglycan biosynthesis and eDNA release. Sci Rep 6:20321. https://doi.org/10.1038/srep20321

Broderick A, Stacy DM, Tal-Gan Y, Kratochvil MJ, Blackwell HE, Lynn DM (2014) Surface coatings that promote rapid release of peptide-based *agrC* inhibitors for attenuation of quorum sensing in *Staphylococcus aureus*. Adv Healthc Mater 3:97–105. https://doi.org/10.1002/adhm.201300119

Cao Y, He S, Zhou Z, Zhang M, Mao W, Zhang H, Yao B (2012) Orally administered thermostable N-Acyl homoserine lactonase from bacillus sp. strain ai96 attenuates aeromonas hydrophila infection in zebrafish. Appl Environ Microbiol 78:1899–1908. https://doi.org/10.1128/AEM.06139-11

Chang CY, Krishnan T, Wang H, Chen Y, Yin W-F, Chong Y-M, Tan LY, Chong TM, Chan K-G (2014) Non-antibiotic quorum sensing inhibitors acting against N -acyl homoserine lactone synthase as drug gable target. Sci Rep 4:7245. https://doi.org/10.1038/srep07245

Cheng G, Hao H, Xie S, Wang X, Dai M, Huang L, Yuan Z (2014) Antibiotic alternatives: the substitution of antibiotics in animal husbandry? Front Microbiol 5:217. https://doi.org/10.3389/fmicb.2014.00217

Cheung GYC, Joo HS, Chatterjee SS, Otto M (2014) Phenol-soluble modulins – critical determinants of staphylococcal virulence. FEMS Microbiol Rev 38:698–719. https://doi.org/10.1111/1574-6976.12057

Christensen LD, Van Gennip M, Jakobsen TH, Alhede M, Hougen HP, Høiby N, Bjarnsholt T, Givskov M (2012) Synergistic antibacterial efficacy of early combination treatment with tobramycin and quorum-sensing inhibitors against Pseudomonas aeruginosa in an intraperitoneal foreign-body infection mouse model. J Antimicrob Chemother 67:1198–1206. https://doi.org/10.1093/jac/dks002

Cirioni O, Ghiselli R, Minardi D, Orlando F, Mocchegiani F, Silvestri C, Muzzonigro G, Saba V, Scalise G, Balaban N, Giacometti A (2007) RNAIII-inhibiting peptide affects biofilm formation in a rat model of staphylococcal ureteral stent infection. Antimicrob Agents Chemother 51:4518–4520. https://doi.org/10.1128/AAC.00808-07

Czajkowski R, Krzyzanowska D, Karczewska J, Atkinson S, Przysowa S, Lojkowska E, Williams P, Jafra S (2011) Inactivation of AHLs by Ochrobactrum Sp. A44 depends on the activity of a novel class of AHL acylase. Environ Microbiol Rep 3:59–68. https://doi.org/10.1111/j.1758-2229.2010.00188

De Lamo Marin S, Xu Y, Meijler MM, Janda KD (2007) Antibody catalyzed hydrolysis of a quorum sensing signal found in Gram-negative bacteria. Bioorg Med Chem Lett 17:1549–1552. https://doi.org/10.1016/j.bmcl.2006.12.118

Dey D, Ghosh S, Ray R, Hazra B (2016) Polyphenolic secondary metabolites synergize the activity of commercial antibiotics against clinical isolates of β-lactamase-producing Klebsiella pneumoniae. Phytother Res 30:272–282. https://doi.org/10.1002/ptr.5527

Dong YH, Wang LH, Xu JL, Zhang HB, Zhang XF, Zhang LH (2001) Quenching quorum-sensing-dependent bacterial infection by an N-acyl homoserine lactonase. Nature 411:813–817. https://doi.org/10.1038/35081101

Fernandes MM, Ivanova K, Francesko A, Rivera D, Burgues J-T, Gedanken A, Mendonza E, Tzanov T (2016) Escherichia coli and Pseudomonas aeruginosa eradication by nanopenicillin G. Nanomed: Nanotechnol Biol Med 12:2061–2069. https://doi.org/10.1016/j.nano.2016.05.018

Fernandes MM, Ivanova K, Hoyo J, Pérez-Rafael S, Francesko A, Tzanov T (2017) Nanotransformation of vancomycin overcomes the intrinsic resistance of Gram-negative bacteria. ACS Appl Mater Interfaces 9:15022–15030. https://doi.org/10.1021/acsami.7b00217

Francesko A, Fernandes MM, Ivanova K, Amorimb S, Reis RL, Pashkuleva I, Mendozad E, Pfeifer A, Heinze T, Tzanov T (2016) Bacteria-responsive multilayer coatings comprising polycationic nanospheres for bacteria biofilm prevention on urinary catheters. Acta Biomater 33:203–212. https://doi.org/10.1016/j.actbio.2016.01.020

Furiga A, Lajoie B, El Hage S, Baziard G, Roques C (2016) Impairment of Pseudomonas aeruginosa biofilm resistance to antibiotics by combining the drugs with a new quorum-sensing inhibitor. Antimicrob Agents Chemother 60:1676–1686. https://doi.org/10.1128/AAC.02533-15

Galloway WRJD, Hodgkinson TJ, Bowden SD, Welch M, Spring DR (2011) Quorum sensing in Gram-negative bacteria: small-molecule modulation of AHL and AI-2 quorum sensing pathways. Chem Rev 111:28–67. https://doi.org/10.1021/cr100109t

Gamby S, Roy V, Guo M, Smith JAI, Wang J, Stewart JE, Wang X, Bentley WE, Sintim HO (2012) Altering the communication networks of multispecies microbial systems using a diverse toolbox of AI-2 analogues. ACS Chem Biol 7:1023–1030. https://doi.org/10.1021/cb200524y

García-Lara B, Saucedo-Mora MA, Roldan-Sanchez JA, Perez-Eretza B, Ramasamy M, Lee J, Coria-Jimenez R, Tapia M, Varela-GuerreroV G-CR (2015) Inhibition of quorum-sensing-dependent virulence factors and biofilm formation of clinical and environmental Pseudomonas aeruginosa strains by ZnO nanoparticles. Lett Appl Microbiol 61:299–305. https://doi.org/10.1111/lam.12456

Geske GD, Mattmann ME, Blackwell HE (2008) Evaluation of a focused library of N-aryl L-homoserine lactones reveals a new set of potent quorum sensing modulators. Bioorg Med Chem Lett 18:5978–5981. https://doi.org/10.1016/j.bmcl.2008.07.089

Grant SS, Hung DT (2013) Persistent bacterial infections, antibiotic tolerance, and the oxidative stress response. Virulence 4:273–283. https://doi.org/10.4161/viru.23987

Gray B, Hall P, Gresham H (2013) Targeting *agr-* and *agr-*like quorum sensing systems for development of common therapeutics to treat multiple Gram-positive bacterial infections. Sensors 13:5130–5166. https://doi.org/10.3390/s130405130

Grover N, Plaks JG, Summers SR, Chado GR, Schurr MJ, Kaar JL (2016) Acylase-containing polyurethane coatings with anti-biofilm activity. Biotechnol Bioeng 113:2535–2543. https://doi.org/10.1002/bit.26019

Gupta P, Chhibber S, Harjai K (2015) Efficacy of purified lactonase and ciprofloxacin in preventing systemic spread of *Pseudomonas aeruginosa* in murine burn wound model Burns. Burns 41:153e162. https://doi.org/10.1016/j.burns.2014.06.009

Hawver LA, Jung SA, Ng W-L (2016) Specificity and complexity in bacterial quorum-sensing systems. FEMS Microbiol Rev 40:738–752. https://doi.org/10.1093/femsre/fuw014

Hema M, AdlinePrincy S, Sridharan V, Vinoth P, Balamurugan P, Sumana MN (2016) Synergistic activity of quorum sensing inhibitor, pyrizine-2-carboxylic acid and antibiotics against multidrug resistant: *V. cholerae*. RSC Adv 6:45938–45946. https://doi.org/10.1039/c6ra04705j

Hentzer M, Givskov M (2003) Pharmacological inhibition of quorum sensing for the treatment of chronic bacterial infections. J Clin Invest 112:1300–1307. https://doi.org/10.1172/JCI200320074

Hong K-W, Koh C-L, Sam C-K, Yin W-F, Chan K-G (2012) Quorum quenching revisited-from signal decays to signalling confusion. Sensors 12:4661–4696. https://doi.org/10.3390/s120404661

Huang JJ, J-In H, Zhang L-H, Leadbetter JR (2003) Utilization of acyl-homoserine lactone quorum signals for growth by a soil pseudomonad and *Pseudomonas aeruginosa* PAO1. Appl Environ Microbiol 69:5941–5949. https://doi.org/10.1128/AEM.69.10.5941

Ilk S, Saglamb N, Özgenc M, Korkusuz F (2017) Chitosan nanoparticles enhances the anti-quorum sensing activity of kaempferol. Int J Biol Macromol 94:653–662. https://doi.org/10.1016/j.ijbiomac.2016.10.068

Ivanova K, Fernandes MM, Tzanov T (2013) In: Mendez-Vilas A (ed) Current advances on bacterial pathogenesis inhibition and treatment strategies. Microbial pathogens and strategies for combating them: science, technology and education, vol 1. Formatex Research Center, Badajoz, pp 322–336

Ivanova K, Fernandes MM, Francesko A, Mendoza E, Guezguez J, Burnet M, Tzanov T (2015a) Quorum-quenching and matrix-degrading enzymes in multilayer coatings synergistically prevent bacterial biofilm formation on urinary catheters. ACS Appl Mater Interfaces 7:27066–27077. https://doi.org/10.1021/acsami.5b09489

Ivanova K, Fernandes MM, Mendoza E, Tzanov T (2015b) Enzyme multilayer coatings inhibit *Pseudomonas aeruginosa* biofilm formation on urinary catheters. Appl Microbiol Biotechnol 99:4373–4385. https://doi.org/10.1007/s00253-015-6378-7

Jabra-Rizk MA, Meiller TF, James CE, Shirtliff ME (2006) Effect of farnesol on *Staphylococcus aureus* biofilm formation and antimicrobial susceptibility. Antimicrob Agents Chemother 50:1463–1469. https://doi.org/10.1128/AAC.50.4.1463–1469.2006

Kai K, Fujii H, Ikenaka R, Akagawa M, Hayashi H (2014) An acyl-SAM analog as an affinity ligand for identifying quorum sensing signal synthases. Chem Commun 50:8586–8589. https://doi.org/10.1039/c4cc03094j

Kalia VC (2015) Microbes: the most friendly beings? In: Kalia VC (ed) Quorum sensing vs quorum quenching: a battle with no end in sight. Springer India, pp 1–5. https://doi.org/10.1007/978-81-322-1982-8_1. ISBN 978-81-322-1981-1

Kasper SH, Hart R, Bergkvist M, Musah RA, Cady NC (2016) Zein nanocapsules as a tool for surface passivation, drug delivery and biofilm prevention. AIMS Microbiol 2:422–433. https://doi.org/10.3934/microbiol.2016.4.422

Kaufmann GF, Park J, Mee JM, Ulevitch RJ, Janda KD (2008) The quorum quenching antibody RS2-1G9 protects macrophages from the cytotoxic effects of the *Pseudomonas aeruginosa* quorum sensing signalling molecule N-3-oxo-dodecanoyl-homoserine lactone. Mol Immunol 45:2710–2714. https://doi.org/10.1016/j.molimm.2008.01.010

Khan BA, Yeh AJ, Cheung GYC, Otto M (2015) Investigational therapies targeting quorum-sensing for the treatment of *Staphylococcus aureus* infections. Informa Healthcare 24:1–16. https://doi.org/10.1517/13543784.2015.1019062

Kim H-S, Lee S-H, Byun Y, Park H-D (2015) 6-gingerol reduces *Pseudomonas aeruginosa* biofilm formation and virulence via quorum sensing inhibition. Sci Rep 5:8656. https://doi.org/10.1038/srep08656

Kiran S, Sharma P, Harjai K, Capalash N (2011) Enzymatic quorum quenching increases antibiotic susceptibility of multidrug resistant *Pseudomonas aeruginosa*. Iran J Microbiol 3:1–12

Koch G, Nadal-Jimeneza P, Reisa CR, Muntendama R, Bokhoveb M, Melilloa E, Dijkstrab BW, Coola RH, Quax WJ (2014) Reducing virulence of the human hathogen *Burkholderia* by altering the substrate specificity of the quorum-quenching Acylase PvdQ. Proc Nat Acad Sci 111:1568–1573. https://doi.org/10.1073/pnas.1311263111

Kuo D, Yu G, Hoch W, Gabay D, Long L, Ghannoum M, Nagy N, Harding CV, Viswanathan R, Shoham M (2015) Novel quorum-quenching agents promote methicillin-resistant *Staphylococcus aureus* (mrsa) wound healing and sensitize mrsa to β-lactam antibiotics. Antimicrob Agents Chemother 59:1512–1518. https://doi.org/10.1128/AAC.04767-14

Lade H, Paul D, Kweon JH (2014) Quorum quenching mediated approaches for control of membrane biofouling. Int J Biol Sci 10:550–565. https://doi.org/10.7150/ijbs.9028

LaSarre B, Federle MJ (2013) Exploiting quorum sensing to confuse bacterial pathogens. Microbiol Mol Biol Rev 77:73–111. https://doi.org/10.1128/MMBR.00046-12

Le KY, Otto M (2015) Quorum-sensing regulation in staphylococci-an overview. Front Microbiol 6:1174. https://doi.org/10.3389/fmicb.2015.01174

Lee J, Zhang L (2014) The hierarchy quorum sensing network in *Pseudomonas aeruginosa*. Protein Cell 6:26–41. https://doi.org/10.1007/s13238-014-0100-x

Lee K, Lee J-H, Kim S-I, Cho MH, Lee J (2014) Anti-biofilm, anti-hemolysis, and anti-virulence activities of black pepper, cananga, myrrh oils, and nerolidol against *Staphylococcus aureus*. Appl Microbiol Biotechnol 98:9447–9457. https://doi.org/10.1007/s00253-014-5903-4

Lee J, Lee I, Nam J, Hwang DS, Yeon K-M, Kim J (2017) Immobilization and stabilization of acylase on carboxylated polyaniline nanofibers for highly effective antifouling application via quorum quenching. ACS Appl Mater Interfaces 9:15424–15432. https://doi.org/10.1021/acsami.7b01528

Li Z, Nair SK (2012) Quorum sensing: how bacteria can coordinate activity and synchronize their response to external signals? Protein Sci 21:1403–1417. https://doi.org/10.1002/pro.2132

Lin YH, Xu J-L, Hu J, Wang L-H, Ong SL, Leadbetter JR, Zhang L-H (2003) Acyl-homoserine lactone acylase from *Ralstonia strain* XJ12B represents a novel and potent class of quorum-quenching enzymes. Mol Microbiol 47:849–860. https://doi.org/10.1046/j.1365-2958.2003.03351.x

Loughlin CTO, Miller LC, Siryaporn A, Drescher K, Semmelhack MF (2013) A quorum-sensing inhibitor blocks *Pseudomonas aeruginosa* virulence and biofilm formation. Proc Natl Acad Sci U S A 110:17981–17986. https://doi.org/10.1073/pnas.1316981110

Miller KP, Wang L, Chen Y-P, Pellechia PJ, Benicewicz BC, Decho AW (2015) Engineering nanoparticles to silence bacterial communication. Front Microbiol 6:189. https://doi.org/10.3389/fmicb.2015.00189

Miquel S, Lagrafeuille R, Souweine B, Forestier C (2016) Anti-biofilm activity as a health issue. Front Microbiol 7:592. https://doi.org/10.3389/fmicb.2016.00592

Monnet V, Juillard V, Gardan R (2016) Peptide conversations in Gram-positive bacteria. Crit Rev Microbiol 42:339–351. https://doi.org/10.3109/1040841X.2014.948804

Montebello NA, Brecht RM, Turner DR, Ghali M, Pu X, Nagarajan R (2014) Acyl-ACP substrate recognition in *Burkholderia mallei* BmaI1 acyl-homoserine lactone synthase. ACS Biochem 53:6231–6242. https://doi.org/10.1021/bi5009529

Munita JM, Arias CA (2016) Mechanism of antibiotic resistance. Microbiol Spectr 4:1–37. https://doi.org/10.1128/microbiolspec.VMBF-0016-2015

Nafee N, Husari A, Maurer CK, Lu C, de Rossi C, Steinbach A, Hartmann RW, Lehr C-M, Schneider M (2014) Antibiotic-free nanotherapeutics: ultra-small, mucus-penetrating solid lipid nanoparticles enhance the pulmonary delivery and anti-virulence efficacy of novel quorum sensing inhibitors. J Control Release 192:131–140. https://doi.org/10.1016/j.jconrel.2014.06.055

Packiavathy AISV, Priya S, Karutha Pandian SK, Ravi AV (2014) Inhibition of biofilm development of uropathogens by curcumin – an anti-quorum sensing agent from *Curcuma longa*. Food Chem 148:453–460. https://doi.org/10.1016/j.foodchem.2012.08.002

Paczkowski JE, Mukherjee S, McCready AR, Cong J-P, Aquino CJ, Kim H, Henke BR, Smith CD, Bassler BL (2017) Flavonoids suppress *Pseudomonas aeruginosa* virulence through allosteric inhibition of quorum-sensing receptors. J Biol Chem 292:4064–4076. https://doi.org/10.1074/jbc.M116.770552

Papenfort K, Bassler B (2015) Quorum-sensing signal-response systems in Gram-negative bacteria. Nat Rev Microbiol 510:84–91. https://doi.org/10.1038/nrmicro.2016.89

Park J, Jagasia R, Kaufmann GF, Mathison JC, Ruiz DI, Moss JA, Meijler MM, Ulevitch RJ, Janda KD (2007) Infection control by antibody disruption of bacterial quorum sensing signaling. Chem Biol 14:1119–1127. https://doi.org/10.1016/j.chembiol.2007.08.013

Percival SL, Suleman L, Vuotto C, Donelli G (2015) Healthcare-associated infections, medical devices and biofilms: risk, tolerance and control. J Med Microbiol 64:323–334. https://doi.org/10.1099/jmm.0.000032

Pollitt EJG, West SA, Crusz SA, Burton-Chellew MN, Diggle SP (2014) Cooperation, quorum sensing, and evolution of virulence in *Staphylococcus aureus*. Infect Immun 82:1045–1051. https://doi.org/10.1128/IAI.01216-13

Pustelny C, Albers A, Buldt-Karentzopoulos K, Parschat K, Chhabra SR, Camara M, Williams P, Fetzner S (2009) Dioxygenase-mediated quenching of quinolone-dependent quorum sensing in *Pseudomonas aeruginosa*. Chem Biol 16:1259–1267. https://doi.org/10.1016/j.chembiol.2009.11.013

Rampioni G, Leoni L, Williams P (2014) The art of antibacterial warfare: deception through interference with quorum sensing-mediated communication. Bioorg Chem 55:60–68. https://doi.org/10.1016/j.bioorg.2014.04.005

Reuter K, Steinbach A, Helms V (2016) Interfering with bacterial quorum sensing. Perspect Medicin Chem 8:1–15. https://doi.org/10.4137/PMC.S13209

Rezzonico F, Smits THM, Duffy B (2012) Detection of AI-2 receptors in genomes of *Enterobacteriaceae* suggests a role of type-2 quorum sensing in closed ecosystems. Sensors (Switzerland) 12:6645–6665. https://doi.org/10.3390/s120506645

Römling U, Balsalobre C (2012) Biofilm infections, their resilience to therapy and innovative treatment strategies. J Intern Med 272:541–561. https://doi.org/10.1111/joim.12004

Scott R, Hasty J (2016) Quorum sensing communication modules for microbial consortia. ACS Synth Biol 5:969–977. https://doi.org/10.1021/acssynbio.5b00286

Singh BR, Singh BN, Singh A, Khan W, Naqvi AH, Singh HB (2015) Mycofabricated biosilver nanoparticles interrupt *Pseudomonas aeruginosa* quorum sensing systems. Sci Rep 5:13719. https://doi.org/10.1038/srep13719

Singh BN, Prateeksha, Upreti DK, Singh BR, Defoirdt T, Gupta VK, De Souza AO, Singh HB, Barreira JCM, Ferreira ICFR, Vahabi K (2017) Bactericidal, quorum quenching and anti-biofilm nanofactories: a new niche for nanotechnologists. Crit Rev Biotechnol 37:525–540. https://doi.org/10.1080/07388551.2016.1199010

Sully EK, Malachowa N, Elmore BO, Alexander SM, Femling JK, Gray BM, DeLeo FR, Otto M, Cheung AL, Edwards BS, Sklar LA, Horswill AR, Hall PR, Gresham HD (2014) Selective chemical inhibition of *agr* quorum sensing in *Staphylococcus aureus* promotes host defense with minimal impact on resistance. PLoS Pathog 10:e1004174. https://doi.org/10.1371/journal.ppat.1004174

Tan CH, Koh KS, Xie C, Zhang J, Tan XH, Lee GP, Zhou Y, Ng WJ, Rice SA, Kjelleberg S (2015) Community quorum sensing signalling and quenching: microbial granular biofilm assembly. NPJ Biofilms Microbiomes 1:15006. https://doi.org/10.1038/npjbiofilms.2015.6

Tang K, Su Y, Brackman G, Cui F, Zhang Y, Shi X, Coenye T, Zhang X-H (2015) MomL, a novel marine-derived N-acyl homoserine lactonase from *Muricauda olearia*. Appl Environ Microbiol 81:774–782. https://doi.org/10.1128/AEM.02805-14

Taraszkiewicz A, Fila G, Grinholc M, Nakonieczna J (2013) Innovative strategies to overcome biofilm resistance. Biomed Res Int 13:150653. https://doi.org/10.1155/2013/150653

Tay SB, Wen Yew WS (2013) Development of quorum-based anti-virulence therapeutics targeting Gram-negative bacterial pathogens. Int J Mol Sci 14:16570–16599. https://doi.org/10.3390/ijms140816570

Tong SYC, Chen LF, Fowler VG Jr (2013) Colonization, pathogenicity, host susceptibility and therapeutics for *Staphylococcus aureus*: what is the clinical relevance? Semin Immunopathol 34:185–200. https://doi.org/10.1007/s00281-011-0300-x

Uroz S, Chhabra SR, Camara M, Williams P, Oger P, Dessaux Y (2005) N-acylhomoserine lactone quorum-sensing molecules are modified and degraded by *Rhodococcus erythropolis* W2 by both amidolytic and novel oxidoreductase activities. Microbiology 151:3313–3322. https://doi.org/10.1099/mic.0.27961-0

Vinoj G, Vaseeharan B, Thomas S, Spiers AJ, Shanthi S (2014) Quorum-quenching activity of the AHL-lactonase from *Bacillus licheniformis* DAHB1 inhibits Vibrio biofilm formation *in vitro* and reduces shrimp intestinal colonisation and mortality. Mar Biotechnol 16:707–715. https://doi.org/10.1007/s10126-014-9585-9

Vinoj G, Patib R, Sonawaneb A, Baskaralingam V (2015) *In vitro* cytotoxic effects of gold nanoparticles coated with functional acyl homoserine lactone lactonase protein from *Bacillus licheniformis* and their antibiofilm activity against *Proteus* species. Antimicrob Agents Chemother 59:763–771. https://doi.org/10.1128/AAC.03047-14

Weiland-Bräuer N, Kisch MJ, Pinnow N, Liese A, Schmitz RA (2016) Highly effective inhibition of biofilm formation by the first metagenome-derived AI-2 quenching enzyme. Front Microbiol 7:1098. https://doi.org/10.3389/fmicb.2016.01098

Xue T, Zhao L, Sun B (2012) LuxS/AI-2 system is involved in antibiotic susceptibility and autolysis in *Staphylococcus aureus* NCTC 8325. Int J Antimicrob Agents 41:85–89. http://dx.doi.org/10.1016/j.ijantimicag.2012.08.016

Yang YX, Xu Z-H, Zhang Y-Q, Tian J, Weng L-X, Wang L-H (2012) A new quorum-sensing inhibitor attenuates virulence and decreases antibiotic resistance in *Pseudomonas aeruginosa*. J Microbiol 50:987–993. https://doi.org/10.1007/s12275-012-2149-7

Yerushalmi SM, Buck ME, Lynn DM, Gabriel N, Meijler MM (2013) Multivalent alteration of quorum sensing in *Staphylococcus aureus*. Chem Commun 49:5177–5179. https://doi.org/10.1039/c3cc41645c

Zang T, Lee BK, Cannonb LM, Ritterb KA, Daia S, Renc D, Wood TK, Zhou ZS (2010) A naturally occurring brominated furanone covalently modifies and inactivates LuxS. Bioorg Med Chem Lett 19:6200–6204. https://doi.org/10.1016/j.bmcl.2009.08.095

Zhang Y, Brackman G, Coenye T (2017) Pitfalls associated with evaluating enzymatic quorum quenching activity: the case of MomL and its effect on *Pseudomonas aeruginosa* and *Acinetobacter baumannii* biofilms. PeerJ 5:e3251. https://doi.org/10.7717/peerj.3251

Chapter 2
Targeting Quorum Sensing Mediated *Staphylococcus aureus* Biofilms: A Proteolytic Approach

Vipin Chandra Kalia, Shikha Koul, Subhasree Ray, and Jyotsana Prakash

Abstract *Staphylococcus aureus* infects human through biofilm formed by the process of quorum sensing. Biofilm confers this human pathogen with high resistance to antibiotics. The inhibition of biofilm forming process or dispersal of already formed biofilm can be the potential targets for treating the diseases. Since, the biofilm is made up of exopolysaccharides, proteins and lipids, the action of proteases can hydrolyse the protein component and disrupt the biofilm matrix. Consequently the bacterium so exposed can be eliminated by low doses of antibacterials.

Keywords Quorum sensing · Proteases · *Bdellovibrio* · *Staphylococcus* · Biofilms

2.1 Introduction

Infectious disease by *Staphylococcus aureus* in humans, is a major cause of concern for Health Departments (Fredheim et al. 2009; Monnappa et al. 2014; Kumar et al. 2016; Fleming and Rumbaugh 2017). The worry emanates from the high level of resistance of this human pathogen to antibiotics like methicillin. In the process of controlling these infectious bacteria, the use of antibiotics has been employed as a first level of defence mechanism (Kalia 2013, 2014). However, the indiscriminate use of antibiotics has led to the evolution of multi-drug resistance in such pathogens. Owing to a strong selection pressure, bacteria develop extraordinary resistance to antibiotics by mechanisms such as high expression of efflux pumps,

V. C. Kalia (✉)
Microbial Biotechnology and Genomics, CSIR – Institute of Genomics and Integrative Biology (IGIB), Delhi, India

Molecular Biology Lab 502; Department of Chemical Engineering, Konkuk University, Seoul, Republic of Korea

S. Koul · S. Ray · J. Prakash
Microbial Biotechnology and Genomics, CSIR – Institute of Genomics and Integrative Biology (IGIB), Delhi, India

© Springer Nature Singapore Pte Ltd. 2018
V. C. Kalia (ed.), *Biotechnological Applications of Quorum Sensing Inhibitors*,
https://doi.org/10.1007/978-981-10-9026-4_2

enzymes for modulating and degrading drugs (Kalia et al. 2014; Koul et al. 2016). This process is further supported by the biofilm formation, which is a direct manifestation of interdependent communal behaviour of bacteria in the ecosystem.

2.2 Quorum Sensing Based Biofilms

Bacteria manifest their unique features, especially those related to pathogenicity including biofilm formation only when they are needed (Archer et al. 2011; Kalia et al. 2017). QS system allows them to regulate the expression of certain genes, which are not beneficial if expressed in an individual bacterium (Kalia and Purohit 2011; Kumar et al. 2013). QS system is mediated through signal molecules, and operates only once bacterial population density has crossed a threshold level. Biofilm formation process involves the following steps: (i) surface attachment, (ii) micro-colony formation, (iii) maturation, and (iv) its dispersal (Boles and Horswill 2011; Chagnot et al. 2013). Biofilm imparts higher tolerance to bacteria against antibiotics and host immune systems (Mootz et al. 2013). In consideration of these challenges, antimicrobial agents with non-microbicidal properties are needed to be researched and utilised against infectious biofilm forming microorganisms. Therefore, the focus is on developing strategies to prevent biofilm formation and on those which are targeted to disintegrate pre-formed biofilms. A direct consequence of disassembly of the biofilm is that it exposes the bacterium, which 'once again' becomes susceptible to lower doses of antimicrobial drugs.

S. aureus has been widely reported to cause infections by forming biofilms. These organisms reside on host tissues and medical implants. Biofilms are surface-attached communities of microbes in a 3-D matrix of exopolysaccharides, proteins, lipids and DNA (Fredheim et al. 2009; Monnappa et al. 2014; Fleming and Rumbaugh 2017). In *S. aureus*, quite a few surface proteins – autolysin, biofilm-associated protein, Fn- and ECM-binding proteins, and clumping factors have been identified (Chagnot et al. 2013; Sugimoto et al. 2013). The proteins – SasC, SasG, Spa and β-toxin, a neutral sphingomyelinase are responsible for the structure of the biofilm matrix, cell and surface interactions (Sugimoto et al. 2013). Gene expression involved in controlling biofilm formation is co ordinated by accessory gene regulator (*agr*) QS system and sigma factor B (*sigB*) acting in response to stress (Archer et al. 2011; Boles and Horswill 2011). In efforts to decipher the role of various factors, which are responsible for *S. aureus* biofilm maturation, mutant strains were generated (Lauderdale et al. 2009). *sigB* deletion mutants lacking sigma factor B activity were identified to significantly affect biofilm maturation process. Here, agr RNAIII levels were higher than wild type strains, and this expression is known to possess antibiofilm activity. The role of proteases was demonstrated by employing protease inhibitors – α-macroglobulin and phenylmethylsulfonyl fluoride, which enabled restoration of biofilm in *sigB* deletion mutants (Lauderdale et al. 2009).

2.3   Disruption of Biofilms by Proteases

The plethora of information on the activities of surface proteins and other protein-aceous compounds during biofilm formation has brought into light the potential role of proteases in inhibiting this process (Chaignon et al. 2007; Flemming and Wingender 2010; Frees et al. 2013; Lister and Horswill 2014; Mukherji et al. 2015; Fleming and Rumbaugh 2017). Among the secretary proteases, SspA (V8 serine protease), ScpA (staphopain) and Aur (auroelysin), have been found to be helpful in disrupting *S. aureus* biofilms (Shaw et al. 2004; Marti et al. 2010; Abraham and Jefferson 2012; Mootz et al. 2013). SspA protease acts by cleaving surface adhesions, especially FnbpAB. It consequently prevents formation of the biofilm structure (O'Neill et al. 2008). Sequenced genome indicates the presence of many proteases in *S. aureus*, some of them having self-cleaving properties, which affect the biofilm integrity. Sigma factor B (ΔsigB) mutants of *S. aureus* were reported to be biofilm negative. Cysteine protease inhibitor E-64 and Staphostatin inhibitors, which act up on extracellular cys-teine proteases SspB and ScpA (also known as termed Staphopains) were identified as those responsible for disrupting already formed biofilm (Mootz et al. 2013).

Esp – a serine protease from *Staphylococcus epidermidis* (a V8 homologue of *S. aureus*) was able to degrade proteins, which were associated with *S. aureus* biofilm formation, especially those responsible for biofilm matrix (Iwase et al. 2010; Sugimoto et al. 2011, 2013). In depth study revealed that Esp protease produced by *S. epidermidis* could degrade about 75 proteins, which also included 11 of those responsible for biofilm formation. The enzyme was effective against human recep-tor proteins of *S. aureus* such as vitronectin, fibrinogen and fibronectin responsible for infection and colonization (Sugimoto et al. 2013). The survival rate of *Caenorhabditis elegans* infected with *S. aureus* was improved with the help of pro-tease produced by *S. epidermidis* (Vandecandelaere et al. 2014).

An interesting case of protease (cysteine proteinase, SpeB) from *Streptococcus pyogenes* – is its ability to degrade a range of substrates: cytokines, extracellular matrix, complement components, immunoglobulins, serum protease inhibitors, and chemokines (Connolly et al. 2011; Nelson et al. 2011). However, the major limitation in its usage is the non-specificity. SpeB can also regulate the streptococcal proteins by releasing them from the bacterial surface (including M protein,) or degrading (such as plasminogen activator streptokinase) them (Nelson et al. 2011). In spite of such a wide range of activities, its ability to inhibit *S. aureus* biofilm has not been reported so far.

Lysostaphin, a glycylglycine endo-peptidase cleaves the pentaglycine linkages of the bacterial peptidoglycan (Wu et al. 2003). It is quite effective in killing *S. aureus* and *S. epidermidis* cells, where MIC90 was found to be 0.001 to 0.064 µg/mL and 12.5 to 64 µg/mL, respectively. Lysostaphin, was found to kill these bacte-rial cells within the biofilm but also could disrupt the extracellular matrix at a low concentration of 1.0 µg/mL. In the absence of Lysostaphin, antibiotics did not shown any biofilm or cellular disruption even at high concentrations (oxacillin, 400 µg/mL; and others like vancomycin and clindamycin at 800 µg/mL) (Wu et al. 2003). Lysostaphin along with the antibiotics – oxacillin showed synergistic effect in inhibiting biofilms formed by MSAA and MRSA strains at sub-biofilm inhibitory concentrations (Walencka et al. 2005).

Biofilm formed by *S. aureus* have been dispersed also through the use of commercially available proteases. Broad range proteases such as Trypsin and Proteinase K resulted in disassembly of biofilms (Boles and Horswill 2008; Mootz et al. 2013). Biofilms could be removed by employing serine proteases – proteinase K and trypsin, which hinder bacterial adherence and formation of the biofilm (Gilan and Sivan 2013; Shukla and Rao 2013; Loughran et al. 2014). The role of serine protease (Esperase HPF, subtilisin) as antifouling agent was shown to be effective against *Microbacterium phyllosphaerae* (a Gram positive bacterium) and a few Gram negative bacteria such as, *Acinetobacter lwoffii, Dokdonia donghaensis* and *Shewanella japonica* (Hangler et al. 2009). Screening of 458 strains belonging to Actinomycetes for restricting *S. aureus* biofilm formation and also for detaching the pre-existing biofilms proved quite productive (Park et al. 2012). The strains – *Streptomyces* sp. BFI250 and *Kribbella* sp. BFI1562 possessing protease (0.1 µg proteinase K/ml) can be exploited as a tool to eradicate staphylococcal infections by more than 80% without causing any bacterial growth inhibition (Park et al. 2012). Biofilms formed by *S. epidermidis* and *S. aureus* were targeted using commercially produced proteases – Alcalase, Flavourzyme and Neutrase. Flavourzyme which had a mixture of endo- and exoproteases obtained from *Aspergillus orizae* turned out to be the most efficient biofilm inhibitor (Elchinger et al. 2014). Neutrase obtained from *Bacillus amyloliquefaciens* had endo-protease activity leading to 72% reduction in bioflm formation by *S. aureus*. Alcalase produced by *Bacillus licheniformis* is a serine endopeptidase, which could disrupt biofilms produced by many bacterial species (Elchinger et al. 2014). Other attempts to inhibit biofilms formed by *Staphylococcus* species involved protease (proteinase K, chymotrypsin and trypsin) and metallo-proteases (serratiopeptidase and carboxypeptidase A) (Artini et al. 2013; Papa et al. 2013). Serratiopeptidase and chymotrypsin showed dramatic reduction in biofilm formation (41 and 64%, respectively) by *S. aureus,* whereas proteinase K was not a strong inhibitor. With *S. epidermidis*, these enzymes resulted in 20–30% reduction in biofilm formation. Serratiopeptidase has been projected as a potential candidate to control *Staphylococcus* biofilm formation process (Artini et al. 2013). Exogenous supplementation of purified proteases – ScpA, SspB and Aur, could disperse pre-formed *S. aureus* biofilms (Loughran et al. 2014).

2.4   Bacteria As Proteolyic Agents

The use of a bacterium as an anti-pathogenic agent is an interesting strategy. *Bdellovibrio bacteriovorus* has the unique capability of attacking Gram-negative group of bacteria (Kadouri and O'Toole 2005; Dwidar et al. 2012, 2013). In order to exploit this predatory behaviour against Gram-positive bacterium such as *S. aureus*, extracellularly produced protease was found to degrade and inhibit biofilms and reduce the pathogenicity of *S. aureus*. The efficiency of the strategy was quite high. An addition of supernatant of *B. bacteriovorus* culture resulted in more than 75% reduction in biofilm formation and could even disperse pre-formed biofilms. LC-MS-MS analysis of supernatant revealed the activities of serine proteases and DNAse (Nijland et al. 2010; Monnappa et al. 2014). These findings opened a new arena for exploiting *B. bacteriovorus* for protecting humans from *S. aureus* infections (Monnappa et al. 2014). Further application of *B. bacteriovorus* has been reported against biofilms produced by *S. aureus* and *Pseudomonas aeruginosa*, which usually infect lungs of cystic fibrosis patients (Iebba et al. 2014). The bacteriolytic behaviour of *B. bacteriovorus* was epibiotic against Gram-positive organisms and periplasmic against Gram-negative organism. It thus acted like a live antibiotic (Iebba et al. 2014). This potential of *Bdellovibrio* has been observed to extend to other biofilm forming pathogens as well: *Aggregatibacter actinomycetemcomitans and Fusobacterium nucleatum*, which cause periodontal disease (Looozen et al. 2015).

2.5 Bacteriophages As Proteolyic Agents

Bacteriophages are known to digest the bacterial cell wall to release its progeny. The lytic enzyme, lysin is effective against gram-positive bacteria in a highly specific manner. These have been demonstrated to control antibiotic resistant bacterial pathogens especially those found on the catheters, mucosal surfaces and infected animal tissues (Curtin and Donlam 2006; Fischetti 2008; Seth et al. 2013). As these bioactive molecules have been envisaged to cause low pressure on these bacteria to generate resistance against them, these are sought after as anti-infective agents. Phage lysins activity against quite a few staphylococci has been demonstrated. Production of tailor-made multi-domain lysins – φ11 lysin and SAL-2, has also been tested for controlling biofilms (Sass and Bierbaum 2007; Son et al. 2010). For maximum hydrolysis of biofilm combined action of the amoidase and endopeptidase domains was necessary (Sass and Bierbaum 2007). The cell wall degrading enzyme SAL-2 could also lyse methicilllin-resistant strains of *S. aureus* (MRSA) (Son et al. 2010). Single-domain lysin – $CHAP_K$ (18.6 kDa), (a cysteine, histidine dependant amidohydrolase/peptidase) a truncated derivative of myoviridae staphylococcal phage K lysis (LysK, 54 kDa) was also employed for controlling biofilms (Fenton et al. 2011 2013). This endopeptidase shows highly specific cleavage in the staphylococcal cell wall peptidoglycan (Becker et al. 2009).

2.6 Conclusion

The process of bacterial infection leading to the manifestation of disease is quite complex. There are numerous possibilities of curtailing this process (Kalia and Purohit 2011; Kalia 2013; Kumar et al. 2013). Here, we have focused on attacking the QS mediated biofilm formation. The QS process involves signal molecules, which interact with receptors and lead to expression of genes responsible for pathogenicity (Kalia et al. 2014; Koul et al. 2016). One can thus prevent the synthesis of signal molecules, inhibit their interaction with receptor or interrupt the gene

Table 2.1 Proteolytic inhibitors to target *Staphylococcus aureus* biofilms

Inhibitor	References
Proteases: SspA (V8 serine protease), ScpA (staphopain) and Aur (auroelysin)	Shaw et al. (2004), Marti et al. (2010), Abraham and Jefferson (2012), Mootz et al. (2013)
Purified proteases – ScpA, SspB and Aur	Loughran et al. (2014)
Esp – a serine protease from *Streptococcus epidermidis* (a V8 homologue of *S. aureus*)	Sugimoto et al. (2013), Vandecandelaere et al. (2014)
Proteases (proteinase K, chymotrypsin and trypsin) and metallo-proteases (serratiopeptidase and carboxypeptidase A)	Artini et al. (2013), Papa et al. (2013)
Endo- and exoproteases: Alcalase, Flavourzyme and Neutrase	Elchinger et al. (2014)
Lysostaphin, a glycylglycine endo-peptidase	Wu et al. (2003), Walencka et al. (2005)
Proteinase K and Trypsin	Boles and Horswill (2008), Gilan and Sivan (2013), Mootz et al. (2013), Shukla and Rao (2013), Loughran et al. (2014)
Protease from Actinomycetes	Park et al. (2012)
Lytic enyzymes from Bacteriophages: φ11 lysin and SAL-2	Sass and Bierbaum (2007), Fischetti (2008), Son et al. (2010)
Lytic enyzymes from Bacteriophages: Single-domain lysin – $CHAP_K$ (18.6 kDa), (a cysteine, histidine dependant amidohydrolase/peptidase	Becker et al. (2009), Fenton et al. (2011, 2013)
Bdellovibrio bacteriovorus	Kadouri and O'Toole (2005), Dwidar et al. (2012, 2013), Iebba et al. (2014), Monnappa et al. (2014)

expression (Begum et al. 2016; Wadhwani et al. 2016; Ahiwale et al. 2017; Azman et al. 2017; Koul and Kalia 2017; Sharma and Lal 2017). However, the manifestation of the disease occurs quite late and the struggle shifts from prevention of biofilm formation to disrupting it and exposing the bacteria. Although, proteins contribute very little to the overall biofilm structure, however, enzyme based hydrolysis of proteins can be effective in curtailing microbial infections. The other possibilities, which can supplement the action of proteases are: DNases, glycoside hydrolases, peptides, dispersal molecules or bioactive molecules (Ma et al. 2012; Gui et al. 2014; Hernández-Saldaña et al. 2016; Jeyanthi and Velusamy 2016; Varsha et al. 2016) (Table 2.1).

Acknowledgments We are thankful to the Director of CSIR – Institute of Genomics and Integrative Biology (CSIR-IGIB), and CSIR-HRD (ES Scheme No. 21(1022)/16/EMR-II) for providing the necessary funds, facilities and moral support. Authors are also thankful to Academy of Scientific & Innovative Research (AcSIR), New Delhi and University Grants Commission (JP). This work was supported by Brain Pool grant (NRF-2018H1D3A2001756) by National Research Foundation of Korea (NRF) to work at Konkuk University.

References

Abraham NM, Jefferson KK (2012) *Staphylococcus aureus* clumping factor B mediates biofilm formation in the absence of calcium. Microbiology 158:1504–1512. https://doi.org/10.1099/mic.0.057018-0

Ahiwale SS, Bankar AV, Tagunde S, Kapadnis BP (2017) A bacteriophage mediated gold nanoparticle synthesis and their anti-biofilm activity. Indian J Microbiol 57:188–194. https://doi.org/10.1007/s12088-017-0640-x

Archer NK, Mazaitis MJ, Costerton JW, Leid JG, Powers ME, Shirtliff ME (2011) *Staphylococcus aureus* biofilms: properties, regulation, and roles in human disease. Virulence 2:445–459. https://doi.org/10.4161/viru.2.5.17724

Artini M, Papa R, Scoarughi GL, Galano E, Barbato G, Pucci P, Selan L (2013) Comparison of the action of different proteases on virulence properties related to the staphylococcal surface. J Appl Microbiol 114:266–277. https://doi.org/10.1111/lam.12038

Azman CA-S, Othman I, Fang C-M, Chan K-G, Goh B-H, Lee L-H (2017) Antibacterial, anti-cancer and neuroprotective activities of rare actinobacteria from mangrove forest soils. Indian J Microbiol 57:177–187. https://doi.org/10.1007/s12088-016-0627-z

Becker SC, Dong S, Baker JR, Foster-Frey J, Pritchard DG, Donovan DM (2009) LysK CHAP endopeptidase domain is required for lysis of live staphylococcal cells. FEMS Microbiol Lett 294:52–60. https://doi.org/10.1111/j.1574-6968.2009.01541.x

Begum IF, Mohankumar R, Jeevan M, Ramani K (2016) GC–MS analysis of bioactive molecules derived from *Paracoccus pantotrophus* FMR19 and the antimicrobial activity against bacterial pathogens and MDROs. Indian J Microbiol 56:426–432. https://doi.org/10.1007/s12088-016-0609-1

Boles BR, Horswill AR (2008) Agr-mediated dispersal of *Staphylococcus aureus* biofilms. PLoS Pathog 4:e1000052. https://doi.org/10.1371/journal.ppat.1000052

Boles BR, Horswill AR (2011) Staphylococcal biofilm disassembly. Trends Microbiol 19:449–445. https://doi.org/10.1016/j.tim.2011.06.004

Chagnot C, Zorgani MA, Astruc T, Desvaux M (2013) Proteinaceous determinants of surface colonization in bacteria: bacterial adhesion and biofilm formation from a protein secretion perspective. Front Microbiol 4:303. https://doi.org/10.3389/fmicb.2013.00303

Chaignon P, Sadovskaya I, Ragunah C, Ramasubbu N, Kaplan JB, Jabbouri S (2007) Susceptibility of staphylococcal biofilms to enzymatic treatments depends on their chemical composition. Appl Microbiol Biotechnol 75:125–132. https://doi.org/10.1007/s00253-006-0790-y

Connolly KL, Roberts AL, Holder RC, Reid SD (2011) Dispersal of group a streptococcal biofilms by the cysteine protease SpeB leads to increased disease severity in a murine model. PLoS One 6:e18984. https://doi.org/10.1371/journal.pone.0018984

Curtin JJ, Donlan RM (2006) Using bacteriophages to reduce formation of catheter-associated biofilms by *Staphylococcus epidermidis*. Antimicrob Agents Chemother 50:1268–1275. https://doi.org/10.1128/AAC.50.4.1268-1275.2006

Dwidar M, Hong S, Cha M, Jang J, Mitchell RJ (2012) Combined application of bacterial predation and carbon dioxide aerosols to effectively remove biofilms. Biofouling 28:671–680. https://doi.org/10.1080/08927014.2012.701286

Dwidar M, Leung BM, Yaguchi T, Takayama S, Mitchell RJ (2013) Patterning bacterial communities on epithelial cells. PLoS One 8:e67165. https://doi.org/10.1371/journal.pone.0067165

Elchinger PH, Delattre C, Faure S, Roy O, Badel S, Bernardi T, Taillefumier C, Michaud P (2014) Effect of proteases against biofilms of *Staphylococcus aureus* and *Staphylococcus epidermidis*. Lett Appl Microbiol 59:507–513. https://doi.org/10.1111/lam.12305

Fenton M, Ross RP, McAuliffe O, O'Mahony J, Coffey A (2011) Characterization of the staphylococcal bacteriophage lysin CHAP$_K$. J Appl Microbiol 111:1025–1035. https://doi.org/10.1111/j.1365-2672.2011.05119.x

Fenton M, Keary R, McAuliffe O, Ross RP, O'Mahony J, Coffey A (2013) Bacteriophage-derived peptidase eliminates and prevents *Staphylococcal* biofilms. Int J Microbiol 2013:625341. https://doi.org/10.1155/2013/625341

Fischetti VA (2008) Bacteriophage lysins as effective antibacterials. Curr Opin Microbiol 11:393–400. https://doi.org/10.1016/j.mib.2008.09.012

Fleming D, Rumbaugh KP (2017) Approaches to dispersing medical biofilms. Microrganisms 5:15. https://doi.org/10.3390/microorganisms5020015

Flemming HC, Wingender J (2010) The biofilm matrix. Nat Rev Microbiol 8:623–633. https://doi.org/10.1038/nrmicro2415

Fredheim EG, Klingenberg C, Rohde H, Frankenberger S, Gaustad P, Flaegstad T, Sollid JE (2009) Biofilm formation by *Staphylococcus haemolyticus*. J Clin Microbiol 47:1172–1180. https://doi.org/10.1099/mic.0.057018-0

Frees D, Brøndsted L, Ingmer H (2013) Bacterial proteases and virulence. Subcell Biochem 66:161–192. https://doi.org/10.1007/978-94-007-5940-4_7

Gilan I, Sivan A (2013) Effect of proteases on biofilm formation of the plastic-degrading actinomycete *Rhodococcus ruber* C208. FEMS Microbiol Lett 342:18–23. https://doi.org/10.1111/1574-6968.12114

Gui Z, Wang H, Ding T, Zhu W, Zhuang X, Chu W (2014) Azithromycin reduces the production of α-hemolysin and biofilm formation in *Staphylococcus aureus*. Indian J Microbiol 54:114–117. https://doi.org/10.1007/s12088-013-0438-4

Hangler M, Burmølle M, Schneider I, Allermann K, Jensen B (2009) The serine protease Esperase HPF inhibits the formation of multispecies biofilm. Biofouling 25:667–674. https://doi.org/10.1080/08927010903096008

Hernández-Saldaña OF, Valencia-Posadas M, de la Fuente-Salcido NM, Bideshi DK, Barboza-Corona JE (2016) Bacteriocinogenic bacteria isolated from raw goat milk and goat cheese produced in the Center of México. Indian J Microbiol 56:301–308. https://doi.org/10.1007/s12088-016-0587-3

Iebba V, Totino V, Santangelo F, Gagliardi A, Ciotoli L, Virga A, Ambrosi C, Pompili M, De Biase RV, Selan L, Artini M, Pantanella F, Mura F, Passariello C, Nicoletti M, Nencioni L, Trancassini M, Quattrucci S, Schippa S (2014) *Bdellovibrio bacteriovorus* directly attacks *Pseudomonas aeruginosa* and *Staphylococcus aureus* cystic fibrosis isolates. Front Microbiol 5:280. https://doi.org/10.3389/fmicb.2014.00280

Iwase T, Uehara Y, Shinji H, Tajima A, Seo H, Takada K, Agata T, Mizunoe Y (2010) *Staphylococcus epidermidis* Esp inhibits *Staphylococcus aureus* biofilm formation and nasal colonization. Nature 465:346–349. https://doi.org/10.1038/nature09074

Jeyanthi V, Velusamy P (2016) Anti-methicillin resistant *Staphylococcus aureus* compound isolation from halophilic *Bacillus amyloliquefaciens* MHB1 and determination of its mode of action using electron microscope and flow cytometry analysis. Indian J Microbiol 56:148–157. https://doi.org/10.1007/s12088-016-0566-8

Kadouri D, O'Toole GA (2005) Susceptibility of biofilms to *Bdellovibrio bacteriovorus* attack. Appl Environ Microbiol 71:4044–4051. https://doi.org/10.1128/AEM.71.7.4044-4051.2005

Kalia VC (2013) Quorum sensing inhibitors: an overview. Biotechnol Adv 31:224–245. https://doi.org/10.1016/j.biotechadv.2012.10.004

Kalia VC (2014) Microbes, antimicrobials and resistance: the battle goes on. Indian J Microbiol 54:1–2. https://doi.org/10.1007/s12088-013-0443-7

Kalia VC, Purohit HJ (2011) Quenching the quorum sensing system: potential antibacterial drug targets. Critic Rev Microbiol 37:121–140. https://doi.org/10.3109/1040841X.2010.532479

Kalia VC, Wood TK, Kumar P (2014) Evolution of resistance to quorum-sensing inhibitors. Microb Ecol 68:13–23. https://doi.org/10.1007/s00248-013-0316-y

Kalia VC, Prakash J, Koul S, Ray S (2017) Simple and rapid method for detecting biofilm forming bacteria. Indian J Microbiol 57:109–111. https://doi.org/10.1007/s12088-016-0616-2

Koul S, Kalia VC (2017) Multiplicity of quorum quenching enzymes: a potential mechanism to limit quorum sensing bacterial population. Indian J Microbiol 57:100–108. https://doi.org/10.1007/s12088-016-0633-1

Koul S, Prakash J, Mishra A, Kalia VC (2016) Potential emergence of multi-quorum sensing inhibitor resistant (MQSIR) bacteria. Indian J Microbiol 56:1–18. https://doi.org/10.1007/s12088-015-0558-0

Kumar P, Patel SKS, Lee JK, Kalia VC (2013) Extending the limits of *Bacillus* for novel biotechnological applications. Biotechnol Adv 31:1543–1561. https://doi.org/10.1016/j.biotechadv.2013.08.007

Kumar R, Koul S, Kumar P, Kalia VC (2016) Searching biomarkers in the sequenced genomes of *Staphylococcus* for their rapid identification. Indian J Microbiol 56:64–71. https://doi.org/10.1007/s12088-016-0565-9

Lauderdale KJ, Boles BR, Cheung AL, Horswill AR (2009) Interconnections between Sigma B, *agr*, and proteolytic activity in *Staphylococcus aureus* biofilm maturation. Infect Immun 77:1623–1635. https://doi.org/10.1128/IAI.01036-08

Lister JL, Horswill AR (2014) *Staphylococcus aureus* biofilms: recent developments in biofilm dispersal. Front Cell Infect Microbiol 2:38. https://doi.org/10.3389/fcimb.2014.00178

Loozen G, Boon N, Pauwels M, Slomka V, Rodrigues Herrero E, Quirynen M, Teughels W (2015) Effect of *Bdellovibrio bacteriovorus* HD100 on multispecies oral communities. Anaerobe 35(Pt A):45–53. https://doi.org/10.1016/j.anaerobe.2014.09.011

Loughran AJ, Atwood DN, Anthony AC, Harik NS, Spencer HJ, Beenken KE, Smeltzer MS (2014) Impact of individual extracellular proteases on *Staphylococcus aureus* biofilm formation in diverse clinical isolates and their isogenic sarA mutants. Microbiology 3:897–909. https://doi.org/10.1002/mbo3.214

Ma Y, Xu Y, Yestrepsky BD, Sorenson RJ, Chen M, Larsen SD, Sun H (2012) Novel inhibitors of *Staphylococcus aureus* virulence gene expression and biofilm formation. PLoS One 7:e47255. https://doi.org/10.1371/journal.pone.0047255

Marti M, Trotonda MP, Tormo-Mas MA, Vergara-Irigaray M, Cheung AL, Lasa I, Penades JR (2010) Extracellular proteases inhibit protein-dependent biofilm formation in *Staphylococcus aureus*. Microbes Infect 12:55–64. https://doi.org/10.1016/micinf.2009.10.005

Monnappa AK, Dwidar M, Seo JK, Hur JH, Mitchell RJ (2014) *Bdellovibrio bacteriovorus* inhibits *Staphylococcus aureus* biofilm formation and invasion into human epithelial cells. Sci Rep 4:3811. https://doi.org/10.1038/srep03811

Mootz JM, Malone CL, Shaw LN, Horswill AR (2013) Staphopains modulate *Staphylococcus aureus* biofilm integrity. Infect Immun 81:3227–3238. https://doi.org/10.1128/IAI.00377-13

Mukherji R, Patil A, Prabhune A (2015) Role of extracellular proteases in biofilm disruption of gram positive bacteria with special emphasis on *Staphylococcus aureus* biofilms. Enz Eng 4:126. https://doi.org/10.4172/2329-6674.1000126

Nelson DC, Garbe J, Collin M (2011) Cysteine proteinase SpeB from *Streptococcus pyogenes* – a potent modifier of immunologically important host and bacterial proteins. Biol Chem 392:1077–1088. https://doi.org/10.1515/BC-2011-208

Nijland R, Hall MJ, Burgess JG (2010) Dispersal of biofilms by secreted, matrix degrading, bacterial DNase. PLoS One 5:e15668. https://doi.org/10.1371/journal.pone.0015668

O'Neill E, Pozzi C, Houston P, Humphreys H, Robinson DA, Loughman A, Foster TJ, O'Gara JP (2008) A novel *Staphylococcus aureus* biofilm phenotype mediated by the fibronectin-binding proteins, FnBPA and FnBPB. J Bacteriol 190:3835–3850. https://doi.org/10.1128/JB.00167-08

Papa R, Artini M, Cellini A, Tilotta M, Galano E, Pucci P, Amoresano A, Selan L (2013) A new antiinfective strategy to reduce the spreading of antibiotic resistance by the action on adhesion-mediated virulence factors in *Staphylococcus aureus*. Microb Pathog 63:44–53. https://doi.org/10.1016/j.micpath.2013.05.003

Park JH, Lee JH, Kim CJ, Lee JC, Cho MH, Lee J (2012) Extracellular protease in Actinomycetes culture supernatants inhibits and detaches *Staphylococcus aureus* biofilm formation. Biotechnol Lett 34:655–661. https://doi.org/10.1007/s10529-011-0825-z

Sass P, Bierbaum G (2007) Lytic activity of recombinant bacteriophage φ11 and φ12 endolysins on whole cells and biofilms of *Staphylococcus aureus*. Appl Environ Microbiol 73:347–352. https://doi.org/10.1128/AEM.01616-06

Seth AK, Geringer MR, Nguyen KT, Agnew SP, Dumanian Z, Galiano RD, Leung KP, Mustoe TA, Hong SJ (2013) Bacteriophage therapy for *Staphylococcus aureus* biofilm-infected wounds: a new approach to chronic wound care. Plast Reconstr Surg 131:225–234. https://doi.org/10.1097/PRS.0b013e31827e47cd

Sharma A, Lal R (2017) Survey of (Meta)genomic approaches for understanding microbial community dynamics. Indian J Microbiol 57:23–38. https://doi.org/10.1007/s12088-016-0629-x

Shaw L, Golonka E, Potempa J, Foster SJ (2004) The role and regulation of the extracellular proteases of *Staphylococcus aureus*. Microbiology 150:217–228. https://doi.org/10.1099/mic.0.26634-0

Shukla SK, Rao TS (2013) Dispersal of Bap-mediated *Staphylococcus aureus* biofilm by proteinase K. J Antibiot (Tokyo) 66:55–60. https://doi.org/10.1038/ja.2012.98

Son JS, Lee SJ, Jun SY, Sj Y, Kang Sh PHR, Kang JO, Choi YJ (2010) Antibacterial and biofilm removal activity of a podoviridae *Staphylococcus aureus* bacteriophage SAP-2 and a derived recombinant cell-wall-degrading enzyme. Appl Microbiol Biotechnol 86:1439–1449. https://doi.org/10.1007/s00253-009-2386-9

Sugimoto S, Iwase T, Sato F, Tajima A, Shinji H, Mizunoe Y (2011) Cloning, expression and purification of extracellular serine protease Esp, a biofilm-degrading enzyme, from *Staphylococcus epidermidis*. J Appl Microbiol 111:1406–1415. https://doi.org/10.1111/j.1365-2672.2011.05167.x

Sugimoto S, Iwamoto T, Takada K, Okuda K, Tajima A, Iwase T, Mizunoe Y (2013) *Staphylococcus epidermidis* Esp degrades specific proteins associated with *Staphylococcus aureus* biofilm formation and host-pathogen interaction. J Bacteriol 195:1645–1655. https://doi.org/10.1128/JB.01672-12

Vandecandelaere I, Depuydt P, Nelis HJ, Coenye T (2014) Protease production by *Staphylococcus epidermidis* and its effect on *Staphylococcus aureus* biofilms. Pathog Dis 70:321–331. https://doi.org/10.1111/2049-632X.12133

Varsha KK, Nishant G, Sneha SM, Shilpa G, Devendra L, Priya S, Nampoothiri KM (2016) Antifungal, anticancer and aminopeptidase inhibitory potential of a phenazine compound produced by *Lactococcus* BSN307. Indian J Microbiol 56:411–416. https://doi.org/10.1007/s12088-016-097-1

Wadhwani SA, Shedbalkar UU, Singh R, Vashisth P, Pruthi V, Chopade BA (2016) Kinetics of synthesis of gold nanoparticles by *Acinetobacter* sp. SW30 isolated from environment. Indian J Microbiol 56:439–444. https://doi.org/10.1007/s12088-016-0598-0

Walencka E, Sadowska B, Rozalska S, Hryniewicz W, Rózalska B (2005) Lysostaphin as a potential therapeutic agent for staphylococcal biofilm eradication. Pol J Microbiol 54:191–200. doi:NA

Wu JA, Kusuma C, Mond JJ, Kokai-Kun JF (2003) Lysostaphin disrupts *Staphylococcus aureus* and *Staphylococcus epidermidis* biofilms on artificial surfaces. Antimicrob Agents Chemother 47:3407–3414. https://doi.org/10.1128/AAC.47.11.3407-3414.2003

Chapter 3
Alternative Strategies to Regulate Quorum Sensing and Biofilm Formation of Pathogenic Pseudomonas by Quorum Sensing Inhibitors of Diverse Origins

P. Sankar Ganesh and V. Ravishankar Rai

Abstract Pathogenic *Pseudomonas* species produce virulence elements and form biofilm through quorum sensing (QS) system. Clinical infections caused by *Pseudomonas aeruginosa* are becoming increasingly tough to treat on account of wide spread drug resistance. The drugs have lost their efficacy due to the virulence elements and biofilms, which are responsible for increased severity of the infection. The search for novel drugs has increased and novel modes of action against Gram-negative pathogenic bacteria are been much more of important. Plants secondary metabolites are widely used for treating the many bacterial diseases. Plant-derived anti-QS and anti-biofilm compounds that do not negatively affect the growth of the bacterial cells, but rather attenuates the QS controlled virulence factors, that might allow the host defense to act more effectively to washout the *P. aeruginosa* infection. In this review mainly focus on overview of pathogenicity, QS controlled virulence factors and biofilm formation of *P. aeruginosa* infection. This review describes a brief account of QS inhibitors and anti-biofilm compounds, which exhibit alternative medicine possible for treating drug resistant *P. aeruginosa*.

Keywords Infections · *Pseudomonas aeruginosa* · Cell to cell communication · Biofilm · Phytochemical · Virulence factors · *Caenorhabditis elegans*

Abbreviations

AHL	N-acyl homoserine lactone
AIDS	Acquired immune diseases
AIs	Autoinducers
C_{12} – HSL	N-dodecanoyl-homoserine lactone
C_4-HSL	N-butyrl –L-homosrine lactone

P. Sankar Ganesh · V. Ravishankar Rai (✉)
Department of Studies in Microbiology, University of Mysore, Mysore, Karnataka, India

© Springer Nature Singapore Pte Ltd. 2018　　　　　　　　　　　　　　　　33
V. C. Kalia (ed.), *Biotechnological Applications of Quorum Sensing Inhibitors*,
https://doi.org/10.1007/978-981-10-9026-4_3

CNS   Cerebrospinal fluid
CO_2 Carbon dioxide
3-Oxo-C_{12}-HSL   N-3-oxo-dodecanoyl-homoserine lactone
EPS   Exopolymeric substance
EVD   External ventricular drainage
HCN   Hydrogen cyanide
ICU   Intensive care unit
QS   Quorum sensing
QSI   Quorum sensing inhibitor
SAM   S-adenosyl methione

3.1 Introduction

Pseudomonas aeruginosa has been recognized as an opportunistic bacterial pathogen. It normally lives in moist environments and has the ability to colonize ecological niches, soil, water, plant and animals. *P. aeruginosa* can infect and proliferate inside host cells such as plants, animals, humans and nematodes. *P. aeruginosa* is normally found in the human body as a normal microflora. It can also cause severe infections in patients (Sadikot et al. 2005). In the hospital environment, *P. aeruginosa* infects immune-compromised patients more rapidly than others (Zawacki et al. 2004). However, healthy people can also develop mild infections especially after exposure to water, contaminated hot tubs and swimming pools (Kiska and Gilligan 2003).

In the health care setting, *P. aeruginosa* cause nosocomial infections leading to significantly high morbidity and mortality, particularly in people with weakened immune system (Kang et al. 2003; Khan et al. 2015). This bacterium is a frequently recovered pathogen from patients in intensive care unit (ICU) and it is mainly transmitted through the hospital coworkers and colonized patients (Chastre and Fagon 2002). *P. aeruginosa* is well adapted and adhered to the respiratory tract, particularly in patients with chronic bronchitis disease or patients who admitted to ICU ward (Bonten et al. 1996). *P. aeruginosa* can infect patients suffering from cystic fibrosis, severe burns and acquired immune diseases syndrome (AIDS) (Sadikot et al. 2005). In addition, *P. aeruginosa* also result in infections of the urinary tract via insertion of urinary catheters, instrumentation and surgery (Nicolle 2014). This organism causes blood stream infections in human and the mortality rate is reported to be greater than 20% as the patients receive inappropriate antimicrobial chemotherapy (Micek et al. 2005). *P. aeruginosa* meningitis is a rare infection, but the organisms are able to infect patients who are immunocompromised or have undergone neurological surgery (Fong and Tomkins 1985). It has the ability to cause meningitis and brain abscesses as it invades through the central nervous system (CNS) from paranasal sinus or inner ear (Dando et al. 2014). Another route of

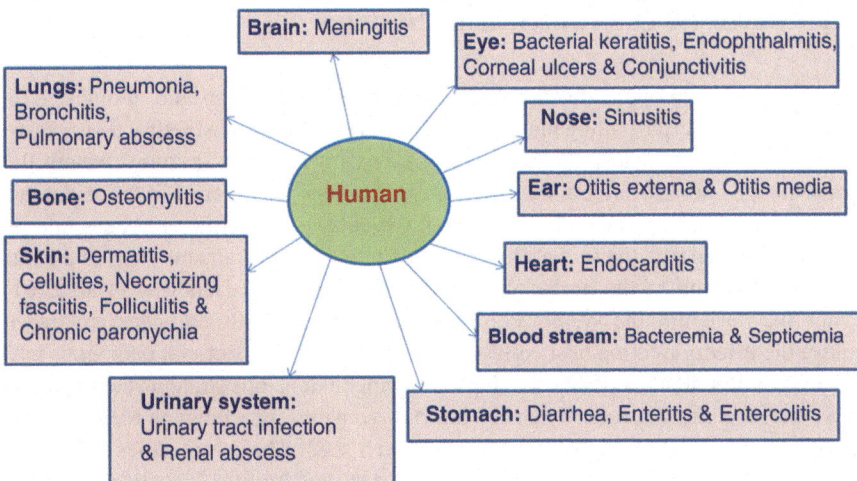

Fig. 3.1 *Pseudonmonas aeruginosa* infection occurs in various parts of the human body

infection is directly by mean of head trauma, head injury and neurological surgery (Dando et al. 2014). *P. aeruginosa* infection is major problem in patients who are undergo external ventricular drainage (EVD) and external cerebrospinal fluid (CSF) via drainage catheters (Fig. 3.1) (Juhi et al. 2009).

P. aeruginosa causes overwhelming infection in the human eye. *P. aeruginosa* can colonize the ocular epithelium and it can proliferate by producing enzymes such as alkaline protease, exotoxin A and elastase, which rapidly lead to destructive infection that may result in permanent loss of vision such as dacroystitis, endophthalmitis, conjunctivitis, infection of corneal ulcers and keratitis (Henry et al. 2012). *P. aeruginosa* rarely infects healthy tissues, but when externally cut it can easily invade the tissues and proliferate to damage the tissue. *P. aeruginosa* is an extracellular pathogen but it also known to invade epithelial cell during infection. After invading, *P. aeruginosa* is able to proliferate and adhere onto the host tissue. It releases virulence factors and toxins resulting in the disease by modifying the immune response, and damaging the host tissue (Moghaddam et al. 2014). *P. aeruginosa* can also colonize medical and surgical devices, such as catheters, implants, prosthetic joints etc. (Donlan 2001). The primary contamination of the medical devices mainly occurs by inoculation from the patient's skin or membranes. Another way of the pathogens may acquire from the surgical instrument by clinical staff and healthcare workers (Donlan 2001). *P. aeruginosa* is notorious for resistance to antibiotics due to adaptive resistance mechanisms. Moreover, the emergence of the *P. aeruginosa* with modified virulent factors makes the treatment difficulties (Fernandez and Hancock 2012). Virulence factors and toxins produced by *P. aeruginosa* are the major causes of infection in human beings. Their production is mediated by small signaling molecules (autoinducer) called quorum sensing (QS) (Bjarnsholt and Givskov 2007).

3.2 Bacterial Quorum Sensing Systems

The bacterial pathogens are normally associated with human and animals. Even though the bacterial pathogens continue contact with the host cells, a successful infection is rare, because the host cell possesses a potential immune system that prevents the cause of infection in such way. The first line order of immune defence mechanisms such as macrophages prevent the entry of bacterial pathogens into the host cells. Similarly, the second line order of immune mechanisms such as an antibody which bind to the bacterial receptors and destroy it (Janeway et al. 2001). Once the host immune system compromised, the bacterial pathogens are free to cause the infection to the host cells. On the other side, the bacterial pathogens have been a counter attack to the host cells; the pathogen develops strategies to overcome host defence mechanisms by using quorum sensing signalling molecules called autoinducers (AIs) (Van Delden and Iglewski 1998). These molecules are largely responsible for producing virulence factors, toxins and biofilm formation in the host cells (Van Delden and Iglewski 1998). Quorum sensing (QS) is a phenomenon, which allows bacteria to communicate and coordinate with the other bacterial cells via a small diffusible signalling molecule called autoinducers (AIs) (LaSarre and Federle 2013). The signaling molecules of Gram-positive bacteria are usually polypeptides which regulate control the bacterial cell density, whereas acyl-homo serine lactones (AHLs) act as signals in the Gram-negative bacteria (LaSarre and Federle 2013). In Gram- positive bacteria, once threshold concentration level is reached, the peptide molecules, bind to a cognate membrane – bound histidine kinase receptor (Ng and Bassler 2009), where was in Gram- negative bacteria, AHLs regulate and coordinate the expression of specific genes (Ng and Bassler 2009) (Fig. 3.2).

The first QS system was reported in the luminescent marine organism – *Vibrio* (Nealson et al. 1970). In Gram-negative bacteria, a number of different QS systems are involved in controlling the population density, virulence factors and toxin

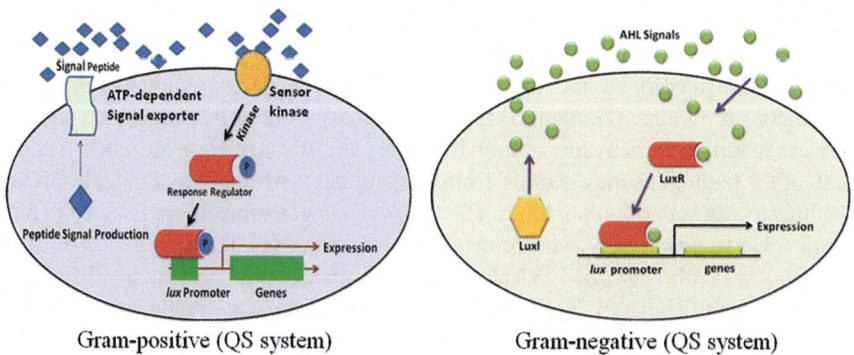

Gram-positive (QS system)   Gram-negative (QS system)

Fig. 3.2 Schematic diagram of Gram-positive and Gram-negative bacterial quorum sensing system

production (Deep et al. 2011; LaSarre and Federle 2013). The archetype QS system in marine bacteria of *Vibrio fisheri* and *V. harveyi* are controlled by homologs of regulatory proteins, LuxI and LuxR (Water and Bassler 2006; Miyashiro and Ruby 2012). QS system mediates bioluminescence, and are products of genes regulated by lux operon (Engebrecht et al. 1983; Miyashiro and Ruby 2012). The prototypes of QS system have been also seen in human pathogens such as *Yersinia pseudotuberculosis, Escherichia coli* and *P. aeruginosa* and a plant associated bacteria such as *Rhizobium leguminosarum, Erwinia carotovora* and *Ralstonia solanacearum* (Fugua et al. 1994; Cha et al. 1998; Atkinson et al. 1999; Sperandio et al. 2002).

3.3 Quorum Sensing in Pseudomonas Aeruginosa

P. aeruginosa operates two QS systems: *las* and *rhl* through AHLs- C4-HSL and N-(3-oxododecanoyl)-L-homoserine lactone (3O-C12-HSL) (Gambello and Iglewski 1991; Pearson et al. 1994, 1995, 1997; Pesci et al. 1997). The QS controls virulence factors involved in cellular toxicity and acute infection (Brint and Ohman 1995; Pearson et al. 1995; Sawa et al. 1998; Wade et al. 2005). Another system regulates amb genes for synthesis of IQS (2-(2-hydroxyphenyl)-thiazole-4-carbaldehyde) molecule, involved in phosphate-stress response encountered during infection leading to cystic fibrosis (Lee et al. 2013) (Fig. 3.3).

3.4 Virulence Factors and Pathogenesis

The QS signaling molecules produce many virulence factors important for infections (Van Delden and Iglewski 1998). Treatment of these infections is becoming difficult due to emergence of drug- resistant strains and biofilms (Tenover 2006). *P. aeruginosa* secretes toxic material and virulence factors such as pyocyanin, pyoverdin, superoxide dismutase, LasA protease, LasB elastase, rhamnolipid, phospholipase C, hydrogen cyanide (HCN), exotoxin A, exoenzyme and colony morphology, swarming motility and biofilms formation to cause diseases and evade the host defenses (Sawa et al. 1998; Hoge et al. 2010; Balasubramanian et al. 2013; Beceiro et al. 2013). Pyocyanin and precursor molecules inhibit the functioning of respiratory cilia in the lungs and alter the expression of immunomodulatory proteins, thus allowing the pathogen to evade host's innate and acquired immunity (Rada and Leto 2013). Lipopolysaccharide endotoxin is a complex glycolipid present in the outer membrane of *P. aeruginsoa.* It responsible for antigenicity, inflammatory response and mediates interactions with conventional antibiotics (Nau and Eiffert 2002). In *P. aeruginosa* infection, considerable damages to host tissue also facilitating their dissemination to distinct sites of infection in blood vessels (Fig. 3.4). The production of these virulence factors limit the treatment options which are currently available as broad-spectrum antibiotics (Wagner et al. 2016).

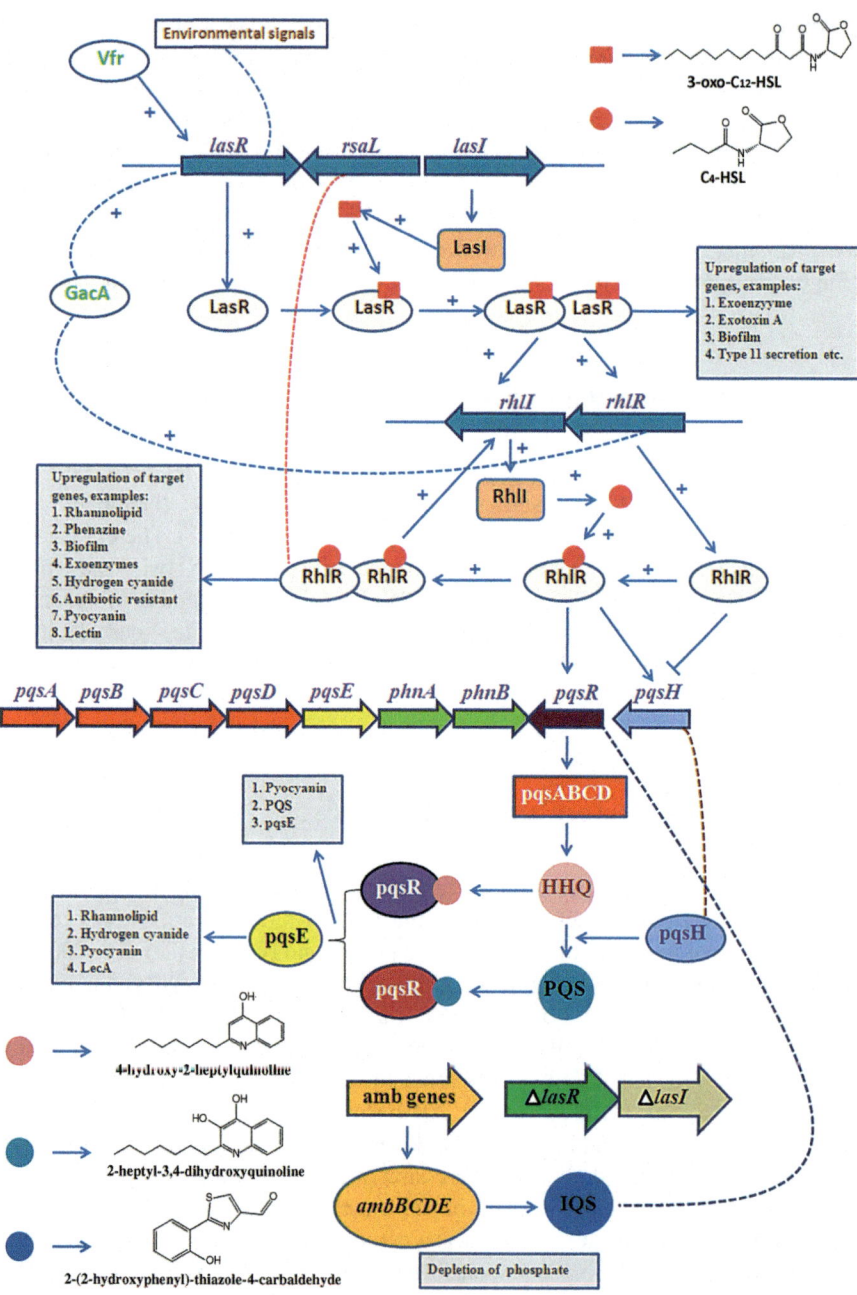

Fig. 3.3 Quorum sensing systems in *Pseudomonas aeruginosa*: (1) Lasl – 3-oxo-C12-homoserine lactone (HSL) (2) Rhll – N-(3-Hydroxybutanoyl) – L -homoserine lactone, (3) PqsABCDH – 2-heptyl-3-hydroxy-4-quinolone (PQS) (4) AmbBCDE – 2-(2-hydroxyphenyl)-thiazole-4-carbaldehyde (IQS)

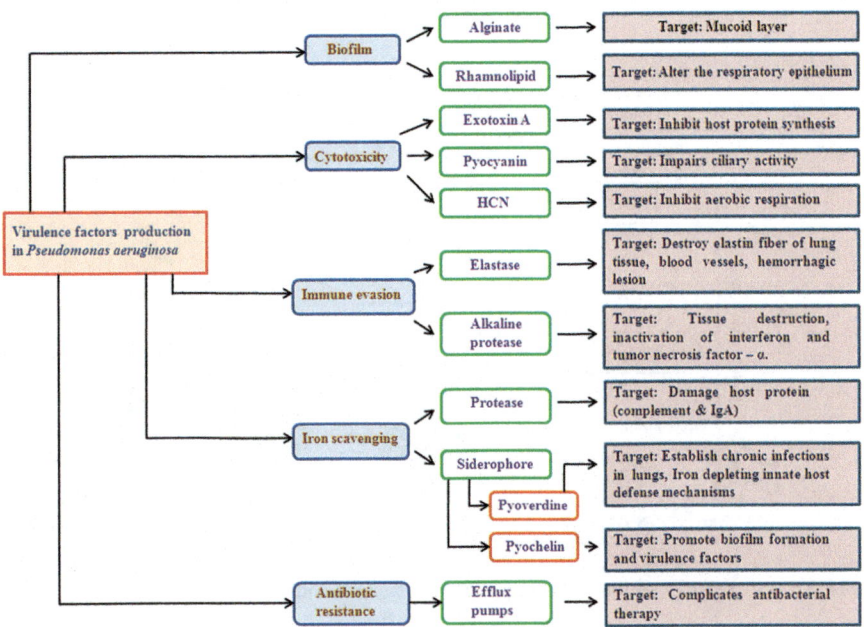

Fig. 3.4 Schematic diagram show virulence factors production and enhancing the ability of *P. aeruginosa* to cause diseases

3.5 Biofilm Formation in *Pseudomonas aeruginosa*

Biofilms are made up of exopolymeric substance (EPS) which hold microbial cells together onto the surface (Donlan 2002). It is made up of polysaccharides, DNAs, proteins, lipids etc., which contribute to the structural scaffold providing the bacterial cell attachment and form a biofilm formation in the host cells (Vu et al. 2009). The EPS is an essential component in the formation of the biofilm matrix. It is mainly involved in the shaping of the cell structure, promoting attachment to surfaces, maintaining the biofilm structure, evading host immune responses and it providing the resistance to antibiotics (Davies 2003; Orgad et al. 2011). Biofilm protects bacteria against immune system and cause acute infections such as pulmonary illnesses (Hoiby et al. 2010; Wang et al. 2014). *P. aeruginosa* has a surface – sensing system – Wsp is a regulatory circuit which consist of membrane-bound methyl accepting chemotoxins protein. WspR regulate the accumulation of cyclic –di-GMP, leading to increased the EPS production (Huangyutitham et al. 2003). Cyclic –di-GMP is another molecule, which regulates the numerous virulence factors, cell cycle, dispersion, motility, differentiation and biofilm formation. In high cell density level, cyclic –di-GMP trigger the formation of biofilm or sessile lifestyle and at low density level are mainly associated with planktonic bacterial cell existence and motility attachment (Hengge 2009). *P. aeruginosa* produces different polysaccharides such as alginate, to stabilize the biofilm (Ryder et al. 2007). Swarming

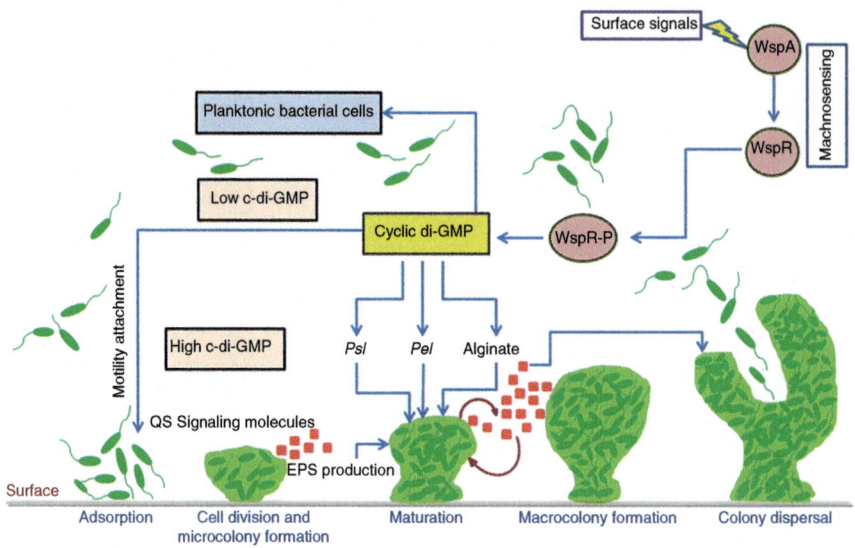

Fig. 3.5 Life cycle of biofilm formation of *Pseudomonas aeruginosa*

motility is under the control of *rhl* system (Ramsamiravaka et al. 2015). The production of rhamnolipids – biosurfactants facilitates *P. aeruginosa* biofilm formation (Pamp and Nielson 2007). These virulence factors results in bacterial persistence and reduced sensitivity to antimicrobials (Wagner et al. 2016) (Fig. 3.5).

3.6   Emergence of Antibiotics Resistance in *Pseudomonas aeruginosa*

Bacteria are naturally resistant to antimicrobial agents, because do not possess molecular target of the antibiotics or acquired resistance occurs through mutation, mobile genetic elements such as plasmids and transposons (Alekshun and Levy 2007). There are four major mechanisms contributing to antibiotics resistance in *P. aeruginosa* includes acquired resistance includes drug inactivation, target modification, reduced permeability and efflux pumps (Sun et al. 2015). *P. aeruginosa* has intrinsic ability to be resistant to quite a few antimicrobial agents and continuous to exposure to many antibiotics results in multi- drug resistant *P. aeruginosa* strains (Morita et al. 2013). It can develop resistance to all conventional antibiotics via different mechanisms. Intrinsic resistant to many antibiotics (e.g. aminoglycosides, β-lactams and fluoroquinoles) is due to low permeability of the outer membrane, which prevents the entry of conventional antimicrobial drugs, due to the presence of porins in this bacterium (Lister et al. 2009). *P. aeruginosa* typically expresses efflux systems, which mainly contributing to antibiotic resistance (Hocquet et al. 2006). This multidrug efflux systems effectively resistances to different classes of

Table 3.1 Antibiotics resistance mechanisms in *Pseudomonas aeruginosa*

Resistance Mechanism	Class of Resistance	Antibiotics	References
Efflux pumps	Intrinsic	MexAB–OprM, MexEF–OprN, MexXY–OprM, MexCD–OprJ, (cephalosporins, quinolones, carbapenems, ureidopenicillins, aminoglycosides)	Hocquet et al. (2006)
β-lactamases	Intrinsic	AmpC (penicillins)	Moya et al. (2009)
Outer membrane impermeability	Intrinsic	OprB, OprD, OprF, (carbapenems, quinolones, aminoglycosides)	Wolter et al. (2009)
Targeted mutation	Acquired	DNA topoisomerase (quinolones)	Ruppe et al. (2015)
		MexZ (quinolones, cefapimes, aminoglycosides), DNA gyrase	
Horizontal transfer	Acquired	ESBLs, Metallo-β-lactamases, (carbapenems, cephalosporins, penicillins)	Shaikh et al. (2015)
Membrane changes	Adaptive	Modification of Lipid A (polymyxins, aminoglycosides)	Lee et al. (2016)

antimicrobial drugs includes β-lactams (and β-lactamase inhibitors), sulfonamides, chloramphenicol, fluoroquinolones, trimethoprim, quinolones, macrolides, tetracycline, fourth-generation cephalosporins, cefpirome, aminoglycosides, and tetracyclines (Nikaido 2010) (Table 3.1).

3.7 Mechanism of *P. aeruginosa* Quorum Sensing (QS) Inhibition

P. aeruginosa has two QS system, which has been extensively studied. AHLs signaling molecules enable *P. aeruginosa* to cause disease (Schuster and Greenberg 2006). The treatment of *P. aeruginosa* infection usually fails due to their host of QS mediated virulence factors expression (Bonte et al. 2007). Interfering with QS system may be a promising strategy to get rid of severe chronic infections (Smith and Iglewski 2003; Bonte et al. 2007).

The main target of QS inhibition in *P. aeruginosa* can be LasR or RhlR gene activation. AHL analogues can act as antagonists for QS signals (Smith and Iglewski 2003). There have been a number of ways to inhibit the QS signaling molecules and virulence factors such as signal binding, degradation of the signaling molecules, competitive inhibition and genetic regulation systems. Catalytic antibodies are hydrolysis of AHLs molecules (QS system) through signal binding. Antibodies are capable of hydrolyzing AHL molecule of $3O-C_{12}$-HSL and it inhibiting the production of pyocyanin in *P. aeruginosa* (Marin et al. 2007). Furanones derivatives isolated from marine algae have successfully inhibited QS regulatory system and cause a reduction in LasR activity (Manefield et al. 1999). QS inhibitors are able to suppress S-adenosyl methione (SAM), which leads to lower the production of C12-AHL (Hoang et al. 2002). Genetic modification of upstream regulators such as Vfr and GacA are involved in the producing many virulence factors (Albus et al. 1997; Reimmann et al. 1997). QS Inhibitor involved genetic modification in upstream regulators, which leads to reducing subsequent production of virulence factors (Smith and Iglewski 2003). Similarly, Lactonases enzymes have significantly decreased production of virulence factors expression in *P. aeruginosa* (Rajesh and Rai 2014).

3.8   Medicinal Plants

Plants have been reported to be the main source of medicinal constituents and have an important role in the maintenance of human health. More than 80% of the world population use active constituents of plant extracts as folk medicine. Similarly, more than 50% of modern drugs have been isolated from the natural source of medicinal plants (Kirbag and Zengin 2009). Antimicrobial screening of medicinal plants' secondary metabolites represents a starting point for antibacterial drug discovery. Some of the plants' secondary metabolites have anti-quorum sensing properties as well. The phytochemical studies have attracted the attention of plant scientists and researchers to develop new drugs for attenuation of drug resistance in clinical strains of Gram-negative and Gram-positive bacteria.

3.8.1   Secondary Metabolites of Medicinal Plants and Targets of Quorum Sensing System

Plants release secondary metabolites which are rich sources of bioactive compounds to protect against pathogen attack and herbivores (Bennett and Wallsgrove 1994). The plants and shrubs release several secondary metabolites: alkaloids, flavonoids, quinines, terpenoids, tannins, steroids and phenylpropanoids to counterattack of bacterial pathogens. These secondary metabolites are the most important source for the discovery of a new formulation of drugs (Verpoorte 1998). Similarly, plant secondary metabolites are economically significant as flavour, fragrances, pesticides and food additives (Husain et al. 2012). Generally, many biologically active compounds are present in plant leaves, because of the favourable storage for desired compounds. Plant fruits also contain a considerable amount of active ingredients which are used in pharmaceutical and food industries (Verpoorte 1998). Similarly, plant parts such as roots, flowers, seeds and stem bark can be employed for extracting therapeutic compounds (Chan et al. 2012). In health care industry, plant secondary metabolites are predominantly used for treating various diseases on account of having antimicrobial, anti-inflammatory, anti-cancer and antioxidant activities

(Wallace 2004; Lu et al. 2013; Yan et al. 2013; Dyduch-Sieminska et al. 2015). Plant secondary metabolites serve as QS antagonists to inhibit the progression of infections. Plants produce various metabolites which can structurally mimic the AHL signaling molecule, and such binding is able to inactivate signal-medicated QS. In addition to this, plant secondary metabolites can degrade the AHLs and consequently inhibit biofilm formation (Koh et al. 2013). Similarly, fruits, roots and seed also show anti-QS activities. The bioactive compound such as rosmarinic acid present in sweet basil plant inhibits the pyocyanin production, elastase, proteases and biofilm formation (Walker et al. 2004). The plant derived secondary metabolites such as phenolic compound, flavonoids, trepans, alkaloids shows the antimicrobial activity. Similarly, at the sub lethal concentration the plant metabolites show anti-QS, anti-virulence and antibiofilm activities, it however does not affect microbial growth (Slobodnikova et al. 2016). The phenolic compound such as esculin, psoralen and nodakenetin isolated from the plant sources potentially prevent formation of biofilm (Zeng et al. 2008; Ding et al. 2011). Flavonoids are a group of natural compounds that exhibit pharmaceutical activities ranging from anti-QS and antibiofilm activities. Recently, flavonoids rich fraction isolated from the herb *Centella asiatica* showed inhibition virulence factors like pyocyanin, elastolytic, proteolytic activities, swimming and swarming motility and biofilm formation (Vasavi et al. 2016). Solenopsin A alkaloid compound purified from *Solenopsis invicta* which inhibits functioning of QS in *P. aeruginosa* (Park et al. 2008).

3.9 Quorum Sensing Inhibitory Compounds

Bacterial QS system regulated toxins and virulence factors enable pathogenic bacteria to evade host immune defence mechanisms and increase the pathogenicity (Brint and Ohman 1995; Ribert and Cossart 2015). In such cases, researchers found QSI compounds which would be attenuate the QS system would reduce the pathogenicity and as a result that is easy to treat the bacterial infections.

3.10 Types of QSIs

3.10.1 Naturally Produced QSIs

Many organisms co-exist in the natural ecosystems. The bacterial community primarily coordinate and compete with other bacterial community. In the natural ecosystem vast number of quorum sensing inhibitor (QSI) compounds can be isolated from the plants and shrubs (LaSarre and Federle 2013). The plants have co-existed with QSI bacteria and do release QSIs in order to reduce the infectivity of pathogenic bacteria. The first QSI such as halogenated furonone compound was recorded from the marine macroalgae (*Delisea pulchra*) (Givskov et al. 1996). The main mode of action of halogenated furanones compounds prevent the bacterial colonization and biofilms formation by interfering with QS mediated genes (e.g., *Serratia liguefaciens* MG1) (Rasmussen et al. 2000; Wu et al. 2004).

3.10.2 Plant Based Quorum Sensing Inhibitors

Plant secondary metabolites also act as quorum-sensing inhibitors because of resemblance in their chemical structure and also ability to degrade signal receptors (LuxR/LasR) in Gram-negative bacteria (Teplitski et al. 2011). Plants produce a number of phenolic bioactive compounds which inhibit growth of pathogenic bacteria. When compared to all other classes of plant bioactive compounds, a vast number of phenolic compounds inhibit the QS system in pathogenic microorganisms (Kalia 2013). QS inhibitors are small molecules which regulate QS controlled gene expression with no toxic effects (Rasmussen and Givskov 2006). Plants and shrubs are able to interfere with the QS systems of Gram-negative organisms of *P. aeruginosa*, secreting signal degrading enzymes and signal mimics (Braeken et al. 2008).

The bioactive compounds of many plants with structural similarity to QS signaling molecules also affect the functioning of signal receptors (Vattem et al. 2007). The Indian system of food habitat, the plants, shrubs and fruits such as garlic, pea, tomato and soybean secrete QSI compounds that can inhibit the AHL receptors (Teplitski et al. 2000). *Vitis vinifera* (grape fruit) extract contain bioactive compound furocoumarins strongly inhibit autoinducer activities, and attenuate the biofilm formation (Girennavar et al. 2008). Similarly, some of the plant extract such as *Prunus armeniaca, Imperata cylindrical, Nelumbo nucifera, Prunella vulgaris, Panax notoginseng*, and *Punica granatum* have potentially inhibited the QS signaling molecules in *Pseudomonas aeruginosa* PAO1 and Gram-negative micoorganisms (Koh and Tham 2011). Isothiocyanate compound produced from *Armoracia rusticana* (horseradish) inhibit the expression of the *lasB-gfp* fusion, which compete with AHL signaling molecules of regulator proteins (Jakobsen et al. 2012). *Emblica officinalis* release secondary metabolites such as pyrogallol is a bioactive compound which exhibit antagonism activity against auto inducer-2 (AI-2) (Ni et al. 2008). Adonizio et al. (2008a) have reported that vescalagin and castalgin compounds from *Conocarpus erectus* show lower AHL production and reduced QS mediated virulence factors. List of plants that are able to inhibit QS system is given in Table 3.2.

Fruits and herbs including blackberry, raspberry, grape, thyme, ginger, turmeric and oregano moderately inhibit QS controlled virulence factors of Gram-negative bacteria and there is no evidence till date (Truchado et al. 2015). But some of the herbs and vegetables were identified to possess anti-QS properties. Similarly, some of the plants derivatives and furanones such as ellagic acids, tannic acids and galic acids have been shown to attenuate the QS system in *P. aeruginosa* (Sarabhai et al. 2013). *Psidium guajava* leaf extract effectively inhibits the AHL production (QS

Table 3.2 Lists of quorum sensing inhibitors isolated from medicinal plants treated against *Pseudomonas aeruginosa*

Plant s name	Parts	Tested organisms	Phytochemical/ compounds	Targets	References
Bucida buceras	Leaves	*Pseudomonas aeruginosa* PAO1	Crude extract	Inhibited the LasA protease and biofilms formation	Adonizio et al. (2008a)
Conocarpus erectus	Leaves	*Pseudomonas aeruginosa* PAO1	Crude extract	Effect of LasA protease, and LasB elastase activity	Adonizio et al. (2008a)
Tetrazygia bicolor	Leaves	*Pseudomonas aeruginosa* PAO1	Crude extract	Inhibited the LasA protease, and biofilms formation	Adonizio et al. (2008a)
Callistemon viminalis	Leaves	*Pseudomonas aeruginosa* PAO1	Crude extract	Effect of LasA protease, LasB elastase and biofilm formation	Adonizio et al. (2008a)
Mangifera indica	Leaves	*Pseudomonas aeruginosa* PAO1	Crude extract	Inhibited the swarming motility	Zahin et al. (2010)
Terminalia catappa	Leaves	*Pseudomonas aeruginosa*	Crude extract	Controlled the biofilm formation	Taganna et al. (2011)
Terminalia bellerica	Leaves	*Pseudomonas aeruginosa*	Crude extract	Inhibited the production of biofilm formation	Ganesh and Rai (2017)
Psidium guajava	Leaves	*Pseudomonas aeruginosa*	Crude extract	Inhibited swarming motility	Ghosh et al. (2014)
Psidium guajava	Leaves	*Pseudomonas aeruginosa*	Quercetin-3- O -arabinoside	Effect of pyocyanin production	Vasavi et al. (2016)

system) in bioreporter strain of *P. aeruginosa* (Ghosh et al. 2014). Similarly, the aqueous extract of *Moringa oleifera* L. consist of bioactive compounds such as gallic acid, quercetin and chlorogenic acid found to inhibit the AHL molecules in Gram-negative bacteria (Singh et al. 2009). *Ocimum sanctum, Ananas comosus* (pineapple), *Musa paradisiaca,* and *Manilkara zapota* provided to be QSI against pyocyanin production, protease, LasA staphylolytic, elastase and biofilms formation (Musthafa et al. 2010). Besides, ethanol extract of *Centella asiatica* found to be inhibitors of QS system in pathogenic *P. aeruginosa* (Vasavi et al. 2016). *Vaccinium oxycoccos* – derived proanthocynidins compound which proved to inhibit AHL synthases and QS transcriptional regulators in *P. aeruginosa* PA14 (Maisuria et al. 2016). Similarly, methanol extract of *Terminalia bellerica* (L) have inhibited the QS and biofilm formation (Ganesh and Rai 2017). Besides, the methanol extracts of *Terminalia chebula* (fruit) decrease the production of virulence factors and biofilms formation in *P. aeruginosa* PAO1 (Sarabhai et al. 2013). Zahin et al. (2010) have reported that, out of 24 medicinal plants, *Hemidesmus indicus* (L.), *Holarrhena antidysenterica, Manggifera indica, Punica granatum, Psoralea corylifolia* shows effectively inhibits violacein production and found to be direct and indirect interference on QS system by *P. aeruginosa* PAO1 (Zahin et al. 2010). Tannins isolated from *Terminalia catappa* inhibits biofilms formation (Taganna et al. 2011).

3.10.3   Essential Oils Antagonist

Essential oils isolated from plant materials are widely used to treat many pathogenic microorganisms, fungi and virus. Essential oils are volatile and hydrophobic nature. Essential oils are generally classified in two groups such as terpenes and terpenoids. Terpenes are volatile unsaturated hydrocarbons which mainly made up of isoprene units are normally found in essential oils of plants (Sterrett 1962). Several compounds of terpenes present in the essential oils such as terpinene, sabinene, p-cymene and pinene. Plant related bioactive compounds derived from isoprene units (Sterrett 1962). Examples of terpenoids present in essential oils are eugenyl acetate, geraniol, citronellol, menthol, carvacrol, geranyl acetate, thymol, linalool, carvone, geranial and 1, 8-cineole. The phenylpropanoids include cinnamaldehyde, phenylpropanoids, chavicol, eugenol, estragole, cinnamyl alcohol, methyl eugenols, and methyl cinnamate (Turek and Stintzing 2013). Essential oils have been shown to be effective antimicrobial agents. Essential oils at the lowest concentration level have shown to be effective anti-QS agents against the pathogenic bacteria. Essential oils such as rosemary, tea tree, juniper and several essential oils showed intensive anti-QS effect. Many compounds such as eugenol (Husain et al. 2013), cinnamaldehyde (Kim et al. 2015b) and vanillin (Plyuta et al. 2013) extracted from plant-derived essential oils inhibit the cell-cell communication and biofilm formation mediated by AHL signaling molecules produced by pathogenic bacteria *P. aeruginosa.* Essential oils of *Syzygium aromaticum* (clove) oil, *Cinnamomum verum* (cinnamon) oil, *Lavandula angustifolia* (lavender) oil, *Mentha balsamea* (peppermint) oil and *Murraya koenigii* essential oil showed high or moderate anti-QS effect (Khan et al. 2009; Ganesh and Rai 2015a). Similarly, essential oils from *Curcuma longa Cuminum cyminum, Cinnamomum verum, Rosmarinus officinalis,* and *Syzygium aromaticum* proved significant anti-QS activity by inhibiting pyocyanin production in *P. aeruginosa* (Ganesh and Rai 2015b). The extracted compounds of essential oils from the plants includes *Ocotea* sp. (α-pine), *Swinglea glutinosa* (β-pinene), *Elettaria cardamomun* (cineol), *Zingiber officinale* (α-zingiberene) and *Minthostachys molis* (pulegone) inhibits the short chain AHL molecules of QS system by *Escherichia coli* (pJBA132) (Jaramillo-Colorado et al. 2012). Ganesh and Rai (2015b) reported that at the lowest concentration essential oils (supercritical CO_2 extraction method) have potentially inhibited the violacein production in pyocyanin production in clinical isolate of *P. aeruginosa.* Sepahi et al. (2015) reported that *Ferula asafetida* (ferula) oil and *Dorema aucheri* (dorema) oil possess anti-QS activity at 25 μg/ml of concentration which resulted in lowering the production of pyocyanin, elastase, pyoveridine and the formation of biofilm *P. aeruginosa* (Fig. 3.6 and Table 3.3).

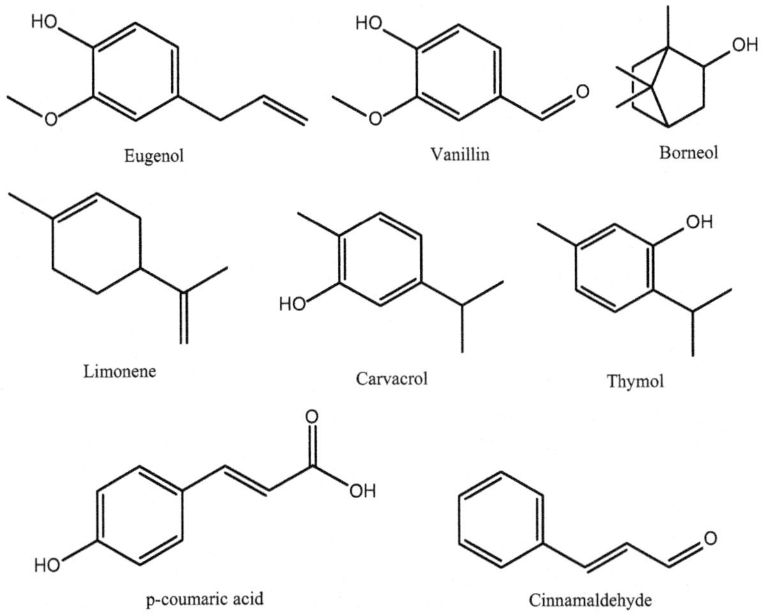

Eugenol   Vanillin   Borneol

Limonene   Carvacrol   Thymol

p-coumaric acid   Cinnamaldehyde

Fig. 3.6 Anti-QS compounds present in plant derived essential oils

Table 3.3 Lists of quorum sensing inhibitors isolated from plant derived essential oils treated against *Pseudomonas aeruginosa*

Plant derived essential oils	Tested organisms	Essential oil/ compounds	Targets	References
Cinnamon verum oil	*Pseudomonas aeruginosa* PAO1	Essential oil	Effects of biofilm formation	Kalia et al. (2015)
Mentha piperita oil	*Pseudomonas aeruginosa* PAO1	Essential oil	Inhibited the biofilm formation	Hussain et al. (2015)
Ferula asafoetida L. oil	*Pseudomonas aeruginosa*	Essential oil	Inhibited the biofilm formation, pyocyanin, pyoverdine and elastase production.	Sepahi et al. (2015)
Dorema aucheri Bioss oil	*Pseudomonas aeruginosa*	Essential oil	Inhibited the biofilm formation, pyocyanin, pyoverdine and elastase production.	Sepahi et al. 2015
Citrus reticulate	*Pseudomonas aeruginosa*	Essential oil	Inhibited the biofilm formation, elastase	Luciardi et al. (2016)
Murraya koenigii oil	*Pseudomonas aeruginosa*	Essential oil	Inhibited the QS and Biofilm formation	Ganesh and Rai (2015a)

3.10.4 Quorum Sensing Inhibitors from Marine Organisms

Marine ecosystem is one of the rich sources of diverse bioactive compounds, QSI discovered from the marine ecosystem especially from marine organisms is scanty. In the recent year more number of bioactive compounds (QSIs) are isolated from various marine sources that includes sponges (*Svenzea tubulosa* and *Ircinia felix*), micro algae (*Asparagopsis taxiformis*), bryozoa, coral (*Eunicea laciniata*) associated bacteria and algae, which inhibit AHL system in Gram-negative bacteria (Peters et al. 2003; Jha et al. 2013; Quintana et al. 2015). Bioactive compounds interfere and completely attenuate the QS system. QSI compounds such as metabolites-manoalide, manoalide monoacetate, and secomanolaide obtained from marine sponge of *Luffariella variabilis* which shows anti-QS activity and antibiofilm activities in *P. aeruginosa* (Skindersoe et al. 2008). Similarly, *Ahnfeltiopsis flabelliformis* (macroalga) synthesize non-lactone QS inhibitors (floridoside, betonicine and isethionic acid), which inhibit QS system (Kim et al. 2007).

3.10.5 Synthetic Quorum Sensing Inhibitors

The highest numbers of synthetic phenolic compounds are reported to inhibit the AHL molecules and biofilms formation when compared to all other classes. Synthetic furanone such as (Z)-5 (bromomethylene)-2(5H)-furanone and (Z)-4-bromo-5-(bromomethylene)-2(5H)-furanone have been found to be inhibiting AHL mediated virulence factors in *P. aeruginosa* infected lungs of mice (Wu et al. 2004). Similarly, the plant derived synthetic compounds such as eugenol and cinnamaldehyde also inhibit the AHL mediated biofilm formation in *P. aeruginosa* (Niu and Afre Gilbert 2006; Zhou et al. 2013). A synthetic analogue of methyl anthranilate compound, which repressed the *Pseudomonas* quinolone signal (PQS) production, has also reduced the elastase production in *P. aeruginosa*. Mustafa et al. (2012a) reported that 2,5-piperazinedione compound interrupt the binding of 3-oxo-C12-HSL to its LasR receptor protein. It resulted in reduced in the production of pyocyanin, extracellular polymeric substance (EPS) and swimming motility in *P. aeruginosa* PAO1 (Fig. 3.7). A synthetic analogue of phenylacetic acid inhibited autoinducer (AI-1) regulated pathogenicity in *P. aeruginosa*. This compound was effective inhibiting QS dependent production of pyocyanin, swarming motility, protease, and elastase in *P. aeruginosa* PAO1 (Mustafa et al. 2012b). Similarly, QSIs such as rosmarinic acid, naringin and morin compounds significantly inhibited the production of elastase, proteases, hemolysin and biofilms formation in *P. aeruginosa*. *In-slico* docking studies these compounds to inhibit the QS system in *P. aeruginosa* (Annapoorani et al. 2012) (Table 3.4).

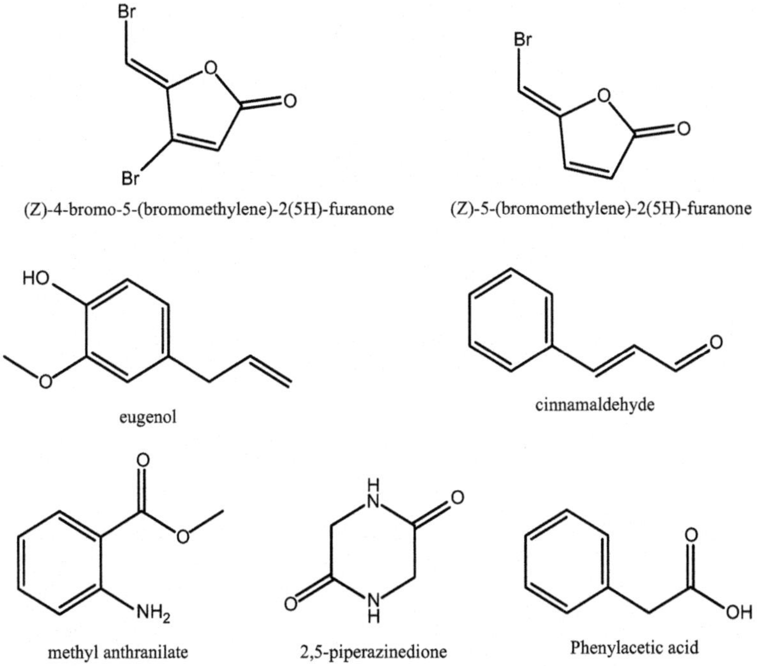

(Z)-4-bromo-5-(bromomethylene)-2(5H)-furanone   (Z)-5-(bromomethylene)-2(5H)-furanone

eugenol   cinnamaldehyde

methyl anthranilate   2,5-piperazinedione   Phenylacetic acid

Fig. 3.7 Synthetic compounds effectively inhibited the quorum sensing system and biofilm formation in Gram-negative bacteria

3.10.6 Natural and Synthetic Inhibitors of Bacterial Biofilms

Nature continues to inspire the discovery of novel bioactive compounds with interesting structure and biological activity. These compounds have served as scaffold for developing synthetic therapeutic agents. Various plants compounds and plant derived compounds can also inhibit the biofilm formation. The methanol extract of *Andrographis paniculata* has potentially inhibited biofilms formation in clinical isolate of *P. aeruginosa* (Murugan et al. 2011). Ginger extract effectively inhibited the biofilms formation in *P. aeruginosa* as a result indicated that the ginger extract reduced the c-di-GMP concentration (Kim and Park 2013). Natural analogue such as manoalide, penicillic acid, and patulin have found to be inhibit biofilms formation in *P. aeruginosa*. Natural halogenated furanones (C-30 and C-56) have exhibit biofilms reduction and target the AHLs systems in *P. aeruginosa* (Wu et al. 2004). Azithromycin antibiotic is derived from *Saccharopolyspora erythraea* which has significant QS and biofilm inhibitory effect (Tateda et al. 2001). Ursolic acid is a plant-derived biofilm inhibitor, which was identified and purified from 13,000 plants and their various parts. At the lowest concentration of ursolic acid (10 μg/ml)

Table 3.4 Lists of quorum sensing inhibitors (synthetic compounds) treated against *Pseudomonas aeruginosa*

Synthetic compound	Tested organisms	Targets	References
Methyl anthranilate	*Pseudomonas aeruginosa*	Effect of *Pseudomonas* quinolone signal PQS	Calfee et al. (2001)
Furanones (C-30 and C-56)	*Pseudomonas aeruginosa*	Inhibited the biofilm formation and AHL system	Wu et al. (2004)
Cinnamaldehyde	*Pseudomonas aeruginosa*	Inhibited the biofilm formation	Niu and Afre Gilbert (2006)
2,5-Piperazinedione	*Pseudomonas aeruginosa* PAO1	Inhibited the production of pyocyanin, protease, elastase, exopolysaccharide and swimming motility	Mustafa et al. (2012a)
Phenylacetic acid	*Pseudomonas aeruginosa* PAO1	Inhibited the production of pyocyanin and protease	Mustafa et al. (2012b)
Eugenol	*Pseudomonas aeruginosa*	Inhibited the biofilm formation	Zhou et al. (2013)
Tannic acid	*Pseudomonas aeruginosa*	Inhibited the AHL production by RhlI,	Chang et al. (2014)
Salicylic acid	*Pseudomonas aeruginosa*	Attenuated the biofilms formation, swarming motility	Chang et al. (2014)
(z)-5-octylidenethiazolidine-2,4-dione	*Pseudomonas aeruginosa*	Inhibited the biofilm formation and swarming motility	Lidor et al. (2015)

reduced biofilm formation in *P. aeruginosa* PAO1 was observed (Kalia 2013). Similarly, salicylic acid inhibits AHL signaling molecule and biofilm formation by and Pseudomonas sp. (Bandara et al. 2006).

The natural phenol compounds such as anacartic acid, tannic acid and cardanol can suppress biofilm formation (Jagani et al. 2009). Ellagic acid compound derived from *Terminalia chebula* decrease lasIR and rh1IR genes expression leading to the reduction in production of virulence factors (Sarabhai et al. 2013). ε-viniferin compound isolated from *Carex pumila*, has inhibits biofilm formation of enterohemorrhagic *E. coli* O157:H7 and *P. aeruginosa* (Cho et al. 2013). Cinnamon bark oil derived from *Cinnamomum verum*, have been found to decrease the production of biofilm, toxins and virulence factors at a concentration of 0.01%, v/v (Kim et al. 2015b). Similarly, 5-hydroxy-1-(4-hydroxy-3-methoxyphenyl)-3-decanone) reduces exoprotease, rhamnolipid and biofilm formation via attenuate the AHL system (Kim et al. 2015a). *Croton nepetaefolius* derived compound (casbane diterpene) significantly inhibits biofilm formation in clinical strain of *P. aeruginosa* without affecting the planktonic cell growth (Carneiro et al. 2010). 2-aminoimidazoles compound conjugated with triazole that has antibiofilm and antibiofouling properties (Rogers et al. 2010). Peppermint (*Mentha piperita*) oil at below the minimum inhibitory concentrations attenuated virulence factors and biofilm formation in *Aeromonas hydrophila* and *P. aeruginosa* (Hussain et al. 2015). Pimenta et al. (2013) have synthesized a library of ethyl gallate, hexyl gallete, octadecyl gallete, (*E*)-*N'*-(4-bromo-benzylidene) benzohydrazide and (2*E*)-1-(3′, 4'-dimethoxyphenyl)-3-(2-naphthyl)-2-propen-1-one compounds which inhibit QS system and biofilm formation in *Enterococcus faecalis, P. aeruginosa, Staphylococcus aureus, Staphylococcus epidermidis* and *Streptococcus mutans*. Similarly, S-phenyl-L-cysteine sulfoxide and its breakdown product of diphenyl disulfide have been found to be potent antibiofilm agents. This compound did not affect planktonic cell growth but influenced the biofilm production (Cady et al. 2012). Pejin et al. (2015) have reported that phytol compound at a sub-MIC level found to inhibit the biofilms formation in *P. aeruginosa*. Similarly, Qu et al. (2016) have reported that norspermidine synthetic compound inhibited QS related gene expression, swarming motility and biofilms formation in *P. aeruginosa* and also norspermidine has potentially prevented the mature biofilms formation on medical devices.

3.10.7   *Host – Pathogen Interaction (Caenorhabditis elegans – Pseudomonas Aeruginosa)*

Caenorhabditis elegans used for host-pathogen interaction and bacterial virulence as well as pharmaceutical drug delivery (Kaletta and Hengartner 2006) (Fig. 3.8). Two major models of infection such as paralysis and fast killing methods have been developed for checking the infection of *P. aeruginosa* on *C. elegans* (Gallagher and Manoil 2001). Under rich physiological medium, *P. aeruginosa* can kill the nematodes rapidly. The fast killing is mediated by the synthesis of phenazines (pyocyanin pigment) (Gallagher and Manoil 2001) which may act through the production of active oxygen species (Mahajan-miklos et al. 1999). In paralytic killing, *C. elegans* has grown on brain heart infusion medium with *P. aeruginosa* become paralyzed

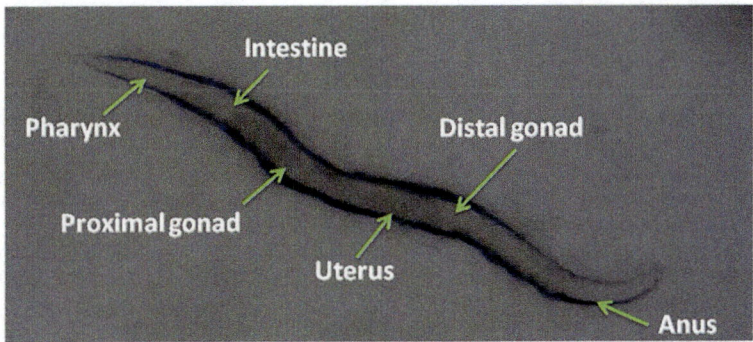

Fig. 3.8 *Caenorhabditis elegans* N2 infection model

and subsequently die (Darby et al. 1999; Gallagher and Manoil 2001). The *P. aeruginosa* mediated killing of *C. elegans* mainly depends on QS-controlled virulence factors (Gallagher and Manoil 2001). Essential oils, plant extracts and purified compounds that attenuate *P. aeruginosa* QS system, can also be expected to reduce the production of virulence factors, toxins and subsequent death of *C. elegans* (Adonizio et al. 2008b; Yu et al. 2014; Hussain et al. 2015; Ganesh and Rai 2016).

3.11 Conclusion and Opinion

In healthcare settings, treatment of infections by *P. aeruginosa* infections are major hurdles as it produces a wide variety of QS signaling molecules, virulence factors and toxins which can influence disease progression. Moreover, this organism has the ability to resist many conventional antibiotics. The development of effective strategies to combat QS and biofilm are challenging task. There are some anti-pathogenic compounds which have been used for the treatment of the *P. aeruginosa* infections. The currently used synthetic anti-pathogenic drugs are not stable and toxic to the host cells. Anti-QS and anti-biofilm agents from natural sources could be less toxic, stable and more specific compared to the synthetic inhibitors. Therefore, there is an increasing need to explore new anti-QS and anti-biofilm compounds of natural origins. Natural products are a rich source of novel antibiotic and anti-pathogenic agents. Moreover, they do not disturb the host cell system. The natural agents could reduce the resistance and suppress the QS system. In alternative therapy, plant-based active compounds can potentially inactivate the QS system and reduce the production of biofilm by infectious bacteria. In our opinion, the anti-pathogenic plant-based drugs need to be investigated for their the mode of action, molecular mechanisms and drug targets of inhibition of QS and biofilm pathway both *in vitro* and *in vivo* level for exploring the possibility of developing them as better drugs of tomorrow to combat *P. aeruginosa* infections in healthcare system.

References

Adonizio A, Kong KF, Mathee K (2008a) Inhibition of quorum sensing-controlled virulence factor production in *Pseudomonas aeruginosa* by South Florida plant extracts. Antimicrob Agents Chemother 52:198–203. https://doi.org/10.1128/AAC.00612-07

Adonizio A, Leal SM, Ausubel FM, Mathee K (2008b) Attenuation of *Pseudomonas aeruginosa* virulence by medicinal plants in a *Caenorhabditis elegans* model system. J Med Microbiol 57:809–813. https://doi.org/10.1099/jmm.0.47802-0

Albus A, Pesci E, Runyen-Janecky L, West S, Iglewski B (1997) Vfr controls quorum sensing in *Pseudomonas aeruginosa*. J Bacteriol 179:3928–3935. https://doi.org/10.1128/jb.179.12.3928-3935.1997

Alekshun MN, Levy SB (2007) Molecular mechanisms of antibacterial multidrug resistance. Cell 128:1037–1050. https://doi.org/10.1016/j.cell.2007.03.004

Annapoorani A, Umamageswaran V, Parameswari R, Pandian SK, Ravi AV (2012) Computational discovery of putative quorum sensing inhibitors against LasR and RhlR receptor proteins of *Pseudomonas aeruginosa*. J Comput Aided Mol Des 26:1067–1077. https://doi.org/10.1007/s10822-012-9599-1

Atkinson S, Throup JP, Stewart GS, Williams P (1999) A hierarchical quorum-sensing system in *Yersinia pseudotuberculosis* is involved in the regulation of motility and clumping. Mol Microbiol 33:1267–1277. https://doi.org/10.1046/j.1365-2958.1999.01578.x

Balasubramanian D, Schneper L, Kumari H, Mathee K (2013) A dynamic and intricate regulatory network determines *Pseudomonas aeruginosa* virulence. Nucleic Acids Res 41:1–20. https://doi.org/10.1093/nar/gks1039

Bandara MB, Zhu H, Sankaridurg PR, Wilcox MD (2006) Salicylic acid reduces the production of several potential virulence factors of *Pseudomonas aeruginosa* associated with microbial keratitis. Invest Ophthalmol Vis Sci 47:4453–4460. https://doi.org/10.1167/iovs.06-0288

Beceiro A, Tomas M, Bou G (2013) Antimicrobial resistance and virulence: a successful or deleterious association in the bacterial world. Clin Microbiol Rev 28:185–230. https://doi.org/10.1128/CMR.00059-12

Bennett RN, Wallsgrove RM (1994) Secondary metabolites in plant defense mechanisms. New Phytol 127:617–633. https://doi.org/10.1111/j.1469-8137.1994.tb02968.x

Bjarnsholt T, Givskov M (2007) The role of quorum sensing in the pathogenicity of the cunning aggressor *Pseudomonas aeruginosa*. Anal Bioanal Chem 387:409–414. https://doi.org/10.1007/s00216-006-0774-x

Bonte SF, Chamot E, Kohler T, Romand JA, van Deldon C (2007) Autoinducer production and quorum-sensing dependent phenotypes of *Pseudomonas aeruginosa* vary according to isolation site during colonization of incubated patients. BMC Microbiol 7:33. https://doi.org/10.1186/1471-2180-7-33

Bonten MJ, Gaillard CA, van der Hulst R, de Leeuw PW, van der Geest S, Stobberingh EE, Soeters PB (1996) Intermittent enteral feeding: the influence on respiratory and digestive tract colonization in mechanically ventilated intensive-care-unit patients. Am J Respir Crit Care Med 154:394–399. https://doi.org/10.1164/ajrccm.154.2.8756812

Braeken K, Daniels R, Ndayizeye M, Vanderleyden J, Michiels J (2008) Quorum sensing bacteria-plants interactions. In: Nautiyal CS, Dion P (ed) Molecular mechanisms of plants and microbe coexistence. Springer, Berlin/Heidelberg, pp 265–289. ISBN: 978-3-540-75574-6. https://doi.org/10.1007/978-3-540-75575-3

Brint JM, Ohman DE (1995) Synthesis of multiple exoproducts in *P. aeruginosa* is under the control of Rh1R-Rh1I, another set of regulators in strain PAO1with homology to the autoinducer-responsive LuxR-LuxI family. J Bacteriol 24:7155–7163. https://doi.org/10.1128/jb.177.24.7155-7163.1995

Cady NC, McKean KA, Behnke J, Kubec R, Mosier AP, Kasper SH, Burz DS, Musah RA (2012) Inhibition of biofilm formation, quorum sensing and infection in *Pseudomonas aeruginosa* by natural products-inspired organosulfur compounds. PLoS One 7:e38492. https://doi.org/10.1371/journal.pone.0038492

Calfee MW, Coleman JP, Pesci EC (2001) Interference with *Pseudomonas* quinolone signal synthesis inhibits virulence factors expression by *Pseudomonas aeruginosa*. Proc Natl Acad Sci U S A 98:11633–11637. https://doi.org/10.1073/pnas.201328498

Carneiro VA, Santos HS, Arruda FV, Bandeira PN, Albuquerque MR, Pereira MO, Henriques M, Cavada BS, Teixeira EHC (2010) *Casbane diterpene* as a promising natural antimicrobial agent against biofilm-associated infections. Molecules 16:190–201. https://doi.org/10.3390/molecules16010190

Cha C, Gao P, Chen YC, Shaw PD, Farrand SK (1998) Production of acyl-homoserine lactone quorum-sensing signals by gram-negative plant-associated bacteria. Mol Plant-Microbe Interact 11:1119–1129. https://doi.org/10.1094/MPMI.1998.11.11.1119

Chan CH, Ngoh GC, Yusoff R (2012) A brief review on anti diabetic plants: global distribution, active ingredients, extraction techniques and acting mechanisms. Pharmacogn Rev 6:22–28. https://doi.org/10.4103/0973-7847.95854

Chang CY, Krishnan T, Wang H, Chen Y, Yin WF, Chong YM, Tan LY, Chong TM, Chan KG (2014) Non-antibiotic quorum sensing inhibitors acting against N-acyl homoserine lactone synthase as druggable target. Sci Rep 4:7245. https://doi.org/10.1038/srep07245

Chastre J, Fagon J-Y (2002) Ventilator-associated pneumonia. Am J Respir Crit Care Med 165:867–903. https://doi.org/10.1164/rccm.2105078

Cho HS, Lee JH, Ryu SY, Joo SW, Cho MH, Lee J (2013) Inhibition of *Pseudomonas aeruginosa* and Escherichia coli O157:H7 biofilm formation by plant metabolite ε-viniferin. J Agric Food Chem 61:7120–7126. https://doi.org/10.1021/jf4009313

Dando SJ, Mackay-Sim A, Norton R, Currie BJ, John JA, Ekberg JA, Batzloff M, Ulett GC, Beacham IR (2014) Pathogens penetrating the central nervous system: infection pathways and the cellular and molecular mechanisms of invasion. Clin Microbiol Rev 27:691–726. https://doi.org/10.1128/CMR.00118-13

Darby C, Cosma CL, Thomas JH, Manoil C (1999) Lethal paralysis of *Caenorhabditis elegans* by *Pseudomonas aeruginosa*. Proc Natl Acad Sci U S A 96:15202–15207. https://doi.org/10.1073/pnas.96.26.15202

Davies D (2003) Understating biofilms resistance to antibacterial agents. Nat Rev Drug Discov 2:114–122. https://doi.org/10.1038/nrd1008

Deep A, Chaudhary U, Gupta V (2011) Quorum sensing and bacterial pathogenicity: from molecules to disease. J Lab Physicians 3:4–11. https://doi.org/10.4103/0974-2727.78553

Ding X, Yin B, Qian L, Zeng Z, Li H, Lu Y, Zhou S (2011) Screening of novel quorum-sensing inhibitors to interfere with the formation of *Pseudomonas aeruginosa* biofilm. J Med Microbiol 60:1827–1834. https://doi.org/10.1099/jmm.0.024166-0

Donlan RM (2001) Biofilms and device-associated infections. Emerg Infect Dis 7:277–281. https://doi.org/10.3201/eid0702.700277

Donlan RM (2002) Biofilms: microbial life on surfaces. Emerg Infect Dis 8:881–890. https://doi.org/10.3201/eid0809.020063

Dyduch-Sieminska M, Najda A, Dyduch J, Gantner M, Klimek K (2015) The content of secondary metabolites and antioxidant activity of wild strawberry fruit (*Fragaria vesca* L.) J Anal Methods Chem 2015:831238. https://doi.org/10.1155/2015/831238

Engebrecht J, Nealson K, Silverman M (1983) Bacterial bioluminescence: isolation and genetic analysis of functions from *Vibrio fischeri*. Cell 32:773–781. https://doi.org/10.1016/0092-8674(83)90063-6

Fernandez L, Hancock REW (2012) Adaptive and mutational resistance: role of porins and efflux pumps in drug resistance. Clin Microbiol Rev 25:661–681. https://doi.org/10.1128/CMR.00043-12

Fong IW, Tomkins KB (1985) Review of *Pseudomonas aeruginosa* meningitis with special emphasis on treatment with ceftazidime. Rev Infect Dis 7:604–612

Fugua WC, Winans SC, Greenberg EP (1994) Quorum sensing in bacteria: the LuxR-LuxI family of cell density-responsive transcriptional regulators. J Bacteriol 176:269–275. https://doi.org/10.1128/jb.176.2.269 275.1994

Gallagher LA, Manoil C (2001) *Pseudomonas aeruginosa* PAO1 kills *Caenorhabditis elegans* by cyanide poisoning. J Bacteriol 183:6207–6214. https://doi.org/10.1128/JB.183.21.6207-6214.2001

Gambello MJ, Iglewski BH (1991) Cloning and characterization of the *Pseudomonas aeruginosa* lasR gene, a transcriptional activator of elastase expression. J Bacteriol 173:3000–3009

Ganesh PS, Rai VR (2015a) *In vitro* antibiofilm activity of *Murraya koenigii* essential oil extracted using supercritical CO_2 method against *Pseudomonas aeruginosa* PAO1. Nat Prod Res 29:2295–2298. https://doi.org/10.1080/14786419.2015.1004673

Ganesh PS, Rai VR (2015b) Evaluation of anti-bacterial and anti-quorum sensing potential of essential oils extracted by supercritical CO_2 method against *Pseudomonas aeruginosa*. J Essent Oil Bear Pl 18:264–275. https://doi.org/10.1080/0972060X.2015.1025295

Ganesh PS, Rai VR (2016) Inhibition of quorum-sensing-controlled virulence factors of *Pseudomonas aeruginosa* by *Murraya koenigii* essential oil: a study in a *Caenorhabditis elegans* infectious model. J Med Microbiol 65:1528–1535. https://doi.org/10.1099/jmm.0.000385

Ganesh PS, Rai VR (2017) Attenuation of quorum-sensing-dependent virulence factors and biofilm formation by medicinal plants against antibiotic resistant *Pseudomonas aeruginosa*. J Tradit Complement Med. https://doi.org/10.1016/j.jtcme.2017.05.008

Ghosh R, Tiwary BK, Kumar A, Chakraborty R (2014) Guava leaf extract inhibits quorum –sensing and *Chromobacterium violaceum* induced lysis of human hepatoma cells: whole transcriptome analysis reveals differential gene expression. PLoS One 9:e107703. https://doi.org/10.1371/journal.pone.0107703

Girennavar B, Cepeda ML, Soni KA, Vikram A, Jesudhasan P, Jayaprakasha GK, Pillai SD, Patil BS (2008) Grapefruit juice and its furocoumarins inhibits autoinducer signaling and biofilm formation in bacteria. Int J Food Microbiol 125:204–208. https://doi.org/10.1016/j.ijfoodmicro.2008.03.028

Givskov M, de Nys R, Manefield M, Gram L, Maximilien R, Eberl L, Molin S, Steinberg P, Kjelleberg S (1996) Eukaryotic interference with homoserine lactone-mediated prokaryotic signaling. J Bacteriol 178:6618–6622. https://doi.org/10.1128/jb.178.22.6618-6622.1996

Hengge R (2009) Principles of c-di-GMP signaling in bacteria. Nat Rev Microbiol 7:263–273. https://doi.org/10.1038/nrmicro2109

Henry CR, Flynn HW, Miller D, Forster RK, Alfonso EC (2012) Infectious keratitis progressing to endophthalmitis: a 15-year-study of microbiology, associated factors, and clinical outcomes. Ophthalmology 119:2443–2449. https://doi.org/10.1016/j.ophtha.2012.06.030

Hoang TT, Sullivan SA, Cusick JK, Schweizer HP (2002) {Beta}-Ketoacyl acyl carrier protein reductase (FabG) activity of the fatty acid biosynthetic pathway is a determining factor of 3-oxo-homoserine lactone acyl chain lengths. Microbiology 148:3849–3856. https://doi.org/10.1099/00221287-148-12-3849

Hocquet D, Nordmann P, Garch FEI, Cabanne L, Plesiat P (2006) Involvement of the MexXY-OprM efflux system in emergence of cefepime resistance in clinical strains of *Pseudomonas aeruginosa*. Antimicr Agents Chemother 50:1347–1351. https://doi.org/10.1128/AAC.50.4.1347-1351.2006

Hoge R, Pelzer A, Rosenau F, Wilhelm S (2010) Weapons of a pathogen: proteases and their role in virulence of *Pseudomonas aeruginosa*. FORMATEX 2:383–395

Hoiby N, Bjarnsholt T, Givskov M, Molin S, Ciofu O (2010) Antibiotic resistance of bacterial biofilms. Int J Antimicrob Agents 35:322–332. https://doi.org/10.1016/j.ijantimicag.2009.12.011

Huangyutitham V, Guvener ZT, Harwood CS (2013) Subcellular clustering of the phosphorylated WspR response regulator protein stimulates its diguanylate cyclase activity. MBio 4:00242–00213. https://doi.org/10.1128/mBio.00242-13

Husain MS, Fareed S, Ansari S, Rahman MA, Ahmad IZ, Saeed M (2012) Current approaches toward production of secondary metabolites. J Pharm Bioallied Sci 4:10–20. https://doi.org/10.4103/0975-7406.92725

Husain FM, Ahmad M, Asif M, Tahseen Q (2013) Influence of clove oil on certain quorum-sensing regulated functions and biofilm of *Pseudomonas aeruginosa* and *Aeromonas hydrophila*. J Biosci 38:835–844

Hussain FM, Ahmad I, Khan MS, Ahmad E, Tahseen Q, Khan MS, Alshabib NA (2015) Sub-MICs of *Mentha piperita* essential oil and menthol inhibits AHL mediated quorum sensing and biofilm of Gram-negative bacteria. Front Microbiol 6:420. https://doi.org/10.3389/fmicb.2015.00420

Jagani S, Chelikani R, Kim DS (2009) Effects of phenol and natural phenolic compounds on biofilm formation by *Pseudomonas aeruginosa*. Biofouling 25:321–324. https://doi.org/10.1080/08927010802660854

Jakobsen TH, Bragason SK, Phipps RK, Christensen LD, van Gennip M, Alhede M, Skindersoe M, Larsen TS, Hoiby N, Bjarnsholt T, Givskov M (2012) Food as a source for quorum sensing inhibitios: Iberin from horseradish revealed as a quorum sensing inhibitor of *Pseudomonas aeruginosa*. Appl Environ Microbiol 78:2410–2421. https://doi.org/10.1128/AEM.05992-11

Janeway CA Jr, Travers P, Walport M, Shlomchik MJ (2001) The front line of host defense. In: Immunology: The immune system in health and disease. Garland Science, New York, pp 295–340

Jaramillo-Colorado B, Olivero-Verbal J, Stashenko EE, Wagner-Dobler I, Kunze B (2012) Anti-quorum sensing activity of essential oils from Colombian plants. Nat Prod Res 26:1075–1086. https://doi.org/10.1080/14786419.2011.557376

Jha B, Kavita K, Westphal J, Hartmann A, Kopplin PS (2013) Quorum sensing inhibition by *Asparagopsis taxiformis,* a marine macro alga: separation of the compound that interrupts bacterial communication. Mar Drugs 11:253–265. https://doi.org/10.3390/md11010253

Juhi T, Bibhabati M, Archana T, Poonam L, Dogra Vinita D (2009) *Pseudomonas aeruginosa* meningitis in post neurological patients. Neurol Asia 14:95–100

Kaletta T, Hengartner MO (2006) Finding function in targets: *C. elegans* as a model organism. Nat Rev Drug Discov 5:387–399. https://doi.org/10.1038/nrd2031

Kalia VC (2013) Quorum sensing inhibitors: an overview. Biotechnol Adv 31:224–245. https://doi.org/10.1016/j.biotechadv.2012.10.004

Kalia M, Yadav VK, Singh PK, Sharma D, Pandey H, Narvi SS, Agarwal V (2015) Effect of cinnamon oil on quorum sensing –controlled virulence factors and biofilms formation in *Pseudomonas aeruginosa.* PLoS One 10:e0135495. https://doi.org/10.1371/journal.pone.0135495

Kang CI, Kim SH, Kim HB, Park SW, Choe YJ, Oh MD, Kim EC, Choe KW (2003) *Pseudomonas aeruginosa* Bacteremia: risk factors for mortality and influence of delayed receipt of effective antimicrobial therapy on clinical outcome. Clin Infect Dis 37:745–751. https://doi.org/10.1086/377200

Khan MSA, Zahin M, Hasan S, Husain FM, Ahmad I (2009) Inhibition of quorum sensing regulated bacterial functions by plant essential oils with special reference to clove oil. Lett Appl Microbiol 49:354–360. https://doi.org/10.1111/j.1472-765X.2009.02666.x

Khan HA, Ahmad A, Mehboob R (2015) Nosocomial infections and their control strategies. Asian Pac J Trop Dis 5:509–514. https://doi.org/10.1016/j.apjtb.2015.05.001

Kim HS, Park HD (2013) Ginger extract inhibits biofilm formation by *Pseudomonas aeruginosa* PA14. PLoS One 8:e76106. https://doi.org/10.1371/journal.pone.0076106

Kim JS, Kim YH, Seo YW, Park S (2007) Quorum sensing inhibitors from the red alga, *Ahnfeltiopsis flabelliformis.* Biotechnol Bioprocess Eng 12:308–311. https://doi.org/10.1007/BF02931109

Kim HS, Lee SH, Byun Y, Park HD (2015a) 6-Gingerol reduces *Pseudomonas aeruginosa* biofilm formation and virulence via quorum sensing inhibition. Sci Rep 5:8656. https://doi.org/10.1038/srep08656

Kim YG, Lee JH, Kim S, Baek KH, Lee J (2015b) Cinnamon bark oil and its components inhibit biofilm formation and toxin production. Int J Food Microbiol 195:30–39. https://doi.org/10.1016/j.ijfoodmicro.2014.11.028

Kirbag S, Zengin F (2009) Antimicrobial activities of extract of some plants. Pakistan. J Bot 41:2067–2070

Kiska DL, Gilligan PH (2003) Pseudomonas. In: Murray PR, Baron EJ, Pfaller MA, Tenover FC, Yolken RH (eds) Manual of clinical microbiology. American Society for Microbiology Press, Washington, DC, pp 517–525

Koh KH, Tham FY (2011) Screening of traditional Chinese medicinal plants for quorum-sensing inhibitors activity. J Microbiol Immunol Infect 44:144–148. https://doi.org/10.1016/j.jmii.2009.10.001

Koh KH, Sam CK, Yin WF, Tan LY, Krishnan T, Chong YM, Chang KG (2013) Plant-derived natural products as sources of anti-quorum sensing compounds. Sensor (Basel) 13:6217–6228. https://doi.org/10.3390/s130506217

LaSarre B, Federle MJ (2013) Exploiting quorum sensing to confuse bacterial pathogens. Microbiol Mol Biol Rev 77:73–111. https://doi.org/10.1128/MMBR.00046-12

Lee J, Wu J, Deng Y, Wang J, Wang J, Wang J, Chang C, Dong Y, Williams P, Zhang L (2013) A cell-cell communication signal integrates quorum sensing and stress response. Nat Chem Biol 9:339–343. https://doi.org/10.1038/nchembio.1225

Lee JY, Park YK, Chung EE, Na IY, Ko KS (2016) Evolved resistance to colistin and its loss due to genetic reversion in *Pseudomonas aeruginosa.* Sci Rep 6:25543. https://doi.org/10.1038/srep25543

Lidor O, Quntar A, Pesci EC, Steinberg D (2015) Mechanistic analysis of a synthetic inhibitor of the *Pseudomonas aeruginosa* LasI quorum-sensing signal synthase. Sci Rep 5. https://doi.org/10.1038/srep16569

Lister PD, Wolter DJ, Hanson ND (2009) Antibacterial-resistant *Pseudomonas aeruginosa*: clinical impact and complex regulation of chromosomally encoded resistance mechanisms. Clin Microbiol Rev 22:582–610. https://doi.org/10.1128/CMR.00040-09

Lu JJ, Bao JL, Wu GS, Xu WS, Huang MQ, Chen XP, Wang YT (2013) Quinones derived from plant secondary metabolites as anti-cancer agents. Anti Cancer Agents Med Chem 13:456–463. https://doi.org/10.2174/1871520611313030008

Luciardi MC, Blazquez MA, Cartagena E, Bardon A, Arena ME (2016) Mandarin essential oils inhibit quorum sensing and virulence factors of *Pseudomonas aeruginosa*. LWT-Food Sci Technol 68:373–380. https://doi.org/10.1016/j.lwt.2015.12.056

Mahajan-miklos S, Tan MW, Rahme LG, Ausubel FM (1999) Molecular mechanisms of bacterial virulence elucidated using a *Pseudomonas aeruginosa–Caenorhabditis elegans* pathogenesis model. Cell 96:47–56. https://doi.org/10.1016/S0092-8674(00)80958-7

Maisuria VB, de Los Santos YL, Tufenkji N, Deziel E (2016) Cranberry-derived proanthocyanidins impair virulence and inhibit quorum sensing of *Pseudomonas aeruginosa*. Sci Rep 6:30169. https://doi.org/10.1038/srep30169

Manefield M, de Nys R, Kumar N, Read R, Givskov M, Steinberg P, Kjelleberg S (1999) Evidence that halogenated furanones from *Delisea pulchra* inhibit acylated homoserine lactones (AHL)-mediated gene expression by displacing the AHL signal from its receptor protein. Microbiol 145:283–291. https://doi.org/10.1099/13500872-145-2-283

Marin SD, Xu Y, Meijler MM, Janda KD (2007) Antibody catalyzed hydrolysis of a quorum sensing signal found in Gram-negative bacteria. Bioorg Med Chem Lett 17:1549–1552. https://doi.org/10.1016/j.bmcl.2006.12.118

Micek ST, Lioyd AE, Ritchie DJ, Reichley RM, Fraser VJ, Koller MH (2005) *Pseudomonas aeruginosa* bloodstream infection: importance of appropriate initial antimicrobial treatment. Antimicrob Agents Chemother 49:1306–1311. https://doi.org/10.1128/AAC.49.4.1306-1311.2005

Miyashiro T, Ruby EG (2012) Shedding light on bioluminescence regulation in *Vibrio fischeri*. Mol Microbiol 84:795–806. https://doi.org/10.1111/j.1365-2958.2012.08065.x

Moghaddam MM, Khodi S, Mirhosseini A (2014) Quorum sensing in bacteria and a glance on *Pseudomonas aeruginosa*. Clin Microbial 3:4. https://doi.org/10.4172/2327-5073.1000156

Morita Y, Tomida J, Kawamura Y (2013) Responses of *Pseudomonas aeruginosa* to antimicrobials. Front Microbiol 4:422. https://doi.org/10.3389/fmicb.2013.00422

Moya B, Dotsch A, Juan C, Blazquez J, Zamorana L, Haussler S, Oliver A (2009) β-lactam resistance response triggered by inactivation of a nonessential penicillin-binding protein. PLoS Pathog 5:e1000353. https://doi.org/10.1371/journal.ppat.1000353

Murugan K, Selvanayaki K, Sohaibani SAI (2011) Antibiofilm activity of *Andrographis paniculata* against cystic fibrosis clinical isolates *Pseudomonas aeruginosa*. World J Microbiol Biotechnol 27:1661–1668. https://doi.org/10.1007/s11274-010-0620-3

Mustafa KS, Balamurugan K, Pandian SK, Ravi AV (2012a) 2,5-piperazinedione inhibits quorum sensing-dependent factor production in *Pseudomonas aeruginosa* PAO1. J Basic Microbiol 52:1–8. https://doi.org/10.1002/jobm.201100292

Mustafa KS, Sivamaruthi BS, Pandian SK, Ravi AV (2012b) Quorum sensing inhibition in *Pseudomonas aeruginosa* PAO1 by antagonistic compound phenylacetic acid. Curr Microbiol 65:475–480. https://doi.org/10.1007/s00284-012-0181-9

Musthafa KS, Ravi AV, Annapoorani A, Packiavathy ISV, Pandian S (2010) Evaluation of anti-quorum-sensing activity of edible plants and fruits through inhibition of the N-acyl-homoserine lactone system in *Chromobacterium violaceum* and *Pseudomonas aeruginosa*. Chemotherapy 56:333–339. https://doi.org/10.1159/000320185

Nau R, Eiffert H (2002) Modulation of release of proinflammatory bacterial compounds by anti-bacterials: potential impact on course of inflammation and outcome in sepsis and meningitis. Clin Microbiol Rev 15:95–110. https://doi.org/10.1128/CMR.15.1.95-110.2002

Nealson KH, Platt T, Hastings JW (1970) Cellular control of the synthesis and activity of the bacterial luminescent. J Bacteriol 104:313–322

Ng W-L, Bassler BL (2009) Bacterial quorum-sensing network architectures. Annu Rev Genet 43:197–222. https://doi.org/10.1146/annurev-genet-102108-134304

Ni N, Choudhary G, Li M, Wang B (2008) Pyrogallol and analogs can antagonize bacterial quorum sensing in *Vibrio harveyi*. Bioorg Med Chemist Lett 18:1567–1572. https://doi.org/10.1016/j.bmcl.2008.01.081

Nicolle LE (2014) Catheter associated urinary tract infections. Antimicrob Resist Infect Control 3:23. https://doi.org/10.1186/2047-2994-3-23

Nikaido H (2010) Multidrug resistance in bacteria. Annu Rev Biochem 78:119–146. https://doi.org/10.1146/annurev.biochem.78.082907.145923

Niu C, Afre Gilbert ES (2006) Sub-inhibitory concentrations of cinnamaldehyde interfere with quorum sensing. Lett Appl Microbiol 43:489–494. https://doi.org/10.1111/j.1472-765X.2006.02001.x

Orgad O, Oren Y, Walker SL, Herzberg M (2011) The role of alginate in *Pseudomonas aeruginosa* EPS adherence, viscoelastic properties and cell attachment. Biofouling 27:787–798. https://doi.org/10.1080/08927014.2011.603145

Pamp SJ, Nielson TT (2007) Multiple roles of biosurfactants in structural biofilm development by *Pseudomonas aeruginosa*. J Bacteriol 189:2531–2539. https://doi.org/10.1128/JB.01515-06

Park J, Kaufmann GF, Bowen JP, Arbiser L, Janda KD (2008) Solenopsin A, a venom alkaloid from the fire ant *Solenopsis invicta*, inhibits quorum-sensing signaling in *Pseudomonas aeruginosa*. J Infect Dis 198:1198–1201. https://doi.org/10.1086/591916

Pearson JP, Gray KM, Passador L, Tucker KD, Eberhard A, Iglewski BH, Greenberg EP (1994) Structure of the autoinducer required for expression of *Pseudomonas aeruginosa* virulence genes. Proc Natl Acad Sci U S A 91:197–201

Pearson JP, Passador L, Iglewski BH, Greenberg EP (1995) A second *N*-acylhomoserine lactone signal produced by *Pseudomonas aeruginosa*. Proc Natl Acad Sci U S A 92:1490–1494. https://doi.org/10.1073/pnas.92.5.1490

Pearson JP, Pesci EC, Iglewski BH (1997) Role of *Pseudomonas aeruginosa las* and *rhl* quorum sensing systems in control of elastase and rhamnolipid biosynthesis genes. J Bacteriol 179:5756–5767

Pejin B, Ciric A, Glamoclija J, Nikolic M, Sokovic M (2015) *In vitro* anti-quorum sensing activity of phytol. Nat Prod Res 29:374–377. https://doi.org/10.1080/14786419.2014.945088

Pesci EC, Pearson JP, Seed PC, Iglewski BH (1997) Regulation of *las* and *rhl* quorum sensing in *Pseudomonas aeruginosa*. J Bacteriol 179:3127–3132

Peters L, Konig AD, Wright AD, Pukall R, Stackebrandt E, Eberl L (2003) Secondary metabolites of *Flustra foliacea* and their influence on bacteria. Appl Environ Microbiol 69:3469–3475. https://doi.org/10.1128/AEM.69.6.3469-3475.2003

Pimenta AL, Delatorre LDC, Mascarello A, Oliveira KAO, Leal PC, Yunes RA, Nedel CB, Aguiar M, Tasca CI, Nunes RJ, Smania AS Jr (2013) Synthetic organic compounds with potential for bacterial biofilm inhibition, a path for the identification of compounds interfering with quorum sensing. Int J Antimicrob Agents 42:519–523. https://doi.org/10.1016/j.ijantimicag.2013.07.006

Plyuta V, Zaitseva J, Lobakova E, Zagoskina N, Kuznetsov A, Khml I (2013) Effect of plant phenolic compounds on biofilm formation by *Pseudomonas aeruginosa*. APMIS 121:1073–1081. https://doi.org/10.1111/apm.12083

Qu L, She P, Wang Y, Liu F, Zhang D, Chen L, Luo Z, Xu H, Qi Y, Wu Y (2016) Effects of norspermidine on *Pseudomonas aeruginosa* biofilm formation and eradication. Microbiology 5:402–412. https://doi.org/10.1002/mbo3.338

Quintana J, Brango-Vanegas J, Costa GM, Castellanos L, Arevalo C, Duque C (2015) Marine organisms as source of extracts to disrupt bacterial communication: bioguided isolation and identification of quorum sensing inhibitors from *Ircinia felix*. Rev Bras Farmacogn 25:199–207. https://doi.org/10.1016/j.bjp.2015.03.013

Rada B, Leto TL (2013) Pyocyanin effects on respiratory epithelium: relevance in *Pseudomonas aeruginosa* airway infections. Trends Microbiol 21:73–81. https://doi.org/10.1016/j.tim.2012.10.004

Rajesh PS, Rai VR (2014) Quorum quenching activity in cell-free lysate of endophytic bacteria isolated from *Pterocarpus santalinus* Linn., and its effect on quorum sensing regulated biofiln in *Pseudomonas aeruginosa* PAO1. Microbiol Res 169:561–569. https://doi.org/10.1016/j.micres.2013.10.005

Ramsamiravaka T, Labtani Q, Duez P, Jaziri MEI (2015) The formation of biofilms by *Pseudomonas aeruginosa:* a review of the natural and synthetic compounds interfering with control mechanisms. Biomed Res Int 2015:1–17. https://doi.org/10.1155/2015/759348

Rasmussen TB, Givskov M (2006) Quorum sensing inhibitors: a bargain of effects. Microbiology 152:895–904. https://doi.org/10.1099/mic.0.28601-0

Rasmussen TB, Manefield M, Andersen JB, Ebert L, Anthoni U, Christophersen C, Steinberg P, Kjelleberg S, Givskov M (2000) How *Delisea pulchra* furanones affect quorum sensing and swarming motility in *Serratia liquefaciens* MG1. Microbiology 146:3237–3244. https://doi.org/10.1099/00221287-146-12-3237

Reimmann C, Beyeler M, Latifi A, Winteler H, Foglino M, Lazdunski A, Haas D (1997) The global activator GacA of *Pseudomonas aeruginosa* PAO1 positively controls the production of the autoinducer N-butyryl-homoserine lactone and the formation of the virulence factors pyocyanin, cyanide, and lipase. Mol Microbiol 24:309–319. https://doi.org/10.1046/j.1365-2958.1997.3291701.x

Ribert D, Cossart P (2015) How bacterial pathogens colonize their hosts and invade deeper tissues. Microbes Infect 17:173–183. https://doi.org/10.1016/j.micinf.2015.01.004

Rogers SA, Huigens RW, Cavanagh J, Melander C (2010) Synergetic effects between conventional antibiotics and 2-aminoimidazole-derived antibiofilm agents. Antimicrob Agents Chemother 54:2112–2118. https://doi.org/10.1128/AAC.01418-09

Ruppe E, Woerther PL, Barbier F (2015) Mechanisms of antimicrobial resistance in gram-negative bacilli. Ann Intensive Care 5:21. https://doi.org/10.1186/s13613-015-0061-0

Ryder C, Byrd M, Wozniak DJ (2007) Role of polysaccharides in *Pseudomonas aeruginosa* biofilm development. Curr Opinion Microbiol 10:644–648

Sadikot RT, Blackwell TS, Christman JW, Prince AS (2005) Pathogen-host interactions in *Pseudomonas aeruginosa* pneumonia. Am J Respir Crit Care Med 17:1209–1223. https://doi.org/10.1164/rccm.200408-1044SO

Sarabhai S, Sharma P, Capalash N (2013) Ellagic acid derivatives from *Terminalia chebula* Retz. downregulate the expression of quorum sensing genes to attenuate *Pseudomonas aeruginosa* PAO1 virulence. PLoS One 8:e53441. https://doi.org/10.1371/journal.pone.0053441

Sawa T, Ohara M, Kurahashi K, Twining SS, Frank DW, Doroques DB, Long T, Michael A, Wiener-kronish JP, Gropper MA (1998) *In vitro* cellular toxicity predicts *Pseudomonas aeruginosa* virulence in lung infections. Infect Immun 66:3242–3249

Schuster M, Greenberg EP (2006) A network of networks: quorum-sensing gene regulation in *Pseudomonas aeruginosa*. Int J Med Microbiol 296:73–81. https://doi.org/10.1016/j.ijmm.2006.01.036

Sepahi E, Tarighi S, Ahmadi FS, Bagheri A (2015) Inhibition of quorum sensing in *Pseudomonas aeruginosa* by two herbal essential oils from *Apiaceae* family. J Microbiol 53:176–180. https://doi.org/10.1007/s12275-015-4203-8

Shaikh S, Fatima J, Shakil S, Rizvi SMD, Kamal MA (2015) Antibiotic resistance and extended spectrum beta-lactamases: types, epidemiology and treatment. Saudi J Biol Sci 22:90–101. https://doi.org/10.1016/j.sjbs.2014.08.002

Singh BN, Singh BR, Singh RL, Prakash D, Dhakarey R, Upadhyay G, Singh HB (2009) Oxidative DNA damage protective activity, antioxidant and anti-quorum sensing potentials of *Moringa oleifera*. Food Chem Toxicol 47:1109–1116. https://doi.org/10.1016/j.fct.2009.01.034

Skindersoe ME, Epstein PE, Rasmussen TB, Bjarnsholt T, de Nys R, Givskov M (2008) Quorum sensing antagonism from marine organisms. Mar Biotechnol 10:56–63. https://doi.org/10.1007/s10126-007-9036-y

Slobodnikova L, Fialova S, Rendekova K, Kovac J, Mucaji P (2016) Antibiofilm activity of plants polyphenols. Molecules 21:1717. https://doi.org/10.3390/molecules21121717

Smith RS, Iglewski BH (2003) *Pseudomonas aeruginosa* quorum sensing as a potential antimicrobial target. J Clin Invest 112:1460–1465. https://doi.org/10.1172/JCI20364

Sperandio V, Torres AG, Kaper JB (2002) Quorum sensing *Escherichia coli* regulators B and C (QseBC): a novel two-component regulatory system involved in the regulation of flagella and motility by quorum sensing in *E. coli*. Mol Microbiol 43:809–821. https://doi.org/10.1046/j.1365-2958.2002.02803.x

Sterrett FS (1962) The nature of essential oils.11. Chemical constituents, analysis. J Chem Edu 39:246–251

Sun J, Deng Z, Yen A (2015) Bacterial multidrug efflux pumps: mechanisms, physiology and pharmacological exploitations. Biochem Biophys Res Commun 453:254–267. https://doi.org/10.1016/j.bbrc.2014.05.090

Taganna JC, Quanico JP, Perono RMG, Amor EC, Rivera WL (2011) Tannin-rich fraction from *Terminalia catappa* inhibits quorum sensing (QS) in *Chromobacterium violaceum* and the QS-controlled biofilm maturation and LasA staphylolytic activity in *Pseudomonas aeruginosa*. J Ethanopharmocol 134:865–871. https://doi.org/10.1016/j.jep.2011.01.028

Tateda K, Comte R, Pechere J-C, Köhler T, Yamaguchi K, van Delden C (2001) Azithromycin inhibits quorum sensing in *Pseudomonas aeruginosa*. Antimicrob Agents Chemother 45:1930–1933. https://doi.org/10.1128/AAC.45.6.1930-1933.2001

Tenover FC (2006) Mechanisms of antimicrobial resistance in bacteria. Am J Infect Control 119:3–10. https://doi.org/10.1016/j.ajic.2006.05.219

Teplitski M, Robinson JB, Bauer WD (2000) Plants secrete substances that mimic bacterial N-acyl-homoserine lactone signal activities and affect population density-dependent behaviors in associated bacteria. Mol Plant-Microbe Interact 13:637–646. https://doi.org/10.1094/MPMI.2000.13.6.637

Teplitski M, Mathesius U, Rumbaugh KP (2011) Perception and degradation of N-acyl homoserine lactone quorum sensing signals by mammalian and plant cells. Chem Rev 111:100–116. https://doi.org/10.1021/cr100045m

Truchado P, Larrosa M, Castro-Ibanez I, Allende A (2015) Plant food extracts and phytochemicals: Their role as quorum sensing inhibitors. Trends Food Sci Technol 43:189–204. https://doi.org/10.1016/j.tifs.2015.02.009

Turek C, Stintzing FC (2013) Stability of essential oils: a review. Compr Rev Food Sci Food Saf 12:40–53. https://doi.org/10.1111/1541-4337.12006

Van Delden C, Iglewski BH (1998) Cell-to-cell signaling and *Pseudomonas aeruginosa* infections. Emerg Infect Dis 4:551–560. https://doi.org/10.3201/eid0404.980405

Vasavi HS, Arun AB, Rekha PD (2016) Anti-quorum sensing activity of flavonoid-rich fraction from *Centella asiatica* L. against *Pseudomonas aeruginosa* PAO1. J Microbiol Immunol Infect 49:8–15. https://doi.org/10.1016/j.jmii.2014.03.012

Vattem DA, Mihali KK, Crixell SH, Mclean RJ (2007) Dietary phytochemicals as quorum sensing inhibitors. Fitoterapia 78:302–310. https://doi.org/10.1016/j.fitote.2007.03.009. 10.1016/S1359-6446(97)01167-7

Verpoorte R (1998) Exploration of natures chemodiversity: the role of secondary metabolites as leads in drug development. Drug Discov Today 3:232–238. https://doi.org/10.1016/S1359-6446(97)01167-7

Vu B, Chen M, Crawford RJ, Ivanova EP (2009) Bacterial extracellular polysaccharides involved in biofilms formation. Molecules 14:2535–2554. https://doi.org/10.3390/molecules14072535

Wade DS, Calfee MW, Rocha ER, Ling EA, Engstrom E, Coleman JP, Pesci EC (2005) Regulation of *Pseudomonas* quinolone signal synthesis in *Pseudomonas aeruginosa*. J Bacteriol 187:4372–4380. https://doi.org/10.1128/JB.187.13.4372-4380.2005

Wagner S, Sommer R, Hinsberger S, Lu C, Hartmann RW, Empting M, Titz A (2016) Novel strategies for the treatment of *Pseudomonas aeruginosa* infections. J Med Chem 59:5929–5969. https://doi.org/10.1021/acs.jmedchem.5b01698

Walker TS, Bais HP, Deziel E, Schweizer HP, Rahme LG, Fall R, Vivanco JM (2004) *Pseudomonas aeruginosa* –plant root interactions. Pathogenicity, biofilm formation, and root exudation. Plant Physiol 134:320–331. https://doi.org/10.1104/pp.103.027888

Wallace RJ (2004) Antimicrobial properties of secondary metabolites. Proc Nutr Soc 63:621–629. https://doi.org/10.1079/PNS2004393

Wang S, Yu S, Zhang Z, Wei Q, Yan L, Ai G Liu H, Ma LZ (2014) Coordination of swarming motility, biosurfactant synthesis, and biofilms matrix exopolysaccharide production in *Pseudomonas aeruginosa*. Appl Environ Microbiol 80:6724–6732. https://doi.org/10.1128/AEM.01237-14

Water CM, Bassler BL (2006) The *Vibrio harveyi* quorum-sensing system uses shared regulatory components to discriminate between multiple autoinducers. Genes Dev 20:2754–2767. https://doi.org/10.1101/gad.1466506

Wolter DJ, Black JA, Liter PD, Hanson ND (2009) Multiple genotypic changes in hyper susceptible strains of *Pseudomonas aeruginosa* isolated from cystic fibrosis patients do not always correlate with the phenotype. J Antimicrob Chemother 64:294–300. https://doi.org/10.1093/jac/dkp185

Wu H, Song Z, Hentzer M, Andersen JB, Molin S, Givskov M, Hoiby N (2004) Synthetic furanones inhibit quorum –sensing and enhance bacterial clearance in *Pseudomonas aeruginosa* lung infection mice. J Antimicrob Chemother 53:1054–1061. https://doi.org/10.1093/jac/dkh223

Yan M, Zhu Y, Zhang HJ, Jiao WH, Han BN, Liu ZX, Qiu F, Chen WS, Lin HW (2013) Anti-inflammatory secondary metabolites from the leaves of *Rosa laevigata*. Bioorg Med Chem 21:3290–3297. https://doi.org/10.1016/j.bmc.2013.03.018

Yu CW, Li WH, Hsu FL, Yen PL, Chang ST, Liao VH (2014) Essential oil alloaromadendrene from mixed –type *Cinnamomum osmophloeum* leaves prolongs the lifespan in *Caenorhabditis elegans*. J Agric Food Chem 62:6159–6165. https://doi.org/10.1021/jf500417y

Zahin M, Hasan S, Aqil F, Khan MSA, Husain FM, Ahmad I (2010) Screening of certain medicinal plants from India for their anti-quorum sensing activity. Indian J Exp Biol 48:1219–1224

Zawacki A, O'Rourke E, Potter-Bynoe G, Macone A, Harbarth S, Goldmann D (2004) An outbreak of *Pseudomonas aeruginosa* pneumonia and bloodstream infection associated with intermittent Otitis externa in a healthcare worker. Infect Control Hosp Epidemiol 25:1083–1089. https://doi.org/10.1086/502348

Zeng Z, Qian L, Cao L, Tan H, Huang Y, Xue X, Shen Y, Zhou S (2008) Virtual screening for novel quorum sensing inhibitors to eradicate biofilm formation of *Pseudomonas aeruginosa*. Appl Microbiol Biotechnol 79:119–126. https://doi.org/10.1007/s00253-008-1406-5

Zhou L, Zheng H, Tang Y, Yu W, Gong Q (2013) Eugenol inhibits quorum sensing at sub-inhibitory concentrations. Biotechnol Lett 35:631–637. https://doi.org/10.1007/s10529-012-1126-x

Chapter 4
Quorum Quenching and Biofilm Inhibition: Alternative Imminent Strategies to Control the Disease Cholera

Lekshmi Narendrakumar, Bhaskar Das, Balasubramanian Paramasivan, Jayabalan Rasu, and Sabu Thomas

Abstract *Vibrio cholerae*, the causative agent of the disease cholera still threatens a large proportion of world's population and is considered as a top priority enteric pathogen. Role of biofilm in *V. cholerae* pathogenesis is well established as it provides the bacterium with enhanced transmission ability during epidemics and also enhanced tolerance to antimicrobial agents. The clinical efficacy of many existing antibiotics is being threatened by the emergence of multi-drug resistant *V. cholerae*. The rapidly increasing number of cholera outbreaks in several developing countries and emergence of multidrug resistant *V. cholerae* necessitates the development of an alternative strategy rather than the existing antibiotic therapy to control the pathogen. In the present chapter, we discuss the different quorum sensing pathways in *V. cholerae*, the common quorum quenching molecules that targets these pathways and a novel strategy of biofilm inhibition in *V. cholerae* using antibiofilm compounds in combination with antibiotics to control the disease. Co-dosing strategy reduce the dosage of antibiotics and such a combination therapy can in turn be used to control the spread of antibiotic resistance.

Keywords Antibiofilm compounds · Antibiotic resistance · Combinatorial therapy · Quorum sensing · Quorum quenching · *Vibrio cholerae*

L. Narendrakumar · S. Thomas (✉)
Cholera and Biofilm Research Laboratory, Rajiv Gandhi Centre for Biotechnology (National Institute under the Department of Biotechnology, Government of India), Thiruvananthapuram, Kerala, India
e-mail: sabu@rgcb.res.in

B. Das · B. Paramasivan
Agriculture and Environmental Laboratory, Department of Biotechnology and Medical Engineering, National Institute of Technology, Rourkela, Odisha, India

J. Rasu
Food Microbiology and Bioprocess Laboratory, Department of Life Science, National Institute of Technology, Rourkela, Odisha, India

© Springer Nature Singapore Pte Ltd. 2018
V. C. Kalia (ed.), *Biotechnological Applications of Quorum Sensing Inhibitors*,
https://doi.org/10.1007/978-981-10-9026-4_4

4.1   Introduction

Vibrio cholerae is the causative agent of the acute diarrheal illness cholera, once the most feared of all pandemics. Today, cholera is managed by oral rehydration therapy and broad spectrum antibiotics, yet the disease is insuperable during epidemics. In 2015, case fatality ratio (CFR) of the disease reported from 42 countries was 0.8%, which includes a total of 172,454 cases with 1304 deaths (World Health Organization 2016). Cholera outbreaks were reported from 16 countries in Africa, 13 in Asia, 1 in Oceania, 6 each in Europe and America. Globally it has been estimated that a total of 21,000–143,000 deaths occur every year from 1.3–4.0 million cholera cases (Ali et al. 2015). However, exact number of cases remains unknown because of the lack in a strong surveillance programs to monitor the disease in developing countries.

V. cholerae is commonly found in saline water bodies, but act as a facultative pathogen when sufficient infective dose reach the human host. There are about 206 serogroups of *V. cholerae* out of which O1 and O139 are known to cause outbreaks. *V. cholerae* O1 are further divided as classical and El Tor biotypes based on phenotypic and genotypic differences. The other serogroups of the bacteria are collectively known as non O1/O139 *V. cholerae* as they do not agglutinate either O1 or O139 polysera. These non O1/O139 *V. cholerae* were considered to be non-toxigenic until it caused a colossal outbreak in1992 (Dutta et al. 2013). However, though improper sanitation has been found to be an important factor for sudden cholera outbreaks, detailed understanding about the role of other factors that play role in sudden outbreaks are still unclear. There have been two hypotheses on the occurrence of sudden cholera outbreaks, the first being that the onset of infection in the population is by transmission of the pathogen from asymptomatic carriers to healthy individuals. This hypothesis explicates that the asymptomatic individuals infect local water bodies leading to explosive sporadic outbreaks of cholera (Frerichs et al. 2012; Eppinger et al. 2014). The second hypothesis assumes that the sporadic sudden cholera cases are initiated when a healthy individual acquires the pathogen from environmental autochthonous toxigenic vibrios (Huq et al. 1983; Alam et al. 2006). Temperature, salinity, nutrients and precipitation are important factors which influences the persistence of vibrios in environmental waters. Also, the pathogen comprises a major component of the commensal flora of phyto-zooplankton persisting as biofilms on them (Jutla et al. 2013).

Biofilms are surface attached, matrix enclosed, multicellular communities of bacteria found in association with both biotic and abiotic surfaces. *V. cholerae* exist in both planktonic and biofilm state during intestinal and aquatic phase of its life cycle. Role of biofilm in *V. cholerae* pathogenesis is well established as it provides the bacterium with enhanced transmission ability during epidemics and also enhanced tolerance to antimicrobial agents. *V. cholerae* biofilms, in its environmental phase provides protection from environmental stresses such as grazing protozoa, bacteriophages and nutrient limitation (Matz et al. 2005). Also, *V. cholerae* biofilms on chitin surfaces induces its natural competence whereby the cells acquire new genetic materials such as resistance genes (Meibom et al. 2005). During inter epidemic periods, metabolically quiescent *V. cholerae* persist in the environment in biofilm state known as Viable but Non Culturable vibrios (VBNC) which cannot be cultured by normal traditional culturing methods. These cells become active only after a passage through the host or from signals produced by active cells present in the environment (Colwell et al. 1996; Bari et al. 2013). The pathogen biofilms in environment have important biological relevance as it provide protection to harsh environmental stresses and also increase the infective dose of cells entering a host capable of initiating the infection. Though development of *V. cholerae* biofilms within host is poorly understood, recent studies using rabbit ileal loop infection models have identified clusters of *V. cholerae* in microcolonies, supporting the fact that *V. cholerae* forms biofilms *in vivo* which will be subsequently excreted in stool thereby increasing the transmission of the disease cholera (Faruque et al. 2006). In the present chapter, we discuss the role of biofim in *V. cholerae* antibiotic resistance, the different quorum sensing systems in the bacteria and different quorum quenching mechanisms. The chapter highlights on the quorum quenching/antibiofilm therapy as an alternative imminent strategy to control the increasing antibiotic use and its probable advantages in decreasing the burden of antibiotic resistance in the pathogen.

4.2   Association of Antibiotic Resistance and Biofilm

Lately, *V. cholerae* strains resistant to these commonly used antibiotics have appeared with increasing frequency in India, as well in as other countries (Bilecen et al. 2015; Gupta et al. 2016). The pathogen develops resistance to antibiotics through several mechanisms. Biofilm formation in *V. cholerae* has been associated with its resistance gene acquisition. Horizontal gene transfer promotes evolution and genetic diversity of the pathogen. Gene transfer among the environmental and clinical strains in the natural environments has led to the emergence of multidrug-resistant bacteria (Martínez 2012). Also, previous studies have demonstrated that EPS matrix prevents diffusion of antimicrobial agents thus providing a protective niche to the bacterial pathogen which is best known to increase transmission via biofilm microcolonies (Teschler et al. 2015). The common antibiotics taken against cholera such as the quinolone antibiotics, tetracyclines and erythromycin does not reach the bacteria that resides within the biofilm. However these cells get exposed to sub-inhibitory concentrations of the antibiotics leading to adaptive evolution of the bacteria to these antibiotics (Bengtsson-Palme and Larsson 2016). Draft genome sequence analysis of an evolved Haitian variant *V. cholerae* strain isolated from a recent outbreak in south India revealed the presence of aminoglycoside gene, *strB* and *strA*, sulphonamide resistance gene, *sul2* and phenicol resistance genes, *floR* and *catB9* suggesting the real time evolution of *V. cholerae* to the commonly used antibiotics (Narendrakumar et al. 2017).

4.3   Genetic Determinants of *V. cholerae* Biofilm Formation

V. cholerae biofilm formation is initiated by the attachment of cells to biotic or abiotic surfaces by Type IV pili (TFP). The bacteria have three types of TFP namely, toxin co-regulated pili (TCP), mannose –sensitive hemagglutinin (MSHA) pilin and chitin- regulated pili (chiRP) of which MSHA pili play a major role in biofilm development. After surface attachment, the pathogen produces an extrapolymeric matrix composed of proteins, nucleic acids and Vibrio exopolysaccharide (VPS). Matrix proteins such as RbmA, RbmC, Bap1 along with VPS maintain the integrity of these biofilms. The VPS genes are clustered as *vpsI* and *vpsII* and located at two different positions in the large chromosome of *V. cholerae* separated by the rbm gene cluster. *vpsI* cluster comprises of *vpsA-K* genes (VC0917-27) and *vpsII* cluster comprises of *vpsL-Q* (VC0934-9) (Fong et al. 2010). Several studies have reported that the VPS genes are important for the colonization and biofilm development both *in-vitro* and *in-vivo*. Expression of genes such as *vpsA* (VC0917), *vpsB* (VC0918), *vpsC* (VC0919) and *vpsN* (VC0936) are found to be up-regulated during the initial colonization stages and final stages of infection in animal model experiments. VpsH (VC0924) was identified to be at detectable levels in cholera patients on *In vivo*-induced antigen technology (IVIAT) (Hang et al. 2003). In frame deletion mutation of vps gene clusters significantly reduced the ability of *V. cholerae* to produce biofilm. However, not all genes in the *vpsI* and *vpsII* gene clusters were important for its biofilm formation. *V. cholerae vps* clusters are positively and negatively regulated by VpsR, VpsT and HapR, CytR respectively (Teschler et al. 2015). VpsR and VpsT were identified to be important for the maximal expression of *vps* genes and mutation studies in either of the genes significantly reduced the biofilm formation in *V. cholerae* (Yildiz et al. 2001). The positive regulators VpsR and VpsT have been identified to be homologous to the two-component regulatory systems that are involved in sensing and responding to environmental stimuli with a sensor histidine kinase that regulates *vps* genes expression.

4.4   Quorum Sensing and Biofilm Formation

Cell to cell communication in bacteria for the regulation and expression of specific genes is known as bacterial quorum sensing (QS). Quorum sensing is dependent on the bacterial cell density and concentration of chemical signal molecules known as autoinducers (AIs) produced by the bacteria. These AIs are released into the environment which accumulate and signals other bacteria of the same kind for collective expression of specific genes for definite functions such as virulence, biofilm formation, bioluminescence etc (Williams and Camara 2009). Many of the molecular mechanisms for intracellular signaling in bacteria are now well studied. Different quorum sensing small molecules produced by microbes include Acyl Homoserine Lactones (AHL), Peptide auto-inducers and Auto-inducer 2.

4.4.1   Acyl Homoserine Lactones (AHL)

AHLs are produced within the bacterial cells and released into the environment. AHLs produced by different bacteria differ in the R-group side chain length which can vary from 4 to 18 carbon atoms and the carbonyl group at the third carbon. AHL signals contains a homoserine lactone linked by a amide bond to the acyl side chain. The first identified AHL molecule in QS was in the marine bacterium *Vibrio fischeri* which were responsible for the bioluminescence in the light organs of Hawaiian bobtail squid, *Euprymna scolopes* (Ruby 1996). After this discovery, AHL signal molecules were identified from different Gram- negative bacteria including *V. cholerae*. AHL signal molecules are chiefly catalyzed by LuxI enzyme family and perceived by LuxR cytoplasmic DNA binding proteins. Based on the length of the side chain, AHL molecules can be distinguished as short side-chain AHL moleculaes and long side- chain AHL molecules. Short chain AHLs diffuse freely across the cell membrane whereas the long chain AHLs requires active efflux pumps to export them from within the cell. Different bacteria are known to produce different AHL molecules and few of them produce diverse AHL signals (Huma et al. 2011). There are also reports of bacteria like *Pseudomonas simiae* and *Pseudomonas brenneri* isolated from Ny-Alesund, Arctic at 79°N producing varied AHL molecules at different temperatures (Kalia et al. 2011; Dharmaprakash et al. 2016).

4.4.2   Peptide Auto-inducers

Many Gram-positive bacteria utilize peptides as signal molecules for QS. They are usually secreted oligopeptides which results from post-translational modifications. Peptide signals can differ in size from 5 to 87 amino acids and also can contain modifications like lactone or thiolactone linkages. These peptide auto-inducers require special export mechanisms like ATP binding cassette transporters. After the accumulation of sufficient peptide auto-inducer in the environment, the signal molecules are perceived by histidine sensor kinase protein of a two-component regulatory system in the bacteria leading to expression of specific genes. Competence signal peptides (CSP) produced by streptococcal species is an example for peptide auto-inducer (De Spiegeleer et al. 2015). Bacteriocins are also categorized to be a peptide AI (Zhao and Kuipers 2016). ComX bacterial peptide AI induces sporulation in *Bacillus subtilis* by activating the ComP/ComA two-component phosphorylation cascade. Activation of the ComP/ComA component in turn leads to increased gene expression of the transcriptional activator ComK which in turn activates the expression of genes required for sporulation (van Sinderen et al. 1995). In *Staphylococcus aureus,* peptide AI system *agr* play an important role in virulence and pathogenicity of the organism (Baldry et al. 2016). The *S. aureus* peptide AI is detected by AgrC/AgrA sensor kinase/response regulator which gets phosphorylated. The phosphorylated ArgA upregulates RNAIII which positively regulates the expression of virulence genes of *S. aureus.*

4.4.3 Auto-inducer 2

Auto-inducer 2 is a furanosyl borate diester signal molecule. AI-2 is produced by many Gram positive and Gram negative bacteria and is believed to be an evolutionary link between the two QS systems. AI-2 facilitates cross species communication in bacteria. The AI-2 system was also first reported from the marine bacterium *V. fischeri*. For bacteria like *E. coli* and *Salmonella typhimurium*, a group of genes called *lsr* gene cassette are induced that encode AI-2 components of the machinery (Pereira et al. 2013). Periplasmic protein LsrB act as the receptor for AI-2 signal molecules which will be delivered inside to the cell cytoplasm by an ABC transporter. This transportation of the signal molecule into the cytoplasm is coupled to its phosphorylation by LsrK kinase (Xavier and Bassler 2005). The prime function of the phosphorylated AI-2 is to deactivate the transcriptional repressor LsrR which in turn activates the positive feedback of AI-2. The synthesis of AI-2 is facilitated by the enzyme LuxS which converts S- ribosyl homocysteine to homocystein and 4,5-dihydroxy-2,3-pentanedione (DPD), the precursor compound of AI-2 (Marques et al. 2011).

4.4.4 Quorum Sensing Signals in Vibrionaceae Family and Biofilm Formation

Vibrionaceae comes under the Gammaproteobacteriae comprising facultative anaerobic Gram negative bacteria such as *V. cholerae*, *V. parahaemolitycus*, *V. harveyii*, *V. anguillarum*, *V. vulnificus*, *V. fischeri* etc. The family includes *Vibrio* and *Photobacterium* genera. Of these genera are many top priority human diarrheal pathogens such as *V. cholerae* and *V. parahaemolyticus* and fish pathogens such as *V. anguillarum* and *V. vulnificus*. There are also many bacterial symbionts such as *V. pomeroyi*, *V. aestuarianus*, and *V. fischeri* in this family (Yang et al. 2011). QS systems have been identified in many bacteria of the Vibrionaceae family such as *V. harveyi* (Bassler et al. 1997), *V. cholerae* (Zhu et al. 2002), *V. anguillarum* (Milton et al. 1997; Buchholtz et al. 2006), *A. salmonicida* (Bruhn et al. 2005), *V. vulnificus* (Valiente et al. 2009) and *Photobacterium phosphoreum* (Flodgaard et al. 2005). In a study to determine global phylogenetic distribution of AHL molecules in bacteria belonging to Vibrionaceae family using biological monitors and LC-MS identification, it was identified that acyl homoserine lactones, including AHLs with odd numbers of carbon, was the most abundant signaling molecule. The study also revealed an AHL fingerprint correlated to specific phylogenetic subclades in Vibrionaceae family (Rasmussen et al. 2014). However, quorum sensing systems rapidly diverge in nature and signal orthogonality and mutual inhibition frequently occur among closely related diverging systems of Vibrionaceae. Different homologues of LuxI/ LuxR proteins have been identified that respond to each of the different AHL molecules. Also, the degree of sensitivity of LuxI/LuxR to different AHL molecules differ (Tashiro et al. 2016).

Most of the members in the Vibrionaceae family are also good biofilm formers. QS controls biofilm formation in Vibrionaceae by different mechanisms such as regulating coordinated behavior and synchronized environmental response (Mireille Aye et al. 2015; Okutsu et al. 2015), regulating the synthesis of biofilm matrix (Tseng et al. 2016), indirectly upregulating biofilm formation by increasing bacterial motility (Yang et al. 2014) and also regulate dispersal of matured biofilm to initiate a new developmental cycle of biofilm formation (Emerenini et al. 2015). A recent study has proved the ability of *Vibrio parahaemolyticus* to regulate biofilm formation and enhancement of colonization by QS (Vinoj et al. 2014). Also studies on genetic basis of QS mediated regulation of virulence related gene Hfq, motility/extracellular protein Pep and colony phenotype intermediated protein valR was studied in biofilm forming *V. alginolyticus* (Chang et al. 2010; Cao et al. 2011; Liu et al. 2011). VanT, a homologue of *Vibrio harveyi* LuxR, is known to regulate biofilm formation in *V. anguillarum* (Croxatto et al. 2002). AphA, DNA binding regulators in vibrios belonging to padR family proteins which is important for the QS in vibrios at low cell density is known to regulate biofilm formation, motility and virulence in pandemic *V. parahaemolyticus* (Wang et al. 2013). It is also interesting to note that there is about 85% homology between the AphA protein of *V. parahaemolyticus* and *V. cholerae* suggesting a possibility of similar regulation in *V. cholerae* also.

4.4.5   V. cholerae *Quorum Sensing Systems*

V. cholerae posseses two well characterized and one predicted quorum sensing mechanism as compared to *V. fischeri*. The well-established systems of *V. cholerae* includes the CqsA/CAI-1/CqsS system and LuxS/AI2/LuxPQ system (Ke et al. 2014). In *V. cholerae,* there are two AIs such as the Cholera Auto-inducer (CAI), S-3-hydroxy-tridecan-4-one and Auto-inducer-2 (2S, 4S- 2- methyl-2,3,3,4-tetrahydroxy tetrahydrofuran-borate) and two cognate receptors. The CAI-1 signals are synthesized by CqsA and sensed by CqsS system sensor (Ng and Bassler 2009). AI-2 is synthesized by LuxS and the cognate sensor is LuxP/Q. The components of the third predicted QS system in *V. cholerae* is unknown.

At low bacterial cell density when the AIs are at minimum concentration, the QS system sensors CqsS and LuxP/Q transfers a phosphate group to the cytoplasmic integral protein LuxU which in turn phosphorylates LuxO (Milton 2006). The phosphorylated LuxO in association with the σ^{54} initiate the repression of *hapR* via the activation of a putative repressor. HapR is a transcriptional repressor of the genes which have specific functions in *V. cholerae* biofilm formation and virulence (Amy et al. 2009). Previous studies have reported that a deletion mutant of *hapR* produces strong biofilm compared to its wild type counterpart. At a higher cell density, the AIs exported out reaches the maximum threshold level converting the LuxQ and CqsS kinases to phosphatases. This backflow of phosphate group from the LuxO destabilizes the repression of putative protein on *hapR* activating its expression. HapR activates Haemagglutinin Protease A (HapA) which disperses the *V. cholerae* in biofilm to planktonic cells.

Another major component that determines the biofilm formation of *V. cholerae* is the concentration of a secondary messenger, c-di-GMP (Cotter and Stibitz 2007). c-di- GMP messengers are synthesized by proteins containing specific GGDEF motifs and its degradation is carried out by proteins containing domains with EAL or HD-GYP motifs. Previous studies reports the presence of 62 genes that encode proteins governing c-di-GMP levels in *V. cholerae* (Galperin 2004). Increase in the concentration of the secondary messenger within the bacterial cell induces the activation of the transcriptional activator *vpsT*. VpsT in turn activates the genes in the vps clusters leading to the increase in extracellular matrix production and increased biofilm. However, the master transcriptional repressor gene HapR is identified to regulate the c-di-GMP mediated biofilm formation at two levels. HapR has been identified to repress 14 genes that encodes for proteins that synthesis c-di-GMP. Also, HapR directly represses the *vpsT* gene which prevents the downstream activation of *vps* genes.

4.5   Quorum Quenching

Quorum quenching (QQ) or QS interference is a strategy wherein the cell to cell communication of bacteria is interrupted by specific compouds. QQ strategy came up with the understanding that bacterial cells have the ability to communicate with each other to collectively regulate various important traits such as virulence, biofilm formation, antimicrobial resistance (Sharma and Jangid 2015; Shiva Krishna et al. 2015). This communication occurs via various signaling molecules which are recognized by specific receptors. The signal molecule binding to the receptor further activates a cascade of molecular signaling which ultimately results in the specific gene regulation. Theoretically, mechanisms that can effactually interfere this signaling/communication between the bacteria can be used as a QQ molecule. A QQ molecule can act on either the signaling molecule, receptor to the signal, regulatory proteins in the signaling cascade.

4.5.1   Quorum Quenching by Signal Molecule Degradation

There are QQ enzymes like AHL-lactonase, AHL-acylase and paraoxonases (PONs), which degrade AHL signals.

4.5.1.1 Quorum Quenching by AHL-Lactonase

AHL- lactonase act on the QS signal, AHL by hydrolyzing the homoserine lactone ring (Dong and Zhang 2005). The first AHL-lactonase, encoded by the *aiiA* gene was identified from a *Bacillus* sp. isolate 240B1. The enzyme has two zinc ions at their active site which initiates a nucleophilic attack at the substrate's carbonyl carbon group. The lactone ring interact with the enzymes Zn^1 and the substrate's carbonyl oxygen interact with Zn^2 thereby weakening the carbonyl bond breaking the lactone ring to yield an open-ring product. Due to its specificity in action, AHL-lactonase is the most precise AHL-degradation enzyme which can hydrolyse both short chain and long chain AHLs efficiently. The ability of AiiA to disrupt QS in *V. harveyii* and *V. cholerae* and thereby inhibit their biofilm formation has been well documented (Bai et al. 2008; Augustine et al. 2010). AiiA enzyme have been successfully cloned into plants such as potato, tobacco, eggplant, cabbage, carrot and celery plant to produce *Erwenia carotovora* infection resistant crops (Dong et al. 2001). Also, *Escherichia coli* containing cloned *Bacillus thuringiensis aiiA* was identified to express greater amount of the enzyme compared to the parent strain increasing its industrial applications (Lee et al. 2002). Another AHL lactonase AttM identified from Agrobacterium tumefaciens with only 24.8% identity to AiiA was characterized to have bacterial fitness properties within the plant tumor (Haudecoeur et al. 2009).

4.5.1.2 Acylases

AHL-Acylases, like lactonases interfere with bacterial QS to attenuate major functions such as virulence, motility and biofilm production. Acylases have been identified to decrease 3OC12HSL and C4HSL accumulation within the bacterial cells which drives the virulence factor production machinery. Acylases have been identified both in Gram-positive and Gram- negative bacteria. AHL-acylase AhlM identified from Streptomyces strain M664 could cleave both medium and long chain AHL signal molecules. AHL acylases AiiC and Aac identified from Anabena strain PCC7120 and Shewanella respectively were proved effective in disrupting biofilm formation of fish pathogen *V. anguillarum in-vitro* (Morohoshi et al. 2008).

4.5.1.3 Oxidoreductases

A third class of AHL- degrading enzymes are the oxidoreductases that inactivates AHL molecules via oxidation or reduction of the acyl chain of the signal molecule. P450 monooxygenase and the NADH-dependent enzyme BpiB09 are the two well-studied AHL signal molecule degrading oxidoreductases. Heterologous expression of BpiB09 have been identified to decrease AHL accumulation in *Pseudomonas aeruginosa* subsequently inhibiting swarming motility, biofilm formation and pyocyanin production (Bijtenhoorn et al. 2011; Kumar et al. 2015).

4.5.2 Quorum Quenching by Inhibition of Signal Molecule Synthesis

Apart from degrading signal molecules that play important role in bacterial cell to cell communication, there have been many compounds that inhibit or decrease the signal molecule synthesis thereby interfering QS. Most of the signal molecule synthesis inhibitors work by indirectly inhibiting precursor molecules which are important for signal molecule synthesis. For example, small molecule triclosan act on enoyl-ACP reductase, an important precursor of AHL synthesis and immucillin A (ImmA) inhibits 5-MAT/S-adenosyl-homocysteine nucleosidase (MTAN) which is crucial for both AHL and AI-2 synthesis (Hoang and Schweizer 1999; Singh et al. 2005). Very low concentration of nucleoside analogues have been found effective to inhibit MTAN activity in virulent *V. cholerae* strains (Gutierrez et al. 2009).

Recent studies on *V. cholerae* QS precursor molecule (S)-4,5-dihydroxy-2,3-pentanedione (DPD) showed that nucleoside analoges of DPD delineate QS in *V. cholerae, V. harveyi, V. anguillarum, V. vulnificus and S. typhimurium* (Meijler et al. 2004; Lowery et al. 2008; Smith et al. 2009; Brackman et al. 2009). Adenosine analogues have been found to have potent antibiofilm activity by blocking AI-2 based QS.

4.5.3 Quorum Quenching by Inhibition of AHL Receptor

The CqsS receptor of *V. cholerae* and *V. harveyi* share extensive homology. However, when it comes to inhibition of CqsS receptors in both the bacteria, possess different overall stringencies for ligands. Many small molecule inhibitors that block the binding of signal molecule to the receptor thereby cutting off the QS signaling cascade has been identified recently. High throughput screening studies have identified many small molecule inhibitors of QS receptors in *V. cholerae* (Peach et al. 2011).

4.6 Antibiofilm Activity of Natural Compounds Against *V. cholerae*

Natural products are a good source of compounds that have various biological activities including antimicrobial and antibiofilm properties. These products are considered safe to administer and are believed to cause lower degree of antimicrobial resistance unlike antibiotics. Many of these compounds are identified to inhibit bacterial virulence or biofilm formation by interfering with the QS of bacterial pathogens. List of important QQ molecules have been listed in Table 4.1.

Table 4.1 List of important quorum quenching molecules in Vibrionaceae family

Quorum quencher	Active against	References
Hexyl-4,5-dihydroxy-2,3-pentaedione	*V. harveyi*	Schaefer et al. (1996)
		Lowery et al. (2009)
4-hydroxy cis or trans analogs	*V. fischeri*	Olsen et al. (2002)
N-sulfonyl-HSL	*V. fischeri*	Castang et al. (2004)
Furanone C-30	*V. anguillarum*	Rasch et al. (2004)
N-(heptyl-sulfanyl acetyl)-L-HSL (HepS-AHL)	*V. fischeri*	Persson et al. (2005)
AiiA enzyme	*V. cholerae*	Augustine et al. (2010)
Oceanobacillu/ Halomonas extract	*V. cholerae* and *V. cholerae. parahaemolyticus*	Nithya et al. (2010)
Artctic actinomycetes extract	*V. cholerae*	Augustine et al. (2012)
Resveratrol	*V. cholerae*	Augustine et al. (2014)

4.6.1   Antibiofilm Activity of Phytochemicals

Plant extracts and plant compounds have been used since time immemorial to treat diarrheal diseases and other stomach ailments. Traditional medicines of India and China use plant extracts for treating bronchitis, asthma, gastric ailments, phlegm, dysentery, leukorrhea, kidney trouble, urethritis, and dropsy (Mitra et al. 2007).

Resveratrol, a phytochemical present mostly in the skin of grapes, blueberries, raspberries, mulberries etc have been identified to target upstream targets of QS system, reducing *V. cholerae* biofilm production upto 80% (Fig. 4.1). Molecular docking studies, AphB, a LysR-type regulator important in the QS of *V. cholerae* was identified to be the putative target for resveratrol (Augustine et al. 2014).

Vikram et al. (2011) demonstrated that flavonoid compounds such as naringenin and quercetin isolated from citrus fruits have been identified to reduce biofilm formation in *V. harveyii* and *V. cholerae* by acting as antagonists of AHL and AI-2 (Vikram et al. 2011). Also, several studies are available showing the anti-virulence activity of plants like 'neem', apple, hop, green tea and elephant garlic via inhibition of bacterial growth or the secreted cholera toxin (CT). Mangrove plant extracts have been known for their mosquito larvicidal, antifungal, antiviral, anti-cancer and anti-diabetic activity. Leaves and bark extracts of mangrove plants *Ceriops tagal* and *Pemphis acidula* revealed to have potent antibiofilm ability against a wide set of microorganisms such as *P. aeruginosa*, *Klebsiella pneumonia*, *V. parahaemolyticus*, *S. aureus* and *V. cholerae* (Arivuselvan et al. 2011). Cinnamaldehyde and its derivatives have been found effective as an antibiofilm compound against all Vibrio sp., by interfering AI-2 mediated QS pathway by decreasing the DNA-binding ability of

Fig. 4.1 Isolation of active compounds from natural sources and analysis of antibiofilm activity of the compound against *V. cholerae*. CLSM images of *V. cholerae* biofilm inhibition using resveratrol. (*a*) untreated culture. (*b, c*) Biofilm treated with 15 & 20 μg/ml of resveratrol. (*d*) 3D view of biofilm thickness of untreated. (*e, f*) 3D view of biofilm thickness of cultures treated with 15 and 20 μg/ml resveratrol

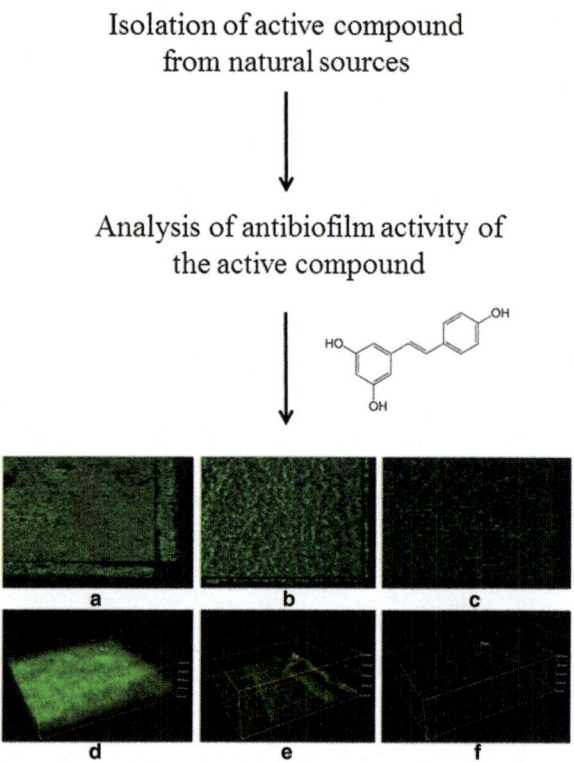

LuxR (Brackman et al. 2008). Also, in *V. cholerae*, cinnamaldehyde analogues have been proved to have antibiofilm (Niu et al. 2006; Brackman et al. 2011). Water soluble extract of Cranberry has found to have potent antibiofilm property against *V. cholerae* acting via down-regulating the *vps* operon by modulating the level of c-di-GMP in the QS pathway. Also it has been shown that Cranberry extracts can block the initial attachment of *V. cholerae* into the host enterocytes during an infection and also inhibit cholera toxin production by regulating LuxO-HapR pathway (Dinh et al. 2014).

4.6.2 Antibiofilm Activity of Marine Compounds

Compounds from marine sources have been a rich source of bioactive compounds. Marine organisms, plants and even sediment compounds have shown antibacterial and antibiofilm activity against a wide array of bacterial pathogens. Quorum quenching from marine sources gained its importance from early discoveries of antibiofilm activity of these compounds against biofouling bacteria (Kalia et al. 2014). Marine bacterial exopolysaccharaides have been able to disrupt Vibrio sp., QS and thereby inhibit virulence gene expression and biofilm formation. Many marine bacteria such as *Oceanobacillus and Halomonas* have been identified to produce small molecule QQ enzymes that are able to disrupt *V. cholerae* and *V. parahaemolyticus* biofilm. Bacteria such as *Bacillus indicus, B. pumilus* and *Bacillus* sp. SS4 isolated from Palk Bay (Bay of Bengal) were shown to cause substantial inhibition of QS based biofilm formation in Gram-negative bacteria such as *Vibrio* species, *Serratia marcescens* and *P. aeruginosa* PAO1 (Nithya et al. 2010, 2011; Musthafa et al. 2011). Other major sources of antibiofilm compounds are halogenated furanones produced by *Delisea pulchra,* secretions from *Chlamydomonas reinhardtii* which mimic bacterial QS signals, bromoperoxidase produced by algae *Laminaria digitata* which deactives AHL molecules, *Ahnfeltiopsis flabelliformis,* red algae that produces α-D-galactopyranosyl-(1 → 2)-glycerol (Floridoside), betonicine and isoethionic acid which produces QS analogues (Gram et al. 1996; Manefield et al. 2000; Borchardt et al. 2001; Teplitski et al. 2004; Kim et al. 2007; Musthafa et al. 2011).

4.6.3 Activity of Synthetic QS Analogues

Many synthetic analogues have been identified to target QS molecules of Gram negative bacteria including *V. cholerae*. Though most of the QQ studies of synthetic quorum sensing inhibitors have been carried out in *V. fischeri*, the compound being translated to target QS systems of *V. cholerae* is not very difficult. Probable QQ compounds that could be used against *V. cholerae* are 4-hydroxy cis or trans analogs of HSL ring of signal molecule 3OC8HSL which have been found effective in inhibiting *V. fischeri* LuxR based QS reporter system, N-(heptyl-sulfanyl acetyl)-L-HSL (HepS-AHL) identified as potent LuxR inhibitor in *V. fischeri*, N-sulfonyl-HSL (with a pentyl chain), identified as an potent LuxR analogue in *V. fischeri*, Hexyl-4,5-dihydroxy-2,3-pentaedione identified to interfere QS in *V. harveyi* and thereby reduce its bioluminescence and Furanone C-30 which reduce virulence gene expression of *V. anguillarum* (Schaefer et al. 1996; Olsen et al. 2002; Castang et al. 2004; Rasch et al. 2004; Persson et al. 2005; Lowery et al. 2009). Nanoparticle therapy has also been proved to be effective against bacterial biofilms (Dobrucka and Długaszewska 2015; Szweda et al. 2015; Ahiwale et al. 2017). Quest for a single quorum quenching molecule that can limit quorum sensing in multiple pathogens is underway (Koul and Kalia 2017).

4.7   Advantage of Antibiofilm Drugs Over Antibiotics

Biofilms are generally insensitive to antibiotics. However, use of antibiotics or anti-microbial agents is effective against planktonic bacteria. Planktonic bacterial cells are known to be 1000- times sensitive to antibiotics compared to their biofilm counterparts (Rasmussen and Givskov 2006). Antibiotics and other antimicrobial agents generally act by inhibiting the growth of bacterial cells by interfering with their major metabolic or biosynthesis pathway killing them. Biofilm mode of life helps bacteria to survive the harsh antibiotic treatment. Within the biofilm, bacteria have the ability to exist as heterogeneous population and have a reduced metabolic activity (Mah et al. 2003). Bacterial cells with reduced metabolic activity have been identified to be inherently more resistant to antimicrobial therapies (Coates and Hu 2008).

Prolonged persistence of such antimicrobial compounds can induce toxicity to not just the target microorganism but also to non-target beneficial microbes existing in the particular environment. Another major problem in using antimicrobial agents over long period of time is the emerging antibiotic resistance. Multidrug resistant bacteria have now days become a global problem aggravating mortality due to infections (Neu 1992). This escalation of antimicrobial resistance among bacterial pathogens worldwide has necessitated an urgent need to look for alternative strategies to combat bacterial infections by reducing virulence rather than by killing the bacteria.

4.8   Combinatorial Therapy of Antibiofilm Agents and Antibiotics

Since the discovery of persistent antibiotic resistant cells in bacterial biofilms, alternative approaches to control bacterial diseases by using a combination of quorum sensing inhibitors that inhibit or disperse the biofilm and antibiotics that kill the dispersed bacterial cells have gained importance. Co-dosing of QSI/bio-film inhibitors with sub-inhibitory concentration of an antibiotic that can kill the bacteria that become sensitive to antibiotics upon release from biofilms in turn reduces the chance of antimicrobial resistance as these compounds target bio-film/QS pathways and does not affect bacterial growth. A pictorial representation of the combinatorial therapy that could be used to treat cholera has been depicted in Fig. 4.2.

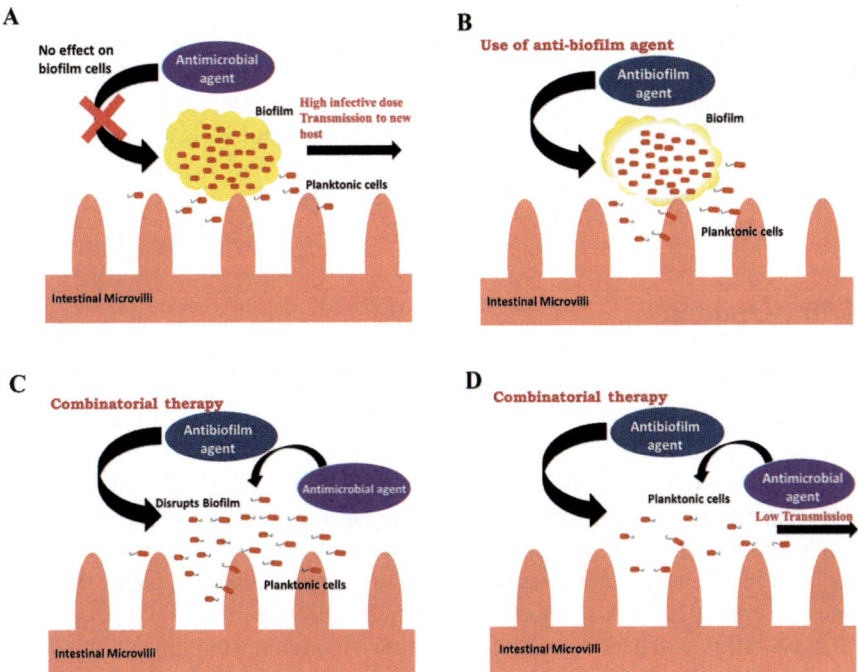

Fig. 4.2 (**a**) Conventional therapy of antibacterial agent alone (**b**) Use of anti-biofilm agent (**c** & **d**) Combinatorial therapy using anti-biofilm agent and antibacterial agents

4.9   Conclusion

The escalation of antimicrobial resistance among bacterial pathogens worldwide is becoming a critical concern. This necessitates an urgent need to look for alternative strategies to combat bacterial infections by reducing virulence rather than by killing the bacteria. In this context, scientists from all over the world are trying to develop novel therapeutic strategies that give importance to bacteriostatic compounds rather than bacteriocidal drugs. Similar to many opportunistic pathogens *V.cholerae* also rely on Quorum sensing, a bacterial cell to cell communication system for biofilm formation and virulence character expression. Quorum quenching or antipathogenic approach and biofilm inhibition will be the promising alternative strategies to contain the disease in future. Vibrios that reside within mature biofilms are highly resistant to antibiotics and host immune response due to the complex architecture and composition of the extracellular matrix. Recent studies have shown the key role played by the biofilm mode of life adapted by vibrios in the emergence of resistant strains, pathogenicity, host colonization and survival in the natural as well as human niches of vibrio species. Small molecules that can inhibit biofilm formation of *V.cholerae* and/or disrupt biofilms will greatly reduce transmission potential of the bacteria especially in epidemic situations. Inhibiting quorum sensing mechanism appears to be an ideal alternative to conventional therapy as *V.cholerae* suppresses both virulence and biofilm formation at high cell densities.

4.10   Opinion

Previous studies have strongly associated biofilm formation to the virulence of *V.cholerae*. Thus targeting biofilm formation is considered as a potential anti-virulent strategy to treat infections caused by bacterial pathogens. To the best of our knowledge, there is no antibiofilm compound that is clinically approved thus far. In this context, discovering potential Quorum Quenching/anti-biofilm compounds against vibrios is highly warranted. This can be achieved by employing two strategies: (1) Screening natural sources especially from anti-diarrheal plants (formulation of herbal extracts), actinomycetes (small molecules) and well characterised phytochemicals and peptides. (2) Identification of conserved biofilm inhibiting/quorum quenching targets by employing multi-omic approaches such as transcriptomics and proteomics. The identified targets could be used in modern computational based drug discovery approaches for accelerating the development of broad spectrum antibiofilm compounds against the major pathogens in the Vibrionaceae family. It has also been suggested that targeting the quorum sensing system and biofilm forming ability, which could disarm the bacteria may offer a affirming avenue to fray both pathogenesis as well as antibiotic resistance. Hence, by taking it as a top priority research agenda will definitely help to move forward to control the disease. Studies in some bacterial species have successfully demonstrated that these quorum sensing inhibitors can specifically bind to target proteins and inhibit virulence gene expression. However, the efficacy of these anti-quorum sensing drug targets for the inhibition of *V.cholerae* quorum sensing has not been extensively evaluated. A strengthened international networking and improved coordination among researchers active in cholera research program will accelerate the program at large.

Acknowledgement Lekshmi. N is obliged to Department of Science and Technology, Govt. of India for providing INSPIRE fellowship (Fellow code: IF 140851). Authors from NIT Rourkela acknowledge the PhD scholar institute fellowship given by NIT Rourkela, Odisha, India.

References

Ahiwale SS, Bankar AV, Tagunde S, Kapadnis BP (2017) A bacteriophage mediated gold nanoparticle synthesis and their anti-biofilm activity. Indian J Microbiol 57:188–194. https://doi.org/10.1007/s12088-017-0640-x

Alam M, Hasan NA, Sadique A, Bhuiyan NA, Ahmed KU, Nusrin S, Nair GB, Siddique AK, Sack RB, Sack DA, Huq A, Colwell RR (2006) Seasonal cholera caused by *Vibrio cholerae* serogroups O1 and O139 in the coastal aquatic environment of Bangladesh. Appl Environ Microbiol 72:4096–4104. https://doi.org/10.1128/AEM.00066-06

Ali M, Nelson AR, Lopez AL, Sack DA (2015) Updated global burden of Cholera in endemic countries. PLoS Negl Trop Dis 9:e0003832. https://doi.org/10.1371/journal.pntd.0003832

Amy TM, cai T, Lui Z, Zhu J, Kulkarni RV (2009) Regulatory targets of Quorum sensing in *Vibrio cholerae*: evidence for two distict HapR- binding motifs. Nucleic Acids Res 8:2747–2756. https://doi.org/10.1093/nar/gkp121

Arivuselvan N, Silambarasan D, Govindan T, Kathiresan K (2011) Antibacterial activity of Mangrove Leave and bark extracts against Human pathogens. Adv Biol Res 5:251–254. doi: Not Available

Augustine N, Kumar P, Thomas S (2010) Inhibition of *Vibrio cholerae* biofilm by AiiA enzyme produced from Bacillus spp. Arch Microbiol 192:1019–1022. https://doi.org/10.1007/s00203-010-0633-1

Augustine N, Wilson Peter A, Kerkar S, Thomas S, (2012) Arctic actinomycetes as potential inhibitors of Vibrio cholerae biofilm. Curr Microbiol 64(4):338–342

Augustine N, Goel AK, Sivakumar KC, Kumar RA, Thomas S (2014) Resveratrol—a potential inhibitor of biofilm formation in *Vibrio cholerae*. Phytomedicine 21:286–289. https://doi.org/10.1016/j.phymed.2013.09.010

Bai F, Han Y, Chen J, Zhang XH (2008) Disruption of quorum sensing in *Vibrio harveyi* by the AiiA protein of *Bacillus thuringiensis*. Aquaculture 274:36–40. https://doi.org/10.1007/s12088-013-0415-y

Baldry M, Kitir B, Frøkiær H, Christensen SB, Taverne N, Meijerink M, Franzyk H, Olsen CA, Wells JM, Ingmer H (2016) The *agr* inhibitors Solonamide B and analogues alter immune responses to *Staphylococccus aureus* but do not exhibit adverse effects on immune cell functions. PLoS One 11:e0145618. https://doi.org/10.1371/journal.pone.0145618

Bari SMN, Roky MK, Mohiuddin M, Kamruzzaman M, Mekalanos JJ, Faruque SM (2013) Quorum-sensing autoinducers resuscitate dormant *Vibrio cholerae* in environmental water samples. Proc Natl Acad Sci U S A 110:9926–9231. https://doi.org/10.1073/pnas.1307697110

Bassler BL, Greenberg EP, Stevens AM (1997) Cross-species induction of luminescence in the quorum-sensing bacterium *Vibrio harveyi*. J Bacteriol 179:4043–4045. https://doi.org/10.1128/jb.179.12.4043-4045.1997

Bengtsson-Palme J, Larsson DGJ (2016) Concentrations of antibiotics predicted to select for resistant bacteria: proposed limits for environmental regulation. Environ Int 86:140–149. https://doi.org/10.1016/j.envint.2015.10.015

Bijtenhoorn P, Mayerhofer H, Muller-Dieckmann J, Utpatel C, Schipper C, Hornung C, Szesny M, Grond S, Thurmer A, Brzuszkiewicz E, Daniel R, Dierking K, Schulenburg H, Streit WR (2011) A novel metagenomic short-chain dehydrogenase/reductase attenuates *Pseudomonas aeruginosa* biofilm formation and virulence on *Caenorhabditis elegans*. PLoS One 6:e26278. https://doi.org/10.1371/journal.pone.0026278

Bilecen K, Fong JCN, Cheng A, Jones CJ, Zamorano-Sánchez D, Yildiz FH (2015) Polymyxin B resistance and biofilm formation in *Vibrio cholerae* are controlled by the response regulator CarR. Infect Immun 83:1199–1209. https://doi.org/10.1128/IAI.02700-14

Borchardt SA, Allain EJ, Michels JJ, Stearns GW, Kelly RF, McCoy WF (2001) Reaction of acylated homoserine lactone bacterial signaling molecules with oxidized halogen antimicrobials. Appl Environ Microbiol 67:3174–3179. https://doi.org/10.1128/AEM.67.7.3174-3179.2001

Brackman G, Defoirdt T, Miyamoto C, Bossier P, Van Calenbergh S, Nelis H, Coeyne T (2008) Cinnamaldehyde and cinnamaldehyde derivatives reduce virulence in *Vibrio* spp. by decreasing the DNA-binding activity of the quorum sensing response regulator LuxR. BMC Microbiol 8:149. https://doi.org/10.1186/1471-2180-8-149

Brackman G, Celen S, Baruah K, Bossier P, Van Calenbergh S, Nelis HJ, Coenye T (2009) AI-2 quorum-sensing inhibitors affect the starvation response and reduce virulence in several Vibrio species, most likely by interfering with LuxPQ. Microbiology 155:4114–4122. https://doi.org/10.1099/mic.0.032474-0

Brackman G, Celen S, Hillaert U, Calenbergh SV, Cos P, Maes L, Nelis HS, Coenye T (2011) Structure-activity relationship of cinnamaldehyde analogs as inhibitors of AI-2 based quorum sensing and their effect on virulence of *Vibrio* spp. PLoS One 6:e16084. https://doi.org/10.1371/journal.pone.0016084

Bruhn JB, Dalsgaard I, Nielsen KF, Buchholtz C, Larsen JL, Gram L (2005) Quorum sensing signal molecules (acylated homoserine lactones) in gram-negative fish pathogenic bacteria. Dis Aquat Org 65:43–52. https://doi.org/10.3354/dao065043

Buchholtz C, Nielsen KF, Milton DL, Larsen JL, Gram L (2006) Profiling of acylated homoserine lactones of *Vibrio anguillarum in vitro* and *in vivo*: influence of growth conditions and serotype. Syst Appl Microbiol 29:433–445. https://doi.org/10.1016/j.syapm.2005.12.007

Cao X, Wang Q, Liu Q, Rui H, Liu H, Zhang Y (2011) Identification of a luxO-regulated extracellular protein Pep and its roles in motility in *Vibrio alginolyticus*. Microb Pathog 50:123–131. https://doi.org/10.1016/j.micpath.2010.12.003

Castang S, Chantegrel B, Deshayes C, Dolmazon R, Gouet P, Haser R, Reverchon S, Nasser W, Hugouvieux- Cotte- Pattat N, Doutheau A (2004) N-Sulfonyl homoserine lactones as antagonists of bacterial quorum sensing. Bioorg Med Chem Lett 14:5145–5149. https://doi.org/10.1016/j.bmcl.2004.07.088

Chang C, Jing-Jing Z, Chun-Hua R, Chao-Qun H (2010) Deletion of *valR*, a homolog of *Vibrio harveyis luxR* generates an intermediate colony phenotype between opaque/rugose and translucent/smooth in *Vibrio alginolyticus*. Biofouling 26:595–601. https://doi.org/10.1080/08927014.2010.499511

Coates AR, Hu Y (2008) Targeting non-multiplying organisms as a way to develop novel antimicrobials. Trends Pharmacol Sci 29:143–150. https://doi.org/10.1016/j.tips.2007.12.001

Colwell RR, Brayton P, Herrington D, Tall B, Huq A, Levine MM (1996) Viable but non-culturable *Vibrio cholerae* O1 revert to a cultivable state in the human intestine. World J Microbiol Biotechnol 12:28–31. https://doi.org/10.1007/BF00327795

Cotter PA, Stibitz S (2007) c-di-GMP-mediated regulation of virulence and biofilm formation. Curr Opin Microbiol 10:17–23. https://doi.org/10.1016/j.mib.2006.12.006

Croxatto A, Chalker VJ, Lauritz J, Jass J, Hardman A, Williams P, Cámara M, Milton DL (2002) VanT, a homologue of *Vibrio harveyi* LuxR, regulates serine, metalloprotease, pigment, and biofilm production in *Vibrio anguillarum*. J Bacteriol 184:1617–1629. https://doi.org/10.1128/JB.184.6.1617-1629.2002

De Spiegeleer B, Verbeke F, D'Hondt M, Hendrix A, Van De Wiele C, Burvenich C, Peremans K, De Wever O, Bracke M, Wynendaele E (2015) The quorum sensing peptides PhrG, CSP and EDF promote angiogenesis and invasion of breast cancer cells in vitro. PLoS One 10:e0119471. https://doi.org/10.1371/journal.pone.0119471

Dharmaprakash A, Reghunathan D, Sivakumar KC, Prasannakumar M, Thomas S (2016) Insights into the genome sequences of an *N*-acyl homoserine lactone molecule producing two *Pseudomonas* spp. isolated from the Arctic. Genome Announc 4:e00767–e00716. https://doi.org/10.1128/genomeA.00767-16

Dinh J, Angeloni JT, Pederson DB, Wang X, Cao M, Dong Y (2014) Cranberry extract standardized for Proanthocyanidins promotes the immune response of *Caenorhabditis elegans* to *Vibrio cholerae* through the p38 MAPK pathway and HSF-1. PLoS ONE 9:e103290. https://doi.org/10.1371/journal.pone.0103290

Dobrucka R, Długaszewska J (2015) Antimicrobial activities of silver nanoparticles synthesized by using water extract of *Arnicae anthodium*. Indian J Microbiol 55:168–174. https://doi.org/10.1007/s12088-015-0516-x

Dong YH, Zhang LH (2005) Quorum sensing and quorum-quenching enzymes. J Microbiol 43:101–109. doi: Not Available

Dong YH, Wang LH, Xu JL, Zhang HB, Zhang XF, Zhang LH (2001) Quenching quorum-sensing-dependent bacterial infection by an N-acyl homoserine lactonase. Nature 411:813–817. https://doi.org/10.1038/35081101

Dutta D, Chowdhury G, Pazhani GP, Guin S, Dutta S, Ghosh S, Rajendran K, Nandy RK, Mukhopadhyay AK, Bhattacharya MK, Mitra U, Takeda Y, Nair GB, Ramamurthy T (2013) *Vibrio cholerae* non-O1, non-O139 serogroups and cholera-like diarrhea, Kolkata, India. Emerg Infect Dis 19:464–467. https://doi.org/10.3201/eid1903.121156

Emerenini BO, Hense BA, Kuttler C, Eberl HJ (2015) A mathematical model of quorum sensing induced biofilm detachment. PLoS One 10:e0132385. https://doi.org/10.1371/journal.pone.0132385

Eppinger M, Pearson T, Koenig SSK, Pearson O, Hicks N, Agrawal S, Sanjar F, Galens K, Daugherty S, Crabtree J, Hendriksen RS, Price LB, Upadhyay BP, Shakya G, Fraser CM, Ravel J, Keim PS (2014) Genomic epidemiology of the Haitian cholera outbreak: a single introduction followed by rapid, extensive, and continued spread characterized the onset of the epidemic. MBio 5:e01721–e01714. https://doi.org/10.1128/mBio.01721-14

Faruque SM, Biswas K, Udden SMN, Ahmad QS, Sack DA, Nair GB, Mekalanos JJ (2006) Transmissibility of cholera: in vivo-formed biofilms and their relationship to infectivity and persistence in the environment. Proc Natl Acad Sci U S A 103:6350–6355. https://doi.org/10.1073/pnas.0601277103

Flodgaard LR, Dalgaard P, Andersen JB, Nielsen KF, Givskov M, Gram L (2005) Non-bioluminescent strains of *Photobacterium phosphoreum* produce the cell-to-cell communication signal *N*-(3-Hydroxyoctanoyl) homoserine lactone. Appl Environ Microbiol 71:2113–2120. https://doi.org/10.1128/AEM.71.4.2113-2120.2005

Fong JC, Syed KA, Klose KE, Yildiz FH (2010) Role of Vibrio polysaccharide (vps) genes in VPS production, biofilm formation and *Vibrio cholerae* pathogenesis. Microbiology 156:2757–2769. https://doi.org/10.1099/mic.0.040196-0

Frerichs RR, Keim PS, Barrais R, Piarroux R (2012) Nepalese origin of cholera epidemic in Haiti. Clin Microbiol Infect 18:E158–E163. https://doi.org/10.1111/j.1469-0691.2012.03841.

Galperin MY (2004) Bacterial signal transduction network in a genomic perspective. Environ Microbiol 6:552–567. https://doi.org/10.1111/j.1462-2920.2004.00633.x

Gram L, de Nys R, Maximilien R, Givskov M, Steinberg P, Kjelleberg S (1996) Inhibitory effects of secondary metabolites from the red alga *Delisea pulchra* on swarming motility of *Proteus mirabilis*. Appl Environ Microbiol 62:4284–4287. doi: Not Available

Gupta PK, Pant ND, Bhandari R, Shrestha P (2016) Cholera outbreak caused by drug resistant *Vibrio cholerae* serogroup O1 biotype El Tor serotype Ogawa in Nepal; a cross-sectional study. Antimicrob Resist Infect Control 5:23. https://doi.org/10.1186/s13756-016-0122-7

Gutierrez JA, Crowder T, Rinaldo-Matthis A, Ho M-C, Almo SC, Schramm VL (2009) Transition state analogs of 5'-methylthioadenosine nucleosidase disrupt quorum sensing. Nat Chem Biol 5:251–257. https://doi.org/10.1038/nchembio.153.

Hang L, John M, Asaduzzaman M, Bridges EA, Vanderspurt C, Kirn TJ, Taylor RK, Hillman JD, Progulske-Fox A, Handfield M, Ryan ET, Calderwood SB (2003) Use of in vivo-induced antigen technology (IVIAT) to identify genes uniquely expressed during human infection with *Vibrio cholerae*. Proc Natl Acad Sci U S A 100:8508–8513. https://doi.org/10.1073/pnas.1431769100

Haudecoeur E, Tannieres M, Cirou A, Raffoux A, Dessaux Y, Faure D (2009) Different regulation and roles of lactonases AiiB and AttM in *Agrobacterium tumefaciens* C58. Mol Plant Microbe Interact 22:529–537. https://doi.org/10.1094/MPMI-22-5-0529

Hoang TT, Schweizer HP (1999) Characterization of *Pseudomonas aeruginosa* enoyl-acyl carrier protein reductase (FabI): a target for the antimicrobial triclosan and its role in acylated homoserine lactone synthesis. J Bacteriol 181:5489–5497. doi: Not Available

Huma N, Shankar P, Kushwah J, Bhushan A, Joshi J, Mukherjee T, Raju SC, Purohit HJ, Kalia VC (2011) Diversity and polymorphism in AHL-lactonase gene (*aiiA*) of *Bacillus*. J Microbiol Biotechnol 21:1001–1011. https://doi.org/10.4014/jmb.1105.05056

Huq A, Small EB, West PA, Huq MI, Rahman R, Colwell RR (1983) Ecological relationships between *Vibrio cholerae* and planktonic crustacean copepods. Appl Environ Microbiol 45:275–283. doi: Not Available

Jutla A, Whitcombe E, Hasan N, Haley B, Akanda A, Huq A, Alam M, Sack RB, Colwell RR (2013) Environmental factors influencing epidemic cholera. Am J Trop Med Hyg 89:597–607. https://doi.org/10.4269/ajtmh.12-0721

Kalia VC, Raju SC, Purohit HJ (2011) Genomic analysis reveals versatile organisms for quorum quenching enzymes: acyl-homoserine lactone-acylase and –lactonase. Open Microbiol J 5:1–13. https://doi.org/10.2174/1874285801105010001

Kalia VC, Kumar P, Pandian SK, Sharma P (2014) Biofouling control by quorum quenching. In: Kim SK (ed) Hb_25 Springer handbook of marine biotechnology chapter 15. Springer, Berlin, pp 431–440. https://doi.10.1007/978-3-642-53971-8_15

Ke X, Miller LC, Ng WL, Bassler BL (2014) CqsA-CqsS quorum-sensing signal-receptor specificity in *Photobacterium angustum*. Mol Microbiol 91:821–833. https://doi.org/10.1111/mmi.12502

Kim JS, Kim YH, Seo YW, Park S (2007) Quorum sensing inhibitors from the red alga, *Ahnfeltiopsis flabelliformis*. Biotechnol Bioprocess Eng 12:308–311. https://doi.org/10.1007/s12257-008-0131-3

Koul S, Kalia VC (2017) Multiplicity of quorum quenching enzymes: a potential mechanism to limit quorum sensing bacterial population. Indian J Microbiol 57:100–108. https://doi.org/10.1007/s12088-016-0633-1

Kumar P, Koul S, Patel SKS, Lee JK, Kalia VC (2015) Heterologous expression of quorum sensing inhibitory genes in diverse organisms. In: Kalia VC (ed) Quorum sensing vs quorum quenching: a battle with no end in sight. Springer, New Delhi, pp 343–356. https://doi.org/10.1007/978-81-322-1982-8_28

Lee SJ, Park SY, Lee JJ, Yum DY, Koo BT, Lee JK (2002) Genes encoding the N-acyl homoserine lactone-degrading enzyme are wide spread in many subspecies of *Bacillus thuringiensis*. Appl Environ Microbiol 68:3919–3924. https://doi.org/10.1128/AEM.68.8.3919-3924.2002

Liu H, Wang Q, Liu Q, Cao X, Shi C, Zhang Y (2011) Roles of Hfq in the stress adaptation and virulence in fish pathogen *Vibrio alginolyticus* and its potential application as a target for live attenuated vaccine. Appl Microbiol Biotechnol 91:353–364. https://doi.org/10.1007/s00253-011-3286-3

Lowery CA, Park J, Kaufmann GF, Janda KD (2008) An unexpected switch in the modulation of AI-2-based quorum sensing discovered through synthetic 4,5-Dihydroxy-2,3-pentanedione analogues. J Am Chem Soc 130:9200–9201. https://doi.org/10.1021/ja802353

Lowery CA, Abe T, Park J, Eubanks LM, Sawada D, Kaufmann GF, Janda KD (2009) Revisiting AI-2 quorum sensing inhibitors: direct comparison of alkyl-DPD analogues and a natural product fimbrolide. J Am Chem Soc 131:15584–15585. https://doi.org/10.1021/ja9066783

Mah TF, Pitts B, Pellock B, Walker GC, Stewart PS, O'Toole GA (2003) A genetic basis for *Pseudomonas aeruginosa* biofilm antibiotic resistance. Nature 426:306–310. https://doi.org/10.1038/nature02122

Manefield M, Harris L, Rice SA, de Nys R, Kjelleberg S (2000) Inhibition of luminescence and virulence in the black tiger prawn (*Penaeus monodon*) pathogen *Vibrio harveyi* by intercellular signal antagonists. Appl Environ Microbiol 66:2079–2084. doi: Not Available

Marques JC, Lamosa P, Russell C, Ventura R, Maycock C, Semmelhack MF, Miller ST, Xavier KB (2011) Processing the interspecies quorum-sensing signal autoinducer-2 (AI-2): characterization of phospho-(S)-4,5-dihydroxy-2,3-pentanedione isomerization by LsrG protein. J Biol Chem 286:18331–18343. https://doi.org/10.1074/jbc.M111.230227

Martínez JL (2012) Natural antibiotic resistance and contamination by antibiotic resistance determinants: the two ages in the evolution of resistance to antimicrobials. Front Microbiol 3:1. https://doi.org/10.3389/fmicb.2012.00001

Matz C, McDougald D, Moreno AM, Yung PY, Yildiz FH, Kjelleberg S (2005) Biofilm formation and phenotypic variation enhance predation-driven persistence of *Vibrio cholerae*. Proc Natl Acad Sci U S A 102:16819–16824. https://doi.org/10.1073/pnas.0505350102

Meibom KL, Blokesch M, Dolganov NA, Wu C-Y, Schoolnik GK (2005) Chitin induces natural competence in *Vibrio cholerae*. Science 310:1824–1827. https://doi.org/10.1126/science.1120096

Meijler MM, Hom LG, Kaufmann GF, McKenzie KM, Sun CZ, Moss JA, Matsushita M, Angew JKD (2004) Synthesis and biological validation of a ubiquitous quorum-sensing molecule. Chem Int Ed 43:2106–2108. https://doi.org/10.1002/anie.200353150

Milton DL (2006) Quorum sensing in vibrios: complexity for diversification. Int J Med Microbiol 296:61–71. https://doi.org/10.1016/j.ijmm.2006.01.044

Milton DL, Hardman A, Camara M, Chhabra SR, Bycroft BW, Stewart GS, Williams P (1997) Quorum sensing in *Vibrio anguillarum*: characterization of the *vanI/vanR* locus and identification of the autoinducer *N*-(3-oxodecanoyl)-L-homoserine lactone. J Bacteriol 179:3004–3012. https://doi.org/10.1128/jb.179.9.3004-3012.1997

Mireille Aye A, Bonnin-Jusserand M, Brian-Jaisson F, Ortalo-Magne A, Culioli G, Koffi Nevry R, Rabah N, Blache Y, Molmeret M (2015) Modulation of violacein production and phenotypes associated with biofilm by exogenous quorum sensing N-acylhomoserine lactones in the marine bacterium *Pseudoalteromonas ulvae* TC14. Microbiology 161:2039–2051. https://doi.org/10.1099/mic.0.000147

Mitra R, Orbell J, Muralitharan MS (2007) Agriculture- medicinal plants of Malaysia. Asia-Pac Biotech News 11:105–110. https://doi.org/10.1142/S0219030307000110

Morohoshi T, Nakazawa S, Ebata A, Kato N, Ikeda T (2008) Identification and characterization of N-acylhomoserine lactone-acylase from the fish intestinal Shewanella sp. strain MIB015. Biosci Biotechnol Biochem 72:1887–1893. https://doi.org/10.1271/bbb.80139

Musthafa KS, Saroja V, Pandian SK, Ravi AV (2011) Antipathogenic potential of marine *Bacillus* sp. SS4 on N-acyl-homoserine-lactone-mediated virulence factors production in *Pseudomonas aeruginosa* (PAO1). J Biosci 36:55–67. https://doi.org/10.1007/s12038-011-9011-7

Narendrakumar L, Suryaletha K, Reghunathan D, Prasannakumar M, Thomas S (2017) Insights into the draft genome sequence of a Haitian variant *Vibrio cholerae* strain isolated from a clinical setting in Kerala, South India. Genome Announc 5:e00843–e00817. https://doi.org/10.1128/genomeA.00843-17

Neu HC (1992) The crisis in antibiotic resistance. Science 257:1064–1073. https://doi.org/10.1126/science.257.5073.1064

Ng WL, Bassler BL (2009) Bacterial quorum-sensing network architectures. Annu Rev Genet 43:197–222. https://doi.org/10.1146/annurev-genet-102108-134304

Nithya C, Arvindraja C, Pandian SK (2010) *Bacillus pumilus* of Palk Bay origin inhibits quorum-sensing-mediated virulence factors in gram-negative bacteria. Res Microbiol 161:293–304. https://doi.org/10.1016/j.resmic.2010.03.002

Nithya C, Devi MG, Pandian SK (2011) A novel compound from the marine bacterium *Bacillus pumilus* S6-15 inhibits biofilm formation in gram-positive and gram-negative species. Biofouling 27:519–528. https://doi.org/10.1080/08927014.2011.586127

Niu C, Afre S, Gilbert ES (2006) Subinhibitory concentrations of cinnamaldehyde interfere with quorum sensing. Lett Appl Microbiol 43:489–494. https://doi.org/10.1111/j.1472-765X.2006.02001.x

Okutsu N, Morohoshi T, Xie X, Kato N, Ikeda T (2015) Characterization of N-acyl-homoserine lactones produced by bacteria isolated from industrial cooling water systems. Sensors 16:E44. https://doi.org/10.3390/s16010044

Olsen JA, Severinsen R, Rasmussen TB, Hentzer M, Givskov M, Nielsen J (2002) Synthesis of new 3- and 4-substituted analogues of acyl homoserine lactone quorum sensing autoinducers. Bioorg Med Chem Lett 12:325–328. https://doi.org/10.1016/S0960-894X(01)00756-9

Peach KC, Bray WM, Shikuma NJ, Gassner NC, Lokey RS, Yildiz FH, Linington RG (2011) An image-based 384-well high-throughput screening method for the discovery of biofilm inhibitors in *Vibrio cholerae*. Mol BioSyst 7:1176–1184. https://doi.org/10.1039/c0mb00276c

Pereira CS, Thompson JA, Xavier KB (2013) AI-2-mediated signalling in bacteria. FEMS Microbiol Rev 37:156–181. https://doi.org/10.1111/j.1574-6976.2012.00345.x

Persson T, Givskov M, Nielsen J (2005) Quorum sensing inhibition: targeting chemical communication in gram-negative bacteria. Curr Med Chem 12:3103–3115. https://doi.org/10.2174/092986705774933425

Rasch M, Buch C, Austin B, Slierendrecht WJ, Ekmann KS, Larsen JL, Johansen C, Riedel K, Ebery L, Givskov M, Gram L (2004) An inhibitor of bacterial quorum sensing reduces mortyalities caused by Vibriosis in Rainbow trout (*Oncorhynchus mykiss, Walbaum*). Syst Appl Microbiol 27:350–359. https://doi.org/10.1078/0723-2020-00268

Rasmussen TB, Givskov M (2006) Quorum-sensing inhibitors as anti-pathogenic drugs. Int J Med Microbiol 296:149–161. https://doi.org/10.1016/j.ijmm.2006.02.005

Rasmussen BB, Nielsen KF, Machado H, Gram L, Sonnenschein E (2014) Global and phylogenetic distribution of quorum sensing signals, acyl homoserine lactones, in the family of Vibrionaceae. Mar Drugs 12:5527–5546. https://doi.org/10.3390/md12115527

Ruby EG (1996) Lessons from a cooperative, bacterial- animal association: the *Vibrio fischeri-Euprymna scolopes* light organ symbiosis. Annu Rev Microbiol 50:591–624. https://doi.org/10.1146/annurev.micro.50.1.591

Schaefer AL, Hanzelka BL, Eberhard A, Greenberg EP (1996) Quorum sensing in *Vibrio fischeri*: probing autoinducer-LuxR interactions with autoinducer analogs. J Bacteriol 178:2897–2901. doi: Not Available

Sharma R, Jangid K (2015) Fungal quorum sensing inhibitors. In: Kalia VC (ed) Quorum sensing vs quorum quenching: a battle with no end in sight. Springer, New Delhi, pp 237–257

Shiva Krishna P, Sudheer Kumar B, Raju P, Murty MSR, Prabhakar Rao T, Singara Charya MA, Prakasham RS (2015) Fermentative production of pyranone derivate from marine *Vibrio* sp. SKMARSP9: isolation, characterization and bioactivity evaluation. Indian J Microbiol 55:292–301. https://doi.org/10.1007/s12088-015-0521-0

Singh V, Evans GB, Lenz DH, Mason JM, Clinch K, Mee S, Painter GF, Tyler PC, Furneaux RH, Lee JE, Howell PL, Schramm VL (2005) Femtomolar transition state analogue inhibitors of 5′-methylthioadenosine/S-adenosylhomocysteine nucleosidase from *Escherichia coli*. J Biol Chem 280:18265–18273. https://doi.org/10.1074/jbc.M414472200

Smith JAI, Wang J, Nguyen-Mau S-M, Leeb V, Sintim HO (2009) Biological screening of a diverse set of AI-2 analogues in *Vibrio harveyi* suggests that receptors which are involved in synergistic agonism of AI-2 and analogues are promiscuous. Chem Commun 45:7033–7035. https://doi.org/10.1039/b909666c

Szweda P, Gucwa K, Kurzyk E, Romanowska E, Dzierżanowska-Fangrat K, Jurek AZ, Kuś PM, Milewski S (2015) Essential oils, silver nanoparticles and propolis as alternative agents against fluconazole resistant *Candida albicans, Candida glabrata* and *Candida krusei* clinical isolates. Indian J Microbiol 55:175–183. https://doi.org/10.1007/s12088-014-0508-2

Tashiro Y, Kimura Y, Furubayashi M, Tanaka A, Terakubo K, Saito K, Kawai-Noma S, Umeno D (2016) Directed evolution of autoinducer selectivity of *Vibrio fischeri* LuxR. J Gen Appl Microbiol 62:240–247. https://doi.org/10.1093/nar/gkq1070

Teplitski M, Chen H, Rajamani S, Gao M, Merighi M, Sayre RT, Robinson JB, Rolfe BG, Bauer WD (2004) *Chlamydomonas reinhardtii* Secretes compounds that mimic bacterial signals and interfere with quorum sensing regulation in bacteria. Plant Physiol 134:137–146. https://doi.org/10.1104/pp.103.029918

Teschler JK, Zamorano-Sánchez D, Utada AS, Warner CJA, Wong GCL, Linington RG, Yildiz FH (2015) Living in the matrix: assembly and control of *Vibrio cholerae* biofilms. Nat Rev Microbiol 13:255–268. https://doi.org/10.1038/nrmicro3433

Tseng BS, Majerczyk CD, Passos da Silva D, Chandler JR, Greenberg EP, Parsek MR (2016) Quorum sensing influences *Burkholderia thailandensis* biofilm development and matrix production. J Bacteriol 198:2643–2650. https://doi.org/10.1128/JB.00047-16

Valiente E, Bruhn JB, Nielsen KF, Larsen JL, Roig FJ, Gram L, Amaro C (2009) *Vibrio vulnificus* produces quorum sensing signals of the AHL-class. FEMS Microbiol Ecol 69:16–26. https://doi.org/10.1111/j.1574-6941.2009.00691.x

van Sinderen D, Luttinger A, Kong L, Dubnau D, Venema G, Hamoen L (1995) comK encodes the competence transcription factor, the key regulatory protein for competence development in *Bacillus subtilis*. Mol Microbiol 15:455–462. https://doi.org/10.1111/j.1365-2958.1995.tb02259.x

Vikram A, Jesudhasan PR, Jayaprakasha GK, Pillai SD, Patil BS (2011) Citrus limonoids interfere with *Vibrio harveyi* cell-cell signaling and biofilm formation by modulating the response regulator LuxO. Microbiology 157:99–110. https://doi.org/10.1099/mic.0.041228-0

Vinoj G, Vaseeharan B, Thomas S, Spiers AJ, Shanthi S (2014) Quorum-quenching activity of the AHL-lactonase from *Bacillus licheniformis* DAHB1 inhibits Vibrio biofilm formation *in vitro* and reduces shrimp intestinal colonisation and mortality. Mar Biotechnol 16:707–715. https://doi.org/10.1007/s10126-014-9585-9

Wang L, Ling Y, Jiang H, Qiu Y, Qiu J, Chen H, Yang R, Zhou D (2013) AphA is required for biofilm formation, motility, and virulence in pandemic *Vibrio parahaemolyticus*. Int J Food Microbiol 160:245–251. https://doi.org/10.1016/j.ijfoodmicro.2012.11.004

Williams P, Camara M (2009) Quorum sensing and environmental adaptation in *Pseudomonas aeruginosa*: a tale of regulatory networks and multifunctional signal molecules. Curr Opin Microbiol 12:182–191. https://doi.org/10.1016/j.mib.2009.01.005

World Health Organization (2016) Weekly epidemiological record, vol 91, 38:433–440. doi: Not Available

Xavier KB, Bassler BL (2005) Interference with AI-2-mediated bacterial cell-cell communication. Nature 437:750–753. https://doi.org/10.1038/nature03960

Yang Q, Han Y, Zhang XH (2011) Detection of quorum sensing signal molecules in the family Vibrionaceae. J Appl Microbiol 110:1438–1448. https://doi.org/10.1111/j.1365-2672.2011.04998.x

Yang K, Meng J, Huang YC, Ye LH, Li GJ, Huang J, Chen HM (2014) The role of the QseC quorum-sensing sensor kinase in epinephrine-enhanced motility and biofilm formation by *Escherichia coli*. Cell Biochem Biophys 70:391–398. https://doi.org/10.1007/s12013-014-9924-5

Yildiz FH, Dolganov NA, Schoolnik GK (2001) VpsR, a member of the response regulators of the two-component regulatory systems, is required for expression of vps biosynthesis genes and EPS^{ETr}-associated phenotypes in Vibrio cholerae O1 El Tor. J Bacteriol 183:1716–1726. https://doi.org/10.1128/JB.183.5.1716-1726.2001

Zhao X, Kuipers OP (2016) Identification and classification of known and putative antimicrobial compounds produced by a wide variety of Bacillales species. BMC Genomics 17:882. https://doi.org/10.1186/s12864-016-3224-y

Zhu J, Miller MB, Vance RE, Dziejman M, Bassler BL, Mekalanos JJ (2002) Quorum-sensing regulators control virulence gene expression in *Vibrio cholerae*. Proc Natl Acad Sci U S A 99:3129–3134. https://doi.org/10.1073/pnas.052694299

Chapter 5
Anti-biofilm Peptides: A New Class of Quorum Quenchers and Their Prospective Therapeutic Applications

Akanksha Rajput and Manoj Kumar

Abstract Biofilms are the major concerns to the researchers, due to their universal distribution among prokaryotes and involvement in antibiotic drug resistance towards conventional drugs. It led the bacteria to become up to 1000 times resistant towards antibiotics. Therefore, diverse types of anti-biofilm agents are continuously designed to target them namely (phyto) chemicals, peptides, enzymes, biosurfactants, microbial extracts, nanoparticles, and many more. Antibiofilm peptides have demonstrated high potential in targeting biofilm due to their low toxicity, and off-target effects. These peptides are experimentally validated to disrupt most of the biofilms developed on medical devices like catheters, stents, dentures, etc. implicated in nosocomial infections by ESKAPE pathogens. However, one of the important reasons for the peptides, to emerge as a new hope against biofilms, is their wide mode of action against different stages and microbial species. In the present chapter, we are focusing to explore various aspects of this important class of antibiofilm therapeutics.

Keywords Anti-biofilm peptides · nosocomial infections · ESKAPE pathogen · antibiotic resistance

5.1 Introduction

Biofilms are the consortium of microbes encapsulated in the self-secreted cocoon composed mainly of extracellular matrix. Biofilm mode of growth is an alternate lifestyle of microbes, where they mimic multicellular behavior rather than unicellular (Kostakioti et al. 2013). Among the microcolonies, the bacteria exhibits difference in their physiological state as compared to their planktonic form. The colonization process is an adaptive mechanism that arose in response to various

A. Rajput · M. Kumar (✉)
Bioinformatics Centre, Institute of Microbial Technology, Council of Scientific and Industrial Research, Chandigarh, India
e-mail: manojk@imtech.res.in

© Springer Nature Singapore Pte Ltd. 2018
V. C. Kalia (ed.), *Biotechnological Applications of Quorum Sensing Inhibitors*,
https://doi.org/10.1007/978-981-10-9026-4_5

environmental stresses including nutrient deficiency, introduction to sub-inhibitory concentration of antibiotics, etc. (Flemming and Wingender 2010). It is ubiquitous in nature and is the dweller of numerous living and non-living surfaces. It was firstly discussed in two seminal papers published in 1930s by Arthur Henrici and Claude Zobell, whereas, the term was coined by Bill Costerton in 1978.

In general, there are five developmental stages of biofilms: reversible attachment, irreversible attachment, proliferation, maturation, and dispersal (Kostakioti et al. 2013). At the very first stage of reversible attachment the planktonic bacteria starts attaching to the surface through weak Vander waal forces or their appendages like flagella. In the next stage the microbes convert the reversible connection to irreversible by overcoming the physical repulsive forces. It is accomplished by increasing the hydrophobic forces among bacteria and extracellular matrix due to secretion of extracellular polymeric substances (EPS) like polysaccharides, proteins, DNA, and many more. After attachment the microbes start proliferating within the biofilm to intensify the colonization process *via* cell division and cell recruitment. In the final stage of development, the biofilm undergoes specialization through certain physiological modifications like efflux pumps, oxygen gradients, division of labor, etc (Dang and Lovell 2016). At the end, the fully developed biofilm starts dispersing with the help of various enzymes that degrades matrix like dispersin B, deoxyribonucleases, etc.

The presence of *cell-to-cell* communication inside the biofilm emerged as an important parameter in shaping the biofilms (Toyofuku et al. 2015). The structure of the biofilms depends on the interaction among the species that can be either competitive or cooperative based on the type of microbial species within them (Li and Tian 2012). Moreover, the linking of the quorum sensing and biofilms was termed as "*sociomicrobiology*" (Parsek and Greenberg 2005; Rajput et al. 2015, 2016).

Various anti-biofilm agents are continuously being designed to impede the biofilm growth namely chemicals, phytochemicals, peptides, nanoparticles, biosurfactants, enzymes, etc as catalogued in ***aBiofilm*** resource by our group (Rajput et al. 2018). In this book chapter we will focus to decipher various important aspects of anti-biofilm peptides (ABPs), which emerged as a new hope to the researchers in the era of antibiotic drug resistance. The chapter will cover the detail of biofilms, its composition and consequences, along with the role of ABPs for targeting the biofilms, its source, chemical modifications, targets and antibiotic drug resistance.

5.2 Characteristics of Biofilms

Overall architecture of the biofilm comprised of self-secreted matrix of hydrated EPS, which is responsible to adhere on the surface and cohesion within them (Branda et al. 2005; Flemming and Wingender 2010). However, the composition of EPS varies according to the microbial species present in the biofilms. Although, polysaccharides are amongst the major constituents of biofilms followed by water, proteins (or enzymes), extracellular DNA (eDNA) that are hydrophilic molecules (Wingender et al. 2001; Conrad et al. 2003). Moreover, the few hydrophobic molecules like lipids and biosurfactants were also reported, which help in adherence and initial microcolony formation among few species like *Thiobacillus ferrooxidans*, *Serratia marcescens*, *P. aeruginosa*, etc (Davey et al. 2003; Sand and Gehrke 2006).

The *polysaccharides* secreted by the microbes in the biofilms are homo and/or hetero in nature. They are composed of monosaccharide units, which can be O- or N- acylated. The EPS of the biofilm are generally species-specific e.g. poly-β-1,6-N-acetylglucosamine is secreted by *E. coli*, *S. aureus*, and *A. pleuropneumoniae*; the polymerization of α-D-galactose, β-D-galactose, and β-D-glucose monomers is the main constituent of *E. persicina* biofilms; the *E. coli* and *Salmonella* biofilms are mainly composed of cellulose; *P. aeruginosa* biofilms comprised of alginate (α-L- guluronic acid and α-L-mannuronic acid), etc (Batoni et al. 2016).

Water channels are amongst the main constituent of biofilm and responsible for nutrient transport. The water channels are considered as homologs to the circulating system of the multicellular organisms, therefore the biofilms are often considered as primitive multicellular organisms. They are able to transport the nutrients in and out from the depths of biofilms (Stewart 2003).

The *protein* portion of biofilms comprised of extracellular proteins, cell surface adhesins, subunits of appendages like flagella and pili, and outer vesicle protein covering. It helps in maintaining structure and stability of the biofilms. Moreover, few proteins with enzymatic properties are responsible for catalyzing matrix components like dispersin B hydrolyzes polysaccharides (Kaplan et al. 2003), DNases to disintegrates extracellular nucleic acids (Nijland et al. 2010), and proteases degrades matrix proteins (Fong and Yildiz 2015).

The *extracellular DNA* is considered as an important stabilizer and maintainer of biofilm architecture in bacteria and fungus. The extracellular DNA derives from the lysis of bacterial cell, and is a hot spot for the horizontal gene transfer within the polyspecies in biofilm. Moreover, they are also considered as one of the factor for transfering the antibiotic drug resistance genes. Intriguingly, the eDNA is also the nutrient source, and cation chelator (Montanaro et al. 2011).

Apart from the hydrophilic molecules, the *hydrophobic molecules* like lipids and biosurfactants are also the constituents of biofilms especially of *Rhodococcos* and *Mycobacterium* genus. The lipids are known to assist in cell adhesion, biofilm formation and development. For example, rhamnolipids helps in modulating the biofilm architecture by retaining the water channels accessible during maturation phase of biofilm (Branda et al. 2005).

The universality of the biofilms and antibiotic resistant behavior is the major concern for the researchers worldwide (Koul et al. 2016; Koul and Kalia 2017). According to the Centre for Disease Control and Prevention (CDC), biofilms are the major cause of nosocomial (hospital-acquired) infections (Davey and O'Toole 2000). However, they are also involving in human health threatening infections e.g. lungs, heart, gastrointestinal tract, oral cavity, urogenital tract, and many more (Li and Tian 2012). Therefore, there is an emergent need to target them through various anti-biofilm agents.

5.3   Anti-biofilm Peptides

Diverse biofilm targeting agents are being designed to impede the dreadful effect of bacteria in biofilm mode. These anti-biofilm agents are varied in nature and ranges from chemicals (Gui et al. 2014; Balamurugan et al. 2015), phytochemicals (Bhargava et al. 2015), peptides, phages (Ahiwale et al. 2017), antibody, biosurfactants, nanoparticles, etc (Kalia 2013; Agarwala et al. 2014; Kalia 2014). Among all the anti-biofilm agents, ABPs emerged as a novel and efficient quencher (Pletzer et al. 2016). Most of the ABPs are ribosomally synthesized and post translationally modified. Moreover, they are also possessing low toxicity as they can be broken down upon ingestion. Mostly, ABPs are short with 5–50 amino acids with cationic and amphipathic in nature (Batoni et al. 2016). The peptide structure of some important ABPs was predicted using PEPstrMOD software (Singh et al. 2015) is provided in Table 5.1.

Source of Anti-biofilm Peptides ABPs are anti-microbial peptides (AMPs) with biofilm targeting activity and are of natural, semi-synthetic and synthetic in nature. Naturally occurring peptides are secreted from humans, plants, animals, and microbes for example Magainin-II, Defensins, Histatins, LL-37, etc (Pletzer and Hancock 2016). While semi-synthetic or synthetic are derivative from natural ABPs or through peptide synthesis methods IDR-1018; C16G2; L-K6; F2,5,12W; R-FV-I16; etc (Batoni et al. 2016).

Chemical Modifications in Anti-biofilm Peptides The chemical modifications in the ABPs make them highly efficient towards the target and increase their half-life. The modifications include post-translational modifications (amidation, carboxylations, etc) (Zhou et al. 2016), physicochemical modifications (deletion and/or substitution of amino acids), sequence truncations (Nagant et al. 2012), designing of retro-inverso peptides (D-enantiomers) (de la Fuente-Nunez et al. 2015), cyclization, hybrids construction (Gopal et al. 2014), and many more (de la Fuente-Nunez et al. 2016).

5.4   Mode of Action of Anti-biofilm Peptides

ABPs target the biofilms *via* different mechanisms both at molecular and physiological level. Several ways by which the ABPs target the biofilms are: degradation of signals within biofilms, permeabilize within cytoplasmic membrane/EPS, modulating EPS production, downregulating the biofilm associated genes, disrupt the adhesion of biofilm, killing of metabolically active cells, and interferes with the motility. The diagrammatic details of mode of action of the ABPs are shown on Fig. 5.1.

Table 5.1 Table showing tertiary structure of important anti-biofilm peptides

Anti-biofilm peptides	Peptide sequence	Structure
1018	VRLIVAVRIWRR-NH2	
Human α-defensin 1	DCYCRIPACIAGEKKYGT CIYQGKLWAFCC	
Amphotericin B	KKVVFWVKFK-NH2	
Competence Stimulating Peptide	SGSLSTFFRLFNRSFTQALGK	
DJK5	VQWRAIRVRVIR	

(continued)

Table 5.1 (continued)

Anti-biofilm peptides	Peptide sequence	Structure
G H12	GLLWHLLHHLLH-NH2	
Histatin 5	DSHAKRHHGYKRKFHEKHHSHRGY	
Lactoferricin B	FKCRRWQWRMK KLGAPSITCVRRAF	
Magainin II	GIGKFLHSAGKFGKAFVGEIMKS	
Pleurocidin	GWGSFFKKAAHVGKHVGKAALTHYL-NH2	
Tachyplesin I	KWCFRVCYRGICYRKCR-NH2	

Degradation of Signals Within Biofilms (QS) The ABPs are known to interfere with the binding and causes the degradation of QS signals and/or secondary messengers (ppGpp) that are important for biofilm formation and maintenance for example, DJK5, DJK6 (de la Fuente-Nunez et al. 2015).

Permeabilize Within Cytoplasmic Membrane The peptides are able to permeabilise and/or form pores inside the cytoplasmic membranes like LL-37, LL-31 (Kanthawong et al. 2012). Moreover, the ABPs are also responsible for the pore formation ability within the lipid component of EPS e.g. pleurocidin (Choi and Lee 2012).

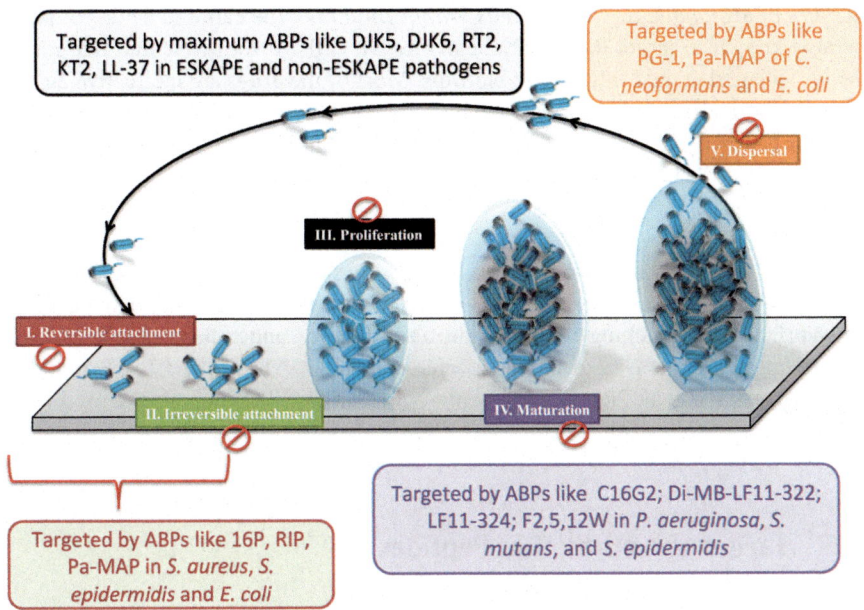

Fig. 5.1 Diagrammatic representation of the mode of action of Anti-biofilm peptides against biofilms

Modulating EPS Production The major components of biofilms are the EPS secreted *via* microbes, which provides a protective sheath and maximally are negatively charged due to presence of eDNA. The ABPs like CSPs used to regulate EPS production in *C. albicans* (Jack et al. 2015).

Downregulating the Biofilm Associated Genes Some of the essential biofilm associated genes were also being targeted with ABPs that are involved in cell adhesion, extracellular matrix hyphal growth, etc like ZAP1, CSH1, ALS3. Whereas, hLF1-11 was used against *C. albicans* biofilm to impede the growth by down regulating various genes (Morici et al. 2016).

Disrupt the Adhesion of Biofilm The initial and important stage of biofilms is adhesion. The ABPs are validated to reduce the adhesion of biofilms with the surface like catheters, stents, dentures, etc. For example, Magainin I, Histatin 5, etc (Pusateri et al. 2009).

Killing of Metabolically Active Cells Within Biofilms The cationic nature of ABPs is also known to target the metabolically active cells (most active and marginally active) located at the centre of the biofilms. Most of the drugs are ineffective against the metabolically active cells as compared to inactive ones due to the modifications in their Lipopolysachharides (LPS). The Cationic Antimicrobial peptides (CAMPs) like GL13K, GH12, hLF1- 11 (Hirt and Gorr 2013) are experimentally validated to target the same.

Interferes with Motility The motility is one of the important parameter for biofilm functionality. It is pre-adhesion stage, when the bacteria started coming to the adherent surface *via* twitching and/or swarming motility under the influence of any chemo-attractant. It is also being interfered through ABPs like LL-37, modified 1037 (Overhage et al. 2008; de la Fuente-Nunez et al. 2012) impede the motility of bacteria.

5.5 Target of Anti-biofilm Peptides

Various ABPs were tested against numerous important pathogens i.e. ESKAPE (*Enterococcus faecium, Staphylococcus aureus, Klebsiella pneumoniae, Acinetobacter, Pseudomonas aeruginosa* and *Enterobacter* spp.) and non-ESKAPE ones. We have extracted the details of various ABPs targeting important pathogens as described below:

5.5.1 Enterococcus faecium

It is a Gram-positive and commensal bacterium of human intestine, responsible to cause nosocomial bacteremia. It is a multidrug resistant ESKAPE pathogen causes biofilm–associated infection through medical devices. Though, various anti-biofilm peptides emerged as a hope to target it, despite being antibiotic drug resistant. For example, ABPs like Siamycin I, Magainin-I, Coprisin, Pleurocidin, etc are reported to inhibit the biofilm of *E. faecium* (Hwang et al. 2013; Winfred et al. 2014).

5.5.2 Staphylococcus aureus

It is a Gram-positive and facultative anaerobic bacterium. Some of the strains developed antibiotic resistance against commonly used antibiotics. They are amongst the most frequent dwellers of medical instruments, and are also called as opportunistic pathogens. According to Nosocomial Infections Surveillance System, *S. aureus* are amongst the frequently occurring nosocomial pathogens isolated from intensive care unit (ICU). Various ABPs are designed to target its stubborn biofilm like Human β-defensin 3, HE12, Myxinidin, etc (de la Fuente-Nunez et al. 2014; Li et al. 2015).

5.5.3 Klebsiella pneumoniae

It is a Gram-negative ESKAPE pathogen, and often considered as "superbug" due to its efficiency to cause range of diseases. It is related to extreme drug resistance (Vuotto et al. 2017), as it is resistant to most of the antibiotics including carbapenems, and a typical example of nosocomial pathogen. Recently, some ABPs were designed to target *K. pneumoniae namely* Pa-MAP, Bac8C, etc (Ding et al. 2014; Cardoso et al. 2016).

5.5.4 Acinetobacter baumanii

It is a Gram-negative, opportunistic and nosocomial pathogen that possess the ability to survive even in unfavorable condition on the hospital instruments. The important factor that led the cocoon of *A. baumanii* so protective is the pili which are expressed from *csuA/BABCDE* operon followed by QS signals (Rajput and Kumar 2017a, b) i.e. Acyl homoserine lactone secreted by *abaI* gene. Now, a days the multidrug resistant (MDR) problem of *A. baumanii* is being targeted by the use of ABPs like LL-37, DJK5, IDR-1018 (Feng et al. 2013; de la Fuente-Nunez et al. 2015).

5.5.5 Pseudomonas aeruginosa

It is a well-known opportunistic pathogen and majorly responsible to cause lung infection in cystic fibrosis patients through biofilms. However, it is also known to cause infections in both plants and animals. Its MDR strains had been proved to be more dreadful and cause nosocomial infections like sepsis and ventilator-associated pneumoniae. The MDR problem of *P. aeruginosa* are treated with various ABPs like Cathelicidin, Indolicidin, Melittin, RIP, etc (Overhage et al. 2008; Gopal et al. 2013).

5.5.6 Escherichia coli

It is a Gram-negative and non-ESKAPE bacteria that is also responsible for nosocomial infection e.g. prostatitis, and urinary tract infections. It forms a thick biofilm with matrix enriched in polysaccharides, and varies according to environmental conditions. The presence of large number of polysaccharides is the main factor for its MDR behavior. However, the ABPs like AGE-RK1, KT2, RT2, Lactoferricin B (Anunthawan et al. 2015) are also being used to target its biofilms.

5.5.7 Candida albicans

It is a fungus that normally found inside the human body, but cause serious illness in immunocompromised patients. It is popularly known to form biofilm on medical instruments like pacemakers, catheters, prosthetic joints, dentures and thus responsible for nosocomial infections. The *C. albicans* biofilms are resistant to conventional anti-fungal drugs. It is responsible to form highly structured biofilms with multiple cells e.g. budding, round, oval pseudohyphal, yeast-form, cylindrical, elongated hyphal cells. Along with the anti-fungal drugs various ABPs like BMAP-28, OSIP108, Amphotericin B are being tested against various srains of *C. albicans* (Theberge et al. 2013; Delattin et al. 2014; Scarsini et al. 2015).

Some other important non-ESKAPE pathogens that are experimentally validated to be targeted with ABPs are *P. putida* targets Putisolvin I and II (Dubern et al. 2006); *S. epidermidis* biofilm is known to be targetted with R- Thanatin, Dalbavancin, Hepcidin 20 (Brancatisano et al. 2014; Knafl et al. 2017) etc. Moreover, the examples of major microbes targeted with ABPs, extracted from the literature are enlisted in Table 5.2.

5.6 Role of Anti-biofilm Peptides in Antibiotic Drug Resistance

Antibiotic drug resistance is a colossal problem due to the microbes residing within biofilms. As compared to the conventional drugs the ABPs are the efficient tool to target biofilm due to their pervasive mode of action (Maisetta et al. 2006) namely killing of multispecies bacteria within biofilm; showing synergistic effect with anti biotics; penetration ability within biofilms, etc. The *S. aureus* (MRSA) can be inhibited with LL-37 peptides (Haisma et al. 2014). The 1018 peptide acts synergistically with antibiotics like ceftazidime, tobramycin, and ciprofloxacin to target the biofilms of ESKAPE pathogens (Reffuveille et al. 2014). The human β-defensins are efficient against various Gram-positive and Gram-negative bacteria involved in nosocomial infections (Maisetta et al. 2006). The DLK-5 and 6 are known to eradicate *P. aeruginosa* infections (Kanthawong et al. 2010). RNAIII-inhibiting peptides (RIP) abolish the *S. aureus* biofilms by interfering their adhesion (Kiran et al. 2008). Despite the presence of more than 600 AMPs at various stages of clinical trials (Preclinical, Phase 1, 2, 3), the status of ABPs in clinical trials is lagging. Notwithstanding, the prevalent mode of action and efficiency to tackle antibiotics drug resistance, the research in the field of ABPs need to be enhanced.

Table 5.2 List of the anti-biofilm peptides used against important pathogens along with the information of peptide sequence, concentration, % inhibition, mode of action, stage of biofilm targeted, and references

Anti-biofilm Peptides	Peptide sequence	Organism	Concentration	Biofilm inhibition activity (%)	Quantification assay	QQA mode of action against biofilm-target	Stage of biofilm targeted	References
16P	YKPVTNF-ST-YKPVTNF-CONH2	*Staphylococcus aureus* ATCC29213	50 µg/mL	58	Crystal violet staining assay	Not Specified	Adhesion and Formation	Zhou et al. (2016)
2C-4	RWWRWF	*Streptococcus mutans* UA159	25 µg/ml	12	OD at 600 nm	Killing of bacterial cells	Formation	He et al. (2010)
6-MO-LF11-322	PFWRIRIRR	*Pseudomonas aeruginosa* PAO1	64 µg/ml	38	MTT assay	Dysregulation of genes related to biofilm formation	Maturation	Sanchez-Gomez et al. (2015)
C16-33	TRRRLFNRSFTQALGKSGGGFKK FWKWFRRF	*Streptococcus mutans* UA159	2.5 ± 2.1 µM	53	Not Specified	Targeted killing of S. mutans	Formation	Eckert et al. (2006)
C16-33	TRRRLFNRSFTQALGKSGGGFK KFWKWFRRF	*Streptococcus sanguinis* NY101	13.3 ± 5.8 µM	43	Not Specified	Targeted killing of S. mutans	Formation	Eckert et al. (2006)
C16G2	TFFRLFNRSFTQALGKGGGKNL RIIRKGIHIIKKY	*Streptococcus mutans* UA140	100 µM	95	Not Specified	Not specified	Maturation	Sullivan et al. (2011)
Coprisin	VTCDVLSFEAKGIAVNHSACALHC IALRKKGGSCQNGVCVCRN-NH2	*Pseudomonas aeruginosa* ATCC 27853	16 µg/mL	92	Crystal violet staining assay	Not Specified	Formation	Hwang et al. (2013)

(continued)

Table 5.2 (continued)

Anti-biofilm Peptides	Peptide sequence	Organism	Concentration	Biofilm inhibition activity (%)	Quantification assay	QQA mode of action against biofilm-target	Stage of biofilm targeted	References
Coprisin	VTCDVLSFEAKGIAVNHSACALH CIALRKKGGSCQNGVCVCRN-NH2	*Escherichia coli* O-157 ATCC 43895	8 µg/mL	85	Crystal violet staining assay	Not Specified	Formation	Hwang et al. (2013)
Coprisin	VTCDVLSFEAKGIAVNHSACALH CIALRKKGGSCQNGVCVCRN-NH2	*Streptococcus mutans* KCTC 3065	8 µg/mL	74	Crystal violet staining assay	Not Specified	Formation	Hwang et al. (2013)
Coprisin	VTCDVLSFEAKGIAVNHSACALH CIALRKKGGSCQNGVCVCRN-NH2	*Staphylococcus aureus* ATCC 25923	16 µg/mL	45	Crystal violet staining assay	Not Specified	Formation	Hwang et al. (2013)
Coprisin	VTCDVLSFEAKGIAVNHSACALH CIALRKKGGSCQNGVCVCRN-NH2	*Enterococcus faecium* ATCC 19434	8 µg/mL	43	Crystal violet staining assay	Not Specified	Formation	Hwang et al. (2013)
Di-MB-LF11-322	PFWRIRIRR	*Pseudomonas aeruginosa* PAO1	32 µg/ml	80	MTT assay	Dysregulation of genes related to biofilm formation	Maturation	Sanchez-Gomez et al. (2015)
DJK5	VQWRAIRVRVIR	*Pseudomonas aeruginosa* PA14	1 µg/ml	50	Crystal violet staining assay	Binds to and promote degradation of the signal for biofilm formation and maintenance i.e. (p)ppGpp	Formation	de la Fuente-Nunez et al. (2015)

DJK5	VQWRAIRVRVIR	Escherichia coli O157	0.8 µg/ml	50	Crystal violet staining assay	Binds to and promote degradation of the signal for biofilm formation and maintenance i.e. (p)ppGpp	Formation	de la Fuente-Nunez et al. (2015)
DJK5	VQWRAIRVRVIR	Klebsiella pneumoniae ATTC 13883	1.6 µg/ml	50	Crystal violet staining assay	Binds to and promote degradation of the signal for biofilm formation and maintenance i.e. (p)ppGpp	Formation	de la Fuente-Nunez et al. (2015)
DJK5	VQWRAIRVRVIR	Salmonella enterica Serovar Typhimurium isolate 14028S	0.8 µg/ml	50	Crystal violet staining assay	Binds to and promote degradation of the signal for biofilm formation and maintenance i.e. (p)ppGpp	Formation	de la Fuente-Nunez et al. (2015)

(continued)

Table 5.2 (continued)

Anti-biofilm Peptides	Peptide sequence	Organism	Concentration	Biofilm inhibition activity (%)	Quantification assay	QQA mode of action against biofilm-target	Stage of biofilm targeted	References
DJK6	VQWRRIRVWVIR	*Pseudomonas aeruginosa* PA14	0.5 µg/ml	50	Crystal violet staining assay	Binds to and promote degradation of the signal for biofilm formation and maintenance i.e. (p)ppGpp	Formation	de la Fuente-Nunez et al. (2015)
DJK6	VQWRRIRVWVIR	*Escherichia coli* O157	8 µg/ml	50	Crystal violet staining assay	Binds to and promote degradation of the signal for biofilm formation and maintenance i.e. (p)ppGpp	Formation	de la Fuente-Nunez et al. (2015)
DJK6	VQWRRIRVWVIR	*Klebsiella pneumoniae* ATTC 13883	2 µg/ml	50	Crystal violet staining assay	Binds to and promote degradation of the signal for biofilm formation and maintenance i.e. (p)ppGpp	Formation	de la Fuente-Nunez et al. (2015)

DJK6	VQWRRIRVWVIR	1 µg/ml	50	Crystal violet staining assay	Binds to and promote degradation of the signal for biofilm formation and maintenance i.e. (p)ppGpp	Formation	de la Fuente-Nunez et al. (2015)
F2,5,12W	RWGRWLRKIRRWRPK	40 µM	83	Crystal violet staining assay	Eradication of formed biofilm by dettachment from surface	Maturation	Molhoek et al. (2011)
IDR-1018	VRLIVAV- RIWRR-NH2	10 µg/ml	100	Crystal violet staining assay, SYT0 9-staining/ Propidium iodide-staining	Inhibition and dispersal of biofilm formation	Formation	de la Fuente-Nunez et al. (2014b)
IDR-1018	VRLIVAV- RIWRR-NH2	10 µg/ml	100	Crystal violet staining assay, SYT0 9-staining/ Propidium iodide-staining	Inhibition and dispersal of biofilm formation	Formation	de la Fuente-Nunez et al. (2014b)

Organisms (Sequence column secondary):
- *Salmonella enterica* Serovar Typhimurium isolate 14028S
- *Staphylococcus epidermidis* BM185
- *Pseudomonas aeruginosa* PA01
- *Escherichia coli* O157

(continued)

Table 5.2 (continued)

Anti-biofilm Peptides	Peptide sequence	Organism	Concentration	Biofilm inhibition activity (%)	Quantification assay	QQA mode of action against biofilm-target	Stage of biofilm targeted	References
IDR-1018	VRLIVAV- RIWRR-NH2	*Acinetobacter baumannii* SENTRY C8	10 µg/ml	100	Crystal violet staining assay, SYTO 9-staining/ Propidium iodide-staining	Inhibition and dispersal of biofilm formation	Formation	de la Fuente-Nunez et al. (2014b)
IDR-1018	VRLIVAV- RIWRR-NH2	*Klebsiella pneumoniae* ATTC13883	2 µg/ml	100	Crystal violet staining assay, SYTO 9-staining/ Propidium iodide-staining	Inhibition and dispersal of biofilm formation	Formation	de la Fuente-Nunez et al. (2014b)
IDR-1018	VRLIVAV- RIWRR-NH2	*Salmonella enterica* Serovar Typhimurium 14028S	10 µg/ml	100	Crystal violet staining assay, SYTO 9-staining/ Propidium iodide-staining	Inhibition and dispersal of biofilm formation	Formation	de la Fuente-Nunez et al. (2014b)
IDR-1018	VRLIVAV- RIWRR-NH2	*Staphylococcus aureus* MRSA #SAP0017	2.5 µg/ml	100	Crystal violet staining assay, SYTO 9-staining/ Propidium iodide-staining	Inhibition and dispersal of biofilm formation	Formation	de la Fuente-Nunez et al. (2014b)

Name	Sequence	Organism	Concentration	Value	Assay	Activity	Type	Reference
IDR-1018	VRLIVAV- RIWRR-NH$_2$	*Burkholderia cenocepacia* IIIa 4813	10 μg/ml	100	Crystal violet staining assay, SYTO 9-staining/Propidium iodide-staining	Inhibition and dispersal of biofilm formation	Formation	de la Fuente-Nunez et al. (2014b)
LF11-324	PFFWRIRIRR	*Pseudomonas aeruginosa* PAO1	8 μg/ml	50	MTT assay	Dysregulation of genes related to biofilm formation	Maturation	Sanchez-Gomez et al. (2015)
LL7-37	RKSKEKIGKEFKRIVQRIKDFLRNLVPRTES	*Pseudomonas aeruginosa* PAO1	10 μM	60	Crystal violet staining assay	Inhibition of biofilm formation	Formation	Nagant et al. (2012)
M8-33	TFFRLFNRSGGGFKKFWKWFRRF	*Streptococcus mutans* UA159	2.5 ± 2.0 μM	64	Not Specified	Targeted killing of S. mutans	Formation	Eckert et al. (2006)
M8-33	TFFRLFNRSGGGFKKFWKWFRRF	*Streptococcus sanguinis* NY101	20 ± 2.0 μM	34	Not Specified	Targeted killing of S. mutans	Formation	Eckert et al. (2006)
M8G2	TFFRLFNRGGGKNLRIIRKGIHIIKKY	*Streptococcus mutans* UA159	25 μM	96	Not Specified	Targeted killing of S. mutans	Formation	Eckert et al. (2006)
M8G2	TFFRLFNRGGGKNLRIIRKGIHIIKKY	*Streptococcus sanguinis* NY101	25 μM	21	Not Specified	Targeted killing of S. mutans	Formation	Eckert et al. (2006)

(continued)

Table 5.2 (continued)

Anti-biofilm Peptides	Peptide sequence	Organism	Concentration	Biofilm inhibition activity (%)	Quantification assay	QQA mode of action against biofilm-target	Stage of biofilm targeted	References
Myxinidin3	RIRWILRYWRWS	*Pseudomonas aeruginosa* ATCC 27853	4 µM	92	Crystal violet staining assay	Through membrane disruption and/or cell penetration.	Formation	Han et al. (2016)
Myxinidin3	RIRWILRYWRWS	*Staphylococcus aureus* ATCC 25923	8 µM	84	Crystal violet staining assay	Through membrane disruption and/or cell penetration.	Formation	Han et al. (2016)
Myxinidin3	RIRWILRYWRWS	*Listeria monocytogenes* KCTC 3710	16 µM	84	Crystal violet staining assay	Through membrane disruption and/or cell penetration.	Formation	Han et al. (2016)
PG-1	RGGRLCYCRRRFCVCVGR	*Cryptococcus neoformans* B3501	8 µM	41	XTT assay	Not specified	Pre-formed	Martinez and Casadevall (2006)
RK-31	RKSKEKIGKEFKRIVQRIK DFLRNLVPRTES	*Burkholderia pseudomallei* 1026b	100 µM	93	Not Specified	Permeabilize/form pores within cytoplasmic membrane	Formation	Kanthawong et al. (2012)

RNAIII-inhibiting Peptide	YKPVTNF-CONH2	Staphylococcus epidermidis XJ75284 (MRSE)	50 μg/mL	60	Crystal violet staining assay	Not Specified	Adhesion and Formation	Zhou et al. (2016)
RNAIII-inhibiting Peptide	YKPVTNF-CONH2	Staphylococcus aureus ATCC29213	50 μg/mL	45	Crystal violet staining assay	Not Specified	Adhesion and Formation	Zhou et al. (2016)
RNAIII-inhibiting Peptide	YKPLTNF-CONH2	Staphylococcus epidermidis XJ75284 (MRSE)	50 μg/mL	40	Crystal violet staining assay	Not Specified	Adhesion and Formation	Zhou et al. (2016)
SMAP-29	RGLRRLGRKIAHGVKK YGPTVLRIIRIAG	Burkholderia thailandensis E264	30 μg/ml	70	Crystal violet staining assay	Not specified	Formation	Blower et al. (2015)

5.7   Conclusion

This book chapter is focused on the new class of therapeutics named ABPs. The ABPs are recently emerged as important and efficient anti-biofilm agents to target MDR biofilms. They are introduced alone and/or in combination with antibiotics. They are proved to be more efficient in inhibiting biofilms involved in nosocomial infections (Pletzer and Hancock 2016). Their importance is gradually increasing due to their broad ranged specificity towards stages and components of biofilms. Moreover, they are proved as a new ray of hope to the world struggling with the problem of antibiotic drug resistance.

References

Agarwala M, Choudhury B, Yadav RN (2014) Comparative study of antibiofilm activity of copper oxide and iron oxide nanoparticles against multidrug resistant biofilm forming uropathogens. Ind J Microbiol 54:365–368. https://doi.org/10.1007/s12088-014-0462-z

Ahiwale SS, Bankar AV, Tagunde S, Kapadnis BP (2017) A bacteriophage mediated gold nanoparticles synthesis and their anti-biofilm activity. Ind J Microbiol 57:188–194. https://doi.org/10.1007/s12088-017-0640-x

Anunthawan T, de la Fuente-Nunez C, Hancock RE, Klaynongsruang S (2015) Cationic amphipathic peptides KT2 and RT2 are taken up into bacterial cells and kill planktonic and biofilm bacteria. Biochim Biophys Acta 1848:1352–1358. https://doi.org/10.1016/j.bbamem.2015.02.021

Balamurugan P, Hema M, Kaur G, Sridharan V, Prabu PC, Sumana MN, Princy SA (2015) Development of a biofilm inhibitor molecule against multidrug resistant Staphylococcus aureus associated with gestational urinary tract infections. Front Microbiol 6:832. https://doi.org/10.3389/fmicb.2015.00832

Batoni G, Maisetta G, Esin S (2016) Antimicrobial peptides and their interaction with biofilms of medically relevant bacteria. Biochim Biophys Acta 1858:1044–1060. https://doi.org/10.1016/j.bbamem.2015.10.013

Bhargava N, Singh SP, Sharma A, Sharma P, Capalash N (2015) Attenuation of quorum sensing-mediated virulence of Acinetobacter baumannii by Glycyrrhiza glabra flavonoids. Future Microbiol 10:1953–1968. https://doi.org/10.2217/fmb.15.107

Blower RJ, Barksdale SM, van Hoek ML (2015) Snake Cathelicidin NA-CATH and Smaller Helical Antimicrobial Peptides Are Effective against Burkholderia thailandensis. PLoS Negl Trop Dis 9: e0003862. doi:https://doi.org/10.1371/journal.pntd.0003862

Brancatisano FL, Maisetta G, Di Luca M, Esin S, Bottai D, Bizzarri R, Campa M, Batoni G (2014) Inhibitory effect of the human liver-derived antimicrobial peptide hepcidin 20 on biofilms of polysaccharide intercellular adhesin (PIA)-positive and PIA-negative strains of Staphylococcus epidermidis. Biofouling 30:435–446. https://doi.org/10.1080/08927014.2014.888062

Branda SS, Vik S, Friedman L, Kolter R (2005) Biofilms: the matrix revisited. Trends Microbiol 13:20–26. https://doi.org/10.1016/j.tim.2004.11.006

Cardoso MH, Ribeiro SM, Nolasco DO, de la Fuente-Nunez C, Felicio MR, Goncalves S, Matos CO, Liao LM, Santos NC, Hancock RE, Franco OL, Migliolo L (2016) A polyalanine peptide derived from polar fish with anti-infectious activities. Sci Rep 6: 21385. doi:https://doi.org/10.1038/srep21385

Choi H, Lee DG (2012) Antimicrobial peptide pleurocidin synergizes with antibiotics through hydroxyl radical formation and membrane damage, and exerts antibiofilm activity. Biochim Biophys Acta 1820:1831–1838. https://doi.org/10.1016/j.bbagen.2012.08.012

Conrad A, Suutari MK, Keinanen MM, Cadoret A, Faure P, Mansuy-Huault L, Block JC (2003) Fatty acids of lipid fractions in extracellular polymeric substances of activated sludge flocs. Lipids 38:1093–1105

Dang H, Lovell CR (2016) Microbial surface colonization and biofilm development in marine environments. Microbiol Mol Biol Rev 80:91–138. https://doi.org/10.1128/mmbr.00037-15

Davey ME, O'Toole GA (2000) Microbial biofilms: from ecology to molecular genetics. Microbiol Mol Biol Rev 64:847–867

Davey ME, Caiazza NC, O'Toole GA (2003) Rhamnolipid surfactant production affects biofilm architecture in Pseudomonas aeruginosa PAO1. J Bacteriol 185:1027–1036

de la Fuente-Nunez C, Korolik V, Bains M, Nguyen U, Breidenstein EB, Horsman S, Lewenza S, Burrows L, Hancock RE (2012) Inhibition of bacterial biofilm formation and swarming motility by a small synthetic cationic peptide. Antimicrob Agents Chemother 56:2696–2704. https://doi.org/10.1128/aac.00064-12

de la Fuente-Nunez C, Mansour SC, Wang Z, Jiang L, Breidenstein EB, Elliott M, Reffuveille F, Speert DP, Reckseidler-Zenteno SL, Shen Y, Haapasalo M, Hancock RE (2014) Anti-biofilm and immunomodulatory activities of peptides that inhibit biofilms formed by pathogens isolated from cystic fibrosis patients. Antibiotics (Basel) 3:509–526. https://doi.org/10.3390/antibiotics3040509

de la Fuente-Nunez C, Reffuveille F, Mansour SC, Reckseidler-Zenteno SL, Hernandez D, Brackman G, Coenye T, Hancock RE (2015) D-enantiomeric peptides that eradicate wild-type and multidrug-resistant biofilms and protect against lethal Pseudomonas aeruginosa infections. Chem Biol 22:196–205. https://doi.org/10.1016/j.chembiol.2015.01.002

de la Fuente-Nunez C, Cardoso MH, de Souza Candido E, Franco OL, Hancock RE (2016) Synthetic antibiofilm peptides. Biochim Biophys Acta 1858:1061–1069. https://doi.org/10.1016/j.bbamem.2015.12.015

Delattin N, De Brucker K, Craik DJ, Cheneval O, Frohlich M, Veber M, Girandon L, Davis TR, Weeks AE, Kumamoto CA, Cos P, Coenye T, De Coninck B, Cammue BP, Thevissen K (2014) Plant-derived decapeptide OSIP108 interferes with Candida albicans biofilm formation without affecting cell viability. Antimicrob Agents Chemother 58:2647–2656. https://doi.org/10.1128/aac.01274-13

Ding Y, Wang W, Fan M, Tong Z, Kuang R, Jiang W, Ni L (2014) Antimicrobial and anti-biofilm effect of Bac8c on major bacteria associated with dental caries and Streptococcus mutans biofilms. Peptides 52:61–67. https://doi.org/10.1016/j.peptides.2013.11.020

Dubern JF, Lugtenberg BJ, Bloemberg GV (2006) The ppuI-rsaL-ppuR quorum-sensing system regulates biofilm formation of Pseudomonas putida PCL1445 by controlling biosynthesis of the cyclic lipopeptides putisolvins I and II. J Bacteriol 188:2898–2906. https://doi.org/10.1128/jb.188.8.2898-2906.2006

Eckert R, He J, Yarbrough DK, Qi F, Anderson MH, Shi W (2006) Targeted killing of Streptococcus mutans by a pheromone-guided "smart" antimicrobial peptide. Antimicrob Agents Chemother 50:3651–3657. https://doi.org/10.1128/aac.00622-06

Feng X, Sambanthamoorthy K, Palys T, Paranavitana C (2013) The human antimicrobial peptide LL-37 and its fragments possess both antimicrobial and antibiofilm activities against multidrug-resistant Acinetobacter baumannii. Peptides 49:131–137. https://doi.org/10.1016/j.peptides.2013.09.007

Flemming HC, Wingender J (2010) The biofilm matrix. Nat Rev Microbiol 8:623–633. https://doi.org/10.1038/nrmicro2415

Fong JN, Yildiz FH (2015) Biofilm matrix proteins. Microbiol Spectr 3. https://doi.org/10.1128/microbiolspec.MB-0004-2014

Gopal R, Lee JH, Kim YG, Kim MS, Seo CH, Park Y (2013) Anti-microbial, anti-biofilm activities and cell selectivity of the NRC-16 peptide derived from witch flounder, Glyptocephalus cynoglossus. Mar Drugs 11:1836–1852. https://doi.org/10.3390/md11061836

Gopal R, Kim YG, Lee JH, Lee SK, Chae JD, Son BK, Seo CH, Park Y (2014) Synergistic effects and antibiofilm properties of chimeric peptides against multidrug-resistant Acinetobacter baumannii strains. Antimicrob Agents Chemother 58:1622–1629. https://doi.org/10.1128/aac.02473-13

Gui Z, Wang H, Ding T, Zhu W, Zhuang X, Chu W (2014) Azithromycin reduces the production of alpha-hemolysin and biofilm formation in *Staphylococcus aureus*. Ind J Microbiol 54:114–117. https://doi.org/10.1007/s12088-013-0438-4

Haisma EM, de Breij A, Chan H, van Dissel JT, Drijfhout JW, Hiemstra PS, El Ghalbzouri A, Nibbering PH (2014) LL-37-derived peptides eradicate multidrug-resistant Staphylococcus aureus from thermally wounded human skin equivalents. Antimicrob Agents Chemother 58:4411–4419. https://doi.org/10.1128/aac.02554-14

Han HM, Gopal R, Park Y (2016) Design and membrane-disruption mechanism of charge-enriched AMPs exhibiting cell selectivity, high-salt resistance, and anti-biofilm properties. Amino Acids 48:505–522. https://doi.org/10.1007/s00726-015-2104-0

He J, Yarbrough DK, Kreth J, Anderson MH, Shi W, Eckert R (2010) Systematic approach to optimizing specifically targeted antimicrobial peptides against Streptococcus mutans. Antimicrob Agents Chemother 54:2143–2151. https://doi.org/10.1128/aac.01391-09

Hirt H, Gorr SU (2013) Antimicrobial peptide GL13K is effective in reducing biofilms of Pseudomonas aeruginosa. Antimicrob Agents Chemother 57:4903–4910. https://doi.org/10.1128/aac.00311-13

Hwang IS, Hwang JS, Hwang JH, Choi H, Lee E, Kim Y, Lee DG (2013) Synergistic effect and anti-biofilm activity between the antimicrobial peptide coprisin and conventional antibiotics against opportunistic bacteria. Curr Microbiol 66:56–60. https://doi.org/10.1007/s00284-012-0239-8

Jack AA, Daniels DE, Jepson MA, Vickerman MM, Lamont RJ, Jenkinson HF, Nobbs AH (2015) Streptococcus gordonii comCDE (competence) operon modulates biofilm formation with Candida albicans. Microbiology 161:411–421. https://doi.org/10.1099/mic.0.000010

Kalia VC (2013) Quorum sensing inhibitors: an overview. Biotechnol Adv 31:224–245. https://doi.org/10.1016/j.biotechadv.2012.10.004

Kalia VC (2014) In search of versatile organisms for quorum-sensing inhibitors: acyl homoserine lactones (AHL)-acylase and AHL-lactonase. FEMS Microbiol Lett 359:143. https://doi.org/10.1111/1574-6968.12585

Kanthawong S, Bolscher JG, Veerman EC, van Marle J, Nazmi K, Wongratanacheewin S, Taweechaisupapong S (2010) Antimicrobial activities of LL-37 and its truncated variants against Burkholderia thailandensis. Int J Antimicrob Agents 36:447–452. https://doi.org/10.1016/j.ijantimicag.2010.06.031

Kanthawong S, Bolscher JG, Veerman EC, van Marle J, de Soet HJ, Nazmi K, Wongratanacheewin S, Taweechaisupapong S (2012) Antimicrobial and antibiofilm activity of LL-37 and its truncated variants against Burkholderia pseudomallei. Int J Antimicrob Agents 39:39–44. https://doi.org/10.1016/j.ijantimicag.2011.09.010

Kaplan JB, Ragunath C, Ramasubbu N, Fine DH (2003) Detachment of Actinobacillus actinomycetemcomitans biofilm cells by an endogenous beta-hexosaminidase activity. J Bacteriol 185:4693–4698

Kiran MD, Adikesavan NV, Cirioni O, Giacometti A, Silvestri C, Scalise G, Ghiselli R, Saba V, Orlando F, Shoham M, Balaban N (2008) Discovery of a quorum-sensing inhibitor of drug-resistant staphylococcal infections by structure-based virtual screening. Mol Pharmacol 73:1578–1586. https://doi.org/10.1124/mol.107.044164

Knafl D, Tobudic S, Cheng SC, Bellamy DR, Thalhammer F (2017) Dalbavancin reduces biofilms of methicillin-resistant Staphylococcus aureus (MRSA) and methicillin-resistant Staphylococcus epidermidis (MRSE). Eur J Clin Microbiol Infect Dis 36:677–680. https://doi.org/10.1007/s10096-016-2845-z

Kostakioti M, Hadjifrangiskou M, Hultgren SJ (2013) Bacterial biofilms: development, dispersal, and therapeutic strategies in the dawn of the postantibiotic era. Cold Spring Harb Perspect Med 3:a010306. https://doi.org/10.1101/cshperspect.a010306

Koul S, Kalia VC (2017) Multiplicity of quorum quenching enzymes: A potential mechanism to limit quorum sensing bacterial population. Indian J Microbiol 57:100–108. https://doi.org/10.1007/s12088-016-0633-1

Koul S, Prakash J, Mishra A, Kalia VC (2016) Potential emergence of multi-quorum sensing inhibitor resistant (MQSIR) bacteria. Indian J Microbiol 56:1–18. https://doi.org/10.1007/s12088-015-0558-0

Li YH, Tian X (2012) Quorum sensing and bacterial social interactions in biofilms. Sensors (Basel) 12:2519–2538. https://doi.org/10.3390/s120302519

Li S, Zhu C, Fang S, Zhang W, He N, Xu W, Kong R, Shang X (2015) Ultrasound microbubbles enhance human beta-defensin 3 against biofilms. J Surg Res 199:458–469. https://doi.org/10.1016/j.jss.2015.05.030

Maisetta G, Batoni G, Esin S, Florio W, Bottai D, Favilli F, Campa M (2006) In vitro bactericidal activity of human beta-defensin 3 against multidrug-resistant nosocomial strains. Antimicrob Agents Chemother 50:806–809. https://doi.org/10.1128/aac.50.2.806-809.2006

Martinez LR, Casadevall A (2006) Cryptococcus neoformans cells in biofilms are less susceptible than planktonic cells to antimicrobial molecules produced by the innate immune system. Infect Immun 74:6118–6123. https://doi.org/10.1128/iai.00995-06

Molhoek EM, van Dijk A, Veldhuizen EJ, Haagsman HP, Bikker FJ (2011) A cathelicidin-2-derived peptide effectively impairs Staphylococcus epidermidis biofilms. Int J Antimicrob Agents 37: 476-479. doi:https://doi.org/10.1016/j.ijantimicag.2010.12.020

Montanaro L, Poggi A, Visai L, Ravaioli S, Campoccia D, Speziale P, Arciola CR (2011) Extracellular DNA in biofilms. Int J Artif Organs 34:824–831. https://doi.org/10.5301/ijao.5000051

Morici P, Fais R, Rizzato C, Tavanti A, Lupetti A (2016) Inhibition of *Candida albicans* biofilm formation by the synthetic lactoferricin derived peptide hLF1-11. PLoS One 11:e0167470. https://doi.org/10.1371/journal.pone.0167470

Nagant C, Pitts B, Nazmi K, Vandenbranden M, Bolscher JG, Stewart PS, Dehaye JP (2012) Identification of peptides derived from the human antimicrobial peptide LL-37 active against biofilms formed by *Pseudomonas aeruginosa* using a library of truncated fragments. Antimicrob Agents Chemother 56:5698–5708. https://doi.org/10.1128/aac.00918-12

Nijland R, Hall MJ, Burgess JG (2010) Dispersal of biofilms by secreted, matrix degrading, bacterial DNase. PLoS One 5:e15668. https://doi.org/10.1371/journal.pone.0015668

Overhage J, Campisano A, Bains M, Torfs EC, Rehm BH, Hancock RE (2008) Human host defense peptide LL-37 prevents bacterial biofilm formation. Infect Immun 76:4176–4182. https://doi.org/10.1128/iai.00318-08

Parsek MR, Greenberg EP (2005) Sociomicrobiology: the connections between quorum sensing and biofilms. Trends Microbiol 13:27–33. https://doi.org/10.1016/j.tim.2004.11.007

Pletzer D, Hancock RE (2016) Antibiofilm peptides: potential as broad-spectrum agents. J Bacteriol 198:2572–2578. https://doi.org/10.1128/jb.00017-16

Pletzer D, Coleman SR, Hancock RE (2016) Anti-biofilm peptides as a new weapon in antimicrobial warfare. Curr Opin Microbiol 33:35–40. https://doi.org/10.1016/j.mib.2016.05.016

Pusateri CR, Monaco EA, Edgerton M (2009) Sensitivity of Candida albicans biofilm cells grown on denture acrylic to antifungal proteins and chlorhexidine. Arch Oral Biol 54:588–594. https://doi.org/10.1016/j.archoralbio.2009.01.016

Rajput A, Kumar M (2017a) Computational exploration of putative LuxR solos in archaea and their functional implications in quorum sensing. Front Microbiol 8:798. https://doi.org/10.3389/fmicb.2017.00798

Rajput A, Kumar M (2017b) In silico analyses of conservational, functional and phylogenetic distribution of the LuxI and LuxR homologs in Gram-positive bacteria. Sci Rep 7:6969. https://doi.org/10.1038/s41598-017-07241-5

Rajput A, Gupta AK, Kumar M (2015) Prediction and analysis of quorum sensing peptides based on sequence features. PLoS One 10:e0120066. https://doi.org/10.1371/journal.pone.0120066

Rajput A, Kaur K, Kumar M (2016) SigMol: repertoire of quorum sensing signaling molecules in prokaryotes. Nucleic Acids Res 44:D634–D639. https://doi.org/10.1093/nar/gkv1076

Rajput A, Thakur A, Sharma S, Kumar M (2018) aBiofilm: a resource of anti-biofilm agents and their potential implications in targeting antibiotic drug resistance. Nucleic Acids Res 46:D894–d900. https://doi.org/10.1093/nar/gkx1157

Reffuveille F, de la Fuente-Nunez C, Mansour S, Hancock RE (2014) A broad-spectrum anti-biofilm peptide enhances antibiotic action against bacterial biofilms. Antimicrob Agents Chemother 58: 5363–5371. doi:https://doi.org/10.1128/aac.03163-14

Sanchez-Gomez S, Ferrer-Espada R, Stewart PS, Pitts B, Lohner K, Martinez de Tejada G (2015) Antimicrobial activity of synthetic cationic peptides and lipopeptides derived from human lactoferricin against Pseudomonas aeruginosa planktonic cultures and biofilms. BMC Microbiol 15:137. https://doi.org/10.1186/s12866-015-0473-x

Sand W, Gehrke T (2006) Extracellular polymeric substances mediate bioleaching/biocorrosion via interfacial processes involving iron(III) ions and acidophilic bacteria. Res Microbiol 157:49–56. https://doi.org/10.1016/j.resmic.2005.07.012

Scarsini M, Tomasinsig L, Arzese A, D'Este F, Oro D, Skerlavaj B (2015) Antifungal activity of cathelicidin peptides against planktonic and biofilm cultures of Candida species isolated from vaginal infections. Peptides 71:211–221. https://doi.org/10.1016/j.peptides.2015.07.023

Singh S, Singh H, Tuknait A, Chaudhary K, Singh B, Kumaran S, Raghava GP (2015) PEPstrMOD: structure prediction of peptides containing natural, non-natural and modified residues. Biol Direct 10:73. https://doi.org/10.1186/s13062-015-0103-4

Stewart PS (2003) Diffusion in biofilms. J Bacteriol 185:1485–1491

Sullivan R, Santarpia P, Lavender S, Gittins E, Liu Z, Anderson MH, He J, Shi W, Eckert R (2011) Clinical efficacy of a specifically targeted antimicrobial peptide mouth rinse: targeted elimination of Streptococcus mutans and prevention of demineralization. Caries Res 45:415–428. https://doi.org/10.1159/000330510

Theberge S, Semlali A, Alamri A, Leung KP, Rouabhia M (2013) C. albicans growth, transition, biofilm formation, and gene expression modulation by antimicrobial decapeptide KSL-W. BMC Microbiol 13:246. https://doi.org/10.1186/1471-2180-13-246

Toyofuku M, Inaba T, Kiyokawa T, Obana N, Yawata Y, Nomura N (2015) Environmental factors that shape biofilm formation. Biosci Biotechnol Biochem 80:7–12. https://doi.org/10.1080/09168451.2015.1058701

Vuotto C, Longo F, Pascolini C, Donelli G, Balice MP, Libori MF, Tiracchia V, Salvia A, Varaldo PE (2017) Biofilm formation and antibiotic resistance in Klebsiella pneumoniae urinary strains. J Appl Microbiol 123:1003–1018. https://doi.org/10.1111/jam.13533

Winfred SB, Meiyazagan G, Panda JJ, Nagendrababu V, Deivanayagam K, Chauhan VS, Venkatraman G (2014) Antimicrobial activity of cationic peptides in endodontic procedures. Eur J Dent 8:254–260. https://doi.org/10.4103/1305-7456.130626

Wingender J, Strathmann M, Rode A, Leis A, Flemming HC (2001) Isolation and biochemical characterization of extracellular polymeric substances from Pseudomonas aeruginosa. Methods Enzymol 336:302–314

Zhou Y, Zhao R, Ma B, Gao H, Xue X, Qu D, Li M, Meng J, Luo X, Hou Z (2016) Oligomerization of RNAIII-inhibiting peptide inhibits adherence and biofilm formation of methicillin-resistant Staphylococcus aureus in vitro and in vivo. Microb Drug Resist 22:193–201. https://doi.org/10.1089/mdr.2015.0170

Chapter 6
Quorum Sensing Inhibition: A Target for Treating Chronic Wounds

Lahari Das and Yogendra Singh

Abstract Chronic wounds are serious medical problem which sometimes become fatal. Bacterial infections are one of the major causes for chronic wounds and delay in their healing. Pathogenic bacteria form biofilms on the surface of wounds. Biofilms are organized polymicrobial structures where bacteria are encased in exopolysaccharide layer and are present in metabolically quiescent state, thus making the wounds resistant to antimicrobial treatment. This delays the wound healing process by slowing down tissue repair and by inducing chronic inflammation at the site of the wound. The cell to cell communication system also known as the quorum sensing (QS) system is required for biofilm formation and coordinated virulence activities. QS inhibitors have emerged as important candidates for inhibiting biofilm formation, maintenance and expression of virulence factors. Use of these compounds alone or in combination with antibiotics may aid in rapid healing of chronic wounds and tissue regeneration. This chapter focuses on the role of QS in chronic wounds and the use of QS inhibitors for treating such wounds and facilitating wound healing.

Keywords Chronic wounds · Wound healing · Quorum sensing · Biofilms · Antibiotic resistance · Quorum sensing inhibitors

6.1 Introduction

Chronic wounds are worldwide problem causing high rates of morbidity and mortality. Wounds are generally healed in the body in an orchestrated manner. The wound healing process involves various steps *viz.* inflammation, proliferatory phase and finally the tissue restructuring phase. However, some wounds due to certain reasons fail to heal and the functional integrity of the tissue is not restored on time.

L. Das
Department of Zoology, University of Delhi North Campus, Delhi, New Delhi, India

Y. Singh (✉)
Department of Zoology, University of Delhi, Delhi, India

Academy of Scientific & Innovative Research (AcSIR), New Delhi, India

© Springer Nature Singapore Pte Ltd. 2018   111
V. C. Kalia (ed.), *Biotechnological Applications of Quorum Sensing Inhibitors*,
https://doi.org/10.1007/978-981-10-9026-4_6

Such wounds are known as chronic wounds. Diabetic foot ulcers, pressure wounds, venous leg ulcers and surgical site infections are some examples of chronic wounds. There are many reasons like venous insufficiency, diabetes, immunosuppression and stress impaired tissue healing that cause chronic wounds. Bacterial infection of the wound area is one of the major cause for making a wound persistent or chronic. Due to the presence of bacterial biofilms on chronic wounds, they are difficult to treat and pose severe problems for antibiotic therapy (Guo and Dipietro 2010; Haji Zaine et al. 2014; Gonzalez et al. 2016).

Bacterial biofilm is a highly organized structure of bacteria encased in exopoly-saccharide substance which adheres to the wound area. Biofilms are polymicrobial in nature and may contain fungi and viruses along with proteins, extracellular DNA and some biogenic substances. Cell to cell communication is indispensable for bio-film formation and this cellular communication is known as quorum sensing (QS). The QS mechanism is governed by signalling molecules that bind to the response regulators and bring about regulation of many genes mostly those which are involved in encoding bacterial virulence factors. Bacterial cells in biofilms are non-motile and metabolically quiescent or show reduced metabolic activity. As a result of which antibiotic resistant phenotypes emerge. The presence of protective exopolysaccha-ride layer and a population of metabolically quiescent bacteria result in decreased effectiveness of antimicrobial compounds (Kalia 2014a). The bacterial infection constantly stimulates the immune system which causes immense damage to the wound and the surrounding tissue. This aggravates inflammation and retards the wound healing process. *Pseudomonas aeruginosa*, *Staphylococcus aureus* and Methicillin resistant *S. aureus* and *Acinetobacter baumannii* are biofilm forming pathogenic bacteria that are responsible for causing most of the chronic wound infec-tions (Castillo-Juarez et al. 2015; Clinton and Carter 2015; De Ryck et al. 2015).

In 1998, Davies et al. were the first to link QS with biofilm formation by showing that QS negative mutants of *P. aeruginosa* were impaired in biofilm formation and were susceptible to antibiotic treatment (Davies et al. 1998). Later, 39 genes were found to be responsible for the QS system in this bacterium. QS mutants were reported to be highly attenuated and caused less infection in mouse pneumonia and burn models (Clinton and Carter 2015).

QS inhibitors (QSI) are gaining importance as potential treatment option for chronic wounds. Small molecules that mimic the QS signalling molecules or com-pounds that quench the signal molecules are being explored for developing novel therapeutic interventions for treating chronic wounds. It is also believed that the small molecules which control the bacterial behaviour and work differently from antibiotics will give the bacterial cells less chance to acquire resistant phenotype. Thus, more studies need to be focussed on synthesizing novel compounds that can inhibit QS and biofilm formation. These will serve as promising candidates for treating chronic wounds by speeding up the wound healing process (Kalia 2013, 2014b; Kalia 2015a, b; Kumar et al. 2013; Agarwala et al. 2014; Brackman and Coenye 2015; Abbas and Shaldam 2016; Begum et al. 2016).

The wound healing process and roles of QS system and biofilms in delaying this process has been explained in the following sections. Also, focus has been laid on the inhibitors of QS and their potential use as alternate therapy for treating chronic wounds.

6.2 Wound Healing Process and Chronic Wound Infections

Wound is a lesion that can be caused due to a trauma, any specific pathological condition or surgery. These lesions cause tissue destruction and can cause damage to specific organelles or to cells as a whole (Shaw and Martin 2009). Healing of wounds is an essential and normal physiological process that involves a complex interplay of many cells and their products. Tissue regeneration and repair processes come into play soon after a wound occurs. The wound healing process comprises of many overlapping cellular and biochemical events that are divided into various stages *viz.*, inflammatory phase in which inflammatory response sets in, proliferative phase characterized by cell division and synthesis of molecules that build up the extracellular matrix occurs, followed by the last phase known as the posterior period or the remodelling phase (Nayak et al. 2009).

Within the first 24 hours of wounding the inflammatory stage sets in wherein, influx of leukocytes occurs in the damaged tissue area (Gonzalez et al. 2016). Erythema and edema are the key signs of inflammation that are observed at the location of the lesion. Immune activation leads to the secretion of chemokines and cytokines by cellular species that aggravate tissue repair (Medrado et al. 2003).

The endothelial transmigration and recruitment of large number neutrophils to the lesion is concurrent with the release of a huge concentration of pro-inflammatory cytokines, reactive oxygen species, antimicrobial peptides and proteases. All of these products lead to the clearance of cellular debris in the wound along with inhibiting microbial colonization of the wound (Gurtner et al. 2008). Another key player in the inflammatory response during wound healing are the blood derived monocytes which are recruited and get differentiated into macrophages in the wounded tissue. These macrophages skew the exudative inflammatory response towards proliferative stage which is a result of the phagocytotic activity of the macrophages coupled to the expression of pro-angiogenic and fibrogenic factors by these cells. The proliferative stage of tissue repair begins within the first 48 hours of injury. This stage is characterized by the contraction and fibroplasia in the tissue area that is aimed towards closure of the wound (Tidball 2005; Eming et al. 2007; Thuraisingam et al. 2010).

Fibroblasts secrete type III collagen that is essential for the formation of the basal membrane and ultimately restoration of tissue integrity (Li et al. 2007). The wound contraction extends from the outer edges of the wound towards its centre, a process mediated by myofibroblasts which are alpha smooth muscle actin rich fibroblasts residing in the border of the wound (Medrado et al. 2010). After wound closure, the type III actin degrades into type I collagen (Gonzalez et al. 2016). These fibres are thicker and are parallely placed thus increasing the tensile strength of the tissue. This stage is marked by the restoration of tissue integrity, appearance and functionality.

Mesenchymal stem cells (MSC) play a keystone role in wound healing. These cells secrete a milieu of growth factors and matrix proteins that aid in tissue restructuring and wound repair (Maxson et al. 2012). MSCs have shown to be promising tool for treating wounds to which these are transplanted to aggravate the wound healing process (Hocking 2015). It has been recently demonstrated taking non-obese-diabetic mice model that not MSCs themselves but the trophic factors released by these cells are critical for tissue regeneration, remodelling and wound healing. Another recent report establishes a tight relationship between hypoxia and MSC function in wound repair. Hypoxia induces the production of Hypoxia-inducible factor - 1α (HIF-1α) which in turn is responsible for regulating the expression of a number of genes involved in wound healing (Hong et al. 2014). The HIF-1α is inhibited by the accumulation of reactive oxygen species. This phenomenon was studied in case of chronic kidney disease (CKD) which is characterized by uremia, anemia, fluid overload and reduced ability to heal wounds. The MSCs derived from CKD have increased concentration of ROS and an oxygen dependent prolyl-hydroxylase enzyme (which negatively regulates HIF-1α function) as well. However, inhibition of above mentioned factors lead to the restoration of HIF-1α activity and aggravated wound healing (Khanh et al. 2017).

The tissue repair or wound healing process is strongly influenced by some external and internal factors. Physiological conditions like diabetes, immunosuppression, metabolic disorders and microbial colonization can hinder the wound healing process. Some of the other factors that are responsible for retarding this process include venous sufficiency, age, gender, sex hormones, stress, ischemia, jaundice, uremia, obesity, use of glucocorticosteroids, non-steroidal anti-inflammatory drugs, chemotherapy, alcoholism, smoking and nutrition. Delay in wound healing leads to the formation of chronic wounds which are severe and fail to heal normally. One of the most common reasons behind chronic wound is microbial infection of the wounded area or tissue. Microorganisms from the environment can get into the damaged tissue and can replicate in that area. If the microorganisms do not replicate at the tissue site, the wound is said to be contaminated. However, if there is presence of replicating microorganisms the wound is classified to be colonized. When the microbial replication is accompanied with the advent of responses in the local tissue, a local infection or critical colonization is said to have occurred (Guo and Dipietro 2010). A wound is classified to be under invasion infection if the microbial replication is resulting in host injury. Bacterial growth in the wound is often accompanied with endotoxin release and both of these are responsible for constantly triggering the inflammatory response. Release of pro-inflammatory cytokines like IL-1 and TNF alpha are responsible for prolonging the inflammatory phase of the wound healing that makes the wound chronic in nature. The prolonged inflammatory reactions lead to the secretion of matrix metalloproteases (MMPs) that can cause the growth factors responsible for angiogenesis and wound restructuring to be rapidly degraded thus, delaying the wound healing process (Menke et al. 2007). Table 6.1 lists the cytokines which are differentially expressed in chronic wounds as compared to the acute wounds.

Treatment of chronic wounds however is possible and involves a multistep treatment regimen. The treatment approach is known as TIME wherein the dead tissue (T) in and around the wound is removed followed by treatment with antibiotics and anti-inflammatory drugs to control infection and inflammation (I). Imbalance in

Table 6.1 Cytokine profile in chronic wounds

Increased expression		Decreased expression	
Cytokines	Cells	Cytokines	Cells
IL-1	Neutrophils, monocytes, keratinocytes	Epidermal growth factor, EGF	Platelets, macrophages, fibroblasts
IL-6	Neutrophils, macrophages	Fibroblast growth factor-2, FGF-2	Keratinocytes, mast cells, fibroblasts, endothelial cells, smooth muscle cells, chondrocytes
TNF-α	Neutrophils, macrophages	Transforming growth factor-beta, TGF-β and Platelet derived growth factor, PDGF	Platelets, keratinocytes, macrophages, lymphocytes, fibroblasts
		Vascular endothelial growth factor, VEGF	Platelets, neutrophils, macrophages, endothelial cells, smooth muscle cells, fibroblasts

moisture (M) is then corrected by specific dressing and growth factors are adminis-tered to promote epithelisation (E) and tissue formation (Clinton and Carter 2015). However, bacterial biofilms in the wound pose a problem for chronic wound healing as they are resistant to antibiotics and phagocytosis by polymorphonuclear neutro-phils (Bjarnsholt et al. 2013).

6.3 Role of Quorum Sensing Mechanism in Inhibiting Wound Healing

QS system is a two-component system consisting of an enzyme that is responsible for catalyzing the signal molecule (autoinducer) synthesis reaction and a receptor molecule that binds to the signal molecule (for e.g. acyl-homoserine lactone, AHL and cyclic peptides) and regulate the transcription of many genes along with that encoding the signal molecule (Gama et al. 2012) (Fig. 6.1).

In bacterial infections, the virulence factors regulated by the QS system are secreted when there is sufficient bacterial load. Researchers have studied QS system in many pathogenic bacteria. *P. aeruginosa* and *S. aureus* are the major pathogens responsible for nosocomial and chronic wound infections (Castillo-Juarez et al. 2015).

6.3.1 Bacterial Biofilms

A bacterial biofilm is formed when a planktonic bacterium swarms its way on an exposed film coated surface. The bacterium needs to overcome the electrostatic forces between the two surfaces, the substratum and the bacterial cell surface in

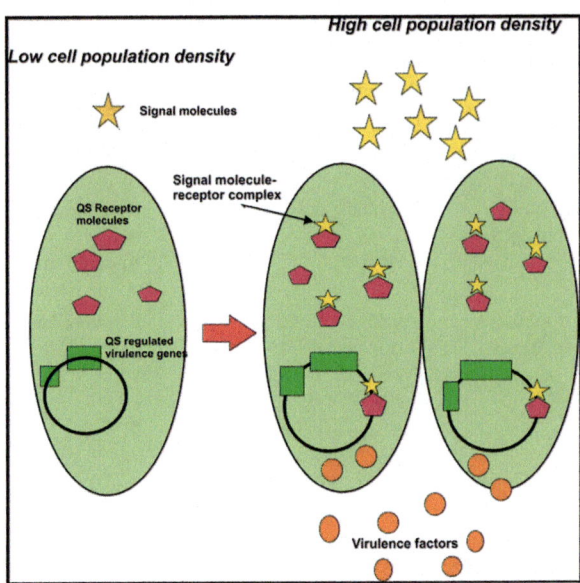

Fig. 6.1 Quorum sensing mechanism and regulation of virulence. The signal molecules bind to the signal-receptor molecules and regulate the transcription of many virulence genes. When present in low population density the signal molecules are not in enough concentration to bind to the receptors and upregulate transcription of the virulence genes. However after a particular population density is reached the signal molecules increase in concentration and carry out the upregulation of the QS dependent virulence genes

order to attach itself. Upon attachment, a microcolony develops that is covered in a protective matrix and marks the expression of biofilm phenotype (Lejeune 2003). A bacterial biofilm is resistant to antimicrobial treatment and host immunity. QS molecules are released at the microcolony stage and expression of many genes are modulated that lead to the maturation of biofilms (Fig. 6.2). Biofilms are source of planktonic bacteria that move to a new location to colonize and establish biofilms (Omar et al. 2017).

Bacterial biofilms exhibit co-existence of diverse bacterial species, both aerobic and anaerobic. Reports indicate the presence of 17 genera and 12–20 different species of microorganisms per wound. Anaerobic bacteria also inhabit the biofilms and are concentrated in the deeper areas of the biofilms. A biofilm contains microbial cells at all stages of growth and physiology. Many microenvironments arise within a biofilm as a result of concentration gradients that confer the antibiotic resistance phenomenon to the biofilm members. Some cells adapt to the microenvironment by becoming metabolically quiescent and are called persister cells which are resistant to antibiotics as well (Kalia et al. 2017; Omar et al. 2017).

P. aeruginosa has been extensively studied as model organism for QS dependent synthesis of virulence factors (proteases), biofilm formation and cell adherence factor (lipopolysaccharide) secretion. Researchers have studied the presence of QS system *in vivo*, wherein autoinducer molecules have been found to be in signifi-

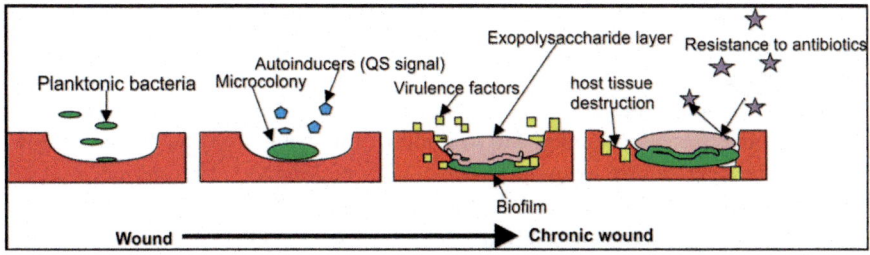

Fig. 6.2 Role of QS and biofilm in chronic wounds. Planktonic bacteria colonize the wound tissue resulting in formation of a microcolony. QS signal molecules accentuate the formation and maturation of biofilm, exopolysaccharide layer formation and virulence factor secretion. Virulence factors prolong inflammation in and around the wound area leading to massive host tissue destruction with increased antibiotic resistance. This leads to enhanced chronicity of wounds which are difficult to treat

cantly increased concentrations in cystic fibrosis patients colonized with *P. aeruginosa* (Keays et al. 2009).

There are many reasons which make the biofilms recalcitrant to antimicrobial therapy. Antibiotics have reduced penetration in biofilms and the negatively charged exopolymer repulse some of the negatively charged antimicrobial compounds. Some microorganisms release antibiotic sequestering molecules as well, thereby making the treatment of chronic biofilm associated wounds difficult. Biofilms due to their highly organized and compact structure increase the propensity of horizontal gene transfer of antibiotic resistance genes between bacteria either belonging to same species or different. Acquisition of such genes leads to irreversible genotypic changes in the bacteria and increase the chance of virulent and persistent infections. As seen with the persister cell population, once the antibiotic therapy is withdrawn the biofilm can resume growth and regenerate itself (James et al. 2008; Clinton and Carter 2015).

6.3.2   Role of P. aeruginosa and S. aureus in Chronic Wounds

Generally, skin infections are caused by *S. aureus*, *P. aeruginosa*, *E.coli* and *Acinetobacter spp.* and coagulase negative *Staphylococcus epidermidis* and *Staphylococcus lugdunensis*. These bacteria produce many virulence factors like enterotoxins, hemolysins, metalloproteases and hyaluronidase. *S. aureus* and *P. aeruginosa* have been implicated in 93.5% and 52.2% chronic wound patients, respectively (Gui et al. 2014;Serra et al. 2015; Jeyanthi and Velusamy 2016).

S. aureus mostly invades the top layer of the wound but *P. aeruginosa* is found to be associated with deeper areas of the wound (Forster et al. 2013; Serra et al. 2015). Bacterial threshold of 10^5 bacteria per gram of tissue causes tissue damage by release of inflammatory cytokines (Arya et al. 2014; Haji Zaine et al. 2014). In this way bacterial infection contribute to the chronicity of the wound. *S. aureus* itself is unable to make strong biofilms however; its co-infection with *P. aeruginosa* creates a strong pathogenic group that enhances the severity of the wound and delays the healing process.

Matrix metalloproteases (MMPs) play very important role in inflammatory responses and thus are considered to be important in wound healing process. MMPs catalyze the degradation of extracellular matrix making the transmigration of the immune cells from bloodstream to the area of inflammation easier. MMPs are secreted in the form of zymogen (inactive form) which is later converted into active form (Serra et al. 2015). However, *P. aeruginosa* produces elastase, an enzyme that converts zymogen form of MMP to active form, resulting into activation of several MMPs that cause hemorrhagic tissue damage and necrosis in wound area thus increasing the severity of wound and prolonging its healing period. Excessive activation of MMPs degrades laminin and type IV collagen which leads to membrane destruction, alteration of capillaries and tissue damage (Beaufort et al. 2013).

S. aureus also aggravates the activation of several MMPs and cause purulent lesions which are result of neutrophil infiltration. Alpha-hemolysin secreted by *S. aureus* interacts with disintegrin and metalloproteinase domain of protein-10 to bring about the disruption of host cell membrane junctions and necrosis of epidermis and dermis layers of skin (Nishifuji et al. 2008; Wilke and Bubeck Wardenburg 2010; Kim et al. 2014).

6.4 Inhibition of QS System for Treating Chronic Wounds

Treatment of chronic wounds involves aggressive debriment of the tissue which dislodges the biofilms and treatment with antibiotics. Use of anti-biofilm therapy for treating chronic wounds is gaining importance (Omar et al. 2017). Several targets have been exploited to disrupt the biofilm integrity and accentuate wound healing process. Some of the biofilm inhibitors are small molecules that; (i) modify cell envelope: D-tyrosine and D-leucine which interfere amyloid fibres required in holding the biofilms together and zaragozic acid which perturbs cell membranes and bacterial lipid rafts, (ii) disrupt matrix: Norspermidine targets extracellular polymeric substance; AA-861 and parthenolide which targets protein component (TasA) present in the extracellular matrix and Rhamnolipid, a surfactant that reduces surface interactions, and (iii) induce apoptosis; Nitric oxide induces apoptosis and cause dismemberment of bacterial biofilms (Oppenheimer-Shaanan et al. 2013).

Bacterial QS system is essential for biofilm formation in chronic wounds so it provides a major target for anti-biofilm therapy (Kalia and Kumar 2015b; Kumar et al. 2015). Since, the QS system is based on signalling by autoinducer molecules, inhibiting these would hinder coordinated virulence activity. QS systems are diverse in Gram-positive and Gram-negative bacteria. In Gram-negative bacteria Autoinducer-1 and furanone based signalling system is present whereas, Gram-positive bacteria rely on oligopeptide and furanone based systems. Apart from these *Pseudomonas* quinolone signal (PQS), diffusible signal factor (DSF) and autoinducer-3 (AI-3) systems exist in bacteria. However, the key steps involving signal supply and response remain conserved in all QS systems thus dividing the QS inhibitors (QSI) into two major classes; the signal-supply inhibitors and the signal-response inhibitors (Rutherford and Bassler 2012; LaSarre and Federle 2013). Quorum quenching is a process by which QS signal molecules are enzymatically cleaved to inhibit their activity (Kalia et al. 2011; Kalia and Purohit 2011; Sharma and Jangid 2015). One such example is the degradation of AHL molecule by AHL lactonases and AHL acylases which cleave the homoserine lactone rings and the amide bonds of AHL molecule, respectively (Huma et al. 2011). Other two enzymes, AHL oxidase and AHL reductase do not cleave the AHL molecule but modify its activity. AHL degrading enzymes are synthesized and secreted by many bacterial species including *P. aeruginosa*, *Acinetobacter* spp. and *Klebsiella pneumonia*, which are mainly associated with chronic wounds in the form of biofilms (Koul et al. 2016; Koul and Kalia 2017). Since biofilms contain bacteria that are resistant to antibiotics, QS inhibition can serve as an alternative to antibiotics. QSIs block the sensing regulated virulence factor production without hindering the bacterial growth. The immune system can then eradicate bacteria which no longer exhibit drug resistance phenomenon (Brackman and Coenye 2015).

Glyceryl trinitrate (GTN) was evaluated for its ability to inhibit QS based biofilm formation in *P. aeruginosa* burn infections (Abbas and Shaldam 2016). GTN is an FDA approved agent that has wound healing and antimicrobial activity. It is used in ointments for treating anal fissures and for inhibiting planktonic *Candida albicans* when used at a concentration 0.15% to 0.3%. GTN was found to inhibit biofilm formation by *P. aeruginosa*. The production of QS molecule violacein was inhibited along with reduction in pyocyanin and protease production. It interferes with the binding of autoinducers to their receptors and hence can aid in the inhibition of biofilm formation and function (Abbas and Shaldam, 2016).

QSIs can be co-administered with antibiotics to get a better response. RNAIII inhibiting peptide (RIP) inhibits biofilm formation and toxin production by Methicillin resistant *S. aureus* (MRSA). In an *in vivo* study, an RIP derived peptide FS10 was administered with an antibiotic tigecycline to treat experimentally induced MRSA associated wound infection in mice. An enhanced therapeutic benefit was obtained by co-administering QSI FS10 with tigecycline (Simonetti et al. 2016). It not only inhibited biofilms signalling but stimulated wound healing process as well. Marine environment serves as a wide source of QSIs and many researchers have exploited marine organisms to isolate QSIs that can function as antibacterial and anti-biofilm agents (Dobrucka and Długaszewska 2015; Shiva Krishna et al. 2015; Saurav et al. 2017). Halogenated furanones and Cinnamaldehyde isolated from red algae *Deliseapulchea* and cinnamon tree respectively, can block the QS dependent signalling by binding to AHL receptor and displacing AHL from it. These compounds can hinder biofilm forming ability of *P. aeruginosa* (Brackman and Coenye 2015).

6.5 Disadvantages of QS Inhibition

QS inhibition exerts a negative effect on biofilm formation and related expression of virulence factors but it can also exert a positive effect on the expression of virulence factors by the biofilm population. The type III secretion system of *P. aeruginosa*, a key determinant of virulence is negatively regulated by the QS system. As a result of which QS inhibition can promote virulence (Kong et al. 2009). Type VI secretion system of *P. aeruginosa* is negatively regulated by QS itself (Sana et al. 2012). Sometimes inhibition of QS accentuates the colonization of more virulent wild type along with the eradication of the less virulent mutated strains that give rise to the increased prevalence of virulent genotypes in nosocomial infections.

QSIs like furanones and synthetic HSLs work in a concentration dependent manner and can activate QS system at a particular concentration rather than inhibiting them. *P. aeruginosa* has been reported to acquire resistance against a quorum quenching furanone C-30. The organism activates multidrug efflux pump MexAB-Opr which is a result of mutation in the transcription repressors MexR and NalC which disrupts their regulatory activity. Many such mutations can be found in clinical isolates as they are selected under immense antibiotic pressure. Another quorum quenching compound 5-fluorouracil has faced resistance in cellular uptake in case of some clinical isolates and multidrug resistant strains. Instead these strains produce increased concentration of virulence factors in presence of QS inhibitor like furanon. Therefore, QS inhibitors should be used judicially as their indiscriminate use may result in selection of QS inhibitor resistant strains and worsen the biofilm mediated virulence (Maeda et al. 2012; Kalia 2013; Garcia-Contreras 2016).

6.6 Conclusion

Wound healing is a complex but normal biological process that involves majorly the inflammation, proliferation and remodelling steps. This process requires complex interplay of cells and their products to repair the wounded tissue and restore its structure and functionality. The normal wound healing process is often hindered in presence of certain physiological factors like diabetes, immunosuppression, stress, oxygen availability etc. and due to microbial infection (Guo and Dipietro 2010; Gonzalez et al. 2016). Microorganisms from the environment get access to the wounded tissue. Some microbial cells are able to establish microcolonies where the QS system comes into play and helps the microcolony to divide and grow to form a biofilm. Biofilms are organized structures consisting of polymicrobial population held together by an exopolysaccharide layer synthesized by the microbes themselves. Biofilms pose an immense problem to the wound healing process as there is upregulation of the QS regulated virulence genes, which are mostly toxins and adhesins. These molecules interact with the host and extend the inflammatory phase causing enhanced tissue damage which takes a long time to heal, resulting in chronic and persistent wounds. QS system is indispensable for biofilm formation and function (Clinton and Carter 2015). It works in a coordinated manner with a basic two component system consisting of signal molecule and receptor molecule interaction when a particular threshold population is reached. *P. aeruginosa* and *S. aureus* are the model pathogens that have been extensively studied to understand QS coordinated expression of virulence factors (Serra et al. 2015; Omar et al. 2017).

Biofilms are resistant to inhibition by antibiotic treatment. The exopolymer layer makes the antibiotic penetration into the biofilms nearly impossible. Also, within biofilms exist microbial cells at different growth stages and some persister cells that are metabolically quiescent and resistant to antibiotics. Thus, chronic wounds are difficult to treat and wound debriment processes temporarily clear off the biofilms but leaves behind some persistent cells that re-establish the biofilms with antibiotic resistant population (Rutherford and Bassler 2012; Serra et al. 2015; Omar et al. 2017). Therefore, there is a need to explore alternate therapy for treating biofilms associated chronic wounds that are antibiotic resistant. Biofilm inhibitors are aimed at inhibiting the matrix molecules or the extracellular polymeric substance, perturbs bacterial cell membranes and induce apoptosis which lead to dismemberment of the bacterial biofilms.

QS system is a promising target for inhibiting biofilms associated virulence factors. The QSIs target the signalling molecule or are diverted towards blocking the signal response (Kalia et al. 2014b; Kalia and Kumar 2015a). Many studies have focussed on blocking the signalling molecule for example AHL inhibitors that include peptides and some small compounds. QSIs have also been exploited in co-therapy with antibiotics. All these studies have indicated that QSIs are indeed efficient in disrupting biofilms and inhibiting the coordinated virulence activities (Maeda et al. 2012; Kalia 2013; Kalia et al. 2014a; Oppenheimer-Shaanan et al. 2013; Arasu et al. 2015; Bose and Chatterjee 2015; Siddiqui et al. 2015; Szweda et al. 2015; Reuter et al. 2016; Wadhwani et al. 2016; Ahiwale et al. 2017; Sharma and Lal 2017). However, there are few reports where resistance to QSIs have been reported to emerge in clinical isolates (Maeda et al. 2012; Garcia-Contreras 2016). Thus, it is important to study the QS system of different pathogens in detail. Extensive research is needed to understand the role of QS in biofilm formation and maintenance along with developing new-toxic but highly active QSIs as anti-biofilm agents.

6.7   Opinion

Biofilm formation is a highly coordinated process employed by pathogenic and non-pathogenic microorganisms to adapt to a particular environment by adhering to any available substratum. When present in a particular population threshold the microbes participating in biofilms become resistant to environmental stress conditions. Biofilms are polymicrobial in nature i.e. several species and genera of microorganisms co-exist in an exopolymer entrapped structure. Microbial cellular communication or QS plays a pivotal role in formation and maintenance of biofilms. Pathogenic bacteria colonize wound area and forms biofilms which is essential for the bacteria to express virulence factors. These virulence factors are responsible for modulating the host immune response and aggravating tissue damage making the wounds chronic. Moreover, the biofilms exhibit antibiotic resistance making treatment of chronic wounds difficult. QSIs block the signalling molecules and disrupt biofilms formation, maintenance and expression of virulence factors. These serve as promising candidates for treating antibiotic resistant chronic wound biofilms. However, the emerging resistance against QSI pose some problem in application of these molecules as alternate anti-biofilm therapy. However, search of novel QS inhibitors and their judicial use would help in treating chronic wounds and aggravating the wound healing process.

References

Abbas H, Shaldam MA (2016) Glyceryl trinitrate is a novel inhibitor of quorum sensing in *Pseudomonas aeruginosa*. Afr Health Sci 16:1109–1117. https://doi.org/10.4314/ahs.v16i4.29

Agarwala M, Choudhury B, Yadav RN (2014) Comparative study of antibiofilm activity of copper oxide and iron oxide nanoparticles against multidrug resistant biofilm forming uropathogens. Indian J Microbiol 54:365–368. https://doi.org/10.1007/s12088-014-0462-z

Ahiwale SS, Bankar AV, Tagunde S, Kapadnis BP (2017) A bacteriophage mediated gold nanoparticle synthesis and their anti-biofilm activity. Indian J Microbiol 57:188–194. https://doi.org/10.1007/s12088-017-0640-x

Arasu MV, Al-Dhabi NA, Rejiniemon TS, Lee KD, Huxley VAJ, Kim DH, Duraipandiyan V, Karuppiah P, Choi KC (2015) Identification and characterization of *Lactobacillus brevis* P68 with antifungal, antioxidant and probiotic functional properties. Indian J Microbiol 55:19–28. https://doi.org/10.1007/s12088-014-0495-3

Arya AK, Tripathi R, Kumar S, Tripathi K (2014) Recent advances on the association of apoptosis in chronic non healing diabetic wound. World J Diabetes 5:756–762. https://doi.org/10.4239/wjd.v5.i6.756

Beaufort N, Corvazier E, Mlanaoindrou S, de Bentzmann S, Pidard D (2013) Disruption of the endothelial barrier by proteases from the bacterial pathogen *Pseudomonas aeruginosa*: implication of matrilysis and receptor cleavage. PLoS One 8:e75708. https://doi.org/10.1371/journal.pone.0075708

Begum IF, Mohankumar R, Jeevan M, Ramani K (2016) GC–MS analysis of bioactive molecules derived from *Paracoccuspantotrophus*FMR19 and the antimicrobial activity against bacterial pathogens and MDROs. Indian J Microbiol 56:426–432. https://doi.org/10.1007/s12088-016-0609-1

Bjarnsholt T, Ciofu O, Molin S, Givskov M, Hoiby N (2013) Applying insights from biofilm biology to drug development - can a new approach be developed? Nat Rev Drug Discov 12:791–808. https://doi.org/10.1038/nrd4000

Bose D, Chatterjee S (2015) Antibacterial activity of green synthesized silver nanoparticles using Vasaka (*Justiciaadhatoda* L.) leaf extract. Indian J Microbiol 55:163–167. https://doi.org/10.1007/s12088-015-0512-1

Brackman G, Coenye T (2015) Quorum sensing inhibitors as anti-biofilm agents. Curr Pharmaceutical Ddesign 21:5–11. doi: Not available

Castillo-Juarez I, Maeda T, Mandujano-Tinoco EA, Tomas M, Perez-Eretza B, Garcia-Contreras SJ, Wood TK, Garcia-Contreras R (2015) Role of quorum sensing in bacterial infections. World J Clinic Cases 3:575–598. https://doi.org/10.12998/wjcc.v3.i7.575

Clinton A, Carter T (2015) Chronic wound biofilms: pathogenesis and potential therapies. Lab Med 46:277–284. https://doi.org/10.1309/LMBNSWKUI4JPN7SO

Davies DG, Parsek MR, Pearson JP, Iglewski BH, Costerton JW, Greenberg, EP (1998) Theinvolvement of cell-to-cell signals in the development of a bacterial biofilm. Science 280:295-298. doi:Not available

De Ryck T, Vanlancker E, Grootaert C, Roman BI, De Coen LM, Vandenberghe I, Stevens CV, Bracke M, Van de Wiele T, Vanhoecke B (2015) Microbial inhibition of oral epithelial wound recovery: potential role for quorum sensing molecules? AMB Express 5:27. https://doi.org/10.1186/s13568-015-0116-5

Dobrucka R, Długaszewska J (2015) Antimicrobial activities of silver nanoparticles synthesized by using water extract of *Arnicae anthodium*. Indian J Microbiol 55:168–174. https://doi.org/10.1007/s12088-015-0516-x

Eming SA, Kaufmann J, Lohrer R, Krieg T (2007) Chronic wounds. Novel approaches in research-and therapy. Hautarzt 58:939–944. https://doi.org/10.1007/s00105-007-1402-1

Forster AJ, Oake N, Roth V, Suh KN, Majewski J, Leeder C, van Walraven C (2013) Patient-level factors associated with methicillin-resistant *Staphylococcus aureus* carriage at hospital admission: a systematic review. Am J Inf Cntrol 41:214–220. https://doi.org/10.1016/j.ajic.2012.03.0

Gama JA, Abby SS, Vieira-Silva S, Dionisio F, Rocha EP (2012) Immune subversion andquorum-sensing shape the variation in infectious dose among bacterial pathogens. PLoS Pathog 8:e1002503. https://doi.org/10.1371/journal.ppat.1002503

Garcia-Contreras R (2016) Is quorum sensing interference a viable alternative to treat *Pseudomonas aeruginosa* infections? Front Microbiol 7:1454. https://doi.org/10.3389/fmicb.2016.01454

Gonzalez AC, Costa TF, Andrade ZA, Medrado AR (2016) Wound healing – a literature review. Anais Brasileiros de Dermat 91:614–620. https://doi.org/10.1590/abd1806-4841.20164741

Gui Z, Wang H, Ding T, Zhu W, Zhuang X, Chu W (2014) Azithromycin reduces the production of α-hemolysin and biofilm formation in *Staphylococcus aureus*. Indian J Microbiol 54:114–117. https://doi.org/10.1007/s12088-013-0438-4

Guo S, Dipietro LA (2010) Factors affecting wound healing. J Dental Res 89:219–229. https://doi.org/10.1177/0022034509359125

Gurtner GC, Werner S, Barrandon Y, Longaker MT (2008) Wound repair and regeneration. Nature 453:314–321. https://doi.org/10.1038/nature07039

Haji Zaine N, Burns J, Vicaretti M, Fletcher JP, Begg L, Hitos K (2014) Characteristics of diabetic foot ulcers in Western Sydney, Australia. J Foot Ankle Res 7:39. https://doi.org/10.1186/s13047-014-0039-4

Hocking AM (2015) The role of chemokines in mesenchymal stem cell homing to wounds. Adv Wound Care 4:623–630. https://doi.org/10.1089/wound.2014.0579

Hong WX, Hu MS, Esquivel M, Liang GY, Rennert RC, McArdle A, Paik KJ, Duscher D, Gurtner GC, Lorenz HP (2014) The role of hypoxia-inducible factor in wound healing. Adv Wound Care 3:390–399. https://doi.org/10.1089/wound.2013.0520

Huma N, Shankar P, Kushwah J, Bhushan A, Joshi J, Mukherjee T, Raju SC, Purohit HJ, Kalia VC (2011) Diversity and polymorphism in AHL-lactonase gene (*aiiA*) of *Bacillus*. J Microbiol Biotechnol 21:1001–1011. https://doi.org/10.4014/jmb.1105.05056

James GA, Swogger E, Wolcott R, Pulcini E, Secor P, Sestrich J, Costerton JW, Stewart PS (2008) Biofilms in chronic wounds. Wound Repair Regen 16:37–44. https://doi.org/10.1111/j.1524-475X.2007.00321.x

Jeyanthi V, Velusamy P (2016) Anti-methicillin resistant *Staphylococcus aureus* compound isolation from halophilic *Bacillus amyloliquefaciens*MHB1 and determination of its mode of action using electron microscope and flow cytometry analysis. Indian J Microbiol 56:148–157. https://doi.org/10.1007/s12088-016-0566-8

Kalia VC (2013) Quorum sensing inhibitors: an overview. Biotechnol Adv 31:224–245. https://doi.org/10.1016/j.biotechadv.2012.10.00

Kalia VC (2014a) Microbes, antimicrobials and resistance: the battle goes on. Indian J Microbiol 54:1–2. https://doi.org/10.1007/s12088-013-0443-7

Kalia VC (2014b) In search of versatile organisms for quorum-sensing inhibitors: acyl homoserinelactones (AHL)-acylase and AHL-lactonase. FEMS Microbiol Letts 359:143. https://doi.org/10.1111/1574-6968.12585

Kalia VC (2015a) Quorum sensing vs quorum quenching: a battle with no end in sight. In: Kalia VC (ed). Springer India, New Delhi. http://link.springer.com/book/10.1007/978-81-322-1982-8

Kalia VC (2015b) Microbes: the most friendly beings?In: Kalia VC (ed) Quorum sensing vs quorum quenching: A battle with no end in sight. Springer India, New Delhi, pp 1–5. doi:https://doi.org/10.1007/978-81-322-1982-8_1

Kalia VC, Kumar P (2015a) Potential applications of quorum sensing inhibitors in diverse fields. In: Kalia VC (ed) Quorum sensing vs Quorum quenching: a battle with no end in sight. Springer India, New Delhi, pp 359–370. doi:https://doi.org/10.1007/978-81-322-1982-8_29

Kalia VC, Kumar P (2015b) The battle: Quorum-sensing inhibitors versus evolution of bacterialresistance. In: Kalia VC (ed) Quorum sensing vs Quorum quenching: a battle with no end in sight. Springer India, New Delhi, pp 385–391. doi:https://doi.org/10.1007/978-81-322-1982-8_31

Kalia VC, Purohit HJ (2011) Quenching the quorum sensing system: potential antibacterial drugtargets. Critic Rev Microbiol 37:121–140. https://doi.org/10.3109/1040841X.2010.532479

Kalia VC, Raju SC, Purohit HJ (2011) Genomic analysis reveals versatile organisms for quorumquenching enzymes: acyl-homoserine lactone-acylase and lactonase. Open Microbiol J 5:1–13. https://doi.org/10.2174/1874285801105010001

Kalia VC, Wood TK, Kumar P (2014a) Evolution of resistance to quorum-sensing inhibitors. Microb Ecol 68:13–23. https://doi.org/10.1007/s00248-013-0316-y

Kalia VC, Kumar P, Pandian SK, Sharma P (2014b) Biofouling control by quorum quenching. In: Kim SK (ed) Hb_25 Springer handbook of marine biotechnology chapter 15. Springer, Berlin, pp 431–440

Kalia VC, Prakash J, Koul S, Ray S (2017) Simple and rapid method for detecting biofilm formingbacteria. Indian J Microbiol 57:109–111. https://doi.org/10.1007/s12088-016-0616-2

Keays T, Ferris W, Vandemheen KL, Chan F, Yeung SW, Mah TF, Ramotar K, Saginur R, Aaron SD (2009) A retrospective analysis of biofilm antibiotic susceptibility testing: a betterpredictor of clinical response in cystic fibrosis exacerbations. J Cystic Fibrosis 8:122–127. https://doi.org/10.1016/j.jcf.2008.10.005

Khanh VC, Ohneda K, Kato T, Yamashita T, Sato F, Tachi K, Ohneda O (2017) Uremic toxins affectthe imbalance of redox state and overexpression of prolyl hydroxylase 2 inhumanadipose tissue-derived mesenchymal stem cells involved in wound healing. Stem Cells Dev 26:948–963. https://doi.org/10.1089/scd.2016.0326

Kim HK, Missiakas D, Schneewind O (2014) Mouse models for infectious diseases caused by Staphylococcus aureus. J Immunol Methods 410:88–99. https://doi.org/10.1016/j.jim.2014.04.007

Kong W, Liang H, Shen L, Duan K (2009) Regulation of type III secretion system by Rhl and PQS quorum sensing systems in Pseudomonas aeruginosa. Acta Microbiol Sin 49:1158–1164

Koul S, Kalia VC (2017) Multiplicity of quorum quenching enzymes: a potential mechanism to limitquorum sensing bacterial population. Ind J Microbiol 57:100–108. https://doi.org/10.1007/s12088-016-0633-1

Koul S, Prakash J, Mishra A, Kalia VC (2016) Potential emergence of multi-quorum sensinginhibitor resistant (MQSIR) bacteria. Indian J Microbiol 56:1–18. https://doi.org/10.1007/s12088-015-0558-0

Kumar P, Patel SKS, Lee JK, Kalia VC (2013) Extending the limits of Bacillus for novelbiotechnological applications. Biotechnol Adv 31:1543–1561. https://doi.org/10.1016/j.biotechadv.2013.08.007

Kumar P, Koul S, Patel SKS, Lee JK, Kalia VC (2015) Heterologous expression of quorum sensinginhibitory genes in diverse organisms. In: Kalia VC (ed) Quorum sensing vs quorum quenching: a battlewith no end in sight. Springer India, New Delhi, pp 343–356. doi:https://doi.org/10.1007/978-81-322-1982-8_28

LaSarre B, Federle MJ (2013) Exploiting quorum sensing to confuse bacterial pathogens. Microbiol Mol Biol Rev 77:73–111. https://doi.org/10.1128/MMBR.00046-12

Lejeune P (2003) Contamination of abiotic surfaces: what a colonizing bacterium sees and how to blur it. Trends Microbiol 11:179–184. https://doi.org/10.1016/S0966-842X(03)00047-7

Li J, Chen J, Kirsner R (2007) Pathophysiology of acute wound healing. Clin Dermatol 25:9–18. https://doi.org/10.1016/j.clindermatol.2006.09.007

Maeda T, Garcia-Contreras R, Pu M, Sheng L, Garci LR, Tomas M, Wood TK (2012) Quorum quenching quandary: resistance to antivirulence compounds. ISME J 6:493–501. https://doi.org/10.1038/ismej.2011.122

Maxson S, Lopez EA, Yoo D, Danilkovitch-Miagkova A, Leroux MA (2012) Concise review: role of mesenchymal stem cells in wound repair. Stem Cells Trans Med 1:142–149. https://doi.org/10.5966/sctm.2011-0018

Medrado AR, Pugliese LS, Reis SR, Andrade ZA (2003) Influence of low level laser therapy on wound healing and its biological action upon myofibroblasts. Lasers Surg Med 32:239–244. https://doi.org/10.1002/lsm.10126

Medrado A, Costa T, Prado T, Reis S, Andrade Z (2010) Phenotype characterization of pericytes during tissue repair following low-level laser therapy. Photodermatol Photoimmunol Photomed 26:192–197. https://doi.org/10.1111/j.1600-0781.2010.00521.x

Menke NB, Ward KR, Witten TM, Bonchev DG, Diegelmann RF (2007) Impaired wound healing. Clin Dermatol 25:19–25. https://doi.org/10.1016/j.clindermatol.2006.12.005

Nayak BS, Sandiford S, Maxwell A (2009) Evaluation of the wound-healing activity of ethanolicextract of *Morinda citrifolia* L. leaf. Evid Compl Alter Med 6:351–356. https://doi.org/10.1093/ecam/nem127

Nishifuji K, Sugai M, Amagai M (2008) Staphylococcal exfoliative toxins: "molecular scissors" of bacteria that attack the cutaneous defense barrier in mammals. J Dermatol Sci 49:21–31. https://doi.org/10.1016/j.jdermsci.2007.05.007

Omar A, Wright JB, Schultz G, Burrell R, Nadworny P (2017) Microbial biofilms and chronicwounds. Microorganisms 5:9. https://doi.org/10.3390/microorganisms5010009

Oppenheimer-Shaanan Y, Steinberg N, Kolodkin-Gal I (2013) Small molecules are natural triggers for the disassembly of biofilms. Trends Microbiol 21:594–601. https://doi.org/10.1016/j.tim.2013.08.005

Reuter K, Steinbach A, Helms V (2016) Interfering with bacterial quorum sensing. Persp Med Chem 8:1–15. https://doi.org/10.4137/PMC.S13209

Rutherford ST, Bassler BL (2012) Bacterial quorum sensing: its role in virulence and possibilitiesfor its control. Cold Spring Harbor Perspect Med 2:pii:a012427. doi:https://doi.org/10.1101/cshperspect.a012427

Sana TG, Hachani A, Bucior I, Soscia C, Garvis S, Termine E, Engel J, Filloux A, Bleves S (2012) The second type VI secretion system of *Pseudomonas aeruginosa* strain PAO1 is regulated by quorum sensing and Fur and modulates internalization in epithelial cells. J Biol Chem 287:27095–27105. https://doi.org/10.1074/jbc.M112.376368

Saurav K, Costantino V, Venturi V, Steindler L (2017) Quorum sensing inhibitors from the sea discovered using bacterial N-acyl-homoserine lactone-based biosensors. Mar Drugs 15. doi: https://doi.org/10.3390/md15030053

Serra R, Grande R, Butrico L, Rossi A, Settimio UF, Caroleo B, Amato B, Gallelli L, de Franciscis S (2015) Chronic wound infections: the role of *Pseudomonas aeruginosa* and *Staphylococcus aureus*. Expert Rev Anti-Infect Ther 13:605–613. https://doi.org/10.1586/14787210.2015.1023291

Sharma R, Jangid K (2015) Fungal Quorum sensing inhibitors. In: Kalia VC (ed) Quorum sensing vs Quorum quenching: a battle with no end in sight. Springer India, New Delhi, pp 237–257

Sharma A, Lal R (2017) Survey of (Meta) genomic approaches for understanding microbial community dynamics. Indian J Microbiol 57:23–38. https://doi.org/10.1007/s12088-016-0629-x

Shaw TJ, Martin P (2009) Wound repair at a glance. J Cell Sci 122:3209–3213. https://doi.org/10.1242/jcs.031187

Shiva Krishna P, Sudheer Kumar B, Raju P, Murty MSR, Prabhakar Rao T, Singara Charya MA, Prakasham RS (2015) Fermentative production of pyranone derivate from marine *Vibrio* sp. SKMARSP9: isolation, characterization and bioactivity evaluation. Indian J Microbiol 55:292–301. https://doi.org/10.1007/s12088-015-0521-0

Siddiqui MF, Rzechowicz M, Harvey W, Zularisam AW, Anthony GF (2015) Quorum sensing based membrane biofouling control for water treatment: a review. J Water Proc Eng 30:112–122. https://doi.org/10.1016/j.jwpe.2015.06.003

Simonetti O, Cirioni O, Cacciatore I, Baldassarre L, Orlando F, Pierpaoli E, Lucarini G, Orsetti E, Provinciali M, Fornasari E (2016) Efficacy of the quorum sensing inhibitor FS10alone and in combination with Tigecycline in an animal model of Staphylococcal infected wound. PLoS One 11:e0151956. https://doi.org/10.1371/journal.pone.0151956

Szweda P, Gucwa K, Kurzyk E, Romanowska E, Dzierżanowska-Fangrat K, Jurek AZ, Kuś PM, Milewski S (2015) Essential oils, silver nanoparticles and propolis as alternative agents against fluconazole resistant *Candida albicans, Candida glabrata* and *Candida krusei* clinical isolates. Indian J Microbiol 55:175–183. https://doi.org/10.1007/s12088-014-0508-2

Thuraisingam T, Xu YZ, Eadie K, Heravi M, Guiot MC, Greemberg R, Gaestel M, Radzioch D (2010) MAPKAPK-2 signaling is critical for cutaneous wound healing. J Invest Dermatol 130:278–286. https://doi.org/10.1038/jid.2009.209

Tidball JG (2005) Inflammatory processes in muscle injury and repair. Am J Physiol Regul Integr Comp Physiol 288:R345–R353. https://doi.org/10.1152/ajpregu.00454.2004

Wadhwani SA, Shedbalkar UU, Singh R, Vashisth P, Pruthi V, Chopade BA (2016) Kinetics of synthesis of gold nanoparticles by *Acinetobacter* sp. SW30 isolated from environment. Indian J Microbiol 56:439–444. https://doi.org/10.1007/s12088-016-0598-0

Wilke GA, Bubeck Wardenburg J (2010) Role of a disintegrin and metalloprotease 10 in *Staphylococcus aureus* alpha-hemolysin-mediated cellular injury. Proc Natl Acad Sci U S A 107:13473–13478. https://doi.org/10.1073/pnas.1001815107

Chapter 7
Efflux Pump-Mediated Quorum Sensing: New Avenues for Modulation of Antimicrobial Resistance and Bacterial Virulence

Manjusha Lekshmi, Ammini Parvathi, Sanath Kumar, and Manuel F. Varela

Abstract Bacterial pathogenesis is frequently enhanced by virulence mechanisms that facilitate growth. Such virulence factors include biofilm formation and antimicrobial resistance mechanisms. One primary resistance mechanism involves the active efflux of antimicrobial agents from cells of pathogenic bacteria. It has been established that quorum sensing serves as a line of communication between the environment and mechanisms of antimicrobial resistance, such as antimicrobial efflux pump systems from bacterial pathogens. This chapter covers several well documented antimicrobial efflux transporter resistance mechanisms and their relationships to key aspects of quorum sensing. Knowledge of these critical relationships may enhance their biotechnological applications.

Keywords Antimicrobial resistance · Bacteria · Biotechnology · Drug resistance · Efflux pump · Multidrug resistance · Pathogenesis · Quorum sensing

M. Lekshmi · S. Kumar
QC Laboratory, Harvest and Post Harvest Technology Division, ICAR-Central Institute of Fisheries Education (CIFE), Mumbai, Maharashtra, India
e-mail: sanathkumar@cife.edu.in

A. Parvathi
CSIR-National Institute of Oceanography (NIO), Regional Centre, Kochi, Kerala, India

M. F. Varela (✉)
Department of Biology, Eastern New Mexico University, Portales, NM, USA
e-mail: manuel.varela@enmu.edu

© Springer Nature Singapore Pte Ltd. 2018
V. C. Kalia (ed.), *Biotechnological Applications of Quorum Sensing Inhibitors*,
https://doi.org/10.1007/978-981-10-9026-4_7

Abbreviations

acyl-HSL	Acyl homoserine lactone
C4-HSL	*N*-butyryl-L-homoserine lactone
DKP	Diketopiperazine
DSF	Diffusible signal factor
HSL	Homoserine lactone
MFS	Major Facilitator superfamily
PMF	Proton motive force
QQ	Quorum quenching enzymes
QS	Quorum sensing
QSI	Quorum sensing signal inhibitors
RND	Resistance-Nodulation-Cell Division
TF	Trifluoromethyl ketone
3OC12-HSL	N-(3-oxododecanoyl)-L-HSL

7.1 Introduction

Throughout their evolution, bacterial pathogens have developed a variety of virulence factors that serve to enhance the morbidity and mortality rates as these microorganisms move through human populations. In recent decades bacterial resistance mechanisms have emerged that involve multiple antimicrobial agents, thus confounding chemotherapeutic efficacy against infectious diseases (Kumar and Varela 2013).

Bacteria have acquired a set of sophisticated integral membrane transporter systems for extruding antibacterial substances from the interiors of their cells (Kumar and Varela 2012; Delmar et al. 2014). In particular, quorum sensing systems have been shown to be involved in certain antimicrobial resistance mechanisms called drug efflux pumps (Singh et al. 2017). This chapter addresses the roles of antimicrobial efflux pumps during quorum sensing, inhibitions of quorum sensing systems via efflux pumps and the biotechnological applications that may result from modulation of quorum sensing.

7.2 Bacterial Antimicrobial Resistance Systems

In order to survive in the face of selective pressure exerted by myriad inhibitory substances present in the environment, bacteria have evolved an array of mechanisms to resist antimicrobials (Kumar and Varela 2013; Blair et al. 2015). Intrinsic antimicrobial resistance involves the lack of a biological target for the antimicrobial agent, and resistance to triclosan, β-lactams and the fluoroquinolones are examples of this type of resistance (Cox and Wright 2013). Acquired resistance involves the development of resistance mechanisms or transfer of genetic elements that transform, transduce or conjugate the bacteria to become antimicrobial resistant (Roberts 2005; van Hoek et al. 2011).

One key and widespread mechanism involves the enzymatic degradation of antimicrobial agents (Sacha et al. 2008). In this resistance mechanism bacteria metabolize antimicrobial agents into inactive products (Wright 2005). Some of these bacterial enzymatic systems have extended spectrums and high catalytic activities (Ghafourian et al. 2015).

Another mechanism involves bacterial alteration of the antimicrobial targets (Lambert 2005). Bacteria will modify the cellular targets of the antimicrobial agents, such as gyrase (Jacoby et al. 2015), RNA polymerase (Floss and Yu 2005), the protein synthetic machinery such as the ribosome (Lambert 2012), and the cell wall synthesizing machinery (Nikolaidis et al. 2014). In these and other cases, the antimicrobial agents will be unable to bind to their bacteriological targets, allowing bacteria with altered targets to survive and grow.

One resistance mechanism involves the prevention of the antimicrobial agents to their internal cellular targets (Kumar and Schweizer 2005). Two sub-categories have been established in this area. The first category involves a reduction in the permeability of the antimicrobial agent into the bacterium, a mechanism found primarily in Gram-negative bacteria with outer membranes (Delcour 2009). The second category involves an enhancement in the active efflux of antimicrobial agents from the interior of bacterial cells (Levy 1992, 2002). Both of these resistance mechanisms prevent the access of growth inhibitory substances into bacteria where their antimicrobial targets are located. This chapter will focus on antimicrobial transport systems that mediate efflux from the cytoplasm of the bacterial cell in order to confer resistance (Kumar and Varela 2012; Floyd et al. 2013; Kumar et al. 2013a, b, 2016; Varela et al. 2013; Andersen et al. 2015; Ranaweera et al. 2015) and their relationships to quorum sensing systems (Dickschat 2010; Yufan et al. 2016).

7.3 Quorum Sensing

Cell-to-cell communication is critical to the survival of bacteria in communities, which enables them to sense their abundance and perform diverse, mutually benefitting activities. Bacteria have a robust communication system in the form of quorum sensing (QS), a key mechanism of communication within the community and with the host (Parsek and Greenberg 2000; Kalia 2015).

7.3.1 Mechanisms

QS regulates myriad functions including metabolism, toxin secretion, biofilm formation, horizontal gene transfer, virulence, growth, luminescence, persistence in the environment, gene expression, toxin and antibiotic production (Kalia 2015; Grandclement et al. 2016).

Bacteria use diverse mechanisms of cell-to-cell communication, but all of these involve small molecules such as peptides and chemicals. Bacteria use more than one signaling molecule and the complex interplay among diverse signals and their cellular targets constitutes a QS apparatus (Papenfort and Bassler 2016). Numerous signaling molecules and their cellular targets have been identified. Acyl homoserine lactone (acyl-HSL) is one such signaling molecule widely employed by Gram negative bacteria to interact with each other and with their hosts (Bassler 1999). HSL molecules are enzymatically synthesized and secreted into the bacterial environment. Binding of the HSL molecules to their receptors regulates gene expression (Miller and Bassler 2001). HSL molecules vary in length depending on the species of bacteria. Short chain HSL molecules diffuse through the membrane easily, while the transport of long chain HSL molecules are mediated by membrane transporters. The receptors of HSL molecules belong to the family of LuxR transcriptional regulators (Parsek and Greenberg 2000). HSL molecules are essential for the interaction of bacteria with their eukaryotic hosts such as in the case of luminescent *Vibrio fischeri* which colonizes the lightorgan of squids and the plant pathogens such as *Erwinia carotovora* and *Pseudostellaria heterophylla* (Holm and Vikström 2014; Zhang et al. 2016). Other known QS signal molecules include AI-2 responsible for bioluminescence in *V. harveyi* (Surette et al. 1999), fatty acid-derived diffusible signal factor (DSF) of *Xanthomonas* (He and Zhang 2008), amino acid-derived diketopiperazines (DKPs) of *Pseudomonas aeruginosa, Enterobacter agglomerans, Citrobacter freundii* etc. (Jimenez et al. 2012), and cyclo-peptides of Gram-positive bacteria (Monnet and Gardan 2015).

Quorum-sensing mediated signaling is necessary for the expression of a host of traits essential for colonization, expression of virulence factors, biofilm formation and antibiotic resistance. The essential feature of quorum sensing in pathogenicity makes it an ideal target for inhibition to disrupt the pathogenic mechanisms of bacteria. This strategy, known as quorum quenching, involves the use of quorum quenching enzymes (QQ) and inhibitors of quorum sensing signals (QSI) for the destruction of QS circuit and the related pathways (Kalia et al. 2011; LaSarre and Federle 2013; Kalia 2014a; Kalia and Kumar 2015a; Koul and Kalia 2017). Unlike antibiotic treatment, quorum quenching does not inhibit or kill bacteria, but interferes in their quorum sensing-dependent virulence mechanisms (Grandclement et al. 2016). As a consequence of this, bacteria will not be able to survive and establish in the host and get cleared by the host immune system. Unlike antibiotics, QSIs do not exert selective pressure and hence do not promote resistance development. Because of this feature, QSIs are being viewed as promising alternatives to antibiotics (Kalia and Purohit 2011; Kalia 2014b).

Bacteria living in a community compete for scarce resources such as nutrients and space by way of secretion of inhibitory substances which include quorum sensing inhibitors. Bacteria producing QQ enzymes such as *Afipia* sp., *Acinetobacter* sp., *Pseudomonas* sp. and *Micrococcus* sp. have been isolated from waste water sludge (Kim et al. 2004). Extracellularly acting QQ enzymes such as lactonases are promising as inhibitors of QS-mediated bacterial activities (Bzdrenga et al. 2017). Several compounds of natural origin possess anti-QS bioactivities (Table 7.1). Coumarin, a

Table 7.1 Examples of quorum sensing inhibitors from natural sources

Inhibitor	Source	References
Dibromohemibastadin	Marine sponge	Le Norcy et al. (2017)
Baicalin	Plant, *Scutellaria baicalensis*	Luo et al. (2017)
Dihydrocoumarin	Coumarin	Hou et al. (2017)
1,5-dihydropyrrol-2-ones	Marine algae, *Delisea pulchra*	Goh et al. (2015)
Coumarin	Plant phenolic compound	Gonzalez-Lamothe et al. (2009)
Ajoene	Garlic	Jakobsen et al. (2012b)
Iberin	Horseradish	Jakobsen et al. (2012a)
Flavonoids	*Piper delineatum*	Martin-Rodriguez et al. (2015)
Glycosylated flavanones	Orange extract	Truchado et al. (2012)
Patulin and penicillic acid	*Penicillium* species	Rasmussen et al. (2005)
Epigallocatechin gallate	*Camellia sinensis* L.	Taganna and Rivera (2008) and Zhu et al. (2015)
Cinnamaldheyde	*Cinnamomum* spp.	Niu et al. (2006) and Brackman et al. (2011)
Ursolic acid	*Diospyros dendo*	Ren et al. (2005)
Salycilic acid	Plants	Yuan et al. (2007)
Urolithin A and B	Pomegranate	Giménez-Bastida et al. (2012)
Polyphenol	Green tea	Zhu et al. (2015)

plant phenolic compound, is a potent QS inhibitor being active against short, medium and long chain N-acyl-homoserine lactones and also acts as an anti-virulence factor against a broad spectrum of pathogens (Gutierrez-Barranquero et al. 2015).

Plant-derived extracts have been widely tested for QS inhibition activities, although in majority of the cases the nature of the bioactive compounds is unknown (Kalia 2013).

In the face of pathogenic bacteria rapidly gaining resistance to multiple antibiotics, the search for new antimicrobials has become imperative to restore the efficacy of antimicrobial chemotherapy (Begum et al. 2016; Jeyanthi and Velusamy 2016; Varsha et al. 2016; Azman et al. 2017). Since QS regulates virulence and several aspects of bacterial pathogenicity, QS inhibitors are potentially useful in medicine. Studies have shown that QSIs enhance the susceptibility of biofilm bacteria to antimicrobials (Brackman et al. 2011). Naturally occurring compounds as well as their structural analogs have been shown to be effective QSIs (Cady et al. 2012; Kalia 2013). QS in *Burkholderia cepacia* is associated with siderophore ornibactin biosynthesis, extracellular protease, swarming and biofilm formation (Lewenza and Sokol 2001; Aguilar et al. 2003). Biofilm-forming bacteria such as *P. aeruginosa* associated with conditions such as cystic fibrosis are responsible for serious infections. QS plays critical role in biofilm formation by this bacterium and interfering with the QS signaling cascade results in the attenuation of its virulence (Hentzer et al. 2003). Decreased virulence facilitates rapid clearance of the bacterium by the immune system, while also potentiating the activities of antibiotics.

7.4   Role of Efflux Pumps in Quorum Sensing

Very few studies have explored the possibility of the involvement of efflux pumps in quorum sensing. Efflux pumps are involved in the transport of HSL molecules. In *P. aeruginosa*, MexA-MexB-OprM efflux pump has been shown to have a role in the transport of 3OC12-HSL (Soto 2013). HSL controls several aspects of *P. aeruginosa* virulence including expression of virulence factors, biofilm formation and survival in the host (Smith and Iglewski 2003). Efflux pump inhibition is therefore a potentially viable method of controlling the pathogenicity of *P. aeruginosa*. The role of efflux pump in the modulation of QS signals was speculated in many early studies. Evans et al. (1998) reported that in *P. aeruginosa*, hyper expression of MexAB-OprM efflux pump resulted in reduced expression of QS-regulated virulence factors possibly due to the extrusion of ASL molecules from the intracellular environment. Further, the expression of MexAB-OprM was increased with the exogenous addition of *N*-butyryl-L-homoserine lactone (C4-HSL), but not *N*-(3-oxododecanoyl)-L-homoserine lactone (Maseda et al. 2004). On the other hand, MexT, a positive regulator of *the mexEF-oprN* repressed the C4-HSL-mediated enhancement of MexAB-OprM expression (Maseda et al. 2004). However, all QS signals are not substrates of efflux pumps. In *P. aeruginosa*, 3-oxo-C12HSL is extruded by the efflux pumps while *N*-butyryl-L-homoserine lactone (C4HSL) passively diffuses out of the cell (Pearson et al. 1999). Microarray analysis of QS-regulated genes in *P. aeruginosa* identified several putative efflux proteins responsive to QS (Wagner et al. 2003).

Initial studies with the pathogen *Burkholderia pseudomallei* showed that extracellular secretion of several homoserine lactones was entirely dependent on the function of the BpeAB-OprB efflux pump (Chan and Chua 2005). In *B. pseudomallei*, the RND efflux pump *BpeAB-OprB* is quorum regulated and the exogenous addition of the autoinducers N-octanoyl-homoserine lactone (C8HSL) and N-decanoyl-homoserine lactone (C10HSL) resulted in its enhanced expression (Chan and Chua 2005). This was further substantiated by the observation in *bpeAB* null mutant as well in *bpeR* (regulator of *bpeAB*) over expressing strains that the production of virulence factors such as siderophore and phospholipase C and biofilm formation were affected in addition to the attenuation of cell invasion and cytotoxic abilities (Chan and Chua 2005). These observations have strengthened the hypothesis that inhibition of the BpeAB-OprB efflux pump could restore the susceptibility of *B. pseudomallei* to aminoglycosides and macrolides and result in attenuation of QS-mediated virulence (Chan et al. 2004).

SdiA, a protein which controls density-dependent cell division in *Escherichia coli*, positively regulates AcrAB efflux pump and its over production leads to increased expression of AcrAB and increased resistance to antibiotics (Rahmati et al. 2002). Since efflux pumps are generally non-specific and extrude structurally diverse substrates, it is possible that QS molecules form substrates of efflux pumps resulting in their over production. Alternatively, transport of QS signals could be the actual functions of efflux pumps in bacteria (Rahmati et al. 2002). This hypothesis remains as an exciting area for future research.

An elegant study (Yang et al. 2006) on quorum sensing and multidrug transporters in *E. coli* found that structurally similar QS communication molecules form substrates for efflux pumps. Bacteria use quinolones as communication signals in QS and quinolones are also powerful chemotherapeutic agents against pathogenic bacteria. Surprisingly, bacteria have multiple quinolone efflux pumps which incidentally efflux structurally similar QS signal molecules. Their study showed that overexpression of quinolone efflux pumps AcrAB and NorE, but not MdfA, resulted in reduced growth rate and early induction of RpoS. Efflux pumps are thought to extrude QS signals at a rate much faster than their passive diffusion causing their rapid accumulation in the extracellular environment and when the signals in sufficient quantities are sensed by the bacterial community, the entire population of bacteria assume stationary phase (Yang et al. 2006).

Further in *P. aeruginosa*, pyocyanin stimulates up regulation of RND efflux pump MexGHI-OpmD and a putative MFS efflux pump PA3718 along with a number of virulence genes which in turn are under the control of QS network (Dietrich et al. 2006). Overexpression of MDR pumps is disadvantageous to the bacterium in terms of reduced virulence and fitness. The production of virulence and fitness factors are controlled by QS signals and the overexpression of efflux pumps results in the extrusion of signaling molecules from the cellular environment making them unavailable for triggering gene expression (Linares et al. 2005). However, all efflux pumps may not have the same effect on QS-regulated virulence gene expression as has been shown in *P. aeruginosa* (Linares et al. 2005). These investigators found that overexpression of MexEF-OprN and MexCD-OprJ led to the reduction in the transcription of the T3SS genes, while the overexpression of either MexAB-OprM or MexXY did not have any detectable effect on T3SS (Linares et al. 2005). In similar lines, the production of phytotoxin in *Burkholderia glumae* and its transport involving putative RND efflux pumps is controlled by QS-regulated transcriptional activator suggesting the involvement of a network comprising of QS and efflux pumps in the pathogenicity of this bacterium (Kim et al. 2014).

7.5 Inhibition of Efflux Pump Mediated Quorum Sensing

Inhibition of quorum sensing offers powerful avenues to control bacterial communication and thereby control them without the use of antimicrobial agents. Since efflux pumps depend on proton motive force (PMF) for their activities, proton scavengers are attractive candidate molecules to inhibit efflux pumps. Trifluoromethyl ketones (TFs) are known inhibitors of PMF and a study has shown that TFs can effectively inhibit QS and efflux pumps in *Chromobacterium violaceum and E. coli* (Varga et al. 2012).

The susceptibility of methicillin resistant *Staphylococcus aureus* to β-lactam antibiotics and tetracycline was significantly enhanced by a thyme plant-derived compound baicalein (Mahmood et al. 2016). Similarly, berberine and 5′-methoxy-hydnocarpin derived from berberis, *N*-trans-feruloyl 4′-O-methyldopamine from *Mirabilis jalapa* modulate the *S. aureus* efflux pump NorA and potentiate the activity of fluoroquinolones (Stermitz et al. 2000; Michalet et al. 2007). Numerous plant extracts have been descried to be efflux pump inhibitors, but in very few studies the bioactive component has been purified (Stavri et al. 2007). Isolation and identification of the active compound is necessary to determine the exact mode of action of the compound and to screen structural homologs for more effective inhibitors using bioinformatics tools.

A recent study showed that vanillin (4-hydroxy-3-methoxybenzaldehyde) could inhibit long chain and short chain ASL molecules and significantly reduce biofilm formation by *Aeromonas hydrophila* on polystyrene surface (Ponnusamy et al. 2009, 2013). It has been shown that *N*-butyryl-L-homoserine lactone induces the expression of *mexAB-oprM* in *P. aeruginosa* establishing a direct link between quorum sensing and efflux mechanisms (Sawada et al. 2004).

With more information on the interplay between QS and efflux activities emerging, the inhibitor strategy should inevitably analyze the effect of efflux pump inhibition from the QS perspective. The involvement of efflux pumps in QS activity has further consolidated the idea that the efflux pump inhibition is a promising and powerful approach for countering pathogenic bacteria and their antibacterial resistance.

Advances in computation techniques have facilitated structure-based ligand screening by molecular docking methods for discovery of potential QS inhibitors, both natural and synthetic (Nandi 2016). Virtual screening using high throughput techniques enormously reduces the cost of drug discovery and testing (Goh et al. 2015).

In *Vibrio cholerae*, the EmrD-3 efflux pump confers resistance to multiple antibacterials (Smith et al. 2009), and a recent study showed that garlic extract and its active component allyl sulfide inhibited EmrD-3-mediated ethidium bromide efflux and lowered the MICs of multiple antibacterials to *V. cholerae* harboring EmrD-3 (Bruns et al. 2017). This finding is interesting and opens up new avenues to further research if the inhibition of EmrD-3 also impacts quorum sensing and the related genes regulated by quorum sensing, and also affects virulence and biofilm formation in *V. cholerae*.

Similarly, in *S. aureus*, the drug transport activities of the multidrug efflux pump LmrS was found to be modulated by the cumin extract (*Cuminum cyminum*) (Kakarla et al. 2017). The new insight into the interplay among quorum sensing, efflux activities, antibiotic resistance and virulence has changed the earlier perspective of treating these traits of a pathogen individually. With the powerful tools such as microarray and transcriptome sequencing, it is expected that the role of efflux pumps will be understood beyond substrate transport and antibiotic resistance, and make them attractive targets for pathogen control in future. However, recent reports on the emergence of QSI-resistant bacteria emphasize the need for careful evaluation of QSIs for resistance development and its implications on the control of harmful bacteria (Kalia and Kumar 2015b; Kalia et al. 2014a; Koul et al. 2016).

7.6   Biotechnological Applications of Quorum Sensing Inhibition

Since QS inhibition affects a wide range of physiological activities in bacteria, this approach has potential biotechnological applications. For example, biofilm formation and biofouling are a serious problem in membrane filtration systems in which quorum sensing plays a critical role. Inhibition of quorum sensing using HSL inhibitors such as vanillin is promising in enhancing the life of membranes used in reverse osmosis (Evans et al. 1998; Ponnusamy et al. 2009; Siddiqui et al. 2015). Environmental friendly antifungal and antifouling agents have been discovered, and their mechanisms of action involve disruption of QS (Kalia et al. 2014b; Sharma and Jangid 2015; Le Norcy et al. 2017).

Since cell-cell-signaling involving HSL and related molecules is essential for the virulence and interactions with the host by many pathogenic bacteria, inhibition of quorum sensing is a powerful means of controlling them. A natural furan (5Z)-4-bromo-5-(bromomethylene)-3-butyl-2(5H)-furanone derived from the red alga *Delisea pulchra* blocks QS-regulated bioluminescence in *V. harveyi* (Defoirdt et al. 2007). This has potential application in shrimp aquaculture systems in which *V. harveyi* is a serious pathogen of all life stages of shrimp. Since quorum sensing is linked to the virulence in *V. harveyi*, its inhibition is expected to attenuate its virulence against the shrimp (Defoirdt and Sorgeloos 2012; Ruwandeepika et al. 2015). Studies on the role of efflux pumps in the virulence, environmental and host adaptation, as well as antibiotic resistance in *V. harveyi* is lacking. If the role of efflux pumps in QS in *V. harveyi* could be established, its inhibition is an attractive approach towards controlling this pathogen.

Food pathogens such as *Salmonella enterica, Listeria monocytogenes, V. parahemolyticus*, Shiga toxin-producing *E.coli, S. aureus* and others use QS to coordinate their survival, multiplication, toxin production and horizontal gene transfer in the food environment. Disruption of QS apparatus using phytochemical additives provides exciting avenues to make food safer, especially those which are minimally processed and have a greater chances of pathogen multiplication when temperature abused.

7.7 Future Directions

Biofilm-forming bacteria are a major problem in food processing industries and such bacteria often contaminate finished products. Biofilm bacteria are resistant to cleaning and disinfectants, and are very difficult to eliminate. Disruption of QS network might lead to poor ability to adhere and form biofilms on food processing surfaces by pathogenic bacteria making them susceptible to antibacterial cleaning agents and facilitate their easy removal. Bacteria are used in food production as well food preservation processes and since their activities are dependent on QS-regulated population density, modulation of these activities for optimal process outcomes using QS inhibition approach is promising.

Acknowledgements This publication and work in our laboratory have been supported in part by an Internal Research Grant (ENMU), a grant from the National Institute of General Medical Sciences (P20GM103451) of the National Institutes of Health, and by a grant from the US Department of Education, HSI STEM program (P031C110114).

References

Aguilar C, Friscina A, Devescovi G, Kojic M, Venturi V (2003) Identification of quorum-sensing-regulated genes of *Burkholderia cepacia*. J Bacteriol 185:6456–6462. https://doi.org/10.1128/JB.185.21.6456-6462.2003

Andersen JL, He GX, Kakarla P, CR K, Kumar S, Lakra WS, Mukherjee MM, Ranaweera I, Shrestha U, Tran T, Varela MF (2015) Multidrug efflux pumps from Enterobacteriaceae, *Vibrio cholerae* and *Staphylococcus aureus* bacterial food pathogens. Int J Environ Res Public Health 12:1487–1547. https://doi.org/10.3390/ijerph120201487

Azman CA-S, Othman I, Fang C-M, Chan K-G, Goh B-H, Lee L-H (2017) Antibacterial, anti-cancer and neuroprotective activities of rare actinobacteria from mangrove forest soils. Indian J Microbiol 57:177–187. https://doi.org/10.1007/s12088-016-0627-z

Bassler BL (1999) How bacteria talk to each other: regulation of gene expression by quorum sensing. Curr Opin Microbiol 2:582–587. https://doi.org/10.1016/S1369-5274(99)00025-9

Begum IF, Mohankumar R, Jeevan M, Ramani K (2016) GC–MS analysis of bioactive molecules derived from *Paracoccus pantotrophus* FMR19 and the antimicrobial activity against bacterial pathogens and MDROs. Indian J Microbiol 56:426–432. https://doi.org/10.1007/s12088-016-0609-1

Blair JM, Webber MA, Baylay AJ, Ogbolu DO, Piddock LJ (2015) Molecular mechanisms of antibiotic resistance. Nat Rev Microbiol 13:42–51. https://doi.org/10.1038/nrmicro3380

Brackman G, Cos P, Maes L, Nelis HJ, Coenye T (2011) Quorum sensing inhibitors increase the susceptibility of bacterial biofilms to antibiotics *in vitro* and *in vivo*. Antimicrob Agents Chemother 55:2655–2661. https://doi.org/10.1128/AAC.00045-11

Bruns MM, Kakarla P, Floyd JT, Mukherjee MM, Ponce RC, Garcia JA, Ranaweera I, Sanford LM, Hernandez AJ, Willmon TM, Tolson GL, Varela MF (2017) Modulation of the multidrug efflux pump EmrD-3 from *Vibrio cholerae* by *Allium sativum* extract and the bioactive agent allyl sulfide plus synergistic enhancement of antimicrobial susceptibility by *A. sativum* extract. Arch Microbiol. https://doi.org/10.1007/s00203-017-1378-x

Bzdrenga J, Daude D, Remy B, Jacquet P, Plener L, Elias M, Chabriere E (2017) Biotechnological applications of quorum quenching enzymes. Chem Biol Interact 267:104–115. https://doi.org/10.1016/j.cbi.2016.05.028

Cady NC, McKean KA, Behnke J, Kubec R, Mosier AP, Kasper SH, Burz DS, Musah RA (2012) Inhibition of biofilm formation, quorum sensing and infection in *Pseudomonas aeruginosa* by natural products-inspired organosulfur compounds. PLoS One 7:e38492. https://doi.org/10.1371/journal.pone.0038492

Chan YY, Chua KL (2005) The *Burkholderia pseudomallei* BpeAB-OprB efflux pump: expression and impact on quorum sensing and virulence. J Bacteriol 187:4707–4719. https://doi.org/10.1128/JB.187.14.4707-4719.2005

Chan YY, Tan TM, Ong YM, Chua KL (2004) BpeAB-OprB, a multidrug efflux pump in *Burkholderia pseudomallei*. Antimicrob Agents Chemother 48:1128–1135. https://doi.org/10.1128/AAC.48.4.1128-1135.2004

Cox G, Wright GD (2013) Intrinsic antibiotic resistance: mechanisms, origins, challenges and solutions. Int J Med Microbiol 303:287–292. https://doi.org/10.1016/j.ijmm.2013.02.009

Defoirdt T, Sorgeloos P (2012) Monitoring of *Vibrio harveyi* quorum sensing activity in real time during infection of brine shrimp larvae. ISME J 6:2314–2319. https://doi.org/10.1038/ismej.2012.58

Defoirdt T, Miyamoto CM, Wood TK, Meighen EA, Sorgeloos P, Verstraete W, Bossier P (2007) The natural furanone (5Z)-4-bromo-5-(bromomethylene)-3-butyl-2(5H)-furanone disrupts quorum sensing-regulated gene expression in *Vibrio harveyi* by decreasing the DNA-binding activity of the transcriptional regulator protein *luxR*. Environ Microbiol 9:2486–2495. https://doi.org/10.1111/j.1462-2920.2007.01367.x

Delcour AH (2009) Outer membrane permeability and antibiotic resistance. Biochim Biophys Acta 1794:808–816. https://doi.org/10.1016/j.bbapap.2008.11.005

Delmar JA, Su CC, Yu EW (2014) Bacterial multidrug efflux transporters. Annu Rev Biophys 43:93–117. https://doi.org/10.1146/annurev-biophys-051013-022855

Dickschat JS (2010) Quorum sensing and bacterial biofilms. Nat Prod Rep 27:343–369. https://doi.org/10.1039/b804469b

Dietrich LE, Price-Whelan A, Petersen A, Whiteley M, Newman DK (2006) The phenazine pyocyanin is a terminal signalling factor in the quorum sensing network of *Pseudomonas aeruginosa*. Mol Microbiol 61:1308–1321. https://doi.org/10.1111/j.1365-2958.2006.05306.x

Evans K, Passador L, Srikumar R, Tsang E, Nezezon J, Poole K (1998) Influence of the MexAB-OprM multidrug efflux system on quorum sensing in *Pseudomonas aeruginosa*. J Bacteriol 180:5443–5447. doi: Not Available

Floss HG, Yu TW (2005) Rifamycin-mode of action, resistance, and biosynthesis. Chem Rev 105:621–632. https://doi.org/10.1021/cr030112j

Floyd JT, Kumar S, Mukherjee MM, He G, Varela MF (2013) A review of the molecular mechanisms of drug efflux in pathogenic bacteria: a structure-function perspective. In: Pandalai SG (ed) Recent research developments in membrane biology, vol 3. Research Signpost, Inc, Thiruvananthapuram, pp 15–66. ISBN: 978-81-308-0529-0

Ghafourian S, Sadeghifard N, Soheili S, Sekawi Z (2015) Extended spectrum beta-lactamases: definition, classification and epidemiology. Curr Issues Mol Biol 17:11–21. https://doi.org/10.21775/cimb.017.011

Giménez-Bastida J, Truchado P, Larrosa M, Espín J, Tomás-Barberán F, Allende A, García-Conesa M (2012) Urolithins, ellagitannin metabolites produced by colon microbiota, inhibit quorum sensing in *Yersinia enterocolitica*: phenotypic response and associated molecular changes. Food Chem 132:1465–1474. https://doi.org/10.1021/jf301365a

Goh WK, Gardner CR, Chandra Sekhar KV, Biswas NN, Nizalapur S, Rice SA, Willcox M, Black DS, Kumar N (2015) Synthesis, quorum sensing inhibition and docking studies of

1,5-dihydropyrrol-2-ones. Bioorg Med Chem 23:7366–7377. https://doi.org/10.1016/j. bmc.2015.10.025

Gonzalez-Lamothe R, Mitchell G, Gattuso M, Diarra MS, Malouin F, Bouarab K (2009) Plant antimicrobial agents and their effects on plant and human pathogens. Int J Mol Sci 10:3400–3419. https://doi.org/10.3390/ijms10083400

Grandclement C, Tannieres M, Morera S, Dessaux Y, Faure D (2016) Quorum quenching: role in nature and applied developments. FEMS Microbiol Rev 40:86–116. https://doi.org/10.1093/femsre/fuv038

Gutierrez-Barranquero JA, Reen FJ, McCarthy RR, O'Gara F (2015) Deciphering the role of coumarin as a novel quorum sensing inhibitor suppressing virulence phenotypes in bacterial pathogens. Appl Microbiol Biotechnol 99:3303–3316. https://doi.org/10.1007/s00253-015-6436-1

He YW, Zhang LH (2008) Quorum sensing and virulence regulation in *Xanthomonas campestris*. FEMS Microbiol Rev 32:842–857. https://doi.org/10.1111/j.1574-6976.2008.00120.x

Hentzer M, Wu H, Andersen JB, Riedel K, Rasmussen TB, Bagge N, Kumar N, Schembri MA, Song Z, Kristoffersen P, Manefield M, Costerton JW, Molin S, Eberl L, Steinberg P, Kjelleberg S, Hoiby N, Givskov M (2003) Attenuation of *Pseudomonas aeruginosa* virulence by quorum sensing inhibitors. EMBO J 22:3803–3815. https://doi.org/10.1093/emboj/cdg366

Holm A, Vikström E (2014) Quorum sensing communication between bacteria and human cells: signals, targets, and functions. Front Plant Sci 5:309. https://doi.org/10.3389/fpls.2014.00309

Hou HM, Jiang F, Zhang GL, Wang JY, Zhu YH, Liu XY (2017) Inhibition of *Hafnia alvei* H4 biofilm formation by the food additive dihydrocoumarin. J Food Prot:842–847. https://doi.org/10.4315/0362-028X.JFP-16-460

Jacoby GA, Corcoran MA, Hooper DC (2015) Protective effect of Qnr on agents other than quinolones that target DNA gyrase. Antimicrob Agents Chemother 59:6689–6695. https://doi.org/10.1128/AAC.01292-15

Jakobsen TH, Bragason SK, Phipps RK, Christensen LD, van Gennip M, Alhede M, Skindersoe M, Larsen TO, Hoiby N, Bjarnsholt T, Givskov M (2012a) Food as a source for quorum sensing inhibitors: iberin from horseradish revealed as a quorum sensing inhibitor of *Pseudomonas aeruginosa*. Appl Environ Microbiol 78:2410–2421. https://doi.org/10.1128/AEM.05992-11

Jakobsen TH, van Gennip M, Phipps RK, Shanmugham MS, Christensen LD, Alhede M, Skindersoe ME, Rasmussen TB, Friedrich K, Uthe F, Jensen PO, Moser C, Nielsen KF, Eberl L, Larsen TO, Tanner D, Hoiby N, Bjarnsholt T, Givskov M (2012b) Ajoene, a sulfur-rich molecule from garlic, inhibits genes controlled by quorum sensing. Antimicrob Agents Chemother 56:2314–2325. https://doi.org/10.1128/AAC.05919-11

Jeyanthi V, Velusamy P (2016) Anti-methicillin resistant *Staphylococcus aureus* compound isolation from halophilic *Bacillus amyloliquefaciens* MHB1 and determination of its mode of action using electron microscope and flow cytometry analysis. Indian J Microbiol 56:148–157. https://doi.org/10.1007/s12088-016-0566-8

Jimenez PN, Koch G, Thompson JA, Xavier KB, Cool RH, Quax WJ (2012) The multiple signaling systems regulating virulence in *Pseudomonas aeruginosa*. Microbiol Mol Biol Rev 76:46–65. https://doi.org/10.1128/MMBR.05007-11

Kakarla P, Floyd J, Mukherjee M, Devireddy AR, Inupakutika MA, Ranweera I, KC R, Shrestha U, Cheeti UR, Willmon TM, Adams J, Bruns M, Gunda SK, Varela MF (2017) Inhibition of the multidrug efflux pump LmrS from *Staphylococcus aureus* by cumin spice *Cuminum cyminum*. Arch Microbiol 199:465–474. https://doi.org/10.1007/s00203-016-1314-5

Kalia VC (2013) Quorum sensing inhibitors: an overview. Biotechnol Adv 31:224–245. https://doi.org/10.1016/j.biotechadv.2012.10.004

Kalia VC (2014a) In search of versatile organisms for quorum-sensing inhibitors: acyl homoserine lactones (AHL)-acylase and AHL-lactonase. FEMS Microbiol Lett 359:143. https://doi.org/10.1111/1574-6968.12585

Kalia VC (2014b) Microbes, antimicrobials and resistance: the battle goes on. Indian J Microbiol 54:1–2. https://doi.org/10.1007/s12088-013-0443-7

Kalia VC (2015) Microbes: the most friendly beings? In: Kalia VC (ed) Quorum sensing vs quorum quenching: a battle with no end in sight. Springer, New Delhi, pp 1–5. ISBN: 978-81-322-1981-1. https://doi.org/10.1007/978-81-322-1982-8_1

Kalia VC, Kumar P (2015a) Potential applications of quorum sensing inhibitors in diverse fields. In: Kalia VC (ed) Quorum sensing vs Quorum quenching: a battle with no end in sight. Springer, New Delhi, pp 359–370. ISBN: 978-81-322-1981-1. https://doi.org/10.1007/978-81-322-1982-8_29

Kalia VC, Kumar P (2015b) The Battle: quorum-sensing inhibitors versus evolution of bacterial resistance. In: Kalia VC (ed) Quorum sensing vs Quorum quenching: a battle with no end in sight. Springer, New Delhi, pp 385–391. ISBN: 978-81-322-1981-1. doi:https://doi.org/10.1007/978-81-322-1982-8_31

Kalia VC, Purohit HJ (2011) Quenching the quorum sensing system: potential antibacterial drug targets. Crit Rev Microbiol 37:121–140. https://doi.org/10.3109/1040841X.2010.532479

Kalia VC, Raju SC, Purohit HJ (2011) Genomic analysis reveals versatile organisms for quorum quenching enzymes: acyl-homoserine lactone-acylase and –lactonase. Open Microbiol J 5:1–13. https://doi.org/10.2174/1874285801105010001

Kalia VC, Wood TK, Kumar P (2014a) Evolution of resistance to quorum-sensing inhibitors. Microb Ecol 68:13–23. https://doi.org/10.1007/s00248-013-0316-y

Kalia VC, Kumar P, Pandian SK, Sharma P (2014b) Biofouling control by quorum quenching. In: Kim SK (ed) Springer handbook of marine biotechnology. Springer, Berlin, pp 431–440. ISBN: 978-3-642-53971-8

Kim J, Kim JG, Kang Y, Jang JY, Jog GJ, Lim JY, Kim S, Suga H, Nagamatsu T, Hwang I (2004) Quorum sensing and the LysR-type transcriptional activator ToxR regulate toxoflavin biosynthesis and transport in *Burkholderia glumae*. Mol Microbiol 54:921–934. https://doi.org/10.1111/j.1365-2958.2004.04338.x

Kim AL, Park SY, Lee CH, Lee CH, Lee JK (2014) Quorum quenching bacteria isolated from the sludge of a wastewater treatment plant and their application for controlling biofilm formation. J Microbiol Biotechnol 24:1574–1582. https://doi.org/10.3109/01443610903506198

Koul S, Kalia VC (2017) Multiplicity of quorum quenching enzymes: a potential mechanism to limit quorum sensing bacterial population. Indian J Microbiol 57:100–108. https://doi.org/10.1007/s12088-016-0633-1

Koul S, Prakash J, Mishra A, Kalia VC (2016) Potential emergence of multi-quorum sensing inhibitor resistant (MQSIR) bacteria. Indian J Microbiol 56:1–18. https://doi.org/10.1007/s12088-015-0558-0

Kumar A, Schweizer HP (2005) Bacterial resistance to antibiotics: active efflux and reduced uptake. Adv Drug Deliv Rev 57:1486–1513. https://doi.org/10.1016/j.addr.2005.04.004

Kumar S, Varela MF (2012) Biochemistry of bacterial multidrug efflux pumps. Int J Mol Sci 13:4484–4495. https://doi.org/10.3390/ijms13044484

Kumar S, Varela MF (2013) Molecular mechanisms of bacterial resistance to antimicrobial agents. In: Méndez-Vilas A (ed) Microbial pathogens and strategies for combating them: science, technology and education. Formatex Research Center, Inc., Badajoz, pp 522–534. ISBN: 978-84-939843-9-7

Kumar S, Floyd JT, He G, Varela MF (2013a) Bacterial antimicrobial efflux pumps of the MFS and MATE transporter families: a review. In: Pandalai SG (ed) Recent research developments in antimicrobial agents & chemotherapy, vol 7. Research Signpost, Thiruvananthapuram, pp 1–21. ISBN: 978-81-308-0465-1

Kumar S, Mukherjee MM, Varela MF (2013b) Modulation of bacterial multidrug resistance efflux pumps of the major facilitator superfamily. Int J Bacteriol 2013. https://doi.org/10.1155/2013/204141

Kumar S, He G, Kakarla P, Shrestha U, Ranjana KC, Ranaweera I, Willmon TM, Barr SR, Hernandez AJ, Varela MF (2016) Bacterial multidrug efflux pumps of the major facilitator superfamily as targets for modulation. Infect Disord Drug Targets 16:28–43. https://doi.org/10.2174/1871526516666160407113848

Lambert PA (2005) Bacterial resistance to antibiotics: modified target sites. Adv Drug Deliv Rev 57:1471–1485. https://doi.org/10.1016/j.addr.2005.04.003

Lambert T (2012) Antibiotics that affect the ribosome. Rev Sci Tech 31:57–64

LaSarre B, Federle MJ (2013) Exploiting quorum sensing to confuse bacterial pathogens. Microbiol Mol Biol Rev 77:73–111. https://doi.org/10.1128/MMBR.00046-12

Le Norcy T, Niemann H, Proksch P, Tait K, Linossier I, Rehel K, Hellio C, Fay F (2017) Sponge-inspired dibromohemibastadin prevents and disrupts bacterial biofilms without toxicity. Mar Drugs 15:pii: E222. https://doi.org/10.3390/md15070222

Levy SB (1992) Active efflux mechanisms for antimicrobial resistance. Antimicrob Agents Chemother 36:695–703

Levy SB (2002) Active efflux, a common mechanism for biocide and antibiotic resistance. Symp Ser Soc Appl Microbiol 92:65S–71S. https://doi.org/10.1046/j.1365-2672.92.5s1.4.x

Lewenza S, Sokol PA (2001) Regulation of ornibactin biosynthesis and N-acyl-L-homoserine lactone production by CepR in *Burkholderia cepacia*. J Bacteriol 183:2212–2218. https://doi.org/10.1128/JB.183.7.2212-2218.2001

Linares JF, Lopez JA, Camafeita E, Albar JP, Rojo F, Martinez JL (2005) Overexpression of the multidrug efflux pumps MexCD-OprJ and MexEF-OprN is associated with a reduction of type III secretion in *Pseudomonas aeruginosa*. J Bacteriol 187:1384–1391. https://doi.org/10.1128/JB.187.4.1384-1391.2005

Luo J, Dong B, Wang K, Cai S, Liu T, Cheng X, Lei D, Chen Y, Li Y, Kong J, Chen Y (2017) Baicalin inhibits biofilm formation, attenuates the quorum sensing-controlled virulence and enhances *Pseudomonas aeruginosa* clearance in a mouse peritoneal implant infection model. PLoS One 12:e0176883. https://doi.org/10.1371/journal.pone.0176883

Mahmood HY, Jamshidi S, Sutton JM, Rahman KM (2016) Current advances in developing inhibitors of bacterial multidrug efflux pumps. Curr Med Chem 23:1062–1081. https://doi.org/10.2174/0929867323666160304150522

Martin-Rodriguez AJ, Ticona JC, Jimenez IA, Flores N, Fernandez JJ, Bazzocchi IL (2015) Flavonoids from *Piper delineatum* modulate quorum-sensing-regulated phenotypes in *Vibrio harveyi*. Phytochemistry 117:98–106. https://doi.org/10.1016/j.phytochem.2015.06.006

Maseda H, Sawada I, Saito K, Uchiyama H, Nakae T, Nomura N (2004) Enhancement of the mexAB-oprM efflux pump expression by a quorum-sensing autoinducer and its cancellation by a regulator, MexT, of the *mexEF-oprN* efflux pump operon in *Pseudomonas aeruginosa*. Antimicrob Agents Chemother 48:1320–1328. https://doi.org/10.1128/AAC.48.4.1320-1328.2004

Michalet S, Cartier G, David B, Mariotte AM, Dijoux-franca MG, Kaatz GW, Stavri M, Gibbons S (2007) N-caffeoylphenalkylamide derivatives as bacterial efflux pump inhibitors. Bioorg Med Chem Lett 17:1755–1758. https://doi.org/10.1016/j.bmcl.2006.12.059

Miller MB, Bassler BL (2001) Quorum sensing in bacteria. Annu Rev Microbiol 55:165–199. https://doi.org/10.1146/annurev.micro.55.1.165

Monnet V, Gardan R (2015) Quorum-sensing regulators in Gram-positive bacteria: 'cherchez le peptide'. Mol Microbiol 97:181–184. https://doi.org/10.1111/mmi.13060

Nandi S (2016) Recent advances in ligand and structure based screening of potent quorum sensing inhibitors against antibiotic resistance induced bacterial virulence. Recent Pat Biotechnol 10:195–216. https://doi.org/10.2174/1872208310666160728104450

Nikolaidis I, Favini-Stabile S, Dessen A (2014) Resistance to antibiotics targeted to the bacterial cell wall. Protein Sci 23:243–259. https://doi.org/10.1002/pro.2414

Niu C, Afre S, Gilbert ES (2006) Subinhibitory concentrations of cinnamaldehyde interfere with quorum sensing. Lett Appl Microbiol 43:489–494. https://doi.org/10.1111/j.1472-765X.2006.02001.x

Papenfort K, Bassler BL (2016) Quorum sensing signal-response systems in Gram-negative bacteria. Nat Rev Microbiol 14:576–588. https://doi.org/10.1038/nrmicro.2016.89

Parsek MR, Greenberg EP (2000) Acyl-homoserine lactone quorum sensing in gram-negative bacteria: a signaling mechanism involved in associations with higher organisms. Proc Natl Acad Sci U S A 97:8789–8793. https://doi.org/10.1073/pnas.97.16.8789

Pearson JP, Van Delden C, Iglewski BH (1999) Active efflux and diffusion are involved in transport of *Pseudomonas aeruginosa* cell-to-cell signals. J Bacteriol 181:1203–1210

Ponnusamy K, Paul D, Kweon JH (2009) Inhibition of quorum sensing mechanism and *Aeromonas hydrophila* biofilm formation by vanillin. Environ Eng Sci 26:1359–1363. https://doi.org/10.1089/ees.2008.0415

Ponnusamy K, Kappachery S, Thekeettle M, Song JH, Kweon JH (2013) Anti-biofouling property of vanillin on *Aeromonas hydrophila* initial biofilm on various membrane surfaces. World J Microbiol Biotechnol 29:1695–1703. https://doi.org/10.1007/s11274-013-1332-2

Rahmati S, Yang S, Davidson AL, Zechiedrich EL (2002) Control of the AcrAB multidrug efflux pump by quorum-sensing regulator SdiA. Mol Microbiol 43:677–685. https://doi.org/10.1046/j.1365-2958.2002.02773.x

Ranaweera I, Shrestha U, Ranjana KC, Kakarla P, Willmon TM, Hernandez AJ, Mukherjee MM, Barr SR, Varela MF (2015) Structural comparison of bacterial multidrug efflux pumps of the major facilitator superfamily. Trends Cell Mol Biol 10:131–140. Not Available

Rasmussen TB, Skindersoe ME, Bjarnsholt T, Phipps RK, Christensen KB, Jensen PO, Andersen JB, Koch B, Larsen TO, Hentzer M, Eberl L, Hoiby N, Givskov M (2005) Identity and effects of quorum-sensing inhibitors produced by *Penicillium* species. Microbiology 151:1325–1340. https://doi.org/10.1099/mic.0.27715-0

Ren D, Zuo R, Gonzalez Barrios AF, Bedzyk LA, Eldridge GR, Pasmore ME, Wood TK (2005) Differential gene expression for investigation of *Escherichia coli* biofilm inhibition by plant extract ursolic acid. Appl Environ Microbiol 71:4022–4034. https://doi.org/10.1128/AEM.71.7.4022-4034.2005

Roberts MC (2005) Update on acquired tetracycline resistance genes. FEMS Microbiol Lett 245:195–203. https://doi.org/10.1016/j.femsle.2005.02.034

Ruwandeepika HA, Karunasagar I, Bossier P, Defoirdt T (2015) Expression and quorum sensing regulation of type III secretion system genes of *Vibrio harveyi* during infection of gnotobiotic brine shrimp. PLoS One 10:e0143935. https://doi.org/10.1371/journal.pone.0143935

Sacha P, Wieczorek P, Hauschild T, Zorawski M, Olszanska D, Tryniszewska E (2008) Metallo-beta-lactamases of *Pseudomonas aeruginosa*–a novel mechanism resistance to beta-lactam antibiotics. Folia Histochem Cytobiol 46:137–142. https://doi.org/10.2478/v10042-008-0020-9

Sawada I, Maseda H, Nakae T, Uchiyama H, Nomura N (2004) A quorum-sensing autoinducer enhances the *mexAB-oprM* efflux-pump expression without the MexR-mediated regulation in *Pseudomonas aeruginosa*. Microbiol Immunol 48:435–439. https://doi.org/10.1111/j.1348-0421.2004.tb03533.x

Sharma R, Jangid K (2015) Fungal quorum sensing inhibitors. In: Kalia VC (ed) Quorum sensing vs quorum quenching: a battle with no end in sight. Springer, New Delhi, pp 237–257. ISBN: 978-81-322-1981-1

Siddiqui MF, Rzechowicz M, Harvey W, Zularisam AW, Anthony GF (2015) Quorum sensing based membrane biofouling control for water treatment: a review. J Water Proc Eng 30:112–122. https://doi.org/10.1016/j.jwpe.2015.06.003

Singh S, Singh SK, Chowdhury I, Singh R (2017) Understanding the mechanism of bacterial biofilms resistance to antimicrobial agents. Open Microbiol J 11:53–62. https://doi.org/10.2174/1874285801711010053

Smith RS, Iglewski BH (2003) *P. aeruginosa* quorum-sensing systems and virulence. Curr Opin Microbiol 6:56–60. https://doi.org/10.1016/S1369-5274(03)00008-0

Smith KP, Kumar S, Varela MF (2009) Identification, cloning, and functional characterization of EmrD-3, a putative multidrug efflux pump of the major facilitator superfamily from *Vibrio cholerae* O395. Arch Microbiol 191:903–911. https://doi.org/10.1007/s00203-009-0521-8

Soto SM (2013) Role of efflux pumps in the antibiotic resistance of bacteria embedded in a biofilm. Virulence 4:223–229. https://doi.org/10.4161/viru.23724

Stavri M, Piddock LJ, Gibbons S (2007) Bacterial efflux pump inhibitors from natural sources. J Antimicrob Chemother 59:1247–1260. https://doi.org/10.1093/jac/dkl460

Stermitz FR, Lorenz P, Tawara JN, Zenewicz LA, Lewis K (2000) Synergy in a medicinal plant: antimicrobial action of berberine potentiated by 5′-methoxyhydnocarpin, a multidrug pump inhibitor. Proc Natl Acad Sci U S A 97:1433–1437. https://doi.org/10.1073/pnas.030540597

Surette MG, Miller MB, Bassler BL (1999) Quorum sensing in *Escherichia coli, Salmonella typhimurium,* and *Vibrio harveyi*: a new family of genes responsible for autoinducer production. Proc Natl Acad Sci U S A 96:1639–1644. https://doi.org/10.1073/pnas.96.4.1639

Taganna JC, Rivera WL (2008) Epigallocatechin gallate from *Camellia sinensis* L.(Kuntze) is a potential quorum sensing inhibitor in *Chromobacterium violaceum*. Sci Diliman 20:24–30. Not Available

Truchado P, Gimenez-Bastida JA, Larrosa M, Castro-Ibanez I, Espin JC, Tomas-Barberan FA, Garcia-Conesa MT, Allende A (2012) Inhibition of quorum sensing (QS) in *Yersinia enterocolitica* by an orange extract rich in glycosylated flavanones. J Agric Food Chem 60:8885–8894. https://doi.org/10.1021/jf301365a

van Hoek AH, Mevius D, Guerra B, Mullany P, Roberts AP, Aarts HJ (2011) Acquired antibiotic resistance genes: an overview. Front Microbiol 2:203. https://doi.org/10.3389/fmicb.2011.00203

Varela MF, Kumar S, He G (2013) Potential for inhibition of bacterial efflux pumps in multidrug-resistant *Vibrio cholerae*. Indian J Med Res 138:285–287. Not Available

Varga ZG, Armada A, Cerca P, Amaral L, Mior Ahmad Subki MA, Savka MA, Szegedi E, Kawase M, Motohashi N, Molnar J (2012) Inhibition of quorum sensing and efflux pump system by trifluoromethyl ketone proton pump inhibitors. In Vivo 26:277–285. Not Available

Varsha KK, Nishant G, Sneha SM, Shilpa G, Devendra L, Priya S, Nampoothiri KM (2016) Antifungal, anticancer and aminopeptidase inhibitory potential of a phenazine compound produced by *Lactococcus* BSN307. Indian J Microbiol 56:411–416. https://doi.org/10.1007/s12088-016-097-1

Wagner VE, Bushnell D, Passador L, Brooks AI, Iglewski BH (2003) Microarray analysis of *Pseudomonas aeruginosa* quorum-sensing regulons: effects of growth phase and environment. J Bacteriol 185:2080–2095. https://doi.org/10.1128/JB.185.7.2080-2095.2003

Wright GD (2005) Bacterial resistance to antibiotics: enzymatic degradation and modification. Adv Drug Deliv Rev 57:1451–1470. https://doi.org/10.1016/j.addr.2005.04.002

Yang S, Lopez CR, Zechiedrich EL (2006) Quorum sensing and multidrug transporters in *Escherichia coli*. Proc Natl Acad Sci U S A 103:2386–2391. https://doi.org/10.1073/pnas.0502890102

Yuan ZC, Edlind MP, Liu P, Saenkham P, Banta LM, Wise AA, Ronzone E, Binns AN, Kerr K, Nester EW (2007) The plant signal salicylic acid shuts down expression of the *vir* regulon and activates quormone-quenching genes in *Agrobacterium*. Proc Natl Acad Sci U S A 104:11790–11795. https://doi.org/10.1073/pnas.0704866104

Yufan C, Shiyin L, Zhibin L, Mingfa L, Jianuan Z, Lianhui Z (2016) Quorum sensing and microbial drug resistance. Yi Chuan 38:881–893. https://doi.org/10.16288/j.yczz.16-141

Zhang L, Guo Z, Gao H, Peng X, Li Y, Sun S, Lee JK, Lin W (2016) Interaction of pseudostellaria heterophylla with quorum sensing and quorum quenching bacteria mediated by root exudates in a consecutive monoculture system. J Microbiol Biotechnol 26:2159–2170. https://doi.org/10.4014/jmb.1607.07073

Zhu J, Huang X, Zhang F, Feng L, Li J (2015) Inhibition of quorum sensing, biofilm, and spoilage potential in *Shewanella baltica* by green tea polyphenols. J Microbiol 53:829–836. https://doi.org/10.1007/s12275-015-5123-3

Chapter 8
CRISPR-Cas Systems Regulate Quorum Sensing Genes and Alter Virulence in Bacteria

Qinqin Pu and Min Wu

Abstract Although understanding how CRISPR and its associated systems are controlled is at the infant stages, recent studies present exciting discoveries in this fast-moving field. Studies find that CRISPR-Cas systems regulate quorum sensing (QS) genes by targeting and degrading *lasR* mRNA, while QS systems can also modulate the CRISPR-Cas as revealed in more recent reports. The importance of the QS system for bacterial pathogenicity is well recognized and the indispensable features of CRISPR-Cas in adaptive immunity and biotechnology application help gaining great attention. Analyzing interaction between QS and CRISPR-Cas systems represents an interesting field as CRISPR-Cas systems are not only the adaptive immunity of bacteria, but also the regulators of their own genes. Undoubtedly, the continued understanding of molecular basis of CRISPR-Cas action and regulation may indicate novel strategies for treatment of bacterial infections.

Keywords CRISPR-Cas systems · Quorum sensing · Endogenous gene targeting

8.1 Introduction

The quorum sensing (QS) system in many bacterial species plays a significant role in cell growth, development and differentiation as an intercellular signaling system that relies on cell density (Hurley and Bassler 2017). Recent studies found that QS regulates a spectrum of critical functions of bacteria, such as synthesis of

Q. Pu
Department of Biomedical Sciences, University of North Dakota, Grand Forks, ND, USA

State Key Laboratory of Biotherapy, West China Hospital, Sichuan University, Chengdu, China

M. Wu (✉)
Department of Biomedical Sciences, University of North Dakota, Grand Forks, ND, USA
e-mail: min.wu@med.und.edu

© Springer Nature Singapore Pte Ltd. 2018
V. C. Kalia (ed.), *Biotechnological Applications of Quorum Sensing Inhibitors*,
https://doi.org/10.1007/978-981-10-9026-4_8

extracellular enzymes and toxins from pathogens, formation of biofilm and production of drug resistance (Atkinson and Williams 2009; Das et al. 2015; Tan et al. 2015; Vuotto et al. 2017). Hence, research into QS signaling may indicate targets to design novel and effective anti-microbial therapeutics. Despite a great deal of efforts in understanding QS regulation, the detailed mechanism of QS gene expression and function has remained largely unknown.

Clustered regularly interspaced short palindromic repeats (CRISPR) and CRISPR-associated (Cas) systems are the essential adaptive immunity to microbes. It is commonly believed that more than 90% archaea and 40% eubacteria possess one or more types of CRISPR-Cas systems (Marraffini and Sontheimer 2010). CRISPR loci are a special DNA sequence located in the bacterial and archaea genomes, often consisting of a leader, a plurality of highly conserved repeats and multiple spacers that were derived from the predators, bacteriophages. Analyzing the flanking sequence of the CRISPR site revealed that there was a polymorphic family gene in the vicinity thereof. The family encodes a group of proteins that contain a functional domain that interacts with the nucleic acid and acts together with the CRISPR region and is named CRISPR associated gene, abbreviated as Cas (Song 2017). When the phage containing the DNA that matches with any of the spacer invades again, the CRISPR-Cas system will cleave the nucleic acids to protect the bacterium itself from killing (Puschnik et al. 2017). CRISPR-Cas systems are categorized to two main classes, 6 types and 19 subtypes, based on the signature Cas protein and other nomenclature features. Currently, there is outstanding interest in understanding the CRISPR-Cas9 biology and function of bacterial *S. pyogenes* due to the powerful gene-editing function and potential application in medicine and biotechnology; however, the original roles of CRISPR-Cas in various aspects of bacterial physiology and immunity are relatively understudied.

Although the best-known Type II CRISPR-Cas (Cas9) was reported to play roles in mammalian host defense (Heidrich and Vogel 2013), it remains unknown whether other CRISPR-Cas systems, such as Type I and Type III CRISPR-Cas, are also involved in host immunity by targeting bacterial endogenous genes. We have recently revealed a new function for Type I-F CRISPR-Cas system which consists of 6 important proteins (Csy1, Csy2, Csy3, Csy4 and Cas1/Cas3 survey complex) in the control of QS associated gene expression in *Pseudomonas aeruginosa*. This study shows that the Type I-F CRISPR immune system of PA14 (a strain of *P. aeruginosa*) modulates the master QS regulator *LasR* by degrading its mRNA, thus protecting the bacteria to evade recognition by host Toll-like receptor 4 (TLR4), and inhibiting host pro-inflammatory response. Hence, targeting the CRISPR-Cas system which regulates quorum-sensing and alters pathogenesis may open a new avenue to tackle the drug resistance of *P. aeruginosa* (Li et al. 2016). Here, we discuss the potential role of CRISPR-Cas systems in regulating endogenous genes to alter virulence.

8.2 CRISPR-Cas in Regulating Virulence and Altering Mammalian Defense

The first study reporting that Type II CRISPR-Cas, namely Cas9, is involved in regulating bacterial endogenous genes was observed in *Francisella novicida* (Sampson et al. 2013). This study demonstrated that Cas9 is necessary for the bacterium to evade detection by Toll-like receptor 2 (TLR2) and cause serious disease. In particular, the Cas9 functions were associated with tracrRNA and crRNA. The authors show that Cas9 alters the stability of endogenous transcript encoding bacterial lipoprotein (BLP) that is essential for its virulence and can be recognized by TLR2. CRISPR-Cas mediates repression of BLP expression by degrading the mRNA and decreasing transcript levels. Hence for the first time, scientists may recognize the CRISPR-Cas components play a key role in pathogenesis of bacteria in causing disease in mammalian systems through self-regulation of their gene expression (Heidrich and Vogel 2013).

Since then, several studies demonstrated that different types of CRISPR-Cas systems were involved in bacterial physiology by targeting the endogenous genes besides silencing of foreign nucleic acids (Yosef et al. 2012; Heussler et al. 2016; Fu et al. 2017). One of the most remarkable examples is the CRISPR-Cas system of *P. aeruginosa* was reported to regulate its biofilm formation (Cady and O'Toole 2011; Heussler et al. 2016; Li et al. 2016). The detailed mechanism is not yet understood but conventional wisdom is that the CRISPR-Cas systems interact with the target genes in the chromosomally integrated prophage to abolish the generation of biofilm. *P. aeruginosa* is lysogenized by bacteriophages and it is clear that the process requires the Cas proteins (Zegans et al. 2009). Biofilm and QS have inseparable relationship. They are two important information departments for bacterial communication. One is for surface-associated communities and the other is for intercellular signaling. How they communicate with each other is known thanks to many recent intense studies but there is much to be learned. In addition to *Vibrio harveyi* and yellow *Myxococcus xanthus*, the population effect of Gram-negative bacteria is generally regulated by LuxR/I-type information systems (Nicola and Vanessa 2005). Studies reported that QS systems may play key roles in regulating the biofilm formation for many bacterial species (Shao et al. 2012; Yu et al. 2012). Recently, Lan and his collaborators found a cascade regulatory pathway to regulate the Rhl population induction system, Crc-Hfq/Lon/RhlI in *P. aeruginosa*, which added knowledge of QS in regulating biofilm (Cao et al. 2014). In recent years, the discoveries of microbial QS system and its relationship with some drug resistance via biofilm provide new perspective and means for the study of drug resistance mechanism. Altogether, these studies provide good theoretical basis for the hypothesis that CRISPR-Cas regulates QS and biofilm to alter pathogenesis. CRISPR-Cas systems may be a new target for bacterial resistant treatments.

Studies reveal that the link between the CRISPR-Cas and QS systems are increasing. A recent paper reported that the QS regulation leads to enhanced expression of the CRISPR-Cas systems in *Serratia* especially for high cell density situation (Patterson et al. 2016; Semenova and Severinov 2016). On the contrary, Zuberi's team found that CRISPR interference (CRISPRi) inhibited biofilm by repressing the luxS QS gene expression in *E. coli* (Zuberi et al. 2017). Consistently, Li et al., demonstrate that self-targeting CRISPR spacers bear sequences for degrading transcription factor mRNA of *lasR* and that CRISPR-Cas systems can control the QS response in some cases (Li et al. 2016). The LasR/LasI system consists of transcriptional activator LasR and acetyl homoserine lactone (AHL) synthase LasI protein in the Gram-negative bacteria QS system with AHL as a self-inducing agent. LasI guides the synthesis of 3-OXO-C-HSL and is secreted into the extracellular spaces by active transport, which binds to LasR at a certain threshold and activates gene transcription, including alkaline protease, exotoxin A, elastase, and other virulence factors. Hence, LasR plays a key role in increasing the expression of *P. aeruginosa* virulence related genes (Lee and Zhang 2015). Li et al. revealed mechanistically that CRISPR-mediated mRNA degradation needs the "5′-GGN-3′" (protospacer adjacent motif [PAM]) sequence and the HD and DExD/H domains of Cas3 protein for recognition in *lasR* mRNA (Li et al. 2016). As the consequence, LasR is decreased, the PA14 strain with CRISPR-Cas shows decreased bacterial phagocytosis by host alveolar macrophages and lower mouse survival than the CRISPR-Cas deleted one. This implicates that the CRISPR-Cas regulates innate immunity, which is exerted via TLR4-initiated signaling as upstream events. These studies have opened up new fields to elucidate the interaction between QS and CRISPR-Cas system and develop new drugs for treating infection (Fig. 8.1).

Once CRISPR-Cas components are activated, the crRNA12 structure and Cascade (Csy1-4 complex) will interact with *lasR* mRNA (or other potential genes) through a sequence matching with crRNA12. Then *lasR* mRNA will be cleaved by Cas3 protein. The degradation of *lasR* mRNA and the changes of downstream genes alter bacterial behaviors and subsequent host inflammatory responses.

A potential function of CRISPR-Cas systems in endogenous gene regulation as well as in pathogenesis may be to acquire self-targeting crRNAs with spacer sequences complementary to chromosomally encoded genes. By analyzing CRISPRs from 330 organisms, Stern et al., found that only approximately 0.4% spacers are potentially self-targeting and that frequent targeting non-moving genes occurs in 18% of all bacteria containing a type of CRISPR-Cas (Stern et al. 2010). The result of self-targeting is likely deleterious chromosomal cleavage and deletion, thus is considered to be detrimental effects to the body of bacteria, hence termed "autoimmunity". In terms of existing knowledge, the Cas proteins, such as Cas9, bind with target genes and inhibit transcription capacity (Qi et al. 2013), while other Cas proteins like Cas1 protein and Cas2 protein are reported to prevent the acquisition of new crRNAs to protect the loss of previously acquired crRNAs (Westra and Brouns 2012). This suggests that the inhibition and destruction of Cas proteins or the process of autoimmunity may be involved in endogenous gene regulation, which may also potentially impact virulence and pathogenesis in some aspects. crRNAs are originally known to target DNA, but a recent study reported that RNA also can be targeted in *F. novicida*

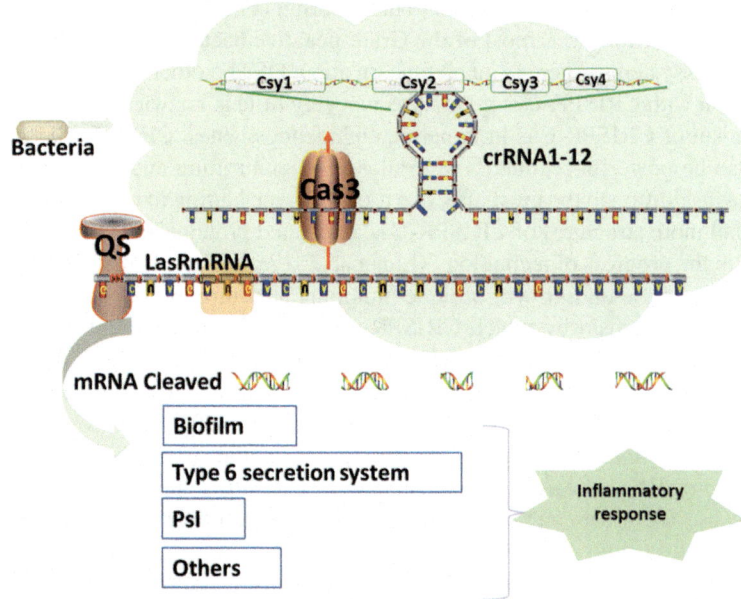

Fig. 8.1 Model of type I-F CRISPR-Cas modulates inflammatory response by temporal mRNA repression of QS genes (*lasR*)

(Sampson et al. 2013). The mechanisms and consequences of autoimmunity as well as their types and substrate specificities in bacteria are totally unknown. Much needs to be done to clearly elucidate how autoimmunity and CRISPR-Cas systems contribute to the regulation of bacterial genes (such as virulence, biofilm and proliferation associated genes) and pathogenesis. Accomplishing these pursues will be very helpful for designing new strategies for anti-infective therapy.

8.3   Discussion

CRISPR-Cas systems as the emerging genome editing tool to identify and edit target genes have built an extremely efficient platform for gene manipulation technology. It is well publicized in scientific research and application due to its high efficiency, simple and economic features, indicating enormous biotechnology and therapeutic values. However, as an adaptive immune system for prokaryotes and archaea, understanding of CRISPR-Cas' own superior functions in virulence regulation is at much slower paces and require strong efforts to further explore. The function of CRISPR-Cas systems in regulating endogenous genes and altering pathogenesis is undoubtedly the great entry point of research. The release of virulence factors is the main form of bacterial infection especially for Gram-negative bacteria that have up to six (type I, II, III, IV, V and VI) types of special secretion systems to transfer virulence factors to neighboring cells or animal host (Green and Mecsas 2016). Moreover, most of the Gram-negative bacteria possess more than 2 types of QS systems. There is much to learn about QS and other virulence factors in interaction with CRISPR-Cas systems as just very little is known about the role and mechanism of CRISPR-Cas in targeting endogenous genes. CRISPR-Cas systems may also be powerful regulators in virulence by controlling endogenous genes in bacteria in addition to the most effective tools for gene editing in mammanlian cells. More and more structures of CRISPR-Cas associated proteins have been character-ized with the progress of technology (Liu et al. 2017a, b; Pausch et al. 2017; Wright et al. 2017).These finding will provide important structural biology to delve in the molecular mechanism by which CRISPR-Cas systems function and interact with other genes. The characterization of structure will greatly broaden our understand-ing about how the CRISPR-Cas systems target to QS or other endogenous genes and improve the transformation and utilization value of CRISPR-Cas systems. For example, the revelation of the Cas13a structure (Liu et al. 2017a, b) may add value in developing RNA research tools and extending CRISPR's application in gene edit-ing because Cas13a is one of the few proteins that can degrade RNA in Class 2 CRISPR-Cas system.

Besides the mechanism that CRISPR-Cas systems degrade QS mRNA, a recent study indicates that small RNA ReaL regulates *P. aeruginosa* QS networks due to the activity of RpoS providing a new perspective to explore the relationship between QS and CRISPR-Cas systems that may be regulated through small RNAs (Carloni et al. 2017). The role of small non-coding RNAs in mammals has been recognized while its function in prokaryote is almost unknown. The research prospect of small RNAs for bacterial physiology is immense and much less is known about their inter-actions with CRISPR-Cas systems. The roles of QS in diverse fields, especially human health and disease, are being dissected. Inhibiting the bacterial QS by deac-tivating their chemical signaling molecules or by producing competitors may help in designing approaches to treating infectious diseases. On contrary, bacteria con-stantly evolve new strategies to battle with drugs and antibiotics. Like the CRISPR-Cas systems, the QS is diverse with strains and species, therefore unravelling the molecular detail in the cross-roads may open new avenues for fighting bacteria.

References

Atkinson S, Williams P (2009) Quorum sensing and social networking in the microbial world. J R Soc Interface 6:959–978. https://doi.org/10.1098/rsif.2009.0203

Cady KC, O'Toole GA (2011) Non-identity-mediated CRISPR-bacteriophage interaction medi-ated via the Csy and Cas3 proteins. J Bacteriol 193:3433–3445. https://doi.org/10.1128/JB.01411-10

Cao Q, Wang Y, Chen F, Xia Y, Lou J, Zhang X, Yang N, Sun X, Zhang Q, Zhuo C, Huang X, Deng X, Yang CG, Ye Y, Zhao J, Wu M, Lan L (2014) A novel signal transduction pathway that modulates rhl quorum sensing and bacterial virulence in *Pseudomonas aeruginosa*. PLoS Pathog 10:e1004340. https://doi.org/10.1371/journal.ppat.1004340

Carloni S, Macchi R, Sattin S, Ferrara S, Bertoni G (2017) The small RNA ReaL: a novel regulatory element embedded in the *Pseudomonas aeruginosa* quorum sensing networks. Environ Microbiol. https://doi.org/10.1111/1462-2920.13886

Das T, Kutty SK, Tavallaie R, Ibugo AI, Panchompoo J, Sehar S, Aldous L, Yeung AW, Thomas SR, Kumar N, Gooding JJ, Manefield M (2015) Phenazine virulence factor binding to extracellular DNA is important for *Pseudomonas aeruginosa* biofilm formation. Sci Rep 5:8398. https://doi.org/10.1038/srep08398

Fu Q, Su Z, Cheng Y, Wang Z, Li S, Wang H, Sun J, Yan Y (2017) Clustered, regularly interspaced short palindromic repeat (CRISPR) diversity and virulence factor distribution in avian Escherichia coli. Res Microbiol 168:147–156. https://doi.org/10.1016/j.resmic.2016.10.002

Green ER, Mecsas J (2016) Bacterial secretion systems: an overview. Microbiol Spectr 4:1–19. https://doi.org/10.1128/microbiolspec.VMBF-0012-2015

Heidrich N, Vogel J (2013) CRISPRs extending their reach: prokaryotic RNAi protein Cas9 recruited for gene regulation. EMBO J 32:1802–1804. https://doi.org/10.1038/emboj.2013.141

Heussler GE, Miller JL, Price CE, Collins AJ, O'Toole GA (2016) Requirements for *Pseudomonas aeruginosa* Type I-F CRISPR-Cas adaptation determined using a biofilm enrichment assay. J Bacteriol 198:3080–3090. https://doi.org/10.1128/JB.00458-16

Hurley A, Bassler BL (2017) Asymmetric regulation of quorum-sensing receptors drives autoinducer-specific gene expression programs in *Vibrio cholerae*. PLoS Genet 13:e1006826. https://doi.org/10.1371/journal.pgen.1006826

Lee J, Zhang L (2015) The hierarchy quorum sensing network in *Pseudomonas aeruginosa*. Protein Cell 6:26–41. https://doi.org/10.1007/s13238-014-0100-x

Li R, Fang L, Tan S, Yu M, Li X, He S, Wei Y, Li G, Jiang J, Wu M (2016) Type I CRISPR-Cas targets endogenous genes and regulates virulence to evade mammalian host immunity. Cell Res 26:1273–1287. https://doi.org/10.1038/cr.2016.135

Liu L, Li X, Ma J, Li Z, You L, Wang J, Wang M, Zhang X, Wang Y (2017a) The molecular architecture for RNA-guided RNA cleavage by Cas13a. Cell 170(714–726):e710. https://doi.org/10.1016/j.cell.2017.06.050

Liu L, Li X, Wang J, Wang M, Chen P, Yin M, Li J, Sheng G, Wang Y (2017b) Two distant catalytic sites are responsible for C2c2 RNase activities. Cell 168(121–134):e112. https://doi.org/10.1016/j.cell.2016.12.031

Marraffini LA, Sontheimer EJ (2010) CRISPR interference: RNA-directed adaptive immunity in bacteria and archaea. Nat Rev Genet 11:181–190. https://doi.org/10.1038/nrg2749

Nicola R, Vanessa S (2005) Quorum sensing: the many languages of bacteri. FEMS Microbiol Lett 254:1–11. https://doi.org/10.1111/j.1574-6968.2005.00001.x

Patterson AG, Jackson SA, Evans TC, Salmond GB, Przybilski GP, R Staals RH, Fineran PC (2016) Quorum sensing controls adaptive immunity through the regulation of multiple CRISPR-Cas systems. Mol Cell 64:1102–1108. https://doi.org/10.1016/j.molcel.2016.11.012

Pausch P, Muller-Esparza H, Gleditzsch D, Altegoer F, Randau L, Bange G (2017) Structural variation of Type I-F CRISPR RNA guided DNA surveillance. Mol Cell 67:622–632. https://doi.org/10.1016/j.molcel.2017.06.036

Puschnik AS, Majzoub K, Ooi YS, Carette JE (2017) A CRISPR toolbox to study virus-host interactions. Nat Rev Microbiol 15:351–364. https://doi.org/10.1038/nrmicro.2017.29

Qi LS, Larson MH, Gilber LA, Doudna JA, Weissman JS, Arkin AP, Lim WA (2013) Repurposing CRISPR as an RNA-guided platform for sequence-specific control of gene expression. Cell 152:1173–1183. https://doi.org/10.1016/j.cell.2013.02.022

Sampson TR, Saroj SD, Llewellyn AC, Tzeng YL, Weis DS (2013) A CRISPR/Cas system mediates bacterial innate immune evasion and virulence. Nature 497:254–257. https://doi.org/10.1038/nature12048

Semenova E, Severinov K (2016) Come together: CRISPR-Cas immunity senses the quorum. Mol Cell 64:1013–1015. https://doi.org/10.1016/j.molcel.2016.11.037

Shao C, Shang W, Yang Z, Sun Z, Li Y, Guo J, Wang X, Zou D, Wang S, Lei H, Cui Q, Yin Z, Li X, Wei X, Liu W, He X, Jiang Z, Du S, Liao X, Huang L, Wang Y, Yuan J (2012) LuxS-dependent AI-2 regulates versatile functions in *Enterococcus faecalis* V583. J Proteome Res 11:4465–4475. https://doi.org/10.1021/pr3002244

Song M (2017) The CRISPR/Cas9 system: their delivery, in vivo and ex vivo applications and clinical development by startups. Biotechnol Prog 33:1035–1043. https://doi.org/10.1002/btpr.2484

Stern A, Keren L, Wurtze O, Amitai G, Sorek R (2010) Self-targeting by CRISPR: gene regulation or autoimmunity? Trends Genet 26:335–340. https://doi.org/10.1016/j.tig.2010.05.008

Tan CH, Koh KS, Xie C, Zhang J, Tan XH, Lee GP, Zhou Y, Ng WJ, Rice SA, Kjelleberg S (2015) Community quorum sensing signalling and quenching: microbial granular biofilm assembly. NPJ Biofilms Microbiomes 1:15006. https://doi.org/10.1038/npjbiofilms.2015.6

Vuotto C, Longo F, Pascolini C, Donell G, Balice MP, Libori MF, Varaldo PE (2017) Biofilm formation and antibiotic resistance in *Klebsiella pneumoniae* urinary strains. J Appl Microbiol. https://doi.org/10.1111/jam.13533

Westra ER, Brouns SJ (2012) The rise and fall of CRISPRs – dynamics of spacer acquisition and loss. Mol Microbiol 85:1021–1025. https://doi.org/10.1111/j.1365-2958.2012.08170.x

Wright AV, Liu JJ, Knott GJ, Doxzen KW, Nogales E, Doudna JA (2017) Structures of the CRISPR genome integration complex. Science 357:1113–1118. https://doi.org/10.1126/science.aao0679

Yosef I, Goren MG, Qimron U (2012) Proteins and DNA elements essential for the CRISPR adaptation process in *Escherichia coli*. Nucleic Acids Res 40:5569–5576. https://doi.org/10.1093/nar/gks216

Yu D, Zhao L, Xue T, Sun B (2012) *Staphylococcus aureus* autoinducer-2 quorum sensing decreases biofilm formation in an icaR-dependent manner. BMC Microbiol 12:288. https://doi.org/10.1186/1471-2180-12-288

Zegans ME, Wagner JC, Cady KC, Murphy DM, Hammond JH, O'Toole GA (2009) Interaction between bacteriophage DMS3 and host CRISPR region inhibits group behaviors of *Pseudomonas aeruginosa*. J Bacteriol 191:210–219. https://doi.org/10.1128/JB.00797-08

Zuberi A, Misb L, Khan AU (2017) CRISPR interference (CRISPRi) inhibition of luxS gene expression in *E. coli*: an approach to inhibit biofilm. Front Cell Infect Microbiol 7:214. https://doi.org/10.3389/fcimb.2017.00214

Chapter 9
Developing Anti-virulence Chemotherapies by Exploiting the Diversity of Microbial Quorum Sensing Systems

Basit Yousuf, Keika Adachi, and Jiro Nakayama

Abstract Quorum sensing is a process of chemical communication that adjusts the genetic expression of certain important biological functions in a cell-density—dependent manner. Quorum sensing (QS) regulates important bacterial behaviors, such as production of virulence factors, formation of biofilm, and antibiotic resistance, via signaling molecules called autoinducers (AIs). Gram-negative bacteria typically use *N*-acyl homoserine lactones (AHLs) and *S*-adenosyl methionine (SAM) molecules as signaling molecules for communication with neighboring cells, while Gram-positive bacteria use oligopeptides for the same purpose. Autoinducer-2 (AI-2) is used as a signaling molecule by both Gram-negative and Gram-positive bacteria. These QS molecules are often involved in the expression of pathogenicity in different bacterial species. Because of the rapid emergence and dissemination of drug-resistant pathogens worldwide, antivirulence chemotherapies may be potential alternatives to the use of antibiotics, which kill bacteria but allow the emergence of resistance. Recently, successful strategies, employed for inhibition/manipulation of QS signaling, has brought real excitement for identifying novel advances to combat these life-threatening pathogens. This review highlights the QS systems used by Gram-negative and Gram-positive bacterial species and discusses promising QS inhibitor (QSI) molecules that may aid in designing novel antimicrobial therapeutics. Utilization of different QS components in the design and development of novel biotechnological products, such as biosensors, engineered microbial consortia, and anticancer molecules, is also addressed.

Keywords Quorum-sensing · Quorum-sensing inhibitor · Autoinducer · Autoinducing peptide · Anti-pathogenic chemotherapy · *N*-acylhomoserine lactone · AI-2

B. Yousuf · K. Adachi · J. Nakayama (✉)
Laboratory of Microbial Technology, Department of Bioscience and Biotechnology, Faculty of Agriculture, Graduate School, Kyushu University, Fukuoka, Japan
e-mail: nakayama@agr.kyushu-u.ac.jp

© Springer Nature Singapore Pte Ltd. 2018
V. C. Kalia (ed.), *Biotechnological Applications of Quorum Sensing Inhibitors*,
https://doi.org/10.1007/978-981-10-9026-4_9

Abbreviations

3OC12-HSL	*N*-3-oxo-dodecanoyl-homoserine lactone
3OC6-HSL	*N*-(3-oxohexanoyl)-*L*-homoserine lactone
3OH-C4HSL	*N*-((R)-3-hydroxybutanoyl)-*L*-homoserine lactone
4-quinolone	2-heptyl-3-hydroxy-4 (1H)-quinolone ACP: acyl carrier protein
AHL	N-acyl homoserine lactone
AI-2	AI-2
AI-3	AI-3
AIP	autoinducing peptides
AMR	antimicrobial resistance
C8-CPA	*N*-octanoyl cyclopentylamide
C8-HSL	*N*-octanoyl-*L*-homoserine lactone
CAI-1	*V. cholerae* AI-1
Cn-CPA	*N*-acyl cyclopentylamide
CSP	competence stimulating peptide
DPD	4,5-dihydroxy-2,3-pentanedione
HAI-1	*V. harveyi* AI-1
HHQ	2-heptyl-4-quinolone
HTS	high-throughput screening
IC_{50}	half maximal (50%) inhibitory concentration
LTTR	LysR-type transcriptional regulator
mBTL	meta-bromo-thiolactone
MDR	multidrug resistant
MRSA	methicillin-resistant *Staphylococcus aureus*
mvfR	multiple virulence factor regulator
PHL	phenylacetanoyl homoserine lactone
PQS	*Pseudomonas* quinolone signal
QQ	quorum quenching
QS	quorum sensing
QSI	quorum sensing inhibitor
RAP	RNA-III activating peptide
RIP	RNA-III inhibiting peptide
R-THMF	(2R,4S)-2-methyl-2,3,3,4- tetrahydroxytetrahydrofuran
SAH	S-adenosylhomocysteine
SAM	S-adenosyl methionine
SAR	structure activity relationship
SRH	S-ribosylhomosysteine
S-THMF-borate	(2S, 4S)-2-methyl-2,3,3,4-tetrahydroxytetrahydrofuran-borate
TCS	two-component regulatory system
TEP	thienopyridine
trAIP	truncated derivative of AIP
TRAP	target of RNA-III activating peptide
TZD-C8	(z)-5-octylidenethiazolidine-2,4-dione

VISA   vancomycin-intermediate and resistant *Staphylococcus aureus*
VRE   vancomycin resistant enterococci
WHO   world health organization

9.1 Introduction

In February 2017, the World Health Organization (WHO) released a list of multi-drug resistant (MDR) "priority pathogens" that pose the greatest threat to healthcare patients. The pathogens were categorized into three groups: critical, high, and medium priority, according to the urgency of need for new antibiotics. The top of the critical group includes *Acinetobacter*, *Pseudomonas,* and various enterobacteriaceae such as *Klebsiella* and *Escherichia coli*. These bacteria cause severe and deadly infections and are becoming resistant to best available antibiotics such as carbapenems and third-generation cephalosporins. The high and medium priority groups include vancomycin-resistant enterococci (VRE), methicillin-resistant *Staphylococcus aureus* (MRSA), vancomycin-intermediate and resistant *Staphylococcus aureus* (VISA), fluoroquinolone-resistant *Salmonella*, and penicillin-non-susceptible *Streptococcus pneumoniae*. The WHO is urging global and immediate action to tackle multi-drug resistant pathogens. If no action is taken by 2050, antimicrobial resistance (AMR) will kill more people each year than cancer, with ten million estimated annual deaths at a cost of $100 trillion to the global economy (O'Neill 2014). Despite great advancements in the developments of new antibiotics, drug-resistant pathogens remain a pressing issue. The use of antivirulence drugs is a potential approach for controlling MDR pathogens (Dickey et al. 2017).

Although bacteria are unicellular organisms, bacterial cells often communicate with each other to behave and survive as a group. To orchestrate this group behavior, bacteria have evolved autocrine signaling to communicate with other cells, and established the so-called quorum sensing (QS), which is a cell-density—dependent regulatory system (Bassler and Losick 2006; Doğaner et al. 2016). Cell-to-cell communication in a QS system is usually mediated by small signaling molecules called autoinducers (AIs). After reaching a particular threshold level, AIs activate their receptors and trigger a signal transduction cascade that eventually results in changes in expression of target genes (Ng and Bassler 2009; Boyer and Wisniewski-Dyé 2009). QS often plays important roles in pathogenic and symbiotic bacteria and their interactions with the host (Boyer and Wisniewski-Dyé 2009). QS is responsible for the phenotypic changes in symbiotic bacteria, such as bioluminescence and root nodulation (Bassler et al. 1994; Sanchez-Contreras et al. 2007), as well as biofilm formation and tolerance to antimicrobial agents, which are the hallmarks of pathogenic bacteria (Davies et al. 1998; Bjarnsholt et al. 2005). Bacteria also use QS to regulate important cellular processes necessary for survival, such as adaptation and concurrent competition in the surrounding environment (Bassler and Vogel

2013; Shanker and Federle 2017). Furthermore, QS plays a critical role in antibiotic resistance, which occurs via overuse of antimicrobial agents, and the spread of which is facilitated by horizontal gene transfer (Rahmati et al. 2002; Schroeder et al. 2017). The overuse of antibiotics has generated a selective pressure for drug-resistant mutations and stimulated bacterial evolution (Rodríguez-Rojas et al. 2013). Owing to accumulation of different resistance genes on plasmids or transposons, many bacteria have developed resistance toward numerous antimicrobial agents; these bacteria are called MDR pathogens (Nikaido 2009). Theoretically, anti-QS strategy does not target cell growth and proliferation and does not induce selective pressure for drug-resistant strains. Therefore, it is considered an alternative, or additive, to antibiotic chemotherapies, particular for infections with MDR pathogens (Hentzer and Givskov 2003; Rasmussen and Givskov 2006; Bhardwaj et al. 2013; O'Loughlin et al. 2013; Allen et al. 2014; Khan et al. 2015; Todd et al. 2017).

Numerous studies have addressed the design and synthesis of small QS inhibitor (QSI) molecules capable of modulating different QS systems. Some researchers emphasize novel strategies for design and discovery of QSIs, which can render these inhibitors potential targeted drugs (Scutera et al. 2014). In recent years, this research area has demonstrated substantial achievements in developing potent anti-microbial agents as attractive alternative for controlling MDR pathogens (Rasmussen and Givskov 2006; Ni et al. 2009; Bhardwaj et al. 2013; Guo et al. 2013; Allen et al. 2014; Dickey et al. 2017). Manipulation of QS, to prevent progression of serious bacterial infections and towards development of future medicines has also been reviewed (González and Keshavan 2006; Galloway et al. 2010; LaSarre and Federle 2013; Bassler 2015; Reuter et al. 2016). The discovery of novel antagonists, with robust and potent activity against QS systems, will help develop broad-spectrum antivirulence agents as alternatives to antibiotics.

This review discusses the current knowledge of the major QS systems, and structure and mechanism of action of inhibitors of MDR pathogens, which is important for advancing the rational design of novel and robust anti- or pro-QS agents. This review will be helpful for further elucidating mechanisms of QS signaling systems and will be cornerstone toward development of potential QSIs, and their importance in regulation of MDR infections, and in developing potential strategies to avoid resistance acquisition.

9.2   Molecular Mechanism of Cell-Cell Communication in Bacteria

Bacteria talk to each other through QS, which involves production, release, and detection of AIs for the orchestration of bacterial behaviors in cell-density—dependent manner. As bacterial density increases, the concentration of AIs also increases and accumulates in the surroundings. The structure, size, and nature of AIs vary greatly across different bacterial species. Bacteria perceive the AI concentration and

Table 9.1   Major QS pathways in bacteria

Bacteria	Pathway	Signal molecules
Gram-negative	AI-1 pathway	Various AHLs
	4QS pathway	PQS and HHL
	AI-3 pathway	AI-3 (unknown structure)
	CAI-1pathway	Hydroxyketones
Gram-negative/positive	AI-2 pathway	R-THMF/S-THMF-borate/unknown
Gram-positive	AIP-TCS pathway	Various oligopeptides
	RNPP/RGG pathway	

respond by amending bacterial behaviors via altering the global pattern of gene expression. Numerous studies have shown the use of at least six QS pathways (Table 9.1), which concomitantly control various phenotypes in numerous bacterial species (Bassler and Losick 2006; Ng and Bassler 2009). Although the QS systems of Gram-negative and Gram-positive bacteria have distinctive differences in regulatory elements and molecular mechanisms, there are several common considerations: (1) bacteria produce AI signaling molecules, which are detectable only after their concentration crosses a certain threshold (2) AIs are recognized by specific cytoplasmic or membrane receptors; and (3) their interaction activates gene expression and the feed-forward autoinduction loop (Miller and Bassler 2001). Here, we provide the detailed overview of the typical communication systems prevalent in the diverse bacterial species of Gram-negative and Gram-positive pathogens.

9.2.1   In Gram-Negative Bacteria

9.2.1.1   Diverse Autoinducers Used by Gram-Negative Bacteria

Numerous diverse compounds are used as AIs by Gram-negative bacteria (Fig. 9.1 and Table 9.2). AHLs are the predominant class of AIs. AHL-mediated QS was first identified in *Vibrio fischeri*, a bioluminescent marine bacterium. AHLs consist of a core homoserine lactone ring and an acyl chain of 4–18 carbon atoms. The acyl chain is derived from fatty acids and occasionally from non-fatty acids such as *p*-coumarate (Schaefer et al. 2008). The acyl chain may have modifications and its length can affect stability, hydrophobicity, and consequently, signaling dynamics (von Bodman et al. 2008; Galloway et al. 2010). The acyl chain has different degrees of saturation and modifications via hydroxyl and carbonyl groups (Fuqua et al. 2001; Whitehead et al. 2001). The different AHL AIs are synthesized by a cognate synthase from LuxI family, which catalyzes the reaction between the SAM of the activated methyl cycle and an acyl carrier protein (Val and Cronan 1998; Ng and Bassler 2009). Gram-negative bacteria also use α-hydroxyketones and their derivatives, alkylquinolones (PQS), and thiazole compounds (IQS) as AIs (Fig. 9.2) (Lee et al. 2013; Papenfort and Bassler 2016). Additionally, furanose derivatives, the

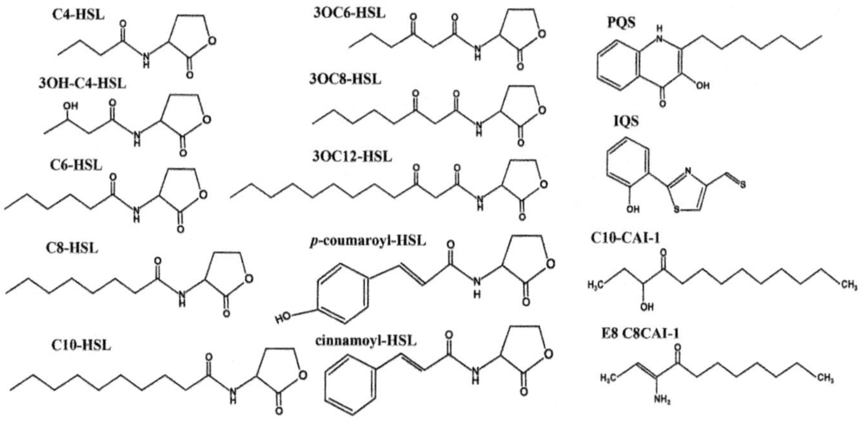

Fig. 9.1 Structure of known AIs used by Gram-negative bacteria

Table 9.2 Representative bacterial QS systems

Autoinducer	Autoinducer synthases	Receptors	Bacteria
3OC6-HSL	LuxI	LuxR	*Vibrio. fischeri*
C8-HSL	AinS	AinR	*V. fischeri*
3OH C4-HSL	LuxM	LuxN	*Vibrio harveyi*
3OC12-HSL	LasI	LasR	*Pseudomonas aeruginosa*
C4-HSL	Rh1I	Rh1R	*P. aeruginosa*
3OC8-HSL	TraI	TraR	*Agrobacterium tumefaciens*
C6-HSL	CviI	CviR	*Chromobacterium violaceum*
Isovaleryl-HSL	BjaI	BjaR	*Bradyrhizobium japonicum*
Cinnamoyl-HSL	BraI	BraR	*Bradyrhizobium spp.*
p-coumaroyl-HSL	RpaI	RpaR	*Rhodopseudomonas palustris*
S-THMF-borate	LuxS	LuxP	*Vibrio* spp.
R-THMF	LuxS	LsrB	*Salmonella*
Ea-C8-CAI-1	CqsA	CqsS	*V. harveyi*
C10-CAI-1	CqsA	CqsS	*Vibrio cholerae*
PQS	PqsABCDH	PqsS	*P. aeruginosa*
IQS	ambBCDE	Unknown	*P. aeruginosa*
AI-3	Unknown	QseEC/ QseEF	*E. coli* and *Salmonella*
AIP (thiolactone/ lactone)	AgrB	AgrC	*Staphylococcus* spp.
AIP (thiolactone)	AgrB	VirS	*Clostridium perfringens*
GBAP (lactone)	FsrA	FsrC	*Enterococcus faecalis*
ComX (linear)	ComC	ComD	*Streptococcus*
CSP (linear)	ComS	ComR	*Streptococcus*

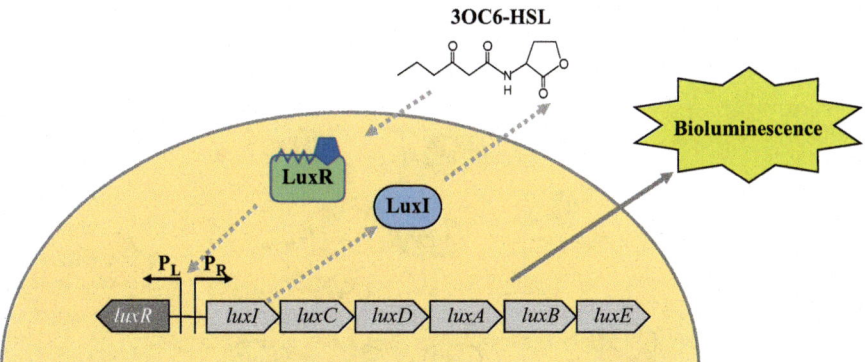

Fig. 9.2 The LuxI/LuxR QS system in *Vibrio fischeri*. 3OC6-HSL is synthesized by LuxI and secreted as AI outside cells. When the concentration of 3OC6-HSL reaches a threshold level, 3OC6-HSL promotes the expression of the luciferase operon, luxICDABE, by binding to the cytoplasmic transcription factor, LuxR

so-called "AI-2," are commonly used as AIs by Gram-negative and Gram-positive bacteria (Galloway et al. 2010).

9.2.1.2 Signaling Intricacies in *Vibrio* spp.

Vibrio fischeri uses the LuxI/LuxR QS system, in which *N*-(3-oxohexanoyl)-homoserine lactone (3OC6-HSL) is synthesized by LuxI and is thereafter perceived by its receptor LuxR; LuxR receptors are cytoplasmic transcription factors (Fig. 9.2) (Schaefer et al. 1996). Similarly, *V. fischeri* also uses the AinS/R type QS system, in which AinS synthase catalyzes the reaction to produce the signal *N*-octanoyl-L-homoserine lactone (C8-HSL) (Hanzelka et al. 1999). C8-HSL interacts with its cognate receptor kinase AinR to initiate a phosphorylation signaling cascade and affect bacterial behavior accordingly (Pérez et al. 2011). Once the AHL-receptor binding complex is formed, it recognizes the binding sequence on the *luxICDABE* operon and activates the expression of luciferase, concomitantly synthesizing AHLs via positive feedback loop (Meighen 1991).

In *V. harveyi*, three types of AIs, namely HAI-1 (harveyi AI-1, 3OH-C4-HSL), AI-2 (*S*-THMF-borate), and CAI-1 (Ea-C8-CAI-1, (Z)-3-aminoundec-2-en-4-one) are used for inter- and intra-species, and intra-genera, communication, consequently controlling approximately 600 genes (Fig. 9.3) (van Kessel et al. 2013; Papenfort and Bassler 2016). These AIs are not recognized by the commonly used LuxIR-type systems of Gram-negative bacteria. The harveyi AI-1 (HAI-1) synthesized by LuxM synthase, is a *N*-((R)-3-hydroxybutanoyl)-L-homoserine lactone (3OH-C4-HSL) (Fig. 9.1) (Bassler et al. 1994; Ke et al. 2015). The LuxN is a membrane histidine kinase that functions as the receptor for HAI-1 (Freeman et al. 2000). LuxM is not the homolog of LuxI, although it uses the same substrates as those used by LuxI viz. SAM and fatty-acid intermediates (Hanzelka and Greenberg 1996). The second AI

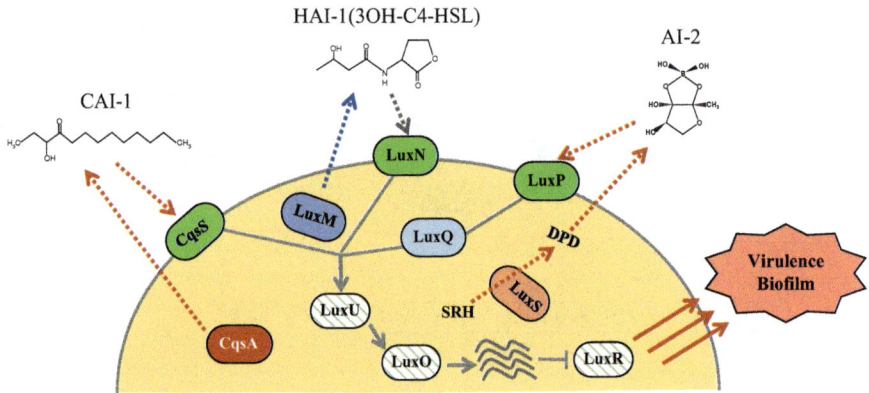

Fig. 9.3 LuxM/LuxN, CqsA/CqsS, and LuxS/LuxPQ QS systems in *V. harveyi*. *V. harveyi* uses three types of AIs, namely HAI-1, CAI-1, and AI-2. These AIs are synthesized by their respective synthases, LuxM, CqsA, and LuxS, perceived by the corresponding sensor kinases, LuxN, CsqS, and LuxP. When bound to corresponding AIs at high cell density, LuxN, CqsS, and, LuxPQ acts as phosphatases; then, LuxO is dephosphorylated and halts the expression of small regulatory RNAs (Orrs), leading to the activation of LuxR expression. LuxR directly or indirectly controls the expression of numerous genes associated with phenotypes

molecule, used by *V. harveyi,* is (Z)-3-aminoundec-2-en-4-one, which is closely related to (*S*)-3-hydroxytriecan-4-one, was identified as AI in *Vibrio cholerae* and designated as cholerae AI-1 (CAI-1 or C10-CAI-1) (Higgins et al. 2007). The CAI-1s of these two species are detected by cognate CqsS receptors and implicated in intra-species QS communication (Higgins et al. 2007; Ng et al. 2011). The LuxM/LuxN and CqsA/CqsS systems are specific to *V. harveyi* and *V. cholerae*, respectively. The AI-2 molecule, (2S,4S)-2-methyl-2,3,3,4-tetrahydroxytetrahydrofuran-borate (S-THMF-borate), is synthesized by LuxS synthase. This is a very active AI in the *Vibrio* spp. and is detected by LuxPQ, which regulates interspecies QS signaling (Surette et al. 1999; Schauder et al. 2001).

These AIs are produced throughout the bacterial growth, but at the basal level; therefore, they do not bind to the receptor protein, which is unstable and degrades rapidly (Zhu and Winans 2001; Henke and Bassler 2004). At this low AI-2 level, LuxN, CqsS, and LuxPQ autophosphorylates their conserved residues and generate a phosphorylation cascade, which subsequently phosphorylates the DNA binding response element, LuxO (Freeman and Bassler 1999; Henke and Bassler 2004). Phosphorylated LuxO promotes the expression of genes encoding small regulatory RNAs, which bind to and block the translation of LuxR mRNA (Lenz et al. 2004). Thus, at the low cell density status, the LuxR protein is not able to activate transcription of luciferase and other target genes (Freeman and Bassler 1999; Waters and Bassler 2006). When bound to corresponding AIs at high cell density, LuxN, CqsS, and LuxPQ act as phosphatases; then, LuxO is dephosphorylated and halts the expression of small regulatory RNAs (Orrs), leading to the activation of LuxR expression (Henke and Bassler 2004; Freeman et al. 2000). LuxR directly or

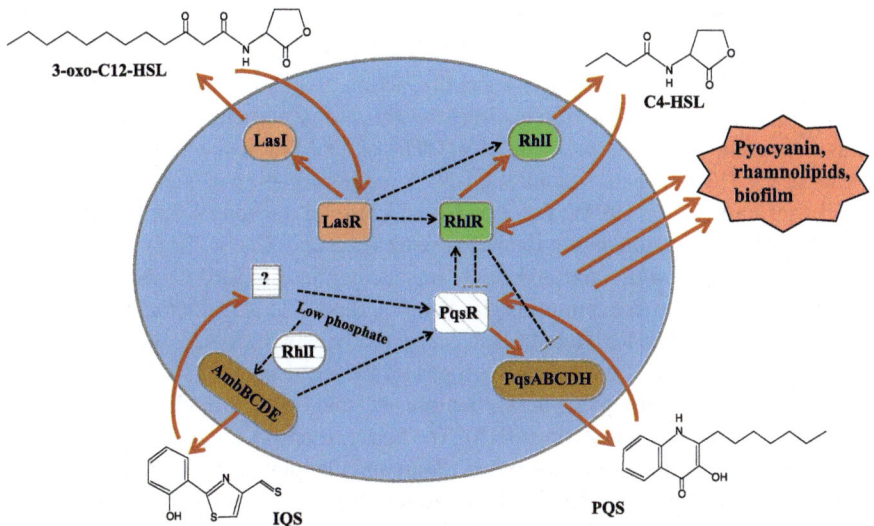

Fig. 9.4 LasI/R, RhlI/R, and Pqs QS systems in *P. aeruginosa*. Four AIs, namely 3OC12-HSL, C4-HSL, PQS, and IQS, were synthesized by their respective synthases, LasI, RhlI, PqsABCDH, and AmbBCDE, and were then perceived by LasR, RhlR, PqsR, and unknown receptors, respectively. These distinct pathways operate in an interconnected manner, thereby inducing a large set of virulence genes in a cell-density—dependent manner

indirectly controls the expression of numerous genes associated with phenotypes. Then, the expression of target genes is activated, and AIs are autoinduced via the feed-forward loop (Ng and Bassler 2009; Papenfort and Bassler 2016). This signaling cascade triggers or blocks the expression of hundreds of genes, thereby regulating diverse physiological functions particularly controlling different bacterial behaviors such as bioluminescence, biofilm formation, and virulence (González and Keshavan 2006; Rutherford and Bassler 2012; LaSarre and Federle 2013).

9.2.1.3 QS Systems Closely Knitted in *Pseudomonas* Signaling Architecture

The cell-cell communication in *Pseudomonas aeruginosa* is regulated by two AHL-mediated QS systems, namely the LasR-LasI and RhlR-RhlI systems, and two non-AHL-based systems, namely the PQS and IQS systems (Fig. 9.4) (Williams 2007; Lee and Zhang 2015; Papenfort and Bassler 2016). The LasR-LasI and RhlR-RhlI systems are mediated by 3-oxo-dodecanoyl-homoserine lactone (3OC12-HSL) and *N*-butanoyl-*L*-homoserine lactone butanoyl (C4-HSL), respectively, and are intimately connected to each other via the signal transduction pathway (Whitehead et al. 2001; Fuqua and Greenberg 2002). These two QS pathways control various bacterial behaviors, including virulence and biofilm formation, by regulating the expression of hundreds of genes (De Kievit et al. 2001; Smith and Iglewski 2003; Rutherford and Bassler 2012).

In the LasR-LasI QS system, LasI acts as synthase toward 3OC12-HSL, which parallels the action of LuxI in the *Vibrio* species (Pearson et al. 1994). The *lasR* gene encodes AI-dependent transcription factors, acting as a global regulator of virulence genes (*lasB, aprA, lasA,* and *toxA*) in *P. aeruginosa* (Gambello and Iglewski 1991; Passador et al. 1993; Pearson et al. 1994). The other AI molecule used by *P. aeruginosa* is C4-HSL, which is controlled by a regulatory protein identified as RhlR (Ochsner and Reiser 1995). The LasB/LasI complex promotes the expression of RhlI/RhlR systems, indicating their close link (Lee and Zhang 2015). RhlR, which shares similarities with LuxI and LasI, acts as synthase toward C4-HSL (Ochsner and Reiser 1995). At concentrations higher than threshold, AIs interact with the corresponding receptors, Rh1R and LasR, respectively, which triggers the activation of receptor proteins and, consequently triggering the expression of several genes such as those for elastase, pyocyanin, hemolysin, and rhamnolipids (Schuster and Greenberg 2007; Lee and Zhang 2015). The *lasB* promotors are activated only when both RhlI and RhlR are fully expressed (Brint and Ohman 1995; Ochsner and Reiser 1995). The positive and negative regulators of Las and Rhl QS systems, such as QscR and VqsR, the homologs of LuxR, have been identified. QscR and VqsR forms heterodimers with LasR/3OC12-HSL and RhlR/C4-HSL, respectively, and prevent their binding with the promoter, thereby suppressing the LasI/LasR, RhlI/RhlR signal transduction cascade (Chugani et al. 2001; Ledgham et al. 2003).

The third type of QS system, which is integrated with the LasI/LasR and RhlI/RhlR signal transduction pathways, employs a quinolone molecule, 4-hydroxy-2-alkylquinolines (HAQs), as the AI (Fig. 9.4) (Deziel et al. 2004). This is designated as the *Pseudomonas* quinolone signaling (PQS) pathway, and the AI can strongly induce the expression LasB in the LasR mutants of *P. aeruginosa* (Pesci et al. 1999; Deziel et al. 2004). mvfR/PqsR functions as the receptor for PQS and regulates the production of PQS (Wade et al. 2005; Xiao et al. 2006a, b; Williams and Cámara 2009). mvfR, a LysR-type transcriptional regulator (LTTR), directs the synthesis of low molecular weight HAQ molecules, such as 4-hydroxy-2-heptylquinoline (HHQ), 3,4-dihydroxy-2-heptylquinoline (PQS), and the non-HAQ, 2-aminoacetophenone (2-AA) (Deziel et al. 2004; Xiao et al. 2006a, 2006b). PqsA, an anthranilate-coenzyme A ligase, conjugates anthranilate with β-keto dodecano-ate, in conjunction with an acyl carrier protein synthase complex, PqsBCD, and synthesizes HHQ (Deziel et al. 2004; Dulcey et al. 2013). Then, the monooxygen-ase, PqsH, converts HHQ into PQS (Dubern and Diggle 2008; Schertzer et al. 2010). PqsE is involved in the production of HHQ (Drees and Fetzner 2015). The expression of PqsH is activated by LasR-3OC12HSL, indicating a close link between the LasR-LasI and Pqs systems (Schertzer et al. 2010; Liu et al. 2015). PQS controls the production of multiple virulence factors, including elastase, rhamnolipids, galacto-philic lectin, pyocyanin, exotoxin A, and formation of biofilm (Dubern and Diggle 2008; Jimenez et al. 2012; Lee and Zhang 2015; de Almeida et al. 2017).

The IQS pathway uses a structurally different signaling molecule, 2-(2-hydroxyphenyl)-thiazole-4-carbaldehyde, which is synthesized by the *ambBCDE* gene cluster (Lee and Zhang 2015). Attenuation of the IQS pathway hinders the production of PQS and C4-HSL as well as various virulence factors

(Lee and Zhang 2015). This signaling molecule plays a critical role in pathogenesis, as indicated by mutational studies, in which mutants restored their phenotypes upon addition of IQS (Lee et al. 2013). Interestingly, the IQS pathway controls the functions of the LasI system during phosphate depletion stress, indicating a close link between stress and IQS QS (Lee et al. 2013; Lee and Zhang 2015). Moreover, the occurrence of mutated LasI/LasR, in *P. aeruginosa* clinical isolates, may be explained by the IQS pathway (Hoffman et al. 2009; Lee and Zhang 2015).

The antibiotic and anticancer activities of quinolones have also demonstrated the multi-functionality of these AIs (Heeb et al. 2011). The different QS systems are interconnected and integrated in a complex way to ensure robust cell-cell communication under diverse conditions. RhlR is key to controlling and containing the expression of virulence genes in *P. aeruginosa* because RhII can be activated by either LasR or PqsR together with RhlR (Jimenez et al. 2012; Papenfort and Bassler 2016). Additionally, PQS and HHQ activate PqsR and indirectly induce pathogenicity. Therefore, LasR, PqsR, and RhlR may be key targets when we search for the QSIs of *P. aeruginosa*.

9.2.1.4 QS in *Acinetobacter* spp.

The AHL-based QS has been recently reported in the *Acinetobacter* spp., an emerging pathogen designated as "Gram-negative MRSA" because of its prevalence and rapid development of resistance similar to *S. aureus* (Taccone et al. 2006; Valencia et al. 2009; Stacy et al. 2012). This bacterium can transfer various antibiotic resistance determinants to other species regulated by QS, which renders it a great challenge to clinical microbiologists (Devaud et al. 1982). *Acinetobacter* regulates biofilm formation and surface motility by using the AbaI/AbaR system, a homolog of the typical LuxR/LuxI QS (Niu et al. 2008; Kang and Park 2010; Bhargava et al. 2015). This involves the AbaR receptor protein, which recognizes the 3-OH-C12 HSL synthesized by AbaI, an AI synthase (Niu et al. 2008; Tomaras et al. 2008). *Acinetobacter* produces five different QS signaling molecules that vary considerably across the *Acinetobacter* species; this poses difficulties in recognizing the virulent and non-virulent strains based on the type of AIs (González et al. 2001).

The homolog of the AbaIR QS system, referred to as AnoIR, has recently been characterized in *Acinetobacter noscomialis* (Oh and Choi 2015). This homolog shows 94% similarity to AbaIR in its amino acid sequence. *Acinetobacter* with an AnoI mutation cannot synthesize 3-OH-C12 HSL, clearly indicating the critical role of AnoI in 3-OH-C12 HSL synthesis. Furthermore, AnoR regulates activity of AnoI, consequently controlling biofilm formation and surface motility (Oh and Choi 2015).

Recently, it has been reported that 3-OH-C12 HSL, produced by *A. baumannii*, can regulate the expression of drug-resistance genes (Dou et al. 2017). Therefore, disrupting QS signaling, referred to as quorum quenching (QQ), can be used for containment of multidrug-resistant *Acinetobacter* infections (Stacy et al. 2012; Dou et al. 2017).

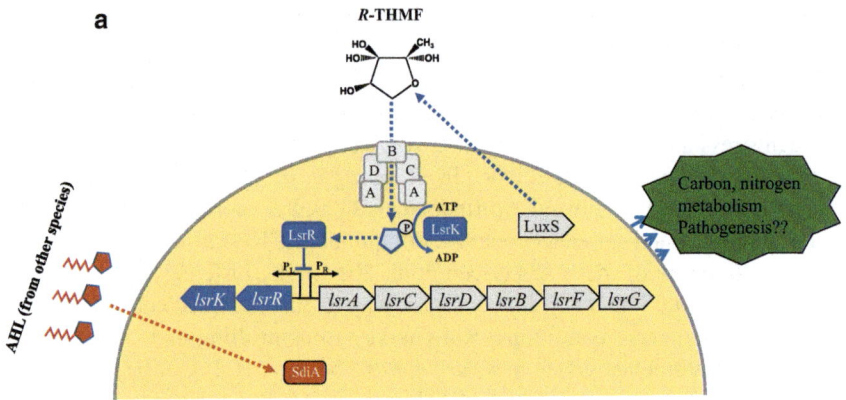

Fig. 9.5a The AI-2/LsrR QS system in *Salmonella typhimurium*. AI-2 is synthesized via a key reaction mediated by LuxS. AI-2, accumulated outside cells, is imported into the cytoplasm by the ABC transporter of LsrABCD, then phosphorylated by LsrK, after which it activates the transcriptional regulator LsrR

9.2.1.5 Non-AHL-Based Communication Network in *E. coli* and *Salmonella*

QS in *E. coli* and *Salmonella* is complex and includes three different signal transduction systems: LuxS/AI-2, SdiA, and AI-3 (Figs. 9.5a and 9.5b) (Ahmer 2004; Walters and Sperandio 2006). These bacteria do not have a gene corresponding to *luxI*, indicating no AHL production (Michael et al. 2001). However, LuxR homolog, SdiA, can sense the diverse AHLs produced by other bacterial species, suggesting that SdiA plays a crucial role in interspecies communication (Michael et al. 2001). Exogenous AIs are perceived by SdiA; the AHLs containing six to eight carbon acyl chains, such as 3O-AHLs, oxoC6, oxoC8, and oxoC10, at 1 to 50 nM concentrations consequently leads to the activation of downstream cascade (Kendall and Sperandio 2014). SdiA does not play any roles in the virulence of *Salmonella* in the mouse, chicken, or bovine models of disease (Ahmer 2004). The other signaling system, which facilitates the degradation of amino acids to pyruvate or succinate in these bacteria, is mediated by indole as the bacterial signal (Wang et al. 2001). It is speculated that indole signaling may play a role in bacterial adaptation in a nutrient-poor environment. Indole signaling, along with SdiA, has been reported to regulate biofilm development and motility (Lee et al. 2007, 2009).

In *Salmonella* and *E. coli*, AI-2 activates the expression of *lsr* operon, consisting of seven genes, *lsrACDBFGE*, which encodes the ABC transporter complex (LsrACDB) to import the AI-2 molecule and LsrEG to process the imported AI-2 (Fig. 9.5a). The primary role of AI-2 in these bacteria has been suggested to be in carbon and nitrogen metabolism, however its role in pathogenesis is not certain and needs to be further investigated (Kendall and Sperandio 2014).

The third type of signaling, used by enterohemorrhagic *E. coli* (EHEC), is mediated by a two-component regulatory system (TCS), QseC/QseB, which senses an autoinducer with unknown structure (AI-3) and mammalian hormones, epinephrine

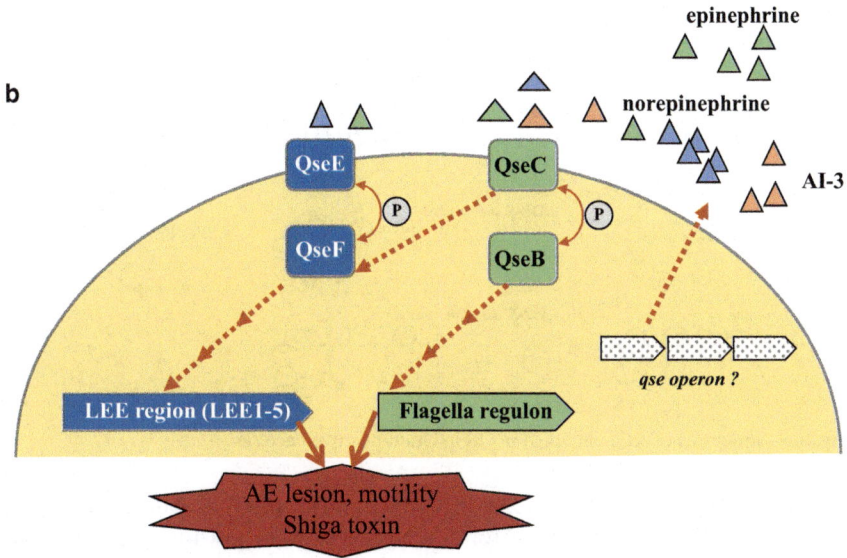

Fig. 9.5b The QseC/QseB and QseE/QseF QS systems of enterohemorrhagic *E. coli* (EHEC). Histidine kinase QseC senses mammalian hormones, epinephrine and norepinephrine, as well as a structure-unknown autoinducer (AI-3). Then, QseC activates QseB via phosphorylation. Through this signal transduction cascade, the flagellum regulon is induced. Another two-component system (TCS), QseE/QseF, also senses these mammalian hormones, but not AI-3, and eventually promotes the *LEE* gene cluster involved attaching-effacing (AE) lesion. These TCSs cross-talk with each other and ultimately tune the virulence of EHEC

and norepinephrine; this indicates communication between host and bacteria (Fig. 9.5b) (Sperandio et al. 1999, 2002). In this signal transduction pathway, the activated histidine kinase, QseC, phosphorylates the transcription factor QseB and eventually promotes the expression of the flagellum regulon (Fig. 9.5b) (Qin et al. 2002; Sperandio et al. 2002; Clarke et al. 2006; Bearson and Bearson 2008). Another TCS, QseE/QseF, also senses these mammalian hormones, but not AI-3, and eventually promotes the *LEE* gene cluster involved attaching-effacing (AE) lesion. These TCSs cross-talk with each other and ultimately tune the virulence of EHEC.

9.2.2 In Gram-Positive Bacteria

9.2.2.1 Variations in the Mode of Cell-Cell Communication and Its Signaling in Gram-Positive Bacteria

Gram-positive bacteria often use oligopeptides as communication signaling molecules. These signaling molecules include autoinducing peptides (AIPs), which are involved with intra-strain communication, as well as autocrine signals and peptides, which mediate inter-strain communication. These peptide molecules vary in

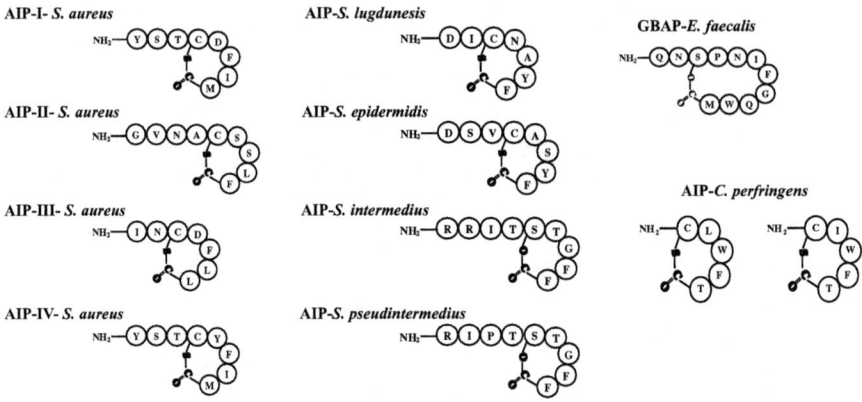

Fig. 9.6 Structures of AIPs involved in QS in Gram-positive bacteria

structure and size, ranging from 5 to 34 amino acids in linear or cyclic form (Fig. 9.6) (Mayville et al. 1999). The cyclic forms are generated via post-translational modification with thiolactone, lactone, lanthionine, cysteine, and geranyl groups (Nakayama et al. 2001; Ansaldi et al. 2002). Based on the structure of signal peptide molecules and the mode of signal transduction, Gram-positive signaling pathways can be classified into three types.

The first is the Agr-type QS system, mediated by thiolactone/lactone AIP. This was originally discovered in *S. aureus* and subsequently found in a wide-range of low-GC Gram-positive bacteria, such as enterococci, clostridia, and listeria as well as staphylococci (Wuster and Babu 2008; Gray et al. 2013). Biosynthesis and perception of AIPs are processed through the functions encoded by cognate *agr*-like gene clusters; AIPs are generated from the AgrD propeptide by the processing enzyme, AgrB, and are the transduced by a TCS consisting of histidine kinase receptor, AgrC, and the response regulator AgrA (Novick and Geisinger 2008; Le and Otto 2015).

The second type is often found in QS involved in the production of Class I or II bacteriocins. The AIPs in this QS system are larger in size compared with that of other types of AIPs. In Class I lantibiotic bacteriocins, bacteriocins themselves often act as AIPs such as nisin and subtilin; while Class II bacteriocins are often under the control of AIPs that are similar in structure but different in size and sequence from, bacteriocins (Kleerebezem et al. 1997). Similar to the generation of Class II bacteriocins, their AIPs are generated by cleavage from precursor peptides at the C-terminal side of double-glycine residues. In streptococci, this type of QS system is used to regulate genetic competence and bacteriocin production. AIPs in this type of QS are transduced through the cognate TCS consisting of histidine kinase receptor and response regulator, each belonging to the same protein family as that of the Agr system.

The third type is mediated by small oligopeptides and their cytosolic receptors. Enterococcal sex pheromones, which activate bacterial conjugation, as well as oligopeptide signals in bacilli, involved in the development of genetic competence, sporulation, virulence, and biofilm formation, have been extensively studied. These signal peptides are internalized into cells and directly bound to the cytosolic receptors belonging to the RNPP family represented by Rap, NprR, PlcR, and PrgX (Declerck et al. 2007). The Rgg-like regulatory family also represents this group as a global transcriptional regulator of metabolism and virulence in streptococci (Cook and Federle 2014).

9.2.2.2 Agr QS System in *Staphylococcus*

The *Staphylococcus* genera include several potent pathogens known to cause a wide array of infections such as boils, pneumonia, toxic shock, endocarditis, and osteomyelitis (Chan et al. 2004). These infections are mediated by a number of virulence factors such as exotoxins, proteases, lipases, and superantigens (Gordon et al. 2013). The Agr QS system plays a central role in *S. aureus*, modulating the expression of several hundreds of genes and consequently controlling virulence and biofilm formation (Fig. 9.7) (Gordon et al. 2013). The Agr system positively regulates the expression of a series of toxins, such as α- and β-haemolysins and enterotoxins, thus promoting virulence (Kong et al. 2006). However, simultaneously, it reduces the expression of several surface adhesins, thereby hampering the formation of biofilm (Kong et al. 2006). Some extracellular proteases, which are critical for disrupting biofilms, are controlled by the Agr QS system (Boles and Horswill 2008).

The Agr QS system consists of four gene products, AgrB, AgrD, AgrC, and AgrA, encoded by one operon (*agrBDCA*). AIP is translated as a prepeptide from *agrD* and secreted as a mature form with the aid of transmembrane endopeptidase,

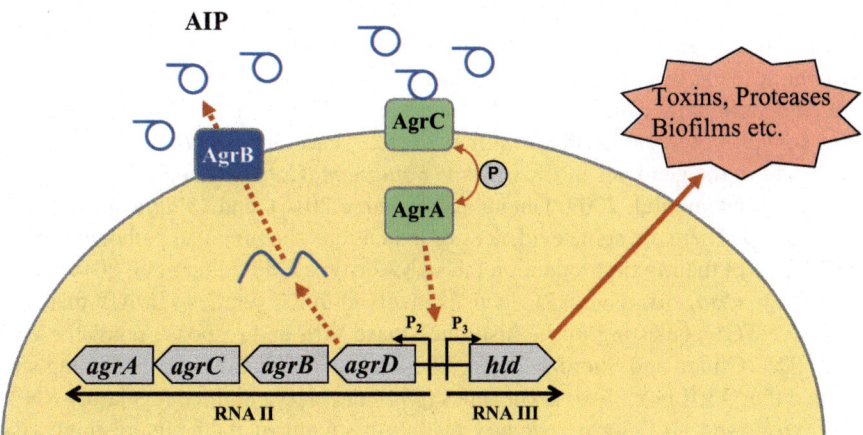

Fig. 9.7 Agr QS system in *Staphylococcus aureus*. Thiolactone/lactone AIP is translated from agrD, processed by AgrB, and then secreted outside cells. A TCS, consisting of histidine kinase AgrC and response regulator AgrB, senses the AIP concentration and promotes the expression of two transcripts, RNAII and RNAIII. RNAIII functions as a regulatory RNA as well as mRNA of delta hemolysin. A series of virulence genes are controlled by the RNAIII molecule

encoded by *agrB* (Zhang et al. 2002). The TCS, composed of membrane histidine kinase and cytoplasmic response regulator, are respectively encoded by *agrC* and *agrA*, each responsible for signal detection and signal transduction to modulate the expression of target genes (Fig. 9.7) (Lina et al. 1998). The *agr* circuit is operated by two intergenic divergent promoters, P2 and P3. The activated AgrA directly promotes the transcription of RNAII, encoding *agrBDCA* operon, and RNAIII, encoding delta hemolysin, via P2 and P3, respectively (Fig. 9.7) (Koenig et al. 2004). In addition to the mRNA of delta hemolysin, RNAIII functions as a regulatory RNA, acting as an effector for the expression of downstream genes. The Agr-based QS system has been extensively studied in staphylococci, and homologous systems have been found in a wide range of low-GC Gram-positive bacteria (Kleerebezem et al. 1997; Wuster and Babu 2008; Gray et al. 2013).

Agr systems in *S. aureus* are classified into four agr groups, *agr* I-*agr* IV, because of polymorphisms within *agrB*, *agrD*, and *agrC* genes (Lina et al. 1998; Jarraud et al. 2011). Similarly, *Staphylococcus epidermidis* has three different agr groups (I – III) (Olson et al. 2014). Each of these groups has a specific AIP of 7–12 amino acids, consisting of macrocyclic thiolactones or lactones (Thoendel and Horswill 2009; Thoendel et al. 2011). The conserved central cysteine or serine covalently links to the C-terminal α-carboxylate to form a 5-amino acids membered thiolactone or lactone structure (Novick and Geisinger 2008). Structure-activity relationship (SAR) studies have revealed that only the intragroup AIP-AgrC interaction leads to the activation of the *agr* response, while cross-group interactions are inhibitory and can block activation by cognate AIPs (Lina et al. 1998; Yang et al. 2016).

9.2.2.3 Agr-Like QS Systems in Clostridia

The clostridia group includes several important pathogens such as *Clostridium botulinum, Clostridium perfringens, Clostridium difficile,* and *Clostridium tetani,* which occasionally cause life-threatening diseases (Carter et al. 2014). These species contain homologs of the *S. aureus agrBD* genes. The *agrBD* loci have been identified in the genomes of *Clostridium botulinum* (Cooksley et al. 2010), *C. perfringens* (Ohtani et al. 2009; Ohtani and Shimizu 2016), and *C. difficile* (Sebaihia et al. 2006). There are some evidences or indications showing that pathogenic clostridia control their toxin production through Agr-type QS (Carter et al. 2014).

In *C. perfringens,* the *agrBD* locus is involved in the synthesis of AIP that activates the TCS consisting of the histidine kinase VirS and response regulator VirR (Fig. 9.8) (Ohtani and Shimizu 2016). The activated VirR binds to its binding site, termed the "VirR box." Two toxin genes (*pfoA* and *ccp*) and three regulatory RNAs (*vrr, virU,* and *virT*) have *virR* box in the up-stream of them. Interestingly, the *agrBD* and *virSR* loci are located at a far location on genomic DNA (Ohtani et al. 2009, 2010). Construction of an *agrB* mutant of *C. perfringens* has unveiled the critical role of Agr QS in the expression of different toxins (Ohtani et al. 2009; Vidal et al. 2012; Yu et al. 2017). Toxin production was restored when *agrB* mutants

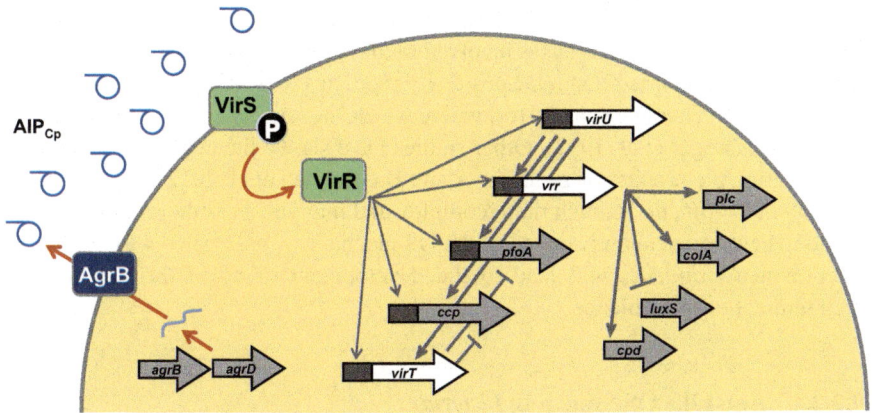

Fig. 9.8 The VirSR QS system in *Clostridium perfringens*. Thiolactone AIP is translated from agrD, processed by AgrB, and then secreted outside cells. A TCS, consisting of histidine kinase VirS and response regulator VirR, senses AIP concentration. The phosphorylated VirR activates the promoters via VirR box, thereby inducing a series of toxin and enzyme genes involved in virulence

of *C. perfringens* were treated with synthetic AIP, further confirming the role of Agr-type QS in this clostridial species (Vidal et al. 2012; Singh et al. 2015).

C. botulinum is well known for botulism poisoning in infants. *C. botulinum* possesses *agrBD* homologs and also two AIPs have been identified. The role of *agrBD1* and *agrBD2* was reported to be involved in the modulation of sporulation and toxin production, respectively (Cooksley et al. 2010). The role of TCSs in toxinogenesis of *C. botulinum* is not well established, but they do play some important roles in the direct or indirect regulation of toxin production. Until now, only the positive regulatory sigma factor, and six TCSs, have been described in *C. botulinum* (Connan et al. 2012; Zhang et al. 2013). Understanding the molecular mechanism of TCSs in toxin regulation of this bacterium will aid in the development of new strategies to control botulism.

The infections, caused by *C. difficile,* are attributed to toxins A and B, encoded by the *tcdA* and *tcdB* genes, respectively (Lyras et al. 2009; Kuehne et al. 2010). The cytotoxicity of toxin B is believed to be stronger than that of toxin A (Lyras et al. 2009). The occurrence of QS in *C. difficile* has not been thoroughly established because of the lack of *agrBD* mutants of this bacterium. However, comparative RNA sequencing of wild type and isogenic *agrA* null mutants of the *C. difficile* strain R20291, suggests that the expression of toxin genes in *C. difficile* are regulated by Agr-type QS (Martin et al. 2013). The *C. difficile* R20291 strain carries *agr1* and *agr2* loci, which are well separated from each other, whereas most *C. difficile* strains carry only the *agr1* locus (Stabler et al. 2009). The *agr1* locus contains only the AIP-encoding genes (*agrB1* and *agrD1*), while the *agr2* locus harbors both AIP and response genes (*agrB2D2* and *agrC2A2*, respectively) (Darkoh and DuPont 2017). Recently, Darkoh et al. has confirmed that some novel thiolactone AIPs trigger toxin production in the *C. difficile* strains 630 and R20291 (Darkoh et al. 2015).

Further, Darkoh et al. performed a study, using allelic exchange-based deletion, and revealed that the *agr1* QS locus is involved in toxin production (Darkoh et al. 2016). Initially, it was reported that *tcdR* positively regulates toxin expression (Mani et al. 2002; Martin-Verstraete et al. 2016), whereas *tcdC* negatively controls toxin expression (Hundsberger et al. 1997). Other studies have shown the negative role of *tcdC* in regulating the production of toxins A and B (Cartman et al. 2012). This indicates that QS in *C. difficile* is much more complex and may involve other regulatory elements. The detailed understanding of the QS mechanism, associated with toxin production in *C. difficile*, will lead to the development of novel therapeutics and containment of its virulence.

9.2.2.4 Agr-Like QS System in *Listeria*

Listeria monocytogenes is a food-borne opportunistic pathogen that causes life-threatening human listeriosis in patients with low immunity (Vázquez-Boland et al. 2001). *L. monocytogenes* contains an Agr-like QS system encoded by the *agrBDCA* operon (Autret et al. 2003). The two components, AgrC and AgrA, act as sensor kinase and response regulator, respectively, transducing the thiolactone AIP signal to induce gene activation (Garmyn et al. 2009). The role of Agr QS in *L. monocytogenes* is critical for infection and biofilm formation (Autret et al. 2003; Riedel et al. 2009). However, *agr* genes are highly conserved among the listeria species, which differs from *S. aureus* in which four *agr* groups are present and interfere with each other (Schmid et al. 2005). Additionally, the RNAIII-like gene, which encodes a regulatory RNA, has not been found (Wuster and Babu 2008). Recent evidence indicates that *L. monocytogenes* positively controls the expression of chitinase genes, *chiA* and *chiB*, via the Agr QS system (Paspaliari et al. 2014).

9.2.2.5 Fsr System and Sex Pheromone Plasmids in Enterococci

Although *Enterococcus faecalis* is a common inhabitant in the animal gastro-intestinal tract, controlling enterococcal nosocomial infections is gaining tremendous attention due to the emergence of drug resistant strains (Gilmore et al. 2013). Strains of enterococci frequently carry genes encoding virulence factors such as cytolysin, gelatinase, serine protease, enterococcal surface protein, and aggregation substance (Haas and Gilmore 1999; Hass et al. 2002). Opportunistic pathogens of enterococcal species, such as *E. faecalis* and *E. faecium,* often harbor mobile genetic elements such as plasmids or transposons. The fact that these elements carry virulence as well as antibiotic resistance genes is worrisome in terms of ecology and medicine.

Among the virulence factors, the two proteases, gelatinase and serine protease, are controlled by the Fsr QS system, which is homologous to the staphylococcal Agr system, while the biosynthesis of cytolysin is controlled by a unique QS system autoregulated by the subunit peptide of cytolysin (Fig. 9.9) (Haas and Gilmore 1999;

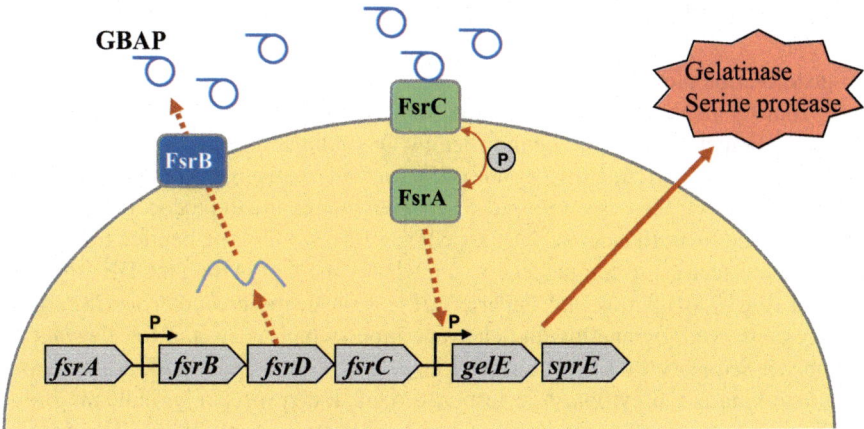

Fig. 9.9 Fsr QS system in *Enterococcus faecalis*. In this system, a lactone AIP, GBAP, functions as AI. AgrD, a propeptide of GBAP, is translated and processed by FsrB and then secreted. A histidine kinase, FsrC, senses the concentration of extracellular GBAP and activates a response regulator FsrA via phosphorylation. The activated FsrA induces the transcription of two operons, fsrBDC, encoding the GBAP precursor and the TCS, and gelE-sprE, encoding two pathogenicity-related extracellular proteases, gelatinase and serine protease

Coburn et al. 2004). The Fsr QS system is mediated by gelatinase biosynthesis-activating pheromone (GBAP), which possesses a lactone instead of a thiolactone, common in the Agr AIPs (Nakayama et al. 2001). GBAP precursor is encoded by *fsrD* which is then matured, cyclized and cleaved enzymatically by FsrB, a membrane transporter protein (Nakayama et al. 2001; Nakayama et al. 2006). Extracellular GBAP is recognized by the TCS system, which is similar to AgrCA in *S. aureus,* and phosphorylated FsrA activates the transcription of target genes.

Some enterococcal strains harbor unique genetic mobile element, which is a plasmid whose transfer is regulated by a peptide sex pheromone belonging to the RNPP family (Dunny et al. 1979; Dunny 2013). Interestingly, a series of machinery, involved in the conjugative transfer, are encoded by the plasmid and are induced by the sex pheromone. Induction of aggregation substance on the surface of plasmid-donor cells is a drastic event, which leads to clumping of donor and recipient cells and enables plasmid transfer at high frequency. Once the recipient acquires the plasmid, it shuts off the activity of the sex pheromone by secreting a sex pheromone inhibitor, which is an antagonistic peptide of sex pheromone. There are different sex pheromone plasmids in *E. faecalis* and *E. faecium,* such as pAD1, pPD1, pAM373, pCF10, and pOB1; whose transfers are respectively induced by the sex pheromones such as cAD1, cPD1, cAM373, cCF10, and cOB1, and suppressed by the corresponding inhibitors such as iAD1, iPD1, iAM373, iCF10, and iOB1 (Nakayama et al. 1995; Cook and Federle 2014). The sex pheromone peptides are imported into the cytosol, wherein they form complexes with the RNPP family regulators, such as PrgX for cCF10 and TraA for cAD1 and cPD1 (Fujimoto et al. 1995; Nakayama et al. 1998; Bae et al. 2004). The complexes eventually determine the conjugation state of the cell, depending on the concentration of pheromone and inhibitor.

9.2.2.6 ComAB/ComCDE and ComRS Systems in Streptococci

Transformation along with conjugation and transduction are involved in horizontal gene transfer and are the reasons behind the spread of antibiotic resistance. In streptococci, competence is tightly controlled by QS and mediated by two types of peptides: competence stimulating peptide (CSP) and alternative sigma factor, *sigX* (also known as *comX*); these are referred to as pheromones in streptococci (Mashburn-Warren et al. 2010; Reck et al. 2015). CSP is a 17-residues long peptide that acts as an inducing factor for competence and mediates the ComAB/ComCDE signaling system (Fig. 9.10) (Cook and Federle 2014). ComC, the precursor to CSP, has a highly conserved glycine-glycine cleavage motif which is critical for cleavage of the leader sequence by ComA (Ishii et al. 2006). ComD and ComE act as sensor histidine kinase and cytoplasmic response regulator, respectively, indicating high similarity to the AgrCA system of *S. aureus* (Pestova et al. 1996). The peptide sequence of CSP is not highly conserved across streptococcal species and sometimes even varies within the species. CSP interacts with ComD, which in turn phosphorylates ComE, thereby activating transcription of competence genes and consequently positive feedback loop as described in Fig. 9.10 (Ween et al. 1999). In addition to competence, CSP stimulates biofilm formation and bacteriocin production (van der Ploeg 2005; Perry et al. 2009). ComCDE signaling is prevailing in mitis and anginosus bacterial groups of streptococcus and is particularly well studied in *S. pneumoniae*.

The other important signaling system in streptococcus, prevalent in pyogenic, salivarius, bovis and mutans groups of streptococcus, is ComRS (Fig. 9.10) (Fontaine et al. 2010, 2013). ComS acts as a signal and varies in length in different

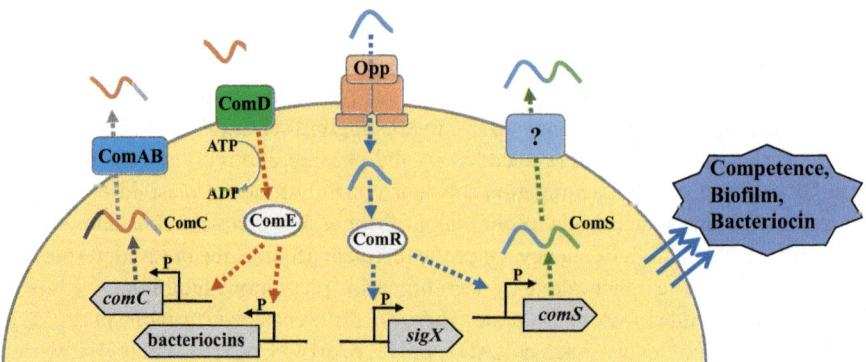

Fig. 9.10 The ComAB/ComCDE QS systems in anginosus and mitis groups of streptococci and ComRS QS system in pyogenic, bovis, salivarius, and mutans group of strepotococci. In former system, a competence stimulating peptide (CSP) is translated from comC and is then processed and secreted by the ComAB complex. Extracellular CSP is sensed by ComD-ComE TCS and eventually promotes the expression of bacteriocin genes. In the second system, ComS is translated, and processed to XIP, and then secreted outside the cells. The extracellular XIP is imported into the cytoplasm and binds to a transcription regulator ComR. The XIP-ComR complex regulates the expression of competence genes

species (Cook and Federle 2014). ComR acts as both a signal receptor and Rgg-type transcriptional regulator. The ComRS system is categorized into two types; type I, found in the salivarius group, and type II found in pyogenic, mutans, and bovins species; the difference is whether a double tryptophan motif is present or not present on the sequence. The alternative sigma factor, ComX, is essential for coupling of competence development and bacteriocin synthesis (Reck et al. 2015). It also regulates bacterial autolysis (the lysing and release of virulence factors) (Cook and Federle 2014). Recently, a Phr-peptide-based signaling system was found in *S. pneumoniae*; this system, called TprA/PhrA, was previously found only in *Bacillus* and appears highly conserved among the different pneumococcal serotype strains (Hoover et al. 2015). This signaling system regulates the synthesis of lantibiotic biosynthesis gene cluster, which provides it with a competitive advantage (Hoover et al. 2015).

9.3 AI-2/LuxS Signaling, a Universal Language for Bacterial Communication

9.3.1 AI-2/LuxS QS System

AI-2/LuxS QS was first discovered in *V. harveyi* as a regulator of bioluminescence (Surette et al. 1999). AI-2-mediated QS is shared by both Gram-negative and Gram-positive bacteria (Sun et al. 2004; Camilli and Bassler 2006). LuxS (S-ribosylhomocysteine lyase) acts as key enzyme in the activated methyl cycle (Fig. 9.11). During the methyl cycle, S-adenosylmethionine (SAM), the key methyl donor for all bacterial cells, donates a methyl group to the substrates; one of the formed byproducts is the toxic *S*-adenosylhomocysteine (SAH) (Pereira et al. 2013). SAH is detoxified by nucleoside Pfs, yielding *S*-ribosylhomocysteine (SRH). The LuxS enzymes then catalyzes the conversion of SRH to homocysteine with the formation of 4,5-dihydroxy-2,3-pentanedione (DPD) (Pereira et al. 2013). DPD undergoes spontaneous cyclization reactions to form a family of interconverting compounds, called furanones, which includes AI-2 that accumulates in culture supernatant (Vendeville et al. 2005). AI-2 is one of the few biomolecules having boron in its structure (Chen et al. 2002). AI-2 acts as a signaling molecules and mediates inter- and intra-species communication across the diverse bacterial groups (Miller et al. 2004; Waters and Bassler 2005; Pereira et al. 2013). The production of AI-2 signaling molecules is regulated by the LuxS gene. The homologs of LuxS gene have been recognized in many bacterial species, suggesting that AI-2 acts as a universal language in bacteria (Surette et al. 1999; Schauder et al. 2001; Miller et al. 2004; Galloway et al. 2010; Pereira et al. 2013).

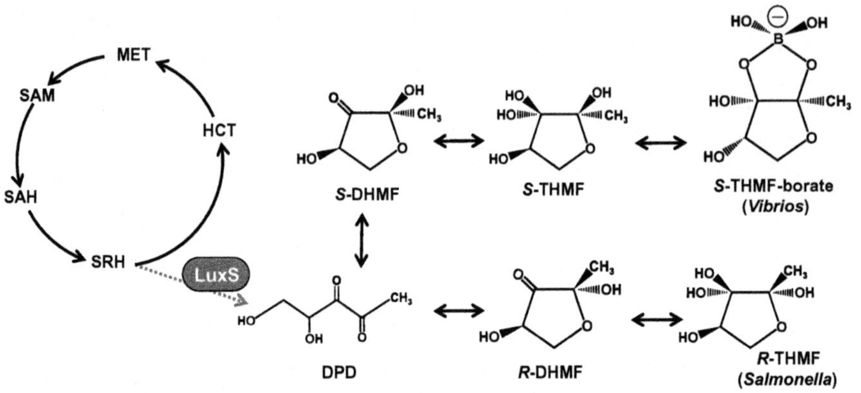

Fig. 9.11 Biosynthesis of AI-2 signaling molecules. The activated methyl cycle and AI-2 biosynthesis are coupled together. During this pathway, SAM is converted to SAH which is subsequentially detoxified by the Pfs enzyme to SRH. LuxS enzyme then catalyzes the conversion of SRH in to DPD and homocysteine. DPD undergoes spontaneous rearrangements to form AI-2. The AI-2 are used as signaling molecules by both Gram-negative and Gram-positive bacteria

9.3.2 AI-2/LuxS Quorum Sensing in Gram-Negative Bacteria

The AI-2 signaling system was identified by observing luciferase activity although at low level in the absence of the HAI-1 AI in *V. harveyi* mutant defective in AHL synthesis (Bassler et al. 1994). The *V. harveyi* mutants were able to perceive the AI-2 signal from even unrelated bacterial species, indicating their efficient functioning of LuxS/AI-2 QS system (Bassler et al. 1994; Bassler 1999). Two AI-2 molecules have been characterized as (2S,4S)-2-methyl-2,3,3,4-tetrahydroxytetrahydrofuryl borate (S-THMF-borate) (Chen et al. 2002), the AI-2 signal of *Vibrionaceae*, and (2R,4S)-2-methyl-2,3,3,4-tetrahydroxytetrahydrofuran (R-THMF), the AI-2 signal of *Enterobacteriaceae* such as *S. typhimurium*, (Fig. 9.5a) (Miller et al. 2004; Xavier et al. 2007). Interestingly, the mode of signal transduction for these two AI-2s differs considerably. *Vibrionaceae* uses a LuxPQ TCS. At low concentrations of AI-2, LuxQ acts as a kinase and generates no luciferase activity. However, at high concentrations, AI 2 binds to LuxP, which in turn changes LuxQ from kinase to phosphatase; thus, no phosphorylation of LuxO, and ultimately LuxR is activated and light is produced (De Keersmaecker et al. 2006; Pereira et al. 2013). *Enterobacteriaceae* imports the AI-2 molecules into cytosol via the LsrABC ABC transporter. After AI-2 is internalized, the LsrK kinase phosphorylates AI-2, and subsequentially LsrG and LsrF enzymes process AI-2 (Taga et al. 2003; Xavier et al. 2007). All these genes, except *LsrK*, operate as a single operon and are regulated by the repressor LsrR (Fig. 9.5a). The phosphorylated AI-2 binds to the repressor LsrR, causing derepression of the operon and vice versa (Pereira et al. 2013). In order to communicate with, and confuse, other bacteria living in the same environment, the Lsr transporter maintains the level of AI-2 in the surrounding environment via positive feedback. The Lsr transporter has been extensively studied in *S. typhimurium*, *E. coli*, and *S. meliloti* (Taga et al. 2003; Xavier et al. 2007; Pereira et al. 2013). Although its homologs have been reported in the

Yersinia spp., *Klebsiella pneumoniae*, and even in Gram-positive *Bacillus* spp. but it is not well characterized. The understanding of AI-2 receptors in other bacterial species is important for determining their molecular mechanisms of AI-2 signaling.

9.3.3 LuxS/AI-2 Communication System in Gram-Positive Bacteria

In addition to AIP-mediated QS, Gram-positive bacteria use the AI-2 signaling for both intra- and interspecies communication (De Keersmaecker et al. 2006; Pereira et al. 2013). The molecular mechanism of AI-2 signaling in Gram-positive bacteria has not been thoroughly characterized. Notably, the receptors for AI-2 signals has not been unraveled and requires more attention. Phenotypes, regulated by AI-2 signaling in Gram-positive bacteria, have been reported to be critical for toxin production and biofilm formation (Rickard et al. 2006; Fitts et al. 2016). It has been indicated that AI-2 upregulates the expression of alpha, kappa, and theta toxin genes in *C. perfringens* (Ohtani et al. 2002). The *C. difficile* toxin genes (*tcdA*, *tcdB*, and *tcdE*) are positively regulated by the LuxS/AI-2 QS (Lee and Song 2005) but only observed before 24-h cell growth (Carter et al. 2005). Shao et al. provided important cues regarding the role of LuxS/AI-2 signaling in vancomycin-resistant *E. faecalis*. Also, externally added AI-2 increased biofilm formation in *E. faecalis* (Shao et al. 2012). It was recently reported that AI-2 QS regulates biofilm formation in *S. aureus* (Ma et al. 2017). The AI-2/LuxS has a signaling function in *S. epidermidis* and *S. intermedius*, regulating antibiotic susceptibility and biofilm formation (Li et al. 2008a; Ahmed et al. 2009). In *S. pneumoniae*, LuxS and AI-2 have also been shown to control biofilm formation (Vidal et al. 2011). These pathogens cause numerous serious infections through biofilm formation, consequently increasing their tolerance toward antimicrobial agents and host defenses.

Although, AI-2 is one of the best studied signaling molecules, however, there are many mysteries revolving around this QS system. How AI-2 signals are perceived where no receptors have been recognized, how AI-2 molecules are exported, regulation of AI-2 molecule synthesis and how its interference affects bacterial behaviors in monocultures and polycultures remains to be answered. Finding the answers to these questions will provide us with better strategies to exploit or inhibit bacterial behaviors for the control of life threatening infectious pathogens.

9.4 Inhibition of Quorum Sensing in Gram-Negative/Gram-Positive Bacteria

9.4.1 Concept of Anti-QS Strategy

In the current cat and mouse game between antibiotics and drug-resistant bacteria, alternatives or additives to antibiotic chemotherapies is considerably prompted (Hentzer and Givskov 2003; Rasmussen and Givskov 2006; Bhardwaj et al. 2013;

O'Loughlin et al. 2013; Allen et al. 2014; Brackman and Coenye 2015; Khan et al. 2015; Todd et al. 2017). Notably, anti-QS strategy is gaining much interest as a novel therapeutic option, because it does not target cell growth and proliferation, and is relieved from the selective pressure for drug resistant strains. QSIs can target different points in QS signaling. Inhibition of AI biosynthesis may be effective in terms of QQ. If a sufficient concentration of a potent AI antagonist is provided, it will effectively interfere with the receptor-AI interaction. Sequestration or degradation of AIs may be easy to control by using an enzyme or antibody. Some studies emphasize on novel strategies for the designing and discovery of QSIs to render these inhibitors potential targeted drugs; other potential inhibitors have been screened from compound libraries or natural resources (Scutera et al. 2014). The QS inhibitors thus far developed are summarized according to category in Table 9.3.

9.4.2 Inhibition of AHL Synthase

Gram-negative bacteria produce a range of signaling molecules among which AHLs are the most prominent and major ones (Ng and Bassler 2009; LaSarre and Federle 2013). The repression of signaling molecules, or interrupting their interactions with receptor proteins will hamper QS-regulated gene expression; hence, bacterial virulence can be attenuated (Dong et al. 2000). In AHLs, the lactone ring is derived from SAM, while the acyl chain is derived from the acyl carrier protein (ACP). For example, RhlI is used to catalyze the synthesis of C4-HSL, a secondary messenger in *P. aeruginosa*, from crotonyl-ACPs and SAM (Hoang and Schweizer 1999). Two analogs of SAM, S-adenosylhomocysteine and sinefungin, were found to inhibit the RhlI-catalyzed synthesis of C4-HSL and QS (Fig. 9.12a) (Musk et al. 2006). However, SAM is involved in both metabolic and signaling pathways, and its inhibition can lead to undesirable repercussions.

In *P. aeruginosa*, LasI is an AHL synthase that regulates the production of numerous determinants of virulence. Tannic acid and trans-cinnamaldehyde have been reported to inhibit LasI and, thus, can efficiently repress AHL production mediated by RhlI (Chang et al. 2014). Furthermore, the authors also demonstrated the potential of trans-cinnamaldehyde, to inhibit QS-regulated production of pyocyanin in *P. aeruginosa* PAO1 (Chang et al. 2014). A novel inhibitory compound, (z)-5-octylidenethiazolidine-2, 4-dione (TZD-C8), has recently been found to inhibit gene expression via inhibition of LasI activity, consequentially affecting the swarming and biofilm formation by *P. aeruginosa* (Fig. 9.12a) (Lidor et al. 2016). Some antibiotics were witnessed to attenuate QS systems. For example, azithromycin, at sub-inhibitory concentrations has been found to inhibit AHL synthase in *P. aeruginosa*; this inhibition was attributed to the arrest of protein synthesis (Tateda et al. 2001). Similarly, sub-inhibitory concentrations of streptomycin have been found to reduce the transcript level of AbaI, the synthase for 3-OH-C12-HSL in *A. baumannii* (Saroj and Rather 2013). However, gentamicin and myomycin had no

Table 9.3 List of described QSIs, their activity, and mechanism of action

QSI	IC50	Activity	References
Inhibitors of Gram-negative bacteria			
Halogenated furanones (C30)		Inhibit ligand binding/promote receptor degradation (LasR)	Ren et al. (2001) and Hentzer et al. (2002)
N-(heptylsulfanylacetyl)-l-homoserine lactone		Competitive inhibition of transcriptional regulators LuxR and LasR.	Persson et al. (2005)
Immucillin A		Inhibition of adenosyl homocysteine nucleosidase	Singh et al. (2005)
PD12	30 nM	Inhibiting production of virulence factor	Müh et al. (2006)
6FABA; 6CABA; 4CABA		Inhibition of 4QS biosynthesis and MvfR-dependent gene expression	Lesic et al. (2007)
N-phenylacetanoyl-*L*-homoserine lactone	0.3 µM	Inhibition of signal receptors; LasR, TraR and LuxR	Geske et al. (2007)
N-mercaptoacetyl-Phe-Tyr-amide		Inhibition of LasB mediated immunomodulation and biofilm formation	Cathcart et al. (2011)
Bromoageliferin; 3-indolacetonitrile; resveratrol; 2-aminoimidazole (2-AI) derivatives		Anti-biofilm activity against *P. aeruginosa*	Huigens et al. (2007) and Frei et al. (2012)
Meta-bromo-thiolactone (mBTL)		Prevents virulence factor expression and biofilm formation	O'Loughlin et al. (2013)
N-[2-(1H-indol-3-yl)ethyl]-urea; *N*-(2-phenethyl)-urea		Suppress QS signaling in *P. aeruginosa*	Chu et al. (2013)
AHL- lactonase, −acylases, −oxidases and -reductases		Degradation of AHLs	Amara et al. (2010)
M64	Nanomolar concentrations	Suppressing MvfR-dependent virulent gene expression	Starkey et al. (2014)
Triclosan		Hindering synthesis of AHL precursor	Hoang and Schweizer (1999) and Chan et al. (2015)
V-06-18	10 µM	Inhibiting production of virulence factor	Müh et al. (2006)and Moore et al. (2015)
6-gingerol		Antagonistic inhibition of virulence factors and biofilm formation	Kim et al. (2015)

(continued)

Table 9.3 (continued)

QSI	IC50	Activity	References
AHL with aryl β-keto ester	23–53 µM	Inhibition of AHL receptors	Forschner-Dancause et al. (2016)
Flavonoids		Noncompetitive inhibition to LasR/RhlRT	Paczkowski et al. (2017)
Inhibitors of Gram-positive bacteria			
RWJ-49815	1.6 µM	Inhibition of histidine kinase of MRSA, VRE and penicillin resistant *S. pneumoniae*	Barrett et al. (1998)
AIP-II;	2.9–3.2 nM	Inhibition of β-lactamase	Mayville et al. (1999)
AIP-II-N3A	180 nM		
trAIP-II	10–272 nM	Inhibition of AgrC receptor-histidine kinase	Lyon et al. (2000)
RNA III inhibiting peptide (RIP) (YSPWTNF)		Inhibition of TRAP in agr QS system	Giacometti et al. (2003), Dell'Acqua et al. (2004), and Balaban et al. (2005)
Thienopyridine (TEP)		Inhibition of histidine kinase	Gilmour et al. (2005)
RIP loaded "poly (methyl methacrylate)" beads		Inhibition of biofilm formation of MRSA	Anguita-Alonso et al. (2007)
Siamycin I	100 nM	Inhibiting biosynthesis of GBAP, and AIPs	Nakayama et al. (2009)
Savirin	<1.0 µM	Interfering with AgrA-DNA binding	Sully et al. (2014)
Benzbromarone	0.13 µM	Interfering with AgrA-DNA binding	Gordon et al. (2013)
AIP-I D5A/AIP-IV Y5A	0.3–8.0 nM	Inhibiting production of toxic shock syndrome toxin (TSST-1) and enterotoxin C3	MDowell et al. (2001) and Gordon et al. (2013)
AIP-III D4A	0.035–0.485 nM	Inhibition of AIP receptor	Tal-Gan et al. (2013)
Truncated RIP YSPWTNF; SPWT		Inhibition of TRAP in agr QS system	Baldassarre et al. (2013)
RIP derivative FS8 (YSPWTNA)		Inhibition of TRAP in agr QS system	Simonetti et al. (2013)
ZBzl-YAA5911	26.2 nM	Antagonistic activity against GBAP receptor (FsrC) of *E. faecalis*	Nakayama et al. (2013)
2,5-di-O-galloyl-d-hamamelose (hamamelitannin)		RNA III production, cell attachment and biofilm formation of *Staphylococci*	Kiran et al. (2008)and Kuo et al. (2015)

(continued)

Table 9.3 (continued)

QSI	IC50	Activity	References
Z-AIPCp-L2A/T5A;	0.32 μM	Inhibition of *pfoA* gene transcription in *C. perfringens*	Singh et al. (2015)
Z-AIPCp-F4A/T5S	0.72 μM		
WS9326A;	0.88–19 μM	Inhibition of agr-, fsr- and VirSR QS signal receptor in *S. aureus* and *Enterococcus*	Desouky et al. (2015)
WS9326B	2.1–23 μM		
Cochinmicin II/III	0.2 μM	Inhibition of agr QS signal receptor	Desouky et al. (2015)
ω-hydroxyemodin; colostrum hexasaccharide		Inhibition of agr QS of *S. aureus*	Daly et al. (2015) and Srivastava et al. (2015)
Avellanin C	4.4 μM	Inhibition of agr QS of *S. aureus*	Igarashi et al. (2015)
AIP-1 of *S. epidermidis* and *S. lugdunensis*	13–419 nM	Competitively inhibits *S. aureus* AgrC-2 and − 3	Gordon et al. (2013) and Tal-Gan et al. (2016)
Oligomerization of RIP peptides		Inhibition of adherence and biofilm formation of MRSA	Zhou et al. (2016)
Ambuic acid	2.5 μM	Inhibiting AIP signal biosynthesis (FsrB) of *E. faecalis*	Todd et al. (2017)
Inhibitors of LuxS/AI-2 quorum sensing			
S-anhydroribosyl-*L*-homocysteine; *S*-homoribosyl-*L*-cysteine	>1 mM	LuxS inhibitors	Alfaro et al. (2004)
KM-03009;	35 μM	Antagonize AI-2-mediated quorum sensing in *V. harveyi*	Li et al. (2008a, b)
SPB-02229	55 μM		
Thiazolidinediones		Inhibition of LuxR and DNA binding in *V. harveyi*	Brackman et al. (2013)
Alkyl-4,5-dihydroxy-2,3-pentanedione analogs; C4-alkoxy-5-hydroxy-2,3-pentanediones; alkyl-DPD analogs	0.15–50 μM	Modulating AI-2 based signaling in *S. typhimurium* and *V. harveyi*	Lowery et al. (2009)
Phenyl-DPD; isobutyl-DPD		Inhibiting Lsr mediated biofilm formation in *E coli*; suppressing toxin production and biofilm formation in *P. aerugenosa*	Roy et al. (2013)

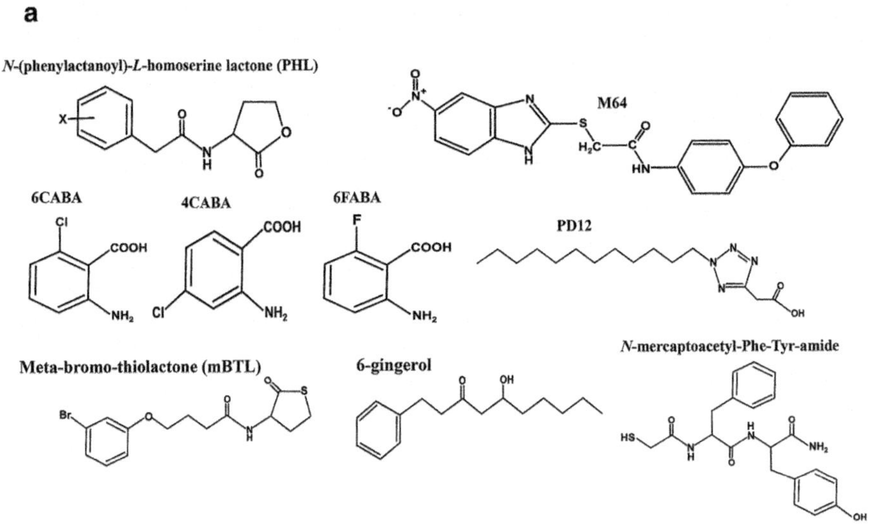

a

N-(phenylactanoyl)-L-homoserine lactone (PHL)

6CABA   4CABA   6FABA   PD12

Meta-bromo-thiolactone (mBTL)   6-gingerol

N-mercaptoacetyl-Phe-Tyr-amide

M64

Fig. 9.12a Structure of potent QSIs targeting Gram-negative bacteria

effect on QS signaling in *A. baumannii*; thus, attenuation of signaling seems not to be as the result of protein synthesis inhibition in this bacterium.

Some compounds have been found to hinder the synthesis of AHL precursors or simply obstruct their synthesis. An AHL precursor, synthesized from enoyl-ACP reductase, is reduced by the small molecule triclosan (Hoang and Schweizer 1999; Chan et al. 2015). Other compounds, such as immucillin A and its derivatives, were identified to inhibit adenosyl-homocysteine nucleosidase, which is typically essential because it is involved in the synthesis of AHL and AI-2 (Singh et al. 2005). The limitation of these inhibitors is that they also affect the central metabolism of amino and fatty acids.

9.4.3   AHL Antagonists

The AHL receptors are the potential target for novel bacterial infection therapies due to their key roles in bacterial population behaviors and pathogenicity (Rasmussen and Givskov 2006; LaSarre and Federle 2013). The modifications in the acyl chain of AHL led to the discovery of numerous potent antagonists, such as N-(heptylsulfanylacetyl)-L-homoserine   lactone   and   N-(phenylacetanoyl)-L-homoserine lactone, with the IC_{50} values of 6 μM and 0.3 μM, respectively (Fig. 9.12a) (Persson et al. 2005; Geske et al. 2007). Recently, Forschner-Dancause et al. modified the phenyl ring at carbon 4 using halo or methoxy groups which generated different QSI compounds ($IC_{50} = 23$–53 μM) that can act as competitive antagonists to AHLs in LuxR-mediated QS (Fig. 9.12a) (Forschner-Dancause et al. 2016).

O'Loughlin and colleagues have reported a potential small molecule QS modulator, meta-bromo-thiolactone (mBTL) (Fig. 9.12a). This compound is effective in thwarting LasR and RhlR, consequently averts the production of pyocyanin and formation of biofilm both *in vitro* and *in vivo* (O'Loughlin et al. 2013). This compound also safeguards *Caenorhabditis elegans* and human lung epithelial cells from QS mediated killing by *Pseudomonas aeruginosa* (O'Loughlin et al. 2013).

Moore et al. comparatively evaluated a library of most potent and efficacious LasR modulators. Their study identified two compounds that exhibit potent LasR modulator activity: a triphenyl compound acting as an agonist and a V-06-018 compound acting as an antagonist (Fig. 9.12a) (Moore et al. 2015). This is an important development toward finding efficient QSI; however, these compounds need to be further evaluated experimentally. Two QQ compounds, (*N*-[2-(1H-indol-3-yl) ethyl]-urea and *N*-(2-phenethyl)-urea), have been identified in Gram-positive bacteria *Staphyloococus delphini* (Chu et al. 2013). These compounds show QQ activity against *P. aeruginosa*. The co-culture of *S. delphini* with *P. aeruginosa* does not inhibit the growth of either. These findings indicate that inhibitors of QS can play important roles in self-protection and competitiveness in natural environments.

Muh and colleagues identified two compounds, tetrazole (PD12) and V-06-018, which showed good inhibitory activity against the expression of genes controlled by LasR and production of virulence factors, such as elastase and pyocyanin in *P. aeruginosa*, with IC_{50} values of 30 nM and 10 μM, respectively (Fig. 9.12a) (Müh et al. 2006). *In silico* studies by Kim et al. demonstrated that pungent oil of ginger, called 6-gingerol, binds antagonistically to the LasR receptor. This research group used transcriptomics to demonstrate that, 6-gingerol reduced production of virulence factors (pyocyanin, rhamnolipid and exoprotease) and formation of biofilm by intercepting the QS of *Pseudomonas* (Kim et al. 2015). Initially, LasR was considered critical for the spread of infection; however, RhlR is now known to play a prominent role in infection, even superseding that of LasR in some cases. Eibergen and colleagues synthesized a series of antagonists targeting RhlR and identified one potent antagonist, E22, having strong activity at the IC_{50} of 17.3 μM (Eibergen et al. 2015).

Recently, naturally produced flavonoids have been reported to act as potent noncompetitive inhibitors of LasR/RhlR. On the basis of SAR analysis, it was observed that two hydroxyl moieties in the flavone A-ring backbone are essential for potent inhibition of LasR/RhlR (Paczkowski et al. 2017). The flavonoids suppressed the production of virulence factors without any bactericidal or bacteriostatic activity in *P. aeruginosa*. Suneby et al. recently synthesized LasR antagonists closely resembling native 3O-C12-HSL and varying only in the length of acyl chain and presence of the 3-oxo group. This research group identified four potent antagonists that form a complex with LasR (the LasR-antagonist complex), which no longer binds to the DNA in *P. aeruginosa* (Suneby et al. 2017). Structure-based virtual screening was performed to identify natural compounds as potential inhibitors of LasR. Using this analysis, six novel and potent inhibitors were identified; these inhibitors can target LasR-mediated QS in *P. aeruginosa* (Kalia et al. 2017).

9.4.4   Enzymes That Degrade AHL

Some bacteria naturally produce enzymes that degrade AHL molecules (Amara et al. 2010; Kalia and Kumar 2015). Lactonases and acylases have gained interest as AHL degrading enzymes that hydrolyze the ester bond in the lactone ring or the amide bond (Fig. 9.12a). These enzymes break down autoinducers, thereby hampering LasR/LasI and RhlI/Rh1R-type QS systems in *P. aeruginosa* (Reimmann et al. 2002). Indeed, lactonase reduces *P. aeruginosa* infections in rats (Hraiech et al. 2014). AHL oxidase and reductase have been reported to catalyze the modification of AHLs (Amara et al. 2010). Because various bacteria share AHLs as their AI molecules, this enzymatic degradation of AIs may be useful in broad-spectrum control of virulence.

9.4.5   Inhibitors Targeting Non-AHL Pathway

PQS signaling is one of the prominent communication pathways in *P. aeruginosa*. This QS pathway can be regulated by inhibiting the synthesis of PQS. Three halogenated anthranilic acid analogs, 6FABA, 6CABA, and 4CABA (Fig. 9.12a) specifically inhibit the 4QS biosynthesis and interfere with MvfR-dependent gene expression in *P. aeruginosa* (Lesic et al. 2007). These compounds can prevent the spread of *P. aeruginosa* infection. Further study of the molecular mechanisms involved in the QS signaling system may provide the basis for a new class of QS-inhibition-based antimicrobial drugs. One of the potential virulence factors in *P. aeruginosa* is elastase encoded by the LasB gene; elastase severely affects the immune response and initiates biofilm formation. Cathcart et al. developed a novel and potent inhibitor of LasB, *N*-mercaptoacetyl-Phe-Tyr-amide (Ki = 41 nM), and demonstrated its ability to completely prevent immunomodulation and biofilm formation initiated by the LasB virulence factor. This LasB inhibitor has also been used in combination with antibiotics to completely eradicate *Pseudomonas* biofilms (Cathcart et al. 2011). Another potent compound, identified using whole-cell high-throughput screening (HTS) and SAR analysis, has been designated as M64; M64 inhibits the MvfR virulence regulon of MDR *P. aeruginosa* (Starkey et al. 2014). M64 is considered a 2nd generation inhibitor that attenuated acute and persistent mammalian infections during *in vivo* assessment, and also can prevent the formation of antibiotic-tolerant bacterial cells (Starkey et al. 2014). Since, homologs of LTTRs occur in diverse virulence regulons across both Gram-negative and Gram-positive pathogens (Schell 1993), these inhibitory compounds are expected to have broad clinical applications against a wide range of bacterial pathogens.

9.4.6   Natural Compounds Inhibiting QS of Gram-Negative Bacteria

The most persuasive natural products with antagonistic activity against *P. aeruginosa* are halogenated furanones (C30), which are structurally similar to AHLs and can be isolated from macroalgae *Delisea pulchra* (Ren et al. 2001; Hentzer et al. 2002). The other important natural compounds, shown effective in inhibiting biofilm formation by *P. aeruginosa*, include bromoageliferin, 3-indolylacetonitrile, and resveratrol (Huigens et al. 2007). These compounds are structurally different from known inhibitors of biofilm formation by *P. aeruginosa*; their mechanism of action is unknown and needs to be studied for future drug development (Huigens et al. 2007). The 2-aminoimidazole (2-AI) derivatives are also reported to strongly inhibit biofilm formation by *P. aeruginosa* (Fig. 9.12a) (Frei et al. 2012).

9.4.7   Histidine Kinase Inhibitors

Histidine kinase, a common bacterial element of the two-component regulatory system (TCS), acts as a sensor for extracellular signals, including environmental signals, and for QS communication signals. This protein family is absent in animal cells, rendering histidine kinases attractive as a drug target. Inhibitors, targeting the QS sensor kinase, may work differently from conventional antibiotics and are expected to be effective against drug-resistant bacteria (Velikova et al. 2016; Gordon et al. 2016; Velikova and Wells 2017).

Matsushita and Janda have thoroughly reviewed inhibition of histidine kinase as potential antimicrobial therapy (Matsushita and Janda 2002). RWJ-49815 is a potent compound, shown to strongly inhibit the autophosphorylation of KinA::Spo0F, a TCS, *in vitro* with an IC_{50} of 1.6 µM; RWJ-49815 belongs to a family of hydrophobic tyramines (Fig. 9.12b) (Barrett et al. 1998). This compound has shown inhibitory activity against MRSA, VRE, and penicillin-resistant *S. pneumoniae*, and is also bactericidal (Barrett et al. 1998). Another potential compound, thienopyridine (TEP), competitively inhibits histidine kinase with respect to ATP without inhibiting the growth of bacterial cells and may be a promising candidate for novel drugs (Gilmour et al. 2005). The highly conserved ATP-binding region in histidine kinase may be a novel target in overcoming multi-drug resistant bacteria (Bem et al. 2015). Igarashi et al. identified a novel secondary metabolite of *Streptomyces*, Waldiomycin, as an inhibitor of WalK. WalK is an essential histidine kinase involved in cell wall metabolism of numerous Gram-positive bacteria including the virulent groups of staphylococci, streptococci, and enterococci (Eguchi et al. 2011; Igarashi et al. 2013). Waldiomycin blocks the autophosphorylation of WalK in *B. subtilis* and *S. aureus* at micromolar levels. Nakayama et al. identified siamycin I as a histidine kinase inhibitor that attenuates GBAP-mediated QS in *E. faecalis* at submicromolar

b

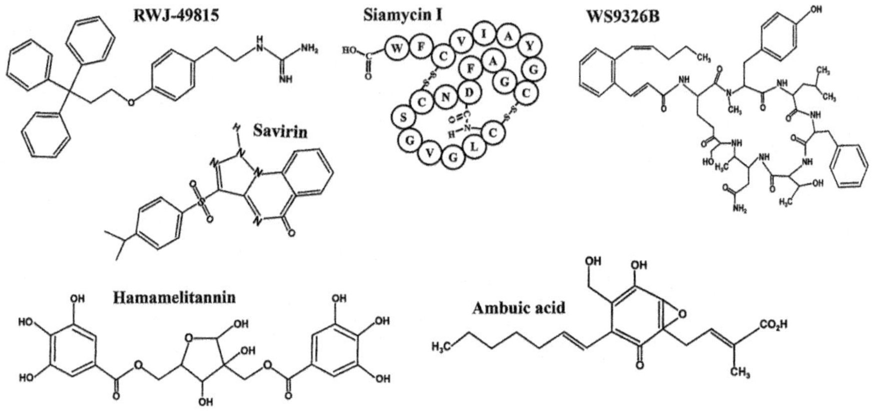

Fig. 9.12b Structure of potent QSIs targeting Gram-positive bacteria

concentrations (IC_{50} = 100 nM) (Nakayama et al. 2007). Siamycin I inhibits the autophosphorylation of FsrC, which is a sensor of GBAP (Ma et al. 2011).

Inhibitors, targeting the sensory domains of QS sensor kinases, have also been developed. Rasko et al. screened 150,000 compounds for inhibitors of QseC, which is a histidine kinase of AI-3 mediated QS system in enterohemorrhagic *E. coli* (EHEC). This study generated LED209 [N-phenyl-4-{[(phenylamino) thioxomethyl]amino}-benzenesulfonamide], a highly potent, non-toxic, and specific QseC inhibitor that inhibits the binding of signals to QseC, preventing its autophosphorylation and consequently inhibiting EHEC virulence traits *in vitro* and *in vivo* (Rasko et al. 2008).

9.4.8 AIP Antagonists

Peptide design of AIP antagonists have been extensively performed for gaining specific inhibitors to block QS signal sensing at histidine kinase. In the case of staphylococci, there is natural interference of QS between different strains and species; this has provided insights into SARs of AIP agonists and antagonists (Ji et al. 1997; Fleming et al. 2006). The SAR study using synthetic peptides, revealed that the exocyclic region of AIP plays a role in the activation of AgrC (Lyon et al. 2000). In that study, the tail was removed from the group-II AIP (AIP-II), yielding a truncate AIP (trAIP-II). The results of an AIP antagonist assay, indicated that trAIP-II blocks Agr signaling at nanomolar concentrations. Subsequently trAIP-II has been used as a lead peptide for further development of AIP antagonists (Lyon et al. 2000; Lyon and Muir 2003). Other potent antagonists, targeting all four AgrC receptors in nM range (IC_{50} = 0.1–5.0 nM), have been synthesized from AIP-I and AIP-II (Fig. 9.12b)

(Scott et al. 2003). Tal-Gan et al. designed and synthesized 30 non-native peptides, based on the AIP of group III *S. aureus,* using an alanine scanning approach. The AIP D5A variant was found to have potent inhibitory activity against AgrC-1, with $IC_{50} = 20$ nM. Furthermore, AIP-III D4A was found to be a picomolar global AgrC inhibitor (Tal-Gan et al. 2013; Gordon et al. 2013). Finally, Tal-Gan et al. have high-lighted the strong potential of truncated analogs, such as tr-(Ala2/Trp3)-AIP-III and tr-(Ala2/Tyr5)-AIP-III, as pan-group inhibitors displaying activity at sub-nanomolar concentrations (Fig. 9.12b) (Tal-Gan et al. 2013). These studies have opened up new avenues to identify novel therapeutic alternatives by interfering with AIP signaling, which is vital for the reduction of the various staphylococci infections (Mayville et al. 1999; Canovas et al. 2016; Wang and Muir 2016).

AIP antagonists have been developed for other staphylococcal species. Truncated AIP analogs, from *S. epidermidis* and *S. lugdunensis,* display potent inhibitory activity against the *S. lugdunensis* group I QS reporter strain. The antagonistic activi-ties of *S. lugdunensis* AIP-1 and AIP-2 are much stronger, with IC_{50} values of 0.2 ± 0.01 µM and 0.3 ± 0.01 µM, respectively, while *S. epidermidis* shows lesser activity at IC_{50} of 2.7 ± 0.1 µM (Fig. 9.12b) (Gordon et al. 2016). Recently it was found that *S. epidermidis* and *S. lugdunensis* AIP-1 competitively inhibits *S. aureus* AgrC-2 and -3, with IC_{50} values ranging between 13 to 419 nM, respectively (Tal-Gan et al. 2016).

AIP antagonists have also been developed for other Gram-positive pathogens, such as *E. faecalis, C. difficile, C. botulinum,* and *L. monocytogenes,* because these bacteria also harbor the Agr-like QS system (Thoendel et al. 2011; Gordon et al. 2016). Singh et al. designed two AIP antagonists, AIPCp-L2A/T5A and Z-AIPCp-F4A/T5S; these inhibitors significantly downregulated perfringolysin production in *C. perfringens,* with $IC_{50} = 0.32$ and 0.72 µM, respectively. However, AIPCp-L2A/T5A acts as a partial agonist, attenuating the production of perfringolysin at a lower concentration but promoting it at higher concentration (Singh et al. 2015). Nakayama et al. have developed a potent GBAP antagonist, ZBzl-YAA5911 [Ala(4,5,6,8,9,11)] Z-GBAP, by using a reverse alanine scanning approach (Nakayama et al. 2013). ZBzl-YAA5911 attenuates gelatinase expression controlled by GBAP-mediated Fsr QS at $IC_{50} = 26$ nM. Subsequently, a pool of GBAP analogs was synthesized, using entirely solid-phase peptide synthesis, to gain in-depth understanding of the SAR trends of GBAP molecules (McBrayer et al. 2017). That study identified a potent GBAP analogs which acts as an activator for FsrC in the picomolar range.

Screening approaches, using natural compound libraries, have offered some interesting histidine kinase-based AIP antagonists. Desouky et al. established a high-throughput screening system using *S. aureus* reporter strains 8325–4. They screened 1000 culture extracts of actinomycetes and identified three cyclodepsipep-tides (WS9326A, WS9326B, and cochinmicin II/III) having anti-QS activity in a micromolar range against pathogenic species of Gram-positive bacteria (Desouky et al. 2013, 2015). These three compounds attenuate Fsr QS-mediated gelatinase expression in *E. faecalis* in a competitive manner, suggesting that these compounds act as antagonists. Indeed, these compounds specifically inhibit the binding of fluorescence-labeled GBAP to FsrC. WS9326A and WS9326B significantly mitigate hemolysis induced by types I, II and III of *S. aureus*, suggesting that these two inhibitors are less specific. WS9326A also significantly inhibits the expression of *pfoA* toxin gene in *C. perfringens,* with IC_{50} of 0.88 µM. The efficacy of WS9326B was demonstrated with a cytotoxicity assay, using *S. aureus* on corneal epithelial cells; the results indicated reduced cytotoxicity at IC_{50} of 20 µM (Desouky et al. 2015). Avellanin C, the cyclic peptide obtained from the fungus *Hamigera ingelheimensis,* was found to attenuate agr-based QS in the *S. aureus* reporter strain (8325–4), with an IC_{50} value of 4.4 µM (Igarashi et al. 2015). These cyclic peptides seem to act as antagonists of AIPs because of their structural analogy with AIP molecules and their binding activity against the histidine kinase receptor. However, it warrants further study to explain their broad range of anti-QS activity.

9.4.9 RNAIII Inhibiting Peptides

Another signal transduction pathway, in staphylococci, is the RAP-TRAP system, which co-regulates the expression of virulence genes with the Agr QS system (Balaban et al. 2005). In this system, RAP (RNAIII activating protein) induces the expression of RNAIII through phosphorylation of TRAP (target of RAP). Interestingly, non-pathogenic staphylococci produces a RAP analog peptide, termed (RNAIII inhibiting peptide) which inhibits the RAP-TRAP system (Balaban et al. 2005). Non-natural RIPs have been extensively developed, notably for the control of biofilm formation by pathogenic staphylococci (Giacometti et al. 2003; Dell'Acqua et al. 2004; Balaban et al. 2005). Different strategies, such as amino acid substitution, modification and oligomerization have been used to improve the efficacy and stability of RIP analogs. The results of alanine scanning and truncation of the RIP peptide (YSPWTNF) have led to the discovery of a novel truncated peptide with the sequence YSPWT; Thr-5 is critical for its activity and shows the best inhibitory activity against *Staphylococcus* (Fig. 9.12b) (Baldassarre et al. 2013). Similarly, another modified RIP derivative, FS8 (YSPWTAA), synthesized by substituting the phenylalanine residue with an alanine at position 7, in conjunction with the antibiotic, tigecycline, prevented prosthesis biofilm in a rat model (Simonetti et al. 2013). Oligomerization of RIP peptides has been shown to significantly suppress adherence and biofilm formation by MRSA, both *in vitro* and *in vivo*, and may be a potential drug against MRSA infections (Zhou et al. 2016). RIP-loaded poly(methyl methacrylate) beads have been very effective in suppressing biofilm formation by MRSA *in vitro* and *in vivo* (Anguita-Alonso et al. 2007). A non-peptide analog of RIP, 2,5-di-O-galloyl-d-hamamelose (hamamelitannin), has also been shown to inhibit RNAIII production, cell attachment, and biofilm formation (Kiran et al. 2008). Hamamelitannin enhances the susceptibility of *S. aureus* to antibiotic treatment by altering cell wall synthesis (Brackman et al. 2016). This compound was also analyzed for *in vivo* applications and has been shown to attenuate MRSA and *S. epidermidis* infections in a rat graft model (Kuo et al. 2015). Biaryl hydroxyketones, in conjunction with nafcillin and cephalothin, were found to significantly reduce the MIC value of these antibiotics to the value equivalent to that of vancomycin when used in murine models (Kuo et al. 2015).

9.4.10   Other Types of Inhibitors Targeting QS of Gram-Positive Bacteria

Two compounds, savirin and benzbromarone, with inhibitory activity against *agr*, were identified during random screening (Sully et al. 2014). Savirin inhibits AIP-induced RNAIII expression, at submicromolar concentrations, by interfering with AgrA binding to DNA (Sully et al. 2014). This inhibitory effect was observed for all four *S. aureus* agr groups but not for *S. epidermidis* strain 12228 (Gordon et al. 2013; Sully et al. 2014). Similarly, other compounds, such as ω-hydroxyemodin and colostrum hexasaccharide, were observed to inhibit agr-mediated pathogenesis by *S. aureus* (Daly et al. 2015; Srivastava et al. 2015).

Nakayama et al. found that a fungal secondary metabolite, ambuic acid, possesses anti-Fsr QS activity (Nakayama et al. 2009). Unlike siamycin, this inhibitory activity was abolished by adding a low dose of synthetic GBAP, suggesting that ambuic acid blocks the biosynthesis of GBAP in *E. faecalis*. The biosynthesis of GBAP is mediated by FsrB, in which the GBAP propeptide is processed and cyclized by the cysteine protease activity of FsrB. Ambuic acid inhibits this processing by blocking protease activity. Interestingly, ambuic acid inhibits not only the biosynthesis of GBAP in *E. faecalis*, but also those in *S. aureus* and *Listeria innocua*. This suggests that ambuic acid targets a common processing mechanism shared among the cyclic AIP-mediated QS systems in Gram-positive bacteria. Recently, the efficacy of ambuic acid was shown *in vitro* as well as *in vivo* using a MRSA strain (Fig. 9.12b) (Todd et al. 2017). With this unique mode of action, ambuic acid may be a model compound for the development of anti-QS drugs.

9.4.11   Inhibition of AI-2 Signaling

Since the LuxS/AI-2 QS system is shared widely among Gram-negative and Gram-positive bacteria, inhibition of AI-2 signaling may be a target for developing novel therapeutic options (Brackman et al. 2013; Guo et al. 2013). The inhibition of AI-2-mediated QS can be achieved by inhibition of the LuxS synthase; sequestration of DPD, which is a precursor to AI-2; or antagonistic inhibition of the AI-2 receptor. Alfaro et al. synthesized two analogs of *S*-ribosyl-*L*-homocysteine: *S*-anhydroribosyl-*L*-homocysteine and *S*-homoribosyl-*L*-cysteine. These two inhibitors were used to suppress the activity of the LuxS enzyme (Fig. 9.12c) (Alfaro et al. 2004). However, these inhibitors exhibit weak activity, with an IC_{50} value of $\approx$ 1 mM. A series of LuxS inhibitors were designed and synthesized by Shen and colleagues which led

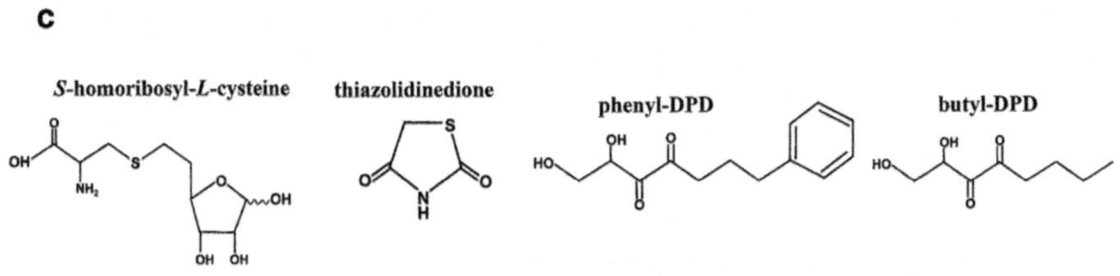

c

S-homoribosyl-L-cysteine   thiazolidinedione   phenyl-DPD   butyl-DPD

Fig. 9.12c Structure of QSIs targeting LuxS/AI-2 signaling pathways

to discovery of some potent and competitive structural analogs of SRH, such as compounds A and B with the Ki values of 0.72 and 0.37 μM, respectively (Shen et al. 2006). The SAR, assessed in this study, suggests that homocysteine and ribose moieties are critical for high-affinity binding to the LuxS catalytic site (Shen et al. 2006). Along with traditional methods of exploring LuxS inhibitors, many researchers have used computer-assisted virtual screening (Li et al. 2008b; Brackman et al. 2013; Shafreen et al. 2014). Two antagonists targeting AI-2 signaling, KM-03009 and SPB-02229, discovered via virtual screening, show inhibitory activity against AI-2 mediated QS at IC_{50} of 35 and 55 μM, respectively, with no cytotoxicity (Fig. 9.12c) (Li et al. 2008b). Cinnamaldehyde and its derivatives were evaluated for their inhibitory activity against AI-2-mediated QS; these agents were consequently proposed for the interception QS in the *Vibrio* spp. because they inhibit LuxR binding to DNA and, thus interfere with biofilm formation and virulence (Niu et al. 2006; Brackman et al. 2008; Brackman et al. 2011). Shafreen et al. further analyzed cinnamaldehyde and its derivatives for antagonistic activity using molecular docking and simulation. That study demonstrated the potent inhibitory activity of these compounds against biofilm formation and virulence factor production by *S. pyogenes* via the LuxS-mediated QS (Shafreen et al. 2014). Brackman et al. designed and synthesized a series of thiazolidinediones and dioxazaborocanes compounds for evaluating obstruction of AI-2 signaling. Thiazolidinediones have structural similarity to furanone compounds, such as *N*-acylaminofuranones signaling molecules, while dioxazaborocanes resemble oxazaborolidine. Among these, thiazolidinediones showed potency, as AI-2 QSIs, in a low micromolar range (Fig. 9.12c) (Brackman et al. 2013). Thiazolidinediones act by suppressing the binding of LuxR to DNA in *V. harveyi*. Although dioxazaborocanes are less potent, their mechanism of action targets LuxPQ.

Several AI-2 antagonists and analogs have been designed and synthesized for intercepting AI-2-mediated QS (Alfaro et al. 2004; Shen et al. 2006; Wnuk et al. 2009). Among the synthesized compounds, alkyl-4,5-dihydroxy-2,3-pentanedione analogs and their modified derivatives, manipulate AI-2-mediated QS in *S. typhimurium* and *V. harveyi* (Lowery et al. 2009). These compounds in combination with antibiotic gentamicin, are highly effective at clearing preexisting biofilms produced by *E. coli* (Roy et al. 2013). The use of phenyl DPD, isobutyl-DPD, and the C1-alkyl AI-2 analog, along with gentamicin, resulted in clearance of preformed

P. aeruginosa biofilms (Roy et al. 2013). These studies suggest that AI-2 inhibitors have broad-spectrum anti-QS properties in conjunction with antibiotics.

CRISPR (Clustered, Regularly Interspaced, Short, Palindromic, Repeat) interference is a new concept in targeting AI-2 signaling; it is used for the knockdown of LuxS gene expression to inhibit biofilm formation by *E. coli* (Zuberi et al. 2017). This approach can potentially attenuate LuxS-mediated QS and inhibit biofilm formation by bacterial populations in nosocomial and environmental settings.

9.5   Emergence of Resistance to QSIs

Various studies have led to the discovery of numerous small chemical compounds, called QSIs, which can disarm pathogens without being bactericidal. However, there is an ongoing debate about QSIs, assuming that development of resistance to novel drugs is inevitable. Some researchers believe that generation of resistance to novel drugs occurs as rapidly as it does to antibiotics (Maeda et al. 2012; Kalia et al. 2014; García-Contreras et al. 2016). However, many reports have confirmed that there are different barriers which can preclude the spread of resistance to QSIs (Rasmussen and Givskov 2006; Rasko and Sperandio 2010; Gerdt and Blackwell 2014). The development of resistance has been examined using competition experiments with QS-deficient vs QS-proficient strain mimics (Gerdt and Blackwell 2014; Ruer et al. 2015). Resistance to QSIs can develop when QQ and growth are tightly dependent on a particular QS system. For example, in *P. aeruginosa*, the LasR QS system controls both virulence and adenosine metabolism (with adenosine as the sole carbon source) (Heurlier et al. 2005). If LasR is targeted under these conditions, consequently growth is also targeted, leading to the development of resistance to these quenchers, albeit at different rates. Similarly, García-Contreras and colleagues reported that QS inhibition, in milieu with stress, leads to bacterial resistance to QSIs (García-Contreras et al. 2016). Infections in animals was successfully treated with QSIs; however, concomitant diminishing of the bacterial pathogen load was observed with treatment, and this can lead to high selective pressure for the selection of resistance (Defoirdt et al. 2010). Moreover, some clinical strains are resistant to certain QQ compounds without any pre-exposure (Maeda et al. 2012; García-Contreras et al. 2013). Potent inhibitors, such as brominated furanone C-30, show higher toxicity and higher potential for inducing resistance, whereas inhibitors with modest activity have less toxicity and less potential for inducing resistance (Gerdt and Blackwell 2014). Inhibitors such as acylases and lactonases, which disrupt AIs, are considered specific and potent; they can be used *in vivo* to determine whether QSIs are inducing resistance.

Some researchers argue that selection for resistance can be reduced, or even reversed, by the careful integration of novel antivirulence drugs and existing traditional therapeutics (Allen et al. 2014). Although QSI therapy may select for resistance as evolution is a natural course of progression, the advantage of evolution can be used to design robust QSIs with minimal resistance-inducing potential (García-

Contreras et al. 2016). The development of novel QS modulators will be critical for understanding the intra- and inter- QS signaling systems at the molecular level, consequently augmenting our ability to manipulate such systems for our benefit.

9.6 Biotechnological Applications

9.6.1 Biosensors

AI biosensors, using reporter strains, can be used for the fast and reliable detection of AI producing bacteria. Numerous biosensors have been developed; some are specific for a certain AI molecule(s), whereas others detect a broad range of AIs (Verbeke et al. 2017). AI sensor strains are basically equipped with an AI transcriptional regulator and a reporter gene under the control of cognate promoter, allowing sensitive and rapid detection of corresponding AIs (Steindler and Venturi 2007; Fletcher et al. 2014). The most commonly used AI biosensor is *Chromobacterium violaceum* VIR07, which is a CviI-knockout mutant producing violacin pigment in response to exogenous AHLs (C10-C16) (McClean et al. 1997). A commonly used biosensor, based on the LuxR receptor in *C. violaceum,* produces antibacterial violacein using the CviI/R AHL QS and is efficient at detecting C6-AHL (McClean et al. 1997). Other biosensors have been constructed to detect a broad range of AHLs. For instance, plasmid pSB401 was constructed using *V. fischeri luxRI* DNA fragment containing a luxR transcriptional activator and *luxI* promoter fused with the *luxCDABE* operon of *Photorhabdus luminescens. E. coli,* carrying pSB401, can detect a wide range of AHLs (C4 to C14) at varying sensitivities (Anbazhagan et al. 2012). The authors also constructed pSB406 and pSB1075 using different reporter genes, *rhlRI:luxCDABE* fusion and *lasRI:luxCDABE* fusion, and introduced these plasmids into *P. aeruginosa* (Winson et al. 1998; Anbazhagan et al. 2012). A dual biosensor was constructed to quantitatively detect PQS and HHQ separately (Fletcher et al. 2007a, b). These biosensors have potential to detect AHLs and their homologs in bacterial culture supernatants and other sources (Fletcher et al. 2007a, b; Müller and Fetzner 2013).

9.6.2 Clinical Applications

Whole-cell biosensors can also be used for clinical and industrial diagnostics to detect the presence of pathogens, or deliver a drug to cancer cells at different sites, such as murine bladder, breast, and brain. The *P. aeruginosa* 3-oxo-C12-HSL was found to impede proliferation of breast cancer cell lines and induce apoptosis (Li et al. 2004). In order to make 3-oxo-C12-HSL efficient for treating cancer, its pro-QS activity needs to be abrogated while retaining its anticancer activity. Anderson et al. engineered bacteria that sense cancerous cells and invade them by releasing a cytotoxic agent (Anderson et al. 2006). QS systems, LasI/LasR and RhlI/RhlR of *V. fischeri* and *P. aeruginosa*, respectively, have been employed to construct the AHL signal transduction system in *E. coli* (Brenner et al. 2008). This artificial AHL QS system can control biofilm formation in a microbial consortium and can be used repeatedly in clinical and industrial settings (Brenner et al. 2008; Hong et al. 2012). Synthetic biology was also used to construct a QS-controllable genetic system, which enabled *E. coli* to efficiently produce toxins capable of killing virulent *P. aeruginosa* and inhibiting biofilm formation (Saeidi et al. 2011; Gupta et al. 2013).

9.6.3 Engineered Microbial Consortia

QS systems can be used to construct engineered microbial consortia (Scott and Hasty 2016). These consortia have potential applications in medicine, human and environmental health, bioenergy, and bio computations (Hays et al. 2015). A strain of *E. coli* was recently engineered to detect and memorize signals in the mouse gut, thereby showing the potential of living therapeutics in the changing microbiome (Kotula et al. 2014). Similarly, QS-based engineered bacterial communication has been shown effective in preventing virulence (Duan and March 2010; Chu et al. 2015). The synchronous use of these systems and synthetic biology will open up new windows towards understanding of naturally occurring microbial consortia and help engineer these consortia for novel therapeutic applications.

9.7 Conclusion

QS is the central part of unicellular bacterial life through which bacteria modulate gene expression in order to adjust themselves according to the surroundings. This post-autocrine signaling cascade enables bacteria to orchestrate numerous phenotypic characteristics many of which are related to virulence and pathogenesis. These QS-controlled bacterial processes significantly impact the health of humans, animals, and the environment. The increasing antibiotic resistance, exhibited by numerous life-threatening pathogens, is alarming. In 2017, the WHO issued a call for immediate global action to tackle the issue of MDR menace.

Considerable research efforts are being dedicated into understanding the mechanism of QS; unique approaches, for interfering with this signaling cascade, are being developed. Numerous potent QSIs have been identified and synthesized with strong antagonistic activity at IC_{50} in nanomolar or submicromolar concentrations. Some of these QSIs, assessed for efficacy *in vitro* and *in vivo*, show stability and can suppress pathogenicity; however, few clinical trials have been conducted. The first clinical trial with human patients, assessing macerated garlic oil as a QS inhibitor of *P. aeruginosa*, was conducted to test this potential therapeutic for the treatment of cystic fibrosis (Smyth et al. 2010).

Although there are concerns that QSIs can exert antibiotic-like effects in bacteria, inducing bacteria to evolve and spread resistance (Kalia et al. 2014; García-Contreras et al. 2016). But QSIs will have lower rates of inducing resistance than do antibiotics as QSIs do not interfere with signaling or growth of normal and beneficial microflora. The synergistic use of QSIs and antibiotics can greatly help control infection, aid in the diagnostics and delivery of drugs, and help engineer microbial consortia. Because resistance to antimicrobial therapeutics cannot be avoided completely, we need to overlook resistance phobia and, instead take advantage of it in developing novel strategies for keeping resistance to a minimum or consider QSIs which are signal-independent.

In order to make QSIs applicable in clinical trials, and as future drugs, we need to better understand the intertwining associations between the multitude of QS pathways, the biological implications of interfering with these pathways, and the functional and mechanistic characteristics of each QS systems within bacteria. Exploring the origin of autocrine signaling; intricacies between autocrine signaling and QS systems, across different bacterial species, can provide new directions toward development of robust QSIs. Furthermore, CRISPER technology is a unique approach for the knockdown of genes involved in QS-signaling mediated pathogenesis.

This review, which presents an overview of QS and QSIs of MDR pathogens, will be a motivation toward in-depth exploration in QS research so that the unraveled questions in this field can be addressed and render this approach a possible additive, or alternative, to antibiotics.

Acknowledgments This work was supported by JSPS KAKENHI Grant Number JP24380050, JP15H04480, and JP16F16101 for J. Nakayama. B. Yousuf gratefully acknowledges the "Japan Society for the Promotion of Science (JSPS)" for providing overseas Postdoctoral Fellowship.

References

Ahmed NA, Petersen FC, Scheie AA (2009) AI-2/LuxS is involved in increased biofilm formation by *Streptococcus intermedius* in the presence of antibiotics. Antimicrob Agents Chemother 53:4258–4263. https://doi.org/10.1128/AAC.00546-09

Ahmer BM (2004) Cell-to-cell signalling in *Escherichia coli* and *Salmonella enterica*. Mol Microbiol 52:933–945. https://doi.org/10.1111/j.1365-2958.2004.04054.x

Alfaro JF, Zhang T, Wynn DP, Karschner EL, Zhou ZS (2004) Synthesis of LuxS inhibitors targeting bacterial cell-cell communication. Org Lett 6:3043–3046. https://doi.org/10.1021/ol049182i

Allen RC, Popat R, Diggle SP, Brown SP (2014) Targeting virulence: can we make evolution-proof drugs? Nature Rev Microbiol 12:300–308. https://doi.org/10.1038/nrmicro3232

Amara N, Krom BP, Kaufmann GF, Meijler MM (2010) Macromolecular inhibition of quorum sensing: enzymes, antibodies, and beyond. Chem Rev 111:195–208. https://doi.org/10.1021/cr100101c

Anbazhagan D, Mansor M, Yan GO, Yusof MY, Hassan H, Sekaran SD (2012) Detection of quorum sensing signal molecules and identification of an autoinducer synthase gene among biofilm forming clinical isolates of *Acinetobacter* spp. PloS One 7:e36696. https://doi.org/10.1371/journal.pone.0036696

Anderson JC, Clarke EJ, Arkin AP, Voigt CA (2006) Environmentally controlled invasion of cancer cells by engineered bacteria. J Mol Biol 355:619–627. https://doi.org/10.1016/j.jmb.2005.10.076

Anguita-Alonso P, Giacometti A, Cirioni O, Ghiselli R, Orlando F, Saba V, Scalise G, Sevo M, Tuzova M, Patel R, Balaban N (2007) RNAIII-inhibiting-peptide-loaded polymethylmethacrylate prevents in vivo Staphylococcus aureus biofilm formation. Antimicrob Agents Chemother 51:2594–2596. https://doi.org/10.1128/AAC.00580-06

Ansaldi M, Marolt D, Stebe T, Mandic-Mulec I, Dubnau D (2002) Specific activation of the Bacillus quorum-sensing systems by isoprenylated pheromone variants. Mol Microbiol 44:1561–1573. https://doi.org/10.1046/j.1365-2958.2002.02977.x

Autret N, Raynaud C, Dubail I, Berche P, Charbit A (2003) Identification of the agr locus of Listeria monocytogenes: role in bacterial virulence. Infect Immun 71:4463–4471. https://doi.org/10.1128/IAI.71.8.4463-4471

Bae T, Kozlowicz BK, Dunny GM (2004) Characterization of cis-acting prgQ mutants: evidence for two distinct repression mechanisms by Qa RNA and PrgX protein in pheromone-inducible enterococcal plasmid pCF10. Mol Microbiol 51:271–281. https://doi.org/10.1046/j.1365-2958.2003.03832.x

Balaban N, Stoodley P, Fux CA, Wilson S, Costerton JW, Dell'Acqua G (2005) Prevention of staphylococcal biofilm-associated infections by the quorum sensing inhibitor RIP. Clin Orthop Relat Res:48–54. https://doi.org/10.1097/01.blo.0000175889.82865.67

Baldassarre L, Fornasari E, Cornacchia C, Cirioni O, Silvestri C, Castelli P, Giocometti A, Cacciatore I (2013) Discovery of novel RIP derivatives by alanine scanning for the treatment of S. aureus infections. Med Chem Comm 4:1114–1117. https://doi.org/10.1039/C3md00122a

Barrett JF, Goldschmidt RM, Lawrence LE, Foleno B, Chen R, Demers JP, Johnson S, Kanojia R, Fernandez J, Bernstein J, Licata L (1998). Antibacterial agents that inhibit two-component signal transduction systems. Proc Natl Acad Sci U S A 95:5317–5322. doi: Not available

Bassler BL (1999) How bacteria talk to each other: regulation of gene expression by quorum sensing. Curr Opin Microbiol 2:582–587. doi: Not available

Bassler B (2015) Manipulating quorum sensing to control bacterial pathogenicity. FASEB J 29:88–101. doi: Not available

Bassler BL, Losick R (2006) Bacterially speaking. Cell 125:237–246. https://doi.org/10.1016/j.cell.2006.04.001

Bassler B, Vogel J (2013) Bacterial regulatory mechanisms: the gene and beyond. Curr Opin Microbiol 16:109–111. https://doi.org/10.1016/j.mib.2013.04.001

Bassler BL, Wright M, Silverman MR (1994) Multiple signaling systems controlling expression of luminescence in Vibrio harveyi: sequence and function of genes encoding a second sensory pathway. Mol Microbiol 13:273–286. https://doi.org/10.1111/j.1365-2958.1994.tb00422.x

Bearson BL, Bearson SM, (2008) The role of the QseC quorum-sensing sensor kinase in colonization and norepinephrine-enhanced motility of Salmonella enterica serovar Typhimurium. Microb Pathog 44:271–278. https://doi.org/10.1016/j.micpath.2007.10.001

Bem AE, Velikova N, Pellicer MT, Baarlen Pv, Marina A, Wells JM (2015) Bacterial histidine kinases as novel antibacterial drug targets. ACS Chem Biol 10:213–224. doi: https://doi.org/10.1021/cb5007135

Bhardwaj AK, Vinothkumar K, Rajpara N (2013) Bacterial quorum sensing inhibitors: attractive alternatives for control of infectious pathogens showing multiple drug resistance. Recent Pat Antiinfect Drug Discov 8:68–83. https://doi.org/10.2174/157489113805290809

Bhargava N, Sharma P, Capalash N (2015) Quorum sensing in Acinetobacter baumannii. In: Kalia VC (ed) Quorum sensing vs quorum quenching: a battle with no end in sight, Springer India, New Delhi, pp 101–113. ISBN 978-81-322-1982-8. https://doi.org/10.1007/978-81-322-1982-8_10

Bjarnsholt T, Jensen PØ, Burmølle M, Hentzer M, Haagensen JA, Hougen HP, Calum H, Madsen KG, Moser C, Molin S, Høiby N (2005) Pseudomonas aeruginosa tolerance to tobramycin, hydrogen peroxide and polymorphonuclear leukocytes is quorum-sensing dependent. Microbiology 151:373–383. https://doi.org/10.1099/mic.0.27463-0

Boles BR, Horswill AR (2008) Agr-mediated dispersal of *Staphylococcus aureus* biofilms. PLoS Pathog 4:e1000052. doi. https://doi.org/10.1371/journal.ppat.1000052

Boyer M, Wisniewski-Dyé F (2009) Cell-cell signalling in bacteria: not simply a matter of quorum. FEMS Microbiol Ecol 70:1–19. doi. https://doi.org/10.1111/j.1574-6941.2009.00745.x

Brackman G, Coenye T (2015) Quorum sensing inhibitors as anti-biofilm agents. Curr Pharm Des 21:5–11. doi:Not available

Brackman G, Defoirdt T, Miyamoto C, Bossier P, Van Calenbergh S, Nelis H, Coenye T (2008) Cinnamaldehyde and cinnamaldehyde derivatives reduce virulence in Vibrio spp. by decreasing the DNA-binding activity of the quorum sensing response regulator LuxR. BMC Microbiol 8:149. https://doi.org/10.1186/1471-2180-8-149

Brackman G, Cos P, Maes L, Nelis HJ, Coenye T (2011) Quorum sensing inhibitors increase the susceptibility of bacterial biofilms to antibiotics *in vitro* and *in vivo*. Antimicrob Agents Chemother 55:2655–2661. https://doi.org/10.1128/AAC.00045-11

Brackman G, Al Quntar AA, Enk CD, Karalic I, Nelis HJ, Van Calenbergh S, Srebnik M, Coenye T (2013) Synthesis and evaluation of thiazolidinedione and dioxazaborocane analogues as inhibitors of AI-2 quorum sensing in *Vibrio harveyi*. Bioorg Med Chem 21:660–667. https://doi.org/10.1016/j.bmc.2012.11.055

Brackman G, Breyne K, De Rycke R, Vermote A, Van Nieuwerburgh F, Meyer E, Van Calenbergh S, Coenye T (2016) The quorum sensing inhibitor hamamelitannin increases antibiotic susceptibility of *Staphylococcus aureus* biofilms by affecting peptidoglycan biosynthesis and eDNA release. Sci Rep 6:20321. https://doi.org/10.1038/srep20321

Brenner K, You L, Arnold FH (2008) Engineering microbial consortia: a new frontier in synthetic biology. Trends Biotech 26:483–489. https://doi.org/10.1016/j.tibtech.2008.05.004

Brint JM, Ohman DE (1995) Synthesis of multiple exoproducts in *Pseudomonas aeruginosa* is under the control of RhlR-RhlI, another set of regulators in strain PAO1 with homology to the autoinducer-responsive LuxR-LuxI family. J Bacteriol 177:7155–7163. https://doi.org/10.1128/jb.177.24.7155-7163

Camilli A, Bassler BL (2006) Bacterial small-molecule signaling pathways. Science 311:1113–1116. https://doi.org/10.1126/science.1121357

Canovas J, Baldry M, Bojer MS, Andersen PS, Grzeskowiak PK, Stegger M, Damborg P, Olsen CA, Ingmer H (2016) Cross-talk between *Staphylococcus aureus* and other *Staphylococcal* species via the *agr* quorum sensing system. Front Microbiol 7:1733. https://doi.org/10.3389/fmicb.2016.01733

Carter GP, Purdy D, Williams P, Minton NP (2005) Quorum sensing in *Clostridium difficile*: analysis of a luxS-type signalling system. J Med Microbiol 54:119–127. https://doi.org/10.1099/jmm.0.45817-0

Carter GP, Cheung JK, Larcombe S, Lyras D (2014) Regulation of toxin production in the pathogenic clostridia. Mol Microbiol 91:221–231. https://doi.org/10.1111/mmi.12469

Cartman ST, Kelly ML, Heeg D, Heap JT, Minton NP (2012) Precise manipulation of the *Clostridium difficile* chromosome reveals a lack of association between the tcdC genotype and toxin production. Appl Environ Microbiol 78:4683–4690. https://doi.org/10.1128/AEM.00249-12

Cathcart GR, Quinn D, Greer B, Harriott P, Lynas JF, Gilmore BF, Walker B (2011) Novel inhibitors of the *Pseudomonas aeruginosa* virulence factor LasB: a potential therapeutic approach for the attenuation of virulence mechanisms in pseudomonal infection. Antimicrob Agents Chemother 55:2670–2678. https://doi.org/10.1128/AAC.00776-10

Chan WC, Coyle BJ, Williams P (2004) Virulence regulation and quorum sensing in *Staphylococcal* infections: competitive AgrC antagonists as quorum sensing inhibitors. J Med Chem 47:4633–4641. https://doi.org/10.1021/jm0400754

Chan KG, Liu YC, Chang CY (2015) Inhibiting *N*-acyl-homoserine lactone synthesis and quenching *Pseudomonas* quinolone quorum sensing to attenuate virulence. Front Microbiol 6:1173. https://doi.org/10.3389/fmicb.2015.01173

Chang CY, Krishnan T, Wang H, Chen Y, Yin WF, Chong YM, Tan LY, Chong TM, Chan KG (2014) Non-antibiotic quorum sensing inhibitors acting against *N*-acyl homoserine lactone synthase as druggable target. Sci Rep 4:7245. https://doi.org/10.1038/srep07245

Chen X, Schauder S, Potier N, Van Dorsselaer A, Pelczer I, Bassler BL, Hughson FM (2002) Structural identification of a bacterial quorum-sensing signal containing boron. Nature 415:545–549. https://doi.org/10.1038/415545a

Chu YY, Nega M, Wölfle M, Plener L, Grond S, Jung K, Götz F (2013). A new class of quorum quenching molecules from *Staphylococcus* species affects communication and growth of gram-negative bacteria. PLoS Pathog 9:e1003654. https://doi.org/10.1371/journal.ppat.1003654

Chu T, Ni C, Zhang L, Wang Q, Xiao J, Zhang Y, Liu Q (2015) A quorum sensing-based *in vivo* expression system and its application in multivalent bacterial vaccine. Microb Cell Fact 14:37. https://doi.org/10.1186/s12934-015-0213-9

Chugani SA, Whiteley M, Lee KM, D'Argenio D, Manoil C, Greenberg EP (2001) QscR, a modulator of quorum-sensing signal synthesis and virulence in *Pseudomonas aeruginosa*. Proc Natl Acad Sci U S A 98:2752–2757. https://doi.org/10.1073/pnas.051624298

Clarke MB, Hughes DT, Zhu C, Boedeker EC, Sperandio V (2006) The QseC sensor kinase: a bacterial adrenergic receptor. Proc Natl Acad Sci U S A 103:10420–10425. https://doi.org/10.1073/pnas.0604343103

Coburn PS, Pillar CM, Jett BD, Haas W, Gilmore MS (2004) *Enterococcus faecalis* senses target cells and in response expresses cytolysin. Science 306:2270–2272. https://doi.org/10.1126/science.1103996

Connan C, Brüggemann H, Mazuet C, Raffestin S, Cayet N, Popoff MR (2012) Two-component systems are involved in the regulation of botulinum neurotoxin synthesis in *Clostridium botulinum* Type A Strain Hall. PloS One 7:10. https://doi.org/10.1371/annotation/c61c1b9e-b406-4057-99c6-ff84d67869bf

Cook LC, Federle MJ (2014) Peptide pheromone signaling in *Streptococcus* and *Enterococcus*. FEMS Microbiol Rev 38:473–492. https://doi.org/10.1111/1574-6976.12046

Cooksley CM, Davis IJ, Winzer K, Chan WC, Peck MW, Minton NP (2010) Regulation of neurotoxin production and sporulation by a putative agrBD signaling system in proteolytic *Clostridium botulinum*. Appl Environ Microbiol 76:4448–4460. https://doi.org/10.1128/AEM.03038-09

Daly SM, Elmore BO, Kavanaugh JS, Triplett KD, Figueroa M, Raja HA, El-Elimat T, Crosby HA, Femling JK, Cech NB, Horswill AR, Oberlies NH, Hall PR (2015) ω-Hydroxyemodin limits *Staphylococcus aureus* quorum sensing-mediated pathogenesis and inflammation. Antimicrob Agents Chemother 59:2223–2235. https://doi.org/10.1128/AAC.04564-14

Darkoh C, DuPont HL (2017) The accessory gene regulator-1 as a therapeutic target for *C. difficile* infections. Expert Opin Ther Targets 21:451–453. https://doi.org/10.1080/14728222.2017.1311863

Darkoh C, DuPont HL, Norris SJ, Kaplan HB (2015) Toxin synthesis by *Clostridium difficile* is regulated through quorum signaling. MBio 6:e02569. https://doi.org/10.1128/mBio.02569-14

Darkoh C, Odo C, DuPont HL (2016) Accessory gene regulator-1 locus is essential for virulence and pathogenesis of *Clostridium difficile*. MBio 7:e01237-16. https://doi.org/10.1128/mBio.01237-16

Davies DG, Parsek MR, Pearson JP, Iglewski BH, Costerton JT, Greenberg EP (1998) The involvement of cell-to-cell signals in the development of a bacterial biofilm. Science 280:295–298. https://doi.org/10.1126/science.280.5361.295

de Almeida FA, de Jesus Pimentel-Filho N, Pinto UM, Mantovani HC, de Oliveira LL, Vanetti MCD (2017) Acyl homoserine lactone-based quorum sensing stimulates biofilm formation by *Salmonella enteritidis* in anaerobic conditions. Arch Microbiol 199:475–486. https://doi.org/10.1007/s00203-016-1313-6

De Keersmaecker SC, Sonck K, Vanderleyden J (2006) Let LuxS speak up in AI-2 signaling. Trends Microbiol 14:114–119. https://doi.org/10.1016/j.tim.2006.01.003

De Kievit TR, Gillis R, Marx S, Brown C, Iglewski BH (2001) Quorum-sensing genes in *Pseudomonas aeruginosa* biofilms: their role and expression patterns. Appl Environ Microbiol 67:1865–1873. https://doi.org/10.1128/AEM.67.4.1865-1873

Declerck N, Bouillaut L, Chaix D, Rugani N, Slamti L, Hoh F, Lereclus D, Arold ST (2007) Structure of PlcR: insights into virulence regulation and evolution of quorum sensing in gram-positive bacteria. Proc Natl Acad Sci U S A 104:18490–18495. https://doi.org/10.1073/pnas.0704501104

Defoirdt T, Boon N, Bossier P (2010) Can bacteria evolve resistance to quorum sensing disruption?. PLoS pathogens. 6:e1000989. https://doi.org/10.1371/journal.ppat.1000989

Dell'Acqua G, Giacometti A, Cirioni O, Ghiselli R, Saba V, Scalise G, Gov Y, Balaban N (2004) Suppression of drug-resistant Staphylococcal infections by the quorum-sensing inhibitor RNAIII-inhibiting peptide. J Infect Dis 190:318–320. https://doi.org/10.1086/386546

Desouky SE, Nishiguchi K, Zendo T, Igarashi Y, Williams P, Sonomoto K, Nakayama J (2013) High-throughput screening of inhibitors targeting Agr/Fsr quorum sensing in *Staphylococcus aureus* and *Enterococcus faecalis*. Bioscience Biotec Biochem 77:923–927. https://doi.org/10.1271/bbb.120769

Desouky SE, Shojima A, Singh RP, Matsufuji T, Igarashi Y, Suzuki T, Yamagaki T, Okubo KI, Ohtani K, Sonomoto K, Nakayama J (2015) Cyclodepsipeptides produced by actinomycetes inhibit cyclic-peptide-mediated quorum sensing in Gram-positive bacteria. FEMS Microbiol Lett 362:109 https://doi.org/10.1093/femsle/fnv109

Devaud M, Kayser FH, Bächi B (1982) Transposon-mediated multiple antibiotic resistance in Acinetobacter strains. Antimicrob Agents Chemother 22:323–329. https://doi.org/10.1128/AAC.22.2.323

Deziel E, Lepine F, Milot S, He J, Mindrinos MN, Tompkins RG, Rahme LG (2004) Analysis of *Pseudomonas aeruginosa* 4-hydroxy-2-alkylquinolines (HAQs) reveals a role for 4-hydroxy-2-heptylquinoline in cell- to-cell communication. Proc Natl Acad Sci U S A 101:1339–1344. https://doi.org/10.1073/pnas.0307694100

Dickey SW, Cheung GY, Otto M (2017) Different drugs for bad bugs: antivirulence strategies in the age of antibiotic resistance. Nat Rev Drug Discov 16:457–471. https://doi.org/10.1038/nrd.2017.23

Doğaner BA, Yan LK, Youk H (2016) Autocrine signaling and quorum sensing: extreme ends of a common spectrum. Trends Cell Biol 26:262–271. https://doi.org/10.1016/j.tcb.2015.11.002

Dong YH, Xu JL, Li XZ, Zhang LH (2000) AiiA, an enzyme that inactivates the acylhomoserine lactone quorum-sensing signal and attenuates the virulence of *Erwinia carotovora*. Proc Natl Acad Sci U S A 97:3526–3531. https://doi.org/10.1073/pnas.060023897

Dou Y, Song F, Guo F, Zhou Z, Zhu C, Xiang J, Huan J (2017) *Acinetobacter baumannii* quorum-sensing signalling molecule induces the expression of drug-resistance genes. Mol Med Rep 15:4061–4068. https://doi.org/10.3892/mmr.2017.6528

Drees SL, Fetzner S (2015) PqsE of *Pseudomonas aeruginosa* acts as pathway-specific thioesterase in the biosynthesis of alkylquinolone signaling molecules. Chem Biol 22:611–618. https://doi.org/10.1016/j.chembiol.2015.04.012

Duan F, March JC (2010) Engineered bacterial communication prevents *Vibrio cholerae* virulence in an infant mouse model. Proc Natl Acad Sci U S A 107:11260–11264. https://doi.org/10.1073/pnas.1001294107

Dubern JF, Diggle SP (2008) Quorum sensing by 2-alkyl-4-quinolones in *Pseudomonas aeruginosa* and other bacterial species. Mol BioSyst 4:882–888. https://doi.org/10.1039/B803796P

Dulcey CE, Dekimpe V, Fauvelle DA, Milot S, Groleau MC, Doucet N, Rahme LG, Lépine F, Déziel E (2013) The end of an old hypothesis: the *Pseudomonas* signaling molecules 4-hydroxy-2-alkylquinolines derive from fatty acids, not 3-ketofatty acids. Chem Biol 20:1481–1491. https://doi.org/10.1016/j.chembiol.2013.09.021

Dunny GM (2013) Enterococcal sex pheromones: signaling, social behavior, and evolution. Annu Rev Genet 47:457–482. https://doi.org/10.1146/annurev-genet-111212-133449

Dunny GM, Craig RA, Carron RL, Clewell DB (1979) Plasmid transfer in *Streptococcus faecalis*: production of multiple sex pheromones by recipients. Plasmid. 2:454–465. https://doi.org/10.1016/0147-619X(79)90029-5

Eguchi Y, Kubo N, Matsunaga H, Igarashi M, Utsumi R (2011) Development of an antivirulence drug against *Streptococcus mutans*: repression of biofilm formation, acid tolerance, and competence by a histidine kinase inhibitor, walkmycin C. Antimicrob Agents Chemother 55:1475–1484. https://doi.org/10.1128/AAC.01646-10

Eibergen NR, Moore JD, Mattmann ME, Blackwell HE (2015) Potent and selective modulation of the RhlR quorum sensing receptor by using non-native ligands: an emerging target for virulence control in *Pseudomonas aeruginosa*. Chembiochem 16:2348–2356. https://doi.org/10.1002/cbic.201500357

Fitts EC, Andersson JA, Kirtley ML, Sha J, Erova TE, Chauhan S, Motin VL, Chopra AK (2016) New insights into autoinducer-2 signaling as a virulence regulator in a mouse model of pneumonic plague. mSphere 1:e00342-16. https://doi.org/10.1128/mSphere.00342-16

Fleming V, Feil E, Sewell AK, Day N, Buckling A, Massey RC (2006) Agr interference between clinical *Staphylococcus aureus* strains in an insect model of virulence. J Bacteriol 188:7686–7688. https://doi.org/10.1128/JB.00700-06

Fletcher MP, Diggle SP, Crusz SA, Chhabra SR, Cámara M, Williams P (2007a) A dual biosensor for 2-alkyl-4-quinolone quorum-sensing signal molecules. Environ Microbiol 9:2683–2693. https://doi.org/10.1111/j.1462-2920.2007.01380.x

Fletcher MP, Diggle SP, Cámara M, Williams P (2007b) Biosensor-based assays for PQS, HHQ and related 2-alkyl-4-quinolone quorum sensing signal molecules. Nat Protoc 2:1254–1262. https://doi.org/10.1038/nprot.2007.158

Fletcher M, Cámara M, Barrett DA, Williams P (2014) Biosensors for qualitative and semiquantitative analysis of quorum sensing signal molecules. Pseudomonas Methods Protoc 245–254. https://doi.org/10.1007/978-1-4939-0473-0_20

Fontaine L, Boutry C, de Frahan MH, Delplace B, Fremaux C, Horvath P, Boyaval P, Hols P (2010) A novel pheromone quorum-sensing system controls the development of natural competence in *Streptococcus thermophilus* and *Streptococcus salivarius*. J Bacteriol 192:1444–1454. https://doi.org/10.1128/JB.01251-09

Fontaine L, Goffin P, Dubout H, Delplace B, Baulard A, Lecat-Guillet N, Chambellon E, Gardan R, Hols P (2013) Mechanism of competence activation by the ComRS signalling system in streptococci. Mol Microbiol 87:1113–1132. https://doi.org/10.1111/mmi.12157

Forschner-Dancause S, Poulin E, Meschwitz S (2016) Quorum sensing inhibition and structure-activity relationships of β-Keto esters. Molecules 21:971. https://doi.org/10.3390/molecules21080971

Freeman JA, Bassler BL (1999) A genetic analysis of the function of LuxO, a two-component response regulator involved in quorum sensing in *Vibrio harveyi*. Mol Microbiol 31:665–677. https://doi.org/10.1046/j.1365-2958.1999.01208.x

Freeman JA, Lilley BN, Bassler BL (2000) A genetic analysis of the functions of LuxN: a two-component hybrid sensor kinase that regulates quorum sensing in *Vibrio harveyi*. Mol Microbiol 35:139–149. https://doi.org/10.1046/j.1365-2958.2000.01684.x

Frei R, Breitbach AS, Blackwell HE (2012) 2-Aminobenzimidazole derivatives strongly inhibit and disperse *Pseudomonas aeruginosa* biofilms. Angewandte Chemie Int Ed 51:5226–5229. https://doi.org/10.1002/anie.201109258

Fujimoto S, Tomita H, Wakamatsu E, Tanimoto K, Ike Y (1995) Physical mapping of the conjugative bacteriocin plasmid pPD1 of *Enterococcus faecalis* and identification of the determinant related to the pheromone response. J Bacteriol 177:5574–5581. https://doi.org/10.1128/jb.177.19.5574-5581

Fuqua C, Greenberg EP (2002) Listening in on bacteria: acyl-homoserine lactone signalling. Nat Rev Mol Cell Biol 3:685. https://doi.org/10.1038/nrm907

Fuqua C, Parsek MR, Greenberg EP (2001) Regulation of gene expression by cell-to-cell communication: acyl-homoserine lactone quorum sensing. Annu Rev Genet 35:439–468. https://doi.org/10.1128/jb.176.10.2796-2806

Galloway WR, Hodgkinson JT, Bowden SD, Welch M, Spring DR (2010) Quorum sensing in gram-negative bacteria: small-molecule modulation of AHL and AI-2 quorum sensing pathways. Chem Rev 111:28–67. https://doi.org/10.1021/cr100109t

Gambello MJ, Iglewski BH (1991) Cloning and characterization of the *Pseudomonas aeruginosa* lasR gene, a transcriptional activator of elastase expression. J Bacteriol 173:3000–3009. https://doi.org/10.1128/jb.173.9.3000-3009

García-Contreras R, Martínez-Vázquez M, Velázquez Guadarrama N, Villegas Pañeda AG, Hashimoto T, Maeda T, Quezada H, Wood TK (2013) Resistance to the quorum-quenching compounds brominated furanone C-30 and 5-fluorouracil in *Pseudomonas aeruginosa* clinical isolates. Pathog Dis 68:8–11. https://doi.org/10.1111/2049-632X.12039

García-Contreras R, Maeda T, Wood TK (2016) Can resistance against quorum-sensing interference be selected?. The. ISME J 10:4–10. https://doi.org/10.1038/ismej.2015.84

Garmyn D, Gal L, Lemaitre JP, Hartmann A, Piveteau P (2009) Communication and autoinduction in the species *Listeria monocytogenes*: A central role for the agr system. Commun Integr Biol 2:371–374. https://doi.org/10.4161/cib.2.4.8610

Gerdt JP, Blackwell HE (2014) Competition studies confirm two major barriers that can preclude the spread of resistance to quorum-sensing inhibitors in bacteria. ACS Chem Biol 9:2291–2299. https://doi.org/10.1021/cb5004288

Geske GD, O'Neill JC, Blackwell HE (2007) *N*-phenylacetanoyl-L-homoserine lactone scan strongly antagonize or superagonize quorum sensing in *Vibrio fischeri*. ACS Chem Biol 2:315–319. https://doi.org/10.1021/cb700036x

Giacometti A, Cirioni O, Gov Y, Ghiselli R, Del Prete MS, Mocchegiani F, Saba V, Orlando F, Scalise G, Balaban N, Dell'Acqua G (2003) RNA III inhibiting peptide inhibits *in vivo* biofilm formation by drug-resistant *Staphylococcus aureus*. Antimicrob Agents Chemother 47:1979–1983. doi: https://doi.org/10.1128/AAC.47.6.1979-1983

Gilmore MS, Lebreton F, van Schaik W (2013) Genomic transition of enterococci from gut commensals to leading causes of multidrug-resistant hospital infection in the antibiotic era. Curr Opin Microbiol 16:10–6. https://doi.org/10.1016/j.mib.2013.01.006

Gilmour R, Foster JE, Sheng Q, McClain JR, Riley A, Sun PM, Ng WL, Yan D, Nicas TI, Henry K, Winkler ME (2005) New class of competitive inhibitor of bacterial histidine kinases. J Bacteriol 187:8196–8200. https://doi.org/10.1128/JB.187.23.8196-8200

González JE, Keshavan ND (2006) Messing with bacterial quorum sensing. Microbiol Mol Biol Rev 70:859–875. https://doi.org/10.1128/MMBR.00002-06

González RH, Nusblat A, Nudel BC (2001) Detection and characterization of quorum sensing signal molecules in *Acinetobacter* strains. Microbiol Res 155:271–277. https://doi.org/10.1016/S0944-5013(01)80004-5

Gordon CP, Williams P, Chan WC (2013) Attenuating *Staphylococcus aureus* virulence gene regulation: a medicinal chemistry perspective. J Med Chem 56:1389–1404. https://doi.org/10.1021/jm3014635

Gordon CP, Olson SD, Lister JL, Kavanaugh JS, Horswill AR (2016) Truncated autoinducing peptides as antagonists of *Staphylococcus lugdunensis* quorum sensing. J Med Chem 59:8879–8888. https://doi.org/10.1021/acs.jmedchem.6b00727

Gray B, Hall P, Gresham H (2013) Targeting agr-and agr-like quorum sensing systems for development of common therapeutics to treat multiple gram-positive bacterial infections. Sensors 13:5130–5166. https://doi.org/10.3390/s130405130

Guo M, Gamby S, Zheng Y, Sintim HO (2013) Small molecule inhibitors of AI-2 signaling in bacteria: state-of-the-art and future perspectives for anti-quorum sensing agents. Int J Mol Sci 14:17694–17728. https://doi.org/10.3390/ijms140917694

Gupta S, Bram EE, Weiss R (2013) Genetically programmable pathogen sense and destroy. ACS Synth Biol 2:715–723. https://doi.org/10.1021/sb4000417

Haas W, Gilmore MS (1999) Molecular nature of a novel bacterial toxin: the cytolysin of *Enterococcus faecalis*. Med Microbiol Immunol 187:183–190. https://doi.org/10.1007/s004300050091

Hanzelka BL, Parsek MR, Val DL, Dunlap PV, Cronan JE, Greenberg EP (1999) Acylhomoserine lactone synthase activity of the *Vibrio fischeri* AinS protein. J Bacteriol 181:5766–5770. doi:Not available

Hanzelka BL, Greenberg EP (1996) Quorum sensing in *Vibrio fischeri*: evidence that S-adenosylmethionine is the amino acid substrate for autoinducer synthesis. J Bacteriol 178:5291–5294. https://doi.org/10.1128/jb.178.17.5291-5294.1996

Hass W, Shepard BD, Gilmore MS (2002) Two-component regulator of *Enterococcus faecalis* cytolysin responds to quorum-sensing autoinduction. Nature 415:84–87. https://doi.org/10.1038/415084a

Hays SG, Patrick WG, Ziesack M, Oxman N, Silver PA (2015) Better together: engineering and application of microbial symbioses. Curr Opin Biotechnol 36:40–49. https://doi.org/10.1016/j.copbio.2015.08.008

Heeb S, Fletcher MP, Chhabra SR, Diggle SP, Williams P, Cámara M (2011) Quinolones: from antibiotics to autoinducers. FEMS Microbiol Rev 35:247–274. https://doi.org/10.1172/JCI200320074

Henke JM, Bassler BL (2004) Three parallel quorum-sensing systems regulate gene expression in *Vibrio harveyi*. J Bacteriol 186:6902–6914. https://doi.org/10.1128/JB.186.20.6902-6914

Hentzer M, Givskov M (2003) Pharmacological inhibition of quorum sensing for the treatment of chronic bacterial infections. J Clin Invest 112:1300–1307. https://doi.org/10.1172/JCI200320074

Hentzer M, Riedel K, Rasmussen TB, Heydorn A, Andersen JB, Parsek MR, Rice SA, Eberl L, Molin S, Høiby N, Kjelleberg S, Givskov M (2002) Inhibition of quorum sensing in *Pseudomonas aeruginosa* biofilm bacteria by a halogenated furanone compound. Microbiology 148:87–102. https://doi.org/10.1099/00221287-148-1-87

Heurlier K, Dénervaud V, Haenni M, Guy L, Krishnapillai V, Haas D (2005) Quorum-sensing-negative (lasR) mutants of *Pseudomonas aeruginosa* avoid cell lysis and death. J Bacteriol 187:4875–4883. https://doi.org/10.1128/JB.187.14.4875-4883.2005

Higgins DA, Pomianek ME, Kraml CM, Taylor RK, Semmelhack MF, Bassler BL (2007) The major *Vibrio cholerae* autoinducer and its role in virulence factor production. Nature 450:883–886. https://doi.org/10.1038/nature06284

Hoang TT, Schweizer HP (1999) Characterization of *Pseudomonas aeruginosa* enoyl-acyl carrier protein reductase (FabI): A target for the antimicrobial triclosan and its role in acylated homoserine lactone synthesis. J Bacteriol 181:5489–5497. doi:Not available

Hoffman LR, Kulasekara HD, Emerson J, Houston LS, Burns JL, Ramsey BW, Miller SI (2009) *Pseudomonas aeruginosa* lasR mutants are associated with cystic fibrosis lung disease progression. J Cyst Fibros 8:66–70. https://doi.org/10.1016/j.jcf.2008.09.006

Hong SH, Hegde M, Kim J, Wang X, Jayaraman A, Wood TK (2012) Synthetic quorum-sensing circuit to control consortial biofilm formation and dispersal in a microfluidic device. Nat Commun 3:613. https://doi.org/10.1038/ncomms1616

Hoover SE, Perez AJ, Tsui HCT, Sinha D, Smiley DL, DiMarchi RD, Winkler ME, Lazazzera BA (2015) A new quorum-sensing system (TprA/PhrA) for *Streptococcus pneumoniae* D39 that regulates a lantibiotic biosynthesis gene cluster. Mol Microbiol 97:229–243. https://doi.org/10.1111/mmi.13029

Hraiech S, Hiblot J, Lafleur J, Lepidi H, Papazian L, Rolain JM, Raoult D, Elias M, Silby MW, Bzdrenga J, Bregeon F (2014) Inhaled lactonase reduces *Pseudomonas aeruginosa* quorum sensing and mortality in rat pneumonia. PloS One 9:e107125. https://doi.org/10.1371/journal.pone.0107125

Huigens RW, Richards JJ, Parise G, Ballard TE, Zeng W, Deora R, Melander C (2007) Inhibition of *Pseudomonas aeruginosa* biofilm formation with Bromoageliferin analogues. J Am Chem Soc 129:6966–6967. https://doi.org/10.1021/ja069017t

Hundsberger T, Braun V, Weidmann M, Leukel P, Sauerborn M, Eichel-Streiber C (1997) Transcription analysis of the genes tcdA-E of the pathogenicity locus of *Clostridium difficile*. Eur J Biochem 244:735–742. https://doi.org/10.1111/j.1432-1033.1997.t01-1-00735.x

Igarashi M, Watanabe T, Hashida T, Umekita M, Hatano M, Yanagida Y, Kino H, Kimura T, Kinoshita N, Inoue K, Sawa R, Nishimura Y, Utsumi R, Nomoto A (2013) Waldiomycin, a novel WalK-histidine kinase inhibitor from Streptomyces sp. MK844-mF10. J Antibiot 66:459–464. https://doi.org/10.1038/ja.2013.33

Igarashi Y, Gohda F, Kadoshima T, Fukuda T, Hanafusa T, Shojima A, Nakayama J, Bills GF, Peterson S (2015) Avellanin C, an inhibitor of quorum-sensing signaling in *Staphylococcus aureus*, from *Hamigera ingelheimensis*. J Antibiot 68:707–710. https://doi.org/10.1038/ja.2015.50

Ishii S, Yano T, Hayashi H (2006) Expression and characterization of the peptidase domain of *Streptococcus pneumoniae* ComA, a bifunctional ATP-binding cassette transporter involved in quorum sensing pathway. J Biol Chem 281:4726–4731. https://doi.org/10.1074/jbc.M512516200

Jarraud S, Lyon GJ, Figueiredo AM, Lina G, Vandenesch F, Etienne J, Muir TW, Novick RP (2011) Exfoliatin-producing strains define a fourth agr specificity group in *Staphylococcus aureus*. J Bacteriol 193:7027. https://doi.org/10.1128/JB.06355-11

Ji G, Beavis R, Novick RP (1997) Bacterial interference caused by autoinducing peptide variants. Science 276:2027–2030. https://doi.org/10.1126/science.276.5321.2027

Jimenez PN, Koch G, Thompson JA, Xavier KB, Cool RH, Quax WJ (2012) The multiple signaling systems regulating virulence in *Pseudomonas aeruginosa*. Microbiol Mol Biol Rev 76:46–65. https://doi.org/10.1128/MMBR.05007-11

Kalia VC, Kumar P (2015) Potential applications of quorum sensing inhibitors in diverse fields. In Quorum sensing vs quorum quenching: a battle with no end in sight. Springer India, New Delhi, pp 359–370. https://doi.org/10.1007/978-81-322-1982-8_29

Kalia VC, Wood TK, Kumar P (2014) Evolution of resistance to quorum-sensing inhibitors. Microb Ecol 68:13–23. https://doi.org/10.1007/s00248-013-0316-y

Kalia M, Singh PK, Yadav VK, Yadav BS, Sharma D, Narvi SS, Mani A, Agarwal V (2017) Structure based virtual screening for identification of potential quorum sensing inhibitors against LasR master regulator in *Pseudomonas aeruginosa*. Microb Pathog 107:136–143. https://doi.org/10.1016/j.micpath.2017.03.026

Kang YS, Park W (2010) Contribution of quorum-sensing system to hexadecane degradation and biofilm formation in *Acinetobacter* sp. strain DR1. J Appl Microbiol 109:1650–1659. https://doi.org/10.1111/j.1365-2672.2010.04793.x

Ke X, Miller LC, Bassler BL (2015) Determinants governing ligand specificity of the *Vibrio harveyi* LuxN quorum-sensing receptor. Mol Microbiol 95:127–142. https://doi.org/10.1111/mmi.12852

Kendall MM, Sperandio V (2014) Cell-to-cell signaling in *E. coli* and *Salmonella*. EcoSal Plus 6. https://doi.org/10.1128/ecosalplus.ESP-0002-2013

Khan BA, Yeh AJ, Cheung GY, Otto M (2015) Investigational therapies targeting quorum-sensing for the treatment of *Staphylococcus aureus* infections. Expert Opin Inv Drug 24:689–704. https://doi.org/10.1517/13543784.2015.1019062

Kim HS, Lee SH, Byun Y, Park HD (2015) 6-Gingerol reduces *Pseudomonas aeruginosa* biofilm formation and virulence via quorum sensing inhibition. Sci Rep 5:8656. https://doi.org/10.1038/srep08656

Kiran MD, Adikesavan NV, Cirioni O, Giacometti A, Silvestri C, Scalise G, Ghiselli R, Saba V, Orlando F, Shoham M, Balaban N (2008) Discovery of a quorum-sensing inhibitor of drug-resistant staphylococcal infections by structure-based virtual screening. Mol Pharmacol 73:1578–1586. https://doi.org/10.1124/mol.107.044164

Kleerebezem M, Quadri LE, Kuipers OP, De Vos WM (1997) Quorum sensing by peptide pheromones and two-component signal-transduction systems in gram-positive bacteria. Mol Microbiol 24:895–904. https://doi.org/10.1046/j.1365-2958.1997.4251782.x

Koenig RL, Ray JL, Maleki SJ, Smeltzer MS, Hurlburt BK (2004) *Staphylococcus aureus* AgrA binding to the RNAIII-agr regulatory region. J Bacteriol 186:7549–7555. https://doi.org/10.1128/JB.186.22.7549-7555.2004

Kong KF, Vuong C, Otto M (2006) *Staphylococcus* quorum sensing in biofilm formation and infection. Int J Med Microbiol 296:133–139. https://doi.org/10.1016/j.ijmm.2006.01.042

Kotula JW, Kerns SJ, Shaket LA, Siraj L, Collins JJ, Way JC, Silver PA (2014) Programmable bacteria detect and record an environmental signal in the mammalian gut. Proceed Nat Acad Sci 111 U S A 4838–4843. doi: https://doi.org/10.1073/pnas.1321321111

Kuehne SA, Cartman ST, Heap JT, Kelly ML, Cockayne A, Minton NP (2010) The role of toxin a and toxin B in *Clostridium difficile* infection. Nature 467:711. https://doi.org/10.1038/nature09397

Kuo D, Yu G, Hoch W, Gabay D, Long L, Ghannoum M, Nagy N, Harding CV, Viswanathan R, Shoham M (2015) Novel quorum-quenching agents promote methicillin-resistant *Staphylococcus aureus* (MRSA) wound healing and sensitize MRSA to beta-lactam antibiotics. Antimicrob Agents Chemother 59:1512–f1518. https://doi.org/10.1128/AAC.04767-14

LaSarre B, Federle MJ (2013) Exploiting quorum sensing to confuse bacterial pathogens. Microbiol Mol Biol Rev 77:73–111. https://doi.org/10.1128/MMBR.00046-12

Le KY, Otto M (2015) Quorum-sensing regulation in staphylococci-an overview. Front Microbiol 6:1174. https://doi.org/10.3389/fmicb.2015.01174

Ledgham F, Ventre I, Soscia C, Foglino M, Sturgis JN, Lazdunski A (2003) Interactions of the quorum sensing regulator QscR: interaction with itself and the other regulators of *Pseudomonas aeruginosa* LasR and RhlR. Mol Microbiol 48:199–210. https://doi.org/10.1046/j.1365-2958.2003.03423.x

Lee AS, Song KP (2005) LuxS/autoinducer 2 quorum sensing molecule regulates transcriptional virulence gene expression in *Clostridium difficile*. Biochem Biophys Res Commun 335:659–666. https://doi.org/10.1016/j.bbrc.2005.07.131

Lee J, Zhang L (2015) The hierarchy quorum sensing network in *Pseudomonas aeruginosa*. Protein Cell 6:26. https://doi.org/10.1007/s13238-014-0100-x

Lee J, Jayaraman A, Wood TK (2007) Indole is an inter-species biofilm signal mediated by SdiA. BMC Microbiol 7:42. https://doi.org/10.1186/1471-2180-7-42

Lee J, Maeda T, Hong SH, Wood TK (2009) Reconfiguring the quorum-sensing regulator SdiA of *Escherichia coli* to control biofilm formation via indole and *N*-acylhomoserine lactone. Appl Environ Microbiol 75:1703–1716. https://doi.org/10.1128/AEM.02081-08

Lee J, Wu J, Deng Y, Wang J, Wang C, Wang J, Chang C, Dong Y, Williams P, Zhang LH (2013) A cell-cell communication signal integrates quorum sensing and stress response. Nature Chem Biol 9:339–343. https://doi.org/10.1038/nchembio.1225

Lenz DH, Mok KC, Lilley BN, Kulkarni RV, Wingreen NS, Bassler BL (2004) The small RNA chaperone Hfq and multiple small RNAs control quorum sensing in *Vibrio harveyi* and *Vibrio cholerae*. Cell 118:69–82. https://doi.org/10.1016/j.cell.2004.06.009

Lesic B, Lépine F, Déziel E, Zhang J, Zhang Q, Padfield K, Castonguay MH, Milot S, Stachel S, Tzika AA, Tompkins RG (2007) Inhibitors of pathogen intercellular signals as selective anti-infective compounds. PLoS pathog 3:e126. https://doi.org/10.1371/journal.ppat.0030126

Li L, Hooi D, Chhabra SR, Pritchard D, Shaw PE (2004) Bacterial N-acylhomoserine lactone-induced apoptosis in breast carcinoma cells correlated with down-modulation of STAT3. Oncogene 23:4894–4902. https://doi.org/10.1038/sj.onc.1207612

Li M, Villaruz AE, Vadyvaloo V, Sturdevant DE, Otto M (2008a) AI-2-dependent gene regulation in *Staphylococcus epidermidis*. BMC Microbiol 8:4. https://doi.org/10.1186/1471-2180-8-4

Li M, Ni N, Chou HT, Lu CD, Tai PC, Wang B (2008b) Structure-based discovery and experimental verification of novel AI-2 quorum sensing inhibitors against *Vibrio harveyi*. ChemMedChem 3:1242–31249. https://doi.org/10.1002/cmdc.200800076

Lidor O, Al-Quntar A, Pesci EC, Steinberg D (2016) Mechanistic analysis of a synthetic inhibitor of the *Pseudomonas aeruginosa* LasI quorum-sensing signal synthase. Sci Rep 6:25257. https://doi.org/10.1038/srep25257

Lina G, Jarraud S, Ji G, Greenland T, Pedraza A, Etienne J, Novick RP, Vandenesch F (1998) Transmembrane topology and histidine protein kinase activity of AgrC, the agr signal receptor in *Staphylococcus aureus*. Mol Microbiol 28:655–662. https://doi.org/10.1046/j.1365-2958.1998.00830.x

Liu YC, Chan KG, Chang CY (2015) Modulation of host biology by *Pseudomonas aeruginosa* quorum sensing signal molecules: messengers or traitors. Front Microbiol 6. https://doi.org/10.3389/fmicb.2015.01226

Lowery CA, Abe T, Park J, Eubanks LM, Sawada D, Kaufmann GF, Janda KD (2009) Revisiting AI-2 quorum sensing inhibitors: direct comparison of alkyl-dpd analogues and a natural product fimbrolide. J Am Chem Soc 131:15584–15585. https://doi.org/10.1021/ja9066783

Lyon GJ, Muir TW (2003) Chemical signaling among bacteria and its inhibition. Chem Biol 10:1007–1021. https://doi.org/10.1016/j.chembiol.2003.11.003

Lyon GJ, Mayville P, Muir TW, Novick RP (2000) Rational design of a global inhibitor of the virulence response in *Staphylococcus aureus*, based in part on localization of the site of inhibition to the receptor-histidine kinase, AgrC. Proc Natl Acad Sci U S A 97:13330–13335. https://doi.org/10.1073/pnas.97.24.13330

Lyras D, O'Connor JR, Howarth PM, Sambol SP, Carter GP, Phumoonna T, Poon R, Adams V, Vedantam G, Johnson S, Gerding DN (2009) Toxin B is essential for virulence of *Clostridium difficile*. Nature 458:1176–1179. https://doi.org/10.1038/nature07822

Ma P, Nishiguchi K, Yuille HM, Davis LM, Nakayama J, Phillips-Jones MK (2011) Anti-HIV siamycin I directly inhibits autophosphorylation activity of the bacterial FsrC quorum sensor and other ATP-dependent enzyme activities. FEBS Lett 585:2660–2664. https://doi.org/10.1016/j.febslet.2011.07.026

Ma R, Qiu S, Jiang Q, Sun H, Xue T, Cai G, Sun B (2017) AI-2 quorum sensing negatively regulates rbf expression and biofilm formation in *Staphylococcus aureus*. Int J Med Microbiol 307:257–267. https://doi.org/10.1016/j.ijmm.2017.03.003

Maeda T, García-Contreras R, Pu M, Sheng L, Garcia LR, Tomás M, Wood TK (2012) Quorum quenching quandary: resistance to antivirulence compounds. ISME J 6:493–501. https://doi.org/10.1038/ismej.2011.122

Mani N, Lyras D, Barroso L, Howarth P, Wilkins T, Rood JI, Sonenshein AL, Dupuy B (2002) Environmental response and autoregulation of *Clostridium difficile* TxeR, a sigma factor for toxin gene expression. J Bacteriol 184:5971–5978. https://doi.org/10.1128/JB.184.21.5971-5978

Martin MJ, Clare S, Goulding D, Faulds-Pain A, Barquist L, Browne HP, Pettit L, Dougan G, Lawley TD, Wren BW (2013) The agr locus regulates virulence and colonization genes in *Clostridium difficile* 027. J Bacteriol 195:3672–3681. https://doi.org/10.1128/JB.00473-13

Martin-Verstraete I, Peltier J, Dupuy B (2016) The regulatory networks that control *Clostridium difficile* toxin synthesis. Toxins 8:153. https://doi.org/10.3390/toxins8050153

Mashburn-Warren L, Morrison DA, Federle MJ (2010) A novel double-tryptophan peptide pheromone controls competence in *Streptococcus* spp. via an Rgg regulator. Mol Microbiol 78:589–606. https://doi.org/10.1111/j.1365-2958.2010.07361.x

Matsushita M, Janda KD (2002) Histidine kinases as targets for new antimicrobial agents. Bioorg Med Chem 10:855–867. https://doi.org/10.1016/S0968-0896(01)00355-8

Mayville P, Ji G, Beavis R, Yang H, Goger M, Novick RP, Muir TW (1999) Structure-activity analysis of synthetic autoinducing thiolactone peptides from *Staphylococcus aureus* responsible for virulence. Proc Natl Acad Sci U S A 96:1218–1223. https://doi.org/10.1073/pnas.96.4.1218

McBrayer DN, Gantman BK, Cameron CD, Tal-Gan Y (2017) An entirely solid phase peptide synthesis-based strategy for synthesis of gelatinase biosynthesis-activating pheromone (GBAP) analogue libraries: investigating the structure-activity relationships of the *Enterococcus faecalis* quorum sensing signal. Org Lett 19:3295–3298. https://doi.org/10.1021/acs.orglett.7b01444

McClean KH, Winson MK, Fish L, Taylor A, Chhabra SR, Camara M, Daykin M, Lamb JH, Swift S, Bycroft BW, Stewart GS (1997) Quorum sensing and Chromobacterium violaceum: exploitation of violacein production and inhibition for the detection of N-acylhomoserine lactones. Microbiology 143:3703–3711. https://doi.org/10.1099/00221287-143-12-3703

Meighen EA (1991) Molecular biology of bacterial bioluminescence. Microbiol Rev 55:123–42. doi: Not available

MDowell P, Affas Z, Reynolds C, Holden MT, Wood SJ, Saint S, Cockayne A, Hill PJ, Dodd CE, Bycroft BW, Chan WC (2001) Structure, activity and evolution of the group I thiolactone peptide quorum-sensing system of *Staphylococcus aureus*. Mol Microbiol 41:503–512.

Michael B, Smith JN, Swift S, Heffron F, Ahmer BM (2001) SdiA of *Salmonella enterica* is a LuxR homolog that detects mixed microbial communities. J Bacteriol 183:5733–5742. https://doi.org/10.1128/JB.183.19.5733-5742

Miller MB, Bassler BL (2001) Quorum sensing in bacteria. Annu Rev Microbiol 55:165–199. https://doi.org/10.1146/annurev.micro.55.1.165

Miller ST, Xavier KB, Campagna SR, Taga ME, Semmelhack MF, Bassler BL, Hughson FM (2004) *Salmonella typhimurium* recognizes a chemically distinct form of the bacterial quorum-sensing signal AI-2. Mol Cell 15:677–687. https://doi.org/10.1016/j.molcel.2004.07.020

Moore JD, Rossi FM, Welsh MA, Nyffeler KE, Blackwell HE (2015) A comparative analysis of synthetic quorum sensing modulators in *Pseudomonas aeruginosa*: new insights into mechanism, active efflux susceptibility, phenotypic response, and next-generation ligand design. J Am Chem Soc 137:14626–14639. https://doi.org/10.1021/jacs.5b06728

Müh U, Schuster M, Heim R, Singh A, Olson ER, Greenberg EP (2006) Novel *Pseudomonas aeruginosa* quorum-sensing inhibitors identified in an ultra-high throughput screen. Antimicrob Agents Chemother 50:3674–3679. https://doi.org/10.1128/AAC.00665-06

Müller C, Fetzner S (2013) A *Pseudomonas putida* bioreporter for the detection of enzymes active on 2-alkyl-4(1*H*)-quinolone signalling molecules. Appl Microbiol Biotechnol 97:751–760. https://doi.org/10.1007/s00253-012-4236-4

Musk J, Dinty J, Hergenrother PJ (2006) Chemical countermeasures for the control of bacterial biofilms: Effective compounds and promising targets. Curr Med Chem 13:2163–2177. https://doi.org/10.2174/092986706777935212

Nakayama J, Yoshida K, Kobayashi H, Isogai A, Clewell DB, Suzuki A (1995) Cloning and characterization of a region of *Enterococcus faecalis* plasmid pPD1 encoding pheromone inhibitor (ipd), pheromone sensitivity (traC), and pheromone shutdown (traB) genes. J Bacteriol 177:5567–5573. https://doi.org/10.1128/jb.177.19.5567-557

Nakayama J, Takanami Y, Horii T, Sakuda S, Suzuki A (1998) Molecular mechanism of peptide-specific pheromone signaling in *Enterococcus faecalis*: functions of pheromone receptor TraA and pheromone-binding protein TraC encoded by plasmid pPD1. J Bacteriol 180:449–456. doi: Not available

Nakayama J, Cao Y, Horii T, Sakuda S, Akkermans AD, de Vos WM, Nagasawa H (2001) Gelatinase biosynthesis- activating pheromone: a peptide lactone that mediates a quorum sensing in *Enterococcus faecalis*. Mol Microbiol 41:145–154. https://doi.org/10.1046/j.1365-2958.2001.02486.x

Nakayama J, Chen S, Oyama N, Nishiguchi K, Azab EA, Tanaka E, Kariyama R, Sonomoto K (2006) Revised model for *Enterococcus faecalis* fsr quorum-sensing system: the small open reading frame *fsrD* encodes the gelatinase biosynthesis-activating pheromone propeptide corresponding to staphylococcal *agrD*. J Bacteriol 188:8321–8326. https://doi.org/10.1128/JB.00865-06

Nakayama J, Tanaka E, Kariyama R, Nagata K, Nishiguchi K, Mitsuhata R, Uemura Y, Tanokura M, Kumon H, Sonomoto K (2007) Siamycin attenuates fsr quorum sensing mediated by a gelatinase biosynthesis- activating pheromone in *Enterococcus faecalis*. J Bacteriol 189:1358–1365. https://doi.org/10.1128/JB.00969-06

Nakayama J, Uemura Y, Nishiguchi K, Yoshimura N, Igarashi Y, Sonomoto K (2009) Ambuic acid inhibits the biosynthesis of cyclic peptide quormones in gram-positive bacteria. Antimicrob Agents Chemother 53:580–586. https://doi.org/10.1128/AAC.00995-08

Nakayama J, Yokohata R, Sato M, Suzuki T, Matsufuji T, Nishiguchi K, Kawai T, Yamanaka Y, Nagata K, Tanokura M, Sonomoto K (2013) Development of a peptide antagonist against fsr quorum sensing of *Enterococcus faecalis*. ACS Chem Biol 8:804–811. https://doi.org/10.1021/cb300717f

Ng WL, Bassler BL (2009) Bacterial quorum-sensing network architectures. Annu Rev Genet 43:197–222. https://doi.org/10.1146/annurev-genet-102108-134304

Ng WL, Perez LJ, Wei Y, Kraml C, Semmelhack MF, Bassler BL (2011) Signal production and detection specificity in *Vibrio* CqsA/CqsS quorum-sensing systems. Mol Microbiol 79:1407–1417. https://doi.org/10.1111/j.1365-2958.2011.07548.x

Ni N, Li M, Wang J, Wang B (2009) Inhibitors and antagonists of bacterial quorum sensing. Med Res Rev 29:65–124. https://doi.org/10.1002/med.20145

Nikaido H (2009) Multidrug resistance in bacteria. Annu Rev Biochem 78:119–146. https://doi.org/10.1146/annurev.biochem.78.082907.145923

Niu C, Afre S, Gilbert ES (2006) Subinhibitory concentrations of cinnamaldehyde interfere with quorum sensing. Lett Appl Microbiol 43:489–494. https://doi.org/10.1111/j.1472-765X.2006.02001.x

Niu C, Clemmer KM, Bonomo RA, Rather PN (2008) Isolation and characterization of an autoinducer synthase from *Acinetobacter baumannii*. J Bacteriol 190:3386–3392. https://doi.org/10.1128/JB.01929-07

Novick RP, Geisinger E (2008) Quorum sensing in *staphylococci*. Annu Rev Genet 42:541–564. https://doi.org/10.1146/annurev.genet.42.110807.091640

O'Loughlin CT, Miller LC, Siryaporn A, Drescher K, Semmelhack MF, Bassler BL (2013) A quorum-sensing inhibitor blocks *Pseudomonas aeruginosa* virulence and biofilm formation. Proc Natl Acad Sci U S A 110:17981–17986. https://doi.org/10.1073/pnas.1316981110

O'Neill J (2014) Antimicrobial resistance: tackling a crisis for the health and wealth of nations. Review on antimicrobial resistance. 11:1–6. doi: Not available

Ochsner UA, Reiser J (1995) Autoinducer-mediated regulation of rhamnolipid biosurfactant synthesis in *Pseudomonas aeruginosa*. Proc Natl Acad Sci U S A 92:6424–6428. doi: Not available

Oh MH, Choi CH (2015) Role of LuxIR homologue AnoIR in *Acinetobacter nosocomialis* and the effect of Virstatin on the expression of anoR Gene. J Microbiol Biotechnol 25:1390–1400. https://doi.org/10.4014/jmb.1504.04069

Ohtani K, Shimizu T (2016) Regulation of toxin production in *Clostridium perfringens*. Toxins 8:207. https://doi.org/10.3390/toxins8070207

Ohtani K, Hayashi H, Shimizu T (2002) The *luxS* gene is involved in cell-cell signalling for toxin production in *Clostridium perfringens*. Mol Microbiol 44:171–179. https://doi.org/10.1046/j.1365-2958.2002.02863.x

Ohtani K, Yuan Y, Hassan S, Wang R, Wang Y, Shimizu T (2009) Virulence gene regulation by the agr system in *Clostridium perfringens*. J Bacteriol 191:3919–3927. https://doi.org/10.1128/JB.01455-08

Ohtani K, Hirakawa H, Tashiro K, Yoshizawa S, Kuhara S, Shimizu T (2010) Identification of a two-component VirR/VirS regulon in *Clostridium perfringens*. Anaerobe 16:258–264. https://doi.org/10.1016/j.anaerobe.2009.10.003

Olson ME, Todd DA, Schaeffer CR, Paharik AE, Van Dyke MJ, Büttner H, Dunman PM, Rohde H, Cech NB, Fey PD, Horswill AR (2014) Staphylococcus epidermidis agr quorum-sensing system: signal identification, cross talk, and importance in colonization. J Bacteriol 196:3482–3493. https://doi.org/10.1128/JB.01882-14

Paczkowski JE, Mukherjee S, McCready AR, Cong JP, Aquino CJ, Kim H, Henke BR, Smith CD, Bassler BL (2017) Flavonoids suppress *Pseudomonas aeruginosa* virulence through allosteric inhibition of quorum-sensing receptors. J Biol Chem 292:4064–4076. https://doi.org/10.1074/jbc.M116.770552

Papenfort K, Bassler BL (2016) Quorum sensing signal-response systems in gram-negative bacteria. Nat Rev Microbiol 14:576–588. https://doi.org/10.1038/nrmicro.2016.89

Paspaliari DK, Mollerup MS, Kallipolitis BH, Ingmer H, Larsen MH (2014) Chitinase expression in *Listeria monocytogenes* is positively regulated by the Agr system. PLoS One 9:e95385. https://doi.org/10.1371/journal.pone.0095385

Passador L, Cook JM, Gambello MJ, Rust L, Iglewski BH (1993) Expression of *Pseudomonas aeruginosa* virulence genes requires cell-to-cell communication. Science 260:1127–1130. doi: Not available

Pearson JP, Gray KM, Passador L, Tucker KD, Eberhard A, Iglewski BH, Greenberg EP (1994) Structure of the autoinducer required for expression of *Pseudomonas aeruginosa* virulence genes. Proc Natl Acad Sci U S A 91:197–201. doi: Not available

Pereira CS, Thompson JA, Xavier KB (2013) AI-2-mediated signalling in bacteria. FEMS Microbiol Rev 37:156–181. https://doi.org/10.1111/j.1574-6976.2012.00345.x

Pérez PD, Weiss JT, Hagen SJ (2011) Noise and crosstalk in two quorum-sensing inputs of *Vibrio fischeri*. BMC Syst Biol 5:153. https://doi.org/10.1186/1752-0509-5-153

Perry JA, Jones MB, Peterson SN, Cvitkovitch DG, Lévesque CM (2009) Peptide alarmone signalling triggers an auto-active bacteriocin necessary for genetic competence. Mol Microbiol 72:905–917. https://doi.org/10.1111/j.1365-2958.2009.06693.x

Persson T, Hansen TH, Rasmussen TB, Skindersø ME, Givskov M, Nielsen J (2005) Rational design and synthesis of new quorum-sensing inhibitors derived from acylated homoserine lactones and natural products from garlic. Org Biomol Chem 3:253–262. https://doi.org/10.1039/b415761c

Pesci EC, Milbank JB, Pearson JP, McKnight S, Kende AS, Greenberg EP, Iglewski BH (1999) Quinolone signaling in the cell-to-cell communication system of *Pseudomonas aeruginosa*. Proc Natl Acad Sci U S A 96:11229–11234. https://doi.org/10.1073/pnas.96.20.11229

Pestova EV, Håvarstein LS, Morrison DA (1996) Regulation of competence for genetic transformation in *Streptococcus pneumoniae* by an auto-induced peptide pheromone and a two-component regulatory system. Mol Microbiol 21:853–862. https://doi.org/10.1046/j.1365-2958.1996.501417.x

Qin X, Singh KV, Weinstock GM, Murray BE (2002) Quorum sensing *Escherichia coli* regulators B and C (QseBC): a novel two-component regulatory system involved in the regulation of flagella and motility by quorum sensing in *E. coli*. Mol Microbiol 43:809–821. https://doi.org/10.1046/j.1365-2958.2002.02803.x

Rahmati S, Yang S, Davidson AL, Zechiedrich EL (2002) Control of the AcrAB multidrug efflux pump by quorum-sensing regulator SdiA. Mol Microbiol 43:677–685. https://doi.org/10.1046/j.1365-2958.2002.02773.x

Rasko DA, Sperandio V (2010) Anti-virulence strategies to combat bacteria-mediated disease. Nat Rev Drug Discov 9:117–128. https://doi.org/10.1038/nrd3013

Rasko DA, Moreira CG, de Li R, Reading NC, Ritchie JM, Waldor MK, Williams N, Taussig R, Wei S, Roth M, Hughes DT, Huntley JF, Fina MW, Falck JR, Sperandio V (2008) Targeting QseC signaling and virulence for antibiotic development. Science 321:1078–1080. https://doi.org/10.1126/science.1160354

Rasmussen TB, Givskov M (2006) Quorum-sensing inhibitors as anti-pathogenic drugs. Int J Med Microbiol 296:149–161. https://doi.org/10.1016/j.ijmm.2006.02.005

Reck M, Tomasch J, Wagner-Döbler I (2015) The alternative sigma factor SigX controls bacteriocin synthesis and competence, the two quorum sensing regulated traits in *Streptococcus mutans*. PLoS Genet 11:e1005353. https://doi.org/10.1371/journal.pgen.1005353

Reimmann C, Ginet N, Michel L, Keel C, Michaux P, Krishnapillai V, Zala M, Heurlier K, Triandafillu K, Harms H, Défago G, Haas D (2002) Genetically programmed autoinducer destruction reduces virulence gene expression and swarming motility in *Pseudomonas aeruginosa* PAO1. Microbiology 148:923–932. https://doi.org/10.1099/00221287-148-4-923

Ren D, Sims JJ, Wood TK (2001) Inhibition of biofilm formation and swarming of *Escherichia coli* by (5Z)-4-bromo-5-(bromomethylene)-3-butyl-2(5H)-furanone. Environ Microbiol 3:731–736. https://doi.org/10.1046/j.1462-2920.2001.00249.x

Reuter K, Steinbach A, Helms V (2016) Interfering with bacterial quorum sensing. Perspect Medicin Chem 8:1–15. https://doi.org/10.4137/PMC.S13209

Rickard AH, Palmer RJ, Blehert DS, Campagna SR, Semmelhack MF, Egland PG, Bassler BL, Kolenbrander PE (2006) Autoinducer 2: a concentration-dependent signal for mutualistic bacterial biofilm growth. Mol Microbiol 60:1446–1456. https://doi.org/10.1111/j.1365-2958.2006.05202.x

Riedel CU, Monk IR, Casey PG, Waidmann MS, Gahan CG, Hill C (2009) AgrD-dependent quorum sensing affects biofilm formation, invasion, virulence and global gene expression profiles in *Listeria monocytogenes*. Mol Microbiol 71:1177–1189. https://doi.org/10.1111/j.1365-2958.2008.06589.x

Rodríguez-Rojas A, Rodríguez-Beltrán J, Couce A, Blázquez J (2013) Antibiotics and antibiotic resistance: a bitter fight against evolution. Int J Med Microbiol 303:293–297. https://doi.org/10.1016/j.ijmm.2013.02.004

Roy V, Meyer MT, Smith JA, Gamby S, Sintim HO, Ghodssi R, Bentley WE (2013) AI-2 analogs and antibiotics: a synergistic approach to reduce bacterial biofilms. Appl Microbiol Biotechnol 97:2627–2638. https://doi.org/10.1007/s00253-012-4404-6

Ruer S, Pinotsis N, Steadman D, Waksman G, Remaut H (2015) Virulence-targeted antibacterials: concept, promise, and susceptibility to resistance mechanisms. Chem Biol Drug Des 86:379–399. https://doi.org/10.1111/cbdd.12517

Rutherford ST, Bassler BL (2012) Bacterial quorum sensing: its role in virulence and possibilities for its control. Cold Spring Harb Perspect Med 2:a012427. https://doi.org/10.1101/cshperspect.a012427

Saeidi N, Wong CK, Lo TM, Nguyen HX, Ling H, Leong SS, Poh CL, Chang MW (2011) Engineering microbes to sense and eradicate *Pseudomonas aeruginosa*, a human pathogen. Mol Syst Biol 7:521. https://doi.org/10.1038/msb.2011.55

Sanchez-Contreras M, Bauer WD, Gao M, Robinson JB, Allan Downie J (2007) Quorum-sensing regulation in rhizobia and its role in symbiotic interactions with legumes. Philos Trans R Soc Lond Ser B Biol Sci 362:1149–1163. https://doi.org/10.1098/rstb.2007.2041

Saroj SD, Rather PN (2013) Streptomycin inhibits quorum sensing in *Acinetobacter baumannii*. Antimicrob Agents Chemother 57:1926–1929. https://doi.org/10.1128/AAC.02161-12

Schaefer AL, Hanzelka BL, Eberhard A, Greenberg EP (1996) Quorum sensing in *Vibrio fischeri*: probing autoinducer-LuxR interactions with autoinducer analogs. J Bacteriol 178:2897–2901. https://doi.org/10.1128/jb.178.10.2897-2901

Schaefer AL, Greenberg EP, Oliver CM, Oda Y, Huang JJ, Bittan-Banin G, Peres CM, Schmidt S, Juhaszova K, Sufrin JR, Harwood CS (2008) A new class of homoserine lactone quorum-sensing signals. Nature 454:595–599. https://doi.org/10.1038/nature07088

Schauder S, Shokat K, Surette MG, Bassler BL (2001) The LuxS family of bacterial autoinducers: biosynthesis of a novel quorum-sensing signal molecule. Mol Microbiol 41:463–476. https://doi.org/10.1046/j.1365-2958.2001.02532.x

Schell MA (1993) Molecular biology of the LysR family of transcriptional regulators. Annu Rev Microbiol 47:597–626. https://doi.org/10.1146/annurev.mi.47.100193.003121

Schertzer JW, Brown SA, Whiteley M (2010) Oxygen levels rapidly modulate *Pseudomonas aeruginosa* social behaviours via substrate limitation of PqsH. Mol Microbiol 77:1527–1538. https://doi.org/10.1111/j.1365-2958.2010.07303.x

Schmid MW, Ng EY, Lampidis R, Emmerth M, Walcher M, Kreft J, Goebel W, Wagner M, Schleifer KH (2005) Evolutionary history of the genus listeria and its virulence genes. Syst Appl Microbiol 28:1–18. https://doi.org/10.1016/j.syapm.2004.09.005

Schroeder M, Brooks BD, Brooks AE (2017) The complex relationship between virulence and antibiotic resistance. Genes 8:39. https://doi.org/10.3390/genes8010039

Schuster M, Greenberg EP (2007) Early activation of quorum sensing in *Pseudomonas aeruginosa* reveals the architecture of a complex regulon. Bmc Genom 8:287. https://doi.org/10.1186/1471-2164-8-287

Scott SR, Hasty J (2016) Quorum sensing communication modules for microbial consortia. ACS Synth Biol 5:969–977. https://doi.org/10.1021/acssynbio.5b00286

Scott RJ, Lian L, Muharram SH, Cockayne A, Wood SJ, Bycroft BW, Williams P, Chan WC (2003) Side-chain- to-tail thiolactone peptide inhibitors of the Staphylococcal quorum-sensing system. Bioorg Med Chem Lett 13:2449–2453. https://doi.org/10.1016/S0960-894X(03)00497-9

Scutera S, Zucca M, Savoia D (2014) Novel approaches for the design and discovery of quorum-sensing inhibitors. Expert Opin Drug Discov 9:353–366. https://doi.org/10.1517/17460441.2014.894974

Sebaihia M, Wren BW, Mullany P, Fairweather NF, Minton N, Stabler R, Thomson NR, Roberts AP, Wang H, Holden MT, Wright A (2006) The multidrug-resistant human pathogen *Clostridium difficile* has a highly mobile, mosaic genome. Nature Genet 38:779–786. https://doi.org/10.1038/ng1830

Shafreen B, Mohmed R, Selvaraj C, Singh SK, Karutha Pandian S (2014) *In silico* and *in vitro* studies of cinnamaldehyde and their derivatives against LuxS in *Streptococcus pyogenes*: effects on biofilm and virulence genes. J Mol Recognit 27:106–116. https://doi.org/10.1002/jmr.2339

Shanker E, Federle MJ (2017) Quorum sensing regulation of competence and bacteriocins in *Streptococcus pneumoniae* and *mutans*. Genes 8:15. https://doi.org/10.3390/genes8010015

Shao C, Shang W, Yang Z, Sun Z, Li Y, Guo J, Wang X, Zou D, Wang S, Lei H, Cui Q, Yin Z, Li X, Wei X, Liu W, He X, Jiang Z, Du S, Liao X, Huang L, Wang Y, Yuan J (2012) LuxS-dependent AI-2 regulates versatile functions in *Enterococcus faecalis* V583. J Proteome Res 11:4465–4475. https://doi.org/10.1021/pr3002244

Shen G, Rajan R, Zhu J, Bell CE, Pei D (2006) Design and synthesis of substrate and intermediate analogue inhibitors of S-ribosylhomocysteinase. J Med Chem 49:3003–3011. https://doi.org/10.1021/jm060047g

Simonetti O, Cirioni O, Mocchegiani F, Cacciatore I, Silvestri C, Baldassarre L, Orlando F, Castelli P, Provinciali M, Vivarelli M, Fornasari E, Giacometti A, Offidani A, Simonetti O, Cirioni O, Mocchegiani F, Cacciatore I, Silvestri C, Baldassarre L, Orlando F, Castelli P, Provinciali M, Vivarelli M, Fornasari E (2013) The efficacy of the quorum sensing inhibitor FS8 and tigecycline in preventing prosthesis biofilm in an animal model of staphylococcal infection. Int J Mol Sci 14:16321–16332. https://doi.org/10.3390/ijms140816321

Singh V, Evans GB, Lenz DH, Mason JM, Clinch K, Mee S, Painter GF, Tyler PC, Furneaux RH, Lee JE, Howell PL, Schramm VL (2005) Femtomolar transition state analogue inhibitors of 5′-methylthioadenosine/S- adenosylhomocysteine nucleosidase from *Escherichia coli*. J Biol Chem 280:18265–18273. https://doi.org/10.1074/jbc.M414472200

Singh RP, Okubo K, Ohtani K, Adachi K, Sonomoto K, Nakayama J (2015) Rationale design of quorum-quenching peptides that target the VirSR system of *Clostridium perfringens*. FEMS Microbiol Lett 362. pii: fnv188. doi: https://doi.org/10.1093/femsle/fnv188

Smith RS, Iglewski BH (2003) *P. aeruginosa* quorum-sensing systems and virulence. Curr Opin Microbiol 6:56–60. https://doi.org/10.1016/S1369-5274(03)00008-0

Smyth AR, Cifelli PM, Ortori CA, Righetti K, Lewis S, Erskine P, Holland ED, Givskov M, Williams P, Cámara M, Barrett DA, Knox A (2010) Garlic as an inhibitor of *Pseudomonas aeruginosa* quorum sensing in cystic fibrosis--a pilot randomized controlled trial. Pediatr Pulmonol 45:356–362. https://doi.org/10.1002/ppul.21193

Sperandio V, Mellies JL, Nguyen W, Shin S, Kaper JB (1999) Quorum sensing controls expression of the type III secretion gene transcription and protein secretion in enterohemorrhagic and enteropathogenic *Escherichia coli*. Proc Natl Acad Sci U S A 96:15196–15201. https://doi.org/10.1073/pnas.96.26.15196

Sperandio V, Torres AG, Kaper JB (2002) Quorum sensing *Escherichia coli* regulators B and C (QseBC): a novel two-component regulatory system involved in the regulation of flagella and motility by quorum sensing in *E. coli*. Mol Microbiol 43:809–821. https://doi.org/10.1046/j.1365-2958.2002.02803.x

Srivastava A, Singh BN, Deepak D, Rawat AK, Singh BR (2015) Colostrum hexasaccharide, a novel *Staphylococcus aureus* quorum-sensing inhibitor. Antimicrob Agents Chemother 59:2169–2178. https://doi.org/10.1128/AAC.03722-14

Stabler RA, He M, Dawson L, Martin M, Valiente E, Corton C, Lawley TD, Sebaihia M, Quail MA, Rose G, Gerding DN (2009) Comparative genome and phenotypic analysis of *Clostridium difficile* 027 strains provides insight into the evolution of a hypervirulent bacterium. Genome Biol 10:R102. https://doi.org/10.1186/gb-2009-10-9-r102

Stacy DM, Welsh MA, Rather PN, Blackwell HE (2012) Attenuation of quorum sensing in the pathogen *Acinetobacter baumannii* using non-native N-acyl homoserine lactones. ACS Chem Biol 7:1719–1728. https://doi.org/10.1021/cb300351x

Starkey M, Lepine F, Maura D, Bandyopadhaya A, Lesic B, He J, Kitao T, Righi V, Milot S, Tzika A, Rahme L (2014) Identification of anti-virulence compounds that disrupt quorum-sensing regulated acute and persistent pathogenicity. PLoS Pathog 10:e1004321. https://doi.org/10.1371/journal.ppat.1004321

Steindler L, Venturi V (2007) Detection of quorum-sensing N-acyl homoserine lactone signal molecules by bacterial biosensors. FEMS Microbiol Lett 266:1–9. https://doi.org/10.1111/j.1574-6968.2006.00501.x

Sully EK, Malachowa N, Elmore BO, Alexander SM, Femling JK, Gray BM, DeLeo FR, Otto M, Cheung AL, Edwards BS, Sklar LA, Horswill AR, Hall PR, Gresham HD (2014) Selective chemical inhibition of agr quorum sensing in *Staphylococcus aureus* promotes host defense with minimal impact on resistance. PLoS Pathog 10:e1004174. https://doi.org/10.1371/journal.ppat.1004174

Sun J, Daniel R, Wagner-Döbler I, Zeng AP (2004) Is autoinducer-2 a universal signal for interspecies communication: a comparative genomic and phylogenetic analysis of the synthesis and signal transduction pathways. BMC Evol Biol 4:36. https://doi.org/10.1186/1471-2148-4-36

Suneby EG, Herndon LR, Schneider TL (2017) *Pseudomonas aeruginosa* LasR·DNA binding is directly inhibited by quorum sensing antagonists. ACS Infect Dis 3:183–189. https://doi.org/10.1021/acsinfecdis.6b00163

Surette MG, Miller MB, Bassler BL (1999) Quorum sensing in *Escherichia coli, Salmonella typhimurium*, and *Vibrio harveyi*: a new family of genes responsible for autoinducer production. Proc Natl Acad Sci U S A 96:1639–1644. https://doi.org/10.1073/pnas.96.4.1639

Taccone FS, Rodriguez-Villalobos H, De Backer D, De Moor V, Devière J, Vincent JL, Jacobs F (2006) Successful treatment of septic shock due to pan-resistant *Acinetobacter baumannii* using combined antimicrobial therapy including tigecycline. Eur J Clin Microbiol Infect Dis 25:257–260. https://doi.org/10.1007/s10096-006-0123-1

Taga ME, Miller ST, Bassler BL (2003) *Lsr*-mediated transport and processing of AI-2 in *Salmonella typhimurium*. Mol Microbiol 50:1411–1427. https://doi.org/10.1046/j.1365-2958.2003.03781.x

Tal-Gan Y, Stacy DM, Foegen MK, Koenig DW, Blackwell HE (2013) Highly potent inhibitors of quorum sensing in *Staphylococcus aureus* revealed through a systematic synthetic study of the group-III autoinducing peptide. J Am Chem Soc 135:7869–7882. https://doi.org/10.1021/ja3112115

Tal-Gan Y, Ivancic M, Cornilescu G, Yang T, Blackwell HE (2016) Highly stable, amide-bridged autoinducing peptide analogues that strongly inhibit the AgrC quorum sensing receptor in *Staphylococcus aureus*. Angew Chem Int Ed Engl 55:8913–8917. https://doi.org/10.1002/anie.201602974

Tateda K, Comte R, Pechere JC, Köhler T, Yamaguchi K, Van Delden C (2001) Azithromycin inhibits quorum sensing in *Pseudomonas aeruginosa*. Antimicrob Agents Chemother 45:1930–1933. https://doi.org/10.1128/AAC.45.6.1930-1933.2001

Thoendel M, Horswill AR (2009) Identification of *Staphylococcus aureus* AgrD residues required for autoinducing peptide biosynthesis. J Biol Chem 284:21828–21838. https://doi.org/10.1074/jbc.M109.031757

Thoendel M, Kavanaugh JS, Flack CE, Horswill AR (2011) Peptide signaling in the staphylococci. Chem Rev 111:117–151. https://doi.org/10.1021/cr100370n

Todd DA, Parlet CP, Crosby HA, Malone CL, Heilmann KP, Horswill AR, Cech NB (2017) Targeting quorum sensing signal biosynthesis to fight antibiotic resistant infections: Ambuic acid as a model inhibitor. Antimicrob Agents Chemother AAC-00263. https://doi.org/10.1128/AAC.00263-17

Tomaras AP, Flagler MJ, Dorsey CW, Gaddy JA, Actis LA (2008) Characterization of a two-component regulatory system from *Acinetobacter baumannii* that controls biofilm formation and cellular morphology. Microbiology 154:3398–3409. https://doi.org/10.1099/mic.0.2008/019471-0

Val DL, Cronan JE (1998) *In vivo* evidence that S-adenosylmethionine and fatty acid synthesis intermediates are the substrates for the LuxI family of autoinducer synthases. J Bacteriol 180:2644–2651. doi: Not available

Valencia R, Arroyo LA, Conde M, Aldana JM, Torres MJ, Fernández-Cuenca F, Garnacho-Montero J, Cisneros JM, Ortiz C, Pachón J, Aznar J (2009) Nosocomial outbreak of infection with pan-drug-resistant *Acinetobacter baumannii* in a tertiary care university hospital. Infect Control Hosp Epidemiol 30:257–263. https://doi.org/10.1086/595977

van der Ploeg JR (2005) Regulation of bacteriocin production in *Streptococcus mutans* by the quorum-sensing system required for development of genetic competence. J Bacteriol 187:3980–3989. https://doi.org/10.1128/JB.187.12.3980-3989

van Kessel JC, Rutherford ST, Shao Y, Utria AF, Bassler BL (2013) Individual and combined roles of the master regulators AphA and LuxR in control of the *Vibrio harveyi* quorum-sensing regulon. J Bacteriol 195:436–443. https://doi.org/10.1128/JB.01998-12

Vázquez-Boland JA, Kuhn M, Berche P, Chakraborty T, Domínguez-Bernal G, Goebel W, González-Zorn B, Wehland J, Kreft J (2001) Listeria pathogenesis and molecular virulence determinants. Clin Microbiol Rev 14:584–640. https://doi.org/10.1128/CMR.14.3.584-640

Velikova N, Wells JM (2017) Rationale and prospects of targeting bacterial two-component systems for antibacterial treatment of cystic fibrosis patients. Curr Drug Targets 18:687–695. https://doi.org/10.2174/1389450117666160208145934

Velikova N, Fulle S, Manso AS, Mechkarska M, Finn P, Conlon JM, Oggioni MR, Wells JM, Marina A (2016) Putative histidine kinase inhibitors with antibacterial effect against multi-drug resistant clinical isolates identified by *in vitro* and *in silico* screens. Sci Rep 6:26085. https://doi.org/10.1038/srep26085

Vendeville A, Winzer K, Heurlier K, Tang CM, Hardie KR (2005) Making sense of metabolism: autoinducer-2, LuxS and pathogenic bacteria. Nat Rev Microbiol 3:383–396. https://doi.org/10.1038/nrmicro1146

Verbeke F, De Craemer S, Debunne N, Janssens Y, Wynendaele E, Van de Wiele C, De Spiegeleer B (2017) Peptides as quorum sensing molecules: measurement techniques and obtained levels *in vitro* and *in vivo*. Front Neurosci 11:183. https://doi.org/10.3389/fnins.2017.00183

Vidal JE, Ludewick HP, Kunkel RM, Zähner D, Klugman KP (2011) The LuxS-dependent quorum-sensing system regulates early biofilm formation by *Streptococcus pneumoniae* strain D39. Infect Immun 79:4050–4060. https://doi.org/10.1128/IAI.05186-11

Vidal JE, Ma M, Saputo J, Garcia J, Uzal FA, McClane BA (2012) Evidence that the Agr-like quorum sensing system regulates the toxin production, cytotoxicity and pathogenicity of *Clostridium perfringens* type C isolate CN3685. Mol Microbiol 83:179–194. https://doi.org/10.1111/j.1365-2958.2011.07925.x

von Bodman SB, Willey JM, Diggle SP (2008) Cell-cell communication in bacteria: united we stand. J Bacteriol 190:4377–4391. https://doi.org/10.1128/JB.00486-08

Wade DS, Calfee MW, Rocha ER, Ling EA, Engstrom E, Coleman JP, Pesci EC (2005) Regulation of *Pseudomonas* quinolone signal synthesis in *Pseudomonas aeruginosa*. J Bacteriol 187:4372–4380. https://doi.org/10.1128/JB.187.13.4372-4380

Walters M, Sperandio V (2006) Quorum sensing in *Escherichia coli* and *Salmonella*. Int J Med Microbiol 296:125–131. https://doi.org/10.1016/j.ijmm.2006.01.041

Wang B, Muir TW (2016) Regulation of virulence in *Staphylococcus aureus*: molecular mechanisms and remaining puzzles. Cell Chem Biol 23:214–224. https://doi.org/10.1016/j.chembiol.2016.01.004

Wang D, Ding X, Rather PN (2001) Indole can act as an extracellular signal in *Escherichia coli*. J Bacteriol 183:4210–4216. https://doi.org/10.1128/JB.183.14.4210-4216

Waters CM, Bassler BL (2005) Quorum sensing: cell-to-cell communication in bacteria. Annu Rev Cell Dev Biol 21:319–346. https://doi.org/10.1146/annurev.cellbio.21.012704.131001

Waters CM, Bassler BL (2006) The *Vibrio harveyi* quorum-sensing system uses shared regulatory components to discriminate between multiple autoinducers. Genes Dev 20:2754–2767. https://doi.org/10.1101/gad.1466506

Ween O, Gaustad P, Håvarstein LS (1999) Identification of DNA binding sites for ComE, a key regulator of natural competence in *Streptococcus pneumoniae*. Mol Microbiol 33:817–827. https://doi.org/10.1046/j.1365-2958.1999.01528.x

Whitehead NA, Barnard AM, Slater H, Simpson NJ, Salmond GP (2001) Quorum-sensing in Gram-negative bacteria. FEMS Microbiol Rev 25:365–404. https://doi.org/10.1111/j.1574-6976.2001. tb00583.x

Williams P (2007) Quorum sensing, communication and cross-kingdom signaling in the bacterial world. Microbiology 153:3923–3938. https://doi.org/10.1099/mic.0.2007/012856-0

Williams P, Cámara M (2009) Quorum sensing and environmental adaptation in *Pseudomonas aeruginosa*: a tale of regulatory networks and multifunctional signal molecules. Curr Opin Microbiol 12:182–191. https://doi.org/10.1016/j.mib.2009.01.005

Winson MK, Swift S, Fish L, Throup JP, Jorgensen F, Chhabra SR, Bycroft BW, Williams P, Stewart GS (1998) Construction and analysis of *luxCDABE*-based plasmid sensors for investigating *N*-acyl homoserine lactone mediated quorum sensing. FEMS Microbiol Lett 163:185–192. https://doi.org/10.1111/j.1574-6968.1998.tb13044.x

Wnuk SF, Robert J, Sobczak AJ, Meyers BP, Malladi VL, Zhu J, Gopishetty B, Pei D (2009) Inhibition of S-ribosylhomocysteinase (LuxS) by substrate analogues modified at the ribosyl C-3 position. Bioorg Med Chem 17:6699–6706. https://doi.org/10.1016/j.bmc.2009.07.057

Wuster A, Babu MM (2008) Conservation and evolutionary dynamics of the *agr* cell-to-cell communication system across firmicutes. J Bacteriol 190:743–746. https://doi.org/10.1128/JB.01135-07

Xavier KB, Miller ST, Lu W, Kim JH, Rabinowitz J, Pelczer I, Semmelhack MF, Bassler BL (2007) Phosphorylation and processing of the quorum-sensing molecule autoinducer-2 in enteric bacteria. ACS Chem Biol 2:128–136. https://doi.org/10.1021/cb600444h

Xiao G, Déziel E, He J, Lépine F, Lesic B, Castonguay MH, Milot S, Tampakaki AP, Stachel SE, Rahme LG (2006a) MvfR, a key *Pseudomonas aeruginosa* pathogenicity LTTR-class regulatory protein, has dual ligands. Mol Microbiol 62:1689–1699. https://doi.org/10.1111/j.1365-2958.2006.05462.x

Xiao G, He J, Rahme LG (2006b) Mutation analysis of the *Pseudomonas aeruginosa* mvfR and pqsABCDE gene promoters demonstrates complex quorum-sensing circuitry. Microbiology 152:1679–1686. https://doi.org/10.1099/mic.0.28605-0

Yang T, Tal-Gan Y, Paharik AE, Horswill AR, Blackwell HE (2016) Structure-function analyses of a *Staphylococcus epidermidis* autoinducing peptide reveals motifs critical for AgrC-type receptor modulation. ACS Chem Biol 11:1982–1991. https://doi.org/10.1021/acschembio.6b00120

Yu Q, Lepp D, Gohari IM, Wu T, Zhou H, Yin X, Yu H, Prescott JF, Nie SP, Xie MY, Gong J (2017) The Agr-like quorum sensing system is required for pathogenesis of necrotic enteritis caused by *Clostridium perfringens* in poultry. Infect Immun 85:e00975-16. https://doi.org/10.1128/IAI.00975-16

Zhang L, Gray L, Novick RP, Ji G (2002) Transmembrane topology of AgrB, the protein involved in the post-translational modification of AgrD in *Staphylococcus aureus*. J Biol Chem 277:34736–34742. https://doi.org/10.1074/jbc.M205367200

Zhang Z, Korkeala H, Dahlsten E, Sahala E, Heap JT, Minton NP (2013) Two-component signal transduction system CBO0787/CBO0786 represses transcription from botulinum neurotoxin promoters in *Clostridium botulinum* ATCC 3502. PLoS Pathog 9:e1003252. https://doi.org/10.1371/journal.ppat.1003252

Zhou Y, Zhao R, Ma B, Gao H, Xue X, Qu D, Li M, Meng J, Luo X, Hou Z (2016) Oligomerization of RNAIII-inhibiting peptide inhibits adherence and biofilm formation of methicillin-resistant *Staphylococcus aureus in vitro* and *in vivo*. Microb Drug Resist 22:193–201. https://doi.org/10.1089/mdr.2015.0170

Zhu J, Winans SC (2001) The quorum-sensing transcriptional regulator TraR requires its cognate signaling ligand for protein folding, protease resistance, and dimerization. Proc Natl Acad Sci U S A 98:1507–1512. https://doi.org/10.1073/pnas.98.4.1507

Zuberi A, Misba L, Khan AU (2017) CRISPR interference (CRISPRi) inhibition of *luxS* gene expression in *E. coli*: an approach to inhibit biofilm. Front Cell Infect Microbiol 7:214. https://doi.org/10.3389/fcimb.2017.00214

Chapter 10
Synergism Between Quorum Sensing Inhibitors and Antibiotics: Combating the Antibiotic Resistance Crisis

Sahana Vasudevan, Shogan Sugumar Swamy, Gurmeet Kaur, S. Adline Princy, and P. Balamurugan

Abstract With the alarming increase in the antibiotic resistance, there is an immediate need for alternative therapeutic strategies to combat this ever-changing bacterial battle. Combinatorial therapies have gained attention owing to their multiple targeted actions. The use of antibiotic is inevitable and antibiotics in combinations have been in use to treat drug resistant infections. Nevertheless, the multidrug resistant strains have found their own mechanisms to surpass such combinatorial treatments. Quorum sensing (QS) inhibition is considered to be the silver lining but is yet to find its way to commercial use. Hence, to combat the antibiotic resistance crisis, the synergy of QSIs and antibiotics is one of the possible revolutionary approaches. In this chapter, we have highlighted the importance and need for the synergy approach with the successful *in vitro* and *in vivo* studies that can possibly be extended to the commercial use.

Keywords Biofilm · Antibiotic · Resistance · Quorum sensing inhibition · Synergy

10.1 Introduction

Antibiotics have been the wonder drugs since 1928 when Sir Alexander Fleming discovered Penicillin, from *Penicillium notatum,* which was a breakthrough in the field of medicines. By 1940, due to the extensive use of antibiotics, there was an accelerated development of antibiotic-resistant strains which C. Lee Ventola describes as 'The Antibiotic resistance crisis'. The major causes were the overuse and misuse of drugs for the up growth of the antibiotic-resistant strains (Ventola 2015).

S. Vasudevan · S. S. Swamy · G. Kaur · S. A. Princy · P. Balamurugan (✉)
Quorum Sensing Laboratory, Centre for Research in Infectious Diseases (CRID),
School of Chemical and Biotechnology, SASTRA Deemed University,
Thanjavur 613401, Tamil Nadu, India
e-mail: adlineprinzy@biotech.sastra.edu; balamurugan@scbt.sastra.edu

© Springer Nature Singapore Pte Ltd. 2018
V. C. Kalia (ed.), *Biotechnological Applications of Quorum Sensing Inhibitors,*
https://doi.org/10.1007/978-981-10-9026-4_10

This is also attributed to improper sanitation, poor public health system and also irregular prescription (amr-review.org). The resistance is developed through bacterial evolution by spontaneous mutations making various drugs incompetent. Humans are affected either by direct contact or indirectly by the consumption of livestock, as antibiotics are used largely in animal feeds to prevent microbial infections. It is supported by the fact that 80% of antibiotics are used in livestock feed, in the US (Ventola 2015). Antibiotic resistance has also become an economic crisis, and many countries have employed various measures to overcome this crisis. It is estimated that by 2050, the death toll will exceed 10 million each year which extends to the cost of 100 trillion USD loss in output (amr-review.org). The main players involved are the pharmaceutical and biotechnology companies who are threatened by the resistance development. This crisis needs to be immediately addressed to avoid the economic loss of a nation.

Bacteria adopt numerous mechanisms to defend against the antimicrobials (Munita and Arias 2016). These mechanisms further ease the way for acquiring phenotypic and genotypic resistance to those antibiotics and make the pathogen to evolve as a superbug. Hence to overcome drug resistance, there is a pressing need for alternative therapeutic approaches. Among them, quorum sensing (QS) inhibition proves to be an encouraging strategy (Kalia and Purohit 2011), as these small molecules will not induce resistance (Gerdt and Blackwell 2014). Quorum sensing is the cell to cell communication, contributes to the defense mechanism and also produces virulence factors. Quorum sensing is attributed not only to the infectious diseases but also to different fields. Subsequently, quorum sensing inhibitors are reported to have varied applications (Kalia et al. 2014a; Kalia and Kumar 2015a). Though many quorum sensing inhibitors (QSIs) have been reported, the use of QSI alone in treatment has not been very successful (Sengupta et al. 2013) because of which commercialization of QSI is not in existence till now. A QSI can attenuate virulence of the pathogenic bacteria by targeting the QS signaling pathway. It is to be noted that QSI will not affect the bacterial growth, rather stop the pathogen to establish a community (Kalia 2013, 2015). A few research works claim that the host immune response will be sufficient enough to clear the bacteria after QS inhibition. However, in a real clinical setting, it may not be sufficient for the complete clearance of the high bacterial load, especially the ones which are in biofilm and may lead to re-emergence of the pathogen as resistant strains. Also to tackle the multidrug resistant strains, resensitization of resistant drugs has become the need of the hour. In spite of being considered as an alternative therapy to antibiotics, a few claims and theories state that bacteria might develop resistance to the QSIs which requires strong experimental evidence (Kalia et al. 2014b; Kalia and Kumar 2015b; Koul et al. 2016). Hence combinatorial treatment of QSI with antibiotics will be a promising approach, as target specific QSI will not pose survival stress and the antibiotic will aid in curbing the pathogenesis, at low concentrations.

10.2   Synergy Between QSIs and Antibiotics

10.2.1   Natural QSIs and Antibiotics

To date, many quorum sensing inhibitors have been tested along with conventionally used antibiotics that are listed in Table 10.1. Many naturally occurring QSIs are
reported previously which were shown to have remarkable synergistic activity with
conventional antibiotics. The following are a few reports which support the combinatorial action. A well-known QSI, furanone C-30, enhanced the susceptibility to
tobramycin against *P. aeruginosa* biofilm (Hentzer et al. 2003). Fujita et al. (2005)
have shown baicalein, a QSI compound from thyme leaves extract reduced the minimum inhibitory concentrations of tetracycline and other β-lactams antibiotics
against Methicillin-resistant *Staphylococcus aureus* (MRSA). Also, they have proposed the possible mechanisms for the synergistic action: inhibition of penicillin-
binding protein 2′(2a) by baicalein, and damage of peptidoglycan. A similar study
was extended to vancomycin-resistant *Enterococcus* where the synergy with gentamicin was reported (Chang et al. 2007). Later, baicalein was shown to interfere with
the transcriptional activator protein (TraR) of *P. aeruginosa* QS system, and also
having combinatorial activity with ampicillin (Zeng et al. 2008).

Farnesol, a quorum signal of *Candida* sp. inhibited *S. epidermidis* biofilm in
synergy with nafcillin and vancomycin (Pammi et al. 2011). Synergism between
tobramycin and baicalin hydrate against *Burkholderia cenocepacia* was shown in a
lung infected animal model (Brackman et al. 2011). It was also shown that the
extracts of *Nymphaea tetragona* and antibiotics could be effective against drug-
resistant Salmonella (Hossain et al. 2014). Synergistic efficacy of sub-MIC concentrations of curcumin with ceftazidime and ciprofloxacin against *P. aeruginosa* QS
system was well documented (Roudashti et al. 2017). The synthetic derivatives of
natural compounds have also been reported for QS inhibition activity. Zeng et al.
(2011) investigated the combinatorial action of 14-alpha-lipoyl andrographolide
(AL-1) and traditionally used antibiotics in inhibiting not only the *P. aeruginosa*
biofilm but also the EPS and pyocyanin.

10.2.2   Synthetic QSIs and Antibiotics

Like natural QSIs, several synthetic small molecule inhibitors have been described
to possess combinatorial action with conventional antibiotics with less or no toxicity. Balaban et al. (2003) have shown the synergistic actions of the RNAIII-inhibiting
peptide (RIP) with conventional antibiotics in 100% clearance of graft-associated *S.
epidermidis* infections *in vivo*, suggesting that RIP may be used to coat medical
devices to prevent staphylococcal infections. The activity of 2-aminoimidazole/tri-
azole conjugate with conventional antibiotics promoted the biofilm dispersion and
along with resensitization of MRSA and multi-drug resistant *Acinetobacter*

Table 10.1 Synergistic action of quorum sensing inhibitors and antibiotics

Quorum sensing inhibitors	Proposed mechanism of quorum sensing inhibition	Concentration of QSI	Antibiotics	Target organism	Effective synergistic action	References
RNAIII-inhibiting peptide	Phosphorylation inhibition of target of RNAIII activating protein (TRAP)	10 µg/ml	Cefazolin Rifampin Imipenem Levofloxacin Teicoplanin Mupirocin Quinupristin Dalfopristin	*Staphylococcus epidermidis*	>10–100-fold bacterial clearance *in vivo*	Balaban et al. (2003)
Furanone C-30	Interference in Acyl-homoserine lactone signaling	10 µM	Tobramycin	*Pseudomonas aeruginosa*	Two to three orders of magnitude more sensitive in bacterial clearance in biofilm	Hentzer et al. (2003)
Baicalein	Efflux pump inhibition	25 µg/ml	β-lactam antibiotics tetracycline	Methicillin-resistant *Staphylococcus aureus*	Reduction in MIC of tetracycline in MRSA strain OM584, from 128 µg/ml to 0.06 µg/ml	Fujita et al. (2005)
	NA	32 µg/ml	Gentamicin	Vancomycin-resistant *Enterococcus*	Two orders of magnitude higher activity in reduction of bacterial growth	Chang et al. (2007)
	Interference in quorum sensing system, transcription activator protein (TraR)	200 mM	Ampicillin	*Pseudomonas aeruginosa*	NA	Zeng et al. (2008)

Compound	Mechanism	Concentration	Antibiotic	Organism	Effect	Reference
Dispersin B	Dispersal of matured biofilm	20–600 µg/ml	Cefamandole nafate	*Staphylococcus epidermidis*	NA	Donelli et al. (2007)
2-Aminoimidazole/triazole conjugates	Inhibition and dispersal of biofilm	2.6–325 nM	Novobiocin; Colistin; Tobramycin; Novobiocin; Tobramycin etc.,	*Acinetobacter baumannii* and *Staphylococcus aureus*	Three orders magnitude increase of biofilm dispersion	Rogers et al. (2010)
Baicalin hydrate	Acyl-homoserine lactone QS signaling	100 µM	Tobramycin	*Burkholderia cenocepacia*, *Burkholderia multivorans* and *Pseudomonas aeruginosa*	Significant decrease in biofilm as compared to antibiotic alone	Brackman et al. (2011)
Cinnamaldehyde		250 µM	Clindamycin	*Pseudomonas aeruginosa* and *Staphylococcus aureus*		
Hamamelitannin		250 µM	Vancomycin	*Staphylococcus aureus*		
14-Alpha-lipoyl andrographolide (AL–1)	Interference in QS system	0.5 mM	Azithromycin; Ciprofloxacin; Fosfomycin; Streptomycin; Gentamicin	*Pseudomonas aeruginosa*	Significant inhibition of biofilm, EPS and pyocyanin compared to antibiotic alone	Zeng et al. (2011)
Lactonase	Increasing antibiotic susceptibility of the biofilm	0.3 units	Ciprofloxacin; Gentamicin	*Pseudomonas aeruginosa*	5–6 folds reduction in minimum biofilm eradication concentration	Kiran et al. (2011)

(continued)

Table 10.1 (continued)

Quorum sensing inhibitors	Proposed mechanism of quorum sensing inhibition	Concentration of QSI	Antibiotics	Target organism	Effective synergistic action	References
Nanosilver	Antibacterial and antibiofilm	0.25 – 2 µg/mL	Ampicillin; Chloramphenicol; Kanamycin	Escherichia coli; Pseudomonas aeruginosa; Staphylococcus aureus; Streptococcus mutans	Reduction of antibiotic concentration	Hwang et al. (2012)
Silver nanocolloids	Antiquorum and antibacterial	10 µg/mL	Amoxicillin; Methicillin; Ampicillin	Escherichia coli; Pseudomonas aeruginosa; Staphylococcus aureus	Increase in antagonistic activity	Arunkumar et al. (2013)
Isobutyl-DPD Phenyl DPD	Inhibiting the genotypic QS responses and dispersal of preformed biofilm	40 and 100 µM	Gentamicin; Gentamicin	Escherichia coli; Pseudomonas aeruginosa	Reduction in biofilm thickness by more than 80% for E. coli and 90% for P. aeruginosa	Roy et al. (2013)
Fimbrolide	Antivirulent and anti-biofilm agents	NA	NO hybrids	Pseudomonas aeruginosa	Increased biofilm inhibition at picomolar to nanomolar levels	Kutty et al. (2013)
Biogenic Silver nanoparticles	Antibacterial and antibiofilm agent	0.1–1 µg/mL	Ampicillin; Chloramphenicol; Erythromycin; Gentamicin; Tetracycline; Vancomycin	Pseudomonas aeruginosa; Shigella flexineri; Staphylococcus aureus; Streptococcus pneumoniae	Enhanced antibacterial activity	Gurunathan et al. (2014)
Citrate-capped silver nanoparticles	Biofilm inhibition and eradication	0.156–10 µg/mL	Aztreonam	Pseudomonas aeruginosa	Synergistic antimicrobial activity	Habash et al. (2014)

Mycofabricated Silver nanoparticles	Antibiofilm agent	25 µg/mL	Tobramycin	*Pseudomonas aeruginosa*	100% killing of biofilm cells	Singh et al. (2015)
4-(Benzylamino) cyclohexyl2-hydroxycinnamate	SarA inhibitor	15 and 65 µg/mL	Gentamicin	*Staphylococcus aureus*	Significant clearance of *Staphylococcus aureus* *in vivo* at lower concentration of gentamicin	Balamurugan et al. (2015)
Biogenic Silver nanoparticles (Fungus derived)	Antimicrobial and antibiofilm	1–128 µg/mL	Amikacin	*Escherichia coli*	Synergistic antibacterial activity	Barapatre et al. (2016)
			Kanamycin	*Pseudomonas aeruginosa*		
			Oxytetracycline	*Staphylococcus aureus*		
			Streptomycin			
2,3-Pyrazine dicarboxylic acid	LuxO modulator	25 µM	Chloramphenicol Doxycycline Erythromycin Tetracycline	*Vibrio cholerae*	Significant reduction in *Vibrio cholerae* growth	Hema et al. (2016)
1, 3-Disubstituted Ureas	ComA inhibitors	0.23–15 µM	Sodium fluoride	*Streptococcus mutans*	Reduced the fluoride concentration to 31.25 ppm	Kaur et al. (2016, 2017)

baumannii (Rogers et al. 2010). Roy et al. (2013) have shown the clearance of pre-formed biofilm by DPD derivatives with a significant increase by 80 and 90% for *E. coli* and *P. aeruginosa* respectively. In another study, the combined use of 2,3-Pyrazine dicarboxylic acid and conventional antibiotics against *V. cholerae* showed a significant reduction in growth (Hema et al. 2016). The combined use of 1,3-disubstituted urea derivatives specifically targeted against the ComA (a bacteriocin associated ABC transporter) of *Streptococcus mutans* and sodium fluoride showed a significant inhibition of growth and biofilm (Kaur et al. 2016). Interestingly, in this case, the results showed a remarkable decrease in the fluoride concentration to 31.25–62.5 ppm (~1000 ppm is presently used in toothpaste formulations). Additionally, 1,3-di-m-tolylurea (DMTU) which is a 1,3-disubstituted ureas derivative along with fluoride was capable of reducing dental caries *in vivo* which was evident from the macroscopic observations and pathological studies (Kaur et al. 2017). Similarly, combinatorial treatments showed promising activity in animal models as shown in Balamurugan et al. (2015). In this study, a significant reduction in gentamicin concentration was reported with SarA (a quorum regulator) targeted QS inhibitor, 4-(benzylamino)cyclohexyl 2-hydroxycinnamate against *Staphylococcus aureus* associated with gestational urinary tract infection.

10.2.3   Quorum Quenching Enzymes and Antibiotics

Enzymes such as acyl homoserine lactone (AHL) lactonases and acylases have also shown potent quorum quenching activities and are found in diverse set of organisms which makes them more versatile to explore (Kalia et al. 2011; Huma et al. 2011; Kalia 2014; Kumar et al. 2015; Koul and Kalia 2017). Interestingly, quorum quenching enzymes also have a significant role in the combinatorial approaches to biofilm treatment. Donelli et al. (2007) have shown the mature biofilm dispersal by dispersin B (β-N-Acetylglucosaminidase) thereby enhancing the antibiotic activity against adherent cells on polyurethane surfaces. Similarly, the lactonase enzyme eradicated biofilm and increased the susceptibility to antibiotics ciprofloxacin and gentamicin against *Pseudomonas aeruginosa* biofilm (Kiran et al. 2011).

10.2.4   Nanoparticles and Antibiotics

In recent times, metal nanoparticles have been explored extensively for antimicrobial activities against multidrug resistant pathogens (Bose and Chatterjee 2015; Dobrucka and Długaszewska 2015; Szweda et al. 2015; Deng et al. 2016; Wan et al. 2016). As an advancement, quorum sensing inhibitory as well as antibiofilm activity of nanoparticles have also been reported (Agarwala et al. 2014; Gurunathan et al. 2014; Ahiwale et al. 2017). The combination of the nanoparticles and antibiotics is proven to have potential synergistic activities enhancing the efficiency of the antibiotic. Silver nanoparticles are the most widely used metal nanoparticle for the antibacterial application which is extended to the combinatorial action with antibiotics. Mycofabricated silver nanoparticles were utilized along with tobramycin which improved the clearing of biofilm cells by facilitating the efficient penetration of the antibiotic (Singh et al. 2015). Gurunathan et al. (2014) have shown the antibiofilm activity of biogenic silver particles with different antibiotics against a wide range of Gram-positive and Gram-negative bacteria. Similar results were obtained for the combination of nano-silver and antibiotics (Hwang et al. 2012). Silver nanocolloids were shown to have effective combinatorial action with the conventionally used antibiotics against *S. aureus*, *E. coli* and *P. aeruginosa* (Arunkumar et al. 2013). The antimicrobial activity of aztreonam has been synergistically enhanced in the presence of citrate-capped silver nanoparticles against *P. aeruginosa* biofilms (Habash et al. 2014). Green synthesized silver nanoparticles from lignin-degrading fungus, *Aspergillus flavus* and *Emericella nidulans*, having antibiofilm activity showed synergistic antimicrobial activity against Gram-positive and Gram-negative bacteria (Barapatre et al. 2016).

Recently, Ilk et al. (2017), have encapsulated the quorum sensing inhibitor, kaempferol, in chitosan nanoparticles, and have shown the increased stability and QSI activity. Hence, the combined use of antibiotics with nanoparticles and QSIs with nanoparticles would be an exciting new choice of treatment in alternative therapies.

10.2.5   Combined Toxicity of Antibiotics and QSI

Even though the combined use of QSIs and antibiotic reduces the concentration of antibiotics, studies on assessing the combined toxicity are limited currently. A recent QSAR-based mechanistic study has assessed the combined toxicity of antibiotics and QSIs against *E. coli*. They have reported the toxicity effects of commonly used antibiotics such as sulfonamides, β–lactams and tetracyclines, and some potential QSIs (including furanone, pyrrolidones, and pyrroles). The eight QSIs taken showed an additive or antagonistic effect in combination with sulphonamides and had antagonistic effects with β-lactams and tetracyclines (Wang et al. 2017).

10.3   Synergy Mechanism

From the above examples, it can be clearly seen that the combinatorial use of QSI and antibiotics work in such a way that the QSI will inhibit/eradicate biofilm formation as well as virulence factors which will favor the effective functioning of antibiotics (Figs. 10.1a and 10.1b). As explained in the mechanism (Figs. 10.1a and 10.1b), QSI targets the quorum sensing pathway thereby reducing the production of QS signals and virulence factors. This interference in the virulence then paves a way for the antibiotics to complete their action in reduced dosage. If the QSI has biofilm

a

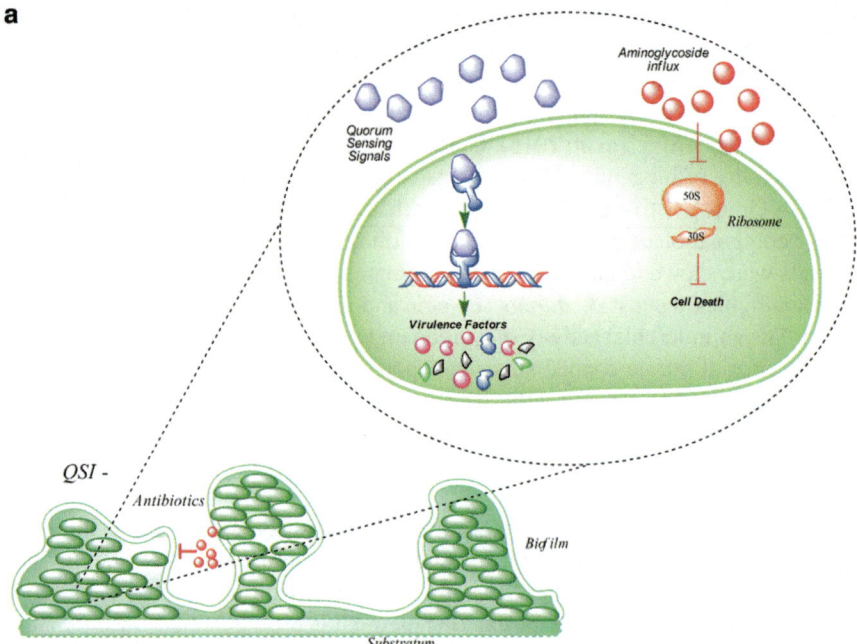

Fig. 10.1a Treatment of biofilm with antibiotics (aminoglycoside) and without quorum sensing inhibitor, QSI−. In the absence of QSI, the mature biofilm impose an antibiotic diffusion barrier for antibiotics. As the antibiotics are unable to enter the biofilm cells and bind to the 30s ribosomal unit, cell death is inhibited even at higher concentrations

disruption activity, then the combination with antibiotics will have an enhanced effect in biofilm dispersal (QSI action) by limiting the antibiotic diffusion barrier as well as the bacterial clearance (antibiotic action).

10.4 Methods of Measurement

In order to evaluate the synergistic action of the drugs, different models have been proposed. When the drugs are given in combination, the resulting effect can be either equal to, greater or less than the corresponding individual drugs' effect, which is termed as additive, synergism or antagonism respectively (Yang et al. 2014). In the case of QSI and antibiotics combination, the desired outcome is synergy rather than additive effect. QSI and antibiotics have independent targets and both the agents should have mutual participation in bringing out the desired effect. Thus, the interaction between a QSI and antibiotics is best explained in terms of synergy.

The most common assays that are carried out to understand synergy are checkerboard assay and time-kill assays (TKA). The checkerboard assay data are analyzed

b

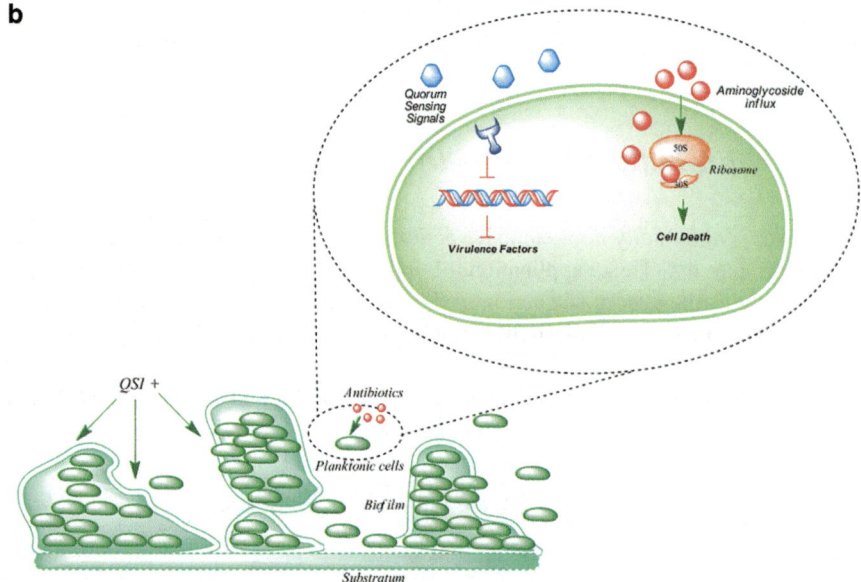

Fig. 10.1b Treatment of biofilm with antibiotics (aminoglycoside) and with quorum sensing inhibitor, QSI+. In the presence of QSI has disrupted the biofilm thereby paving a way for the antibiotic to enter the cell and complete its action. In this case, at low concentrations, the aminoglycoside is able to bind with 30s ribosomal subunit and interfere with the protein translational process and thereby causing cell death

using fractional inhibitory concentration (FIC). The data obtained from TKA gives the rate of killing in addition to the optimum concentration (Doern 2014).

10.4.1 Fractional Inhibitory Concentration Index (FICI)

The effect of the combinatorial treatment can be interpreted as synergistic, indifferent or antagonistic based on the non-parametric method, FICI (Kaur et al. 2016; Subramaniam et al. 2014). The following are the formulae to calculate the FICI of the drug interaction.

Let us consider the two drugs, A and B,

$$FICI = FIC_A + FIC_B$$

Where,

$$FIC_A = (MIC\ of\ A\ in\ combination)/(MIC\ of\ A\ alone)$$

$$FIC_B = (MIC\ of\ B\ in\ combination)/(MIC\ of\ B\ alone)$$

FICI ≤0.5 denotes Synergy; 0.5 < FICI <4denotes indifference or absence of interaction; FICI >4 denotes antagonism.

10.4.2 Bliss Independence Model

Synergism is a mutually non-exclusive action of two drugs leading to the enhanced inhibitory activity. Thus, application of independence probability theory, Bliss model explains the synergistic action of the two drugs be informational (Yang et al. 2014). This model is particularly useful when QSI with antibiofilm activity is given in combination with the antibiotics.

The bliss independence model or BI theory (Goldoni and Johansson 2007; Sun et al. 2008) is described by the following equations.

$$I_i = (I_A + I_B) - (I_A X I_B) \tag{10.1}$$

Where,

Ii = predicted inhibition percentage of A and B.

I_A = experimental inhibition percentage of A (alone).

I_B = experimental inhibition percentage of B (alone).

$$I = 1-E \tag{10.2}$$

Where,

E = growth percentage

Equation 10.3 is obtained by substituting (10.2) in (10.1)

$$E_i = E_A X E_B \tag{10.3}$$

Where,

E_i = predicted growth percentage of A and B.

E_A = observed growth percentage of A

E_B = observed growth percentage of B.

Interaction (ΔE) is given by the formula (10.4):

$$\Delta E = E_{predicted} - E_{observed} \tag{10.4}$$

By the nonparametric approach described by Prichard et al. (1991,1993), E_A and E_B are obtained directly from the experimental data. With the obtained results, the interpretations are as follows: ΔE – positive (synergy) and ΔE – negative (antagonism).

With the previous studies, both the above said models correlated with each other (Barapatre et al. 2016; Kaur et al. 2016; Hema et al. 2016). Response surface methodology can also be used to understand the synergistic pattern of the above synergy. Thus these methods give a mathematical validation to the synergistic activity of the QSIs and antibiotics.

10.5   Conclusion

Combinatorial therapies are currently used in the treatment of complex diseases like cancer. The idea behind the combinatorial therapy is to target different molecular mechanisms thereby disarming the proliferation of the infectious diseases. Antibiotics are broad spectrum and require a very high dosage for the treatment of multidrug resistant strains. On the other hand, QSIs are target specific which works at very low concentrations without inducing a survival stress. With the aforementioned examples, it can be clearly seen that the use of these antibiotics with QSIs potentiates significant synergistic action and is a thoughtful way to combat the overuse of antibiotics.

10.6   Opinion

Currently, a broad range of antibiotics is used to scale down the infections. It is well known that continuous administration of antibiotics leads to resistance development. Therefore, a cocktail of antibiotics will also eventually lead to more resistant strains causing a disastrous epidemic. On the contrary, the use of QSI may not contribute to resistance development but it makes the bacteria more susceptible to antibiotics at low levels. This way the bacteria are exposed to the minimal level of antibiotics such that they are unable to trigger the evolution of resistant strains. Thus, the combinatorial use of antibiotics with quorum sensing inhibitors uplifts the condition of the antibiotics resistance crisis and also the market per se.

References

Agarwala M, Choudhury B, Yadav RN (2014) Comparative study of antibiofilm activity of copper oxide and iron oxide nanoparticles against multidrug resistant biofilm forming uropathogens. Indian J Microbiol 54:365–368. https://doi.org/10.1007/s12088-014-0462-z

Ahiwale SS, Bankar AV, Tagunde S, Kapadnis BP (2017) A bacteriophage mediated gold nanoparticle synthesis and their antibiofilm activity. Indian J Microbiol 57:188–194. https://doi.org/10.1007/s12088-017-0640-x

Arunkumar M, Mahesh N, Balakumar S, Sivakumar R, Priyadharshni S (2013) Antiquorum sensing and antibacterial activity of silver nanoparticles synthesized by mutant *Klebsiella pneumoniae* MTCC 3354. Asian J Chem 25:9961–9964. https://doi.org/10.14233/ajchem.2013.15754

Balaban N, Giacometti A, Cirioni O, Gov Y, Ghiselli R, Mocchegiani F, Viticchi C, Del Prete MS, Saba V, Scalise G, Dell'Acqua G (2003) Use of the quorum-sensing inhibitor RNAIII-inhibiting peptide to prevent biofilm formation in vivo by drug-resistant *Staphylococcus epidermidis*. J Infect Dis 187:625–630. https://doi.org/10.1086/345879

Balamurugan P, Hema M, Kaur G, Sridharan V, Prabu PC, Sumana MN, Princy SA (2015) Development of a biofilm inhibitor molecule against multidrug resistant *Staphylococcus aureus* associated with gestational urinary tract infections. Front Microbiol 6:832. https://doi.org/10.3389/fmicb.2015.00832

Barapatre A, Aadil KR, Jha H (2016) Synergistic antibacterial and antibiofilm activity of silver nanoparticles biosynthesized by lignin-degrading fungus. Bioresour Bioprocess 3:8. https://doi.org/10.1186/s40643-016-0083-y

Bose D, Chatterjee S (2015) Antibacterial activity of green synthesized silver nanoparticles using Vasaka (*Justicia adhatoda* L.) leaf extract. Indian J Microbiol 55:163–167. https://doi.org/10.1007/s12088-015-0512-1

Brackman G, Cos P, Maes L, Nelis HJ, Coenye T (2011) Quorum sensing inhibitors increase the susceptibility of bacterial biofilms to antibiotics in vitro and in vivo. Antimicrob Agents Chemother 55:2655–2661. https://doi.org/10.1128/AAC.00045-11

Chang PC, Li HY, Tang HJ, Liu JW, Wang JJ, Chuang YC (2007) *In vitro* synergy of baicalein and gentamicin against vancomycin resistant *Enterococcus*. J Microbiol Immunol Infect 40:56–61

Deng H, McShan D, Zhang Y, Sinha SS, Arslan Z, Ray PC, Yu H (2016) Mechanistic study of the synergistic antibacterial activity of combined silver nanoparticles and common antibiotics. Environ Sci Technol 50:8840–8848. https://doi.org/10.1021/acs.est.6b00998

Dobrucka R, Długaszewska J (2015) Antimicrobial activities of silver nanoparticles synthesized by using water extract of *Arnicae anthodium*. Indian J Microbiol 55:168–174. https://doi.org/10.1007/s12088-015-0516-x

Doern CD (2014) When does 2 plus 2 equal 5? A review of antimicrobial synergy testing. J Clin Microbiol 52:4124–4128. https://doi.org/10.1128/JCM.01121-14

Donelli G, Francolini I, Romoli D, Guaglianone E, Piozzi A, Ragunath C, Kaplan JB (2007) Synergistic activity of dispersin B and cefamandole nafate in inhibition of staphylococcal biofilm growth on polyurethanes. Antimicrob Agents Chemother 51:2733–2740. https://doi.org/10.1128/AAC.01249-06

Fujita M, Shiota S, Kuroda T, Hatano T, Yoshida T, Mizushima T, Tsuchiya T (2005) Remarkable synergies between baicalein and tetracycline, and baicalein and beta-lactams against methicillin resistant *Staphylococcus aureus*. Microbiol Immunol 49:391–396. https://doi.org/10.1111/j.1348-0421.2005.tb03732.x

Gerdt JP, Blackwell HE (2014) Competition studies confirm two major barriers that can preclude the spread of resistance to quorum-sensing inhibitors in bacteria. ACS Chem Biol 9:2291–2299. https://doi.org/10.1021/cb5004288

Goldoni M, Johansson C (2007) A mathematical approach to study combined effects of toxicants in vitro: evaluation of the Bliss independence criterion and the Loewe additivity model. Toxicol Vitro 21:759–769. https://doi.org/10.1016/j.tiv.2007.03.003

Gurunathan S, Han JW, Kwon DN, Kim JH (2014) Enhanced antibacterial and anti-biofilm activities of silver nanoparticles against Gram-negative and Gram-positive bacteria. Nanoscale Res Lett 9:373–390. https://doi.org/10.1186/1556-276X-9-373

Habash MB, Park AJ, Vis EC, Harris RJ, Khursigara CM (2014) Synergy of silver nanoparticles and aztreonam against *Pseudomonas aeruginosa* PAO1 Biofilms. Antimicrob Agents Chemother 58:5818–5830. https://doi.org/10.1128/AAC.03170-14

Hema M, Princy SA, Sridharan V, Vinoth P, Balamurugan P, Sumana M (2016) Synergistic activity of quorum sensing inhibitor, pyrizine-2-carboxylic acid and antibiotics against multi-drug resistant *V. cholerae*. RSC Adv 6:45938–45946. https://doi.org/10.1039/C6RA04705J

Hentzer M, Wu H, Andersen JB, Riedel K, Rasmussen TB, Bagge N, Kumar N, Schembri MA, Song Z, Kristoffersen P, Manefield M (2003) Attenuation of *Pseudomonas aeruginosa* virulence by quorum sensing inhibitors. EMBO J 22:3803–3815. https://doi.org/10.1093/emboj/cdg366

Hossain MA, Park JY, Kim JY, Suh JW, Park SC (2014) Synergistic effect and antiquorum sensing activity of *Nymphaea tetragona* (water lily) extract. Bio Med Res Int 2014:562173. https://doi.org/10.1155/2014/562173

Huma N, Shankar P, Kushwah J, Bhushan A, Joshi J, Mukherjee T, Raju SC, Purohit HJ, Kalia VC (2011) Diversity and polymorphism in AHL-lactonase gene (*aiiA*) of *Bacillus*. J Microbiol Biotechnol 21:1001–1011. https://doi.org/10.4014/jmb.1105.05056

Hwang IS, Hwang JH, Choi H, Kim KJ, Lee DG (2012) Synergistic effects between silver nanoparticles and antibiotics and the mechanisms involved. J Med Microbiol 61:1719–1726. https://doi.org/10.1099/jmm.0.047100-0

Ilk S, Sağlam N, Özgen M, Korkusuz F (2017) Chitosan nanoparticles enhances the anti-quorum sensing activity of kaempferol. Int J Biol Macromol 94:653–662. https://doi.org/10.1016/j.ijbiomac.2016.10.068

Kalia VC, Purohit HJ (2011) Quenching the quorum sensing system: potential antibacterial drug targets. Crit Rev Microbiol 37:121–140. https://doi.org/10.3109/1040841X.2010.532479

Kalia VC, Raju SC, Purohit HJ (2011) Genomic analysis reveals versatile organisms for quorum quenching enzymes: acyl-homoserine lactone-acylase and –lactonase. Open Microbiol J 5:1–13. https://doi.org/10.2174/1874285801105010001

Kalia VC (2013) Quorum sensing inhibitors: an overview. Biotechnol Adv 31:224–245. https://doi.org/10.1016/j.biotechadv.2012.10.004

Kalia VC, Kumar P, Pandian SK, Sharma P (2014a) Chapter 15: biofouling control by quorum quenching. In: Kim SK (ed) Hb_25 Springer handbook of marine biotechnology. Springer, Berlin, pp 431–440

Kalia VC, Wood TK, Kumar P (2014b) Evolution of resistance to quorum-sensing inhibitors. Microb Ecol 68:13–23. https://doi.org/10.1007/s00248-013-0316-y

Kalia VC (2014) In search of versatile organisms for quorum-sensing inhibitors: acyl homoserine lactones (AHL)-acylase and AHL-lactonase. FEMS Microbiol Lett 359:143. https://doi.org/10.1111/1574-6968.12585

Kalia VC (2015) Microbes: the most friendly beings? In: Kalia VC (ed) Quorum sensing vs quorum quenching: a battle with no end in sight. Springer, New Delhi, pp 1–5. ISBN 978-81-322-1981-1. https://doi.org/10.1007/978-81-322-1982-8_1

Kalia VC, Kumar P (2015a) Potential applications of quorum sensing inhibitors in diverse fields. In: Kalia VC (ed) Quorum sensing vs quorum quenching: a battle with no end in sight. Springer, New Delhi, pp 359–370. https://doi.org/10.1007/978-81-322-1982-8_29

Kalia VC, Kumar P (2015b) The Battle: quorum-sensing inhibitors versus evolution of bacterial resistance. In: Kalia VC (ed) Quorum sensing vs quorum quenching: a battle with no end in sight. Springer India, New Delhi, pp 385–391. https://doi.org/10.1007/978-81-322-1982-8_31

Kaur G, Balamurugan P, Uma Maheswari C, Anitha A, Princy SA (2016) Combinatorial effects of aromatic 1, 3-disubstituted Ureas and fluoride on in vitro inhibition of *Streptococcus mutans* biofilm formation. Front Microbiol 7:861. https://doi.org/10.3389/fmicb.2016.00861

Kaur G, Balamurugan P, Princy SA (2017) Inhibition of the quorum sensing system (ComDE Pathway) by aromatic 1,3-di-m-tolylurea (DMTU): cariostatic effect with fluoride in wistar rats. Front Cell Infect Microbiol 7:313. https://doi.org/10.3389/fcimb.2017.00313

Kiran S, Sharma P, Harjai K, Capalash N (2011) Enzymatic quorum quenching increases antibiotic susceptibility of multidrug resistant *Pseudomonas aeruginosa*. Iran J Microbiol 3:1–12

Koul S, Prakash J, Mishra A, Kalia VC (2016) Potential emergence of multi-quorum sensing inhibitor resistant (MQSIR) bacteria. Indian J Microbiol 56:1–18. https://doi.org/10.1007/s12088-015-0558-0

Koul S, Kalia VC (2017) Multiplicity of quorum quenching enzymes: a potential mechanism to limit quorum sensing bacterial population. Indian J Microbiol 57:100–108. https://doi.org/10.1007/s12088-016-0633-1

Kumar P, Koul S, Patel SKS, Lee JK, Kalia VC (2015) Heterologous expression of quorum sensing inhibitory genes in diverse organisms. In: Kalia VC (ed) Quorum sensing vs quorum quenching: a battle with no end in sight. Springer, New Delhi, pp 343–356. https://doi.org/10.1007/978-81-322-1982-8_28

Kutty SK, Barraud N, Pham A, Iskander G, Rice SA, Black DS, Kumar N (2013) Design, synthesis, and evaluation of fimbrolide-nitric oxide donor hybrids as antimicrobial agents. J Med Chem 56:9517–9529. https://doi.org/10.1021/jm400951f

Munita JM, Arias CA (2016) Mechanisms of antibiotic resistance. Microbiol Spectr 4. https://doi.org/10.1128/microbiolspec.VMBF-0016-2015

Pammi M, Liang R, Hicks JM, Barrish J, Versalovic J (2011) Farnesol decreases biofilms of *Staphylococcus epidermidis* and exhibits synergy with nafcillin and vancomycin. Pediatr Res 70:578–583. https://doi.org/10.1203/PDR.0b013e318232a984

Prichard MN, Prichard LE, Baguley WA, Nassiri MR, Shipman C (1991) Three-dimensional analysis of the synergistic cytotoxicity of ganciclovir and zidovudine. Antimicrob Agents Chemother 35:1060–1065. https://doi.org/10.1128/AAC.35.6.1060

Prichard MN, Prichard LE, Shipman C (1993) Strategic design and three-dimensional analysis of antiviral drug combinations. Antimicrob Agents Chemother 37:540–545. https://doi.org/10.1128/AAC.37.3.540

Rogers SA, Huigens RW, Cavanagh J, Melander C (2010) Synergistic effects between conventional antibiotics and 2-aminoimidazole-derived antibiofilm agents. Antimicrob Agents Chemother 54:2112–2118. https://doi.org/10.1128/AAC.01418-09

Roudashti S, Zeighami H, Mirshahabi H, Bahari S, Soltani A, Haghi F (2017) Synergistic activity of sub-inhibitory concentrations of curcumin with ceftazidime and ciprofloxacin against *Pseudomonas aeruginosa* quorum sensing related genes and virulence traits. World J Microbiol Biotechnol 33:50. https://doi.org/10.1007/s11274-016-2195-0

Roy V, Meyer MT, Smith JA, Gamby S, Sintim HO, Ghodssi R, Bentley WE (2013) AI-2 analogs and antibiotics: a synergistic approach to reduce bacterial biofilms. Appl Microbiol Biotechnol 97:2627–2638. https://doi.org/10.1007/s00253-012-4404-6

Sengupta S, Chattopadhyay MK, Grossart HP (2013) The multifaceted roles of antibiotics and antibiotic resistance in nature. Front Microbiol 4:47. https://doi.org/10.3389/fmicb.2013.00047

Singh BR, Singh BN, Singh A, Khan W, Naqvi AH, Singh HB (2015) Mycofabricated biosilver nanoparticles interrupt *Pseudomonas aeruginosa* quorum sensing systems. Sci Rep 5:13719. https://doi.org/10.1038/srep13719

Subramaniam S, Keerthiraja M, Sivasubramanian A (2014) Synergistic antibacterial action of β-sitosterol-D-glucopyranoside isolated from Desmostachya bipinnata leaves with antibiotics against common human pathogens. Rev Bras Farm 24:44–50. https://doi.org/10.1590/0102-695X20142413348

Sun S, Li Y, Guo Q, Shi C, Yu J, Ma L (2008) *In vitro* interactions between tacrolimus and azoles against *Candida albicans* determined by different methods. Antimicrob Agents Chemother 52:409–417. https://doi.org/10.1128/AAC.01070-07

Szweda P, Gucwa K, Kurzyk E, Romanowska E, Dzierżanowska-Fangrat K, Jurek AZ, Kuś PM, Milewski S (2015) Essential oils, silver nanoparticles and propolis as alternative agents against fluconazole resistant *Candida albicans, Candida glabrata* and *Candida krusei* clinical isolates. Indian J Microbiol 55:175–183. https://doi.org/10.1007/s12088-014-0508-2

Ventola CL (2015) The antibiotic resistance crisis: part 1: causes and threats. Pharm Ther 40:277

Wan G, Ruan L, Yin Y, Yang T, Ge M, Cheng X (2016) Effects of silver nanoparticles in combination with antibiotics on the resistant bacteria *Acinetobacter baumannii.* Int J Nanomedicine 11:3789. https://doi.org/10.2147/IJN.S104166

Wang D, Shi J, Xiong Y, Hu J, Lin Z, Qiu Y (2017) A QSAR-based mechanistic study on the combined toxicity of antibiotics and quorum sensing inhibitors against *Escherichia coli.* J Hazard Mater 341:438–447. https://doi.org/10.1016/j.jhazmat.2017.07.059

Yang H, Novick SJ, Zhao W (2014) Drug combination synergy. In: Zhao W, Yang H (eds) Drug statistical methods in drug combination studies. CRC Press, Boca Raton, pp 17–40. ISBN 978-14-822-1674-5

Zeng Z, Qian L, Cao L, Tan H, Huang Y, Xue X, Shen Y, Zhou S (2008) Virtual screening for novel quorum sensing inhibitors to eradicate biofilm formation of *Pseudomonas aeruginosa.* Appl Microbiol Biotechnol 79:119–126. https://doi.org/10.1007/s00253-008-1406-5

Zeng X, Liu X, Bian J, Pei G, Dai H, Polyak SW, Song F, Ma L, Wang Y, Zhang L (2011) Synergistic effect of 14-alpha-lipoyl andrographolide and various antibiotics on the formation of biofilms and production of exopolysaccharide and pyocyanin by *Pseudomonas aeruginosa.* Antimicrob Agents Chemother 55:3015–3017. https://doi.org/10.1128/AAC.00575-10

Chapter 11
Nanoparticles as Quorum Sensing Inhibitor: Prospects and Limitations

Faizan Abul Qais, Mohammad Shavez Khan, and Iqbal Ahmad

Abstract The emergence and worldwide spread of multi-drug resistant bacterial pathogens and slow pace of drug discovery with novel mode of action has necessitated search for alternative or new strategies to combat bacterial infection. Targeting virulence and pathogenicity of pathogens controlled by quorum sensing (QS) is considered as a promising anti-infective drug target. Several molecules both natural and synthetic were reported to interfere quorum sensing and are potential candidates for anti-infective drugs. The inhibition of QS might successfully attenuate and eradicate the microbial pathogens in combination with host immune system. It is expected that QS inhibition will exert less selection pressure for development of resistance among pathogenic bacteria. The recent progress in nanobiotechnology have given a greater hope for the development of novel anti-QS agents/formulations with improved therapeutic potential, enhanced targeted delivery with lesser toxicity to host system. The improved action of nano-formulations is a fascinating ability compared to their bulk. Recently, nanoparticles such as metal nanoparticles are reported to exhibit promising anti-QS activity both *in vitro* and *in vivo*. Nanomaterials are also been tested as vehicle for targeted delivery of conventionally used antimicrobial agents. There is greater scope of manipulation in nano-based formulations according to desired needs making such therapeutic strategies more efficient. Of note, the risks associated with the application of nanoparticles in drug delivery, diagnostics, production of improved biocompatible material or preventing biofilm formation on medical devices, *etc.* are needed to be scrutinized. In this article, we have made an attempt to review the recent advancements in nanoparticle as anti-QS agents and progress made on nano-based formulations with promising prospects and limitations.

Keywords Quorum sensing · Nanoparticles · Anti-QS agents · Drug delivery · Bacterial infection · Multi-drug resistance

F. A. Qais · M. S. Khan · I. Ahmad (✉)
Department of Agricultural Microbiology, Faculty of Agricultural Sciences,
Aligarh Muslim University, Aligarh, UP, India

© Springer Nature Singapore Pte Ltd. 2018
V. C. Kalia (ed.), *Biotechnological Applications of Quorum Sensing Inhibitors*,
https://doi.org/10.1007/978-981-10-9026-4_11

Abbreviations

AgNPs Silver nanoparticles
AHL Acylhomoserine lactone
CNTs Carbon nanotubes
QS Quorum sensing
QSI Quorum sensing inhibitor
HSL Homoserine lactone
SLNs Solid lipid nanoparticles
MDR Multi-drug resistance

11.1 Introduction

Microbial cell to cell communication is used in microbial system that helps in adaption and monitoring of their surroundings via chemical signalling, contact base chemical exchanges and electric signalling (Galloway et al. 2010; Phelan et al. 2012; Shrestha et al. 2013). One of such system is quorum sensing (QS) in which bacteria examines its local population by monitoring the amount of autoinducers, small chemical signal molecules. For the first time, QS was discovered in *Vibrio fischeri*, a marine bacterium, by Nealson and co-workers (Nealson et al. 1970). The term "quorum sensing" was coined by Fuqua and his group that referred to acylated homoserine lactone (AHL)-mediated luxR/luxI regulated system (Fuqua et al. 1994). In bacteria harbouring QS system, autoinducers interact with transcriptional regulators altering the genetic expression profiles, once their concentration reaches the certain threshold limit (Schaefer et al. 2013; Zhang and Li 2015; Ahmad et al. 2017). Autoinducers also bind to the extracellular domains of histidine kinase receptor, a membrane receptor, leading to autophosphorylation and causes a cognate cytoplasmic response (Ke et al. 2015). Regulation of expression of QS-dependent genes in a population by autoinducers provides an ability to maintain a "society" like structure that controls certain important physiological pathways and produce "co-operative" response such as biofilm formation, pathogenesis, pollutant biodegradation etc. (Ren et al. 2013; Husain et al. 2015; Yong et al. 2015). Recently, considerable amount of literature and reviews are available that highlights the importance of disruption of QS system as a promising strategy for disease control, water treatment systems and biodegradation (Bhardwaj et al. 2013; Díaz et al. 2013; Rampioni et al. 2014; Siddiqui et al. 2015).

In QS, the expression of virulence factors and other proteins are also controlled which are involved in primary metabolic process (Husain et al. 2016; Rajamanikandan et al. 2017). A remarkable portion of bacterial genome *i.e.* 4–10% and more than 20% of bacterial proteome system is influenced by this communication system (Sifri 2008). Apart from above mentioned processes, many other responses are also controlled by QS that includes competence, motility, secretion of virulence factors, sporulation, bioluminescence, and antibiotic production (Roux et al. 2009). Another problem with conventional antimicrobial agents (mainly antibiotics) is that its sub-judicious use had led to the emergence of multi-drug resistant (MDR) strains or "superbugs" (Maheshwari et al. 2016). Virulence and pathogenicity in large number of pathogenic bacteria are regulated by QS. Therefore, it is now considered that disruption of this microbial communication may prove to be an important target for the development of novel anti-infective agents and combating problem of multi-drug resistance (Al-Shabib et al. 2017). It is expected that QS inhibitors may successfully eradicate the microbial infections in combination with host immune system without having any harmful effect on human tissues (Defoirdt et al. 2013). Moreover, microbes are less likely to develop resistance against anti-QS agents since it mainly targets the virulence factors without inhibiting the growth of micro-organisms (Hentzer and Givskov 2003; Ahmad and Husain 2014).

Recent advances in nanotechnology have opened new hope for researchers as it has gained much attention due to its application in medicine, diagnostics, bioremediation, agriculture etc. (Valcárcel and López-Lorente 2016). The underlying reason of nanoparticle's better action compared to their bulk form is their fascinating properties which are superior to those of bulk materials (Wagh et al. 2013). This advancement has attracted the attention of researchers to develop novel antibacterial agents in the form of nanomedicine (Khan et al. 2016). Anticipated application of nanotechnology in healthcare include diagnostics, drug delivery, preventing biofilm formation on medical devices, production of improved biocompatible material *etc.* (Zia et al. 2010). However, risks associated with the application of nanoparticles in medicine is also subject to scrutiny. In this chapter, we have briefly summarised the fundamental concept of nanoparticles, its application in medicine, its current status of knowledge on application of nanoparticles as anti-QS agents and in delivery of anti-QS agents.

11.2 Nanoparticles: Characteristics and Their Interaction with Bacteria

Recently, there has been enormous growth in nanotechnology that has found its vital applications in basic and applied research of biological, chemical, physical and earth sciences (Fernandez-Garcia et al. 2004; Rodríguez and Fernández-García 2007; Raghunath and Perumal 2017). Nanoparticles are usually in the range of 1–100 nm having great versatility in their shapes and size that possess unique chemical and physical characteristics. Till date, nanoparticles have found their application in diagnosis, catalysis, drug delivery, sensing, semiconductors and solid oxide fuel cells (Haddad and Seabra 2012; Corr 2013). Various reports have been found on the application of nanomaterials as antimicrobial agents (Jones et al. 2008; Mahapatra et al. 2008; Tran et al. 2010). The unique chemical and physical properties of nanoparticles compared to their bulk material enable them to differently interact with biological systems and contribute to antimicrobial activity (Singh and Nalwa 2011). The alkaline nature of metal nanoparticles such as magnesium oxide and calcium oxide nanoparticles is the significant component that confers to antimicrobial activity (Sawai et al. 2005). These alkali metal nanoparticles are relatively more soluble that contributes in alkalinity of the medium which is not found in semiconductor metal nanoparticles such as zinc oxide nanoparticles (Zhang et al. 2007). The electrostatic nature of positively charged nanoparticles such as cerium oxide nanoparticles also determines their bacteriostatic and bactericidal property (Thill et al. 2006). Titanium nanoparticles are semiconductor photocatalysts which inhibit the growth of even desiccation tolerant and ultraviolet radiation-resistant bacteria (Sadiq et al. 2010).

Nanoparticles exhibit wide range of action that can serve as broad spectrum antimicrobial agents against micro-organisms including those of multi-drug resistant strains. Different nanoparticles with diverse functional and physicochemical properties makes them good antimicrobial agent and an alternative to the conventionally used antibiotics. Nanoparticles of desired properties can be made as anti-infective agents owing to their high novel electrical, chemical, magnetic, mechanical and optical properties and high surface area-to-volume ratio (Whitesides 2005). The antimicrobial efficacy of nanoparticles is mainly governed by the solubility in aqueous medium, particle size and release of metal ions (Raghunath and Perumal 2017). The mode of action of nanoparticles is quite different from conventional antibiotics that include destruction of enzyme and nucleic acid pathway and alterations of the cell wall (Zhu et al. 2013). Due to limited site of action of antibiotics, bacteria develop resistance against one or more antibiotics. On the other hand, nanoparticle exhibited completely different mode of action including membrane damage and alteration of cellular processes both at molecular and biochemical levels (Kumar et al. 2011a,b). Antibacterial efficacy of nanoparticles is also due to induction of oxidative stress, release of metal ions and non-oxidative stress (Nagy et al. 2011; Gurunathan et al. 2012; Leung et al. 2014). The multiple mode actions require multiple simultaneous genetic changes in bacterial cell to develop antibacterial resistance against nanoparticles (Zaidi et al. 2017) which makes difficult to emergence of early resistance to nanoparticle. Moreover, use of nanomaterials as vehicle for antibiotics for targeted delivery can support and complement traditional antibiotics. The nanomaterials as delivery systems include concurrent delivery of multiple drugs, enhanced drug solubility and prolonged systemic circulation as reported earlier (Davis and Shin 2008; Chetoni et al. 2016). The promising results at research level have endorsed the use of nanoparticles for the treatment of infectious diseases and delivery of vaccines in which many formulations are under various phases of pre-clinical and clinical tests (Raghunath and Perumal 2017; Zaidi et al. 2017).

11.3   Nanoparticles as Anti-QS Agents

In search for novel quorum sensing inhibitors, researchers have tested diverse group of compounds including, phytocompounds and synthetic compounds (Asfour 2017). In recent years, efforts have been made to evaluate various nanoparticles as antimicrobial agents against a number of pathogenic microorganims (Zaidi et al. 2017). Search for novel activities in nanoparticles such as anti-QS properties and QS mediated inhibition of virulence and biofilm have been recently documented (Radzig et al. 2013; Chaudhari et al. 2015; García-Lara et al. 2015; Miller et al. 2015). Some of the relevant literature reports are summarized in Table 11.1. Since most of the studies in this direction are directed towards metallic nanoparticle such as silver nanoparticle. A survey of literature is briefly described below.

11.3.1 Silver Nanoparticles

Silver nanoparticles are nanoscale clusters of silver atoms (Ag^0). The silver nanoparticles is most commonly synthesised by chemical reduction of silver ions with reducing agents (Iravani et al. 2014). However, green synthesis of nanoparticles has now become more common. Various combinations of silver nanoparticles with other metal is used to enhance the availability and activity of metal nanoparticles (Janczak and Aspinwall 2012). Antimicrobial property of silver nanoparticles has been well documented in literature (Agnihotri et al. 2014; Ahmed et al. 2016; Zou et al. 2017). The mechanism of action of silver nanoparticles is not yet fully explored but involves three most common mechanisms:

(i) Free silver ions uptake followed by interruption of ATP production and DNA replication.
(ii) Production of reactive oxygen species(ROS)
(iii) Damage to cytoplasmic membrane.

Silver nanoparticles (AgNPs), one of the most widely studied metal nanoparticles, have been found to inhibit QS controlled virulence factors in both Gram-positive and Gram-negative bacteria (Wagh et al. 2013). A brief study on green synthesized silver nanoparticles from *Cymbopogan citratus* leaf extract demonstrated quorum quenching action and prevented biofilms formation by *Staphylococcus aureus* which was also demonstrated by microscopic data (Masurkar et al. 2012). The research found that silver nanoparticles might be involved in neutralisation of adhesive substances which are required for the initial attachment of microbes and for maintenance of biofilm strength.

A detailed study conducted by Singh and co-workers demonstrated that biosynthesized AgNPs inhibited the violacein production approximately by 100% at 25 μg/ml in *Chromobacterium violaceum* 12472. Furthermore, they examined the effect of AgNPs on *C. violaceum* CV026 and *Pseudomonas aeruginosa* PAO1 both at toxic and non-toxic concentrations. The results demonstrated that AgNPs interfered QS *via* attenuation of AHL production not by its toxic effect. Many QS mediated virulence factors of PAO1 were inhibited by AgNPs such as Las a protease activity (15–86% inhibition), LasB elastase (22–86% inhibition), pyocyanin (18–96% suppression), pyoverdine (14–95% suppression), pyochelin (10–92% suppression) and rhamnolipid (10–70% inhibition) at non-toxic concentrations. The

Table 11.1 Nanoparticles demonstrating anti-QS activity and their mode of action

S. No.	Nanoparticles	Activities found	Test organism	References
1.	Silver nanoparticle	Quorum quenching against *S. aureus* biofilm	*S. aureus*	Masurkar et al. (2012)
2.	Silver nanoparticle	Inhibition of violacein of *C. violaceum*, inhibition of virulence factors such as protease activity, elastase, pyocyanin, pyoverdine, pyochelin and rhamnolipid of *P. aeruginosa*	*C. violaceum* and *P. aeruginosa*	Singh et al. (2015)
3.	Silver nanowires	Inhibition of violacein of *C. violaceum*, inhibition of biofilm of *P. aeruginosa*	*C. violaceum* and *P. aeruginosa*	Wagh et al. (2013)
4.	Silver nanoparticle	Inhibition of violacein of *C. violaceum*	*C. violaceum*	Arunkumar et al. (2013)
5.	Silver nanoparticle	Inhibition of violacein of *C. violaceum*	*C. violaceum*	Anju and Sarada (2016)
6.	Silver nanoparticle	Inhibition of violacein of *C. violaceum*, inhibition of production of pyocyanin, protease, hemolysin and biofilm of *P. aeruginosa*	*C. violaceum* and *P. aeruginosa*	Ali et al. (2017)
7.	Honey polyphenol carrying silver nanoparticle	Inhibition of violacein of *C. violaceum*, inhibition of elastin-degrading elastase, exoprotease, pyocyanin, biofilm, swarming motility and rhamnolipid of *P. aeruginosa* in mice model	*C. violaceum* and *P. aeruginosa*	Prateeksha et al. (2017)
8.	Zinc oxide nanoparticles	Inhibition of violacein of *C. violaceum*, inhibition of elastase, total protease, pyocyanin production, exopolysaccharide production and swarming motility of *L. monocytogenes*, *P. aeruginosa*, *E. coli* and *C. violaceum*, disruption of mature biofilm, down regulation in *pqsA*	*L. monocytogenes*, *P. aeruginosa*, *E. coli* and *C. violaceum*	Al-Shabib et al. (2016)
9.	β-cyclodextrin functionalized silicon dioxide nanoparticles	Inhibition of bioluminescence of *Vibrio fischeri*, down-regulation of *luxA* and *luxR* gene of *V. fischeri*	*Vibrio fischeri*	Miller et al. (2015)
10.	Silver coated carbon nanotubes	Down-regulation of *sdiA* (a quorum sensing gene) and many virulence genes (*safC*, *ychP*, *sseA* and *sseG*) of *S. aureus*	*S. aureus*	Chaudhari et al. (2015)
11.	Silver-titanium nanocomposite	Inhibition of violacein of *C. violaceum*, inhibition of biofilm formation and degradation of homoserine lactone	*C. violaceum*	Naik and Kowshik (2014)
12.	Silver and curcumin nanoparticles	Inhibition of biofilm formation of *P. aeruginosa* and *S. aureus*	*P. aeruginosa* and *S. aureus*	Loo et al. (2015)

expression of QS regulated virulence genes was also significantly reduced. It was revealed by RT-qPCR in planktonic cells of PAO1 that the expression of *lasA*, *lasB*, *phzA1* and *rhlA* were repressed by 79, 84, 68 and 72%, respectively at 25 mg/l of AgNPs. There was also remarkable inhibition (71%) of LasI transcriptional activity and 50% down-regulation of LasR. The level of RhlI and RhlR was also decreased by 64 and 55%, respectively. Many other QS-regulated genes such as *lasI*, *lasR*, *rhlI*, *rhlR*, and *fabH2* were also down-regulated by 71, 51, 63, 36, and 81%, respectively while expression of the *proC* housekeeping gene was not affected. The synthesis of C12-AHL and C4-AHL was also inhibited dose-dependently at tested concentration. Similarly, AgNPs exhibited reduction in biofilm formation at 5–25 mg/l which was evident from confocal laser scanning microscopy (CLSM) and scanning electron microscopy (SEM) data (Singh et al. 2015).

Similarly, silver nanowires (SNWs) synthesized by polyol process was found to inhibit quorum sensing. It was found the nanowires were able to inhibit synthesis of violacein by 60 and 80% at 0.5 and 4 mg/ml respectively, in *C. violaceum* CV026. The concentration above 4 mg/ml was inhibitory to the growth of the tested bacteria. Biofilm of *P. aeruginosa* NCIM 2948 was maximally inhibited by SNWs at 4 mg/ml without interfering its growth (Wagh et al. 2013). The QS-mediated inhibition of biofilm is considered important as at sub-inhibitory growth concentration, there is no selective pressure for the development of resistance against test compound or nanoparticle. Recently, there is growing interest for the development of biomaterials that has shifted from drug molecules to nanomaterials (Knetsch and Koole 2011; Bazaka et al. 2012). Many medical devices have been made using nanomaterials as antibiofilm agents which is aimed to minimize the biofilm formation on their surface (Monteiro et al. 2009).

A preliminary investigation of AgNPs synthesised from double mutant strain *Klebsiella pneumoniae* found the inhibitory effect on formation of purple pigment in CV026. A clear zones of pigment inhibition around wells at varying concentrations (15, 10, 5 μl) of AgNPs were indicative of quorum sensing inhibition. Additionally, antibacterial activity of green synthesized AgNPs and its synergy in combination with antibiotics were also found (Arunkumar et al. 2013). A recent study on anti-QS activity of AgNPs also supported earlier finding and exhibited clear zone of violacein inhibition around the sample followed by a turbid halo zone where indicator organism was not inhibited but depigmented highlighting the anti-QS potential. Not only QS, but bacterial growth was also significantly inhibited of multi-drug resistant pathogens such as *S. aureus* and *P. aeruginosa* (Anju and Sarada 2016). In another investigation, anti-quorum sensing activity was reported by silver nanoparticles at sub-MIC (15 μg/ml) against *C. violaceum*. Similarly, many QS-mediated virulence factors of *P. aeruginosa* were successfully inhibited. The production of pyocyanin was inhibited up to 74.64%, protease production was decreased up to 47.3%, hemolysin activity was decreased to 47.7% in all tested drug resistant clinical isolates of *P. aeruginosa*. The formation of biofilm was also remarkably reduced by 70.9–79.7% as also evident from confocal laser scanning microscopy (Ali et al. 2017).

The inhibition of quorum sensing regulated traits by silver nanoparticle is also attributed to silver. It was found that silver dose dependently attenuated the attachment of *P. aeruginosa* to the cover slip significantly compared to the control group. The biofilm inhibition data at sub-MIC levels was further evaluated by crystal violet assay and total protein assay. The growth cycle of test organism was evaluated at sub-MIC dose of silver to further validate and it was found that after identical period of incubation, silver treated and untreated microorganisms exactly followed the same trend in their growth kinetics (Sharma et al. 2015).

11.3.2 Other Nanoparticles

Prateeksha and colleagues demonstrated that selenium nano-scaffold exhibited enhanced anti-QS activity, anti-virulence potential and anti-biofilm efficacy under *in vitro* and *in vivo* compared to both selenium nanoparticles and honey polyphenols. Preliminary investigation demonstrated that surface conjugated selenium nanoparticles (SeNPs@HP) interfere QS by interacting with AHLs and their receptors. At 4.5 µg/ml of SeNPs@HP, there was decrease in virulence factors elastin-degrading elastase (52.7%), exoprotease (60.2%), pyocyanin (49.6%) and rhamnolipid (59.6%) of *P. aeruginosa* PAO1. A significant reduction in swarming motility was also observed at same tested concentration. As evident from calorimetric and microscopic data, SeNPs@HP reduced the formation of biofilm by more than 90% at sub-MIC concentration. To further validate under *in vivo* condition, the infected mice (with 10^7 cfu/ml of *P. aeruginosa* PAO1) were treated with SeNPs@HP. At early stage of infection (on first day), there was insignificant difference in treated and untreated groups. However, treatment with 4.5 µg/ml of SeNPs@HP, on day 5, 10, 15 and 20 post-infection, there was 31.7, 69.6, 81.6 and 97.3% wound healing respectively. The molecular docking results revealed that interaction of honey polyphenols with *N*-(3-oxododecanoyl)-1-homoserine lactone binding site of LasR might be the reason for successful inhibition of virulence of *P. aeruginosa* PAO1 (Prateeksha et al. 2017).

Zinc oxide nanoparticles (ZnO-NPs) also possess anti-QS potential. Preliminary investigation by disc diffusion assay suggested the QS inhibition in bio-indicator strain *C. violaceum* 12472. It was found that maximum inhibition of violacein was observed at 400 µg/ml (91%) followed by lower concentrations (200–50 µg/ml) in a dose dependent manner. ZnO-NPs inhibited many virulence factors including elastase (35–82%), total protease (20–77%) and pyocyanin production (48–93%). At 10, 20, 40 and 80 µg/ml, there was 35, 55, 78 and 85% inhibition of *lasB* transcriptional activity. Similarly, 41–84% down regulation in *pqsA* was also recorded at varying (10–80 µg/ml) levels of ZnO-NPs. A significant decrease in exopolysaccharide (EPS) production (25–90%) and swarming motility (7–78%) was recorded.

The biofilm formation by *Listeria monocytogenes*, *P. aeruginosa*, *Escherichia coli* and *C. violaceum* was inhibited up to 91, 93, 82 and 83% respectively. Disruption of mature biofilm of different bacterial strains was also achieved at sub-MIC levels (Al-Shabib et al. 2016).

Similarly, β-cyclodextrin functionalized silicon dioxide nanoparticles with 2 μM 3OC6-HSL demonstrated that the β-cyclodextrin moiety was significantly more effective at dimming bioluminescence of *V. fischeri* when functionalized to silicon dioxide nanoparticles than it was as a free-compound. At environmentally-relevant levels of HSLs, bioluminescence of *V. fischeri* was significantly diminished by β-cyclodextrin (P = 0.05). The functionalization of β-cyclodextrin to 50 nm NPs, 133 nM β-cyclodextrin produced similar result as that of 2X concentration of free 250 nM β-cyclodextrin. Quantitative PCR and transcript analysis found that the quantity of transcripts produced by untreated cultures and treated cultures (with 250 nM β-cyclodextrin) were not significantly different. The result confirmed that lower concentrations of free β-cyclodextrin were ineffective in down-regulating *luxA* and *luxR* at environmental levels of *N*-acyl-L-homoserine lactones, however, higher concentration (*i.e.* 2 mM) β-cyclodextrin was able to significantly reduce their expression. The 133 nM -cyclodextrin, in presence of functionalized 50 nm NPs, produced most down-regulation of luxA and luxR transcripts in all treatment groups (Miller et al. 2015).

Chaudhari and co-workers found the anti-QS activity of pegylated silver coated carbon nanotubes (pSWCNTs-Ag). Treatment of pSWCNTs-Ag to *S. aureus* exclusively down regulated the expression of *sdiA* (a quorum sensing gene) and many virulence genes (*safC*, *ychP*, *sseA* and *sseG*) by several folds. It was of noteworthy that pSWCNTs-Ag at bactericidal concentration was found non-toxic to human cells (Chaudhari et al. 2015).

Composite nanoparticles are reported to exhibit anti-QS activity. A qualitative estimation of silver-titanium nanocomposite (AgCl-TiO$_2$NPs) resulted in concentration dependent violacein pigment inhibition of *C. violaceum*. Violacein production was inhibited by 82% at 100 μg/ml and at 300 μg/ml, there was complete (100%) inhibition in nutrient broth. In modified Tris minimal medium, treatment with 50 and 75 μg/ml of AgCl-TiO$_2$NPs inhibited violacein production by 87 and 99% respectively. At 20 μg/ml, there was remarkable decrease in biofilm production and complete inhibition was obtained at 100 μg/ml. One of underlying mechanism of QS-inhibition was found to be degradation of homoserine lactone (HSL) by AgCl-TiO$_2$NPs. HPLC chromatogram exhibited single sharp peak with retention time of 21 min due to HSL. The peak was corresponding to HSL was absent in AgCl-TiO$_2$NPs treated samples and presence of two smaller peaks were attributed to the degradation products/precursors of the signalling molecule (Naik and Kowshik 2014).

Combination of silver and curcumin nanoparticles was also effective in eradication of established mature biofilm as well as inhibition of biofilm formation.

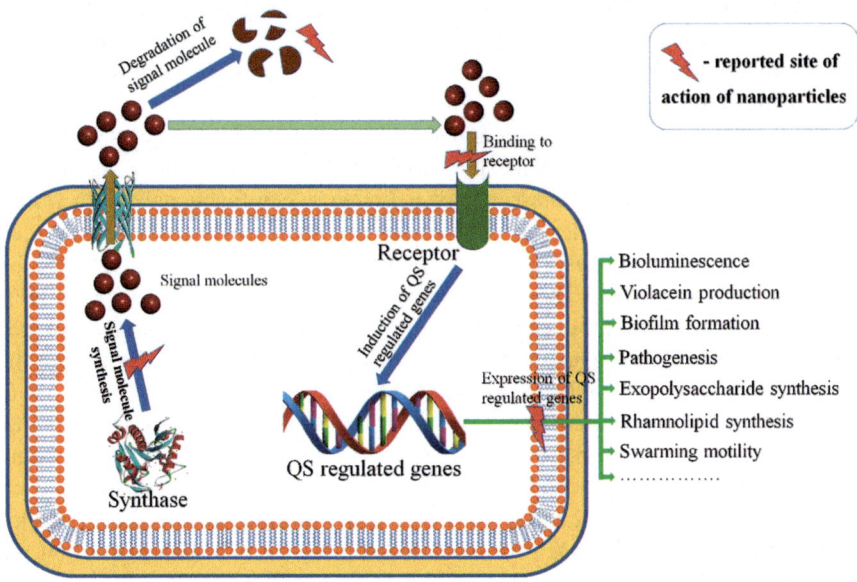

Fig. 11.1 Schematic representation of anti-QS mechanisms of nanoparticles

Treatment of 400 μg/ml of curcumin nanoparticles successfully reduced biofilm biomass of both *P. aeruginosa* and *S. aureus* (Loo et al. 2015). The anti-biofilm activity of curcumin is by attenuation of QS virulence factors and by interfering with the signal molecules (Rudrappa and Bais 2008).

In the beginning of twentieth century, discovery of antibiotics revolutionized the field of medicine by combating a large number of life threatening diseases. The indiscriminate and excessive usage of antibiotics caused the emergence of drug resistance in bacteria (Ciofu et al. 1994). It is important to mention that at least 65% of all infectious diseases are associated with bacterial communities which become virulent by forming biofilms (Lewis 2007). Once in biofilm mode, bacteria become up to 1000 times more resistant to antibiotics compared to their planktonic counterparts (Olson et al. 2002). The behaviour and virulence with in the biofilm is in QS control in which bacteria synthesize chemical signals and express their virulence genes in a cell density dependent manner (Romero et al. 2012). The efforts are being made for the search and synthesis of molecules having tendency to disrupt biofilm formation by quenching the QS system, this phenomenon is called quorum quenching (Huma et al. 2011; Kalia et al. 2011). The quorum sensing inhibitors (QSIs) may target at various sites of QS circuit that can provide an opportunity for the development of new therapeutic agents against pathogens to combat infections (Kalia 2013). Based on the above literature and many other reports, the mode of action nanoparticles in inhibition of QS is summarized in Fig. 11.1.

11.4 Nanoparticles in Delivery of Quorum Sensing Inhibitors (QSI)

The activity of quorum sensing inhibitors (QSI) can also be enhanced by their application and delivery in the form of nano-formulations. Various systems have been developed for the delivery of antibiofilm and anti-QS agents including liposome, noisome, PGLA, dendrimers, chitosan etc. (Sajid et al. 2014). The biodegradable and biocompatible nanoparticles used for controlled delivery of drugs is an effective therapeutic strategy (Daum et al. 2012). In last few years, several nano-based delivery systems like poly(lactic-co-glycolic acid) nanoparticle (PLGA), fusogenic liposomes, solid lipid nanoparticles (SLNs) and lipid-polymer hybrid nano-formulation have proven to be promising vehicles for targeted delivery of drugs (Forier et al. 2014). Many such formulations like protein–polymer conjugates (*e.g.* Intron® A) and liposomes (e.g. AmBisome®) have already reached to market authorization stage. There are many other formulations under preclinical or clinical investigation such as polymeric nanoparticles, dendrimers, lipid nanoparticle, nanosomes, drug–polymer conjugates and complexes (Mohamed-Ahmed et al. 2013). Solid lipid nanoparticles (SLNs) are physiological lipids dispersed in aqueous surfactant solution are one of the attractive class of nanocarriers (Bondì and Craparo 2010). SLNs have advantage of improved drug stability, tendency of readily incorporation of lipophilic drugs, controlled release and a higher safety threshold values due to the evasion of organic solvents (MuÈller et al. 2000; Mehnert and Mäder 2012). There are certain limitations associated with SLNs including risk of drug leakage gelation during storage, low drug loading owing to lipid polymorphism (Müller et al. 2002).

Nafee and co-workers investigated the effect of SLNs incorporated with QSI on pyocyanin production in *P. aeruginosa* PA14. It is interesting to note that not only the biological activity of QSI was maintained but exhibited superior action in its nano-formulation. Further, inhibition of pyocyanin production by QSI encapsulated SLNs was most pronounced at lower concentrations and weaker dependence on dose, while free compound resulted in a clear dose-dependent inhibition. It was also found that free SLNs also inhibited pyocyanin production that strongly suggests an additive inhibitory effect by the QSI and the SLNs. The growth curve data of *P. aeruginosa* confirmed the inhibitory potential of plain SLNs was not due to bacteriostatic or bactericidal effect (Nafee et al. 2014). The result rules out the possibly of QS inhibition by inhibiting microbial growth which is often associated with nanoparticles (Bae et al. 2011).

11.5 Prospects and Limitations

Recent development on efficacy of nanoparticles as antimicrobial and anti-pathogenic agents (anti-biofilm and anti-QS) have indicated the promising prospect in the treatment and prevention of bacterial infections. Due to unique physical and chemical characteristics of nanoparticles, it has also been evaluated and found effective as carrier for a number of therapeutic drugs such as antibiotics, anticancer drugs. Therefore, it is expected that with the increase in discovery of anti-QS agents from natural and synthetic sources, nanoparticles could also be effectively useful in delivery of such anti-QS agents for combating bacterial infection. Different combination or formulations of nanoparticles may be developed to enhance efficacy, safety and availability *in vivo* system.

The *in vitro* reports on efficacy of nanoparticles in interfering QS and its regulated functions are increasing. However, the *in vitro* conditions are different from *in vivo* conditions. Therefore, suitable animal model studies should be conducted to assess the therapeutic efficacy of nanoparticles as anti-QS agents. Anti -QS nanoparticle should not exert toxicity to host cell as well as should not create selection pressure on microbial pathogens especially on MDR bacteria to develop resistance to nanoparticle. Further understanding the role of nanoparticles as anti-QS needs to be evaluated through systemic investigation in suitable animal models.

11.6 Conclusion

Interference of QS is an effective alternate strategy to combat microbial infections. In the present scenario when the efficacy of antibiotics cannot be ensured for longer time, nanoparticles with anti-QS activity may become promising therapeutic agents against bacterial pathogens. Various reports indicated that nanomaterials can also be used as vehicles for the targeted delivery of natural and synthetic compounds which are known for their promising anti-QS activity and reduces the toxicity to non-target tissues. Studies have shown the quorum sensing inhibitory activity of many nanoparticles in biosensor strains and such nanoparticles should be evaluated against clinically relevant pathogens both *in vitro* and in experimental animals to uncover the therapeutic efficacy.

11.7 Opinion

Current antibiotic therapy to combat MDR problem requires immediate attention to develop new anti-infective drugs. QS inhibition is considered a promising drug target to attenuate bacterial pathogenicity and reduce the risk of antibiotic resistance. Recent research indicated that nanoparticles may prove to be better antimicrobial agents compared to conventional antibiotics against MDR bacteria as nanoparticles have multiple targets resulting in broad-spectrum action against pathogenic bacteria. On the other hand, nanoparticle have greater scope for manipulation and can be modified according to specific needs. Although research in the area of medicine is at the early stage and still restricted to the laboratory but the results so far reported is promising. Further research on nanoparticles as QSIs must be continued with special reference to its toxicological impact on host system and *in vivo* efficacy. It is expected that time is not far when inhibition of cellular communication (quorum sensing) by nanoparticles may become a better alternative for treatment of microbial disease in man and animals specially against drug resistant strains.

References

Agnihotri S, Mukherji S, Mukherji S (2014) Size-controlled silver nanoparticles synthesized over the range 5–100 nm using the same protocol and their antibacterial efficacy. RSC Adv 4:3974–3983. https://doi.org/10.1039/C3RA44507K

Ahmad I, Husain FM (2014) Bacterial virulence, biofilm and quorum sensing as promising targets for anti-pathogenic drug discovery and the role of natural products. In: Bhutani KK, Govil JN (ed) Biotechnology. Drug discovery, vol 7. Studium Press LLC, USA, pp 107–149 ISBN:1-62699-015-8

Ahmad I, Khan MS, Altaf MM, Qais FA, Ansari FA, Rumbaugh KP (2017) Biofilms: an overview of their significance in plant and soil health. In: Ahmad I, Husain FM (eds) Biofilms in plant and soil health. Wiley, Hoboken, pp 1–26. ISBN: 978-1-119-24634-3

Ahmed S, Ahmad M, Swami BL, Ikram S (2016) A review on plants extract mediated synthesis of silver nanoparticles for antimicrobial applications: a green expertise. J Adv Res 7:17–28. https://doi.org/10.1016/j.jare.2015.02.007

Ali SG, Ansari MA, Khan HM, Jalal M, Mahdi AA, Cameotra SS (2017) *Crataeva nurvala* nanoparticles inhibit virulence factors and biofilm formation in clinical isolates of *Pseudomonas aeruginosa*. J Basic Microbiol 57:193–203. https://doi.org/10.1002/jobm.201600175

Al-Shabib NA, Husain FM, Ahmad I, Khan MS, Khan RA, Khan JM (2017) Rutin inhibits mono and multi-species biofilm formation by foodborne drug resistant *Escherichia coli* and *Staphylococcus aureus*. Food Control 79:325–332. https://doi.org/10.1016/j.foodcont.2017.03.004

Al-Shabib NA, Husain FM, Ahmed F, Khan RA, Ahmad I, Alsharaeh E, Khan MS, Hussain A, Rehman MT, Yusuf M, Hassan I (2016) Biogenic synthesis of zinc oxide nanostructures from *Nigella sativa* seed: prospective role as food packaging material inhibiting broad-spectrum quorum sensing and biofilm. Sci Rep 6:36761. https://doi.org/10.1038/srep36761

Anju S, Sarada J (2016) Quorum sensing inhibiting activity of silver nanoparticles synthesized by *Bacillus* isolate. Int J Pharm Bio Sci 6:47–53

Arunkumar M, Mahesh N, Balakumar S, Sivakumar R, Priyadharshni S (2013) Antiquorum sensing and antibacterial activity of silver nanoparticles synthesized by mutant *Klebsiella pneumoniae* MTCC 3354. Asian J Chem 25:9961–9964. https://doi.org/10.14233/ajchem.2017.20003

Asfour HZ (2017) Anti-quorum sensing natural compounds. J Micros Ultrastr. https://doi.org/10.1016/j.jmau.2017.02.001

Bae E, Park HJ, Yoon J, Kim Y, Choi K, Yi J (2011) Bacterial uptake of silver nanoparticles in the presence of humic acid and AgNO₃. Korean J Chem Eng 28:267–271. https://doi.org/10.1007/s11814-010-0351-z

Bazaka K, Jacob MV, Crawford RJ, Ivanova EP (2012) Efficient surface modification of biomaterial to prevent biofilm formation and the attachment of microorganisms. Appl Microbiol Biotechnol 95:299–311. https://doi.org/10.1007/s00253-012-4144-7

Bhardwaj AK, Vinothkumar K, Rajpara N (2013) Bacterial quorum sensing inhibitors: attractive alternatives for control of infectious pathogens showing multiple drug resistance. Recent Pat Antiinfect Drug Discov 8:68–83. https://doi.org/10.2174/1574891X11308010012

Bondì ML, Craparo EF (2010) Solid lipid nanoparticles for applications in gene therapy: a review of the state of the art. Expert Opin Drug Deliv 7:7–18. https://doi.org/10.1517/17425240903362410

Chaudhari AA, Jasper SL, Dosunmu E, Miller ME, Arnold RD, Singh SR, Pillai S (2015) Novel pegylated silver coated carbon nanotubes kill *Salmonella* but they are non-toxic to eukaryotic cells. J Nanobiotechnol 13:23. https://doi.org/10.1186/s12951-015-0085-5

Chetoni P, Burgalassi S, Monti D, Tampucci S, Tullio V, Cuffini AM, Muntoni E, Spagnolo R, Zara GP, Cavalli R (2016) Solid lipid nanoparticles as promising tool for intraocular tobramycin delivery: pharmacokinetic studies on rabbits. Eur J Pharm Biopharm 109:214–223. https://doi.org/10.1016/j.ejpb.2016.10.006

Ciofu O, Giwercman B, Høiby N, Pedersen SS (1994) Development of antibiotic resistance in *Pseudomonas aeruginosa* during two decades of antipseudomonal treatment at the Danish CF Center. APMIS 102:674–680. https://doi.org/10.1111/j.1699-0463.1994.tb05219.x

Corr SA (2013) Metal oxide nanoparticles. Nanoscience 2:180–234. https://doi.org/10.1039/9781849734844-00180

Daum N, Tscheka C, Neumeyer A, Schneider M (2012) Novel approaches for drug delivery systems in nanomedicine: effects of particle design and shape. Wiley Interdiscip Rev Nanomed Nanobiotechnol 4:52–65. https://doi.org/10.1002/wnan.165

Davis ME, Shin DM (2008) Nanoparticle therapeutics: an emerging treatment modality for cancer. Nat Rev Drug Discov 7:771–782. https://doi.org/10.1038/nrd2614

Defoirdt T, Brackman G, Coenye T (2013) Quorum sensing inhibitors: how strong is the evidence? Trends Microbiol 21:619–624. https://doi.org/10.1016/j.tim.2013.09.006

Díaz E, Jiménez JI, Nogales J (2013) Aerobic degradation of aromatic compounds. Curr Opin Biotechnol 24:431–442. https://doi.org/10.1016/j.copbio.2012.10.010

Fernandez-Garcia M, Martinez-Arias A, Hanson JC, Rodriguez JA (2004) Nanostructured oxides in chemistry: characterization and properties. Chem Rev 104:4063–4104. https://doi.org/10.1021/cr030032f

Forier K, Raemdonck K, De Smedt SC, Demeester J, Coenye T, Braeckmans K (2014) J Control Release 190:607–623. https://doi.org/10.1016/j.jconrel.2014.03.055

Fuqua WC, Winans SC, Greenberg EP (1994) Quorum sensing in bacteria: the LuxR-LuxI family of cell density-responsive transcriptional regulators. J Bacteriol 176:269–275. https://doi.org/10.1128/jb.176.2.269-275.1994

Galloway WR, Hodgkinson JT, Bowden SD, Welch M, Spring DR (2010) Quorum sensing in gram-negative bacteria: small-molecule modulation of AHL and AI-2 quorum sensing pathways. Chem Rev 111:28–67. https://doi.org/10.1021/cr100109t

García Lara B, Saucedo Mora MÁ, Roldán Sánchez JA, Pérez Eretza B, Ramasamy M, Lee J, Coria Jimenez R, Tapia M, Varela Guerrero V, García Contreras R (2015) Inhibition of quorum sensing dependent virulence factors and biofilm formation of clinical and environmental *Pseudomonas aeruginosa* strains by ZnO nanoparticles. Lett Appl Microbiol 61:299–305. https://doi.org/10.1111/lam.12456

Gurunathan S, Han JW, Dayem AA, Eppakayala V, Kim JH (2012) Oxidative stress-mediated antibacterial activity of graphene oxide and reduced graphene oxide in *Pseudomonas aeruginosa*. Int J Nanomedicine 7:5901–5914. https://doi.org/10.2147/IJN.S37397

Haddad PS, Seabra AB (2012) Biomedical applications of magnetic nanoparticles. In: Martinez AI (ed) Iron oxides: structure, properties and applications, vol 1. Nova Science Publishers, Inc., New York, pp 165–188

Hentzer M, Givskov M (2003) Pharmacological inhibition of quorum sensing for the treatment of chronic bacterial infections. J Clin Invest 112:1300–1307. https://doi.org/10.1172/JCI200320074

Huma N, Shankar P, Kushwah J, Bhushan A, Joshi J, Mukherjee T, Raju SC, Purohit HJ, Kalia VC (2011) Diversity and polymorphism in AHL-lactonase gene (aiiA) of *Bacillus*. J Microbiol Biotechnol 21:1001–1011. https://doi.org/10.4014/jmb.1105.05056

Husain FM, Ahmad I, Baig MH, Khan MS, Khan MS, Hassan I, Al-Shabib NA (2016) Broad-spectrum inhibition of AHL-regulated virulence factors and biofilms by sub-inhibitory concentrations of ceftazidime. RSC Adv 6:27952–27962. https://doi.org/10.1039/C6RA02704K

Husain FM, Ahmad I, Khan MS, Al-Shabib NA (2015) *Trigonella foenum-graceum* (seed) extract interferes with quorum sensing regulated traits and biofilm formation in the strains of *Pseudomonas aeruginosa* and *Aeromonas hydrophila*. Evid Based Complement Alternat Med 27:879540. https://doi.org/10.1155/2015/879540

Iravani S, Korbekandi H, Mirmohammadi SV, Zolfaghari B (2014) Synthesis of silver nanoparticles: chemical, physical and biological methods. Res Pharm Sci 9:385–406

Janczak CM, Aspinwall CA (2012) Composite nanoparticles: the best of two worlds. Anal Bioanal Chem 402:83–89. https://doi.org/10.1007/s00216-011-5482-5

Jones N, Ray B, Ranjit KT, Manna AC (2008) Antibacterial activity of ZnO nanoparticle suspensions on a broad spectrum of microorganisms. FEMS Microbiol Lett 279:71–76. https://doi.org/10.1111/j.1574-6968.2007.01012.x

Kalia VC (2013) Quorum sensing inhibitors: an overview. Biotechnol Adv 31:224–245. https://doi.org/10.1016/j.biotechadv.2012.10.004

Kalia VC, Raju SC, Purohit HJ (2011) Genomic analysis reveals versatile organisms for quorum quenching enzymes: acyl-homoserine lactone-acylase and-lactonase. Open Microbiol J 5:1–13. https://doi.org/10.2174/1874285801105010001

Ke X, Miller LC, Bassler BL (2015) Determinants governing ligand specificity of the Vibrio harveyi LuxN quorum sensing receptor. Mol Microbiol 95:127–142. https://doi.org/10.1111/mmi.12852

Khan MF, Ansari AH, Hameedullah M, Ahmad E, Husain FM, Zia Q, Baig U, Zaheer MR, Alam MM, Khan AM, AlOthman ZA (2016) Sol-gel synthesis of thorn-like ZnO nanoparticles endorsing mechanical stirring effect and their antimicrobial activities: potential role as nano-antibiotics. Sci Rep 6:27689. https://doi.org/10.1038/srep27689

Knetsch ML, Koole LH (2011) New strategies in the development of antimicrobial coatings: the example of increasing usage of silver and silver nanoparticles. Polymers 3:340–366. https://doi.org/10.3390/polym3010340

Kumar A, Pandey AK, Singh SS, Shanker R, Dhawan A (2011a) Cellular response to metal oxide nanoparticles in bacteria. J Biomed Nanotechnol 7:102–103

Kumar A, Pandey AK, Singh SS, Shanker R, Dhawan A (2011b) Engineered ZnO and TiO$_2$ nanoparticles induce oxidative stress and DNA damage leading to reduced viability of *Escherichia coli*. Free Radic Biol Med 51:1872–1881. https://doi.org/10.1016/j.freeradbiomed.2011.08.025

Leung YH, Ng A, Xu X, Shen Z, Gethings LA, Wong MT, Chan C, Guo MY, Ng YH, Djurišić AB, Lee PK (2014) Mechanisms of antibacterial activity of MgO: non-ROS mediated toxicity of MgO nanoparticles towards *Escherichia coli*. Small 10:1171–1183. https://doi.org/10.1002/smll.201302434

Lewis K (2007) Persister cells, dormancy and infectious disease. Nat Rev Microbiol 5:48–56. https://doi.org/10.1038/nrmicro1557

Loo CY, Rohanizadeh R, Young PM, Traini D, Cavaliere R, Whitchurch CB, Lee WH (2015) Combination of silver nanoparticles and curcumin nanoparticles for enhanced anti-biofilm activities. J Agric Food Chem 64:2513–2522. https://doi.org/10.1021/acs.jafc.5b04559

Mahapatra O, Bhagat M, Gopalakrishnan C, Arunachalam KD (2008) Ultrafine dispersed CuO nanoparticles and their antibacterial activity. J Exp Nanosci 3:185–193. https://doi.org/10.1080/17458080802395460

Maheshwari M, Ahmad I, Althubiani AS (2016) Multidrug resistance and transferability of bla CTX-M among extended-spectrum β-lactamase-producing enteric bacteria in biofilm. J Glob Antimicrob Resist 6:142–149. https://doi.org/10.1016/j.jgar.2016.04.009

Masurkar SA, Chaudhari PR, Shidore VB, Kamble SP (2012) Effect of biologically synthesised silver nanoparticles on *Staphylococcus aureus* biofilm quenching and prevention of biofilm formation. IET Nanobiotechnol 6:110–114. https://doi.org/10.1049/iet-nbt.2011.0061

Mehnert W, Mäder K (2012) Solid lipid nanoparticles: production, characterization and applications. Adv Drug Deliv Rev 64:83–101. https://doi.org/10.1016/S0169-409X(01)00105-3

Miller KP, Wang L, Chen YP, Pellechia PJ, Benicewicz BC, Decho AW (2015) Engineering nanoparticles to silence bacterial communication. Front Microbiol 6:189. https://doi.org/10.3389/fmicb.2015.00189

Mohamed-Ahmed AA, Ginn C, Croft S, Brocchini S (2013) Anti-infectives. In: Fundamentals of pharmaceutical nanoscience. Springer, New York, pp 429–464

Monteiro DR, Gorup LF, Takamiya AS, Ruvollo-Filho AC, de Camargo ER, Barbosa DB (2009) The growing importance of materials that prevent microbial adhesion: antimicrobial effect of medical devices containing silver. Int J Antimicrob Agents 34:103–110. https://doi.org/10.1016/j.ijantimicag.2009.01.017

MuÈller RH, MaÈder K, Gohla S (2000) Solid lipid nanoparticles (SLN) for controlled drug delivery–a review of the state of the art. Eur J Pharm Biopharm 50:161–177. https://doi.org/10.1016/S0939-6411(00)00087-4

Müller RH, Radtke M, Wissing SA (2002) Solid lipid nanoparticles (SLN) and nanostructured lipid carriers (NLC) in cosmetic and dermatological preparations. Adv Drug Deliv Rev 1:54. https://doi.org/10.1016/S0169-409X(02)00118-7

Nafee N, Husari A, Maurer CK, Lu C, de Rossi C, Steinbach A, Hartmann RW, Lehr CM, Schneider M (2014) Antibiotic-free nanotherapeutics: ultra-small, mucus-penetrating solid lipid nanoparticles enhance the pulmonary delivery and anti-virulence efficacy of novel quorum sensing inhibitors. J Control Release 192:131–140. https://doi.org/10.1016/j.jconrel.2014.06.055

Nagy A, Harrison A, Sabbani S, Munson RS Jr, Dutta PK, Waldman WJ (2011) Silver nanoparticles embedded in zeolite membranes: release of silver ions and mechanism of antibacterial action. Int J Nanomedicine 6:1833–1852. https://doi.org/10.2147/IJN.S24019

Naik K, Kowshik M (2014) Anti quorum sensing activity of AgCl-TiO$_2$ nanoparticles with potential use as active food packaging material. J Appl Microbiol 117:972–983. https://doi.org/10.1111/jam.12589

Nealson KH, Platt T, Hastings JW (1970) Cellular control of the synthesis and activity of the bacterial luminescent system. J Bacteriol 104:313–322

Olson ME, Ceri H, Morck DW, Buret AG, Read RR (2002) Biofilm bacteria: formation and comparative susceptibility to antibiotics. Can J Vet Res 66:86–92

Phelan VV, Liu WT, Pogliano K, Dorrestein PC (2012) Microbial metabolic exchange [mdash] the chemotype-to-phenotype link. Nat Chem Biol 8:26–35. https://doi.org/10.1038/nchembio.739

Prateeksha BR, Shoeb M, Sharma S, Naqvi AH, Gupta VK, Singh BN (2017) Scaffold of selenium Nanovectors and honey phytochemicals for inhibition of *Pseudomonas aeruginosa* quorum sensing and biofilm formation. Front Cell Infect Microbiol 7:93. https://doi.org/10.3389/fcimb.2017.00093

Radzig MA, Nadtochenko VA, Koksharova OA, Kiwi J, Lipasova VA, Khmel IA (2013) Antibacterial effects of silver nanoparticles on gram-negative bacteria: influence on the growth and biofilms formation, mechanisms of action. Colloids Surf B Biointerfaces 102:300–306. https://doi.org/10.1016/j.colsurtb.2012.07.039

Raghunath A, Perumal E (2017) Metal oxide nanoparticles as antimicrobial agents: a promise for the future. Int J Antimicrob Agents 49:137–152. https://doi.org/10.1016/j.ijantimicag.2016.11.011

Rajamanikandan S, Jeyakanthan J, Srinivasan P (2017) Molecular docking, molecular dynamics simulations, computational screening to design quorum sensing inhibitors targeting LuxP of *Vibrio harveyi* and its biological evaluation. Appl Biochem Biotechnol 181:192–218. https://doi.org/10.1007/s12010-016-2207-4

Rampioni G, Leoni L, Williams P (2014) The art of antibacterial warfare: deception through interference with quorum sensing–mediated communication. Bioorg Med Chem 55:60–68. https://doi.org/10.1016/j.bioorg.2014.04.005

Ren TT, Li XY, Yu HQ (2013) Effect of N-acy-l-homoserine lactones-like molecules from aerobic granules on biofilm formation by *Escherichia coli* K12. Bioresour Technol 129:655–658. https://doi.org/10.1016/j.biortech.2012.12.043

Rodriguez JA, Fernández-García M (2007) Synthesis, properties, and applications of oxide nanomaterials. Wiley, Hoboken

Romero M, Acuña L, Otero A (2012) Patents on quorum quenching: interfering with bacterial communication as a strategy to fight infections. Recent Pat Biotechnol 6:2–12. https://doi.org/10.2174/187220812799789208

Roux A, Payne SM, Gilmore MS (2009) Microbial telesensing: probing the environment for friends, foes, and food. Cell Host Microbe 6:115–124. https://doi.org/10.1016/j.chom.2009.07.004

Rudrappa T, Bais HP (2008) Curcumin, a known phenolic from *Curcuma longa*, attenuates the virulence of *Pseudomonas aeruginosa* PAO1 in whole plant and animal pathogenicity models. J Agric Food Chem 56:1955–1962. https://doi.org/10.1021/jf072591j

Sadiq IM, Chandrasekaran N, Mukherjee A (2010) Studies of effect of TiO_2 nanoparticles on growth and membrane permeability of *Escherichia coli*, *Pseudomonas aeruginosa* and *Bacillus subtilis*. Curr Nanosci 6:381–387. https://doi.org/10.2174/157341310791658973

Sajid M, Khan MS, Cameotra SS, Ahmad I (2014) Drug delivery systems that eradicate and/or prevent biofilm formation. In: Antibiofilm agents. Springer, Berlin/Heidelberg, pp 407–424

Sawai J, Himizu K, Yamamoto O (2005) Kinetics of bacterial death by heated dolomite powder slurry. Soil Biol Biochem 37:1484–1489. https://doi.org/10.1016/j.soilbio.2005.01.011

Schaefer AL, Lappala CR, Morlen RP, Pelletier DA, Lu TY, Lankford PK, Harwood CS, Greenberg EP (2013) LuxR-and LuxI-type quorum-sensing circuits are prevalent in members of the *Populus deltoides* microbiome. Appl Environ Microbiol 79:5745–5752. https://doi.org/10.1128/AEM.01417-13

Sharma BK, Saha A, Rahaman L, Bhattacharjee S, Tribedi P (2015) Silver inhibits the biofilm formation of *Pseudomonas aeruginosa*. Adv Microbiol 5:677–685. https://doi.org/10.4236/aim.2015.510070

Shrestha PM, Rotaru AE, Summers ZM, Shrestha M, Liu F, Lovley DR (2013) Transcriptomic and genetic analysis of direct interspecies electron transfer. Appl Environ Microbiol 79:2397–2404. https://doi.org/10.1128/AEM.03837-12

Siddiqui MF, Rzechowicz M, Harvey W, Zularisam AW, Anthony GF (2015) Quorum sensing based membrane biofouling control for water treatment: a review. J Water Proc Eng 7:112–122. https://doi.org/10.1016/j.jwpe.2015.06.003

Sifri CD (2008) Quorum sensing: bacteria talk sense. Clin Infect Dis 47:1070–1076. https://doi.org/10.1086/592072

Singh BR, Singh BN, Singh A, Khan W, Naqvi AH, Singh HB (2015) Mycofabricated biosilver nanoparticles interrupt *Pseudomonas aeruginosa* quorum sensing systems. Sci Rep 5:13719. https://doi.org/10.1038/srep13719

Singh R, Nalwa HS (2011) Medical applications of nanoparticles in biological imaging, cell labeling, antimicrobial agents, and anticancer nanodrugs. J Biomed Nanotechnol 7:489–503. https://doi.org/10.1166/jbn.2011.1324

Thill A, Zeyons O, Spalla O, Chauvat F, Rose J, Auffan M, Flank AM (2006) Cytotoxicity of CeO_2 nanoparticles for *Escherichia coli*. Physico-chemical insight of the cytotoxicity mechanism. Environ Sci Technol 40:6151–6156. https://doi.org/10.1021/es060999b

Tran N, Mir A, Mallik D, Sinha A, Nayar S, Webster TJ (2010) Bactericidal effect of iron oxide nanoparticles on *Staphylococcus aureus*. Int J Nanomedicine 5:277–283. https://doi.org/10.2147/IJN.S9220

Valcárcel M, López-Lorente ÁI (2016) Recent advances and trends in analytical nanoscience and nanotechnology. Trends Anal Chem 84:1–2. https://doi.org/10.1016/j.trac.2016.05.010

Wagh MS, Patil RH, Thombre DK, Kulkarni MV, Gade WN, Kale BB (2013) Evaluation of anti-quorum sensing activity of silver nanowires. Appl Microbiol Biotechnol 97:3593–3601. https://doi.org/10.1007/s00253-012-4603-1

Whitesides GM (2005) Nanoscience, nanotechnology, and chemistry. Small 1:172–179. https://doi.org/10.1002/smll.200400130

Yong YC, Wu XY, Sun JZ, Cao YX, Song H (2015) Engineering quorum sensing signaling of *Pseudomonas* for enhanced wastewater treatment and electricity harvest: a review. Chemosphere 140:18–25. https://doi.org/10.1016/j.chemosphere.2014.10.020

Zaidi S, Misba L, Khan AU (2017) Nano-therapeutics: a revolution in infection control in post antibiotic era. Nanomedicine 13:2281–2301. https://doi.org/10.1016/j.nano.2017.06.015

Zhang L, Jiang Y, Ding Y, Povey M, York D (2007) Investigation into the antibacterial behaviour of suspensions of ZnO nanoparticles (ZnO nanofluids). J Nanopart Res 9:479–489. https://doi.org/10.1007/s11051-006-9150-1

Zhang W, Li C (2015) Exploiting quorum sensing interfering strategies in gram-negative bacteria for the enhancement of environmental applications. Front Microbiol 6:1535. https://doi.org/10.3389/fmicb.2015.01535

Zhu X, Hondroulis E, Liu W, Li CZ (2013) Biosensing approaches for rapid genotoxicity and cytotoxicity assays upon nanomaterial exposure. Small 9:1821–1830. https://doi.org/10.1002/smll.201201593

Zia Q, Farzuddin M, Ansari MA, Alam M, Ali A, Ahmad I, Owais M (2010) Novel drug delivery systems for antifungal compounds. In: Combating fungal infections, polisher. Springer, Berlin/Heidelberg, pp 485–528

Zou X, Deng P, Zhou C, Hou Y, Chen R, Liang F, Liao L (2017) Preparation of a novel antibacterial chitosan-poly (ethylene glycol) cryogel/silver nanoparticles composites. J Biomater Sci Polym Ed 27:1–4. https://doi.org/10.1080/09205063.2017.1321346

Chapter 12
Nanotechnological Approaches in Quorum Sensing Inhibition

A. Jamuna Bai and V. Ravishankar Rai

Abstract The increasing incidence of drug resistance in pathogenic bacteria has made it essential to explore novel antimicrobials and drug targets. The nanoparticles have been considered as one of the most potential therapeutic agents. Nanomaterials have unique physicochemical properties. In the recent years, nanoparticles have been well characterized for their antimicrobial properties. Apart from their inhibitory effects on pathogens, they are also being increasingly investigated for their effects on biofilm formation and signaling in bacterial cells at sub-inhibitory levels. Quorum sensing (QS) or cell to cell signaling is known to regulate biofilm formation and virulence factor production in pathogenic bacteria. Hence, the QS mechanism offers new drug targets. The nanomaterials at sub-inhibitory concentration can inhibit QS and prevent biofilm formation and virulence development in pathogens. The chapter focuses on the application of nanoparticles as QS inhibitory or quorum quenching agents to attenuate pathogenicity in bacteria and control their recalcitrant biofilms.

Keywords Quorum sensing · Biofilms · Nanoparticles · Nanotechnology · Quorum sensing inhibitors · Anti-biofilm agents

12.1 Introduction

Increased incidence of drug resistance in pathogenic and infectious bacteria is a matter of serious global public health concern. Though there are advances in the study of microbial pathogenesis and developments in modern diagnostics and therapeutic designs, still there is high morbidity and mortality due to microbial infections (Kolar et al. 2001). Hence, novel drug target strategies and new antimicrobials from natural and synthetic techniques are being explored. Silver and copper based antimicrobials have been used to treat microbial infections (Moghimi 2005).

A. Jamuna Bai · V. Ravishankar Rai (✉)
Department of Studies in Microbiology, University of Mysore, Mysore, Karnataka, India

© Springer Nature Singapore Pte Ltd. 2018
V. C. Kalia (ed.), *Biotechnological Applications of Quorum Sensing Inhibitors*,
https://doi.org/10.1007/978-981-10-9026-4_12

However, the advances in nanotechnology has led to the use of inorganic and organic nanomaterials in medicine as therapeutics (Gajjar et al. 2009).

Quorum sensing in bacterial populations generally coordinate communal behavior by cell to cell communication using diffusible signaling molecules (Reading and Sperandio 2006). QS regulates the expression of a number of phenotypes in bacteria such as motility, biofilm formation and production of virulence factors (Davies 2003). Many infectious bacteria use QS regulation to form recalcitrant biofilms during pathogenesis (Whitehead et al. 2001). In clinical settings, biofilms cause persistent infections prolonging treatment period (Davies 2003).

Quorum sensing regulates the expression of virulence factors and biofilm formation in pathogens. Hence, they are promising targets for developing novel therapeutics. The use of QS inhibitory drugs for attenuating bacterial pathogenicity than inhibiting their growth has many benefits in the light of the emergence of drug resistant bacteria (Kalia and Purohit 2011; Kalia 2013). The compounds that can interfere and attenuate pathogenicity are termed anti-pathogenic drugs. So far, numerous synthetic signal analogues and natural products have been screened for QS inhibitory potential. QS inhibitors make recalcitrant biofilms susceptible to antimicrobials, reduce virulence and have shown to increase clearance of bacteria in animal models during infection studies (Rasmussen and Givskov 2006).

QS and biofilm formation enhances the resistance of pathogens to antimicrobials and the host's defences (Blango and Mulvey 2009). Biofilms make it difficult for antimicrobials to reach constituent cells and hence their eradication is challenging in clinical and industrial settings. In the recent years a number of bacteria-resistant surfaces have been shown to prevent biofilm growth (Rana and Matsuura 2010). There are also reports on the use of nanostructures as an antibiofilm agent based on their growth inhibitory property (Lellouche et al. 2009).

Nanomaterials are finding increasing application due to their unique properties compared to their bulk counterparts (Xia et al. 2003). The inherent property of nanoparticles i.e., the large surface-to-bulk ratio makes them an attractive candidate for various applications (Murphy et al. 2006). The nanoparticles due to their unique electrical, optical, magnetic, and thermal properties are used to fabricate nanoscale devices which have varied applications (Zhai et al. 2009). The one dimensional nanoparticles, nanorods and nanowires with desired size and shape have high technological applications. Their characteristic size and shape determines their optical, electronic and chemical properties (Huang et al. 2001; Hu and Chan 2004). The metal nanoparticles with their unique electrical, thermal, optical and catalytic properties are used in fabricating nanoscale electronic sensing devices, surface enhanced Raman scattering, photonic crystals, etc. (Murphy et al. 2006; Chen et al. 2007; Hu et al. 2010)

In the present chapter, the focus is on the use of nanoparticles as QS inhibitors in pathogenic bacteria, their anti-biofilm activity at sub lethal levels and ability to enhance the efficacy of other known QS inhibitory compounds on conjugation and functionalization. The potential application of nanoparticle based QS inhibitors as anti-biofouling agents in industrial and water treatment plants; and as biofilm inhibitory and anti-infectives/anti-pathogenic drugs in clinical settings has also been discussed.

12.2 Properties of Nanoparticles

Nanoparticles have a dimension of 1–100 nanometers (nm) with properties different from the bulk counterpart. At the atomic scale, they have unique physico-chemical, optical and biological properties (Feynman 1991). The nanoparticles can be manipulated for desired applications, they are increasingly used in medicine (Parak et al. 2003). The inorganic nanoparticles are used more than the organic nanoparticles as they withstand adverse processing conditions (Whitesides 2003). The antimicrobial activity of the nanoparticles has been well studied and is attributed to the increased surface available which increases its interaction with microbes. Metal nanoparticles based on their antimicrobial nature have found promising applications in biomedical and surgical devices, water treatment, food processing and packaging and synthetic textiles (Gutierrez et al. 2010). The nanocomposites and functionalized nanomaterials have shown to enhance the biological activity (Eustis and El-Sayed 2006). The nanoparticles have unique physico-chemical, optical and electronic property due to their size and shape. This has led them to have unique properties such as quantum confinement, surface plasmon resonance and super paramagnetism (Jamieson et al. 2007). The enhanced catalytic activity of nanoparticles due to its highly active facets makes it attractive for various industrial applications (Gupta and Gupta 2005). The nanoparticles have been used in biomedical application for delivery of drugs, diagnsotics and imaging, and to design artificial implants. Biocompatibility of some of the nanomaterials i.e. have also made them interesting candidates in medical application (Samia et al. 2006).

12.3 Nanoparticles as Quorum Sensing Inhibitors

In the past decade, nanoparticles have been extensively investigated as broad-spectrum antimicrobial agents. However, recent studies are focused on the ability of nanoparticles to inhibit the bacterial cell to cell communication or quorum sensing. In the following section, we review some of the studies on the quorum sensing inhibitory potential of nanoparticles (Table 12.1).

Silver nanoparticles (20 nm) and single-wall carbon nanotubes (1.4 nm) have been investigated for quorum sensing inhibition. The study showed that nanomaterials affected AHL synthesis and its effect varied with bacterial species. The quorum sensing inhibitory effect of silver nanoparticles and single-wall carbon nanotubes was studied in *Pseudomonas syringae* and *Pantoea stewartii* bacteria (Mohanty et al. 2016).

The QS inhibition activity of nanoparticles has also been shown to be due to adsorption of AHLs to nanoparticles. Functionalized silicon dioxide nanoparticles (SiNP) were loaded with β-cyclodextrin. Cyclodextrins are known to bind AHLs, and reduce the vicinal AHLs. SiNP (15–50 nm) were designed to target AHLs to disrupt quorum sensing. The surface functionalized SiNPs with the β-cyclodextrin

Table 12.1 Quorum Sensing Inhibitory Properties of Nanomaterials

Nanomaterials	Target organism	References
Silver nanoparticles (20 nm) and carbon nanotubes (1.4 nm)	*Pseudomonas syringae* and *Pantoea stewartii*	Mohanty et al. (2016)
Silicon dioxide nanoparticles functionalized with β-cyclodextrin (15–50 nm)	*Vibrio fischeri*	Miller et al. (2015)
Silver nanowires (200–250 nm)	*P. aeruginosa* PA01 and *Chromobacterium violaceum*	Wagh et al. (2013)
Kaempferol loaded chitosan nanoparticles (192.27 ± 13.6 nm)	*C. violaceum*	Ilk et al. (2017)
AgCl-TiO$_2$ nanoparticles (6–7 nm)	*C. violaceum*	Naik and Kowshik (2014)
Spice essential oil nanoemulsions of cumin (52.89 nm), fennel (59.52 nm) and pepper (82.08 nm)	*C. violaceum*	Venkadesaperumal et al. (2016)
Mycofabricated silver nanoparticles (410 nm)	*P. aeruginosa* PA01	Singh et al. 2015
Silver nanoparticles from *Crataeva nurvala* (15 nm)	*P. aeruginosa* PA01	Ali et al. 2017
Zinc oxide nanoparticles (65 ± 17 nm)	*P. aeruginosa* PA01	Garcia-Lara et al. (2015)
SeNPs honey conjugates (12–15 nm)	*P. aeruginosa* PA01	Prateeksha et al. (2017)
ZnO nanoparticles (<50 nm)	*P. aeruginosa* PA01	Lee et al. (2014a, b), Aswathanarayan and Vittal (2017)
Chitosan Nanocapsules (114–155 nm)	*Escherichia coli*	Qin et al. (2017)
Gold nanoemulsions (75.52 nm)	*Pectobacterium carotovorum* sub sp. *Carotovorum*	Joe et al. (2015)

(β-CD), were able to reduce luminescence, a QS mediated process in *Vibrio fischeri*. Gene expression studies by qPCR also confirmed reduced expression of luminescence genes. The SiNPs were capable of binding to AHLs as observed by NMR studies. It was observed that high concentrations of SiNPs engineered with β-CD, a QS inhibitor, was capable of quenching vicinal AHLs. The β-CD bind AHLs and down-regulate bacterial QS genes The engineered NPs were capable of blocking AHL signalling in actively-metabolizing cultures. Thus, SiNPs act as scaffold for loading QS inhibitors. Silica nanoparticles on functionalization can enhance the efficacy of quorum quenching agents. The quenching ability of β-CD increased with functionalized Si-NPs (50 nm). As these have QS inhibitory potential, they can also be explored for biofilm inhibitory activity (Miller et al. 2015).

Silver nanowires (SNWs) of 200–250 nm size were synthesized by polyol reduction method. Polyvinylpyrrolidone (PVP) was used as a capping agent and ethylene glycol was used as the solvent and reductant. In the polyol synthesis approach, the size and shape of the nanoparticles can be controlled by optimizing reactant concentrations. The SNWs were studied for their ability to inhibit the QS-mediated biofilms in *P. aeruginosa* and violacein production in *Chromobacterium violaceum*.

The SNWs inhibited violacein synthesis at 0.5 mg/ml without affecting the growth of the bacterium. It could inhibit violacein production by 60% at 0.5 mg/ml, and by 80% at 4 mg/mL, after which the cell growth ceased. The SNWs were able to reduce biofilm formation in *P. aeruginosa* significantly at 4 mg/ml. It was hypothesized that the antibiofilm activity of SNWs is due to their direct diffusion through the exopolysaccharide layer through the pores. However, SEM studies showed that only the outermost layer of the biofilms was affected by SNWs. This could probably be due to the diffusion limit to SNWs owing to their rod shape (Wagh et al. 2013).

Kaempferol loaded chitosan nanoparticles of size 192.27 ± 13.6 nm and zeta potential of +35 mV, were synthesized. The kaempferol encapsulated chitosan nanoparticles were evaluated for QS inhibitory activity in *C. violaceum* CV026. The novel nanoparticle system was able to significantly inhibit violacein production in *C. violaceum* during the 30 storage days. The kaempferol - chitosan nanoparticles with loading and encapsulation efficiency between 78 and 93% will be a potential QS inhibitor (Ilk et al. 2017).

The anti-quorum sensing (anti-QS) activity of AgCl-TiO$_2$ nanoparticles (ATNPs) and its mechanism was studied. Anti-QS activity of ATNPs (6–7 nm) was evaluated using the *C. violaceum*. Silver present in ATNPs reduced the production of violacein pigment. Anti-QS activity was confirmed by the absence of signalling molecule, oxo-octanoyl homoserine lactone during growth in the presence of ATNPs. TiO$_2$ acted as a good supporting matrix facilitating controlled release of silver with prolonged residual activity. ATNPs are proposed as QS inhibitors with potential for use as an antipathogenic but nontoxic bioactive material. Although silver is well known for its antibacterial, antifungal and antiviral properties, this study confirms its anti-QS activity and its potential use in food packaging industry (Naik and Kowshik 2014).

Nanoemulsions of essential oils derived from spices have been tested for anti-QS activity against food borne pathogens. The nanoemulsions of essential oils were formulated by ultrasonic emulsification of Tween80 and water. Cumin, pepper and fennel oil nanoemulsions of 52.89 nm, 82.08 nm and 59.52 nm, respectively, were tested for inhibition of QSI activity including violacein pigment in *C. violaceum* CV026 and biofilm formation in foodborne pathogens. The three nanoemulsions efficiently inhibited biofilm formation & EPS production in the pathogens. Spice oil nanoemulsion can be used as anti-QS and biofilm inhibitors in food-borne pathogens (Venkadesaperumal et al. 2016).

Silver nanoparticles were synthesized using metabolites from soil fungus *Rhizopus arrhizus* BRS-07. These mycofabricated silver nanoparticles (mfAgNPs) were evaluated for inhibition of QS-regulated virulence and biofilm formation in *P. aeruginosa*. The mfAgNPs (410 nm) were able to downregulate expression of receptors. Further, they reduced AHL synthesis, biofilm formation and virulence factor expression and was confirmed by genomic studies. As mfAgNPs can inhibit *P. aeruginosa* QS signalling and virulence factor expression, these nanoparticles can find application as anti-virulence agents or anti-infectives (Singh et al. 2015).

Similarly, silver nanoparticles were biosynthesized from the bark extract of the medicinal plant, *Crataeva nurvala,* having extensive usage in Ayurveda medicinal system. The AgNPs were 15.2 nm in size and at a concentration of 15 $\mu g\,ml^{-1}$ efficiently inhibited biofilm formation and virulence factor expression in drug-resistant clinical isolates of *P. aeruginosa*. The silver nanoparticles were able to reduce biofilm formation by 79.70%. CLSM and TEM studies showed that the nanoparticles were internalized in the *P. aeruginosa* cell and could prevent colonization of the bacteria on the surface and thus inhibit biofilm formation. The AgNPs were also able to inhibit production of virulence factors such as pyocyanin by 74.64% and hemolysin by 47.7% (Ali et al. 2017).

The ZnO nanoparticles (65.17 nm) previously tested for quorum sensing inhibition in the laboratory strain *P. aeruginosa* PAO1, clinical isolates from cystic fibrosis patients, furanone C-30 resistant UTI *P. aeruginosa*, C 30 and gallium resistant PA14 and environmental isolates. The NPs inhibited virulence factor production such as elastase, pyocyanin, and biofilm formation in *P. aeruginosa* irrespective of its origin and resistance to C-30 or gallium. Thus, ZnO NPs can be used as an alternative therapeutics for *P. aeruginosa* infections (Garcia-Lara et al. 2015).

Honey, a rich source of polyphenolic compounds has been previously ascertained for its QS inhibitory activity. However, its lower water solubility and bioavailability is a limitation for its therapeutic application. Scaffolds of Selenium nanoparticles were designed for conjugation with honey on their surface. The nanoparticles were used as carriers to enhance the QS inhibitory activity of nanoparticles. The SeNPs honey conjugates were 12–15 nm in size. The developed selenium nano-scaffold of honey showed higher anti-QS and anti-biofilm activity, under *in vitro* and *in vivo* condition in comparison to nanoparticle or honey alone. They were capable of inhibiting violacein production in *C. violaceum* wildtype. The SeNPs and honey could inhibit violacein production at 7.5 $\mu g/ml$ and 0.6%, respectively. While the nanoparticle phytochemical conjugate inhibited violacein production at 4.5 $\mu g/ml$. At the same concentration, it was able to reduce biofilm formation in *P. aeruginosa* PA01 by >90% and suppress virulence factor expression including pyocyanin formation by 49.6% and elastase activity by 52.7%. Docking studies showed that the QS receptor LasR is inhibited by selenium nano-scaffold. Binding occurs between honey and the QS receptor LasR through hydrogen bonding and hydrophobic interactions. Thus, selenium-based nano-carriers can be used for delivery of QS inhibitors to improve their activity (Prateeksha et al. 2017).

Thirty-six metal ions were investigated for antivirulence and antibiofilm activity in *P. aeruginosa*. Of all the tested metal ions, Zinc ions and ZnO nanoparticles significantly inhibited biofilms of *P. aeruginosa*. They could also inhibit pyocyanin and pyochelin production. The *Pseudomonas* quinolone signal signalling was also affected. The NPs were also capable of reducing hemolytic activity in *P. aeruginosa* at sub-inhibitory concentrations. ZnO NPs induced zinc cation efflux pump *czc* operon and porin gene *opdT* and type III repressor *ptrA* transcriptional regulators, and repressed pyocyanin-related *phz* operon as seen by transcriptomic studies. Using mutants, it was observed that ZnO NPs controlled the production of virulence factor pyocyanin. They also increased the hydrophilicity of *P. aeruginosa* cells. ZnO NPs (<50 nm) inhibited *P. aeruginosa* biofilm formation at 1 mM by 95% on polystyrene surface. Thus, ZnO NPs have antivirulence activity against *P. aeruginosa* infections (Lee et al. 2014a).

In a similar study, the ZnO NPs were capable of inhibiting pyocyanin virulence factor production in *P. aeruginosa* PA01 at a sub-MIC of 3.125 µg/ ml. They were also studied for the antibiofilm activity at sub-MIC concentrations. The ZnO NPs (<100 nm) had higher biofilm inhibitory activity than NPs of <50 nm size and at 3.125 µg/ml inhibited *P. aeruginosa* PA01 biofilm by 61.08%. Therefore, ZnO NPs have anti-infective potential and biofilm inhibitory activity and can be developed as anti-pathogenic drugs against *P. aeruginosa* PA01 (Aswathanarayan and Vittal 2017).

The interaction between chitosan-based nanocapsules (NC), (114–155 nm) and zeta potential of +50 mV with *E. coli* quorum sensing reporter strain were studied. The stoichiometric ratio of 80 NC per bacterium was inferred. SEM studies showed the aggregation between NC and bacteria. Custom *in silico* platform simulation showed the behavior of NCs on the bacterial surface. Computation studies showed force interactions between NCs and NC-bacteria and a maximum number of 145 particles interacting at the bacterial surface. The stoichiometric ratio of NC and bacteria influenced bacterial behavior and quorum sensing response, particularly due to the bacterial aggregation in presence of NC. It was also observed that only chitosan-coated nanocapsules could influence zeta potential of *E.coli*, whereas nanoemulsions had no such effect. The nanocapsules binds to *E.coli* and promotes their aggregation. Further, chitosan-coated nanocapsules attenuate bacterial quorum sensing response in *E. coli* biosensor strain as seen by reduced expression of GFP. The number of added NC equal to stoichiometric ratio of NC per bacterium compensates electrical charge of the bacterial cell wall and induces aggregation. It was observed in both cases where the AHLs were added to the culture or on loading onto the core of the NC. The QS inhibition is probably due to limited diffusion of AHL due to bacterial aggregation caused by NCs (Qin et al. 2017).

The *in vitro* QS inhibitory activity of nanoemulsions of gold (AUSN1) was studied in plant pathogen *Pectobacterium carotovorum* sub sp. *Carotovorum*. The pathogen is known to cause soft rot disease in horticultural crops. The AUSN1 (75.52 nm) completely eliminated planktonic population and inhibited biofilms by 30–51%. in biofilms and. The AUSN1 apart from biofilm formation also inhibited swarming and swimming motility and AHL production in *P. carotovorum* strain. The exopolysaccharide (EPS) synthesis required for biofilm formation and hydrolytic enzymes production was also affected. AUSN1 has the potential to inhibit the *P. carotovorum* soft root infections in potato tubers (Joe et al. 2015).

12.4 QSI Based Nanotherapeutics

P. aeruginosa is known to cause respiratory infections and is involved in the cystic fibrosis pathogenesis. It uses the quorum sensing mechanism to regulate virulence factor production and pathogenesis. Thus, bacterial pathogenicity in *P. aeruginosa* can be attenuated by targeting the quorum sensing system. Though, QS inhibitors with potent *in vivo* activity have been developed against *P. aeruginosa*, however, their lipophilic nature hinders their penetration in non-cellular barriers such as mucus and bacterial biofilms. Therefore, it has been hypothesized that successful anti-infective inhalation therapy can be developed by using biodegradable nanocarrier. Ultra-small solid lipid nanoparticles (us-SLNs) of <100 nm size were prepared by hot melt homogenization. These had a high encapsulation efficiency of 68–95% and were able to penetrate into artificial sputum. They were efficiently nebulized and their viability was maintained on loading with quorum sensing inhibitors in Calu-3 cells. The quorum sensing inhibitors had a sevenfold superior anti-virulence activity after encapsulation in us-SLNs. The SLNs also exhibited anti-virulence effects. Thus, the use of nanocarriers for the delivery of anti-infectives has various advantages such as high loading efficiency and prolonged release, ability to penetrate mucus and effective delivery to pulmonary and enhanced anti-virulence efficacy of the quorum sensing inhibitors (Nafee et al. 2014).

In *Vibrio cholera*, the quorum sensing signal CAI-1 has been seen as a potential candidate for developing anti-pathogenic drugs for cholera treatment. The oral therapy for active CAI-1 has limitations due to inherent hydrophobicity of CAI-1 and the presence of water and mucus layers in the crypt openings in the intestinal cells. Therefore, nanoparticles have been used for the effective delivery of quorum sensing inhibitors to gastrointestinal environments. Flash NanoPrecipitation (FNP) method was used to develop CAI-1 nanoparticles (CAI-1 NPs), which is a kinetically controlled and block copolymer-directed precipitation process. The technique has been used for developing hydrophobic therapeutics as it is capable of producing PEG layers suitable for mucus penetration. The CAI-1 NPs could inhibit biofilms of *V. cholerae* WN1102 cells under static growth conditions in a dose-responsive manner. At 50 μM CAI-1 concentration, the CAI-1 NPs reduced biofilm production by 71 ± 2%, whereas only CAI-1 reduced biofilms by 32 ± 4%. It was also observed that CAI-1 NPs induced 5x quorum-sensing agonism in comparison to nonencapsulated CAI-1 under *in vitro* conditions. Thus, nanotechnology application can be used to develop quorum sensing inhibitor based therapeutics. The FNPs enhanced the water dispersible and bioactive potential of CAI-1 NPs for effective penetration through mucus films and biofilms to attenuate *V. cholera* virulence (Lu et al. 2015).

12.5 Recent Trends in QS Inhibition

12.5.1 Biological Nanofactories

Biological nanofactories have been engineered to detect and synthesize AI-2 signalling molecules. They contain a fusion protein (enzymes Pfs and LuxS) which synthesizes quorum sensing signals (Autoinducer 2) on binding target bacteria. They have antibody which help in targeting bacteria of interest (Fernandes et al. 2010). These nanofactories have been used in designing novel nano based quorum sensing inhibitors.

In one such study, a biological nanofactory was used in regulating bacterial behaviour by coating the surface of epithelial cells and induce autoinducer 2 (AI-2) quorum sensing signal production. The biological nanofactories was constructed using fusion proteins for AI- synthesis and an antibody for purification. The bionanofactories could modulate QS in *E. coli* in Caco-2 cell models (Hebert et al. 2010).

Nanofactory loaded biopolymer capsules made of chitosan alginate were able to interrupt AI-2 mediated QS signalling. The nanofactory contains fusion protein – antibody complex, wherein the fusion protein is involved in AI-2 synthesis and the antibody is used for permeability control. Biopolymer capsules are prepared by electrostatic interaction between cationic chitosan and anionic sodium alginate. Further, the nanofactories are encapsulated in chitosan alginate capsules. The capsule shell has tripolyphosphate crosslinked to chitosan as it imparts structural integrity. The capsule shell is only permeable to small molecules such as the substrate S-adenosylhomocysteine (SAH). The capsule takes up vicinal SAH by diffusion and the encapsulated nanofactories converts them into AI-2. The synthesized signalling molecule can diffuse out of the cell and modulate QS behaviour in *E. coli* (Gupta et al. 2013).

12.5.2 Nanoparticles as Efflux Pump Inhibitors

One mechanism by which the bacteria show resistance to antimicrobials is by drug extrusion through efflux pumps. Efflux pumps have a wide range of substrate specificity and are able to extrude the drug molecules outside bacterial cell. Recent studies have shown that efflux pumps are involved in the exclusion and inclusion of quorum sensing signals. Thus, by disrupting the efflux pumps function, the signal molecule movement can be effected. Metallic nanoparticles are potential candidates to be used as efflux pump inhibitors. The nanoparticles as efflux pump inhibitors can disrupt quorum sensing signalling and in turn control biofilm formation. Various studies have been performed wherein nanoparticles have been used in synergy with antibiotics for efflux pump inhibition. It has been suggested that by disrupting efflux pump function, cell membrane properties get altered which effects cellular aggregation and biofilm formation (Gupta et al. 2017).

Nanoparticles can disrupt efflux pump functioning by two possible mechanisms. Nanoparticles can be a competitive inhibitor of antibiotic for the efflux pumps binding site. The nanoparticles bind to the active site of efflux pumps and can block antibiotic extrusion (Padwal et al. 2014). In another mechanism, the nanoparticles can disrupt the efflux kinetics and affect the functioning of efflux pumps. Silver nanoparticles have been shown to interfere with efflux kinetics of MDR efflux pump, MexAM-OPrM, in *P. aeruginosa* (Nallathamby et al. 2010). The nanoparticles can cause termination of proton gradient which leads to disruption of membrane potential or loss of proton motive force, ultimately affecting efflux pump activity (Dibrov et al. 2002; Choi et al. 2008).

12.5.3   Quorum Quenching Bionanocatalyst

AiiA is a 28-kDa acylhomoserine lactonase synthesized by *Bacillus* sp. 240B1. It hydrolyzes and inactivates variety of AHLs. AiiA can find application as bio-decontaminating agent for disrupting QS in industrial and environmental samples. However, the commercial application of AiiA has setbacks due to high production cost and lack of enzyme recovery technique for its reuse. To overcome this limitation, cloned, expressed and purified recombinant AiiA (r-AiiA) enzyme was covalently immobilized onto magnetic nanoparticles (MNPs). It was observed that r-AiiA-MNP nanobiocatalyst could hydrolyze 3O-C10 AHL and inhibit QS and recovered by using external magnetic field. Further, the quorum quenching nanocatalysts could be used multiple times for 3O-C10 AHL hydrolysis. Thus, the nanoformulation of AiiA enzyme can find immense industrial applications (Beladiya et al. 2015).

Similary, enzymatic quorum quenching technique was successfully applied to control membrane biofouling in wastewater treatment. The quorum quenching enzyme, an acylase was immobilized on a nanofiltration membrane to control biofouling. The acylase-immobilized membrane with quorum quenching activity inhibited mushroom-shaped mature biofilm formation as a result of reduced EPS secretion. The acylase-immobilized membrane retained 90% of its initial catalytic activity even after 20 times of usage. The acylase-immobilized membrane was able to effectively mitigate biofouling as observed by the spatial distribution of cells and polysaccharides on the membrane surface using CLSM (Kim et al. 2011).

Thus, quorum quenching enzymes have been shown to be a potential approach for biofouling control in the membrane bioreactor (MBR) in wastewater treatment process. Acylase are immobilized on magnetic enzyme carriers (MEC). The MEC showed enzymatic activity after under rigorous test conditions. It was also observed that the enzyme showed enhanced anti-biofouling activity on immobilizing with magnetic particles than in comparison to the free enzyme on recycle as well as stability wise. It was observed that in a continuous operation of MBR with MEC, there was increased membrane permeability than in a conventional MBR without enzyme (Yeon et al. 2009).

By immobilizing and stabilizing an acylase as a cross-linked enzymes in magnetically separable mesoporous silica, a highly effective antifouling technique was developed. The nanoenzyme reactors of acylase enzyme (NER-AC) inhibited *P. aeruginosa* PAO1 biofilm maturation on the membrane surface in a membrane filtration advanced water treatment plant. Thus, NER-AC which is highly stable and recoverable by magnetic separation, can be used to prevent biofouling in membrane filtration units of water treatment plant (Lee et al. 2014b).

Acylase has been immobilized and stabilized on carboxylated polyaniline nanofibers (cPANFs). They were found to be effective antifouling nanobiocatalysts having high enzyme loading and stability. Acylase was immobilized by three different techniques of covalent attachment, enzyme coating, and magnetically separable enzyme precipitate coating (Mag-EPC). It was observed that the enzyme loading and enzyme activity per unit weight of cPANFs with Mag-EPC was 75 and 300 times higher than covalent attachment and enzyme coating. Acylase immobilized on Mag-EPC retained 55% of its initial activity even after incubation under shaking at 200 rpm for 20 days. The highly loaded and stable Mag-EPC showed antifouling activity against the biofouling by *P. aeruginosa* under static- and continuous-flow conditions. The addition of Mag-EPC reduced the production of AHLs in the biofouling bacteria. Thus, the inhibition of biofilm formation and biofouling by Mag-EPC is due to the hydrolysis of AHLs by the immobilized acylase on Mag-EPC (Lee et al. 2017).

The N-acylated homoserine lactonases are well known quorum quenching agents. Gold nanoparticles coated with AHL lactonase proteins (AiiA AuNPs) were studied for QS inhibition. The enzymes were obtained from *Bacillus licheniformis*. The AiiA AuNPs were spherical in shape and sized 10–30 nm. AiiA AuNPs reduced exopolysaccharide production, metabolic activities, and cell surface hydrophobicity and biofilms in multidrug-resistant *Proteus* species. AiiA AuNPs showed antibiofilm activity at 2–8 M concentrations without effecting macrophages. Thus, AuNPs coated with AiiA can attenuate pathogens without affecting host cells. The AiiA AuNPs have potential antibiofilm activity against multidrug-resistant Proteus species (Vinoj et al. 2015).

12.6 Detection of QS Signals Using Nanomaterials

12.6.1 Surface Enhanced Raman Scattering (SERS)

QS signal production by bacteria are detected by using a biosensor organism which is genetically modified to express phenotypes on sensing signalling molecules. The signal response as fluorescence or luminescence is measured and detected by colorimetry or fluorescence technique. However, in the recent years Surface enhanced Raman scattering (SERS) technique has been increasingly used in detection. SERS uses large electric field generated by nanostructured metals on excitation by light, which increases the inelastic scattering of photons by molecules in the vicinity of the metal. It has been shown that enhancements of Raman scattering of molecules is possible by adsorbing them on the metals such as gold or silver. It even allows detection of single molecule. SERS gives a vibrational fingerprint of the target molecules and is non-invasive and label-free Recently, it has been proposed that Surface enhanced Raman scattering (SERS) spectroscopy can be used to detect QS signals without using labels or biosensor organisms (Hill and Liz-Marzan 2017).

SERS has also been used to the detect bacteria and quorum sensing signals in biofilms. SERS was used to detect chemical variations in biofilm matrices of different growth phases of *Escherichia coli*, *Pseudomonas putida*, and *Bacillus subtilis*. SERS sensor contained hydroxylamine hydrochloride-reduced colloidal silver nanoparticles of 20–30 nm. It was observed that carbohydrates, proteins, and nucleic acids contents in the biofilm matrix increased along with the biofilm growth as inferred by the intensities and appearance probabilities of related marker peaks in the SERS spectra. Lipid content increased only in the Gram-negative biofilms *E. coli* and *P. putida*. Thus, SERS can be used to study chemical variations during biofilm formation (Chao and Zhang 2012).

SERS was used to detect a quorum sensing signal molecule at a concentration below 1 nM in both ultrapure water and physiological conditions. SERS is a highly suitable for in situ measurements of low AHLs in biofilms. SERS can be used to detect low levels of AHLs due to structural differences in the corresponding SERS spectra. SERS was measured using Ag colloidal nanoparticles produced by the hydroxylamine reducing method. SERS spectra of the AHL, C12-HSL in ultrapure water and in supplemented minimal medium were measured at different concentrations ranging from 2 μM to 0.2 nM. Thus, AHL molecules ranging from 1 nM to 1 μM could be detected using SERS (Claussen et al. 2013).

The vibrational spectroscopy technique of SERS is also used to study low concentrations of proteins in their active state. Using the high sensitivity of SERS, it has been possible to detect molecular interactions between the ligand-binding domains of LasR (LasR-LBD), the receptor of AHL molecules in *P. aeruginosa*. Thin films of gold nanoparticle fabricated using polyelectrolyte layer-by-layer (LbL) assembly. It was synthesized by alternate deposition of positively charged polyelectrolyte, poly (diallyldimethylammonium chloride) (PDDA), and negatively charged, citrate-stabilized Au NPs (60 nm) on a glass slide. The LbL film comprising three bilayers (PDDA-Au NPs) was used as SERS substrate. It was observed differential SERS fingerprints were produced by QS activators and inhibitors in LasR-LBD. Molecular docking analysis showed signal-specific structural changes in LasR on ligand binding, and confirmed the use of SERS for studying ligand induced conformational changes in proteins (Costas et al. 2015).

Using SERS, nanostructured plasmonic substrates were rationally designed for the *in situ*, label-free detection of AHLs in *P. aeruginosa* biofilms. Thus, SERS can be applied for studying intercellular communication using secreted molecules as signals. Pyocyanin is a heterocyclic nitrogen containing compound of the phenazine family produced by *P. aeruginosa*. It is an intercellular signalling molecule in the QS network of *P. aeruginosa* is involved in the biofilm morphogenesis. SERS detection and imaging of pyocyanin was performed to study QS in biofilms and microcolonies of *P. aeruginosa*. The SERS platform was developed using hybrid materials containing a plasmonic component within a porous matrix for diffusion of only small molecules. Three types of cell-compatible plasmonic platform was fabricated. Macroporous poly-N-isopropylacrylamide (pNIPAM) hydrogels loaded with Au nanorods (Au@pNIPAM), had a highly porous platform. It was used for plasmonic detection of pyocyanin in colonized and non-colonized regions of the substrate. Mesostructured Au@TiO$_2$ containing mesoporous TiO$_2$ thin film coated on monolayer of Au nanospheres was used to generate SERRS maps. It showed difference in QS plasmonic signal in biofilms. A mesoporous silica-coated micropatterned supercrystal arrays of Au nanorods (Au@SiO2), having a high electromagnetic enhancement factor, could detect QS phenotypes in early stages of biofilm formation. Phenazine could also be imaged in microcolonies of bacterial. Thus, using nanostructured hybrid materials and SERS, it is possible for label free detection of QS in biofilms. SERS can used to study QS signaling by the release of pyocyanin from biofilms (Bodelon et al. 2016).

SERS can be used to study biofilms of those bacteria which produce SERS active QS molecules. Specific surface chemistry or surface topography can be designed on SERS-active substrate for monitoring QS signalling in biofilms. Also, the attenuation of bacterial biofilm growth formation will cause different SERS pattern from the QS signal. Quorum quenching molecules can be detected by modification of the SERS substrates due to the presence of these compounds. SERS detecting of QS signals can be performed label-free and non-invasive technique method by constantly observing surface topography and chemistry (Hill and Liz-Marzan 2017).

12.6.2 Förster Resonance Energy Transfer (FRET)

AHL based QS regulates various phenotypes in *Vibrio fischeri*. Hence, quenching of this signal would help in controlling the marine bacterium. Conventional QS signaling molecule detection techniques using biosensors and analytical techniques such as HPLC take time and are less sensitive. As 3OC6HSL binds to LuxR receptor of *V. fischeri*, it changes the receptor conformation. A biosensor was engineered by incorporating LuxR in the Forster resonance energy transfer pair YFP/CFP. This biosensor had a LOD of 100 μM for signal detection (Zhang and Ye 2014).

12.7 Opinion

The nanoparticles are promising as QS inhibitory agents and hence have the potential for application as anti-biofilm/biofouling agents in industrial and clinical settings. Their ability to inhibit QS in bacteria and enhance the activity of other QS inhibitory agents shows that functionalized nanoparticles can be developed as anti-infectives or anti-pathogenic drugs to treat infectious diseases.

References

Ali SG, Ansari MA, Khan HM, Jalal M, Mahdi AA, Cameotra SS (2017) *Crataeva nurvala* nanoparticles inhibit virulence factors and biofilm formation in clinical isolates of *Pseudomonas aeruginosa*. J Basic Microbiol 57(3):193–203. https://doi.org/10.1002/jobm.201600175

Aswathanarayan JB, Vittal RR (2017) Antimicrobial, biofilm inhibitory and anti-infective activity of metallic nanoparticles against pathogens MRSA and *Pseudomonas aeruginosa* PA01. Pharm Nanotechnol 5(2):148–153. https://doi.org/10.2174/2211738505666170424121944

Beladiya C, Tripathy RK, Bajaj P, Aggarwal G, Pande AH (2015) Expression, purification and immobilization of recombinant AiiA enzyme onto magnetic nanoparticles. Protein Expr Purif 113:56–62. https://doi.org/10.1016/j.pep.2015.04.014

Blango MG, Mulvey MA (2009) Bacterial landlines: contact-dependent signaling in bacterial populations. Curr Opin Microbiol 12:177–181. https://doi.org/10.1016/j.mib.2009.01.011

Bodelon G, Montes-Garcia V, Lopez-Puente V, Hill EH, Hamon C, Sanz-Ortiz MN (2016) Detection and imaging of quorum sensing in *Pseudomonas aeruginosa* biofilm communities by surface-enhanced resonance Raman scattering. Nat Mater 15:1203–1211. https://doi.org/10.1038/nmat4720

Chao Y, Zhang T (2012) Surface-enhanced Raman scattering (SERS) revealing chemical variation during biofilm formation: from initial attachment to mature biofilm. Anal Bioanal Chem 404:1465–1475. https://doi.org/10.1007/s00216-012-6225-y

Chen J, Wiley BJ, Xia Y (2007) One-dimensional nanostructures of metals: large-scale synthesis and some potential applications. Langmuir 23:4120–4129. https://doi.org/10.1021/la063193y

Choi O, Deng KK, Kim NJ, Ross L, Surampalli RY, Hu Z (2008) The inhibitory effects of silver nanoparticles, silver ions, and silver chloride colloids on microbial growth. Water Res 42:3066–3074. https://doi.org/10.1016/j.watres.2008.02.021

Claussen A, Abdali S, Berg RW, Givskov M, Sams T (2013) Detection of the quorum sensing signal molecule *N*-Dodecanoyl-DL-homoserine lactone below 1 nanomolar concentrations using surface enhanced Raman spectroscopy. Curr Phys Chem 3:199–210. https://doi.org/10.2174/1877946811303020010

Costas C, Lopez-Puente V, Bodelon G, Gonzalez-Bello C, Perez-Juste, Pastoriza-Santos I, Liz Martin LM (2015) Using surface enhanced Raman scattering to analyze the interactions of protein receptors with bacterial quorum sensing modulators. ACS Nano 9:5567–5576. https://doi.org/10.1021/acsnano.5b01800

Davies D (2003) Understanding biofilm resistance to antibacterial agents. Nat Rev Drug Discov 2:114–122. https://doi.org/10.1038/nrd1008

Dibrov P, Dzioba J, Gosink KK, Häse CC (2002) Chemiosmotic mechanism of antimicrobial activity of Ag^+ in *Vibrio cholerae*. Antimicrob Agents Chemother 46:2668–2670. https://doi.org/10.1128/AAC.46.8.2668-2670.2002

Eustis S, El-Sayed MA (2006) Why gold nanoparticles are more precious than pretty gold: noble metal surface plasmon resonance and its enhancement of the radiative and nonradiative properties of nanocrystals of different shapes. Chem Soc Rev 35:209–217. https://doi.org/10.1039/b514191e

Fernandes R, Roy V, Wu HC, Bentley WE (2010) Engineered biological nanofactories trigger quorum sensing response in targeted bacteria. Nat Nanotechnol 5:213–217. https://doi.org/10.1038/nnano.2009.457

Feynman R (1991) There's plenty of room at the bottom. Science 254:1300–1301. https://doi.org/10.1126/science.254.5036.1300

Gajjar P, Pettee B, Britt DW, Huang W, Johnson WP, Anderson J (2009) Antimicrobial activities of commercial nanoparticles against an environmental soil microbe, *Pseudomonas putida* KT2440. J Biol Eng 3:9–22. https://doi.org/10.1186/1754-1611-3-9

Garcia-Lara B, Saucedo Mora MA, Roldan Sanchez JA, Perez-Eretza B, Ramasamy M, Lee J, Coria-Jimenez R, Tapia M, Varela-Guerrero V, Garcia-Contreras R (2015) Inhibition of quorum-sensing-dependent virulence factors and biofilm formation of clinical and environmental *Pseudomonas aeruginosa* strains by ZnO nanoparticles. Lett Appl Microbiol 61(3):299–305. https://doi.org/10.1111/lam.12456

Gupta A, Terrell JL, Fernandes R, Dowling MB, Payne GF, Raghavan SR, Bentley WE (2013) Encapsulated fusion protein confers "sense and respond" activity to chitosan–alginate capsules to manipulate bacterial quorum sensing. Biotechnol Bioeng 110:552–562. https://doi.org/10.1002/bit.24711

Gupta AK, Gupta M (2005) Synthesis and surface engineering of iron oxide nanoparticles for biomedical applications. Biomaterials 26:3995–4021. https://doi.org/10.1016/j.biomaterials.2004.10.012

Gupta D, Singh A, Khan AU (2017) Nanoparticles as efflux pump and biofilm inhibitor to rejuvenate bactericidal effect of conventional antibiotics. Nanoscale Res Lett 12:454. https://doi.org/10.1186/s11671-017-2222-6

Gutierrez FM, Olive PL, Banuelos A, Orrantia E, Nino N, Sanchez EM, Ruiz F, Bach H, Gay YA (2010) Synthesis, characterization, and evaluation of antimicrobial and cytotoxic effect of silver and titanium nanoparticles. Nanomedicine 6:681–688. https://doi.org/10.1016/j.nano.2010.02.001

Hebert CG, Gupta A, Fernandes R, Tsao CY, Valdes JJ, Bently WE (2010) Biological nanofactories target and activate epithelial cell surfaces for modulating bacterial quorum sensing and interspecies signaling. ACS Nano 4:6923–6931. https://doi.org/10.1021/nn1013066

Hill EH, Liz-Marzan LM (2017) Toward plasmonic monitoring of surface effects on bacterial quorum-sensing. Curr Opin Colloid Interface Sci 32:1–10. https://doi.org/10.1016/j.cocis.2017.04.003

Hu X, Chan CT (2004) Photonic crystals with silver nanowires as a near-infrared superlens. Appl Phys Lett 85:1520–1522. https://doi.org/10.1063/1.1784883

Hu L, Kim HS, Lee J-Y, Peumans P, Cui Y (2010) Scalable coating and properties of transparent, flexible silver nanowire electrodes. ACS Nano 4:2955–2963. https://doi.org/10.1021/nn1005232

Huang Y, Duan X, Cui Y, Lauhon LJ, Kim K-H, Lieber CM (2001) Logic gates and computation from assembled nanowire building blocks. Science 294:1313–1317. https://doi.org/10.1126/science.1066192

Ilk S, Saglam N, Ozgen M, Korkusuzda F (2017) Chitosan nanoparticles enhances the anti-quorum sensing activity of kaempferol. Int J Biol Macromol 94:653–662. https://doi.org/10.1016/j.ijbiomac.2016.10.068

Jamieson T, Bakhshi R, Petrova D, Pocock R, Imani M, Seifalian AM (2007) Biological applications of quantum dots. Biomaterials 28:4717–4732. https://doi.org/10.1016/j.biomaterials.2007.07.014

Joe MM, Benson A, Sarvanan VS, Tongmin S (2015) In vitro antibacterial activity of nanoemulsion formulation on biofilm, AHL production, hydrolytic enzyme activity, and pathogenicity of *Pectobacterium carotovorum sub sp. Carotovorum*. Physiol Mol Plant Pathol 91:46–55. https://doi.org/10.1016/j.pmpp.2015.05.009

Kalia VC, Purohit HJ (2011) Quenching the quorum sensing system: potential antibacterial drug targets. Crit Rev Microbiol 37:121–140. https://doi.org/10.3109/1040841X.2010.532479

Kalia VC (2013) Quorum sensing inhibitors: an overview. Biotechnol Adv 31:224–245. https://doi.org/10.1016/j.biotechadv.2012.10.004

Kim JH, Choi DC, Yeon KM, Kim SR, Lee CH (2011) Enzyme-immobilized nanofiltration membrane to mitigate biofouling based on quorum quenching. Environ Sci Technol 45:1601–1607. https://doi.org/10.1021/es103483j

Kolar M, Urbanek K, Latal T (2001) Antibiotic selective pressure and development of bacterial resistance. Int J Antimicrob Agents 17:357–363

Lee JH, Kim YG, Cho MH, Lee J (2014a) ZnO nanoparticles inhibit *Pseudomonas aeruginosa* biofilm formation and virulence factor production. Microbiol Res 169:888–896. https://doi.org/10.1016/j.micres.2014.05.005

Lee B, Yeon KM, Shim J, Kim SR, Lee CH, Lee J, Kim J (2014b) Effective antifouling using quorum-quenching acylase stabilized in magnetically-separable mesoporous silica. Biomacromolecules 15:1153–1159. https://doi.org/10.1016/j.jconrel.2014.06.055

Lee J, Lee I, Nam J, Hwang DS, Yeon KM, Kim J (2017) Immobilization and stabilization of acylase on carboxylated polyaniline nanofibers for highly effective antifouling application via quorum quenching. ACS Appl Mater Interfaces 9:15424–15432. https://doi.org/10.1021/bm401595q

Lellouche J, Kahana E, Elias S, Gedanken A, Banin E (2009) Antibiofilm activity of nanosized magnesium fluoride. Biomaterials 30:5969–5978. https://doi.org/10.1016/j.biomaterials.2009.07.037

Lu HD, Spiegel A, Hurley A, Perez LJ, Maisel K, Ensign LM, Hanes J, Bassler BL, Semmelhack MF, Prud'homme RK (2015) Modulating *Vibrio cholerae* quorum sensing controlled communication using autoinducer loaded nanoparticles. Nano Lett 15:2235–2241. https://doi.org/10.1021/acs.nanolett.5b00151

Miller KP, Wang L, Chen Y, Pellechia PJ, Benicewicz BC, Decho AW (2015) Engineering nanoparticles to silence bacterial communication. Front Microbiol 6:189. https://doi.org/10.3389/fmicb.2015.00189

Moghimi SM (2005) Nanomedicine: prospective diagnostic and therapeutic potential. Asia Pacific Biotech News 9:1072–1077. https://doi.org/10.1517/14712598.5.1.1

Mohanty A, Tan CH, Cao B (2016) Impacts of nanomaterials on bacterial quorum sensing: differential effects on different signals. Environ Sci Nano 3:351–356. https://doi.org/10.1039/C5EN00273G

Murphy CJ, Gole AM, Hunyadi SE, Orendorff CJ (2006) One-dimensional colloidal gold and silver nanostructures. Inorg Chem 45:7544–7554. https://doi.org/10.1021/ic0519382

Nafee N, Husari A, Maurer CK, Lu C, de Rossi C, Steinbach A, Hartmann RW, Lehr CM, Schneider M (2014) Antibiotic-free nanotherapeutics: ultra-small, mucuspenetrating solid lipid nanoparticles enhance the pulmonary delivery and anti-virulence efficacy of novel quorum sensing inhibitors. J Control Release 192:131–140. https://doi.org/10.1016/j.jconrel.2014.06.055

Naik K, Kowshik M (2014) Anti-quorum sensing activity of AgCl-TiO$_2$ nanoparticles with potential use as active food packaging material. J Appl Microbiol 117:972–983. https://doi.org/10.1111/jam.12589

Nallathamby PD, Lee KJ, Desai T, Xu XH (2010) Study of the multidrug membrane transporter of single living *Pseudomonas aeruginosa* cells using size-dependent plasmonic nanoparticle optical probes. Biochemistry 49:5942–5953. https://doi.org/10.1021/bi100268k

Padwal P, Bandyopadhyaya R, Mehra S (2014) Polyacrylic acid-coated iron oxide nanoparticles for targeting drug resistance in mycobacteria. Langmuir 30:15266–15276. https://doi.org/10.1021/la503808d

Parak WJ, Gerion D, Pellegrino T, Zanchet D, Micheel C, Williams CS, Boudreau R, Le Gros MA, Larabell CA, Alivisatos AP (2003) Biological applications of colloidal nanocrystals. Nanotechnology 14:15–27. https://doi.org/10.1088/0957-4484/14/7/201

Prateeksha SBR, Shoeb M, Sharma S, Naqvi AH, Gupta VK, Singh BN (2017) Scaffold of selenium nanovectors and honey phytochemicals for inhibition of *Pseudomonas aeruginosa* quorum sensing and biofilm formation. Front Cell Infect Microbiol 7:93. https://doi.org/10.3389/fcimb.2017.00093

Qin X, Engwer C, Desai S, Vila-Sanjurjo C, Goycoole FM (2017) An investigation of the interactions between an E. coli bacterial quorum sensing biosensor and chitosan-based nanocapsules. Colloids Surf B: Biointerfaces 149:358–368. https://doi.org/10.1016/j.colsurfb.2016.10.031

Rana D, Matsuura T (2010) Surface modifications for antifouling membranes. Chem Rev 110:2448–2471. https://doi.org/10.1021/cr800208y

Rasmussen TB, Givskov M (2006) Quorum sensing inhibitors: a bargain of effects. Microbiology 152:895–904. https://doi.org/10.1099/mic.0.28601-0

Reading NC, Sperandio V (2006) Quorum sensing: the many languages of bacteria. FEMS Microbiol Lett 254:1–11. https://doi.org/10.1111/j.1574-6968.2005.00001.x

Samia ACS, Dayal S, Burda C (2006) Quantum dot-based energy transfer: perspectives and potential for applications in photodynamic therapy. Photochem Photobiol 82:617–625. https://doi.org/10.1562/2005-05-11-IR-525

Singh BR, Singh A, Khan W, Naqvi AH, Singh HB (2015) Mycofabricated biosilver nanoparticles interrupt *Pseudomonas aeruginosa* quorum sensing systems. Sci Rep 5:13719. https://doi.org/10.1038/srep13719

Venkadesaperumal G, Rucha S, Sundar K, Shetty PH (2016) Anti-quorum sensing activity of spice oil nanoemulsions against food borne pathogens. LWT Food Sci Technol 66:225–231. https://doi.org/10.1016/j.lwt.2015.10.044

Vinoj G, Pati R, Sonawane A, Vaseeharan B (2015) In vitro cytotoxic effects of gold nanoparticles coated with functional acyl homoserine lactone lactonase protein from *Bacillus licheniformis* and their antibiofilm activity against Proteus species. Antimicrob Agents Chemother 59:763–771. https://doi.org/10.1128/AAC.03047-14

Wagh (nee Jagtap) MS, Patil RH, Thombre DK, Kulkarni MV (2013) Evaluation of anti-quorum sensing activity of silver nanowires. Appl Microbiol Biotechnol 97:3593. https://doi.org/10.1007/s00253-012-4603-1

Whitehead NA, Barnard AM, Slater H, Simpson NJ, Salmond GP (2001) Quorum sensing in Gram-negative bacteria. FEMS Microbiol Rev 25:365–404. https://doi.org/10.1111/j.1574-6976.2001.tb00583.x

Whitesides GM (2003) The 'right' size in Nanobiotechnology. Nat Biotechnol 21:1161–1165. https://doi.org/10.1038/nbt872

Xia Y, Yang P, Sun Y, Wu Y, Mayers B, Gates B, Yin Y, Kim F, Yan H (2003) One-dimensional nanostructures: synthesis, characterization and applications. Adv Mater 15:353–389. https://doi.org/10.1002/adma.200390087

Yeon KM, Cheong WS, Oh HS, Lee WN, Hwang BK, Lee CH, Beyenal H, Lewandowski Z (2009) Quorum sensing: a new biofouling control paradigm in a membrane bioreactor for advanced wastewater treatment. Environ Sci Technol 43:380–385. https://doi.org/10.1021/es8019275

Zhai T, Fang X, Liao M, Xu X, Zeng H, Yoshio B, Golberg D (2009) A comprehensive review of one-dimensional metal-oxide nanostructure photodetectors. Sensors 9:6504–6529. https://doi.org/10.3390/s9080650

Zhang C, Ye BC (2014) Real-time measurement of quorum-sensing signal autoinducer 3OC6HSL by a FRET-based nanosensor. Bioprocess Biosyst Eng 37(5):849–855. https://doi.org/10.1007/s00449-013-1055-7

Chapter 13
Bacterial-Mediated Biofouling: Fundamentals and Control Techniques

Soumya Pandit, Shruti Sarode, Franklin Sargunaraj, and Kuppam Chandrasekhar

Abstract Biofouling is a serious drawback found in technological tools exposed in an aqueous medium; it plays a significant role in selecting suitable materials used in a different industrial application like food processing industry, shipping yards, water treatment and desalting tools; consequently, it has huge impact viability and economic feasibility of those systems. Biofouling includes organic fouling, particulate/colloidal fouling and fouling occurred due to microbial/biological entities. The present chapters dealt about bacterial mediated biofouling. Bacterial-mediated biofouling represents the "Achilles heel" due to bacteria's ability to multiply over time; this type of biofouling is potentially more dangerous. An iota of living cells attached to the surface can grow and form biofilms using the dissolve organic substances in the water. Bacterial biofouling is recently gaining much interest due to it's severe economic and environmental adverse effects. In the present book chapter, the different types of biofouling and their causes have been highlighted. A thorough understanding of the fundamental principles of biofilm generation would help to prevent biofouling. Basics of biofilm development causing fouling and the factors affecting fouling have been depicted. A short discussion about the means to mitigate biofouling has been provided. The present chapters also described different anti-biofouling strategies adapted to in various industrial tools. A brief description of the influence of biofouling in water treatment, food processing, and biomedical devices has been provided. Advantages and disadvantages of bacterial-mediated biofouling have been discussed.

Authors Soumya Pandit and Shruti Sarode have contributed equally with all other contributors.

S. Pandit (✉) · S. Sarode
Department of Biotechnology, IIT-Kharagpur, Kharagpur, West Bengal, India

F. Sargunaraj
Faculty of Medicine, Department of Human and Medical Genetics,
Vilnius University, Vilnius, Lithuania

K. Chandrasekhar (✉)
Bio-Engineering and Environmental Science (BEES), CSIR-IICT,
Hyderabad, Telangana, India

School of Applied Bioscience, Kyungpook National University (KNU), Daegu, South Korea

© Springer Nature Singapore Pte Ltd. 2018
V. C. Kalia (ed.), *Biotechnological Applications of Quorum Sensing Inhibitors*,
https://doi.org/10.1007/978-981-10-9026-4_13

Keywords Antimicrobial · Biofilm · Bacterial adhesion · Quorum sensing · Surface charge · Hydrophobicity

13.1 Introduction

The term 'Biofouling' is a non-desirable aggregation of microorganisms (bacteria, algae, fungus) and macroorganisms (arthropods, crustaceans) on a solid surface in an aqueous environment. Any solid surface that is exposed to water is susceptible to biofouling and hence, it is a menace for different manufacturing units and industrial tools such as paper production machines, food processors, wastewater treatment equipment and desalination plants. Biofouling can lead to deterioration in the material strength of an object, loss in heat transfer efficiency and mechanical blockage of fluid transport systems. Consequently, the economic losses associated with biofouling are substantial. In most industrial systems, bacterial biofilms are the cause of biofouling (Singh et al. 2006).

The main focus of this review will be on bacterial biofouling. Biofilm is a term referring to the accumulation of bacterial colony onto a material. Usually, bacteria are embedded to each other by their secretion known as extracellular polymeric substances (EPS) (Sutherland 2001). Bacterial biofilms are highly adaptive (de la Fuente-Núñez et al. 2013) that they are capable of being adhered to any living (like a tooth) surfaces or non-living (membranes) surfaces. Biofilms certainly use material surfaces as a substrate for their growth; in other cases, they are also capable of retrieving nutrition from them (Speranza et al. 2011). These bacteria produce molecules that can cause corrosion, product contamination and threaten public health. When the effect of biofilm formation turns out to be deleterious, then it is known as biofouling caused by a bacterium. The adversity of biofilms is even exacerbated when they developed resistance to drugs, leading to a greater biofouling with the lack of control. Hence, measures have to be taken to prevent biofouling to save billions of dollars spent in repairing damages caused due to biofouling.

13.2 Classification and Reasons

Bacterial-mediated biofouling consists of several steps from the irreversible attachment of seed bacteria to the maturation of biofilm. The adhesion of a bacterial cell to material surface is the initialization of bacterial-mediated biofouling. Initial microbial adhesion on the surface depends on several factors like surface roughness, hydrophobicity, surface charge, etc. The electrokinetic and hydrophobic interaction plays a major role in initial bacterial attachment. The attachment is followed by cell multiplication and EPS generation to build biofilm using the soluble biodegradable organic substances in the aqueous environment. The EPS produced by the bacteria facilitates in anchoring the cells to surface or substratum.

13.2.1   Classification

Biofouling is generally classified into two types based on the size of an organism that causes biofouling. Macrofouling refers to the sticking of larger organisms like mussels, mollusks, and barnacles; whereas microfouling refers to adhesion of bacteria or certain fungal slimes. The deeper classification relies on the type of fouling materials,

(i) Hard Fouling – is caused by organisms that secrete calcareous substances.
(ii) Soft Fouling – refers to less rigidity (not adhesive strength) of the fouling materials like polysaccharides secreted by bacteria (Chaudhury et al. 2005).

Bacterial fouling is micro-level and soft. Nevertheless, bacterial biofilms are harder to mitigate because of the resistance in bacteria against antibiotics (Cortés et al. 2011).

13.2.2   Reasons

13.2.2.1   Adaptive Reasons

Biofilms are formed as an adaptive response arising due to environmental stress. They are formed via a complex regulatory network and molecular mechanisms that make them incapable of being destroyed by antibiotics. They could also adjust their physiological systems to adapt to any ambient conditions (Baker and Dudley 1998).

13.2.2.2   Genetic Reasons

Researchers have honed on genetic principles of biofilm, especially their initiation in recent past (Pratt and Kolter 1999). Genes are the cause for every biological process which includes biofilm. Genes causing flagellar motility contributes significantly to biofilm formation, as bacterial motion (chemotaxis) necessitates initial adhesion and their migration along the surface, and are important for influencing the aggregation of bacteria on the surface through chemotaxis (Johnson 2008), which consequently causes biofouling.

13.3   Factors Affecting Fouling

Several physical and chemical interactions influence organic fouling (Mi and Elimelech 2008) on substrates. They are explained with reasons under this section.

13.3.1   Physical Factors

13.3.1.1   Temperature

An increased temperature in water systems has shown increased biofilm thickness because of enhanced metabolism. Certain bacteria originating from hydrothermal vents are capable of bearing high temperatures. They cause biofouling at devices that even emit heat like chimneys (Stokke et al. 2015), irrespective of their materials. E.g., metallic surfaces are affected by fouling (Guezennec et al. 1998). High temperatures will not be an ideal factor to prevent biofouling. Cleaning performed at high levels of temperature did not remove biofilm formation but even increased the adherent quality of the bacterial films (Marion-Ferey et al. 2003). Therefore, increase in temperature is always positive in adversening the effects of biofouling (Garrett et al. 2008).

13.3.1.2   Gravity

The *Pseudomonas aeruginosa*, grown in microgravity and normal gravity conditions showed that microbes growing in space conditions exhibited excess growth when compared to the earthly conditions, regardless of the amount of nutrition available (Kim et al. 2013). Gravity is a minor factor influencing the initial adhesion of microbes in biofilm formation; this is proved using *Streptococcus mutans* and *Escherichia coli* as model microorganisms. The absence of gravity will not mitigate biofouling; rather can augment it (Walt et al. 1985).

13.3.1.3   Roughness

The roughness of a surface is directly proportional to the biofilm development (Kerr et al. 1999). Although Kerr et al. 1999 examined only for marine biofouling, this concept applies to any material and for any organism that is capable of biofouling (Tang et al. 2009). For example, it has been found through confocal microscopy that on dental implants, the bacteria *S. mutans* severely adheres on resin materials which are rougher than the ceramic type of materials (Aykent et al. 2010).

13.3.1.4   Hydrodynamics

Hydrodynamics has significant effects on biofilm development, the extent of biofouling changes with different flow rate. A patchy and rough cell aggregates were found when biofilm developed under laminar flow condition, on the contrary, the cells in biofilm developed in turbulent flow condition are usually elongated 'streamers'. It was noticed that EPS characteristics also change with different shear force due to variation in flow rate. The extent of microbial colonization also depends on the hydrodynamic conditions. It differs in the laminar and turbulent flow. The formation of biofilms increases as the roughness of the surface increases because the surface area is higher on rough surfaces (Stoodley et al. 1998).

Several other factors affect biofilm formation on a different type of surfaces. The surfaces which are in contact with the aqueous medium are modified by the polymers from the medium to form a conditioning film. This formation of the film causes chemical and physical transformation on the surface to which the microbes attach, it also affects the rate of microbial attachment. The wettability or surface hydrophobicity promotes biofouling; the surface charge also influences bacterial adhesion and therefore biofilm. Usually, the negative charge of the surface has repulsion towards bacteria due to bacterial negative surface charges bestowed by peptidoglycan or different acidic group. However, it depends on solution pH and ionic concentration (Giaouris et al. 2005).

13.3.2 Biological Factors

Quorum sensing (QS), microcolonies, and chemotaxis are the main biological factors that cause biofouling (Weitere et al. 2005).

13.3.2.1 Quorum Sensing

The biofilm development takes place with a minimum number of bacteria (quorum) which coordinate among themselves using specific stimuli and responses accordingly. This determines the effect of biofouling. A bacterium uses QS for communicating to perform various physiological activities and also for forming a biofilm which is of adaptive significance (Li and Tian 2012). The biofilm phenotype (i.e., the expressed nature of biofilm) is contingent upon the social behavior of bacteria which in turn arises due to QS. The QS determines the molecular structure and functions within the biofilm, therefore understanding this type of sense would help us determine the adversity of biofouling (Li 2009). The QS process is controlled via discharge and detection of a specific small chemical substance-auto inducer (AI). Generation of AIs plays a pivotal role in the establishment of biofilm. There is three major type of AIs identified so far which helps in regulating biofilm formation-N-acyl homoserine lactones (AHL), oligopeptides, and autoinducer-2 (AI-2). Expression of LuxS gene is responsible for this AI. AHL and oligopeptide control the intra-species communication for both gram positive and negative bacteria while AI-2 acts as a messenger for the interspecies bacterial community.

13.3.2.2 Chemotaxis

Computational models have indicated that the motile bacterial cells are capable of forming a biofilm that spreads on the entire surface, whereas immotile genetic variants of cells cause round colonies. Therefore, motility is disadvantageous to the material surfaces by causing severe fouling. They (genetically motile bacteria) not only grow in length but also in height, by dethatching and reattaching, which further worsens the degree of biofouling (Picioreanu et al. 2007). In the following image, we can see that yellow (motile cells) migrate and spreads to a larger surface than blue cells (immotile). The images obtained from computational simulations by (Picioreanu et al. 2007) and experiments done by (Klausen et al. 2003) are mere the same. Although chemotaxis is a biophysical motion of bacteria, they are strongly rooted in their genetics (see Sect. 2) for complimentary information.

13.3.2.3 Horizontal Gene Transfer

Horizontal transfer of genetic material is an important phenomenon for the evolution of the microbial community. The genetic diversity increases when the surrounding environment changes. Mobile genetic elements using transposon, conjugative plasmid or bacteriophages are responsible for the horizontal transfer of gene (Madsen et al. 2012). The phenotypic characteristics of bacteria are different from biofilm mode or attached based growth to suspended growth (planktonic).

13.3.2.4 Microcolonies

Quorum sensing instigates the detection of adjacent bacterial cells, leading to their attachment to each other via chemotaxis (Pratt and Kolter 1999) forming microcolonies within biofilm structures (Davey and O'toole 2000). Therefore, these three factors are highly interrelated with each other in causing biofouling. Microcolonies are formed at an initial stage during when the biofilm is stronger against resistant, this phenomenon is known as microcolony mediated protection. However, at the later stages, microcolonies formation is rare, with a decrease in biofilm mass (Weitere et al. 2005). This is correlated with the possibility that because of microcolonies biofouling will be adverse at the beginning and lessens over a period (Fig. 13.1).

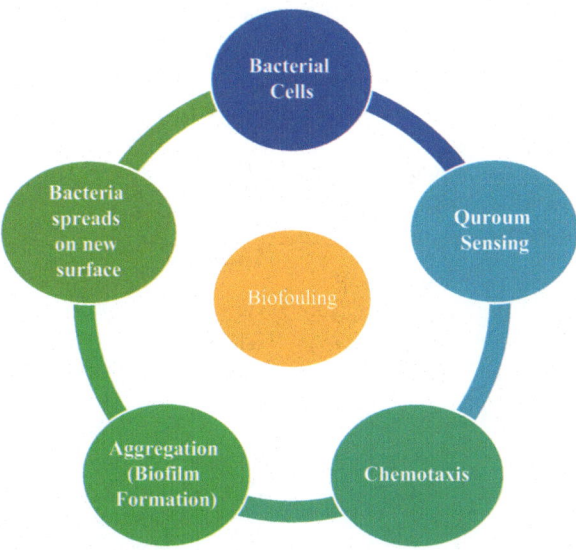

Fig. 13.1 Explanation of causes for biofouling

13.3.3   Chemical Factors

The presence of chemicals such as NaCl, Ca^{+2}, and the pH influences biofouling, the coupled interaction between physical and chemical factors such as binding of Ca^{+2} ions play a greater role in fouling especially in membrane systems (Hong and Elimelech 1997). It has been shown that in *P. aeruginosa* that presence of Calcium ions cross-links components of EPS making their tensile strength to be higher (Körstgens et al. 2001). The Ca ions also play a vital role, in intra- and extra- cellular Ca binding bacterial proteins, which is more evident in biofilm than in cultures, mainly by increasing cell aggregation and thickening of biofilm layer which increases biofouling (Rose and Turner 1998). It has also been shown when EPS and SMP (Soluble Microbial Products) are compared under low and optimal Ca^{+2} concentrations on membranes, that at low Ca conditions biofouling is higher (Kim and Jang 2006) indicating that biofouling reaches a plateau and decrease after certain higher concentrations of Calcium ion. A pH of 6 to 9 is positive for biofouling and the conditions which are more basic or acidic than this limit decreases biofilm growth and efficiency (Patil et al. 2011). Biofilm production is also induced by bacterial genes by recognizing NaCl (Lim et al. 2004) (Table 13.1).

Table 13.1 Essential parameters influencing bacterial adhesion to material surfaces

Bacteria	Surface property	Aqueous condition/ physicochemical factors	References
Species	Surface charge	Temperature	Giaouris et al. (2005)
Growth phase	Roughness	pH	Kerr et al. (1999)
Hydrophobicity of the membrane	Hydrophilicity	Shear force	Stoodley et al. (1998)
Charge	Chemical composition	Viscosity	Stoodley et al. (1998)
Population density	Surface tension/ wettability	Dissolved organic substance	Giaouris et al. (2005)
Composition of the mixed population	Porosity	Gravity	Walt et al. (1985)
Physiological responses (Quoram sensing, chemotaxis)		Concentration of halogens	Giaouris et al. (2005)
		Concentration of calcium	Hong and Elimelech (1997)

13.4 Criteria to Mitigate Biofouling

The qualities of the material, chemical or a biomolecular function could be utilized as criteria to mitigate biofouling. Examples are listed below.

13.4.1 Physical, Material Science and Modification Criteria to Mitigate Biofouling

Gedge et al. 2012 showed that ultrasonic sound waves minimized biofouling (Gedge et al. 2012). The use of Phosphorylcholine-based polymers reduces by biofouling by decreasing cell and protein adhesion on surfaces (Lewis 2000). Coating of nano silver particles onto substrates like filtration membranes reduced biofouling (Yang et al. 2009). Modifying the properties of the surface so as to inhibit the binding of cells and organic polymers to the surface is an effective strategy to prevent the formation of biofilms. Several strategies to decrease the protein absorption to biomaterials were adopted, i.e. the development of strongly hydrophilic and hydrophobic surfaces (Biology and Board 2000). According to a study which compares the protein adsorption of different strongly hydrophilic surfaces, including negatively charged (AA), positively charged (N-vinylpyrolidon), and non-charge (PEG), it was observed that the non-charged strongly hydrophilic surfaces are more advantageous than the other alternatives (Flemming 1993). This is because hydrophobic surface strongly repel water and the presence of non-sterile aqueous medium is essential for the formation of biofilms. Also, the neutral nature of the surface makes it tougher for charged moieties on organic molecules to attach to the solid surface. PEG or polyethylene glycol is most often used for marine biofouling prevention. The inert nature of PEG makes it an ideal candidate for the creation of bio-inert material surfaces. "Dry" chemistry methods are used for the formation of simply cleaned or non-fouling protein repellant material surfaces which can be cleaned easily. Some of the dry methods include plasma treatment in vacuum or at normal pressure, in atmospheres of different gases, as well as ion or electron beam. The formation of exceedingly smooth surfaces can be done by employing the surface topography to improvise the evenness and engineering local surface hydrodynamics. This may be a more credible approach to decreased biofouling. Mechanical agitation of a surface has also been shown to be effective in disrupting the formation of biofilms.

13.4.2 Chemical Approaches to Mitigate Fouling

Ultra-pure water systems also support the development of biofilms, hence inhibiting the formation of biofilms cannot be done by limiting the carbon (Gupta et al. 2016). Biofouling can be controlled by the following three principal approaches: (1) extermination or inactivation of biofouling organisms using antibiotics, biocides, cleaning chemicals, etc., (2) detachment of biofoulers mechanically; and (3) modification of the surface by converting its material into a low-fouling or non-sticking (non-adhesive) one. Such modification generally changes several factors such as the surface chemical composition and morphology, surface topography and roughness, the hydrophilic/hydrophobic balance, in addition to the surface energy and polarity.

Chemical agents used for cleaning and disinfection have an adverse effect on the environment, and hence, there's a need to shift to alternatives. Enzyme-based detergents, also called "green chemicals" are being used in place of traditional chemicals in the food industry. The cleaning efficiency was improved by combining the proteolytic enzymes with surfactants because it increased the wettability of biofilms formed by a thermophilic *Bacillus* (Dufour et al. 2010).

The term "Disinfection" refers to reducing the surface population of viable cells left after cleaning and hence it prevents the growth of microbes on the surfaces before the growth of microbes restarts. Disinfectants are more effective if organic material like carbohydrates, fat and other protein-based materials have been removed beforehand by cleaning. The efficiency of disinfectants depends on several factors such as the temperature pH, water hardness, presence of chemical inhibitors, the concentration of disinfectant used and the contact time (Hall-Stoodley et al. 2004). The prevention of biofilm formation can be mainly approached done by cleaning and regularly disinfecting before bacteria attach firmly to surfaces (Dufour et al. 2010). Cleaning agents help to dislodge and detach the microorganisms from the surface while disinfectants kill the remaining microorganisms and prevent further growth.

Organic material can be removed by using surfactants which decrease the surface tension, emulsifies fats, and denatures proteins. Cleaning agents can remove a maximum of 90% of all the microorganisms and need to be supplemented with biocides to prevent regrowth. In fact, cleaning agents break up the EPS matrix that provides mechanical support and protection to bacteria in biofilm and allows biocides to reach the microorganisms. Also, cleaning procedures can often consume much time and hence, and are associated with equipment downtime and subsequent economic losses. Lastly, chemicals used for cleaning are harsh, and they can also have adverse impacts on the mechanical properties of sensitive and fragile surfaces like membranes.

13.4.3 Microbiology Based Approach to Mitigate Fouling

Quorum quenching will be good criteria for biofouling mitigation as it reverses quorum sensing (Kalia 2014a). Quorum quenching is a term that refers to degradation of quorum sensing signals (Dong et al. 2007). In microfiltration membranes, a quorum quenching enzyme named acylase was immobilized onto it, and flow cell experimental results showed prohibited mushroom shaped biofilm formation and reduced EPS (Kim et al. 2011). Another innovative strategy for prevention of the formation of biofilms is to introduce exogenous microbial species that can outcompete the native species by utilizing the given energy sources in a better way. According to studies microbial molecules, such as lauricidin, nisin, reuterin and pediocin, which are commonly used as biopreservatives are very efficient in controlling the formation of biofilms formed by microorganisms which are commonly found in dairy processing facilities, including *Listeria monocytogenes* (Dufour et al. 2010). Biofilms can also be destroyed in a highly specific and non-toxic way by introducing bacteriophages, viruses that infect bacteria. Studies by Hughes, Sutherland, and Jones (1998) have shown that biofilms formed by *Enterobacter agglomerans*can be disrupted by the use of phages that can effectively lyse the cells. The phages also produced polysaccharide degrading enzymes that digested the EPS and caused biofilm slough off.

13.5 Anti-biofouling Strategies

We propose three distinct strategies (biological, physical and chemical) in theory, and bolstered the usefulness of our proposed strategies with at least one experimental evidence from literature for each of the strategy.

13.5.1 Strategy from Material Physics – Bacteria Resistant Material

Anti-microbial coatings on surfaces would prevent bacterial biofilm formation and biofouling in devices (Reid 1999). There are several materials which were able to resist bacteria because of their physical and toxic nature. This quality could be strategized to prevent biofouling.

13.5.1.1 Metals

The metallic surfaces in the industry which are biofouled can be replaced by bactericidal metals. E.g., Copper is capable of killing a wide variety of microbes including. However, their mechanism of action is quite unknown. It is also worthy to note that bacterium is not only capable of developing a resistive nature against biological and chemical agents but also on physical surfaces. It has been shown that bacterial species slowly develop a resistive trait only against copper surface trying to grow in them; nevertheless, these do not develop resistance to other antibiotics, since it will be too expensive for them to produce resistance (Santo et al. 2010).

13.5.2 Chemical Methods

Using chemicals are a long term conventional strategies in preventing biofouling (Bereschenko et al. 2011) It will be interesting to look upon how these chemicals inhibit biofouling (Mechanism of action), apart from just knowing that they are capable of preventing biofilm formation or growth. *P. aeruginosa* forms biofilm through quorum sensing which is mediated by a signaling molecule called *N-acyl* homoserine lactone. Halogenated furanone compounds of algal origin (Manefield et al. 2001) inhibit quorum sensing by penetrating to the microcolonies. There is a double advantage for using this halogenated furanone because it kills bacteria and disrupts biofilm architecture by bacterial detachments (Hentzer et al. 2002). Thus, the effect of biofouling a surface could be reversed. In summary, halogenated furanones are capacitive of penetrating microcolonies, disturbs cell signals, kills bacteria, and removes the biofilm from surface forming multiple modes of strategies against biofouling.

Halogens would be a key component to mitigate biofouling, Ca^{+2} intensifies biofilm strength; the addition of fluoride will inhibit calcium-mediated cell association, thus stops biofouling (Rose and Turner 1998). Supplementation of biocides/antimicrobial substances in the feed water is a most common way to curb the menace of biofouling. The control of biofouling using biocides depends on several parameters: concentration of antimicrobial agents, type of bacterial populations, type of bioac-

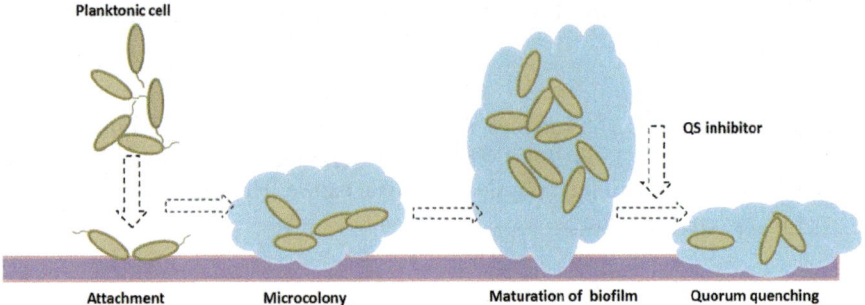

Fig. 13.2 Schematics of biofilm formation and dispersal using QS inhibitor

tivity in the system, solution pH type of dosing (continuous or shock dosing), residence time, etc. Chlorine is the most popular choice for disinfection of contaminated water. It is used both in the form of a gas or hypochlorite as a powder. Chlorine dioxide is also used as a potent antimicrobial substance due to its high bactericidal activity and generation of relatively harmless by-product. Ozone is utilized as a bactericidal agent due to its strong oxidizing ability wastewater treatment industry. It can even kill endospore of bacteria. Peracetic acid, iodine and hydrogen peroxide are other oxidizing chemicals used to mitigate biofouling in industries. The choice of materials depends on health safety, material cost, generation of assimilable organic carbon (AOC) as by-product, etc. The quaternary ammonium compounds formaldehyde, glutaraldehyde is few non-oxidizing antibacterial agents used to avoid biofouling problem.

The QS is a system with which bacteria can use to organize their communal behavior during the development of biofilm (Kalia et al. 2014). The critical phenomena like the development of a biofilm, movement of flagella and pili, swarming, excretion of EPS can be controlled with QS system (Kalia et al. 2017). Disturbing the QS signal was found as an effective approach to preventing biofouling. Chemical substances like 2(5-H) furanone and vanillin (4-hydroxy-3-methoxybenzaldehyde) were applied successfully to inhibit biofouling on membrane surfaces used in water treatment plant (Kalia and Purohit 2011). Nitric oxide (NO) was found efficacious in biofilm dispersal of sessile bacteria (Kalia 2013). However, direct supplementation of NO is restricted in aqueous condition as NO don not dissolve in water and it gets oxidized easily. Researchers used a different type of NO donors like sodium nitrite, diazeniumdiolate, S-nitroso-N-acetylpenicillamine, sodium nitroprusside etc. to disperse biofilm (Kalia 2014b) (Fig. 13.2).

UV irradiation and application of electric field are the two efficient physical processes to disinfect biofouled water. UV irradiation is capable of inactivating and destroying bacteria including endospore. It generates harmful hydroxyl radicals which facilitate in killing bacteria. Bacterial DNA can be fragmented using irradiation at 254 nm which prohibits further bacterial growth. Nevertheless, expensive system and the problem associated with optimizing the magnitude of dosages limit its havoc usage.

Table 13.2 Anti-biofouling strategies and mode of action

S. no	Strategy	Mode of action	Reference
1.	Physics		
	Copper materials	Enzyme inhibition	Agarwala et al. (2014)
	Silver materials	Enzyme inhibition, intracellular reactive oxygen species generation	Bose and Chatterjee (2015)
	Carbon nanomaterials (graphene, carbon nanotube)	Intracellular reactive oxygen species generation	Ji et al. (2016)
2.	Chemical strategy		
	Halogenated furanone	Multiple: Penetrate microcolonies, disturbs signals, kills bacteria, and removes biofilm	Cheng et al. (2016)
	Halogens	Inhibit bacterial protein and cell-cell adhesion	Rose and Turner (1998)
	Quaternary ammonium compound	Disruption of bacterial cell membrane,	
3.	Biological strategies	Quorum quenching	Kalia (2013)
	IgG	Antibody (Immunelysis)	Hedegaard et al. (2016)
	Protozoan	Predation	

13.5.3 Biological Methods

Further, it was proposed that natural antibiotics, natural enemies and their characteristics that resist biofilm would be a better strategy for controlling biofouling, as their role is quite evident in nature. For example, Immunoglobulin G is a protein secreted by the human immune system. This is a glycosylated protein dimer secreted in mucous, found in urogenital tracts, sweat, saliva also in colostrum (mother's milk) and plays a vital role in preventing the accumulation of bacteria (Underdown and Schiff 1986). These IgGs are present in body systems that are more prone to infection and biofilm formation, like urinary tracts (Stamm 1991); gastrointestinal tracts and respiratory airways. Since this immunoglobulin is an external secretion, they should be capable of being efficient *in vitro*.

Protozoans are capable of finding bacterial hot spots and graze over them. Therefore, protozoans would act as key control factors to mitigate biofouling. However, the mode of action of protozoan grazing on biofilm is quite unknown. The bacterial microcolonies can prevent grazing at initial stages (Weitere et al. 2005) and is also capable of developing a resistive nature against protozoan grazing (Hentzer et al. 2002). On the contrary, protozoans are capable of forming a symbiotic relationship with biofilm, by transporting and stabilizing solute flux, which would in turn increase biofouling (Arndt et al. 2003). If protozoan is to be used as a strategy to mitigate biofouling then a careful preliminary experiment is required (Table 13.2).

Nevertheless, prevention of biofilm formation is a more rational option over its treatment. One should take into consideration that, the current successful prevention or control strategies come with adverse side effects. In industrial processes, one of the control methods of biofouling can be the use of bio-dispersants. However, in less controlled environments, the biofilms are treated by coatings of biocides, heat treatments or pulses of energy. Nontoxic mechanical approaches that avoid organisms from attaching take account of choosing a material or coating with a smooth surface, making an ultra-low fouling surface with the usage of zwitterions, or establishment of nanoscale surface topologies alike to the skin of sharks and dolphins which only offer poor anchor points.

Cleaning and disinfecting the surface regularly before bacterial attachment to surfaces is an effective strategy for the prevention of biofilm formation (Gilbert et al. 1997). Biofilm detectors monitor the surface colonization by bacteria and allow the control of biofilms in the early stages of development which makes it easy to control the formation of biofilms. More preventive actions were developed to select materials which facilitate in preventing biofilm formation. Reducing the availability of the carbon source is another strategy suppress biofilm development, nevertheless, inhibiting biofilm in aqueous condition is virtually impossible (Khardori and Yassien 1995). One method involves in blocking the supply of growth factors to the microorganisms so that adhesion to the surface is no longer beneficial for the bacterial population. In the past, several research projects have focused on strategies to avert biofilm development by the introducing of antimicrobial agents into surface materials by modifying the surfaces physicochemical properties or by coating surfaces with antimicrobial agents like polyethylene glycol or hydrogel etc. (Gilbert et al. 1997). Studies have shown that a reduction in infection rate is observed using silicone rubber implants with covalently coupled quaternary ammonium coatings (Gottenbos et al. 2002). Nonionic and strong anionic surfactants were introduced to inhibit the initial attachment of *P. aeruginosa* to stainless steel and glass surfaces. The surfactants demonstrated more than 90% inhibition of adhesion.

13.5.4 Characterization Techniques for Biofouling Study

The on-line image capture and digital image analysis by using microscopy make the system popular for biofilm study in recent past. The epifluorescence microscopy (EFM), confocal laser scanning microscopy (CLSM) and electron microscopy are commonly used microscopy to study morphology, the surface roughness of bio-fouled materials.

Scanning transmission X-ray microscopy (STXM) can be used for examining hydrated biofilms due to the ability of soft X-rays to penetrate water. CLSM helps to find the 3-dimensional structure of biofouled substances; image analysis provides detail parameters related to biofilm. Prior to experiment in CLSM, different staining dye can be used for detection of viable, non-viable bacteria and EPS. Recently, researchers used two noninvasive technology for study related to bacterial adherence and biofilm formation experiments. The quartz crystal microbalance (QCM) is a nanogram sensitive technique that utilizes acoustic waves generated by oscillating a piezoelectric, single crystal quartz plate to measure bacterial mass. QCM with dissipation monitoring (QCM-D) helps us to understand the viscoelastic properties of biomass adhered on the quartz plate or sensor. Optical coherence tomography (OCT) is other industrial nondestructive testing (NDT) imaging technique that uses coherent light to capture micrometer-resolution, two- and three-dimensional images from within optical scattering biofilm. The technology is based on low-coherence interferometry, typically employing near-infrared light.

Different common microbiological techniques like contact killing method, heterotrophic plate counts, cell counts, deposited biomass measurement can be done to evaluate the extent of biofouling. Biochemical assays such as total carbohydrate, protein, Total organic carbon, adenosine triphosphate, EPS and can be determined to measure active biomass responsible for biofouling. In this respect, installation of efficient monitoring systems is imperative to develop and optimize proper implementation of any anti-biofouling strategies.

13.6 Disadvantages of Biofouling

Biofilms can pose threats in some food industry sectors like beverage, poultry, fish, and food processing units. Mostly, the source of problems regarding contamination of dairy products is biofilm-related. In most cases, biofilms are harmful to both economy and ecology. They get attached to object surfaces, as a consequence, the corrosion is increased, and loss in performance is observed. One of its most common drawbacks is decreasing the efficiency of propulsion of ships. Thus the fouling of the underbody of flying boats may result in their inability to get off the water. The growths may interfere with the mechanisms which actuate mines and reduce the efficiency of underwater acoustic devices. They also cause problems in paper mills, since open systems are used in them, which provide favorable conditions for microbial growth. The formation of biofilms not only causes economic losses but also it leads to the deterioration of raw materials, it causes the breakdowns and lowers the quality of products, it also causes problems in wastewater treatment. The operating costs are increased because of the accumulation of microbes on the surfaces of industrial processing equipment (Baker and Dudley 1998).

13.7 Biofouling in Industrial Applications

13.7.1 Biofouling in Water Treatment Industry

Biofouling can cause severe financial losses and environmental damages hence they are undesirable for many applications. In the marine environment, biofouling affects the underwater cables, offshore structures, platforms, seawater cooling systems, sonar devices, etc. Hence they contribute to the reduced speed of ships, increased cost due to more fuel consumption, corrosion, environmental concerns and safety hazards. The ship engine also has to undergo increased stress from the extra drag. Not only marine biofouling, but industrial biofouling also is a major problem which is encountered in a broad spectrum of technical systems from the food industry, water purification, pharmaceuticals to nuclear power plants. Biofilms not only tend to increase the energy needs and the friction between two surfaces, but they also decrease the heat-transfer efficiency. Biofilms can also be a home to dangerous pathogenic microbes in potable water supplies. Most water supplies, even stainless steel pipes with ultrapure water, are susceptible to biofilms (Block et al. 1993).

Biofouling also negatively affects membrane systems by decreasing the membrane flux. The differential and feed pressure is also increased due to biofilm resistance, which in turn leads to increased energy consumption. Hence, the costs are increased which leads to high financial losses (Iorhemen et al. 2016). The biofilms also produce acidic by-products which cause membrane biodegradation. Since biofilms are mostly a threat to the industries which use water, the most common method for anti-fouling is to feed water continuously mixed with antimicrobial substances. Chlorine is the most widely used disinfectant for this purpose. Oxidizing biocides for disinfecting water include ozone, which is very effective for deactivating bacteria, viruses, protozoa and endospore. However, it can produce carcinogenic agents in treated water. The non-oxidizing biocides include glutaraldehyde and formaldehyde, though the long-term use of these biocides may cause the acclimation of the microbes to be resistant to them. UV radiation can also be used for the disinfection of water, but it is relatively more expensive than its other counterparts. Also, it is tough to optimize the dosage at a large scale. Another method to deal with biofouling is by limiting the nutrients available to the microbes. Phosphorus is one of the major requirements for the growth of these surface microbes; its limitation may restrict the microbial growth. Hence, methods such as electrochemical coagulation and usage of adsorbents (like blast furnace slag and dolomite) are used for phosphate removal (Nguyen et al. 2012).

13.7.2   Biofouling in Food Industry

Biofouling is prevalent in food and dairy industries and often leads to contamination because of the microbes that the biofilms harbor on their surface. It causes food spoilage and poses a threat to workers' health. The biofilm growth in food processing unit enhances the possibility of microbial contamination. Owing to the production of EPS, microorganisms are safe from sanitizer and cleaning. Surface tension value of the substratum also plays a major role in bacterial attachment and biofilm formation. The *Listeria momocytogens* was predominantly found in most food processing unit particularly in the dairy industry. The fouling can be controlled by using proper equipment design, controlled temperature, and diminution in nutrient and moisture level. One of the potent ways to reduce fouling or contamination is the efficient cleansing method at potential growth site (Poulsen 1999). A blend of alkali compounds in combination with chelators/sequestrants and anionic wetting agents are commonly utilized as cleaning agents. Acids, halogens, peroxygens and quaternary ammonium compounds are generally used as major components of sanitizers. The quaternary ammonium compounds as cationic surfactants are found effective against a wide range of microorganisms. Hydrogen peroxide generator is also used as sanitizer; peroxide is efficient sanitizer against *L monocytogenes* and salmonella species (Chmielewski and Frank 2003).

13.7.3 Biofouling in Medical Devices

Bacterial contamination or the biofouling or is a serious concern in biomedical instruments. Biofilms are formed by either a single species microorganism or a variety of them; this depends on the type of surface and the amount of time taken for the formation. Biofilm forming bacterial contamination affect the medical devices like the urinary catheter, central venous catheter etc. The organisms that cause these biofilms can be either gram positive or gram negative. The *Staphylococcus aureus*; *Staphylococcus epidermis, Klebsiella pneumoniae* are common biofilm formers in the indwelling central venous catheter. The origin of these bacteria is mostly from patient's own skin microflora or exogenous microflora from medico persons. SEM and TEM images suggested that these microbes are universally present in catheter while inserting them for the purpose of administration of fluid, medications, nutritional substances and hemodynamic monitoring. Migration of bacteria occurs from skin to exterior part of the catheter; colony formation may take 3 days to 5 days from the initial attachment (Percival et al. 2015). Long-term utilization of catheter causes havoc development of biofilm inside the inner lumen of the catheter. To control biofilm growth, researchers applied different antimicrobial agents particularly antibiotics solution to investigate their potency to mitigate contamination. Different approaches like catheter impregnated with cephalosporin bonded cationic surfactant or a blend of minocycline and rifampicin were found effective in removing in eliminating biofilm development (Gilbert et al. 1997).

Prosthetic heart valves are vulnerable to microbial colonization. Microbes like diptheroids, *S. aureus*; streptococcus, enterococci, *S epidermis* etc. were found mostly at the suture site of the prosthetic valve. Silicon and latex materials are utilized for urinary catheters; in open system urinary catheter gets contaminated within 4 days, however, in a closed system, the catheter is not as much susceptible to urinary tract infection (UTI). *Enterococcus faecal, Proteus mirabilis, E. coli* are predominantly detected from the contaminated site. Urease produced by some of these bacteria generates ammonium hydroxide which eventually increases solution of pH to alkaline sight. As a result, minerals in the form of struvite and hydroxyapatite deposited, the forming of encrustation sometimes completely blocked the inner lumen of the catheter and causes havoc infection (Deva et al. 2013). Impregnations of the catheter with silver oxide, bladder irrigation were some strategies found efficient in inhibiting microbial colony formation. Intrauterine devices (IUD) are also prone to microbial contamination. The tail of IUD are often attached with *Lactobacillus, S. aureus; S epidermis*, different species of *Enterobacter, Corynebacterium*, other anaerobes.

Materials for both soft and hard eye contact lens are prone to microbial colonization. Species of *Serratia, Proteus, P. aeruginosa E. coli, Candida* etc. are often responsible for bacterial microcolony development on the lens. Operational parameters like water content on the surface of lens, electrolyte concentration, the polymer composition of lens, hydrophobicity and surface charge of lens surface affect biofilm development (Donlan 2001).

13.8 Advantages of Biofouling

Having seen enough about the disadvantages of biofouling or biofilms, it is best to know some of their advantages. Of course, this relies on the basic principle that bacterium is both harmful and beneficial. Biofouling plays a vital role in bioremediation especially biofilm formed from genetically engineered bacteria. The chemotaxic ability of bacteria aids the growth of biofilm on non-degradable recalcitrant compounds (like hydrocarbons) and they cause bioremediation by immobilizing these compounds (Singh et al. 2006). That is to say, biofouling on an unwanted substrate is advantageous. The other advantage of biofouling mediated bioremediation is that bacterial cells that are capable of fouling are highly adaptive (de la Fuente-Núñez et al. 2013); destructive to adhering materials, and protected within EPS (Decho 2000).

13.9 Conclusion

Biofouling is a complex process which involves a wide diversity of microorganisms and compounds. Consequently, to prevent biofouling, both chemical and physical strategies should be used. Since biofilms are formed by diverse microorganisms which are structurally complex, little information about bacterial attachment, biofilm development and structure, and the physiology of bacteria in biofilms. Further research must be carried out to learn about the effect of antimicrobial products on microbial biofilms and their recovery responses to damage because microorganisms can not only develop resistance, they can successively survive control procedures which were previously effective. In the coming years, our knowledge of biofilms will expand faster because of the development of new technologies like microsensors and CSLM,OCT, QCM etc. Progress in research and development may lead to the discovery of new strategies to control biofouling, as well as we'll get more insight into how biofilms may be controlled and employed for biotechnological applications.

References

Agarwala M, Choudhury B, Yadav RNS (2014) Comparative study of antibiofilm activity of copper oxide and iron oxide nanoparticles against multidrug resistant biofilm forming uropathogens. Indian J Microbiol 54:365–368. https://doi.org/10.1007/s12088-014-0462-z

Arndt H, Schmidt-Denter K, Auer B, Weitere M (2003) Protozoans and biofilms. In: Krumbein WE, Paterson DM, Zavarzin GA (eds) Fossil and recent biofilms. Springer, Dordrecht, pp 161–179. https://doi.org/10.1007/978-94-017-0193-8_10

Aykent F, Yondem I, Ozyesil AG, Gunal SK, Avunduk MC, Ozkan S (2010) Effect of different finishing techniques for restorative materials on surface roughness and bacterial adhesion. J Prosthet Dent 103:221–227. https://doi.org/10.1016/S0022-3913(10)60034-0

Baker JS, Dudley LY (1998) Biofouling in membrane systems – a review. Desalination 118:81–89. https://doi.org/10.1016/S0011-9164(98)00091-5

Bereschenko LA, Prummel H, Euverink GJW, Stams AJM, van Loosdrecht MCM (2011) Effect of conventional chemical treatment on the microbial population in a biofouling layer of reverse osmosis systems. Water Res 45:405–416. https://doi.org/10.1016/j.watres.2010.07.058

Biology NRC (US) B. on, Board NRC (US) O.S (2000) Bacterial biofilms and biofouling: Translational research in marine biotechnology. National Academies Press (US), Washington, DC

Block JC, Haudidier K, Paquin JL, Miazga J, Levi Y (1993) Biofilm accumulation in drinking water distribution systems. Biofouling 6:333–343. https://doi.org/10.1080/08927019309386235

Bose D, Chatterjee S (2015) Antibacterial activity of green synthesized silver nanoparticles using vasaka (*Justicia adhatoda* L.) leaf extract. Indian J Microbiol 55:163–167. https://doi.org/10.1007/s12088-015-0512-1

Chaudhury MK, Finlay JA, Chung JY, Callow ME, Callow JA (2005) The influence of elastic modulus and thickness on the release of the soft-fouling green alga *Ulva linza* (syn. Enteromorphalinza) from poly(dimethylsiloxane) (PDMS) model networks. Biofouling 21:41–48. https://doi.org/10.1080/08927010500044377

Cheng Y, Gao B, Liu X, Zhao X, Sun W, Ren H, Wu J (2016) In vivo evaluation of an antibacterial coating containing halogenated furanone compound-loaded poly(l-lactic acid) nanoparticles on microarc-oxidized titanium implants. Int J Nanomedicine 11:1337–1347. https://doi.org/10.2147/IJN.S100763

Chmielewski RAN, Frank JF (2003) Biofilm formation and control in food processing facilities. Compr Rev Food Sci Food Saf 2:22–32. https://doi.org/10.1111/j.1541-4337.2003.tb00012.x

Cortés ME, Consuegra J, Sinisterra RD (2011) Biofilm formation, control and novel strategies for eradication. Sci Microb Pathog Commun Curr Res Technol Adv 2:896–905

Davey ME, O'toole GA (2000) Microbial biofilms: from ecology to molecular genetics. Microbiol Mol Biol Rev 64:847–867

de la Fuente-Núñez C, Reffuveille F, Fernández L, Hancock RE (2013) Bacterial biofilm development as a multicellular adaptation: antibiotic resistance and new therapeutic strategies. Curr Opin Microbiol 16:580–589. https://doi.org/10.1016/j.mib.2013.06.013

Decho AW (2000) Microbial biofilms in intertidal systems: an overview. Cont Shelf Res 20:1257–1273. https://doi.org/10.1016/S0278-4343(00)00022-4

Deva AK, Adams WP, Vickery K (2013) The role of bacterial biofilms in device-associated infection. Plast Reconstr Surg 132:1319–1328. https://doi.org/10.1097/PRS.0b013e3182a3c105

Dong Y-H, Wang L-H, Zhang L-H (2007) Quorum-quenching microbial infections: mechanisms and implications. Philos Trans R Soc B Biol Sci 362:1201–1211. https://doi.org/10.1098/rstb.2007.2045

Donlan RM (2001) Biofilms and device-associated infections. Emerg Infect Dis 7:277–281. https://doi.org/10.3201/eid0702.700277

Dufour D, Leung V, Lévesque CM (2010) Bacterial biofsilm: structure, function, and antimicrobial resistance. Endod Top 22:2–16. https://doi.org/10.1111/j.1601-1546.2012.00277.x

Flemming H-C (1993) Biofilms and environmental protection. Water Sci Technol 27:1–10

Garrett TR, Bhakoo M, Zhang Z (2008) Bacterial adhesion and biofilms on surfaces. Prog Nat Sci 18:1049–1056. https://doi.org/10.1016/j.pnsc.2008.04.001

Gedge M, Voon L, Glynne-Jones P, Mowlem M, Morgan H, Hill M (2012) The use of ultrasonic waves to minimise biofouling in oceanographic microsensors. AIP Conf Proc 1433:765–768. https://doi.org/10.1063/1.3703293

Giaouris E, Chorianopoulos N, Nychas GJE (2005) Effect of temperature, pH, and water activity on biofilm formation by *Salmonella enterica* enteritidis PT4 on stainless steel surfaces as indicated by the bead vortexing method and conductance measurements. J Food Prot 68:2149–2154

Gilbert P, Das J, Foley I (1997) Biofilm susceptibility to antimicrobials. Adv Dent Res 11:160–167. https://doi.org/10.1177/08959374970110010701

Gottenbos B, van der Mei HC, Klatter F, Nieuwenhuis P, Busscher HJ (2002) In vitro and in vivo antimicrobial activity of covalently coupled quaternary ammonium silane coatings on silicone rubber. Biomaterials 23:1417–1423. https://doi.org/10.1016/S0142-9612(01)00263-0

Guezennec J, Ortega-Morales O, Raguenes G, Geesey G (1998) Bacterial colonization of artificial substrate in the vicinity of deep-sea hydrothermal vents. FEMS Microbiol Ecol 26:89–99. https://doi.org/10.1016/S0168-6496(98)00022-1

Gupta P, Sarkar S, Das B, Bhattacharjee S, Tribedi P (2016) Biofilm, pathogenesis and prevention—a journey to break the wall: a review. Arch Microbiol 198:1–15. https://doi.org/10.1007/s00203-015-1148-6

Hall-Stoodley L, Costerton JW, Stoodley P (2004) Bacterial biofilms: from the natural environment to infectious diseases. Nat Rev Microbiol 2:95–108. https://doi.org/10.1038/nrmicro821

Hedegaard CJ, Strube ML, Hansen MB, Lindved BK, Lihme A, Boye M, Heegaard PMH (2016) Natural pig plasma immunoglobulins have anti-bacterial effects: potential for use as feed supplement for treatment of intestinal infections in pigs. PLoS One 11:e0147373. https://doi.org/10.1371/journal.pone.0147373

Hentzer M, Riedel K, Rasmussen TB, Heydorn A, Andersen JB, Parsek MR, Rice SA, Eberl L, Molin S, Høiby N, Kjelleberg S, Givskov M (2002) Inhibition of quorum sensing in pseudomonas aeruginosa biofilm bacteria by a halogenated furanone compound. Microbiology 148:87–102. https://doi.org/10.1099/00221287-148-1-87

Hong S, Elimelech M (1997) Chemical and physical aspects of natural organic matter (NOM) fouling of nanofiltration membranes. J Membr Sci 132:159–181. https://doi.org/10.1016/S0376-7388(97)00060-4

Iorhemen OT, Hamza RA, Tay JH (2016) Membrane bioreactor (MBR) technology for wastewater treatment and reclamation: membrane fouling. Membranes 6:33. https://doi.org/10.3390/membranes6020033

Ji H, Sun H, Qu X (2016) Antibacterial applications of graphene-based nanomaterials: recent achievements and challenges. Adv Drug Deliv Rev 105:176–189. https://doi.org/10.1016/j.addr.2016.04.009

Johnson LR (2008) Microcolony and biofilm formation as a survival strategy for bacteria. J Theor Biol 251:24–34. https://doi.org/10.1016/j.jtbi.2007.10.039

Kalia VC (2013) Quorum sensing inhibitors: an overview. Biotechnol Adv 31:224–245. https://doi.org/10.1016/j.biotechadv.2012.10.004

Kalia VC (2014a) Microbes, antimicrobials and resistance: the battle goes on. Indian J Microbiol 54:1–2. https://doi.org/10.1007/s12088-013-0443-7

Kalia VC (2014b) In search of versatile organisms for quorum-sensing inhibitors: acyl homoserine lactones (AHL)-acylase and AHL-lactonase. FEMS Microbiol Lett 359:143–143. https://doi.org/10.1111/1574-6968.12585

Kalia VC, Prakash J, Koul S, Ray S (2017) Simple and rapid method for detecting biofilm forming bacteria. Indian J Microbiol 57:109–111. https://doi.org/10.1007/s12088-016-0616-2

Kalia VC, Purohit HJ (2011) Quenching the quorum sensing system: potential antibacterial drug targets. Crit Rev Microbiol 37:121–140. https://doi.org/10.3109/1040841X.2010.532479

Kalia VC, Wood TK, Kumar P (2014) Evolution of resistance to quorum-sensing inhibitors. Microb Ecol 68:13–23. https://doi.org/10.1007/s00248-013-0316-y

Kerr A, Beveridge CM, Cowling MJ, Hodgkiess T, Parr ACS, Smith MJ (1999) Some physical factors affecting the accumulation of biofouling. J Mar Biol Assoc UK 79:357–359. https://doi. org/10.1017/S002.531549800040X

Khardori N, Yassien M (1995) Biofilms in device-related infections. J Ind Microbiol 15:141–147. https://doi.org/10.1007/978-3-540-68119-9

Kim IS, Jang N (2006) The effect of calcium on the membrane biofouling in the membrane bioreactor (MBR). Water Res 40:2756–2764. https://doi.org/10.1016/j.watres.2006.03.036

Kim J-H, Choi D-C, Yeon K-M, Kim S-R, Lee C-H (2011) Enzyme-immobilized nanofiltration membrane to mitigate biofouling based on quorum quenching. Environ Sci Technol 45:1601–1607. https://doi.org/10.1021/es103483j

Kim W, Tengra FK, Young Z, Shong J, Marchand N, Chan HK, Pangule RC, Parra M, Dordick JS, Plawsky JL, Collins CH (2013) Spaceflight promotes biofilm formation by pseudomonas aeruginosa. PLoS One 8:e62437. https://doi.org/10.1371/journal.pone.0062437

Klausen M, Heydorn A, Ragas P, Lambertsen L, Aaes-Jørgensen A, Molin S, Tolker-Nielsen T (2003) Biofilm formation by *Pseudomonas aeruginosa* wild type, flagella and type IV pili mutants. Mol Microbiol 48:1511–1524. https://doi.org/10.1046/j.1365-2958.2003.03525.x

Körstgens V, Flemming H-C, Wingender J, Borchard W (2001) Influence of calcium ions on the mechanical properties of a model biofilm of mucoid *Pseudomonas aeruginosa*. Water Sci Technol 43:49–57

Lewis AL (2000) Phosphorylcholine-based polymers and their use in the prevention of biofouling. Colloids Surf B Biointerfaces 18:261–275. https://doi.org/10.1016/S0927-7765(99)00152-6

Li Y-H (2009) Quorum sensing and signal transduction in biofilms: the impacts of bacterial social behavior on biofilm ecology. In: Schlesinger LS, Jaykus L-A, Wang HH (eds) Food-borne microbes. American Society of Microbiology, Washington, DC, pp 117–133. https://doi. org/10.3390/s120302519

Li Y-H, Tian X (2012) Quorum sensing and bacterial social interactions in biofilms. Sensors 12:2519–2538. https://doi.org/10.3390/s120302519

Lim Y, Jana M, Luong TT, Lee CY (2004) Control of glucose- and NaCl-induced biofilm formation by rbf in staphylococcus aureus. J Bacteriol 186:722–729. https://doi.org/10.1128/JB.186.3.722-729.2004

Madsen JS, Burmølle M, Hansen LH, Sørensen SJ (2012) The interconnection between biofilm formation and horizontal gene transfer. FEMS Immunol Med Microbiol 65:183–195. https:// doi.org/10.1111/j.1574-695X.2012.00960.x

Manefield M, Welch M, Givskov M, Salmond GP, Kjelleberg S (2001) Halogenated furanones from the red alga, *Delisea pulchra*, inhibit carbapenem antibiotic synthesis and exoenzyme virulence factor production in the phytopathogen *Erwinia carotovora*. FEMS Microbiol Lett 205:131–138. https://doi.org/10.1111/j.1574-6968.2001.tb10936.x

Marion-Ferey K, Pasmore M, Stoodley P, Wilson S, Husson GP, Costerton JW (2003) Biofilm removal from silicone tubing: an assessment of the efficacy of dialysis machine decontamination procedures using an in vitro model. J Hosp Infect 53:64–71. https://doi.org/10.1053/jhin.2002.1320

Mi B, Elimelech M (2008) Chemical and physical aspects of organic fouling of forward osmosis membranes. J Membr Sci 320:292–302. https://doi.org/10.1016/j.memsci.2008.04.036

Nguyen T, Roddick FA, Fan L (2012) Biofouling of water treatment membranes: a review of the underlying causes, monitoring techniques and control measures. Membranes 2:804–840. https://doi.org/10.3390/membranes2040804

Patil SA, Harnisch F, Koch C, Hübschmann T, Fetzer I, Carmona-Martínez AA, Müller S, Schröder U (2011) Electroactive mixed culture derived biofilms in microbial bioelectrochemical systems: the role of pH on biofilm formation, performance and composition. Bioresour Technol 102:9683–9690. https://doi.org/10.1016/j.biortech.2011.07.087

Percival SL, Suleman L, Vuotto C, Donelli G (2015) Healthcare-associated infections, medical devices and biofilms: risk, tolerance and control. J Med Microbiol 64:323–334. https://doi. org/10.1099/jmm.0.000032

Picioreanu C, Kreft J-U, Klausen M, Haagensen JAJ, Tolker-Nielsen T, Molin S (2007) Microbial motility involvement in biofilm structure formation – a 3D modelling study. Water Sci Technol 55:337. https://doi.org/10.2166/wst.2007.275

Poulsen LV (1999) Microbial biofilm in food processing. LWT Food Sci Technol 32:321–326. https://doi.org/10.1006/fstl.1999.0561

Pratt LA, Kolter R (1999) Genetic analyses of bacterial biofilm formation. Curr Opin Microbiol 2:598–603. https://doi.org/10.1016/S1369-5274(99)00028-4

Reid G (1999) Biofilms in infectious disease and on medical devices. Int J Antimicrob Agents 11:223–226. https://doi.org/10.1016/S0924-8579(99)00020-5

Rose RK, Turner SJ (1998) Extracellular volume in streptococcal model biofilms: effects of pH, calcium and fluoride. Biochim Biophys Acta Gen Subj 1379:185–190. https://doi.org/10.1016/S0304-4165(97)00098-6

Santo CE, Morais PV, Grass G (2010) Isolation and characterization of bacteria resistant to metallic copper surfaces. Appl Environ Microbiol 76:1341–1348. https://doi.org/10.1128/AEM.01952-09

Singh R, Paul D, Jain RK (2006) Biofilms: implications in bioremediation. Trends Microbiol 14:389–397. https://doi.org/10.1016/j.tim.2006.07.001

Speranza B, Corbo MR, Sinigaglia M (2011) Effects of nutritional and environmental conditions on *Salmonella* sp. biofilm formation. J Food Sci 76:M12–M16. https://doi.org/10.1111/j.1750-3841.2010.01936.x

Stamm WE (1991) Catheter-associated urinary tract infections: epidemiology, pathogenesis, and prevention. Am J Med Proc Third Decennial Int Conf Nosocomial Infect 91:S65–S71. https://doi.org/10.1016/0002-9343(91)90345-X

Stokke R, Dahle H, Roalkvam I, Wissuwa J, Daae FL, Tooming-Klunderud A, Thorseth IH, Pedersen RB, Steen IH (2015) Functional interactions among filamentous Epsilonproteobacteria and bacteroidetes in a deep-sea hydrothermal vent biofilm. Environ Microbiol 17:4063–4077. https://doi.org/10.1111/1462-2920.12970

Stoodley P, Dodds I, Boyle JD, Lappin-Scott HM (1998) Influence of hydrodynamics and nutrients on biofilm structure. J Appl Microbiol 85(Suppl 1):19S–28S. https://doi.org/10.1111/j.1365-2672.1998.tb05279.x

Sutherland IW (2001) Biofilm exopolysaccharides: a strong and sticky framework. Microbiology 147:3–9. https://doi.org/10.1099/00221287-147-1-3

Tang H, Cao T, Liang X, Wang A, Salley SO, McAllister J, Ng KYS (2009) Influence of silicone surface roughness and hydrophobicity on adhesion and colonization of Staphylococcus epidermidis. J Biomed Mater Res A 88:454–463. https://doi.org/10.1002/jbm.a.31788

Underdown BJ, Schiff JM (1986) Immunoglobulin A: strategic defense initiative at the mucosal surface. Annu Rev Immunol 4:389–417. https://doi.org/10.1146/annurev.iy.04.040186.002133

Walt DR, Smulow JB, Turesky SS, Hill RG (1985) The effect of gravity on initial microbial adhesion. J Colloid Interface Sci 107:334–336. https://doi.org/10.1016/0021-9797(85)90185-7

Weitere M, Bergfeld T, Rice SA, Matz C, Kjelleberg S (2005) Grazing resistance of Pseudomonas aeruginosa biofilms depends on type of protective mechanism, developmental stage and protozoan feeding mode. Environ Microbiol 7:1593–1601. https://doi.org/10.1111/j.1462-2920.2005.00851.x

Yang H-L, Lin JC-T, Huang C (2009) Application of nanosilver surface modification to RO membrane and spacer for mitigating biofouling in seawater desalination. Water Res 43:3777–3786. https://doi.org/10.1016/j.watres.2009.06.002

Chapter 14
Technological Developments in Quorum Sensing and Its Inhibition for Medical Applications

Swapnil C. Kamble and Santoshkumar N. Patil

Abstract Recent years have seen great advances in utilizing quorum sensing (QS) and its inhibition, including application of X-ray crystallography, and genetic engineering to improve and develop an array of analytical tools for medical applications. Through utilization of X-ray crystallography to computational approaches, our understanding of QS and quorum sensing inhibition (QSIn) has grown by leaps. Bacterial circuitries are usually characterized by high specificity, selectivity and sensitivity, and thus suggest applications leading to low cost, miniaturization of circuitry for portable equipment and potential multiplexing. This review is intended to give its reader an overview of technologies based on QS and QSIn for medical use. The sections: techniques and technology platforms/protocols towards discovery and development of quorum sensing inhibitor (QSI), the advancement in technology towards theragnostic applications and technology enabled development of QS and QSIn as an anti-virulence strategy are aimed to provide current research insights. Specifically, we examine screening of pathogens using aptamers, biofilm prevention, vaccine development and treatment of infections and cancer.

Keywords Quorum sensing inhibitors · Informatics · Aptamers · Vaccines · Implants · Nanotechnology · Bacteriocin · Cancer therapy · Theragnostics

S. C. Kamble
Department of Technology, Savitribai Phule Pune University, Pune, Maharashtra, India

S. N. Patil (✉)
Sai Life Sciences Ltd, Pune, Maharashtra, India
e-mail: santosh.p@sailife.com

© Springer Nature Singapore Pte Ltd. 2018
V. C. Kalia (ed.), *Biotechnological Applications of Quorum Sensing Inhibitors*,
https://doi.org/10.1007/978-981-10-9026-4_14

Abbreviations

QS	Quorum sensing
QSIn	quorum sensing inhibition
AHL	N –acyl homoserine lactone
SOC	standard of care
*is*CADDIP	*in silico* CADD techniques and informatics protocol
ADME	absorption, distribution, metabolism, and excretion
MEMS	microelectromechanical system
TRAP	target of ribonucleic acid III activating protein
XDR	extensively drug- resistant
3-oxo-C_{12}-HSL	N-3-oxododecanoyl-L-homoserine lactone
VLP	Virus-like particles
MRSA	multiple drug resistant Staphylococcus aureus
C4-HSL	N-butyrylhomoserine lactone
SELEX	Systematic evolution of ligands by exponential enrichment
OM	otitis media
EPS	extracellular polymeric substance
TFP	Type IV pilus
SLC	synchronized lysis circuit
AI	autoinducer
QSI	Quorum sensing inhibitor
CADD	Computer- aided drug design
RIP	RNA III inhibiting peptide
RAP	ribonucleic acid III activating protein
T3SS	type III secretion system
AIP	autoinducing peptides
ALS	amino lactam surrogate
NTHI	non typeable Haemophilus influenzae
IHF	integration host factor
HSL	homoserine lactone
MRSA	methicillin resistant *Staphylococcus aureus*

14.1 Introduction

Microorganism are usually present as mixed population in nature. Intra- and inter-species communication in single-cell organisms become crucial for fundamental functions in this milieu of life. Population density based gene expression was initially described in marine bacteria for bioluminescence. Coordinated activities based on cellular and molecular density as discovered through quorum sensing (QS) has led to our understanding of early lineage of communications. These communications mediated through small signaling molecules called autoinducers (AI), are a complex signaling mechanism. AIs could induce gene expression in cluster of bacteria for the functions that may be otherwise absent in planktonic forms. For example, bioluminescence, biofilm formation, virulence, etc. have been shown to be caused by QS and modulation of AI by bacteria. Hence, an understanding on QS and its functioning in bacterial physiology and pathology underlies the potential to help in development of new systems to tackle it. Natural resistance to QS by bacterial and fungal communities suggest presence of natural QS inhibitors. Like the discovery of QS was contributed with improvement of methods and techniques and eventual application towards development of biomonitor and biosensors, quorum sensing inhibition (QSIn) also evoked technological milestones for biomedical application. The techniques of gene circuit in QS and QSIn potentially connects sensing machinery to different physiological activities. Simple sensor-reporter system thus, become a stepping stone to develop applications. These have found a way for utilization for microbial inhibitions and further has been correlated to various human diseases. In this chapter we classify the technology platforms utilizing QS and QSIn to two different areas: prevention and screening of infection, and disease treatment.

14.1.1 Technology Protocols Towards Quorum Sensing Inhibitor (QSI) Development

QS machinery revolves around production, release into the microenvironment and auto-detections of the signaling molecule that leads to virulence in pathogenic bacteria. Thus, the QS machinery can be inhibited at three stages in case of Gram-negative bacteria (Fig. 14.1).

1. Preventing production of signaling molecule by LasI or LuxI protein (Parsek et al. 1999);
2. Signal scrambling by degradation of signaling molecule (eg. Lactonase, Dong and Zhang 2005) or signal sequestration (Truchado et al. 2015); or
3. Interference of signal reception by blocking of LasR or LuxR protein (competitive or non-competitive antagonist, Hentzer et al. 2002).

Traditionally, medicinal chemist attempted to inhibit the QS circuits based on known natural antifouling agents or inhibitors. For example, halogenated furanone secreted by red marine algae inhibits QS in *P. aeruginosa* (stage 3, Hentzer et al. 2002). Likewise, researchers attempted to screen marine natural products and even food such as garlics (Bjarnsholt et al. 2005), honey (Hussain 2017) and plant based products for QS inhibition activity (Truchado et al. 2015). However, with advancements in understanding of QS phenomenon at molecular level, *de-novo* medicinal chemistry based approaches have evolved that utilizes synthesis of small molecules to prevent virulence. These screening assays involve high-throughput (like 96-well plate format) read out wherein small molecules would decrease light based fluorescence or bioluminescence if it disturbs the QS circuits. Most of them utilize QS

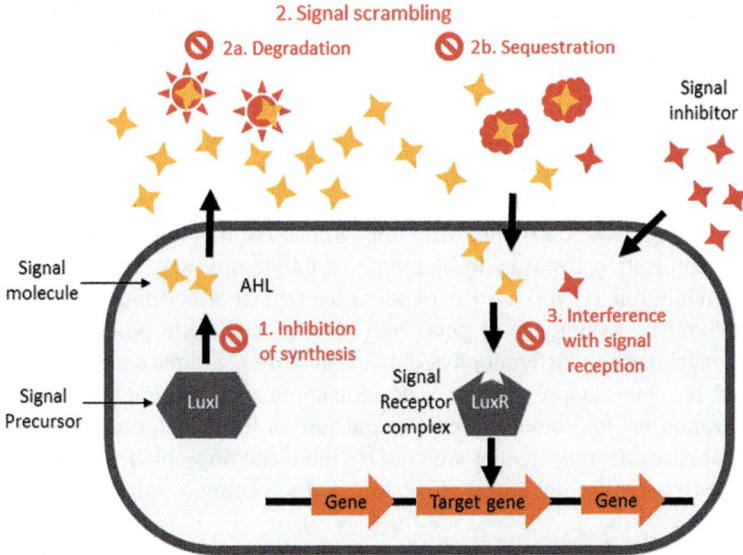

Fig. 14.1 Current quorum sensing inhibition strategies for LuxI/LuxR QS system. The signaling molecule AHL (golden star), produced using LuxI synthase, freely diffuses out of the cell and upon reaching critical concentration diffuses back inside, binds with LuxR and the LuxR-AHL complex triggers QS transcription. The inhibition target stages are *1*. synthesis and release of AHL, *2*. AHL molecule degradation or sequestration, and *3*. interference of AHL mediated LuxR trigger by AHL inhibitor (red star)

reported strains that demonstrate fluorescence in response to bioluminescence in *Vibrio harveyi*. However, the light production (fluorescence or bioluminescence) is an energy expensive endeavor for cell and it involves metabolic reactions. If a test compound also interferes with light producing metabolic machinery of cell, it may lead to false positive results and could be reported as QSI. For example, Pyrogallol has been a reported QSI that prevents *N*-acylhomoserine lactone (AHL) induced bioluminescence in *V. harveyi* without affecting bacterial growth (Ni et al. 2008). Subsequent study revealed the apparent QSI activity of Pyrogallol was due to its toxic side reaction (inhibition of bioluminescence due to peroxide production). Therefore, such assays may mislead and many compounds previously claimed to be QSI could be false positive hits. This apparent urgency for newer methodology was looked through informatics based screening protocols with proper control experiments, to rule out false positive QSI and confirm QSI for medical applications. In this section, we will discuss advanced technologies like X-ray crystallography and informatics based protocols towards finding of QSI.

14.1.2 X-ray Crystallographic Techniques and Informatics Based Protocol

Advanced techniques such as X-ray crystallography in conjunction with traditional cell- based assays could consolidate the QSIn mechanism. X-ray crystallography is an advanced technique wherein researchers are able co-crystallize the inhibitor and protein involved in QS circuits that could gain major insight into binding of QSI and QS protein at molecular level. Some of the recent X-ray co-crystal data is given in Table 14.1.

The binding information can be used towards rational quorum sensing inhibitor development. Towards it, researchers are following computer aided drug discovery (CADD) based tools and informatics protocol (Fig. 14.2). The crystal structure and molecular information of QS protein enabled researchers to design inhibitors using *in silico* approaches. This approach can be used to treat infections caused by fungus and bacteria.

For example, dermatophytes are fungi that require keratin pigment for their growth. Dermatophytes is a common term used for three type of fungi belonging to genera *Microsporum*, *Epidermophyton* and *Trichophyton,* that can cause skin, hair and nail infection in humans to obtain nutrition from keratinized materials. *Trichophyton rubrum* produces squalene epoxidase, an enzyme known to trigger events leading to prolonged dermatophytes infection. The current standard of care (SOC) medicine to treat dermatophytes infection is terbinafine. It can be taken orally or applied to the skin as a cream or ointment. However, nail infections cannot be cured by cream and ointment (Hamilton 2015). The oral administrations of terbinafine, the only choice in certain dermatophytes infection, has been implicated in

Table 14.1 The X-ray crystallographic information of QS macromolecules and QSI

Title	Ligand name	PDB ID
Structure of 5′-methylthionadenosine/ S-Adenosylhomocysteinenucleosidase from *Streptococcus pneumoniae* with a transition-state inhibitor MT-ImmA	(3S,4R)-2-(4-Amino-5h-Pyrrolo[3,2-D] Pyrimidin-7-Yl)-5-[(Methylsulfanyl) Methyl]Pyrrolidine-3,4-Diol	1ZOS
Crystal structure of PqsR co-inducer binding domain of *P. aeruginosa* with inhibitor 3NH2-7Cl-C9QZN	3-amino-7-chloro-2-nonylquinazolin-4(3H)-one	4JVI
QS signal integrator LuxO - catalytic domain in complex with AzaU inhibitor	Acetate ion	5EP2
QS signal integrator LuxO - catalytic domain bound to CV-133 inhibitor	2,2-dimethylpropyl 2-[(3-oxidanylidene-5-sulfanylidene-2~{H}-1,2,4-triazin-6-yl)amino] ethanoate	5EP3
Crystal structure of the complex of the peptidase domain of *Streptococcus mutans* ComA with a small molecule inhibitor	[(1~{S},2~{R},4~{S},5~{R})-5-[5-(4-methoxyphenyl)- 2-methyl-pyrazol-3-yl]-1-azabicyclo[2.2.2]octan- 2-yl] methyl ~{N}-propylcarbamate	5XE9

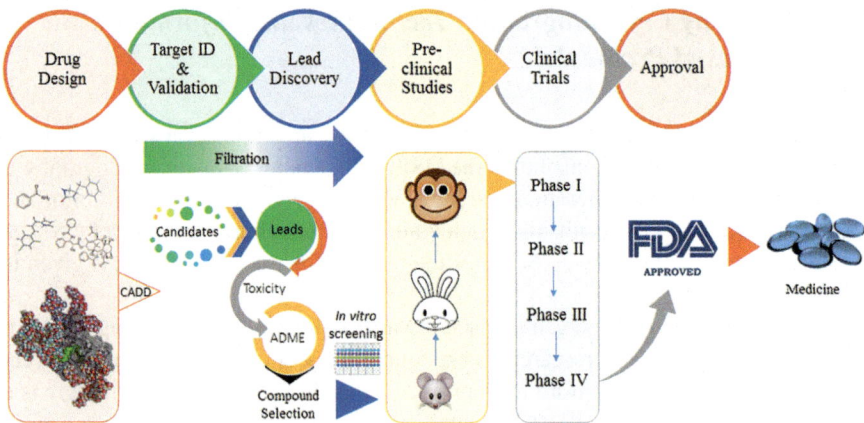

Fig. 14.2 General representation of *is*CADDIP: The informatics and CADD protocol are used to select lead QS inhibitor candidate. CADD tools such as molecular modeling, identify binding interactions of ligand (small molecule) and receptor (a typical macromolecule such as protein and enzymes like LuxR). The virtual screening of various databases such as ZINC, Maybridge, Chembridge, etc. results into identification of hit candidates that are further filtered using various *in silico* toxicity and ADME parameters. The filtration process results in identification of few molecules called lead. The selected lead molecules are then purchased or prepared and screened using *in vitro* assay to confirm the robustness of protocol. These lead molecules are further optimized into QSI with medicinal applications value. This product enters into battery of clinical trials and FDA approvals to become a drug

severe adverse side effects (McGuire 2008). Therefore, these is an urgent need to find better candidate than terbinafine. Towards mitigating side effects and improve pharmacological properties, researchers have attempted to prepare terbinafine analogs. Karumuri et al. developed one such analog of terbinafine for treatment of disease caused by *T. rubrum* (Karumuri et al. 2015). The authors designed analogs and utilized CADD enabled molecular docking approach. The preliminary docking score between terbinafine and squalene epoxidase of −338.75 kcal/mol, suggested tight binding between small molecule inhibitor and squalene epoxidase. These analogs of terbinafine, if pass through stringent FDA and clinical trials could be better alternative of current SOC.

The above mentioned CADD technique can be used to design QSI. The design of QSI involve interfering with QS circuits, one popular target being blockage of LuxR receptor (Fig. 14.1, Stage 3). The LuxR receptor dimerises upon binding of autoinducer homoserine lactone, translocates to nuclei and triggers QS genes (Tsai and Winans 2010). Thus, LuxR is an attractive target towards QS inhibition. The binding of the AHL and LuxR have been well documented. Recently, researchers are coupling CADD technique with informatics driven drug discovery protocol towards finding a drug like QSI. For example, the *in silico* CADD techniques and informatics protocol (*is*CADDIP) (Fig. 14.2) have been used by Rajamanikandan et al. for exploring the selectivity of auto-inducer complex LuxR and discovery of QSI (Rajamanikandan et al. 2017).

Rajamanikandan and Srinivasan reported the mode of binding and molecular interactions at functional amino acid residue level (Rajamanikandan and Srinivasan 2017). The prediction of binding mode was made possible by various computational approaches such as molecular docking, *in silico* mutational studies, molecular dynamics simulations, and free energy calculation. The authors identified amino acids Asn133 and Gln137 residues play an important role in recognizing AI homoserine lactones based on hydrogen bond interactions between protein and carbonyl groups of lactone and amide functionalities of homoserine lactone. The virtual screening parameters such as docking score, binding affinity and mode of interactions with the receptor (LuxR) were analyzed. The best score was obtained with 4-Benzyl-2-pyrrolidinone and N-[2(1-cyclohexen-1-yl) enthyl]-N′(2-ethoxyphenyl). The structures obtained with various *in silico* hits were further screened/filtered through absorption, distribution, metabolism, and excretion (ADME) prediction software to generate lead molecules. The authors proposed that these lead molecules be used as anti-QS drug for medical applications Thus, an informatics centered protocol through a series of systematic CADD based screenings steps, could be used to identify lead molecule.

The same group applied *is*CADDIP protocol to explore selectivity of cinnamaldehyde derivatives complexed with LuxR and did virtual screening based on shape and e-pharmacophore and carried out its biological evaluation (Rajamanikandan et al. 2017). It was reported that cinnamaldehyde derivatives interfere with QS circuit LuxR/LuxI by binding to LuxR. Cinnamaldehyde upon binding to LuxR, decreases the DNA binding ability of LuxR. The authors carried out molecular docking and dynamics simulations to derive parameters such binding mode, binding dynamics and energy. These parameters coupled to *in vitro* screening experiments were used to develop e-pharmacophore model using LuxR-3,4-dichloro-cinnamaldehyde complex (a pharmacophore model is a model that represents abstract description of molecular features that are necessary for molecular recognition between ligand (small molecule) and macromolecule (protein)). Further, the pharmacophore model was used for carrying out virtual screening. The molecules listed in Chembridge database were screened through pharmacophore model for binding affinity for LuxR, resulting into nine hits. These hits when screened through ADMET prediction, prime MM-GBSA calculations and dynamics simulations resulted in a top compound 3-(2,4-dichlorophenyl)-1-(1H-pyrrol-2-yl)-2-propen-1-one ((ChemBridge-7364106), which was then procured and screened for quorum sensing inhibition in *V. harveyi* using *in vitro* bioluminescence assay. In addition to reduction in bioluminescence, ChemBridge-7364106 showed significant biofilm reduction. The author proposes this molecule hold a potential and could be future optimized into a drug. Further, screenings were performed against ChemBridge database to identify suitable ligands for LuxR. Comparisons of these resulted in screenings, 9 best hit molecules, which were further studied for ADMET prediction, dynamics simulations, molecular and Prime MM-GBSA analysis. Among these 9, the top most compound was 3-(2,4-dichlorophenyl)-1-(1H-pyrrol-2-yl)-2-propen-1-one (ChemBridge-7364106). It was selected for *in vitro* assays with *V. harveyi*. It revealed that ChemBridge-7364106 reduced the

Table 14.2 Informatics based approaches towards QS studies

Informatics technique	Quorum sensing organism	Key findings	References
Comparative genome analyses of mycobacteria	*Mycobacterium abscessus*	Specific genes (811 genes) including bacterial QS genes may help bacteria to adapt to harsher environment.	Wee et al. (2017)
Clustering of MS2 spectra done using CluMSID	*P. aeruginosa*	Authors clustered several classes of primary and secondary metabolites and annotated 27 undescribed QS signal molecules. Belonging to canonical classes of alkyl quinolone	Depke et al. (2017)
Network mining utilizing total 110 scientific articles, corresponding to 1004 annotations were analyzed	*P. aeruginosa*	Database containing a knowledge network of potential QSI molecules for *P. aeruginosa* was created using literature mining tool.	Pérez-Pérez et al. (2017)

bioluminescence, whereas ChemBridge-7364106 inhibited biofilm formation and motility, suggesting their usage as QSIs.

Similarly, *is*CADDIP protocol depicted in Fig. 14.2 was followed by Kalia et al. wherein 2603 compounds available with ZINC registry were virtually screened against LasR receptor to obtain hit molecules (Kalia et al. 2017). These hit molecules were further filtered based upon Lipinski rule (rule based upon five physico-chemical properties that predicts drug likeness) and various *in silico* ADME and toxicology filters to obtain six potential QS inhibitors. Likewise, various informatics tools are playing pivotal role in finding medical application of QS inhibitors. The following table summarizes the recent informatics based approaches used towards understanding of QS and could be used for development of potential QS inhibitors/modulators (Table 14.2).

14.1.2.1   Infection Prevention Strategies Using QSI Technologies

Orthopedic Intelligent Implant: Biofilm Detection and Control

Implants have been plagued with the fear of bacterial biofilm formation since they are difficult to contain. One of the most obvious methods is the painful removal of the infected implant. However, considering the cost, effort and pain to be borne by the patient, a need for different approach is required. Prevention or treatment of biofilm that can be targeted towards the biofilm forming microbes could help in long-term solution for implantation. It has been a consensual agreement among surgeons that >60% of arthroplasty infections are caused by *Staphylococcus aureus* and *S. epidermidis*. A *Staphylococcus* sensing mechanism placed within the implant could suggest an intervention using antibiotics. However, treatment of biofilm on implant have limited accessibility for natural immunity or antibiotics to act on. The idea of an external source of antibiotics that can deliver directly to the implant by itself is flawed, since it may become an alternate place for biofilm formation. Hence, a system with an inbuilt mechanism for detection of microorganisms and sensor regulated dispersal of antibiotics seems an apparent solution.

Ehrlich et al. developed electrical current based system for bacterial detection and removal from artificial joints (Ehrlich et al. 2005). This microsystem incorporated a microelectromechanical system (MEMS) biosensor capable of signaling gated reservoirs for controlled release of biofilm inhibitors and antibiotics. MEMS biosensor was designed to detect early stages of staphylococcal interactions leading to biofilm formation that are mediated through ribonucleic acid III activating protein (RAP) – target of RAP (TRAP) communications. The biosensor comprised of a chimeric TRAP molecule which undergoes a conformational switch upon binding with bacteria released RAP. The chimeric TRAP was generated by fusing its extracellular domain with a glucosidase-conjugated transmembrane conformational switching domain. The functioning of the sensor was made on cantilever-based microviscometer sensing unit – i.e. the changes in viscosity are measured. The viscosity is measured for a glucose/dextran binding with concanavalin A molecule (developed by Jeckelmann and Siebold). In absence of glucose, dextran molecules undergo binding with concanavalin A to produce a highly viscous solution. In contrast, glucose molecules when present will compete with dextran for concanavalin A binding sites and hence decrease the viscosity of the solution. The deflection of cantilever is inversely proportional to the measured viscosity.

In an unbound state, glucosidase is in inactive conformation and hence biosensor detects a high viscosity state, suggesting absence of bacteria. In presence of *Staphylococcus*, RAP secreted in the microenvironment conjugates to TRAP of the biosensor and induce its conformational change. This change brings about activation of glucosidase that digests inherently stored glucose polymer to release glucose molecules. The glucose molecules compete with dextran for concanavalin A to decrease viscosity.

These signals are passed to a pair of integral gated reservoirs that contain biofilm inhibitors and antibiotics. RAP-TRAP interactions can be interfered by RNA III inhibiting peptide (RIP) and hence was suitably included in staphylococcal biofilm inhibitor reservoir along with anti-RAP antibodies. The second reservoir had nafcillin and vancomycin to kill planktonic staphylococcus. The discharge of inhibitors and antibiotics was monitored by another MEMS-based sensor to safeguard the controlled release. Both the MEMS-based biosensors and two reservoirs were connected to a memory module and an embedded telemetry system to communicate the *in situ* activities to Bluetooth monitoring device for patient. Thus, an intelligent device to sense presence of *Staphylococcus* and provide an *in situ* treatment was developed.

Preventing Biofilm Formation in Body

Biofilm formation is mechanistically dependent on QS and acts as an armor for bacteria, therefore preventing QS based signaling is expected to prevent biofilm formation. Lung infection caused due to biofilm produced by alginate-producing strains of *P. aeruginosa* has been an important medical concern. Recently, scientists have targeted actin, which serves as a matrix in formation of biofilms, for development of potential therapy for cystic fibrosis (Parks et al. 2009). Similar strategies and technology platforms were developed for treatment of biofilm formation in dental caries, wounds, acne, etc. In parallel, efforts are also ongoing to prevent biofilm formation in medical devices. Researchers are also developing chemical QSIs for impregnating catheters and replacement hips (Veerachamy et al. 2014). Elimination of post-operative infections could reduce the usage of antibiotics and their potential resistance.

Development of synthetic molecules as QS inhibitors seems a viable option for resistant microbes including extensively drug resistant (XDR) *P. aeruginosa*. Kalaiarasan et al. synthesized two anti-QS molecules namely, *N*-(4-{4-fluoroanilno} butanoyl)-L-homoserine lactone (FABHL) and *N*-(4-{4-chlororoanilno}butanoyl)-L-homoserine lactone (CABHL) to counter XDR *P. aeruginosa* biofilm formation (Kalaiarasan et al. 2017). Anti-biofilm property, without inducing cell death, of FABHL or CABHL was confirmed by determining the expression levels of lasR and rhlR using qRT-PCR. Molecular modeling assays showed the binding energy of FABHL (−4.27) and CABHL (−4.51) with LasR protein enables them as potential candidate for potent anti-biofilm agent.

However, some researchers are cautioning in this approach, since dispersing a biofilm could end an infection at that specific site, and it might get distributed to new sites. Another limitation towards biofilm prevention is that it is complex mix of microorganism like cheaters and cooperators (towards QS activity). The cheaters typically get benefit from biofilm but they do not cooperate towards biofilm formation (Czárán and Hoekstra 2009). The QS inhibitor technology would be stymied by cheaters and hence rendering it ineffective. Therefore, before developing QS inhibitor technologies we need to understand more about QS phenomenon.

Numerous companies are exploring various options on biofilm prevention – both in medical and industrial arena. Few of academic spin off and start-up companies working in the field of QSI or companies applying QS as principle for such an application have been discussed below.

(a) Selenium, Ltd., founded in 2004 is a spin-off company based on research performed by Dr. Julian Spallholz and Dr. Ted Reid at Texas Tech University, Texas, USA and supported by venture-capital firm Emergent Technologies. It serves medical segment (implants including catheters, contact lenses and voice prostheses) and industrial companies through two products namely Seldox and SeGuard, respectively by generating superoxides that prevents adhesion of microbes.

(b) Curza, founded in 2016 in Salt Lake City, Utah, USA is developing coatings for prevention of biofilms formation on hip and knee implants.

Even with these excitements, QS inhibitors are yet to create a buzz in market and are yet to entice big pharma companies. Half a dozen of the startups have started, some of which have been described above, by bright scientists and based upon brilliant idea but still failed to bring QS technology to market. The direct reason was the approach to use QS inhibitor as a sole alternative to antibiotics. However, repurposing of existing drugs seems to give a middle path wherein, QS and antibiotic could be used in combination. For example, hamamelitannin – a naturally occurring QSI in the bark of witch hazel – result in Staphylococcus biofilms becoming susceptible to antibiotics (Cobrado et al. 2012).

Bacterial Vaccine

Traditionally, development and usage of live attenuated or part of antigens for eliciting immune (both humoral and cellular) response, in order to develop immunity has worked for different microbes. The strategy to obtain a recombinant antigen and vaccine carrier are crucial steps that determine the success of vaccine development. The safety and efficacy of such a vaccine would suffice the developmental needs. Low copy number vectors have limited expression which is insufficient for immune response, thereby requiring multiple dosing. In contrast, high copy number vector achieves the required antigen expression. However, it may cause over-attenuation of carrier that prevents immunogenic reaction. A middle way that allows enough antigen presentation molecules is required to develop and lead to a multivalent vaccine preparation. One of the lucrative possibilities is use of bacteria for delivery of exogenous antigens produced by introduction of plasmid. However, it may be unstable and involves undesirable metabolic liability on the vector. Another approach is application of *in vivo* inducible promoter with high effectiveness and minimal leakiness of expression. This is a challenging system to have, owing to its ambitious nature.

Chu et al. used this principle to construct a synthetic binary regulation system for enteric Gram-negative intracellular pathogen – *Edwardsiella tarda* (Chu et al. 2015). This system was composed of QS genes originating from *V. fischeri* and it was associated with iron uptake regulons, referred to as "ironQS". IronQS was activated in growth medium upon exhaustion of Fe^{2+} which indicated attainment of threshold cell density. A protective antigen encoding glyceraldehyde-3-phosphate dehydrogenase (GAPDH) of *Aeromonas hydrophila* LSA34, a known fish pathogen was introduced in ironQS and evaluated in *Scophtal musmaximus*, a turbot. Majority of these vaccinated fish survived upon encounter of *A. hydrophila* LSA34 or *E. tarda* EIB202. The key advantage of this research was the high expression efficiency of the designed circuitry that is maintained by ironQS. This work proves the potential of QS towards development of vector vaccines in a controlled and *in vivo-*inducible environment.

In case of people with limited or compromised immunity like wound, infections by *P. aeruginosa* are quite common and the chances of survival are very bleak. The virulence of *P. aeruginosa* is mediated through QS machinery and type III secretion system (T3SS). Alteration in functioning of these two systems may pave a way for development of a candidate vaccine. *P. aeruginosa* predominantly produces two AIs *viz.* *N*-butanoyl-L-homoserine lactone (C4-HSL) and *N*-3-oxododecanoyl-L-homoserine lactone (3-oxo-C_{12}-HSL). In addition, QS is involved in regulation of T3SS which is responsible for delivery of toxins in the host cell. Golpasha et al. targeted 3-oxo-C_{12}-HSL and conjugated it to one of the T3SS (PcrV) proteins to generate a bivalent antigen as a vaccine candidate (Golpasha et al. 2015). Mice were immunized thrice (at an interval of 2 weeks) with individual or combined antigens or with saline (control). Post-one month of last immunization, all animals were burned and challenged with *P. aeruginosa* PAO1. Analyses of survival and bacterial burden on skin, liver and spleen showed survival after 2 days among the bivalent and PcrV immunized group in comparison to control and 3-oxo-C_{12}-HSL group; and significantly high IgG in serum and low bacterial burden in these groups. Thus, a bivalent vaccine to inhibit Pseudomonas infections was developed.

Virus-like particles (VLP) are another route to administer antigens to illicit immune response. Daly et al. have invented a VLP with epitopes of *Staphylococcus aureus* (autoinducing peptides (AIP), toxins and leucocidins) which when injected in mouse model of *S. aureus* dermonecrosis generated required immunity (Daly et al., 2017). This was achieved by expressing AIP1 amino acid sequence on antigenic surface of *P. aeruginosa* RNA phage PP7 coat protein. Skin and soft tissue infections mouse model was administered with this protein and subsequently challenged with virulent strain of *Staphylococcus aureus* (MRSA). In comparison to the control animals, PP7-AIP1 combination vaccinated animals had less *agr*-regulated virulence at the site of infection, less pathogenesis and increased bacterial clearance. Hence, VLP could be potential vaccination tool for similar bacterial infections.

Bacteriocin

QS regulates bacteriocin production and induces competency for survival. QS regulates entrance of a bacteria into a competent state in *Streptococcus mutans* mediated through XIP/ComR pathway (Perez-Pascual et al. 2016). Genome sequencing of Streptococcal genomes have revealed conservation of potential competence and bacteriocin QS pathways. This has been exemplified with documentation of ComRS pathway in *S. mutans* and *Streptococcus thermophilus* and its orthologs in other members of Streptococcal genera. Most members of this genera express SigX in response to treatment except *S. pyogenes*. Interestingly, *S. pyogenes* growing in a biofilm undergo transformation (Marks et al. 2014). This could be initiated by QS pathways that have not been identified yet.

Considering the potential of bacteriocin in QS regulation, discovery of novel bacteriocins from different sources are necessitated. The wide spectrum of bacteria exists in milk and its derivatives that have antibacterial activity. Hernández-Saldaña used raw goat milk and goat cheese to isolate bacteriocinogenic bacteria (Hernández-Saldaña et al. 2016). The group isolated few bacteria from both the sources that had potential antibacterial activities against clinically significant bacteria - *Staphylococcus aureus*, *Escherichia coli*, *Listeria inoccua*, *L. monocytogenes*, *Bacillus cereus*, *Shigella flexneri*, *Pseudomonas aeruginosa*, *Serratia marcescens*, *Klebsiella pneumonieae*, and *Enterobacter cloacae L. lactis* and *S. agalactiae*. The stability of these bacteriocins produced by these isolates was evaluated for pH, temperature and proteinases. It was observed that the bacteriocins were acid stable (pH 2-6), stable till temperature of 100 °C but were prone to degradation by proteinases. Discovery of newer bacteriocins adds to the array of potential QS signaling disruption molecules.

Anti-infective Material

One approach of preventing infections is by designing surfaces that do not permit adhesion of macromolecules and hence avoid biofilm formation. These surfaces may release QSI that restrict development of virulence. However, some material require an intermediate release of QSI over a period of time and those that allow its scalability for multiple applications. Kratochvil utilized AIP analog peptide 1 to prevent *S. aureus* infection on new material keeping these points as objectives (Kratochvil et al. 2017). The researchers developed AIP-loaded and hydrolytically degradable polymer nano-fiber mesh made by electrospinning. These materials were fabricated by applying a strong electrical potential that extruded concentrated solution of polymer mixed with AIP through fine needle on electrically grounded surface. Experimental analysis showed that AIP was released for over 3 weeks in physiologically similar buffer and could disrupt *agr*-based QS signaling. This was evident from lack of production of hemolysin, a *S. aureus* virulence phenotype involved in erythrocyte lysis. This fabricated material exemplifies future applications of various QSIs for medical applications including mats, gauzes or implantable objects.

Screenings of Pathogens Through Aptamers

P. aeruginosa employs QS autoinducers of *las* and *rhl* systems to cause virulence. These systems comprise of transcriptional activators (LasR and RhlR) and AI synthase (LasI and RhlI) that cause synthesis of AIs (3-oxo-C_{12}-HSL and C4-HSL respectively) (Antunes et al. 2010; Dekimpe and Déziel 2009). These two systems are capable of activating each other partially. Hence, inhibition of the two AI by preventing their synthesis, release or AI mediated response is expected to prevent virulence. Such an interference is possible using aptamers which can have high affinity for specific areas of macromolecules.

Zhao et al. used an amino lactam surrogate (ALS) of AHL to obtain eight potential AHL aptamers through systematic evolution of ligands by exponential enrichment (systemic evolution of ligands by exponential enrichment, SELEX) (Zhao et al. 2013). Two of these eight aptamers showcased high inhibition to the *P. aeruginosa* virulence as determined through decrease in biofilm formation, lower secretions of LasA protease, LasB elastase and pyocyanin. The presence of aptamers did not affect the bacterial growth and hence makes them vulnerable to application of antibiotics. Thus, these high-affinity aptamers against AHL present an alternative method for virulence inhibition that has potential for the management of *P. aeruginosa* infection.

A similar method was used by Park and colleagues wherein SELEX was used for selection of 40 aptamers with specificity towards oral pathogens *viz. Porphyromonas gingivalis, Treponema denticola, Streptococcus mutans, S. sanguis* and *S. oralis* (Park et al. 2015).

14.1.2.2   Disease Treatment Through QS Technology

Diseases caused by micro-organisms have only grown more resistant over time due to large-scale and indiscriminate use of antibiotics (WHO 2014). This became evident with identification of penicillin resistant strains of Staphylococcus during 1940s and later erythromycin resistant Streptococcus (1960s) (Kalia 2014). Repeated treatment over longer duration has rendered broad spectrum antibiotic inefficient (Oliver et al. 2000). Microbes evolve faster to given constraints by antibiotics and hence discovery of newer antibiotics or synthesis of novel molecules are bound to 'create' resistant micro-organisms. Pathogenicity of micro-organisms have been identified to be signaled by the QS circuits. Further, immune evasion and activation in *S. aureus* has been shown to be defined by the confinement within lymphocytes and induction of QS response (Qazi et al. 2001). Similarly, biofilm formation by few cells of *P. aeruginosa* has been implicated through QS (Boedicker et al. 2009). Since major group of bacteria utilize different QS mechanisms – oligopeptides (Gram positive) and AHL (Gram negative), treatment for respective class of bacteria will logically require applications based on respective QSI (Kalia et al. 2014). Thus, these discoveries have directly guided research towards discovery and development of QSI based bacterial population disruption, which can be acted upon by regular antibiotics. QSI is ever gaining popularity since current antibiotic based treatment has major issue of antibiotic resistance. Hence, an ideal therapy may be considered wherein bacteria would not develop resistance. QSI is being considered as one such anti-virulence strategy wherein QS inhibitors would not kill or stop bacterial growth. Rather, they would stop bacterial virulence without having to pay evolution penalty. Desired properties of QSI for medical application include host enzymatic stability, low molecule mass, high specificity, longer side chain as compared to native AHL and no adverse reaction on the host system. The validation of anti-virulence concept has been ongoing with various infection experiments and preclinical stage finalization is being attempted. Clinically though, the protocols and new methodologies for microbial diagnosis and follow up treatment needs to be standardized.

QSI for Treating Otitis Media

Recent application of QSI has been attempted in otitis media (OM) – a common middle ear inflammatory disease frequently observed in children. Unencapsulated or non typeable *Haemophilus influenza* (NTHI) is known to cause OM by biofilm formation. This biofilm comprises of the extracellular polymeric substance (EPS), extracellular DNA, integration host factor (IHF), adhesins (including Type IV pilus, TFP), enzymes, lipooligosaccharide and DNABII binding protein among other macromolecules. Removal of IHF from the biofilm destabilizes it and causes exposure of NTHI, thereby allowing their inhibition to therapy (Brockson et al. 2014). This was achieved by transcutaneous immunization of *E. coli* derived IHF in the middle ears of chinchillas with previously NTHI-induced OM. Another target for biofilm dispersal could be NTHI pilin protein since both immature and mature biofilms produced in the middle ears of OM models had antibodies against it. Novotny et al., studied the mechanism involved in NTHI biofilm dispersal and deciphered that expression of TFP and LuxS is co-regulated (Novotny et al. 2015). Further, administration of combination of IHF and rsPilA (recombinant and soluble *N*-terminally truncated form of PilA) by transcutaneous immunization to the test chinchillas led to significant decrease in OM.

QS for Cancer Therapy

Identification of novel anti-cancer molecule is a continuous on-going work worldwide. The biological control of cancer through plant-originating alkaloids (like taxanes, vinca alkaloids and Podophyllotoxins) based chemotherapies, phenazines from *Lactococcus*, extracts from actinobacteria present in mangrove soils (*Microbacterium mangrovi* MUSC 115^T) are indicative of presence of natural molecules for limiting cancer growth and spread (Newman and Cragg 2007; Varsha et al. 2016; Azman et al. 2017). Beyond plantae kingdom, considering the enormous number of bacteria present, the microbial world is apt for looking for potential candidate. Indeed, various anticancer activities with microbial activity have been reported and recently Balhouse et al. have reported interactions of 3-oxo-C_{12}-HSL with breast cancer cell lines (Balhouse et al. 2017). Further, the ever increasing library of QS small molecules and inhibitors may present a possible candidate with application in cancer therapy. Some of these have been mentioned in Table 14.3.

Discovery of limited QS molecules with anti-cancer properties and development of structural deciphering tools gave a platform of making synthetic QS based molecules that can target cancer cells. Different AI have been used to model these synthetic analogs and evaluated in cell lines of cancers of colorectum, prostate, gingiva, tongue, stomach, breast, liver among others (Table 14.4).

Table 14.3 Natural QS molecules with anti-cancer activities

QS molecule	Source	Targeted pathway	Cancers and cell line models	References
3-oxo-C_{12}-HSL	*P. aeruginosa*	Cytoskeletal changes leading to susceptibility to 5-fluoruuracil	Human colorectal cancer (H630)	Dolnick et al. (2005)
3-oxo-C_{12}-HSL	*P. aeruginosa*	Downregulation of STAT3	Breast cancer (MCF-7, BR293 and MDA-MB-468)	Li et al. (2004)
Farnesol	*Candida albicans*	Intrinsic and extrinsic apoptotic pathways	Oral squamous cell carcinoma (OSCC 9 and OSCC 25)	Scheper et al. (2008)

Table 14.4 Chemically synthesized analogs of AI for cancer inhibition

Compound	Structure based on AI	Cancer	References
3-oxo-12-phenyldodecanoyl-L-homoserine lactone	3-oxo-C_{12}-HSL from *P. aeruginosa*	Human colorectal carcinoma cell lines, H630 (parental) and H630-1 (5-fluorouracil resistant), and human prostate carcinoma cell line (PC3)	Oliver et al. (2009)
Compounds 5 and 87	*N-acyl-L-homoserine lactone (AHL)*	Human gingival carcinoma cell line (Ca9-22) and tongue cancer cell line (SAS)	Chai et al. (2012)
Compounds 1 and 2	*HSL-based analogs*	Chronic myeloid leukemia K562	Hazawa et al. (2012)
Compounds 10a-k, 14, 10i, 11e	*HSL*	Human gastric cancer cell line (MGC-803), human breast cancer cell line (MCF-7), human esophageal cancer cell line (EC-9706) and human hepatocellular carcinoma cell line (SMMC-7721)	Ren et al. (2015)
ITC-12, ITC-Cl and Br-Furanone	Multiple	Hodgkin's lymphoma cells (L428)	Nandakumar et al. (2017)

Natural toxins can be utilized for targeted cancer therapy if effectively controlled. The eminent control through QS is an obvious choice. Anderson and colleagues developed a complex QS based circuitry and engineered it in *E. coli* bacteria to target and kill cancer cells (Anderson et al. 2006). Here, the authors utilized invasin encoding *inv* gene from *Yersinia pseudotuberculosis* and engineered it in *E. coli.* to induce binding and invasion of β1-integrins expressing cancer cells. Since cancer community is characterized by high cellular density and hypoxia, the authors placed *inv* gene under QS lux operon derived from *V. fisheri*, with anaerobically induced formate dehydrogenase (*fdhF*) promoter. The *lux* genetic circuit differentiates low (10^3cfu/ml) *versus* high (10^{10}cfu/ml) densities while the selected promoter triggers invasion only under shift of aerobic to anaerobic environment. The functioning was evaluated in metastatic breast cancer (HeLa), hepatocarcinoma (HepG2) and osteosarcoma (U2OS) lines. This approach has potential to be utilized for delivery of specific toxins in the cancer cells. Din et al., synchronized the cycles of engineered bacterial lysis for *in vivo* delivery of the toxins using a similar approach (Din et al. 2016). A synchronized lysis circuit (SLC) was made using a positive (autoinducer, AHL) and negative (bacteriophage lysis, φX174 E) feedback loops under *luxI* promoter. The AI allows the bacteria to attain a specific population size after which they undergo lysis due to bacteriophage lysis activation. The cytotoxic component was mediated through Haemolysin E, encoded by *hlyE* from *E. coli*. This SLC was incorporated in *Salmonella enterica* subsp. enterica serovar Typhimurium and evaluated in a co-culture system with human metastatic breast cancer cell line (HeLa). It was further verified in a sub-cutaneous tumor model wherein the bacterial strains were fed orally. The limited action of chemotherapy on avascular regions of tumors was overcome when the SLC strains were used in combination with 5-fluorouracil (5-FU).

Nanotechnology Assisted QSI

Nanoparticles have multiple application for treatment and prevention of bacterial infections. Biocompatibility and bioavailability can make nanoparticles important deliverables for potential therapeutic usage. Inert inorganic metals such as silver have already been used for various medicinal purposes such as tooth filling in dentistry etc. Silver nanoparticles have been tried in various medical applications and have recently applied to interrupt QS. Singh et al. mycofabricated silver nanoparticles using soil fungus *Rhizopus arrhizus* BRS-07 (Singh et al. 2015). The metabolites were evaluated for QS-regulated virulence in *P. aeruginosa*. Gene expression studies confirmed reduction in cellular levels of LasIR-RhlIR. Further, application of silver nanoparticles prevented biofilm formation and subsequently, the associated virulence and reduced AHLs production. The synthesis of nanoparticles when carried out using green synthesis obviates the toxicity issue. Bose and Chatterjee considered green synthesis of silver nanoparticles for inherent properties of biocompatibility (Bose and Chatterjee 2015). They used leaf extracts of *Justicia adhatoda* for synthesis of silver nanoparticles that had an average size of 20 nm. The antibacterial properties were confirmed by their application against *P. aeruginosa* MTCC 741. Likewise, Dobrucka et al.. performed ecofriendly synthesis of silver nanoparticles using aqueous extract of *Arnicae anthodium* and evaluated their antimicrobial activity against tested bacterial strains (*S. aureus* ATCC 4163, *E. coli* ATCC 25922, *P. aeruginosa* ATCC 6749) (Dobrucka and Długaszewska 2015). Agarwala et al compared antibacterial potential of nanotized metal oxides CuO and Fe_2O_3 against biofilm forming MRSA (Agarwala et al. 2014). It was observed that CuO with maximum antibacterial activity with zone of inhibition of (22 ± 1) mm is better than Fe_2O_3. Ahiwale et al. evaluated gold nanoparticles synthesized using rare bacteriophage belonging to family Podoviridae, as a potential therapeutic agent against biofilm producing *P. aeruginosa* (Ahiwale et al. 2017). In order to produce optimum gold nanoparticles, the author evaluated various physiological parameters like concentration of salts, pH, and temperature and characterized the phase inspired gold nanoparticles using various standard material characterizing techniques such as UV–Vis spectrophotometry, scanning electron microscopy, energy dispersive spectroscopy, X-ray diffraction and dynamic light scattering. It was found that gold nanoparticles were produced with varied sizes (100–200 nM) and varied shapes (spheres, hexagons, triangles, rhomboids and rectangular etc). They observed that 0.2 mM gold nanoparticles concentration inhibited 80% of biofilm produced by *P. aeruginosa*. The AuNPs were also evaluated by Wadhwani et al.., wherein author studied influence of varying cell density of *Acinetobacter* sp. SW30 and gold salt concentrations, on synthesis of nanoparticles and its morphology (Wadhwani et al. 2016). The spherical gold nanoparticles of size ~19 nm were reported at lowest cell density and $HAuCl_4$ salt concentration, whereas increased cell density in gave rise to polyhedral gold nanoparticles (~39 nm). The kinetics of gold nanoparticles synthesis was also reported.

Ivanova et al showed that the combination of nanotised antibacterial gentamicin and quorum quenching enzyme, acylase completely eradicate *P. aeruginosa* in both planktonic and sessile forms (Ivanova 2017). Nanotized gentamicin was produced using ultrasound-assisted nanotransformation process. It has been shown that these nanosphere have capacity to penetrate and disturb bacterial cell membrane. The inhibitory concentration of these hybrid nanobacterial did not show any innocuous effect on human fibroblasts (BJ-5ta cells).

Bioengineered Bacteria to Kill Pathogenic Bacteria

Jayaraman et al. developed a proof of concept model for sensing pathogenic *Vibrio cholerae* in a microenvironment of engineered *Escherichia coli* (Jayaraman et al. 2017). The QS circuitry of *E. coli* was designed to specifically detect *V. cholerae* by its QS signaling molecule CAI-1. Further, in order to kill these pathogenic bacteria, the *E. coli* were made to respond by expressing cell lysis protein (YebF-Art-085) that caused self-lysis that simultaneously released Art-085 in the vicinity to kill *V. cholerae*. This work represents a stepping stone on potential of probiotic mode of treatment in the future.

Since QSI could be applied to myriad of issues related to human health (including treatment of infections), some companies are working in field of development of QSI *viz.*: Health Care Technology – Quonova LLC, Sorretto Therapeutics, etc.

(a) Quonova LLC, founded in 2006 in Melbourne, Florida, USA, develops QSIn platforms that permit discovery of biofilm preventing QSI.
(b) Sorretto Therapeutics, founded in 2006 and based in San Diego, California, USA has obtained an exclusive license from The Scripps Research Institute, California, USA for a technology that is based on targeting AIP produced by (MRSA). AIP is central to QS of *S. aureus* and its targeting suggests an alternative to cellular direct targeting.

14.2 Conclusion

In this chapter we have attempted to present technology perspective towards QS and QSI development for medical use. Recent techniques and technology platforms/protocols towards: a) discovery and development of QSI, b) QS linked disease prevention and c) QS linked disease treatment are reviewed. The development of QSI for intended medical use covers gamut's of activities beyond traditional discovery of QSI and is being attempted using *is*CADDIP. The QS technologies are employed to prevent QS related biofilm formation inside body and on implants. The aptamer based pathogen screenings, bacteriocin and bacterial vaccine are being attempted to prevent QS related infections. The QS technologies could be evolved into anti-virulence treatment strategy. The natural QS signaling molecules and QSI hold promise in cancer treatment. Thus, QS based technology hold promise towards anticancer treatment. Although, still in infancy stage, the nanotechnology based preparations and applications of QSI is also gaining momentum. Overall, technology advancements have catapulted discovery and development of QS system and QSI for medical use.

14.3 Opinion

QS and QSI have generated a lot of enthusiasm in research over a period of time that can be validated by increase in publications over the years. The underlying technology that have been critically designed and executed for medicinal application was underrated. In process of reviewing the available literature, we observed the significance of standard reporter based assay – with any flaw could lead to detection as false positive results. This indicates the necessity to re-examine previously discovered and/or synthesized QS and QSI molecules. We opine amalgamation of the traditional QS/QSI discovery and development with the established technology platforms/protocols for graduating from an initial hit into a drug. The newer QS interference strategies such as signal sequestering could be applied towards development of QSI. Further, protocols like *is*CADDIP hold promise for de-novo discovery of QSI. However, these designs require undergoing complete drug discovery cycle to become a drug. Considering these developments, we believe that the technological advances could propel future of QS and QSIn strategies beyond isolated work on prevention and treatment scenario into the theragnostic paradigm, wherein inbuilt detections and treatment capabilities are expected to be developed. In summation, in this exciting time, a new era of QS awaits that would now further propel real time medical applications.

Acknowledgment Authors would like to thank Sai Life Sciences management and leadership team and Dean, Faculty of Technology, Savitribai Phule Pune University for approval and support. A special thanks to Dr. Sarma BVNBS, VP, Sai Life Sciences, Pune, India for support and encouragement provided during writing this chapter and Dr. Joyita Sarkar for proof-reading. We also thank Prof. Kalia, Emeritus Scientist, CSIR – Institute of Genomics and Integrative Biology, for guidance and an invitation to write book chapter.

References

Agarwala M, Choudhury B, Yadav RNS (2014) Comparative study of antibiofilm activity of copper oxide and iron oxide nanoparticles against multidrug resistant biofilm forming uropathogens. Indian J Microbiol 54:365–368. https://doi.org/10.1007/s12088-014-0462-z

Ahiwale SS, Bankar AV, Tagunde S, Kapadnis BP (2017) A bacteriophage mediated gold nanoparticles synthesis and their anti-biofilm activity. Indian J Microbiol 57:188–194. https://doi.org/10.1007/s12088-017-0640-x

Anderson JC, Clarke EJ, Arkin AP, Voigt CA (2006) Environmentally controlled invasion of cancer cells by engineered bacteria. J Mol Biol 355:619–627. https://doi.org/10.1016/j.jmb.2005.10.076

Antunes LCM, Ferreira RBR, Buckner MMC, Finlay BB (2010) Quorum sensing in bacterial virulence. Microbiology 156:2271–2282. https://doi.org/10.1099/mic.0.038794-0

Azman A-S, Othman I, Fang C-M, Chan K-G, Goh B-H, Lee L-H (2017) Antibacterial, anticancer and neuroprotective activities of rare actinobacteria from Mangrove forest soils. Indian J Microbiol 57:177–187. https://doi.org/10.1007/s12088-016-0627-z

Balhouse BN, Patterson L, Schmelz EM, Slade DJ, Verbridge SS (2017) N-(3-oxododecanoyl)-L-homoserine lactone interactions in the breast tumor microenvironment: implications for breast cancer viability and proliferation in vitro. PLoS One 12:e0180372. https://doi.org/10.1371/journal.pone.0180372

Bjarnsholt T, Jensen PØ, Rasmussen TB, Christophersen L, Calum H, Hentzer M, Hougen H-P, Rygaard J, Moser C, Eberl L, Høiby N, Givskov M (2005) Garlic blocks quorum sensing and promotes rapid clearing of pulmonary Pseudomonas aeruginosa infections. Microbiology 151:3873–3880. https://doi.org/10.1099/mic.0.27955-0

Boedicker JQ, Vincent ME, Ismagilov RF (2009) Microfluidic confinement of single cells of bacteria in small volumes initiates high-density behavior of quorum sensing and growth and reveals its variability. Angew Chemie Int Ed 48:5908–5911. https://doi.org/10.1002/anie.200901550

Bose D, Chatterjee S (2015) Antibacterial activity of green synthesized silver nanoparticles using vasaka (*Justicia adhatoda* L.) leaf extract. Indian J Microbiol 55:163–167. https://doi.org/10.1007/s12088-015-0512-1

Brockson ME, Novotny LA, Mokrzan EM, Malhotra S, Jurcisek JA, Akbar R, Devaraj A, Goodman SD, Bakaletz LO (2014) Evaluation of the kinetics and mechanism of action of anti-integration host factor-mediated disruption of bacterial biofilms. Mol Microbiol 93:1246–1258. https://doi.org/10.1111/mmi.12735

Chai H, Hazawa M, Shirai N, Igarashi J, Takahashi K, Hosokawa Y, Suga H, Kashiwakura I (2012) Functional properties of synthetic N-acyl-L-homoserine lactone analogs of quorum-sensing gram-negative bacteria on the growth of human oral squamous carcinoma cells. Investig New Drugs 30:157–163. https://doi.org/10.1007/s10637-010-9544-x

Chu T, Ni C, Zhang L, Wang Q, Xiao J, Zhang Y, Liu Q (2015) A quorum sensing-based *in vivo* expression system and its application in multivalent bacterial vaccine. Microb Cell Factories 14:37. https://doi.org/10.1186/s12934-015-0213-9

Cobrado L, Azevedo MM, Silva-Dias A, Ramos JP, Pina-Vaz C, Rodrigues AG (2012) Cerium, chitosan and hamamelitannin as novel biofilm inhibitors? J Antimicrob Chemother 67:1159–1162. https://doi.org/10.1093/jac/dks007

Czárán T, Hoekstra RF (2009) Microbial communication, cooperation and cheating: quorum sensing drives the evolution of cooperation in bacteria. PLoS One 4:e6655. https://doi.org/10.1371/journal.pone.0006655

Daly SM, Joyner JA, Triplett KD, Elmore BO, Pokhrel S, Frietze KM, Peabody DS, Chackerian B, Hall PR (2017) VLP-based vaccine induces immune control of Staphylococcus aureus virulence regulation. Sci Rep 7:637. https://doi.org/10.1038/s41598-017-00753-0

Dekimpe V, Déziel E (2009) Revisiting the quorum-sensing hierarchy in Pseudomonas aeruginosa: the transcriptional regulator RhlR regulates LasR-specific factors. Microbiology 155:712–723. https://doi.org/10.1099/mic.0.022764-0

Depke T, Franke R, Brönstrup M (2017) Clustering of MS 2 spectra using unsupervised methods to aid the identification of secondary metabolites from Pseudomonas aeruginosa. J Chromatogr B: pii S1570-0232(17)30999-6. doi: https://doi.org/10.1016/j.jchromb.2017.06.002

Din MO, Danino T, Prindle A, Skalak M, Selimkhanov J, Allen K, Julio E, Atolia E, Tsimring LS, Bhatia SN, Hasty J (2016) Synchronized cycles of bacterial lysis for in vivo delivery. Nature 536:81–85. https://doi.org/10.1038/nature18930

Dobrucka R, Długaszewska J (2015) Antimicrobial activities of silver nanoparticles synthesized by using water extract of Arnicae anthodium. Indian J Microbiol 55:168–174. https://doi.org/10.1007/s12088-015-0516-x

Dolnick R, Wu Q, Angelino NJ, Stephanie LV, Chow K-C, Sufrin JR, Dolnick BJ (2005) Enhancement of 5-fluorouracil sensitivity by an rTS signaling mimic in H630 colon cancer cells. Cancer Res 65:5917–5924. https://doi.org/10.1158/0008-5472.CAN-05-0431

Dong Y-H, Zhang L-H (2005) Quorum sensing and quorum-quenching enzymes. J Microbiol 43 Spec No:101–109. doi: Not Available

Ehrlich GD, Stoodley P, Kathju S, Zhao Y, McLeod BR, Balaban N, Hu FZ, Sotereanos NG, Costerton JW, Stewart PS, Post JC, Lin Q (2005) Engineering approaches for the detection and control of orthopaedic biofilm infections. Clin Orthop Relat Res 437:59–66. doi: Not Available

Golpasha ID, Mousavi SF, Owlia P, Siadat SD, Irani S (2015) Immunization with 3-oxododecanoyl-L-homoserine lactone-r-PcrV conjugate enhances survival of mice against lethal burn infections caused by Pseudomonas aeruginosa. Bosn J Basic Med Sci 15:15–24. https://doi.org/10.17305/bjbms.2015.292

Hamilton RJ (2015) Tarascon pocket pharmacopoeia, 16th edn. Jones & Bartlett Learning, Burlington

Hazawa M, Kudo M, Iwata T, Saito K, Takahashi K, Igarashi J, Suga H, Kashiwakura I (2012) Caspase-independent apoptosis induction of quorum-sensingautoinducer analogs against chronic myeloid leukemia K562. Investig New Drugs 30:862–869. https://doi.org/10.1007/s10637-010-9623-z

Hentzer M, Riedel K, Rasmussen TB, Heydorn A, Andersen JB, Parsek MR, Rice SA, Eberl L, Molin S, Høiby N, Kjelleberg S, Givskov M (2002) Inhibition of quorum sensing in Pseudomonas aeruginosa biofilm bacteria by a halogenated furanone compound. Microbiology 148:87–102. https://doi.org/10.1099/00221287-148-1-87

Hernández-Saldaña OF, Valencia-Posadas M, de la Fuente-Salcido NM, Bideshi DK, Barboza-Corona JE (2016) Bacteriocinogenic bacteria isolated from raw goat milk and goat cheese produced in the center of México. Indian J Microbiol 56:301–308. https://doi.org/10.1007/s12088-016-0587-3

Hussain MB (2017) Role of honey in topical and systemic bacterial infections. J Altern Complement Med 24:15. https://doi.org/10.1089/acm.2017.0017

Ivanova KD (2017) Nanostructured coatings for controlling bacterial biofilms and antibiotic resistance. Universitat Politècnica de Catalunya, Barcelona

Jayaraman P, Holowko MB, Yeoh JW, Lim S, Poh CL (2017) Repurposing a two-component system-based biosensor for the killing of Vibrio cholerae. ACS Synth Biol 6:1403–1415. https://doi.org/10.1021/acssynbio.7b00058

Kalaiarasan E, Thirumalaswamy K, Harish BN, Gnanasambandam V, Sali VK, John J (2017) Inhibition of quorum sensing-controlled biofilm formation in Pseudomonas aeruginosa by quorum-sensing inhibitors. Microb Pathog 111:99–107. https://doi.org/10.1016/j.micpath.2017.08.017

Kalia VC (2014) Microbes, antimicrobials and resistance: the battle goes on. Indian J Microbiol 54:1–2. https://doi.org/10.1007/s12088-013-0443-7

Kalia VC, Wood TK, Kumar P (2014) Evolution of resistance to quorum-sensing inhibitors. Microb Ecol 68:13–23. https://doi.org/10.1007/s00248-013-0316-y

Kalia M, Singh PK, Yadav VK, Yadav BS, Sharma D, Narvi SS, Mani A, Agarwal V (2017) Structure based virtual screening for identification of potential quorum sensing inhibitors against LasR master regulator in Pseudomonas aeruginosa. Microb Pathog 107:136–143. https://doi.org/10.1016/j.micpath.2017.03.026

Karumuri S, Singh PK, Shukla P (2015) In silico analog design for terbinafine against trichophyton rubrum: a preliminary study. Ind J Microbiol 55:333–340. https://doi.org/10.1007/s12088-015-0524-x

Kratochvil MJ, Yang T, Blackwell HE, Lynn DM (2017) Nonwoven polymer nanofiber coatings that inhibit quorum sensing in staphylococcus aureus: toward new nonbactericidal approaches to infection control. ACS Infect Dis 3:271–280. https://doi.org/10.1021/acsinfecdis.6b00173

Li L, Hooi D, Chhabra SR, Pritchard D, Shaw PE (2004) Bacterial N-acylhomoserine lactone-induced apoptosis in breast carcinoma cells correlated with down-modulation of STAT3. Oncogene 23:4894–4902. https://doi.org/10.1038/sj.onc.1207612

Marks LR, Mashburn-Warren L, Federle MJ, Hakansson AP (2014) Streptococcus pyogenes biofilm growth in vitro and in vivo and its role in colonization, virulence, and genetic exchange. J Infect Dis 210:25–34. https://doi.org/10.1093/infdis/jiu058

McGuire, S (2008) Australian regulators issue warning on Novartis' Lamisil – Medical Marketing and Media. http://www.mmm-online.com/channel/australian-regulators-issue-warning-on-novartis-lamisil/article/104983/. Accessed 9 Oct 2017

Nandakumar N, Dandela R, Gopas J, Meijler MM (2017) Quorum sensing modulators exhibit cytotoxicity in Hodgkin's lymphoma cells and interfere with NF-κB signaling. Bioorg Med Chem Lett 27:2967–2973. https://doi.org/10.1016/j.bmcl.2017.05.012

Newman DJ, Cragg GM (2007) Natural products as sources of new drugs over the last 25 years. J Nat Prod 70:461–477. https://doi.org/10.1021/np068054v

Ni N, Choudhary G, Li M, Wang B (2008) Pyrogallol and its analogs can antagonize bacterial quorum sensing in Vibrio harveyi. Bioorg Med Chem Lett 18:1567–1572. https://doi.org/10.1016/j.bmcl.2008.01.081

Novotny LA, Jurcisek JA, Ward MO, Jordan ZB, Goodman SD, Bakaletz LO (2015) Antibodies against the majority subunit of type IV pili disperse nontypeable haemophilus influenzae biofilms in a LuxS-dependent manner and confer therapeutic resolution of experimental otitis media. Mol Microbiol 96:276–292. https://doi.org/10.1111/mmi.12934

Oliver A, Cantón R, Campo P, Baquero F, Blázquez J (2000) High frequency of hypermutable Pseudomonas aeruginosa in cystic fibrosis lung infection. Science 288:1251–1254. https://doi.org/10.1126/science.288.5469.1251

Oliver CM, Schaefer AL, Greenberg EP, Sufrin JR (2009) Microwave synthesis and evaluation of phenacylhomoserine lactones as anticancer compounds that minimally activate quorum sensing pathways in Pseudomonas aeruginosa. J Med Chem 52:1569–1575. https://doi.org/10.1021/jm8015377

Park J-P, Shin HJ, Park S-G, Oh H-K, Choi C-H, Park H-J, Kook M, Ohk S (2015) Screening and development of DNA Aptamers specific to several oral pathogens. J Microbiol Biotechnol 25:393–398. https://doi.org/10.4014/jmb.1407.07019

Parks QM, Young RL, Poch KR, Malcolm KC, Vasil ML, Nick JA (2009) Neutrophil enhancement of Pseudomonas aeruginosa biofilm development: human F-actin and DNA as targets for therapy. J Med Microbiol 58:492–502. https://doi.org/10.1099/jmm.0.005728-0

Parsek MR, Val DL, Hanzelka BL, Cronan JE, Greenberg EP (1999) Acyl homoserine-lactone quorum-sensing signal generation. Proc Natl Acad Sci U S A 96:4360–4365

Perez-Pascual D, Monnet V, Gardan R (2016) Bacterial cell–cell communication in the host via RRNPP peptide-binding regulators. Front Microbiol 7:706. https://doi.org/10.3389/fmicb.2016.00706

Pérez-Pérez M, Jorge P, Pérez Rodríguez G, Pereira MO, Lourenço A (2017) Quorum sensing inhibition in Pseudomonas aeruginosa biofilms: new insights through network mining. Biofouling 33:128–142. https://doi.org/10.1080/08927014.2016.1272104

Qazi SNA, Counil E, Morrissey J, Rees CED, Cockayne A, Winzer K, Chan WC, Williams P, Hill PJ (2001) Agr expression precedes escape of internalized Staphyloccccus aureus from the host endosome. Microbiology 69:7074–7082. https://doi.org/10.1128/IAI.69.11.7074

Rajamanikandan S, Srinivasan P (2017) Exploring the selectivity of auto-inducer complex with LuxR using molecular docking, mutational studies and molecular dynamics simulations. J Mol Struct 1131:281–293. https://doi.org/10.1016/j.molstruc.2016.11.056

Rajamanikandan S, Jeyakanthan J, Srinivasan P (2017) Discovery of potent inhibitors targeting Vibrio harveyi LuxR through shape and e-pharmacophore based virtual screening and its biological evaluation. Microb Pathog 103:40–56. https://doi.org/10.1016/j.micpath.2016.12.003

Ren J-L, Zhang X-Y, Yu B, Wang X-X, Shao K-P, Zhu X-G, Liu H-M (2015) Discovery of novel AHLs as potent antiproliferative agents. Eur J Med Chem 93:321–329. https://doi.org/10.1016/j.ejmech.2015.02.026

Scheper MA, Shirtliff ME, Meiller TF, Peters BM, Jabra-Rizk MA (2008) Farnesol, a fungal quorum-sensing molecule triggers apoptosis in human oral squamous carcinoma cells. Neoplasia 10:954–963. https://doi.org/10.1593/neo.08444

Singh BR, Singh BN, Singh A, Khan W, Naqvi AH, Singh HB (2015) Mycofabricated biosilver nanoparticles interrupt Pseudomonas aeruginosa quorum sensing systems. Sci Rep 5:13719. https://doi.org/10.1038/srep13719

Truchado P, Larrosa M, Castro-Ibáñez I, Allende A (2015) Plant food extracts and phytochemicals: their role as quorum sensing inhibitors. Trends Food Sci Technol 43:189–204. https://doi.org/10.1016/j.tifs.2015.02.009

Tsai C-S, Winans SC (2010) LuxR-type quorum-sensing regulators that are detached from common scents. Mol Microbiol 77:1072–1082. https://doi.org/10.1111/j.1365-2958.2010.07279.x

Varsha KK, Nishant G, Sneha SM, Shilpa G, Devendra L, Priya S, Nampoothiri KM (2016) Antifungal, anticancer and aminopeptidase inhibitory potential of a phenazine compound produced by Lactococcus BSN307. Indian J Microbiol 56:411–416. https://doi.org/10.1007/s12088-016-0597-1

Veerachamy S, Yarlagadda T, Manivasagam G, Yarlagadda PK (2014) Bacterial adherence and biofilm formation on medical implants: a review. Proc Inst Mech Eng Part H J Eng Med 228:1083–1099. https://doi.org/10.1177/0954411914556137

Wadhwani SA, Shedbalkar UU, Singh R, Vashisth P, Pruthi V, Chopade BA (2016) Kinetics of synthesis of gold nanoparticles by Acinetobacter sp. SW30 isolated from environment. Indian J Microbiol 56:439–444. https://doi.org/10.1007/s12088-016-0598-0

Wee WY, Dutta A, Choo SW (2017) Comparative genome analyses of mycobacteria give better insights into their evolution. PLoS One 12:e0172831. https://doi.org/10.1371/journal.pone.0172831

WHO (2014) Antimicrobial resistance global report on surveillance. World Health Organization, Geneva

Zhao ZG, Yu YM, Xu BY, Yan SS, Xu JF, Liu F, Li GM, Ding YL, Wu SQ (2013) Screening and anti-virulent study of N-acyl homoserine lactones DNA aptamers against Pseudomonas aeruginosa quorum sensing. Biotechnol Bioprocess Eng 18:406–412. https://doi.org/10.1007/s12257-012-0556-6

Chapter 15
Combating Staphylococcal Infections Through Quorum Sensing Inhibitors

Nishant Kumar, Hansita Gupta, Neha Dhasmana, and Yogendra Singh

Abstract *Staphylococcus aureus* is a clinically important pathogen mainly causing hospital borne infections. These bacterial infections range from mild skin infections to serious health threats like endocarditis, osteomyelitis, and pneumonia. Few strains have developed resistance against antibiotics used to treat *S. aureus* infections and are termed as Methicillin Resistant *S. aureus* strains. The pathogen releases Auto Inducing Peptides to establish cell density dependent inter-cell communication, also known as quorum sensing (QS). QS results in the expression of accessory gene regulator system. It causes successful biofilm formation and enhanced expression of toxins. QS mediated biofilm formation provides an additional resistance against the antibiotics used. An innovative therapeutic approach has been studied vastly in last decade to deal with severe infections using specific QS inhibitors (QSIs). This chapter comprehensively describes the QSIs studied to control the infections caused by *S. aureus* strains.

Keywords Agr system · *Staphylococcus aureus* · Biofilm · Inhibitors · Quorum sensing · RAP/TRAP

Nishant Kumar and Hansita Gupta have contributed equally with all other contributors.

N. Kumar · N. Dhasmana (✉)
Allergy and Infectious Diseases, CSIR – Institute of Genomics and Integrative Biology, Delhi, India

Academy of Scientific & Innovative Research (AcSIR), New Delhi, India

H. Gupta
Department of Zoology, University of Delhi, Delhi, India

Y. Singh (✉)
Department of Zoology, University of Delhi, Delhi, India

Academy of Scientific & Innovative Research (AcSIR), New Delhi, India

© Springer Nature Singapore Pte Ltd. 2018
V. C. Kalia (ed.), *Biotechnological Applications of Quorum Sensing Inhibitors*,
https://doi.org/10.1007/978-981-10-9026-4_15

15.1 Introduction

Quorum sensing (QS) is reported to be critical for various human pathogens for example *Staphylococcus aureus, Staphylococcus epidermidis, Pseudomonas aeruginosa, Serratia pneumonia, Yersinia pestis, Brucella abortus* and *Burkholderia pseudomallei* (Swift et al. 2001; Williams 2002). The successful establishment of disease is governed through the pathogen's ability to invade and forms biofilm in the host. This also facilitates the pathogen to avoid antibiotic mediated killing in vivo. There are various methods to prevent the formation and disrupt the pathogenic biofilms in the host for examples nanoparticles, azithromycin, etc (Agarwala et al. 2014; Gui et al. 2014; Wadhwani et al. 2016; Ahiwale et al. 2017). Interfering bacterial QS through QS inhibitors (QSIs) is a novel therapeutic approach to curb the bacterial infection (Kumar et al. 2015). *Staphylococcus aureus* is the major cause of nosocomial infections in USA and developing countries as well. *S. aureus* secretes few virulence factors which are under the control of *agr* operon. The *agr* operon encodes AgrB (membrane bound peptidase), AgrD (precursor of AIP), AgrC (membrane bound histidine kinase), and AgrA (response regulator). The promoter region P2 governs polycistronic operon *agrBDCA* while the adjacent promoter P3 encodes mRNA for δ-hemolysin and pleiotropic regulator of other virulence genes (Bronesky et al. 2016). AgrB cleaves AgrD into a thiolactone intermediate which is secreted to undergo subsequent cleavage to yield mature AIP. AIP is then sensed by the receptor histidine kinase AgrC, thus phosphorylating itself and response regulator. The activated response regulator AgrA binds to the P2 and P3 to enhance the expression through these promoters. RNA III encoded RNA which acts as an antisense and interferes with the translation of 'repressor of toxin' Rot, an inhibitor of α-hemolysin. Another QS system RAP/TRAP consists of two proteins, which are RAP (**R**NAIII-**a**ctivating **p**rotein) and TRAP (**t**arget **R**NAIII-**a**ctivating **p**rotein). RAP activates the production of toxins by phosphorylating the histidine amino acid of TRAP, when RAP reaches a certain threshold concentration (Balaban et al. 2001). Emergence of multidrug resistance in *S. aureus* strains is a serious public health issue (Kalia 2014a, 2015). Therefore, an alternative approach of targeting the QS molecules of bacteria is a viable option to effectively treat the infections (Koul and Kalia 2017) (Fig. 15.1).

15.2 QS Inhibitors in Controlling *S. aureus* Infections

QSIs are antimicrobial compounds that interfere with the ability of bacteria to communicate in a colony (Kalia and Purohit 2011, Kumar et al. 2013). They should be specific for the protein to be targeted to avoid killing of host and its microbiome. Specific QSIs have been researched for their application in the prophylaxis of *S. aureus* borne infections. Structurally QSIs can belong to different categories of macromolecules for examples, peptides, sugar, amides or their analogues. Following, we have discussed the inhibitors showing significant potency against the infection caused by methicillin resistant *S. aureus* (Table 15.1).

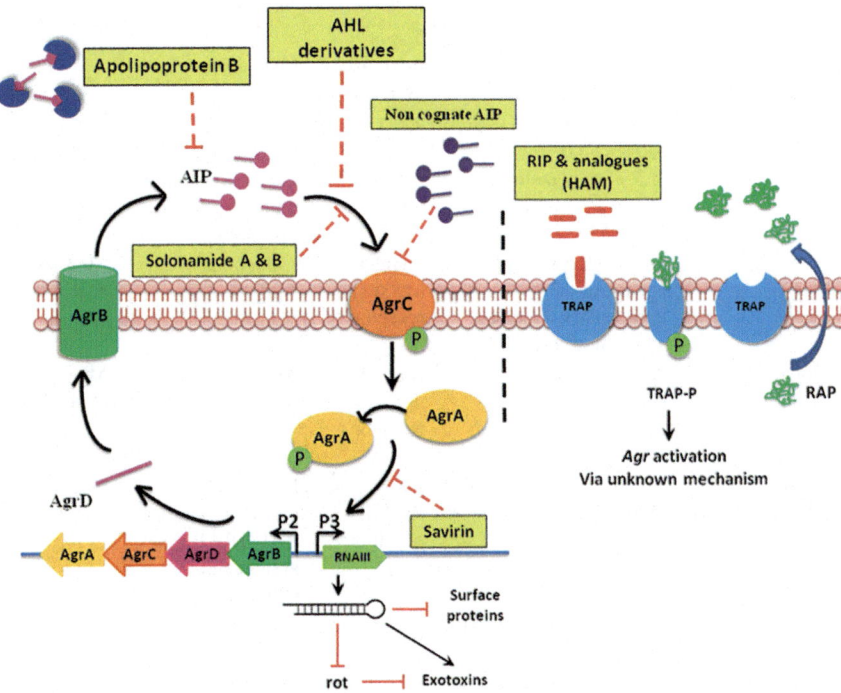

Fig. 15.1 **Inhibition of Agr and RAP/TRAP systems of** *S. aureus* **using quorum sensing inhibitors.** Left panel shows Agr inhibition while Right panel shows RAP/TRAP inhibition. Red Dotted Lines indicate specific inhibition steps in the pathways caused by the inhibitors

15.3   Savirin Inhibits Growth via Interaction with AgrA

High throughput screening of small molecule inhibitors led to the discovery of the Savirin (*Staphylococcus **a**ureus* **vir**ulence **in**hibitor), which specifically inhibits the *agr* mediated signaling in *S. aureus* without affecting the growth of skin commensal *S. epidermidis*. Chemically, Savirin is 3-(4-propan-2-ylphenyl) sulfonyl-1H-triazolo (1,5-a) quinazolin-5-one (Sully et al. 2014). Apart from its molecular weight, the lipophilic nature of savirin makes it an interesting drug candidate for treatment of *S. aureus* infections (Lipinski et al. 2001). *S. aureus* possess a two-component system (TCS) which comprises of AgrC, histidine kinase and AgrA, response regulator. The extracellular autoinducing peptide (AIP) binds to the transmembrane protein AgrC, which in turn phosphorylates AgrA. Thus activated AgrA binds to promoter P2 and P3 encodes AgrB, AgrD, AgrC, AgrA and RNAIII respectively. RNA levels of RNAIII increases dramatically upon AgrA binding (Koenig et al. 2004). RNAIII primarily functions as an antisense and is earlier reported to inhibit transcription of repressor of toxins such as rot (Boisset et al. 2007). RNAIII mediated inhibition of *rot* expression thus increases transcription of downstream virulence factors such as α-hemolysin (Yarwood and Schlievert 2003; Le and Otto 2015). Recently, it was

Table 15.1 Natural and synthetic inhibitors of *S. aureus* quorum sensing

Compound Name	Molecular Details	Target proteins/ systems	Mechanism of action	References
Savirin	(3-(4-propan-2-ylphenyl) sulfonyl-1H-triazolo [1,5-a] quinazolin-5-one)	AgrA, the *agr* response regulator	Inhibitor of AgrA-DNA interaction and thus inhibiting RNAIII synthesis.	Sully et al. (2014)
Solonamide A	A cyclodepsipeptide consist of a 3-hydroxyhexanoic acid and four amino acids	AgrC, the *agr* signal receptor	Antagonist of AgrC.	Mansson et al. (2011)
Solonamide B	A cyclodepsipeptide consist of a 3-hydroxyoctanoic acid and four amino acids	AgrC, the *agr* signal receptor	Antagonist of AgrC.	Mansson et al. (2011)
RNAIII-inhibiting peptide	A heptapeptide (YSPXTNF-NH2)	RNAII and RNAIII, biofilms	Inhibits TRAP phosphorylation of RAP/TRAP QS system and *agr* expression.	Gov et al. (2001)
Apolipoprotein B	4536 amino acid protein	AIP	Sequesters AIP	Elmore et al. (2015)
Hamamelitannin and its analogues	Ester of D- hamamelose (2-hydroxymethyl-D-ribose) with 2 molecules of gallic acid (2′,5-di-O-galloyl-Dhamamelose)	Inhibits *agr* expression by blocking TRAP	Non peptide analogue of RIP	Brackman et al. (2016), Vermote et al. (2017)
Non cognate AIP	7–9 amino acid residues in length with 5 membered ring; C-terminus forms the thiolactone bond with conserved central cysteine.	Blocking AgrC receptor	Inhibition of agr by non cognate binding of AgrC-AIP	Tal-Gan et al. (2013a), Vasquez et al. (2017)
TMA, TOA and TTA derivatives of AHL	Modifications in 3-oxo-12-HSL structure	Inhibitor *agr* signalling	Inhibition of AgrC-AIP interaction	Murray et al. (2014), Zapotoczna et al. (2017)

shown that total activity of α-hemolysin was significantly reduced in savirin-treated bacterial supernatants of MRSA isolated from different sites of infection. Since histidine kinase domain of AgrC is conserved in *S. aureus* and *S. epidermidis* (a skin commensal), therefore AgrA was selected as a target for drug development using high throughput screening. C-terminal DNA binding domain of AgrA was used to identify the drug candidates using swissdock, an online server. The study reveals that savirin binds to the CTD of AgrA from *S. aureus* (SA_AgrA) however was unable to interact with AgrA from *S. epidermidis* (SE_AgrA). DNA binding domain of SA_AgrA and SE_AgrA differs in two positions (229, Tyr to Phe) and (227, His to Asn), which significantly reduces the binding affinity of Savirin (Sully et al. 2014). The crystal structure of LytTR domain of the SA_AgrA was analyzed in a DNA unbound form. At the same time, screening a library of small molecules reveals that the AgrA-DNA interactions might destabilize by targeting an exposed hydrophobic cleft with a small molecule (Leonard et al. 2012). Mechanistic studies involving a novel reporter strain of SA_AgrA activation and electromobility shift assays have demonstrated the efficacy of savirin, both in vivo and in vitro, by inhibiting the binding function of SA_AgrA with DNA in *S. aureus*. These evidences suggested that savirin impedes the function of SA_AgrA, thus preventing the transcription from *agrBDCA* promoter P2 and RNAIII promoter P3 and other *agr*-regulated virulence genes. The clinical isolate of *S. aureus* was studied for few generations for the emergence of resistance to savirin both in vivo and in vitro. Even the persistent exposure to the drug savirin could not lead to the emergence of resistant strain (Sully et al. 2014). Unlike conventional antibiotics, savirin is highly specific and does not foster stress responses and disrupt membrane integrity (Defoirdt et al. 2013). Moreover, AgrA has similar sequence in all four *S. aureus agr* groups, making it more desirable therapeutic target (Wang and Muir 2016).

15.4 Solonamide A and B Act As Antagonist of AgrC

Two of the most important strains of MRSA are "hospital acquired" or HA-MRSA and "community acquired" or CA-MRSA (Gordon and Lowy 2008). In general, HA-MRSA is an opportunistic pathogen unable to infect healthy individuals while in recent years, the most common strain of CA-MRSA, USA300 has emerged as a serious concern due to its capability of infecting healthy individuals (Loughman et al. 2009). Increased resistance to different antibiotics in MRSA has led to the development of new therapeutic strategies. The *agr* system regulates the expression of virulence gene in *S. aureus* (Gordon and Lowy 2008). Therefore, anti-virulence therapy has received an appreciable interest for combating *S. aureus* infections (Wright and Sutherland 2007). Recently, two novel compounds were isolated from the marine bacterium *Photobacterium halotolerans* (strain S2753) named Solonamides A and B. They impede *agr* QS of *S. aureus* and subsequently disrupt the expression of virulence gene. Based on NMR data, solonamides structure was characterized as cyclodepsipeptides consisting of a 3-hydroxy fatty acid and four amino acids (phenylalanine, alanine and two leucines). It was also found that solonamide A contains a 3-hydroxyhexanoic acid (Hha), whereas solonamide B is made up of 3-hydroxyoctanoic acid (Hoa) (Mansson et al. 2011). In vitro, it was shown that signals downstream to the *agr* sensing system upregulates the expression of α-hemolysin encoded by *hla* and downregulates the expression of cell surface protein such as protein A encoded by *spa* at the beginning of the stationary growth phase in *S. aureus* (Vuong et al. 2000). Northern blot analysis examined the amount of mRNA isolated from strain of CA-MRSA, USA300 and *S. aureus* 8325–4 after the treatment with solon-amide, verified the interference of these compounds in virulence gene expression. Solonamide B minimizes the expression of *hla* and *rnaIII* and increases the expression of *spa*. Whereas solonamide A has been shown to increase the expression of *spa* however there were minor reduction in *hla* and *rnaIII* expression in USA300 and 8325-4 strains (Mansson et al. 2011). The primary host defense in opposition to *S. aureus* infections are neutrophils and therefore, lysis of neutrophils is crucial for the virulence of these strains. The PSMs and α-hemolysin are two major virulence factors; both are remarkable at killing immune cells and responsible for an increased virulence of CA-MRSA (Bubeck Wardenburg et al. 2007; Wang et al. 2007). It was reported that solonamide B reduces the expression of virulence factors such as phenol soluble modulins, the PSMs and α-hemolysin in USA300 strain. Additionally the toxicity of supernatants was shown to be minimized when tested against human neutrophils. AgrA, the response regulator of *agr* QS system directly controls the expression of PSMs. Apart from disrupting expression of genes via RNAIII, solonamide B also affects expression of PSMs through AgrA (Nielsen et al. 2014). The QS signal molecules of *S. aureus* are the cyclic thiolactone peptides generally known as autoinducing peptides (AIPs). AIPs activate the *agr* QS system and thus controlling the virulence gene expression via the effector molecule RNAIII (Novick and Geisinger 2008). Depending on the strain, there are four distinct types of AIPs in which AIP of one type specifically binds to its cognate receptor agrC (*agr* signal receptor) but shows antagonistic activity in strains harboring other types of AIPs (George and Muir 2007). It was suggested that solonamides are the competitive inhibitors of the agr system as they have structures similar to the AIPs. Solonamide is a lactone whereas AIP is a thiolactone. However in recent studies, AIP analogues harboring lactone instead of thiolactone have been found to act as competitive inhibitors. It was also found that both solonamides contain hydrophobic phenylalanine and leucine residues that are crucial for the impediment of the *agr* response (Mayville et al. 1999; Mansson et al. 2011). Moreover, Baldry and colleagues chemically synthesized the solonamide analogues to improve its anti-virulence candidacy (Baldry et al. 2016). These findings suggest that inhibition via solonamides is probable alternative therapeutic approach to treat MRSA infections.

15.5   Apolipoprotein Act As Sequester of AIP

In recent times serum lipoproteins (LP) have emerged as a molecule having a dual role of contributing to cholesterol homeostasis as well as host innate defense. It has been established that very low levels of serum lipoprotein (hypolipoproteinemia) is related to increased bacterial infection in critically ill patients (Han 2010; Femling et al. 2013). In this respect Apolipoprotein B (apoB100), a 4536 amino acid protein is essential for the formation of these LPs (LDLs, VLDLs, Chylomicrons, etc.). Recent studies have shown that apoB100 disrupts virulence factor expression of *S. aureus* thus limiting its pathogenesis (Hall et al. 2013). It is done by binding of apoB100 to AIPs and thus disrupting *agr* mediated virulence. While in human intestinal enterocytes, a truncated form of apoB100 is produced, which is apoB48. It is studied that enteral feeding in critically ill patients leads to reduced risk of infection as compared to parenteral feeding, which suggests the importance of apoB48 in host innate immune response, however the mechanism is unknown (Kattelmann et al. 2006). This led to the development of new quorum quenching inhibitor i.e. apoB48 to control *agr* mediated *S. aureus* QS by Bradley and colleague. It was seen that apoB48 and apoB100 antagonizes agr signalling with similar IC_{50} of 3.5 and 2.3 nM, respectively. The IC_{50} values were found to below the reported EC_{50} (28 nM) for activation of agr system via AIP1. This could provide effective protection against *S. aureus* infections. In vivo studies also showed that exogenous apoB48 treated mice infected with *S. aureus* USA300 strain had decreased bacterial burden at site of infection as compared to untreated mice. This data makes apoB48 an important inhibitor of *agr* signalling mediated QS in vivo and providing protection against *S. aureus* infection (Elmore et al. 2015). Thus apolipoprotein can prove to be a global inhibitor of QS and warrants more research for its use as therapeutic agent.

15.6 Non Cognate AIP

QS in *S. aureus* is controlled by the chromosome locus named *agr* (**A**ccessory **G**ene **R**egulator). It is an operon system, genes of which encodes for and also sense a small peptide autoinducer named AIP (**A**utoinducing **P**eptide) (Novick and Geisinger 2008). AIP consists of 7–9 amino acid residues and harbours a five membered ring wherein the C-terminal forms the thiolactone bond with cysteine (central position). This arrangement is crucial for AIP's activity (Ji et al. 1997; Mayville et al. 1999; McDowell et al. 2001). A conserved hydrophobic patch in the C-terminus and few specific contacts aid in binding of AIP to its cognate receptor, AgrC via the hexahelical transmembrane (TM) sensor domain. Thus resulting in activation of downstream signalling cascade (Lyon et al. 2002; Wright et al. 2004; Geisinger et al. 2008). The *agr* locus possesses polymorphism within a single species. This polymorphism is due to the variability in the regions of RNAII, AgrB, AgrD and AgrC, giving rise to four allelic variants of *S. aureus*. This hypervariability guides the generation of four different types of AIPs (I-IV) on the basis of the strain (Ji et al. 1997; Jarraud et al. 2000). Generally, only the cognate interaction of AIP with AgrC guides the expression of *agr* operon whilst the non-cognate interactions of the same lead to the inhibition of the expression, thus causing the inhibition of QS. Owing to this property of inhibition of QS by non-cognate AIPs Lyon and McDowell research groups independently designed hybrid AIPs by altering length or amino acid sequence, by introducing truncations and structural substitutions. The hybrid AIPs thus created have the property to act as universal inhibitors of all the AgrC and thus outcompeting all types of AIP (Lyon et al. 2000, 2002; McDowell et al. 2001). Based on the **S**tructure **a**ctivity **r**elationship (SAR) studies conducted on AIP-I, II and III, a few important points have been revealed. Modifications of these can convert the AIPs to global *agr* inhibitors, for instance, a 16-membered macrocycle important for binding. Playing with the size and stoichiometry of this ring is deleterious to AIP activity (McDowell et al. 2001; Johnson et al. 2015). Second, C-terminal end of AIPs have hydrophobic residues which are important for effective binding to AgrC. Point mutations on alanine at these particular positions destroy the potency of the AIPs (McDowell et al. 2001; Tal-Gan et al. 2013b). Lastly, structural modification of AIPs plays a detrimental role in its activity. Owing to this the second residue within the macrocycle and the exocyclic tail are required for AgrC activation. Modification and truncations of these sites lead to loss of its potency (Tal-Gan et al. 2013a). Owing to its peptidic backbone and its consequent higher immunogenicity and lack of stability in vivo, elaborate research is underway to make the peptidomimetics corresponding to these AIPs. For this purpose, modifications in AIP-III by replacement of amino acid residues with corresponding peptoids or N-methyl mimics has produced new QSIs (Tal-Gan et al. 2014). Further research is needed to completely turn them into peptidomimetics, which would help to bring them in clinical trials.

Recently a group of scientists created a focussed library of 63 peptidomimetic by using standard Fmoc **S**olid **P**hase **P**eptide **S**ynthesis (SPPS) method for evaluating AgrC inhibition in four groups of *S. aureus*. These were the simplified peptidomimetics of the previously reported truncated native AIP, *t*-AIP-II (Lyon et al. 2002; George et al. 2008). Out of these, three peptidomimetics namely *n*7FF, *n*8FF, and *n*7OFF inhibited AgrC activity in the clinically relevant group I: *S. aureus* strain with potencies similar to that of the parent peptide minus their shortcomings like solubility and stability (Vasquez et al. 2017). However, further research is required to test these non cognate AIPs as therapeutic agents to control infections by methicillin resistant *S. aureus* strains.

15.7 Analogues of Signal Molecules

Acyl homoserine lactone (AHL) is a class of QS molecule produced by gram negative bacteria and shows polymorphisms even in the same genera (Huma et al. 2011; Kalia 2014b). Two AHL compounds are produced by *P. aeruginosa*, which are short chain N-butanoyl-L-homoserine lactone (C4-HSL) and long chain N-(3-oxododecanoyl)-L-homoserine lactone (3-oxo-C12-HSL). These compounds regulate virulence and the generation of secondary metabolites. However only 3-oxo-C12-HSL acts on gram positive bacteria by inhibiting their growth. The 3-oxo-C12-HSL is earlier reported to have a killing effect on *S. aureus* (Kaufmann et al. 2005; Qazi et al. 2006). While at subinhibitory concentrations it hinders the release of *S. aureus* exotoxins (α-hemolysin, δ-hemolysin and toxic shock syndrome toxin) and thus acts as a quorum quenching agents (Qazi et al. 2006; Kalia et al. 2011). 3-oxo-C12-HSL undergoes intramolecular changes to give acid product 3-(1-hydroxydecylidene)-5-(2-hydroxyethyl)pyrrolidine-2,4-dione [(S)-5-hydroxyethyl-3-decanoyltetramic acid;8 5-HE-C10-TMA, 5] (Kaufmann et al. 2005). This belongs to TMA family of compounds which have antibacterial activity. Lately Murray and colleagues designed a series of 3-oxo-C12-HSL, TMA, and TOA analogues. This was done by bringing about systematic modifications on the parent compound 3-oxo-12-HSL focusing on (I) homoserine lactone, (II) 3-oxo substituent, (III) acyl side chain and (IV) amide structural units. HSL analogue namely 3-oxo-C12-HSL **1** having modifications in the homoserine lactone ring inhibited AgrC with an IC$_{50}$ of 22 ± 6 μM. TMA analogues (namely **3–13**) created by varying the 3-acyl chain length **3–8**, stereochemistry **9**, and substitution at the 5-position of the heterocyclic ring **12** and **13** were tested for their inhibitory activity against *agr*. It was observed that compound **4** 5-HE-C8-TMA has good inhibitory activity (42 ± 13 μM) against *agr*. It also fully abolished the expression of *agr*-mediated exotoxin α-hemolysin at 100 μM. This makes it a good candidate for future therapeutics however research should be focused on increasing its stability (Murray et al. 2014). Next in line are the TOA compounds (namely **14–18**) synthesized by bringing about variations in TMA structure wherein the ring nitrogen was replaced by oxygen. Upon evaluation of these TOAs against *S. aureus* growth and *agr* inhibition, it was found that C-14 TOA **17** was the most effective having an IC$_{50}$ of 3 ± 1 μM which is approximately 8 times lower than the MIC (25 μM). Another compound C-12 TOA **16** was found to be most potent than any other compound in preventing AIP mediated activation of AgrC by maintaining allosteric interaction with AgrC. Finally C-14 TOA **17** also reduced *S. aureus* colonization of human nasal passage. C-14 TOA **17** also showed its potency in mouse model system without any toxicity to host (Murray et al. 2014). Recently Zapotoczna and colleagues tested antibacterial and anti-biofilm potential along with a new sulphur-containing analogue (3-tetradecanoylthiotetronic acid; C14-TTA) towards MRSA and MSSA strains of *S. aureus*. Their potential clinical use as catheter lock solution was also examined using in vitro and in vivo models of IVC infection (Zapotoczna et al. 2015). Evaluation of biofilm killing activity of these compounds 5HE-C14-TMA killed over 50% of both MSSA and MRSA biofilms at 128 μg/ml with full abolishment at 512–1024 μg/ml. Similar results were obtained in in vivo rat model for IVC infections. However the efficacy of C14-TOA and C14-TTA were far less in killing MSSA and MRSA biofilms. Taking into account of all these observations 5HE-C14-TMA proves to be a compound of therapeutic value against *S. aureus* biofilms (Zapotoczna et al. 2017).

15.8 RNAIII-Inhibiting Peptide (RIP) Binds to TRAP

The key feature in pathogenesis of *S. aureus* is the regulation of toxin production. *S. aureus* produces different toxins during its proliferation that can cause severe disease. At the initiation of growth, when the population of *S. aureus* is scarce, various molecules required for adhesion such as protein A, fibronectin binding-proteins and fibrinogen binding-proteins are expressed and help bacteria to colonize and attach to host cells. Whereas at early stationary phase of growth, bacteria are in greater density, produce toxic molecules such as hemolysins, enterotoxins and Toxic Shock Syndrome Toxin-1 (TSST-1) that help the bacteria to spread, survive and initiate the infection (Lowy 1998). There are two QS mechanisms in *S. aureus* which regulates the production of toxin molecules in greater densities and adhesion molecules expression in lesser densities. The first one is RAP/TRAP QS system, made up of two components, RAP and TRAP (mentioned in introduction). RAP is a protein that activates the production of toxins by phosphorylating the histidine amino acid of TRAP, when RAP reaches a certain threshold concentration (Balaban et al. 2001). With an unknown mechanism, phosphorylation of TRAP causes increased cell attachment to the host and activation of *agr* QS system. The chromosomal locus, *agr* encodes RNAII and RNAIII transcripts. RNAII transcript encodes AgrA, AgrD, AgrC and AgrB, where propeptide AgrD is processed, and secreted in the form of an autoinducer AIP with the help of transmembrane protein, AgrB. In the mid exponential phase of growth, *agr* is activated which results in AIP secretion. The secreted AIP molecules then bind to the AgrC and causes AgrC phosphorylation. In turn, AgrA is activated which leads to RNAIII production. RNAIII upregulates the expression of toxins and downregulates the expression of cell surface proteins (Bronesky et al. 2016). In addition, AIP reduces the phosphorylation of TRAP and thus, leading to decreased cell adhesion (Balaban et al. 2001). RNAIII- inhibiting peptide (RIP) is a heptapeptide that can attenuate the virulence of *S. aureus*. YSPXTNF-NH2 was identified as a sequence of RIP (Balaban et al. 1998). RIP acts as a competitor of RAP on activating TRAP and thus inhibits its phosphorylation, which leads to attenuation of transcription from RNAII and RNAIII promoters and thus inhibiting toxin production. Synthetic analogues of RIP, YSPWTNF was made and shown to effectively inhibits the RNAIII synthesis in vitro and reduces the *S. aureus* infections caused by different strains in vivo, including osteomylitis,cellulitis, mastitis, septic arthritis and keratitis. Theoretically, RIP would lead to increase bacterial adhesion as it inhibits the RNAIII synthesis and RNAIII function is to decrease the cell surface adhesion molecules. But, by using atomic force and fluorescence microscopy, it was shown that RIP decreases attachment of bacterial cells to mammalian cells (HEP2) and to polystyrene. Thus, RIP can be used as a better therapeutic candidate for *S. aureus* infections (Gov et al. 2001). *S. aureus* infections connected to biofilm formation are commonly linked with the implantated medical devices (Costerton et al. 1999). After the removal of devices, the predominant species found on biofilms are *S. aureus* (Marr 2000). Biofilm is the structure formed due to QS or cell-cell communication and highly resistant to antibiotics. A novel way to treat biofilm related *S. aureus* infections is to use RNAIII inhibiting peptide, which disrupts the QS system and decreases bacterial adhesion. In an experiment, RIP was applied systematically and locally in a vascular-graft rat model, suggested that RIP completely inhibits the antibiotic- resistant *S. aureus* infections (Dell'Acqua et al. 2004). Therefore, RIP can thus be used as a coating material for various medical devices to be used during medical procedure. Moreover, antibiotics such as carbapenems (imipenem) and cephalosporins (cefazolin) in combination with RIP, inhibits the infection completely (Giacometti et al. 2003). Therefore, RIP can inhibit QS regulated toxin production and biofilm formation.

15.9 Non-peptide Analogues of RIP

Hamamelitannin (HAM), condensed tannin is a natural product obtained from the bark of the plant witch hazel (*Hamamelis virginiana*). It is the ester of D-hamamelose (2-hydroxymethyl-D-ribose) with 2 molecules of gallic acid (2′, 5-di-O-galloyl-Dhamamelose). Because gallic acid contains three phenolic functional groups, it is considered a polyphenol. Owing to studies on HAM in last decade, it emerged as a candidate of QSI of drug resistant Staphylococcal infection. It works by acting as non-peptide analogue of RIP and thus hinders biofilm formation. Non-peptide analogue of RIP also block the production of RNAIII in vitro as well as in vivo by blocking TRAP phosphorylation and thus affects TRAP mediated *agr* expression (Gov et al. 2004; Kiran et al. 2008). A recent study conducted by Brackman and colleagues showed that HAM increased the antibiotic susceptibility of *S. aureus* biofilms. It was observed that HAM in combination with vancomycin resulted in enhanced killing of *S. aureus* Mu50 biofilm cells compared to vancomycin alone in in vitro models. Similar results were observed for a combination of HAM with clindamycin. The in vivo effect of combined treatment was seen in *C. elegans* model system. HAM and vancomycin together significantly ($p < 0.01$) increased the survival of *S. aureus* Mu50 infected *C. elegans* model system (Brackman et al. 2011). They further elaborated their study to give the mechanistic view about the action of HAM by showing that this increase in susceptibility towards antibiotics is via affecting peptidoglycan biosynthesis and exogenous DNA (eDNA) release. Combintion of HAM with other antibiotics such as vancomycin, cefazolin, cefalonium, cephalexin, cefoxitin, daptomycin, linezolid, tobramycin or fusidic acid also significantly increase the killing of biofilm cells for various *S. aureus* strains. Mutations in gene belonging to QS and RNA sequencing studies showed that HAM has specificity towards TRAP receptor (Brackman et al. 2016).

However the structure of HAM makes it very polar affecting its bioavailability. It is also more prone to oxidation and glucoronidation because of its aromatic hydroxy functional moieties. Formation of ester linkages in vivo also raise an issue related to its stability (Vermote et al. 2016). Based on these observations, Vermote and group worked on making analogues of HAM by improving its stability. Three modifications were made in the HAM structure. These were modification or elimination of the aromatic hydroxy groups, replacement of the ester groups with isosteric linker moieties and lastly removal of the anomeric hydroxy group. This resulted in developing of rigid and structurally well-defined tetrahydrofuran core (position 5). Further changes led to the development of 58 analogues of HAM. Out of these the ortho chloro derivative i.e. **38** came out to be the most potent analogue of HAM. The compound **38** in combination with vancomycin resulted in enhanced killing of *S. aureus* Mu50 biofilm cells. Also it had better stability in vivo and displayed no cytotoxicity towards host cells. Thus giving **38** a better hand over HAM for therapeutic use (Vermote et al. 2016). In a latest study by same group more analogues were created by making changes at C-2′ position and conducting

Structure Activity Relationship (SAR) based studies. This led to the generation of 52 analogues of HAM focussing on benzamides with different substituents at different positions. Three derivatives namely **10u**, **15** and **25** showed promising results when tested for disruption of *S. aureus* biofilm cells in vitro and their susceptibility to vancomycin on these biofilm cells. These products warrant more study for their therapeutic use (Vermote et al. 2017).

15.10 Future Directions

Hospital borne infections are a nuisance to the medical industry. *Staphylococcus aureus* is the causative agent of diseases like endocarditis, osteomyelitis, and pneumonia. Using small molecule inhibitors to combat the infections is therapeutically effective approach in case of various pathogens like *Bacillus anthracis* (Dhasmana et al. 2014). As discussed in this review, various small molecule inhibitors have been tested against *S. aureus* which have proved their efficacy in various in vitro as well as in vivo model systems. However bacterial colonization takes places during the initial phases of disease establishment and hence the implication of QSIs becomes limiting. It is important to take precautionary measures in case of medical devices. These potent QSIs could be used as a coating material on these medical devices, which would help in reducing nosocomial infections by MRSA strains. Currently, there are fewer studies testing these inhibitors on various medical devices and this field should be explored further.

Acknowledgements This work is supported by J C Bose Fellowship (SERB) to YS and Research Grant by University of Delhi. NK is UGC-SRF. HG is Masters of Science in Zoology from University of Delhi. ND is Shyama Prasad Mukherjee Fellow (CSIR-SRF) and Fulbright-Nehru Doctoral Fellow (2015–16) at NIAID NIH.

Author Information The authors declare no competing financial interests. Correspondence and requests for materials should be addressed to YS (ysinghdu@gmail.com).

References

Agarwala M, Choudhury B, Yadav RN (2014) Comparative study of antibiofilm activity of copper oxide and iron oxide nanoparticles against multidrug resistant biofilm forming uropathogens. Indian J Microbiol 54:365–368. https://doi.org/10.1007/s12088-014-0462-z

Ahiwale SS, Bankar AV, Tagunde S, Kapadnis BP (2017) A bacteriophage mediated gold nanoparticle synthesis and their anti-biofilm activity. Indian J Microbiol 57:188–194. https://doi.org/10.1007/s12088-017-0640-x

Balaban N, Goldkorn T, Nhan RT, Dang LB, Scott S, Ridgley RM, Rasooly A, Wright SC, Larrick JW, Rasooly R, Carlson JR (1998) Autoinducer of virulence as a target for vaccine and therapy against *Staphylococcus aureus*. Science 280:438–440. https://doi.org/10.1126/science.280.5362.438

Balaban N, Goldkorn T, Gov Y, Hirshberg M, Koyfman N, Matthews HR, Nhan RT, Singh B, Uziel O (2001) Regulation of *Staphylococcus aureus* pathogenesis via target of RNAIII-activating protein (TRAP). J Biol Chem 276:2658–2667. https://doi.org/10.1074/jbc.M005446200

Baldry M, Kitir B, Frøkiær H, Christensen SB, Taverne N, Meijerink M, Franzyk H, Olsen CA, Wells JM, Ingmer H (2016) The agr inhibitors solonamide B and analogues alter immune responses to *Staphylococccus aureus* but do not exhibit adverse effects on immune cell functions. PLoS One 11:e0145618. https://doi.org/10.1371/journal.pone.0145618

Boisset S, Geissmann T, Huntzinger E, Fechter P, Bendridi N, Possedko M, Chevalier C, Helfer AC, Benito Y, Jacquier A, Gaspin C, Vandenesch F, Romby P (2007) *Staphylococcus aureus* RNAIII coordinately represses the synthesis of virulence factors and the transcription regulator Rot by an antisense mechanism. Genes Dev 21:1353–1366. https://doi.org/10.1101/gad.423507

Brackman G, Cos P, Maes L, Nelis HJ, Coenye T (2011) Quorum sensing inhibitors increase the susceptibility of bacterial biofilms to antibiotics in vitro and in vivo. Antimicrob Agents Chemother 55:2655–2661. https://doi.org/10.1128/aac.00045-11

Brackman G, Breyne K, De Rycke R, Vermote A, Van Nieuwerburgh F, Meyer E, Van Calenbergh S, Coenye T (2016) The quorum sensing inhibitor hamamelitannin increases antibiotic susceptibility of *Staphylococcus aureus* biofilms by affecting peptidoglycan biosynthesis and eDNA release. Sci Rep 6:20321. https://doi.org/10.1038/srep20321

Bronesky D, Wu Z, Marzi S, Walter P, Geissmann T, Moreau K, Vandenesch F, Caldelari I, Romby P (2016) *Staphylococcus aureus* RNAIII and its regulon link quorum sensing, stress responses, metabolic adaptation and regulation of virulence gene expression. J Clin Invest 70:299–316. https://doi.org/10.1146/annurev-micro-102215-095708

Bubeck Wardenburg J, Bae T, Otto M, Deleo FR, Schneewind O (2007) Poring over pores: alpha-hemolysin and Panton-valentine leukocidin in *Staphylococcus aureus* pneumonia. Nat Med 13:1405–1406. https://doi.org/10.1038/nm1207-1405

Costerton JW, Stewart PS, Greenberg EP (1999) Bacterial biofilms: a common cause of persistent infections. Science 284:1318–1322. https://doi.org/10.1126/science.284.5418.1318

Defoirdt T, Brackman G, Coenye T (2013) Quorum sensing inhibitors: how strong is the evidence? Trends Microbiol 21:619–624. https://doi.org/10.1016/j.tim.2013.09.006

Dell'Acqua G, Giacometti A, Cirioni O, Ghiselli R, Saba V, Scalise G, Gov Y, Balaban N (2004) Suppression of drug-resistant staphylococcal infections by the quorum-sensing inhibitor RNAIII-inhibiting peptide. J Infect Dis 190:318–320. https://doi.org/10.1086/386546

Dhasmana N, Singh LK, Bhaduri A, Misra R, Singh Y (2014) Recent developments in anti-dotes against anthrax. Recent Pat Antiinfect Drug Discov 9:83–96

Elmore BO, Triplett KD, Hall PR (2015) Apolipoprotein B48, the structural component of chylomicrons, is sufficient to antagonize *Staphylococcus aureus* quorum-sensing. PLoS One 10:e0125027. https://doi.org/10.1371/journal.pone.0125027

Femling JK, West SD, Hauswald EK, Gresham HD, Hall PR (2013) Nosocomial infections after severe trauma are associated with lower apolipoproteins B and AII. J Trauma Acute Care Surg 74:1067–1073. https://doi.org/10.1097/TA.0b013e3182826be0

Geisinger E, George EA, Muir TW, Novick RP (2008) Identification of ligand specificity determinants in AgrC, the *Staphylococcus aureus* quorum-sensing receptor. J Biol Chem 283:8930–8938. https://doi.org/10.1074/jbc.M710227200

George EA, Muir TW (2007) Molecular mechanisms of agr quorum sensing in virulent staphylococci. Chembiochem 8:847–855. https://doi.org/10.1002/cbic.200700023

George EA, Novick RP, Muir TW (2008) Cyclic peptide inhibitors of staphylococcal virulence prepared by Fmoc-based thiolactone peptide synthesis. J Am Chem Soc 130:4914–4924. https://doi.org/10.1021/ja711126e

Giacometti A, Cirioni O, Gov Y, Ghiselli R, Del Prete MS, Mocchegiani F, Saba V, Orlando F, Scalise G, Balaban N, Dell'Acqua G (2003) RNA III inhibiting peptide inhibits in vivo biofilm formation by drug-resistant *Staphylococcus aureus*. Antimicrob Agents Chemother 47:1979–1983. https://doi.org/10.1128/AAC.47.6.1979-1983.2003

Gordon RJ, Lowy FD (2008) Pathogenesis of methicillin-resistant *Staphylococcus aureus* infection. Clin Infect Dis 46:350–359. https://doi.org/10.1086/533591

Gov Y, Bitler A, Dell'Acqua G, Torres JV, Balaban N (2001) RNAIII inhibiting peptide (RIP), a global inhibitor of *Staphylococcus aureus* pathogenesis: structure and function analysis. Peptides 22:1609–1620. https://doi.org/10.1016/S0196-9781(01)00496-X

Gov Y, Borovok I, Korem M, Singh VK, Jayaswal RK, Wilkinson BJ, Rich SM, Balaban N (2004) Quorum sensing in staphylococci is regulated via phosphorylation of three conserved histidine residues. J Biol Chem 279:14665–14672. https://doi.org/10.1074/jbc.M311106200

Gui Z, Wang H, Ding T, Zhu W, Zhuang X, Chu W (2014) Azithromycin reduces the production of α-hemolysin and biofilm formation in *Staphylococcus aureus*. Indian J Microbiol 54:114–117. https://doi.org/10.1007/s12088-013-0438-4

Hall PR, Elmore BO, Spang CH, Alexander SM, Manifold-Wheeler BC, Castleman MJ, Daly SM, Peterson MM, Sully EK, Femling JK, Otto M, Horswill AR, Timmins GS, Gresham HD (2013) Nox2 modification of LDL is essential for optimal apolipoprotein B-mediated control of agr type III *Staphylococcus aureus* quorum-sensing. PLoS Pathog 9:e1003166. https://doi.org/10.1371/journal.ppat.1003166

Han R (2010) Plasma lipoproteins are important components of the immune system. Microbiol Immunol 54:246–253. https://doi.org/10.1111/j.1348-0421.2010.00203.x

Huma N, Shankar P, Kushwah J, Bhushan A, Joshi J, Mukherjee T, Raju SC, Purohit HJ, Kalia VC (2011) Diversity and polymorphism in AHL-lactonase gene (*aiiA*) of *Bacillus*. J Microbiol Biotechnol 21:1001–1011. https://doi.org/10.4014/jmb.1105.05056

Jarraud S, Lyon GJ, Figueiredo AMS, Gerard L, Vandenesch F, Etienne J, Muir TW, Novick RP (2000) Exfoliatin-producing strains define a fourth agr specificity group in *Staphylococcus aureus*. J Bacteriol 182:6517–6522. https://doi.org/10.1128/JB.182.22.6517-6522.2000

Ji G, Beavis R, Novick RP (1997) Bacterial interference caused by autoinducing peptide variants. Science 276:2027–2030. https://doi.org/10.1126/science.276.5321.2027

Johnson JG, Wang BY, Debelouchina GT, Novick RP, Muir TW (2015) Increasing AIP macrocycle size reveals key features of agr activation in *Staphylococcus aureus*. Chem Bio Chem 16:1093–1100. https://doi.org/10.1002/cbic.201500006

Kalia VC (2014a) Microbes, antimicrobials and resistance: the battle goes on. Indian J Microbiol 54:1–2. https://doi.org/10.1007/s12088-013-0443-7

Kalia VC (2014b) In search of versatile organisms for quorum-sensing inhibitors: acyl homoserine lactones (AHL)-acylase and AHL-lactonase. FEMS Microbiol Letts 359:143. https://doi.org/10.1111/1574-6968.12585

Kalia VC, Purohit HJ (2011) Quenching the quorum sensing system: potential antibacterial drug targets. Critical Rev Microbiol 37:121–140. https://doi.org/10.3109/1040841X.2010.532479

Kalia VC, Raju SC, Purohit HJ (2011) Genomic analysis reveals versatile organisms for quorum quenching enzymes: acyl-homoserine lactone-acylase and –lactonase. Open Microbiol J 5:1–13. https://doi.org/10.2174/1874285801105010001

Kalia VC (2015) Microbes: the most friendly beings? In: Quorum sensing vs quorum quenching: a battle with no end in sight. Springer, New Delhi, pp 1–5. http://dx.doi.org/10.1007/978-81-322-1982-8_1

Kattelmann KK, Hise M, Russell M, Charney P, Stokes M, Compher C (2006) Preliminary evidence for a medical nutrition therapy protocol: enteral feedings for critically ill patients. J Am Diet Assoc 106:1226–1241. https://doi.org/10.1016/j.jada.2006.05.320

Kaufmann GF, Sartorio R, Lee SY, Rogers CJ, Meijler MM, Moss JA, Clapham B, Brogan AP, Dickerson TJ, Janda KD (2005) Revisiting quorum sensing: discovery of additional chemical and biological functions for 3-oxo-N-acylhomoserine lactones. Proc Natl Acad Sci U S A 102:309–314. https://doi.org/10.1073/pnas.0408639102

Kiran MD, Adikesavan NV, Cirioni O, Giacometti A, Silvestri C, Scalise G, Ghiselli R, Saba V, Orlando F, Shoham M, Balaban N (2008) Discovery of a quorum-sensing inhibitor of drug resistant staphylococcal infections by structure-based virtual screening. Mol Pharmacol 73:1578–1586. https://doi.org/10.1124/mol.107.044164

Koenig RL, Ray JL, Maleki SJ, Smeltzer MS, Hurlburt BK (2004) *Staphylococcus aureus* AgrA binding to the RNAIII-agr regulatory region. J Bacteriol 186:7549–7555. https://doi.org/10.1128/JB.186.22.7549-7555.2004

Koul S, Kalia VC (2017) Multiplicity of quorum quenching enzymes: a potential mechanism to limit quorum sensing bacterial population. Indian J Microbiol 57:100–108. https://doi.org/10.1007/s12088-016-0633-1

Kumar P, Patel SKS, Lee J-K, Kalia VC (2013) Extending the limits of Bacillus for novel biotechnological applications. Biotechnol Adv 31(8):1543–1561

Kumar P, Koul S, Patel SKS, Lee JK, Kalia VC (2015) Heterologous expression of quorum sensing inhibitory genes in diverse organisms. In: Quorum sensing vs quorum quenching: a battle with no end in sight. Springer, New Delhi, pp 343–356. http://dx.doi.org/10.1007/978-81-322-1982-8_28

Le KY, Otto M (2015) Quorum-sensing regulation in staphylococci-an overview. Front Microbiol 6:1174. https://doi.org/10.3389/fmicb.2015.01174

Leonard PG, Bezar IF, Sidote DJ, Stock AM (2012) Identification of a hydrophobic cleft in the LytTR domain of AgrA as a locus for small molecule interactions that inhibit DNA binding. Biochemistry 51:10035–10043. https://doi.org/10.1021/bi3011785

Lipinski CA, Lombardo F, Dominy BW, Feeney PJ (2001) Experimental and computational approaches to estimate solubility and permeability in drug discovery and development settings. Adv Drug Deliv Rev 46:3–26. https://doi.org/10.1016/S0169-409X(00)00129-0

Loughman JA, Fritz SA, Storch GA, Hunstad DA (2009) Virulence gene expression in human community acquired *Staphylococcus aureus* infection. J Infect Dis 199:294–301. https://doi.org/10.1086/595982

Lowy FD (1998) *Staphylococcus aureus* infections. N Engl J Med 339:520–532. https://doi.org/10.1056/NEJM199808203390806

Lyon GJ, Mayville P, Muir TW, Novick RP (2000) Rational design of a global inhibitor of the virulence response in *Staphylococcus aureus*, based in part on localization of the site of inhibition to the receptor-histidine kinase AgrC. Proc Natl Acad Sci U S A 97:13330–13335. https://doi.org/10.1073/pnas.97.24.13330

Lyon GJ, Wright JS, Muir TW, Novick RP (2002) Key determinants of receptor activation in the agr autoinducing peptides of *Staphylococcus aureus*. Biochemistry 41:10095–11104. https://doi.org/10.1021/bi026049u

Mansson M, Nielsen A, Kjærulff L, Gotfredsen CH, Wietz M, Ingmer H, Gram L, Larsen TO (2011) Inhibition of virulence gene expression in *Staphylococcus aureus* by novel depsipeptides from a marine photobacterium. Mar Drugs 9:2537–2552. https://doi.org/10.3390/md9122537

Marr KA (2000) *Staphylococcus aureus* bacteremia in patients undergoing hemodialysis. Semin Dial 13:23–29. https://doi.org/10.1046/j.1525-139x.2000.00009.x

Mayville P, Ji G, Beavis R, Yang H, Goger M, Novick RP, Muir TW (1999) Structure-activity analysis of synthetic autoinducing thiolactone peptides from *Staphylococcus aureus* responsible for virulence. Proc Natl Acad Sci U S A 96:1218–1223. https://doi.org/10.1073/pnas.96.4.1218

McDowell P, Affas Z, Reynolds C, Holden MT, Wood SJ, Saint S, Cockayne A, Hill PJ, Dodd CE, Bycroft BW, Chan WC, Williams P (2001) Structure, activity and evolution of the group I thiolactone peptide quorum-sensing system of *Staphylococcus aureus*. Mol Microbiol 41:503–512. https://doi.org/10.1046/j.1365-2958.2001.02539.x

Murray EJ, Crowley RC, Truman A, Clarke SR, Cottam JA, Jadhav GP, Steele VR, O'Shea P, Lindholm C, Cockayne A, Chhabra SR, Chan WC, Williams P (2014) Targeting *Staphylococcus aureus* quorum sensing with nonpeptidic small molecule inhibitors. J Med Chem 57:2813–2819. https://doi.org/10.1021/jm500215s

Nielsen A, Månsson M, Bojer MS, Gram L, Larsen TO, Novick RP, Frees D, Frøkiær H, Ingmer H (2014) Solonamide B inhibits quorum sensing and reduces *Staphylococcus aureus* mediated killing of human neutrophils. PLoS One 9:e84992. https://doi.org/10.1371/journal.pone.0084992

Novick RP, Geisinger E (2008) Quorum sensing in staphylococci. Annu Rev Genet 42:541–564. https://doi.org/10.1146/annurev.genet.42.110807.091640

Qazi S, Middleton B, Muharram SH, Cockayne A, Hill P, O'Shea P, Chhabra SR, Cámara M, Williams P (2006) N-acylhomoserine lactones antagonize virulence gene expression and quorum sensing in *Staphylococcus aureus*. Infect Immun 74:910–919. https://doi.org/10.1128/IAI.74.2.910-919.2006

Sully EK, Malachowa N, Elmore BO, Alexander SM, Femling JK, Gray BM, DeLeo FR, Otto M, Cheung AL, Edwards BS, Sklar LA, Horswill AR, Hall PR, Gresham HD (2014) Selective chemical inhibition of agr quorum sensing in *Staphylococcus aureus* promotes host defense with minimal impact on resistance. PLoS Pathog 10:e1004174. https://doi.org/10.1371/journal.ppat.1004174

Swift S, Downie JA, Whitehead N, Barnard AML, Salmond GPC, Williams P (2001) Quorum sensing as a population density dependent determinant of bacterial physiology. Adv Microb Physiol 45:199–270. https://doi.org/10.1016/S0065-2911(01)45005-3

Tal-Gan Y, Ivancic M, Cornilescu G, Cornilescu CC, Blackwell HE (2013a) Structural characterization of native autoinducing peptides and abiotic analogues reveals key features essential for activation and inhibition of an AgrC quorum sensing receptor in *Staphylococcus aureus*. J Am Chem Soc 135:18436–18444. https://doi.org/10.1021/ja407533e

Tal-Gan Y, Stacy DM, Foegen MK, Koenig DW, Blackwell HE (2013b) Highly potent inhibitors of quorum sensing in *Staphylococcus aureus* revealed through a systematic synthetic study of the group-III autoinducing peptide. J Am Chem Soc 135:7869–7882. https://doi.org/10.1021/ja3112115

Tal-Gan Y, Stacy DM, Blackwell HE (2014) N-methyl and peptoid scans of an autoinducing peptide reveal new structural features required for inhibition and activation of AgrC quorum sensing receptors in *Staphylococcus aureus*. Chem Commun 50:3000–3003. https://doi.org/10.1039/c4cc00117f

Vasquez JK, Tal-Gan Y, Cornilescu G, Tyler KA, Blackwell HE (2017) Simplified AIP-II peptidomimetics are potent inhibitors of *Staphylococcus aureus* AgrC quorum sensing receptors. Chem Bio Chem 18:413–423. https://doi.org/10.1002/cbic.201600516

Vermote A, Brackman G, Risseeuw MDP, Vanhoutte B, Cos P, Van Hecke K, Breyne K, Meyer E, Coenye T, Van Calenbergh S (2016) Hamamelitannin analogues that modulate quorum sensing as potentiators of antibiotics against *Staphylococcus aureus*. Angew Chem Int Ed 55:6551–6555. https://doi.org/10.1002/anie.201601973

Vermote A, Brackman G, Risseeuw MDP, Cappoen D, Cos P, Coenye T, Van Calenbergh S (2017) Novel potentiators for vancomycin in the treatment of biofilm-related MRSA infections via a mix and match approach. ACS Med Chem Lett 8:38–42. https://doi.org/10.1021/acsmedchemlett.6b00315

Vuong C, Götz F, Otto M (2000) Construction and characterization of an agr deletion mutant of *Staphylococcus epidermidis*. Infect Immun 68:1048–1053. https://doi.org/10.1128/IAI.68.3.1048-1053.2000

Wadhwani SA, Shedbalkar UU, Singh R, Vashisth P, Pruthi V, Chopade BA (2016) Kinetics of synthesis of gold nanoparticles by *Acinetobacter* sp. SW30 isolated from environment. Indian J Microbiol 56:439–444. https://doi.org/10.1007/s12088-016-0598-0

Wang B, Muir TW (2016) Regulation of virulence in *Staphylococcus aureus*: molecular mechanisms and remaining puzzles. Cell Chem Biol 23:214–224. https://doi.org/10.1016/j.chembiol.2016.01.004

Wang R, Braughton KR, Kretschmer D, Bach TH, Queck SY, Li M, Kennedy AD, Dorward DW, Klebanoff SJ, Peschel A, DeLeo FR, Otto M (2007) Identification of novel cytolytic peptides as key virulence determinants for community-associated MRSA. Nat Med 13:1510–1514. https://doi.org/10.1038/nm1656

Williams P (2002) Quorum sensing: an emerging target for antibacterial chemotherapy? Expert Opin Ther Targets 6:257–274. https://doi.org/10.1517/14728222.6.3.257

Wright JS 3rd, Lyon GJ, George EA, Muir TW, Novick RP (2004) Hydrophobic interactions drive ligand-receptor recognition for activation and inhibition of staphylococcal quorum sensing. Proc Natl Acad Sci U S A 101:16168–16173. https://doi.org/10.1073/pnas.0404039101

Wright GD, Sutherland AD (2007) New strategies for combating multidrug-resistant bacteria. Trends Mol Med 13:260–267. https://doi.org/10.1016/j.molmed.2007.04.004

Yarwood JM, Schlievert PM (2003) Quorum sensing in *Staphylococcus* infections. J Clin Invest 112:1620–1625. https://doi.org/10.1172/JCI20442

Zapotoczna M, McCarthy H, Rudkin JK, O'Gara JP, O'Neill E (2015) An essential role for coagulase in *Staphylococcus aureus* biofilm development reveals new therapeutic possibilities for device-related infections. J Infect Dis 212:1883–1893. https://doi.org/10.1093/infdis/jiv319

Zapotoczna M, Murray EJ, Hogan S, O'Gara JP, Chhabra S, Chan WC, O'Neil E, Williams P (2017) 5-Hydroxyethyl-3-tetradecanoyltetramic acid represents a novel treatment for intravascular catheter infections due to *Staphylococcus aureus*. J Antimicrob Chemother 72:744–753. https://doi.org/10.1093/jac/dkw482

Part II
Plant Health

Chapter 16
Marine Biodiversity As a Resource for Bioactive Molecules As Inhibitors of Microbial Quorum Sensing Phenotypes

Faseela Hamza and Smita Zinjarde

Abstract The emergence of antibiotic resistant pathogenic strains is well documented and currently available drugs are becoming less effective in combating infections. In addition, pathogens often form biofilms as a survival strategy. This phenomenon is significant in aquaculture, industrial and environmental settings as well. In most pathogens, quorum sensing plays an important role in survival; development of virulence factors and in pathogenicity. An understanding of the molecular basis of this phenomenon has led to the development of new strategies for disease control and a search for molecules interfering with related processes. Marine environments are hotspots for biodiversity and various quorum sensing inhibitors (QSIs) have been reported from this habitat. This chapter describes the salient features of biosensor strains used for screening QSIs from the marine environment and details chemical structures of inhibitors derived from different marine organisms (bacteria, fungi, algae and invertebrates) in a classified manner. Different biotechnological applications of QSIs as antivirulence drugs, antibiofilm agents, antifouling compounds and their use in aquaculture practices have also been highlighted. QSIs may in the future may, to some extent, be useful as replacements for antibiotics.

Keywords Quorum sensing inhibitors · Marine · Bacteria · Fungi · Algae · Invertebrates

F. Hamza
Institute of Bioinformatics and Biotechnology, Savitribai Phule Pune University, Pune, India

S. Zinjarde (✉)
Institute of Bioinformatics and Biotechnology, Savitribai Phule Pune University, Pune, India

Department of Microbiology, Savitribai Phule Pune University, Pune, India
e-mail: smita@unipune.ac.in; hodmicro@unipune.ac.in

© Springer Nature Singapore Pte Ltd. 2018
V. C. Kalia (ed.), *Biotechnological Applications of Quorum Sensing Inhibitors*,
https://doi.org/10.1007/978-981-10-9026-4_16

16.1   Introduction

The indiscriminate and frequent use of antibiotics has led to decreased effectiveness of currently available drugs in combating infectious diseases. Most pathogens have evolved mechanisms of resisting antimicrobial agents. Susceptible strains develop into resistant ones by bringing about alterations in relevant genes or by obtaining them from resistant organisms (Arias and Murray 2008; Yong et al. 2009). Another important survival strategy that pathogens adopt is formation of biofilms (Lewis 2007). Biofilms are aggregates of microbial communities that grow on different biotic and abiotic surfaces and are particularly significant in the medical field. Biofilms are also prevalent in aquaculture, industrial, and environmental settings. Compared to their planktonic counterparts, they are more resistant towards antimicrobial agents and are difficult to eliminate (Høiby et al. 2010; Heidari et al. 2015). Biofilm formation is an important aspect for survival, virulence, and stress resistance in most pathogens (Faruque et al. 2006). On account of the aforementioned factors, there is need to explore alternative means of controlling biofilms and related infections.

Insights in the mechanisms related to bacterial pathogenesis and microbial communication have led to the development of alternative strategies for combating bacterial diseases. Within biofilms, bacterial behavior is regulated by the phenomenon of quorum sensing (QS) wherein virulence genes are expressed in a cell density-dependent manner and small signal molecules are produced (Bhargava et al. 2010; Romero et al. 2012). Interference with this type of intercellular signaling via quorum sensing inhibitors can affect the ability of pathogens to invade host tissues. Such compounds negatively modulate bacterial cell-to-cell-communication and thereby decrease virulence and biofilm formation.

16.2   Quorum Sensing (QS) Inhibition: An Anti-infective Strategy

Quorum sensing is a process of regulating expression of certain genes via small signal molecules. This mechanism was first discovered in *Vibrio fischeri* (Nealson et al. 1970). QS involves (i) production of extracellular signaling molecules referred to as autoinducers (ii) their detection by bacterial populations and (iii) elicitation of appropriate responses. QS systems are known to control diverse functions such as bioluminescence, conjugation, biofilm formation, antibiotic production, swarming, nodulation, sporulation and expression of virulence factors such as toxins, siderophores, lytic enzymes and adhesion molecules (Dunny and Leonard 1997; de Kievit and Iglewski 2000; Jayaraman and Wood 2008). In opportunistic pathogens such as *Pseudomonas aeruginosa*, more than 6% of the genes involved in pathogenesis are regulated by QS. During the initial stages of infection, expression of virulence related genes is low. When sufficiently high population densities are reached, QS mediates high expression of virulence related genes and disease progression occurs (Hentzer et al. 2003; Rasmussen et al. 2005).

Bacterial QS operates via five main types of signal molecules. These are N-acyl homoserine lactones (AHLs), oligopeptides (5–10 amino acid cyclic thiolactone), furanosyl borate (Autoinducer-2/AI-2), methyl dodecanoic acid and hydroxyl-palmitic acid methyl ester. Among these, AHLs produced by more than 70 species of Gram-negative bacteria and peptide based QS systems in Gram-positive bacteria are two most widely studied systems. AHLs diffuse across cell membranes and bind to regulatory proteins within the cell and peptides operate via membrane bound receptor histidine kinases (Kalia 2013). There are a large number of studies that focus on inhibiting bacterial QS and controlling microbial pathogenesis. Since this strategy is neither bactericidal nor bacteriostatic, selective pressures are not imposed and development of resistance is less likely. Identification of compounds inhibiting QS (quorum sensing inhibitors, QSIs) is of considerable interest as they can be used to decrease virulence, pathogenicity and biofilm formation. This strategy is particularly important in recent years when pathogenic bacteria are rapidly becoming resistant towards antibiotics. QSIs can be of biological or chemical origin. The former are preferred over the latter on account of their non toxic nature. Biological compounds can be obtained from a variety of organisms inhabiting different ecosystems including marine ones (Dobretsov et al. 2009). Marine bacteria, fungi, algae, sponges and tunicates to mention a few produce QSIs (Dobretsov et al. 2011a). In order to identify QSIs, a variety of biosensor reporter strains have been developed. These are generally genetically modified strains that express reporter genes when specific QS signals are received (Steindler and Venturi 2007). Some of the main biosensor strains used for screening QSIs from the marine environment are detailed below.

16.3 Biosensor Strains Used to Screen QSIs from the Marine Environment

During the screening of compounds for quorum sensing inhibitory activity, appropriate biosensors need to be employed. These strains allow qualitative and quantitative detection of QS signals. Most of the biosensors are based on AHL and AI-2 reporters. The main biosensor strains employed for screening and identification of QSIs from the marine environment based on their phenotypic expression are described in the following section.

16.3.1 Pigment Based Biosensor Strains

Chromobacterium violaceum is one of the most widely used biosensor strain. This bacterium synthesizes a purple pigment (violacein), which is regulated by CviI/CviR quorum sensing system. An AHL-deficient non pigmented mutant strain (CV026) has been employed as a biosensor to perceive the incidence of AHLs. In the presence of exogenous AHL, this non-pigmented biosensor produces violacein and forms purple colonies. In the presence of QSIs, reduced pigmentation is observed. Two more mutants (CV017 and VIR24) have also has been used for bioactive guided isolation of QSIs (McClean et al. 1997; Chernin et al. 1998; Someya et al. 2009). *Serratia marcescens* based biosensor strains have also been developed.

16.3.2   Bioluminescence Based Biosensor Strains

Screening for QSIs can be performed by using bioluminescence based biosensor strains. Inhibitory activity in such systems can be measured quantitatively (by using a luminometer) or qualitatively (by using bioluminescence microscopy). One of the most widely used bioluminescence-based QS reporter strain of *E. coli* is based on plasmid pSB401. This employs the *luxCDABE* operon from *Photorhabdus lumine-scens* under the control of *PluxI* gene and the *V. fischeri luxR* DNA fragment. Other reporters based on pSB403, pSB1075, pSB536 and pSU2007 have also can be used (Swift et al. 1997; Winson et al. 1998; Rasmussen et al. 2005).

16.3.3   β-Galactosidase Based Biosensors

Agrobacterium tumefaciens strain NT1 bearing plasmid pZLR4 is a regularly applied biosensor in this class. Strain NT1 does not produce native AHLs and harbors a plasmid encoding β-galactosidase. Different type of colonies (blue: when exogenous AHLs are present and colorless: when QSIs are included) are observed. This strain has the ability to respond to a wide range of AHLs at very low concentrations. Another biosensor strain in this category is the *E. coli* strain harboring the plasmid pKDT17 (Farrand et al. 2002).

16.3.4   Green Fluorescent Protein (Gfp) Based Biosensors

The commonly used *gfp*-based AHL sensor plasmids are pKR-C12, pAS-C8 and QSIS3. pKR-C12 is based on *lasB-gfp* (ASV) translational fusion wherein constitutive expression of *lasR* gene occurs under the influence P*lac* and is based on the plasmid pBBR1MCS-5. pAS-C_8 biosensor depends on the quorum sensing method associated with *Burkholderia cepacia* and is responsive to the presence of N-octanoyl-L-homoserine lactone (C_8-HSL). It contains P*cepI-gfp* (ASV) along with the *cepR* regulator gene, which is under the control of P*lac*. Qualitative and quantitative screening can be performed with these strains by using epifluorescence microscope (Andersen et al. 2001). *E. coli* QSIS3 based system (derived from *Vibrio fischeri* LuxR QS) is another popular system. When QSI is present, de-repression of antibiotic resistance leads to growth of the biosensor strain, which in turn produces fluorescence (Rasmussen et al. 2005). The biosensor *E. coli* strain (JB525 harboring the *gfp* plasmid pJBA132) has also been used. Another QS reporter system was constructed by harboring the *lasB-gfp* (ASV) fusion in a *P. aeruginosa* PAO1 Tn5-Las background where the expression of unstable Gfp (ASV) is regulated by the QS-controlled *PlasB*. In the presence of an exogenous QSI, decreased fluorescence that is proportional to the concentration and efficacy of the QS inhibitor is observed (Hentzer et al. 2002). All these indicator strains can be used to screen for molecules with potential QS inhibitory activities.

16.4 QSIs Derived from the Marine Environment

Marine locations are a rich resource of exclusive bioactive compounds, with diverse chemical structures. Prokaryotic and eukaryotic marine microorganisms (Fig. 16.1) and other invertebrates (bryozoans, sponges and cnidarians) have been reported to produce QSIs with varied biotechnological applications. In most of the cases, chemical structures of these compounds have also been elucidated. The following subsections describe QSIs obtained from different marine biological forms.

16.4.1 Marine Bacteria

Marine environments are hot spots for microbial diversity (Blunt et al. 2016). Marine bacteria produce an array of bioactive compounds to protect themselves from competing microorganisms and from bacterivorous Eukaryotic predators (Nasrolahi et al. 2012). Epibiotic bacteria that are associated with higher organisms such as corals, sponges, snails and mollusks often produce such compounds.

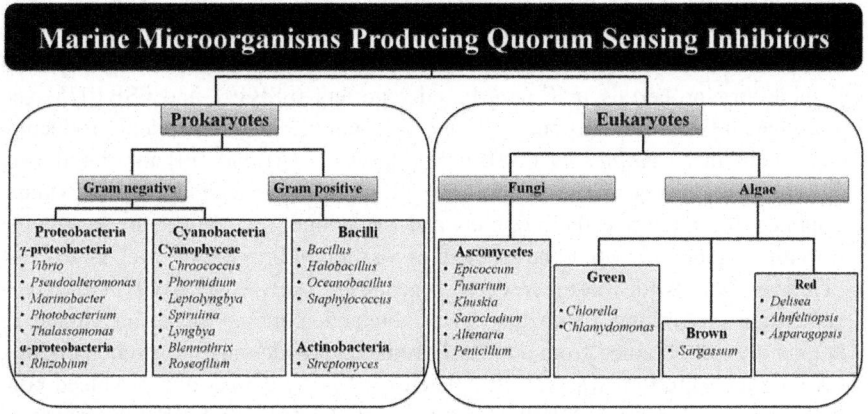

Fig. 16.1 Marine microorganisms as a potential source of QSIs

These bacteria are also potentially significant for bioprospecting different types of QSIs that can be used for various biotechnological applications (Dobretsov et al. 2015; Wu et al. 2015). Bacteria producing QSIs in this chapter have been categorized as per their Gram character and are depicted in the following text.

16.4.1.1   Gram Negative Bacteria

Several genera of Gram negative bacteria can inhibit QS by producing analogues of QS molecules or by degrading QS molecules such as AHL. Most of the reports are related to bacteria belonging to the Phylum Proteobacteria and Cyanobacteria as detailed below.

Proteobacteria

Bacteria belong to two classes of Phylum Proteobacteria namely; γ-proteobacteria and α-proteobacteria produce QSIs. Several medically and ecologically important bacteria are classified under the γ-proteobacteria class. Some members belonging to this class (*Vibrio*, *Pseudoalteromonas*, *Marinobacter* and *Photobacterium*) inhibit QS by synthesizing analogues of QS molecules. *Thalassomonas* on the other hand inhibits QS by degrading AHL molecules.

During a study on the isolation of QSIs from marine mixed populations, a variety of culturable bacteria were found to be bioactive. From the selectively isolated bacteria, 93.7% inhibited quorum sensing-based bioluminescence in *V. harveyi* and 21% displayed activity against quorum sensing-regulated pigment production in *S. marcescens*. These isolates belonged to the genera *Vibrio* and *Pseudoalteromonas*. Extracts from *Pseudoalteromonas* species also interfered with quorum-sensing-governed processes in *C. violaceum* (Linthorne et al. 2015).

Extracts obtained from a *Marinobacter* species (SK-3 obtained from south eastern Oman) repressed QS. Four correlated diketopiperazines (i) cyclo(L-Pro-L-Leu), (ii) cyclo(L-Pro-L-isoLeu), (iii) cyclo(L-Pro-L-Phe), and (iv) cyclo(L-Pro-D-Phe) were mined from this bacterium (Fig. 16.2a–d). These compounds displayed QS inhibitory properties in *E. coli*-based reporters (pSB401 and pSB1075) and *C. violaceum* CV017. Compounds (i) and (iii) inhibited QS-dependent production of violacein in *C. violaceum* CV017. Similarly, (i) (ii) and (iii) interfered with QS-dependent luminescence in *E. coli* pSB401. These isolates were extreme halophiles and moderate thermophiles and could find applications in developing antifouling agents that would be efficient in marine settings (Abed et al. 2013).

During a worldwide marine research expedition, several (over five hundred) bacterial strains inhibiting pathogens were obtained. Among these, an isolate of *Photobacterium* obtained from a mussel produced a wide range of cyclic peptides. This γ-proteobacterium was closely related to *P. halotolerans* and produced non-ribosomal peptides such as holomycin (Fig. 16.2e) that inhibited the growth of *Vibrio anguillarum* and *Staphylococcus aureus* (Wietz et al. 2010).

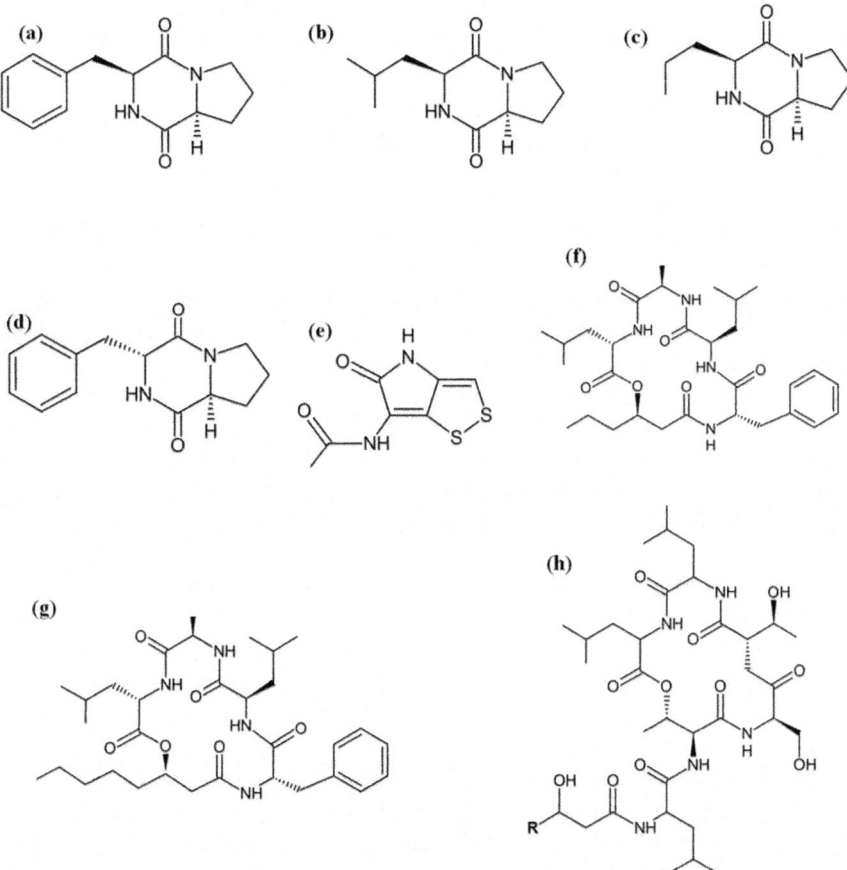

Fig. 16.2 Structural details of QSIs derived from Proteobacteria (**a**) cyclo(L-Pro-L-Phe), (**b**) cyclo(L-Pro-L-Leu), (**c**) cyclo(L-Pro-L-isoLeu), (**d**) cyclo(L-Pro-D-Phe), (**e**) holomycin, (**f**) solonamide A, (**g**) solonamide B, (**h**) ngercheumicin F (R = $C_{11}H_{21}$); G (R = $C_{11}H_{23}$); H (R = $C_{13}H_{25}$); I (R = $C_{13}H_{27}$)

In addition, a wide range of peptides (diketopiperazines and cyclodepsipeptides) have also been isolated. These compounds were tested for their ability to interfere with *agr* quorum sensing system that governs expression of virulence genes in *S. aureus*. It must be noted that the *agr* quorum sensing system depends on the presence of the autoinducing peptide (AIP). When required cell numbers in a population build up, extracellular toxins including *hla* encoding α-hemolysin and expression of surface factors such as the *spa* encoded Protein A is observed (Novick and Geisinger 2008). Among the peptides that were isolated during the study, two novel cyclodepsipeptides, designated solonamide A and B were found to be interfering with the *agr* quorum sensing system (Fig. 16.2f, g). Solonamides A and B were composed of four amino acids (L-Leu-D-Ala-D-Leu-L-Phe) linked to 3-hydroxyhexanoic and 3-hydroxyoctanoic acid, respectively (Mansson et al. 2011). During a continuation of this study, other cyclodepsipeptides (ngercheumicin F, G, H, and I) were described [Fig. 16.2h, ngercheumicin F (R = $C_{11}H_{21}$); G (R = $C_{11}H_{23}$); H (R = $C_{13}H_{25}$); I (R = $C_{13}H_{27}$)]. The four ngercheumicins consisted of six identical amino acids (three leucines, two threonines, and one serine) and a 3-hydroxy fatty acid. The structural difference between the four analogues was observed in the length and saturation of the unbranched fatty acid chain.

These analogues also affected the expression of virulence related genes via the *agr* QS system in *S. aureus*. The ngercheumicins were structural analogues of *S. aureus* associated AIPs (Kjaerulff et al. 2013). In another study, 127 bacterial isolates were obtained from coral species and screened for their QS inhibition properties. Among these isolates, approximately 12, 11, and 24% of the isolates exhibited QS inhibition when *E. coli* pSB1075, *C. violaceum* CV026, and *A. tumefaciens* KYC55 indicator strains, respectively were used. One isolate obtained from the coral *Favia* sp. identified as *V. harveyi* produced a compound that was in the mass range of AHL molecules (Golberg et al. 2013). Another species of *Vibrio* (*V. alginolyticus* G16) isolated from the red alga *Gracilaria gracilis* is reported to produce the QSI namely, phenol, 2,4-bis(1,1-dimethylethyl) abbreviated as PD (Padmavathi et al. 2014).

Some members of γ-Proteobacteria inhibit QS by degrading AHL molecules. *Thalassomonas* sp. PP2-459 isolated from a bivalve hatchery displayed quorum quenching (QQ) activity by degrading AHLs differing in alkyl chain lengths (C_4, C_6, C_8, C_{10} and C_{12}). This activity was tested by the standard agar plate diffusion assays using three biosensors namely, *C. violaceum* (CV026, VIR07) and *A. tumefaciens* NTL4. In addition, this bacterium degraded AHLs produced by *Halomonas anticariensis* FP35 and *Vibrio anguillarum* ATCC 19264 (Torres et al. 2013).

The extracts derived from the α-proteobacterium *Rhizobium* sp. NAO1 collected from Atlantic Ocean displayed QS inhibition in *C. violaceum* ATCC 12472. Secondary metabolites of this bacterium included AHL-based QS analogues that were effective in disrupting biofilm formation by *P. aeruginosa* PAO1 and in downregulating AHL mediated production of virulence factors (Chang et al. 2017).

Cyanobacteria

Memebrs of the Phylum Cyanobacteria are well known for their abilities to produce bioactive compounds (Kleigrewe et al. 2015). Most of the reports on QS inhibition activity are associated with members of Class Cyanophyceae. Extracts derived from cyanobacterial communities present in hot springs located in Oman inhibited QS in reporter strains *C. violaceum* CV017 and *A. tumefaciens* NTL4. These isolates mainly belonged to the genus *Chroococcus*, *Phormidium*, *Leptolyngbya*, *Spirulina* and *Lyngbya* (Dobretsov et al. 2011a).

Different isolates of *Lyngbya majuscula* synthesize a variety of QSIs. For example, extracts derived from *L. majuscula* collected from Florida, USA yielded three QSIs namely, malyngamide C (Fig. 16.3a, R=·······ₗₗOH) 8-epi-malyngamide C (Fig. 16.3a, R=———OH) and lyngbic acid (Fig. 16.3b). All these compounds ham-

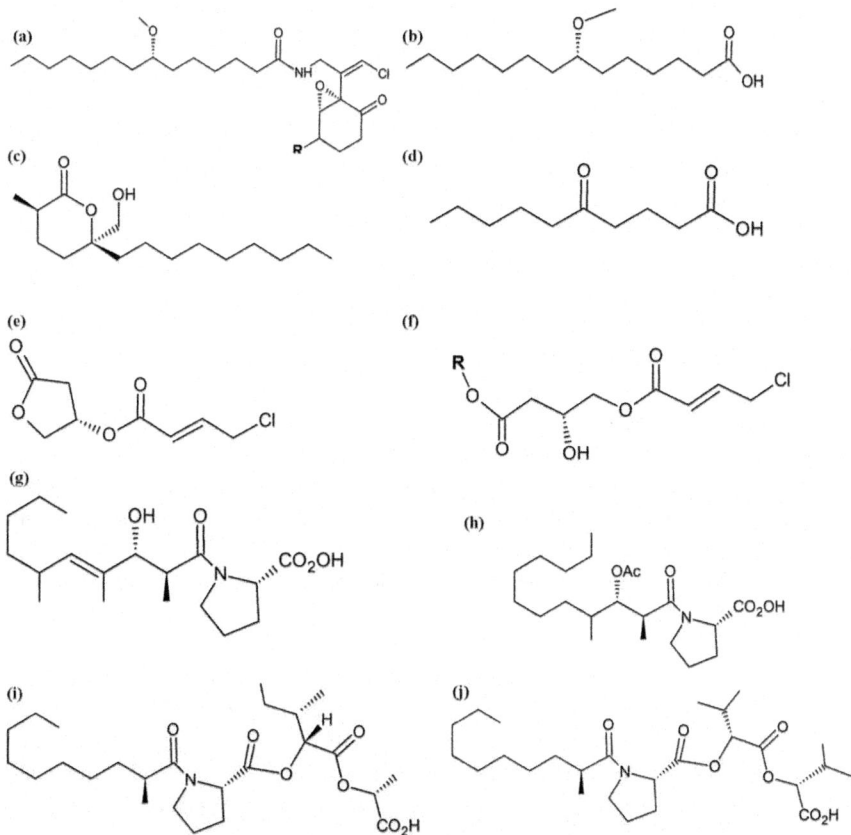

Fig. 16.3 Chemical structures of QSIs derived from marine Cyanobacteria (**a**) malyngamide C (R = ·····ııOH); 8-epi-malyngamide C (R = ▬▬OH) (**b**) lyngbic acid, (**c**) malyngolide (**d**) pitinoic acid A (**e**) honaucin A (**f**) honaucin B (R = CH$_2$CH$_3$); honaucin C (R = CH$_3$), (**g**) tumonoic acid E, (**h**) tumonoic acid F, (**i**) tumonoic acid G, (**j**) tumonoic acid H

pered 3-oxo-C$_{12}$-HSL based signaling in reporter based on pSB1075 (Kwan et al. 2010, 2011). Malyngolide (Fig. 16.3c), another compound obtained from *L. majuscula* decreases QS-dependent elastase production in *P. aeruginosa* PAO1 (Dobretsov et al. 2010). A peptide (lyngbyastatin) and two lipopeptides (microcolins A and B) isolated from *L. majuscula* also inhibited QS in the bacterial reporter *C. violaceum* CV017 (Dobretsov et al. 2011b). Pitinoic acid A (Fig. 16.3d) isolated from a Guamanian *Lyngbya* sp. inhibited QS in *P. aeruginosa*. This compound brought about a significant reduction in LasB and pyocyanin levels (Montaser et al. 2013).

Leptolyngbya crossbyana, Blennothrix cantharidosmum and Roseofilum reptotaenium are other cyanobacteria that synthesize QSIs. Three γ-butyrolactones (honaucins A–C) were obtained from *L. crossbyana* isolated from Hawaiian corals (Fig. 16.3e, f). All these compounds inhibited QS in *V. harveyi* BB120 and *E. coli* JB525. Several synthetic analogues of honaucin A were synthesized and screened for QSI activity. Analogs containing halogen atom at the 4-position of the crotonic acid subunit (4′-bromohonaucin A and 4′-iodohonaucin A) were more potent than the natural honaucins (Choi et al. 2012). Tumonoic acids A, D-I, were isolated from *B. cantharidosmum* samples native to Duke of York Island, Papua New Guinea. Among these compounds, Tumonoic acids E-H (Fig. 16.3g–j) inhibited biolumines-cence in wild-type strain *V. harveyi*, and the tumonoic acid F was most active (Clark et al. 2008). Lyngbic acid [(4E,7S)-7-methoxytetradec-4-enoic acid] obtained from the filamentous cyanobacterium *R. reptotaenium* is also reported to display strong QSI activity against *V. harveyi* (Meyer et al. 2016).

16.4.1.2 Gram Positive Bacteria

Gram positive bacteria belonging to two main classes namely; Bacilli (*Bacillus, Halobacillus, Oceanobacillus* and *Staphylococcus*) and Actinobacteria (*Streptomyces*) produce analogues of QS molecules. *Halobacillus salinus* isolated from a sea grass sample, inhibited bioluminescence in *V. harveyi*, violacein produc-tion in *C. violaceum* CV026 and GFP production in *E. coli* JB525. Two phenethyl-amide metabolites (Fig. 16.4a, b) were identified to inhibit bacterial QS by competing with AHLs (Teasdale et al. 2009). In a further study by these authors, several Gram positive bacterial isolates (mainly *Bacillus* and *Halobacillus*) col-lected from diverse marine habitats were screened for quorum sensing inhibitory activity. *N*- (2′-phenylethyl) isobutyramide and cyclo-L-proline-L-tyrosine and (Fig. 16.4a, c) derived from these isolates interfered with *V. harveyi* biolumines-cence and *C. violaceum* violacein production (Teasdale et al. 2011). In another study, tyrosol (Fig. 16.4d R = H) and tyrosol acetate (Fig. 16.4d R = Ac) obtained from *Oceanobacillus profundus* that was associated with Caribbean soft coral

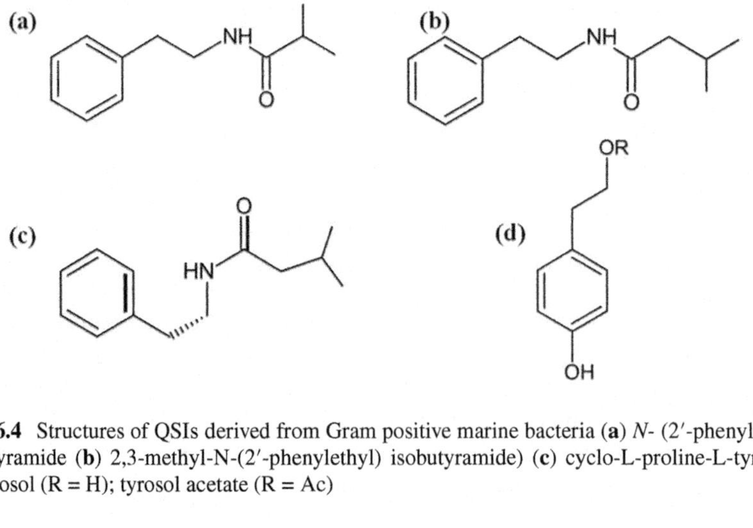

Fig. 16.4 Structures of QSIs derived from Gram positive marine bacteria (**a**) *N*- (2′-phenylethyl) isobutyramide (**b**) 2,3-methyl-N-(2′-phenylethyl) isobutyramide) (**c**) cyclo-L-proline-L-tyrosine (**d**) tyrosol (R = H); tyrosol acetate (R = Ac)

specimens (*Antillogorgia elisabethae*) displayed quorum sensing inhibitory activity when *C. violaceum* ATCC 31532 was used (Martínez-Matamoros et al. 2016). The diketopiperazine cyclo(Pro-Leu), extracted from a marine strain of *Staphylococcus saprophyticus* 108 inhibited QS in reporter strain *C. violaceum* 12,472 (Li et al. 2013).

Culture supernatants derived from *Streptomyces* species collected from marine sediments are also reported to display quorum sensing inhibition. These cultures could inhibit the synthesis of prodigiosin, a quorum-sensing regulated pigment in *S. marcescens*. Ethyl acetate extracts were effective in inhibiting biofilm formation by a clinical isolate of *Proteus mirabilis* and in attenuating QS dependent factors such as motility, hemolysin and urease production (Younis et al. 2016).

16.4.2 Marine Fungi

Marine fungi associated with algae, sponges, invertebrates, and sediments are a rich source for secondary metabolites such as alkaloids, polyketides, terpenoids, isoprenoid and quinines (Hasan et al. 2015). Most of the reports on QSIs are from Ascomycetous fungi. Extracts derived from the genera *Epicoccum, Fusarium, Khuskia* and *Sarocladium* isolated from coral reefs displayed quorum sensing inhibitory activity in *C. violaceum* CVO26 (Martín-Rodríguez et al. 2014). Kojic acid was isolated from *Altenaria* sp. associated with the green alga *Ulva pertusa* (Li et al. 2003). In a later study, the ability of this compound in inhibiting violacein production in *C. violaceum* CV017 and bioluminescence in *E. coli* pSB401 was demonstrated. Another compound, meleagrin obtained from *Penicillum chrysogenium* also inhibited QS in *C. violaceum* CV017 (Dobretsov et al. 2011b). In a recent study, six QSIs namely, aculene C-E, penicitor B, aspergillumarin A and B were isolated from *Penicillium* sp. SCS-KFD08. These compounds were also reduced violacein production in *C. violaceum* CV026 (Kong et al. 2017).

16.4.3 Marine Algae

Marine algae are reported to produce a variety of bioactive compounds including QSIs (Blunt et al. 2016). For example the effect of some microalgae strains on AHL-regulated QS in *V. harveyi* and *C. violaceum* CV026 has been investigated (Natrah et al. 2011). In particular, extracts derived from the green microalga *Chlorella saccharophila* CCAP211/48 inhibited violacein production in CV026 and interfered with bioluminescence in *V. harveyi* without affecting cell densities. *Chlamydomonas reinhardtii* is another green microalga that is reported to inhibit AHL-mediated luminescence. Contents of QSIs were found to be considerably higher in phototrophically cultured algae than in cultures grown on acetate (Teplitski et al. 2004). *Delisea pulchra* a red macroalga of marine origin is documented to produce halogenated furanones that act as antagonists for AHL-mediated QS. These halogenated furanones were structurally similar to AHLs and most probably bound to LuxR type proteins (Manefield et al. 2000). In addition, the red macroalga *Ahnfeltiopsis flabelliformes* produces AHL antagonists. After bioactivity-guided fractionation, three compounds namely, floridoside, betonicine and isethionic acid were isolated (Kim et al. 2007). The effectiveness of commercially available isethionic acid and chemically synthesized floridoside and betonicine individually and in combinations was also evaluated in a later study (Liu et al. 2008). Mixtures of floridoside and isethionic acid inhibited QS in a dose-dependent manner. Another marine red macroalgae *Asparagopsis taxiformis* displayed QSI activity against *C. violaceum* (CV026) and *Serratia liquefaciens* MG44. The active molecule in this study was identified as 2-dodecanoyloxyethanesulfonate (Jha et al. 2013). In a recent study, the extract of *Sargassum muticum* was observed to inhibit QS in the reporter *C. violaceum* CV017. The crude extract displayed strong antifouling activity against larval forms of a bryozoan (*Bugula neritina*) and the diatom *Cylindrotheca closterium* (Schwartz et al. 2017).

16.4.4 Marine Invertebrates

Certain invertebrates present in the marine environment are also significant in producing QSIs. In particular, Bryozoans, Sponges and Cnidarians produce such inhibitors of QS.

Bryozoans (also referred to as sea mosses) are avid producers of bioactive compounds with varied applications (Sharp et al. 2007). With respect to the production of QSIs, two brominated alkaloids were obtained from the North Sea bryozoan *Flustra foliacea* (Peters et al. 2003). These alkaloids exhibited antagonistic activities towards N-acyl-homoserine lactone-dependent QS systems observed in *Pseudomonas putida* (pKR-C12), *E. coli* (pSB403) and *P. putida* (pAS-C8).

Sponges are considered to be a rich source of natural products. Various bioactive compounds such as terpenes, sphingoids, taurinated fatty acids, polyacetylenes, peptides, and alkaloids have been reported from these invertebrates (Blunt et al. 2016). Most of the QSIs are reported from sponges that belong to Class Demospongiae as detailed here. The alkaloid hymenialdisin was originally isolated from the marine sponge *Axinella carteri* (Supriyono et al. 1995). During a study on the screening of several natural products, quorum sensing inhibitory activity associated with this alkaloid was demonstrated in reporter strains pSB401 and pSB1075 (Dobretsov et al. 2011b). Sponge samples collected from Colombian Caribbean Sea and Brazilian Coast were evaluated for QSIs. Crude extracts from *Svenzea tubulosa*, *Ircinia felix* and *Neopetrosia carbonaria* were found to be most promising. *I. felix* produced furanosesterterpenes that displayed were structurally similar to AHL (Quintana et al. 2015). Sponges prevalent in the Mediterranean (*Ircinia variabilis*, *Sarcotragus* species) and Red Sea (*Suberites clavatus* and *Negombata magnifica*) are also reported to produce QSIs. However, lead molecules from these sources have not been characterized (Saurav et al. 2016). Sponge extracts from the Great Barrier Reef were screened for QSIs (Skindersoe et al. 2008). Three related C_{25} sesterterpenes (manoalide, manoalide monoacetate, and secomanolaide) displaying quorum sensing inhibitory properties (in reporter systems QSIS1 and QSIS2) were purified from *Luffariella variabilis*. These compounds were also effective QSIs for *P. aeruginosa* PAO1. Two sponges namely, *Haliclona megastoma* and *Clathria atrasanguinea* were obtained from Palk Bay, India. Methanol extracts of these sponges inhibited violacein production in *C. violaceum* (ATCC 12472) and CV026 (Annapoorani et al. 2012).

Some sponges belonging to other classes Hexactinellida (*Aphrocallistes bocagei*), Homoscleromorpha (*Plakortis lita*) and Calcarea (*Leucetta chagosensis*) also produce QSIs. A γ- lactone plakofuranolactone capable of reducing protease activity (under the control of LasI/R QS system) was extracted from Indonesian specimens of *P. lita*. Quorum sensing inhibitory activity of this compound was evaluated using *E. coli* (pSB1075, pSB401) and *C. violeaceum* CV026 biosensor strains (Costantino et al. 2017). Two alkaloids namely, Isonaamidine A and isonaamine D were isolated from *L. chagosensis* that effectively inhibited quorum sensing pathways of *V. harveyi* (Mai et al. 2015).

Cnidarians are a diverse and ecologically important group of marine invertebrates characterized by specialized cells called cnidocytes or nematocytes that are utilized for capturing prey. This phylum mainly includes jellyfish, sea anemones and corals. More than 3000 bioactive natural products have been described from Cnidarians (Daly et al. 2007; Rocha et al. 2011). Two cembranoid diterpenes namely, knightol and knightal were isolated from the sea whip *Eunicea knighti* collected from the Colombian Caribbean. These exhibited QS inhibitory activity when evaluated with biosensors *E. coli* pS401 and *P. putida* IsoF wild type strain. In addition, semisynthetic compounds (obtained after transformation of natural compounds) displayed antifouling activity against different marine bacteria (Tello et al. 2009). In a later study, *E. knighti* was also shown to produce in low concentrations, three more cembranoid diterpenes (knightine, 11(R)-hydroxy-12(20)-en-knightol acetate and 11(R)-hydroxy-12(20)-en-knightal) (Tello et al. 2012). Three other cembranoid epimers (at the C-8 position) that inhibited production of violacein in *C. violaceum* ATCC 31532 were obtained from the organic extracts of the coral *Pseudoplexaura flagellosa* (Tello et al. 2011).

16.5 Biotechnological Applications of QSI

Most of the bacterial pathogens coordinate gene expression by using quorum sensing signal molecules. These signal molecules control diverse functions such as bioluminescence, conjugation, biofilm formation, antibiotic production, virulence and pathogenicity. Their ability to attenuate the abovementioned bioprocess signals makes them useful for a variety of biotechnological purposes. QSIs derived from the marine environment are mainly used as anti infective agents and some of their specific applications are detailed below and in Fig. 16.5.

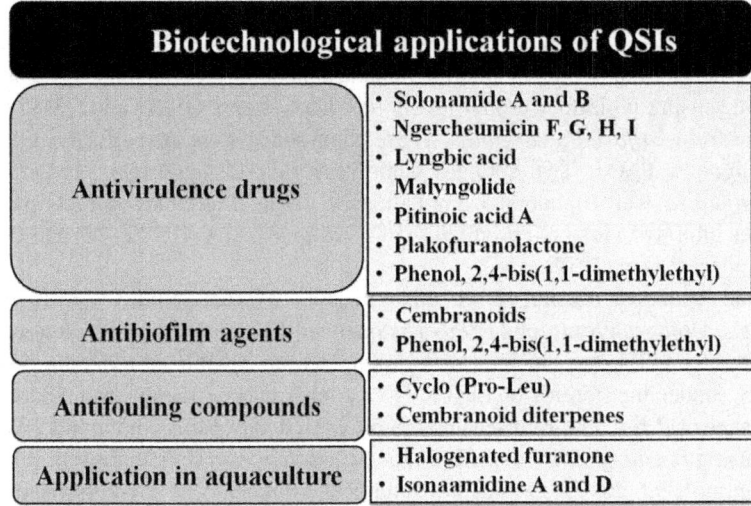

Fig. 16.5 Schematic representation of biotechnological applications of QSIs derived from marine organisms

16.5.1 Antivirulence Drugs

Several compounds from the marine environment are efficient in interfering with expression of virulence factors that are controlled via QS in pathogens such as *S. aureus, P. aeruginosa, S. marcescens* and *P. mirabilis*. For example, cyclodepsipeptides namely solonamide A, B, ngercheumicin F, G, H, and I derived from *P. halotolerans* are reported to interfere with the *agr* QS system in *S. aureus* USA300 (Mansson et al. 2011; Kjaerulff et al. 2013). Solonamide B interfered with the binding of AIPs derived from *S. aureus* to regulatory components (histidine kinase and AgrC) of the *agr* system. In this pathogenic strain, solonamide B also decreased the activity of central virulence factors such as α- hemolysin and the transcription of *psma* gene encoding phenol soluble modulins (PSMs). It must be noted that hypervirulence associated with this strain is linked to increased expression of α -hemolysin and PSMs. In the presence of QSIs, the strain thus exhibited lower toxicity towards human neutrophils and rabbit erythrocytes (Nielsen et al. 2014).

Various QSIs have also been used to attenuate virulence in *P. aeruginosa*. The extract derived from *Rhizobium* species is known to lower pathogenicity of *P. aeruginosa* PAO1 by down-regulating AHL mediated virulence factors such as elastase and siderophore production (Chang et al. 2017). Lyngbic acid, malyngolide and Pitinoic acid A obtained from *Lyngbya* species decrease QS regulated production of pyocyanin and elastase in *P. aeruginosa* (Kwan et al. 2011; Dobretsov et al. 2010; Montaser et al. 2013). Total protease activity (an important virulence factor) associated with *P. aeruginosa* was seen to be lower after treatment with plakofuranolactone a QSI extracted from the sponge *P. lita* (Costantino et al. 2017).

Methanolic extracts of some sponges (*A. bocagei*, *H. megastoma* and *C. atrasanguinea*) were efficient in altering virulence gene expressions in a clinical isolate of *S. marcescens* (PS1). These extracts inhibited AHL-dependent pigment (prodigiosin) production, virulence associated enzymes (protease, hemolysin) and biofilm formation (Annapoorani et al. 2012). Another compound PD [phenol, 2,4-bis(1,1-dimethylethyl)] extracted from *V. alginolyticus* G16 negatively affected the production of QS regulated virulence factors such as protease, haemolysin, lipase, prodigiosin and extracellular polysaccharide in *S. marcescens* without affecting growth (Padmavathi et al. 2014). In addition, culture supernatants derived from marine *Streptomyces* species were found to attenuate QS dependent virulence factors such as hemolysin, urease activity and motility in *P. mirabilis* (Younis et al. 2016). Such compounds capable of attenuating production of virulence factors and pathogenicity can be used for developing novel drugs.

16.5.2   Antibiofilm Agents

Bioactive compounds derived from the marine environment have also been evaluated for their antibiofilm potential. Compound PD, the QSI extracted from *Vibrio alginolyticus* G16 is reported to down-regulate genes involved in biofilm formation (*fimA*, *fimC*, *flhD* and *bsmA*) in *S. marcescens* and increase susceptibility towards gentamicin (Padmavathi et al. 2014). During another study, ethyl acetate extracts obtained from *Streptomyces* species were effective in controlling biofilm formation by an antibiotic resistant clinical isolate of *P. mirabilis* (Younis et al. 2016).

The extract derived from *Rhizobium* species is reported to be effective in inhibiting biofilm formation in *P. aeruginosa* PAO1 thereby increasing its sensitivity towards aminoglycoside antibiotics (Chang et al. 2017). The culture supernatant from *V. harveyi* was found to effectively inhibit biofilm formation by *P. aeruginosa* and *Acinetobacter baumannii* (Golberg et al. 2013). Various cembranoids also inhibit biofilm formation in *P. aeruginosa* ATCC 27853, *S. aureus* ATCC 25923 and *V. harveyi* PHY-2A (Tello et al. 2011, 2012). The ability of QSIs in inhibiting biofilm formation is of relevance not only in the field of medicine but also in industrial and environmental settings where biofilms may be detrimental.

16.5.3   Antifouling Agents

Various QSIs have also been exploited for their antifouling activities towards larger organisms and microorganisms. For example, larvae of the marine fouling organisms *Balanus reticulatus* and *Pinctada martensi* were prevented from settling on surfaces when cyclo (Pro-Leu) derived from *S. saprophyticus* was used (Li et al. 2013). Extracts obtained from the brown alga *S. muticum* displayed strong antifouling activity by disallowing settlement, growth and existence of *Cylindrotheca closterium,* a diatom. The larvae of certain bryozoa (*Bugula neritina*) were also considerably affected by this extract. The activity was also confirmed by field experiments (Schwartz et al. 2017).

QSIs isolated from the sea whip *E. knighti* (cembranoid diterpenes, knightol, knightal) and semisynthetic derivatives (obtained after chemical transformations of these natural compounds) exhibited antifouling activity towards bacteria. These compounds were found to be active against *Ochrobactrum pseudogringnonense,* *Alteromonas* sp., *Oceanobacillus iheyensis, Bacillus* sp., and *Kocuria* sp. isolated from the surface of a heavily fouled marine sponge (Tello et al. 2009).

16.5.4   Applications in Aquaculture

Aquaculture practices can also be improved by using QSIs derived from marine organisms. The use of antibiotics and sanitizers has limited success in preventing aquaculture associated diseases (King et al. 2008). Moreover, the persistent and unrestrained application of antibiotics has resulted in the emergence of resistant strains in such situations (Agersø et al. 2007; Akinbowale et al. 2007). In this regard, two different strategies have been employed. One makes use of microbial communities or pure cultures that are capable of degrading QS molecules and the other involves the introduction of QSIs that are capable of controlling pathogens associated with aquaculture species. Bacteria utilizing AHL molecules as carbon and nitrogen sources have been effective in protecting larvae of turbots and the fresh water prawn *Macrobrachium rosenbergii* from infections (Tinh et al. 2008; Cam et al. 2009a; Nhan et al. 2010). In one such study, AHL degrading enrichment cultures (EC) were obtained from fish samples. Two such cultures EC5(D) (composed of *Bacillus circulans, Bacillus* sp. and *Vibrio* sp) and EC5(L) (including members of the family Enterobacteriaceae) were able to degrade AHLs. These cultures protected prawn larvae from infections and increased their survival under experimental conditions (Cam et al. 2009a). Another enrichment culture capable of degrading homoserine lactone and accumulating poly-β- hydroxybutyrate (PHB) was effective as a biocontrol agent against *Vibrio* infections in *Artemia* (Cam et al. 2009b). Inclusion of such cultures in overcoming infections could prove to be an effective alternative for carrying out aquaculture farming in a sustainable manner. In a similar manner, *Thalassomonas* species (PP2–459) isolated from a bivalve hatchery was capable of degrading AHL produced by *V. anguillarum* ATCC 19264, a known pathogen in aquaculture settings. This bacterium could be introduced as a probiotic strain in bivalve aquaculture farms to enhance productivity and control infections (Torres et al. 2013).

Certain QSIs are reported to control infections in aquaculture farms. The halogenated furanone extracted from *Delisea pulchra* was structurally analogous with AHLs and was found to be an antagonist for AHL-mediated QS. Intramuscular administration of this QSI in the shrimp *Penaeus monodon* decreased disease-associated mortality by 50% (Manefield et al. 2000). Two other quorum sensing inhibitory alkaloids (Isonaamidine A and isonaamine D) isolated from *L. chagosensis* could also effectively inhibit QS pathways in *V. harveyi* an established aquaculture pathogen (Mai et al. 2015).

16.6   Opinion

The emergence of antibiotic resistant strains and biofilm formation are a major challenge in combating infectious diseases. The significance of quorum sensing in biofilm formation and in the production of virulence factors is well-known. An understanding on the molecular basis of quorum sensing has opened up new lines of investigations and a search for novel molecules that interfere with this phenomenon has been initiated. The marine environment is a rich source of bioactive molecules including QSIs. Marine organisms such as bacteria, fungi, algae, other invertebrates produce QSIs. Generally QS inhibition is mediated via the production of compounds that are analogous to QS molecules. In other cases, pathogen derived QS molecules are degraded by marine microorganisms. In most of the investigations, biomolecules involved in this process have been characterized and structural details have been provided. QSIs such as solonamide A and B, ngercheumicin F, G, H, and I, lyngbic acid, malyngolide, pitinoic acid A, plakofuranolactone and phenol, 2,4-bis(1,1-dimethylethyl) have been studied for suitability to use as antivirulence drugs. Compounds such as phenol, 2,4-bis(1,1-dimethylethyl) and cembranoids diterpenes are effective as antibiofilm agents. Similarly, cyclo (Pro-Leu) and cembranoid diterpenes are suitable for antifouling applications. Halogenated furanone, isonaamidine A and D find use in the aquaculture practices. To conclude, the use of QSIs in controlling pathogens may in the future, provide an effective alternative to combat bacterial infections.

References

Abed RM, Dobretsov S, Al-Fori M, Gunasekera SP, Sudesh K, Paul VJ (2013) Quorum-sensing inhibitory compounds from extremophilic microorganisms isolated from a hypersaline cyanobacterial mat. J Indust Microbiol Biotechnol 40:759–772. https://doi.org/10.1007/s10295-013-1276-4

Agersø Y, Bruun MS, Dalsgaard I, Larsen JL (2007) Tetracycline resistance gene *tet*(E) is frequently occurring and present on large horizontally transferable plasmids in *Aeromonas* spp. from fish farms. Aquaculture 266:47–52. https://doi.org/10.1016/j.aquaculture.2007.01.012

Akinbowale OL, Peng H, Barton MD (2007) Diversity of tetracycline resistance genes in bacteria from aquaculture sources in Australia. J Appl Microbiol 103:2016–2025. https://doi.org/10.1111/j.1365-2672.2007.03445.x

Andersen JB, Heydorn A, Hentzer M, Eberl L, Geisenberger O, Christensen BB, Molin S, Givskov M (2001) gfp-based N-acyl homoserine-lactone sensor systems for detection of bacterial communication. Appl Environ Microbiol 67:575–585. https://doi.org/10.1128/AEM.67.2.575-585.2001

Annapoorani A, Jabbar AKKA, Musthafa SKS, Pandian SK, Ravi AV (2012) Inhibition of quorum sensing mediated virulence factors production in urinary pathogen *Serratia marcescens* PS1 by marine sponges. Indian J Microbiol 52:160–166. https://doi.org/10.1007/s12088-012-0272-0

Arias CA, Murray BE (2008) Emergence and management of drug-resistant enterococcal infections. Expert Rev Anti-Infect Ther 6:637–655. https://doi.org/10.1586/14787210.6.5.637

Bhargava N, Sharma P, Capalash N (2010) Quorum sensing in *Acinetobacter*: an emerging pathogen. Crit Rev Microbiol 36:349–360. https://doi.org/10.3109/1040841X.2010.512269

Blunt JW, Copp BR, Keyzers RA, Munroa MH, Prinsepd MR (2016) Natural product reports. Nat Prod Rep 33:382–431. https://doi.org/10.1039/c5np00156k

Cam DTV, Nhan DT, Ceuppens S, Hao NV, Dierckens K, Wille M, Sorgeloos P, Bossier P (2009a) Effect of *N*-acyl homoserine lactone-degrading enrichment cultures on *Macrobrachium rosenbergii* larviculture. Aquaculture 294:5–13. https://doi.org/10.1016/j.aquaculture.2009.05.015

Cam DTV, Hao NV, Dierckens K, Defoirdt T, Boon N, Sorgeloos P, Bossier P (2009b) Novel approach of using homoserine lactone degrading and poly-β-hydroxybutyrate accumulating bacteria to protect *Artemia* from the pathogenic effects of *Vibrio harveyi*. Aquaculture 291:23–30. https://doi.org/10.1016/j.aquaculture.2009.03.009

Chang H, Zhou J, Zhu X, Yu S, Chen L, Jin H, Cai Z (2017) Strain identification and quorum sensing inhibition characterization of marine-derived *Rhizobium* sp. NAO1. R Soc Open Sci 4:170025. https://doi.org/10.1098/rsos.170025

Chernin LS, Winson MK, Thompson JM, Haran S, Bycroft BW, Chet I, Williams P, Stewart GSAB (1998) Chitinolytic activity in *Chromobacterium violaceum*: substrate analysis and regulation by quorum sensing. J Bacteriol 180:4435–4441. doi: Not available

Choi H, Mascuch SJ, Villa FA, Byrum T, Teasdale ME, Smith JE, Preskitt LB, Rowley DC, Gerwick L, Gerwick WH (2012) Honaucins A– C, potent inhibitors of inflammation and bacterial quorum sensing: synthetic derivatives and structure-activity relationships. Chem Biol 19:589–598. https://doi.org/10.1016/j.chembiol.2012.03.014

Clark BR, Engene N, Teasdale ME, Rowley DC, Matainaho T, Valeriote FA, Gerwick WH (2008) Natural products chemistry and taxonomy of the marine cyanobacterium *Blennothrix cantharidosmum*. J Nat Prod 71:1530–1537. https://doi.org/10.1021/np800088a

Costantino V, Della SG, Saurav K, Teta R, Bar-Shalom R, Mangoni A, Steindler L (2017) Plakofuranolactone as a Quorum quenching agent from the Indonesian sponge *Plakortis* cf. lita. Mar Drugs 15:59. https://doi.org/10.3390/md15030059

Daly M, Brugler MR, Cartwright P, Collins AG, Dawson MN, Fautin DG, France SC, McFadden CS, Opresko DM, Rodriguez E, Romano SL (2007) The phylum Cnidaria: a review of phylogenetic patterns and diversity 300 years after Linnaeus. Zootaxa 1668:127–182. www.mapress.com/zootaxa/

de Kievit TR, Iglewski BH (2000) Bacterial quorum sensing in pathogenic relationships. Infect Immun 68:4839–4849. https://doi.org/10.1128/IAI.68.9.4839-4849.2000

Dobretsov S, Teplitski M, Paul V (2009) Mini-review: quorum sensing in the marine environment and its relationship to biofouling. Biofouling 25:413–427. https://doi.org/10.1080/08927010902853516

Dobretsov S, Teplitski M, Alagely A, Gunasekera SP, Paul VJ (2010) Malyngolide from the cyanobacterium *Lyngbya majuscula* interferes with quorum sensing circuitry. Environ Microbiol Rep 2:739–744. https://doi.org/10.1111/j.1758-2229.2010.00169.x

Dobretsov S, Abed RM, Al-Maskari SM, Al-Sabahi JN, Victor R (2011a) Cyanobacterial mats from hot springs produce antimicrobial compounds and quorum-sensing inhibitors under natural conditions. J Appl Phycol 23:983–993. https://doi.org/10.1007/s10811-010-9627-2

Dobretsov S, Teplitski M, Bayer M, Gunasekera S, Proksch P, Paul VJ (2011b) Inhibition of marine biofouling by bacterial quorum sensing inhibitors. Biofouling 27:893–905. https://doi.org/10.1080/08927014.2011.609616

Dobretsov S, Al-Wahaibi AS, Lai D, Al-Sabahi J, Claereboudt M, Proksch P, Soussi B (2015) Inhibition of bacterial fouling by soft coral natural products. Int Biodeter Biodegr 98:53–58. https://doi.org/10.1016/j.ibiod.2014.10.019

Dunny GM, Leonard BAB (1997) Cell-cell communication in Gram-positive bacteria. Annu Rev Microbiol 51:527–564. doi: Not available

Farrand SK, Qin Y, Oger P (2002) Quorum-sensing system of *Agrobacterium* plasmids: analysis and utility. Methods Enzymol 358:452–484. https://doi.org/10.1016/S0076-6879(02)58108-8

Faruque SM, Biswas K, Udden SM, Ahmad QS, Sack DA, Nair GB, Mekalanos JJ (2006) Transmissibility of cholera: *in vivo*-formed biofilms and their relationship to infectivity and persistence in the environment. Proc Natl Acad Sci U S A 103:6350–6355. https://doi.org/10.1073/pnas.0601277103

Golberg K, Pavlov V, Marks RS, Kushmaro A (2013) Coral-associated bacteria, quorum sensing disrupters, and the regulation of biofouling. Biofouling 29:669–682. https://doi.org/10.1080/08927014.2013.796939

Hasan S, Ansari MI, Ahmad A, Mishra M (2015) Major bioactive metabolites from marine fungi: a review. Bioinformation 11:176. https://doi.org/10.6026/97320630011176

Heidari AE, Moghaddam S, Truong KK, Chou L, Genberg C, Brenner M, Chen Z (2015) Visualizing biofilm formation in endotracheal tubes using endoscopic three-dimensional optical coherence tomography. J Biomed Opt 20:126010. https://doi.org/10.1117/1.JBO.20.12.126010

Hentzer M, Riedel K, Rasmussen TB, Heydorn A, Andersen JB, Parsek MR, Rice SA, Eberl L, Molin S, Hoiby N, Kjelleberg S (2002) Inhibition of quorum sensing in *Pseudomonas aeruginosa* biofilm bacteria by a halogenated furanone compound. Microbiology 148:87–102. https://doi.org/10.1099/00221287-148-1-87

Hentzer M, Wu H, Andersen JB, Riedel K, Rasmussen TB, Bagge N, Kumar N, Schembri MA, Song Z, Kristoffersen P, Manefield M (2003) Attenuation of *Pseudomonas aeruginosa* virulence by quorum sensing inhibitors. EMBO J 22:3803–3815. https://doi.org/10.1093/emboj/cdg366

Høiby N, Bjarnsholt T, Givskov M, Molin S, Ciofu O (2010) Antibiotic resistance of bacterial biofilms. Int J Antimicrob Agents 35:322–332. https://doi.org/10.1016/j.ijantimicag.2009.12.011

Jayaraman A, Wood TK (2008) Bacterial quorum sensing: signals, circuits, and implications for *biofilms and disease*. Ann Rev Biomed Eng 10:145–167. https://doi.org/10.1146/annurev.bioeng.10.061807.160536

Jha B, Kavita K, Westphal J, Hartmann A, Schmitt-Kopplin P (2013) Quorum sensing inhibition by *Asparagopsis taxiformis*, a marine macro alga: separation of the compound that interrupts bacterial communication. Mar Drugs 11:253–265. https://doi.org/10.3390/md11010253

Kalia VC (2013) Quorum sensing inhibitors: an overview. Biotechnol Adv 31:224–245. https://doi.org/10.1016/j.biotechadv.2012.10.004

Kim J, Kim Y, Seo Y, Park S (2007) Quorum sensing inhibitors from the red alga, *Ahnfeltiopsis flabelliformis*. Biotechnol Bioprocess Eng 12:308–311. https://doi.org/10.1007/BF02931109

King RK, Flick GJ, Smith SA, Pierson MD, Boardman GD, Coale CW (2008) Response of bacterial biofilms in recirculating aquaculture systems to various sanitizers. J Appl Aquac 20:79–92. https://doi.org/10.1080/10454430802191766

Kjaerulff L, Nielsen A, Mansson M, Gram L, Larsen TO, Ingmer H, Gotfredsen CH (2013) Identification of four new agr quorum sensing-interfering cyclodepsipeptides from a marine Photobacterium. Mar Drugs 11:5051–5062. https://doi.org/10.3390/md11125051

Kleigrewe K, Almaliti J, Tian IY, Kinnel RB, Korobeynikov A, Monroe EA, Duggan BM, Di Marzo V, Sherman DH, Dorrestein PC, Gerwick L (2015) Combining mass spectrometric metabolic profiling with genomic analysis: a powerful approach for discovering natural products from cyanobacteria. J Nat Prod 78:1671–1682. https://doi.org/10.1021/acs.jnatprod.5b00301

Kong FD, Zhou LM, Ma QY, Huang SZ, Wang P, Dai HF, Zhao YX (2017) Metabolites with Gram-negative bacteria quorum sensing inhibitory activity from the marine animal endogenic fungus *Penicillium* sp. SCS-KFD08. Arch Pharm Res 40:25–31. https://doi.org/10.1007/s12272-016-0844-3

Kwan JC, Teplitski M, Gunasekera SP, Paul VJ, Luesch H (2010) Isolation and biological evaluation of 8-epi-malyngamide C from the Floridian marine cyanobacterium *Lyngbya majuscula*. J Nat Prod 73:463–466. https://doi.org/10.1021/np900614n

Kwan JC, Meickle T, Ladwa D, Teplitski M, Paul V, Luesch H (2011) Lyngbyoic acid, a "tagged" fatty acid from a marine cyanobacterium, disrupts quorum sensing in *Pseudomonas aeruginosa*. Mol BioSyst 7:1205–1216. https://doi.org/10.1039/C0MB00180E

Lewis K (2007) Persister cells, dormancy and infectious disease. Nat Rev Microbiol 5:48–56. https://doi.org/10.1038/nrmicro1557

Li X, Jeong JH, Lee KT, Rho JR, Choi HD, Kang JS, Son BW (2003) γ-Pyrone derivatives, kojic acid methyl ethers from a marine-derived fungus *Altenaria* sp. Arch Pharm Res 26:532–534. doi: Not available

Li M, Huiru Z, Biting D, Yun J, Wei J, Kunming D (2013) Study on the anti-quorum sensing activity of a marine bacterium *Staphylococcus saprophyticus* 108. Biotechol Indian J 7:11. doi: Not available

Linthorne JS, Chang BJ, Flematti GR, Ghisalberti EL, Sutton DC (2015) A direct pre-screen for marine bacteria producing compounds inhibiting quorum sensing reveals diverse planktonic bacteria that are bioactive. Mar Biotechnol 17:33–42. https://doi.org/10.1007/s10126-014-9592-x

Liu HB, Koh KP, Kim JS, Seo Y, Park S (2008) The effects of betonicine, floridoside, and isethionic acid from the red alga *Ahnfeltiopsis flabelliformis* on quorum-sensing activity. Biotechnol Bioprocess Eng 13:458–463. https://doi.org/10.1007/s12257-008-0145

Mai T, Tintillier F, Lucasson A, Moriou C, Bonno E, Petek S, Magre K, Al Mourabit A, Saulnier D, Debitus C (2015) Quorum sensing inhibitors from *Leucetta chagosensis* Dendy, 1863. Lett Appl Microbiol 61:311–317. https://doi.org/10.1111/lam.12461

Manefield M, Harris L, Rice SA, De Nys R, Kjelleberg S (2000) Inhibition of luminescence and virulence in the black tiger prawn (*Penaeus monodon*) pathogen *Vibrio harveyi* by intercellular signal antagonists. Appl Environ Microbiol 66:2079–2084. https://doi.org/10.1128/AEM.66.5.2079-2084.2000

Mansson M, Nielsen A, Kjærulff L, Gotfredsen CH, Wietz M, Ingmer H, Gram L, Larsen TO (2011) Inhibition of virulence gene expression in *Staphylococcus aureus* by novel depsipeptides from a marine *Photobacterium*. Mar Drugs 9:2537–2552. https://doi.org/10.3390/md9122537

Martínez-Matamoros D, Fonseca ML, Duque C, Ramos FA, Castellanos L (2016) Screening of marine bacterial strains as source of quorum sensing inhibitors (QSI): first chemical study of *Oceanobacillus profundus* (RKHC-62B). Vitae 23:30–47. https://doi.org/10.17533/udea.vitae.v23n1a04

Martín-Rodríguez AJ, Reyes F, Martín J, Pérez-Yépez J, León-Barrios M, Couttolenc A, Espinoza C, Trigos Á, Martín VS, Norte M, Fernández JJ (2014) Inhibition of bacterial quorum sensing by extracts from aquatic fungi: first report from marine endophytes. Mar Drugs 12:5503–5526. https://doi.org/10.3390/md12115503

McClean KH, Winson MK, Fish L, Taylor A, Chhabra SR, Camara M, Daykin M, Lamb JH, Swift S, Bycroft BW, Stewart GS (1997) Quorum sensing and *Chromobacterium violaceum*: exploitation of violacein production and inhibition for the detection of N-acylhomoserine lactones. Microbiology 143:3703–3711. https://doi.org/10.1099/00221287-143-12-3703

Meyer JL, Gunasekera SP, Scott RM, Paul VJ, Teplitski M (2016) Microbiome shifts and the inhibition of quorum sensing by Black Band disease cyanobacteria. ISME J 10:1204–1216. https://doi.org/10.1038/ismej.2015.184

Montaser R, Paul VJ, Luesch H (2013) Modular strategies for structure and function employed by marine cyanobacteria: characterization and synthesis of pitinoic acids. Org Lett 15:4050–4053. https://doi.org/10.1021/ol401396u

Nasrolahi A, Stratil SB, Jacob KJ, Wahl M (2012) A protective coat of microorganisms on macroalgae: inhibitory effects of bacterial biofilms and epibiotic microbial assemblages on barnacle attachment. FEMS Microbiol Ecol 81:583–595. https://doi.org/10.1111/j.1574-6941.2012.01384.x

Natrah FMI, Kenmegne MM, Wiyoto W, Sorgeloos P, Bossier P, Defoirdt T (2011) Effect of microalgae commonly used in aquaculture on acyl homoserine lactone quorum sensing. Aquaculture 317:53–57. https://doi.org/10.1016/j.aquaculture.2011.04.038

Nealson KH, Platt T, Hastings W (1970) Cellular control of the synthesis and activity of the bacterial biolumionescent system. J Bacteriol 104:313–322. doi: Not available

Nhan DT, Cam DTV, Wille M, Defoirdt T, Bossier P, Sorgeloos P (2010) Quorum quenching bacteria protect *Macrobrachium rosenbergii* larvae from *Vibrio harveyi* infection. J Appl Microbiol 109:1007–1016. https://doi.org/10.1111/j.1365-2672.2010.04728.x

Nielsen A, Månsson M, Bojer MS, Gram L, Larsen TO, Novick RP, Frees D, Frøkiær H, Ingmer H (2014) Solonamide B inhibits quorum sensing and reduces *Staphylococcus aureus* mediated killing of human neutrophils. PLoS One 9:e84992. https://doi.org/10.1371/journal.pone.0084992

Novick RP, Geisinger E (2008) Quorum sensing in staphylococci. Annu Rev Genet 42:541–564. https://doi.org/10.1146/annurev.genet.42.110807.091640

Padmavathi AR, Abinaya B, Pandian SK (2014) Phenol, 2, 4-bis (1, 1-dimethylethyl) of marine bacterial origin inhibits quorum sensing mediated biofilm formation in the uropathogen *Serratia marcescens*. Biofouling 30:1111–1122. https://doi.org/10.1080/08927014.2014.972386

Peters L, König GM, Wright AD, Pukall R, Stackebrandt E, Eberl L, Riedel K (2003) Secondary metabolites of *Flustra foliacea* and their influence on bacteria. Appl Environ Microbiol 69:3469–3475. https://doi.org/10.1128/AEM.69.6.3469-3475.2003

Quintana J, Brango-Vanegas J, Costa GM, Castellanos L, Arévalo C, Duque C (2015) Marine organisms as source of extracts to disrupt bacterial communication: bioguided isolation and identification of quorum sensing inhibitors from *Ircinia felix*. Rev Bras Farmacogn 25:199. https://doi.org/10.1016/j.bjp.2015.03.013

Rasmussen TB, Bjarnsholt T, Skindersoe ME, Hentzer M, Kristoffersen P, Kote M, Nielsen J, Eberl L, Givskov M (2005) Screening for quorum-sensing inhibitors (QSI) by use of a novel genetic system, the QSI selector. J Bacteriol 187:1799–1814. https://doi.org/10.1128/JB.187.5.1799-1814.2005

Rocha J, Peixe L, Gomes NCM, Calado R (2011) Cnidarians as a source of new marine bioactive compounds—an overview of the last decade and future steps for bioprospecting. Mar Drugs 9:1860–1886. https://doi.org/10.3390/md9101860

Romero M, Acuña L, Otero A (2012) Patents on quorum quenching: interfering with bacterial communication as a strategy to fight infections. Recent Pat Biotechnol 6:2–12. https://doi.org/10.2174/187220812799789208

Saurav K, Bar-Shalom R, Haber M, Burgsdorf I, Oliviero G, Costantino V, Morgenstern D, Steindler L (2016) In search of alternative antibiotic drugs: quorum-quenching activity in sponges and their bacterial isolates. Front Microbiol 7:416. https://doi.org/10.3389/fmicb.2016.00416

Schwartz N, Dobretsov S, Rohde S, Schupp PJ (2017) Comparison of antifouling properties of native and invasive *Sargassum* (Fucales, Phaeophyceae) species. Eur J Phycol 52:116–131. https://doi.org/10.1080/09670262.2016.1231345

Sharp JH, Winson MK, Porter JS (2007) Bryozoan metabolites: an ecological perspective. Nat Prod Rep 24:659–673. https://doi.org/10.1039/B617546E

Skindersoe ME, Ettinger-Epstein P, Rasmussen TB, Bjarnsholt T, de Nys R, Givskov M (2008) Quorum sensing antagonism from marine organisms. Mar Biotechnol 10:56–63. https://doi.org/10.1007/s10126-007-9036-y

Someya N, Morohoshi T, Okano N, Otsu E, Usuki K, Sayama M, Sekiguchi H, Ikeda T, Ishida S (2009) Distribution of N-acylhomoserine lactone-producing fluorescent pseudomonads in the phyllosphere and rhizosphere of potato (*Solanum tuberosum*L.) Microbes Environ 24:305–314. https://doi.org/10.1264/jsme2.ME09155

Steindler L, Venturi V (2007) Detection of quorum-sensing N-acyl homoserine lactone signal molecules by bacterial biosensors. FEMS Microbiol Lett 266:1–9. https://doi.org/10.1111/j.1574-6968.2006.00501.x

Supriyono A, Schwarz B, Wray V, Witte L, Müller WEG, Soest RV, Sumaryono W, Proksch P (1995) Bioactive alkaloids from the tropical marine sponge *Axinella carteri*. Z Naturforschung C 50:669–674. https://doi.org/10.1515/znc-1995-9-1012

Swift S, Karlyshev AV, Fish L, Durant EL, Winson MK, Chhabra SR, Williams P, Macintyre S, Stewart GS (1997) Quorum sensing in *Aeromonas hydrophila* and *Aeromonas salmonicida*: identification of the LuxRI homologs AhyRI and AsaRI and their cognate N-acylhomoserine lactone signal molecules. J Bacteriol 179:5271–5281. https://doi.org/10.1128/jb.179.17.5271-5281.1997

Teasdale ME, Liu J, Wallace J, Akhlaghi F, Rowley DC (2009) Secondary metabolites produced by the marine bacterium *Halobacillus salinus* that inhibit quorum sensing-controlled phenotypes in gram-negative bacteria. Appl Environ Microbiol 75:567–572. https://doi.org/10.1128/AEM.00632-08

Teasdale ME, Donovan KA, Forschner-Dancause SR, Rowley DC (2011) Gram-positive marine bacteria as a potential resource for the discovery of quorum sensing inhibitors. Mar Biotechnol 13:722–732. https://doi.org/10.1007/s10126-010-9334-7

Tello E, Castellanos L, Arevalo-Ferro C, Duque C (2009) Cembranoid diterpenes from the Caribbean Sea whip *Eunicea knighti*. J Nat Prod 72:1595–1602. https://doi.org/10.1021/np9002492

Tello E, Castellanos L, Arevalo-Ferro C, Rodríguez J, Jiménez C, Duque C (2011) Absolute stereochemistry of antifouling cembranoid epimers at C-8 from the Caribbean octocoral *Pseudoplexaura flagellosa*. Revised structures of plexaurolones. Tetrahedron 67:9112–9121. https://doi.org/10.1016/j.tet.2011.09.094

Tello E, Castellanos L, Arévalo-Ferro C, Duque C (2012) Disruption in quorum-sensing systems and bacterial biofilm inhibition by Cembranoid Diterpenes isolated from the octocoral *Eunicea knighti*. J Nat Prod 75:1637–1642. https://doi.org/10.1021/np300313k

Teplitski M, Chen H, Rajamani S, Gao M, Merighi M, Sayre RT, Robinson JB, Rolfe BG, Bauer WD (2004) *Chlamydomonas reinhardtii* secretes compound that mimic bacterial signals and interfere with quorum sensing regulation in bacteria. Plant Physiol 134:137–146. https://doi.org/10.1104/pp.103.029918

Tinh NTN, Yen VHN, Dierckens K, Sorgeloos P, Bossier P (2008) An acyl homoserine lactone-degrading microbial community improves the survival of first feeding turbot larvae (*Scophthalmus maximus* L.) Aquaculture 285:56–62. https://doi.org/10.1016/j.aquaculture.2008.08.018

Torres M, Romero M, Prado S, Dubert J, Tahrioui A, Otero A, Llamas I (2013) N-acylhomoserine lactone-degrading bacteria isolated from hatchery bivalve larval cultures. Microbiol Res 168:547–554. https://doi.org/10.1016/j.micres.2013.04.011

Wietz M, Mansson M, Gotfredsen CH, Larsen TO, Gram L (2010) Antibacterial compounds from marine Vibrionaceae isolated on a global expedition. Mar Drugs 8:2946–2960. https://doi.org/10.3390/md8122946

Winson MK, Swift S, Fish L, Throup JP, Jorgensen F, Chhabra SR, Bycroft BW, Williams P, Stewart GS (1998) Construction and analysis of luxCDABE-based plasmid sensors for investigating N-acyl homoserine lactone-mediated quorum sensing. FEMS Microbiol Lett 163:185–192. https://doi.org/10.1111/j.1574-6968.1998.tb13044.x

Wu B, Ohlendorf B, Oesker V, Wiese J, Malien S, Schmaljohann R, Imhoff JF (2015) Acetyl cholinesterase inhibitors from a marine fungus *Talaromyces* sp. strain LF458. Mar Biotechnol 17:110–119. https://doi.org/10.1007/s10126-014-9599-3

Yong D, Toleman MA, Giske CG, Cho HS, Sundman K, Lee K, Walsh TR (2009) Characterization of a new metallo-beta-lactamase gene, bla(NDM-1), and a novel erythromycin esterase gene carried on a unique genetic structure in *Klebsiella pneumoniae* sequence type 14 from India. Antimicrob Agents Chemother 53:5046–5054. https://doi.org/10.1128/AAC.00774-09

Younis KM, Usup G, Ahmad A (2016) Secondary metabolites produced by marine *streptomyces* as antibiofilm and quorum-sensing inhibitor of uropathogen *Proteus mirabilis*. Environ Sci Pollut Res 23:4756–4767. https://doi.org/10.1007/s11356-015-5687-9

Chapter 17
Quorum Sensing in Phytopathogenic Bacteria and Its Relevance in Plant Health

Firoz Ahmad Ansari and Iqbal Ahmad

Abstract Bacteria can regulate expression of certain genes through quorum sensing (QS), a cell density dependent communication system. The signal molecules up-regulate their own synthesis and hence act as auto-inducers. Most of well characterized phytopathogenic bacteria depend on this communication system for virulence and pathogenicity. Although pathogenesis is a multifactorial phenomenon but expression of virulence factors regulated by QS is a perquisite to cause infection. Majority of phytopathogenic bacteria are Gram negative and their signal molecules and QS systems are fairly investigated. Increased understanding on gene expression of QS regulated virulence functions has led to development of QS-disrupting strategies to fight bacterial plant diseases. Although success *in vivo* achieved in combating bacterial diseases are varied but encouraging. In this particular chapter, the role of QS in mediating virulence factors and pathogenicity among important phytopathogenic bacteria are reviewed and summaries various strategies of QS-disruption of plant pathogens to protect plant health.

Keywords Quorum sensing · Signal molecules · Phytopathogenic bacteria · Virulence factors · QS interference · Transgenic plant · Plant health

17.1 Introduction

Plant microbiome associated with rhizosphere and phyllosphere are diverse and vast majority live as saprophytic, symbiotic and endophytic. However, only a few bacterial species have become pathogenic to their host. A majority of the pathogens belong to Gram negative bacteria of certain families. Only few members are Gram positive. Expression of virulence factors in microorganisms defines their pathogenicity. Expression of virulence factors among micro-organisms, may be the structural

F. A. Ansari (✉) · I. Ahmad
Department of Agricultural Microbiology, Faculty of Agricultural Sciences,
Aligarh Muslim University, Aligarh, UP, India

© Springer Nature Singapore Pte Ltd. 2018
V. C. Kalia (ed.), *Biotechnological Applications of Quorum Sensing Inhibitors*,
https://doi.org/10.1007/978-981-10-9026-4_17

component and its products result in pathogenesis of the host. A number of chemical weapons (virulence factors) described in phytopathogenic bacteria is listed in Table 17.1. The expression of quite a few virulence factors are mediated at different level. Quorum sensing (QS) is considered as a global gene regulatory mechanism in microbes. Such regulation of genes through quorum sensing controls essential biological functions in bacteria including several virulence factors. QS signaling molecules are produced and maintain its concentration in proportion to its cell density. Theses signaling molecules at initial small concentration can amplify its production through transcription activators. These signaling molecules above a threshold level, result in expression of genes (Pereira et al. 2013). QS involves a variety of signal molecules: N-acylhomoserine lactones (AHLs) in gram-negative bacteria (Fuqua et al. 2001) and oligopeptides in Gram positive type of bacteria (Kleerebezem et al. 1997). Autoinducer 2 (AI-2) are common to all bacteria (Chen et al. 2002). Signal molecules like quinolones and cyclic dipeptides are reported in *Pseudomonas* species (Holden et al. 1999), where as diffusible molecules are reported among *Xyllela fastidiosa*, *Xanthomonas* spp., and *Burkholderia cenocepacia* (Ryan and Dow 2008). Presence of 3-hydroxy-palmitic acid methyl ester are reported by Flavier et al. (1997) in *Ralstonia solanacearum*. Similarly, γ-butyrolactone is found in *Streptomyces* spp. (Takano 2006). Bacteria coordinate their behavior according to neighboring communities under the influence of discrete responses that these signals receive (Genin and Denny 2012). *Xanthomonas campestris* possess diffusible signal factor (DSF) (Ham 2013), which includes a secondary messenger, cyclic-di-guanosine monophosphate (cyclic-di-GMP) that facilitates the coupling of QS networks (Kai et al. 2015). This system is an associated in the regulation of various bacterial traits. Recently, the research on QS has now gained attention beyond prokaryotic interactions after realizing that the signal produced by bacteria can modulate certain phenotypes in eukaryotic organisms (Hughes and Sperandio 2008). In this chapter, an extensive review on the recent literature and variable on the understanding on quorum sensing in phytopathogenic bacteria and its relevance in plant health.

17.2 Quorum Sensing (QS) in Phyto-Pathogenic Bacteria

Several bacteria causing disease in plants produce virulence factors mediated by quorum sensing. Some of the bacterial functions related to the virulence are induced only when a thresh hold density of the bacterial population reached to cells (Andersen et al. 2010). AHLs, signal molecules are comprised of a lactone ring connected to a fatty acyl side-chain.

Among phytopathogens, AHL-QS based infections, soft rot pectobacteria cause disease in many plants including potato (Põllumaa et al. 2012). In many pectobacteria, production of virulence factors such as plant cell wall degrading enzymes, necrosis inducing factors, secreted virulence factors and type III secreted functions are regulated by QS (Barras et al. 1994; Corbett et al. 2005; Whitehead et al. 2002; Barnard et al. 2007; Mukherjee et al. 1997; Barnard and Salmond 2007; Liu et al.

Table 17.1 Diversity of quorum sensing (QS) signals and QS-dependent traits of plant-pathogenic bacteria

Phytopathogenic bacteria	QS signal molecule	Phenotype	References
Agrobacterium tumefaciens	3-oxo-C8-HSL	Ti plasmid conjugal transfer genes	Tannières et al. (2017)
Burkholderia glumae	C6-HSL, C8-HSL	Toxoflavin biosynthesis and transport	Gao et al. (2015)
Dickeya zeae	3-oxo-C6-HSL, C6-HSL	Cell motility, aggregation	Hussain et al. (2008)
Dickeya dadantii	3-oxo-C6-HSL, 3-oxo-C8-HSL	Role of AHL not known	Crépin et al. (2012)
Dickeya dadantii	VFM signal	Extracellular cell wall-degrading enzymes	Nasser et al. (2013)
Pantoea stewartii ssp. *stewartii*	3-oxo-C6-HS	EPS stewartan, biofilm development, host colonization	Koutsoudis et al. (2006)
Pectobacterium atrosepticum	3-oxo-C6-HSL, C6-HSL, 3-oxoC8-HSL and 3-oxo-C10-HSL	Extracellular cell wall-degrading enzymes, antibiotic carbapenem, virulence factor	Crépin et al. (2012) and Valente et al. (2017)
Pectobacterium carotovorum	3-oxo-C6-HSL, C6-HSL, 3-oxoC8-HS	Extracellular cell wall-degrading enzymes, antibiotic carbapenem, harpin HrpN	Crépin et al. (2012)
Pseudomonas syringae pv. Syringae	3-oxo-C6-HSL	Exopolysaccharide (EPS), oxidative stress tolerance, extracellular degrading enzymes, negative regulator of swarming	Cheng et al. (2016)
Pseudomonas syringae pv. Tabaci	3-oxo-C6-HSL, C6-HSL	Negative regulation of biosurfactant, extracellular polysaccharides, iron acquisition, virulence	Taguchi et al. (2006)
Ralstonia solanacearum	3-Hydroxypalmitic acid methyl ester	EPS, endoglucanase, pectin methyl esterase	Mori et al. (2017)
Xanthomonas oryzae pv. oryzae	DSF, BDSF, CDSF	EPS, extracellular xylanase	Zheng et al. (2016)
Xanthomonas campestris pv. campestris	DF, DSF	Xanthomonadin, EPS, extracellular enzymes, biofilm dispersal, oxidative stress	He et al. (2011) and Kakkar et al. (2015)
Xyllela fastidiosa	DSF (*Xyllela*)	Biofilm formation in insects	Ionescu et al. (2014)

C6-HSL, N-hexanoyl-L-homoserine lactone; C8-HSL, N-octanoyl-L-homoserine lactone; 3-oxo-C6-HSL, N-(3-oxohexanoyl)-L-homoserine lactone; 3-oxo-C8-HSL, N-(3-oxooctanoyl)-L-homoserine lactone; 3-oxo-C10-HSL, N-(3-oxodecanoyl)-L-homoserine lactone; AHL, N-acylhomoserine lactone; DF, dodecenoic acid, 3-hydroxybenzoic acid; BDSF and CDSF, cis-11-methyldodeca-2, 5-dienoic acid; DSF (Xyllela), 12-methyl-tetradecanoic acid; VFM, virulence factor modulating signal—the signal has not yet been identified

2008). The common AHL molecules are produced includes C6HSL and C8HSL. Bacterial phyto-pathogenicity is mostly relies on type II secretion system, which secretes plant cell wall-degrading enzymes (PCWDEs). Pectinases, cellulases, hemicellulases and proteases are the common PCWDEs which disrupt the integrity of cell and cause rotting. However, soft rot enterobacteria *Erwinia amylovora* may cause infection without developing visible disease symptoms. QS in Pectobacterium strictly controls the production of PCWDEs whereas global regulator (PecS) restrict the early expression of PCWDEs in *D. dadantii*. This helps to prevent the premature activation of plant defenses because defense responses are triggered in the host plant with the action of PCWDEs that release cell wall fragment (Davidsson et al. 2013). It has been further observed that PCWDE and QS in *Pectobacterium* also influence the expression of other virulence factors relevant to plant defense suppression such as T3SS and the AvrE-like T3E DspE/A (Davidsson et al. 2013).

There are several plant pathogenic bacteria which are well known for their QS regulated pathogenicity (Table 17.1). Several workers have reviewed plant pathogenic bacteria based on their importance, a brief description of phytopathogens, their pathogenicity and possible role of quorum sensing is described below.

17.2.1 Ralstonia solanacearum

Ralstonia solanacearum belongs to pectobacteria β subdivision is found in soil. The bacterium infects several plants like potato, eggplant and tomato and various dicots crops and few monocots. The mode of entry of this pathogen could be from wounds, crack sites or root tips. It is widely distributed in tropics and subtropics.

The pathogen colonizes the root cortex and can successfully penetrate to the xylem vessels and reaches to the aerial parts of the plants. OE1-1 strain of *Ralstonia solanacearum* produces a signal molecule characterized as methyl 3-hydroxymyristate (3-OH MAME) and contribute to the virulence of bacteria. PhcA is known to be regulate virulence through the QS system through *ralA* that encodes for furanone synthase responsible for the synthesis of arylfuranone and ralfuranone are under regulated by QS. A mutant deficient in Ralfuranone (DralA) showed reduced virulence when inoculated directly into the xylem vessels of tomato. In addition, DralA showed down-regulated expression of QS positively regulated (>90%) genes and 75% of the up regulated genes. ralD and ralA encoded enzymes production, transaminase and furanone synthase respectively depend on PhcA function via phc QS system responsible for the production of biosynthesis of ralfuranone. (Kai et al. 2015). Mori et al. 2017 demonstrated by the using real time PCR that showed down regulated expression of vsrAD and vsrBC as the result of deletion of ralA and provided evidence that integrated signaling through ralfuranones modulate QS and virulence of *R. solanacearum*.

17.2.2   Agrobacterium tumefaciens

Agrobacterium tumefaciens, which causes crown gall disease in plants, affects various crop species worldwide and severely reduced the growth and crop yield. *A. tumefaciens* cause tumor development to plants through infecting plant cell by virulence part of Ti-plasmid (T-DNA) and several virulence proteins (Gohlke and Deeken 2014). It is interesting to note that transfer of oncogenic Ti-plasmid by *A. tumefaciens* is achieved by QS. The QS system belongs to LuxR/LuxI class and AHLs are the signal molecules. TraR gene of *A. tumefaciens* is found homologous to LuxR of *Vibrio fisheri.* There are two version of TraR, found simultaneously in Ti-plasmids coding for nopaline and octapine production. Further, synthesis of OC8HSL was found to be encoded by similar sequence to TraI of Ti-plasmid (Qin et al. 2007). *A. tumefaciens* QS system is similar to other LuxI/LuxR type, has another component responsible for negative modulation of TraR activity and OC8HSL. Ti-plasmid mediated production of TraM protein suppressing transcriptional activity of TraR (Gelencser et al. 2012). *A. tumefaciens* opines and QS signal pathways are interconnected and opines acts as the dominant regulator. Therefore presence of appropriate opines is necessary for TraA-3-oxoC8HSL signaling. In octapine type plasmid, TraR gene is controlled by an OccR (octapine responsive activator) as reported by Seet and Zhang 2011. OccR lead to transcription of traR in the presence of sufficient octapine. The TraR operon is expressed when sufficient 3-oxo-C8HSL is accumulated. However traR gene regulation in nopaline type Ti-plasmid is controlled by opines by the using different routes. Similarly role of agrocinopines A and B have induces expression and regulation of traR gene (Costa et al. 2012).

17.2.3   Pectobacterium *spp.*

Pectobacterium atrosepticum (Pca) is mainly limited to cooler climates (Toth et al. 2003). Soft rot disease of several crops is caused by *Pectobacterium carotovorum (Pcc)* whereas *Pca* is associated with of particular blackleg disease of potato (Pérombelon 2002). Phytopathogenesis of soft rot pectobacteria was first demonstrated the role of PCWDEs as important virulence factor. The important enzymes identified as cellulases, pectinases and proteases. Significant role of PCWDEs in virulence as resulted in the discovery of enzyme secretion system. Further, the significant role of PCWDEs in virulence has further resulted in the discovery of Type I and Type II secretory pathways (Evans et al. 2009). Many virulence factors and PCWDEs are produced and regulated through QS and Gac/Rsm pathway in *P. wasabiae* (Valente et al. 2017). Papenfort and Bassler (2016) reported significant role of RsmB in signal transduction responsible for induction of virulence of *P. wasabiae* in mixed species plant lesions.

Analysis of PCWDEs regulation mechanism in *Pcc* uncovered the role QS in the elaboration of diseases (Liu et al. 2008). The above role of QS in pectobacterial pathogensis was also demonstrated on genetically engineered plant based experimentation (Toth et al. 2004). Similarly production of 1-carbapen-2-em-3-carboxylic acid in *Pcc* was regulated by PCWDEs virulence through QSS. The importance of QS in plant physiology and virulence pathogenicity has now been well established through *in planta* transcriptome studies (Vakulskas et al. 2015).

17.2.4 Xanthomonas *spp.*

Xanthomonas spp., Proteobacteria produce yellow soluble pigments, xanthomonadins and EPSs, which help to protect them from the harsh environments. (Büttner and Bonas 2010). *Xanthomonas oryzae* infects rice plant mainly through the opening like hydathodes, trichomes and wounds in the plant surface. *X. oryzae* multiply in the intercellular spaces and enter the xylem vessels (Park et al. 2010). Similar to *Xanthomonas compestris sub sp. compestris* (*Xcc*), virulence factors and type such as production of EPS, extracellular enzymes and type III factor are considered essential for *Xoo* (An et al. 2014). Two QS factors, DSF and Ax21 and type I secreted proteins are employed by *Xoo* for expression of virulence and pathogenicity (Han et al. 2011). It has been shown that Ax21 play a binary role in the activation of host immune innate response. Biofilm development, motility and virulence are linked to Ax21. Further it was investigated that the *rfp* (regulation of pathogenicity factors) gene cluster is associated with DSF regulated quorum sensing (Yu et al. 2015). In *X. oryzae*, two component signal transduction systems are employed in adapting to the environmental factors. In this QS systems – SreKRS (salt response kinase, regulator, and StoS (stress tolerance-related oxygen sensor) and) are associated with regulation of EPS production and swarming. However absence of above system could not result in attenuation of virulence. EPS and swarming regulation is supposedly regulated by metabolism of carbohydrate and proteins respectively. Moreover, StoS and SreKRS demonstrate production of virulence factors, hypersensitivity response and pathogenicity (Hrp) proteins (Zheng et al. 2016).

17.2.5 Xylella fastidiosa

Xylella fastidiosa is a plant pathogen causing serious diseases in many agricultural crops including grapevine, almond and in citrus plant. Pathogenicity of *X. fastidiosa* is related with the ability of the organism to multiply within vascular system, xylem vessels and resulted in the development of symptoms due to interference in flow of xylem sap (Pierce et al. 2014).

Due to absence of T3SS in *X. fastidiosa*, it is believed that translocation of effectors does not occur in plant cell to elicit host response. However, T1SS and T2SSs are active in *X. fastidiosa*. The expression of virulence genes in *X. fastidiosa* involves an Rpf cell to cell communication system with DSF (Rai et al. 2012).

17.2.6 Erwinia amylovora

Erwinia amylovora belongs to the family Enterobactereaceae and causes disease of blackberry, quince, apple, raspberry and pear, and much wild and cultivated rosaceous ornamental plant (Vanneste 2000). Development of sporadic disease with destructive in nature especially in young fruit trees has been documented.

The important virulence factors of this pathogen have been characterized which includes (i) type III secretaries (ii), production of EPS (iii) motility and biofilm formation. A complex regulatory network is prerequisite to establishment infection by the bacterium to sense, respond and coordinate expression of virulence determinants. This involves two component signal transduction systems, bis-$(3'-5')$-cyclic di-GMP (c-di-GMP) and QS (Piqué et al. 2015). Due to absence of plant cell degrading tools, components like hrp pathogenicity island and the EPS appear to be its most important pathological tools. The role of T3SS secreted proteins – DspA/E and HrpN in pathogenicity has been well recognized (Bocsanczy et al. 2008). EPS, amylovoran and levan are major virulence factors which are crucial role in the process of biofilm establishment and pathogenicity (Koczan et al. 2009).

17.3 Strategies to Inhibit QS Signal and Its Influence on Plant Health

Several strategies are known to interfere the QS system thus reducing plant diseases and promote plant health. Some bacteria mainly species of *Bacillus* produce enzymes like Lactonase and Acylase which degrade the N-acylhomoserine lactone (AHL) by the process of Quorum Quenching (QQ) Fig. 17.1. Disruption of QS signal by volatile organic compounds has also been reported. Atleast three sulfur compounds, dimethyl sulfide, dimethyl disulfide and dimethyl trisulfide are produced by *Serratia plymuthica* was found to effectively inhibiting the growth of *Agrobacterium* (Dandurishvili et al. 2010a, b). A number of plant molecules regulate the QS via the secretion of dozen of molecules for their potential to interfere with bacterial signaling. Many plants secrete AHL mimics which can inhibit or even stimulate QSS systems (Bauer and Mathesius 2004; Teplitski et al. 2011; Tannières et al. 2017).

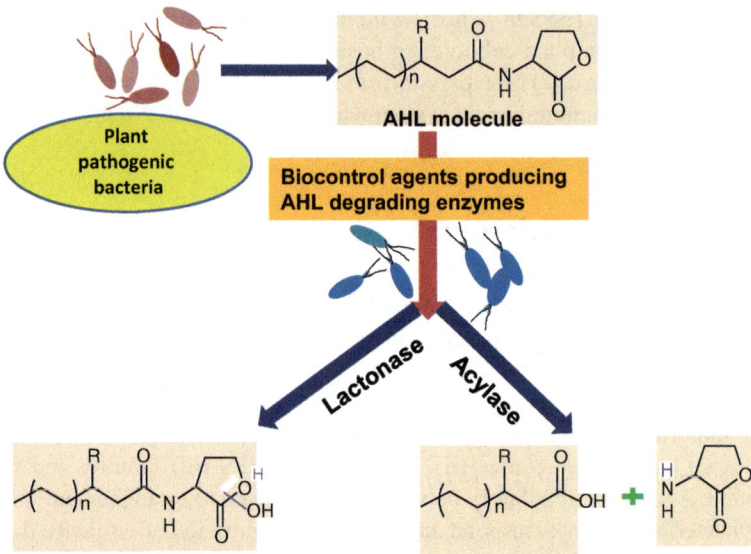

Fig. 17.1 Degradation of *N*-acylhomoserinelactone by AHL degrading enzymes lactonase and acylase

17.3.1 QS Disruption by Enzymatic Degradation of AHL

The inhibition of quorum sensing by QS inhibiting enzymes has been recognized as a promising strategy. The AHL degrading enzymes was first detected and purified from *Bacillus sp.* and the gene responsible was named as aiiA (AI inactivation). The AHL inactivation is achieved by hydrolysis of lactone ring. It is believe that the production of such enzyme by bacteria might be primarily aimed to degrade AHL to use it as source of C and energy rather than QQ (LaSarre and Federleb 2013). Quorum quenching activities were further identified in other bacteria such as *Variovorax paradoxus* (Hanano et al. 2014) and later in *Ralstonia* strain XJ12B. AHL degradation by *Ralstonia* sp. is achieved by the production of acylase enzyme (aiiD) through breaking amide linkage of AHL (Lin et al. 2003). Expression of aiiD acylase in *P. aeruginosa* resulted in the reduction of AHL released from the transformed bacterium as well as reduced production of proteolytic enzymes and swarming. Several workers now have reported AHL degrading enzymes from different bacteria such as *Bacillus spp.* (Han et al. 2010), *Pseudomonas spp.* (Fekete et al. 2010), *Arthrobacter* sp., *Klebsiella pneumonia* (Park et al. 2003) *Rhodococcus spp.* (Park et al. 2006), *Comamonas* spp. (Uroz et al. 2003), *Agrobacterium tumefaciens* (Haudecoeur et al. 2009), *Brucella melitensis* (Terwagne et al. 2013), *Ochrobactrum* sp. (Mei et al. 2010), *Microbacterium* spp. (Wang et al. 2012), and *Ralstonia* sp. (Chen et al. 2009; Han et al. 2010), *Actinobacter sp.* (Kang et al. 2004) and *Lysinibacillus* sp.(Garge and Nerurkar 2016). More recently Torres et al. (2017) have reported a novel QQ enzyme of AHL lactonase family with broad spectrum activity.

17.3.2 AHL- Degrading Bacteria As Biocontrol Agents

Exploitation of AHL degrading bacteria and its gene to control the phytopathogenic bacteria by quenching/ attenuating their virulence and pathogenicity mediated by QS is under scrutiny Molina et al. (2003) have made first attempt in this direction. Two biocontrol strain *Bacillus sp*. A24 and genetically modified *P. fluorescens* carrying aiiA genes were listed against phytopathogenic bacteria, *A. tumefaciens* and *Pe. Carotovorum*. Infected potato plants with *Pe. carotovorum* showed significant reduction in rot symptoms when co-inoculation with *Bacillus* sp. A24. Similar protection to *A. tumefaciens* infection was observed when co-inoculation with *P. fluorescens* β – pME6863 and *Bacillus* sp. in tomato plants.

Molina et al. have revealed biocontrol activity by using bacteria harboring AHL-degrading enzymes. In another study, *Rhodococcus* sp. have been investigated. Theses bacteria possess QQ enzymatic activities such as lactonase, reductase and amidohydrolase (Wu et al. 2016; Cirou et al. 2012). A significant increase in the occurrence of *Rhodococcus* spp. was reported through biostimulation by γ-caprolactones in hydroponic potato plants systems in the potato plants. Hayward et al. (2010) have reported *Lysobacter enzymogenes* as an effective PGPR and capable of controlling pathogenic fungi such as *Rhizoctonia solani, Fusarium graminearum* and *F. solani* and by the oomycete *Phytophthora capsici*. However it is effective against *Pectobacterium* spp. The efficacy of this bacterium has been improved by introducing aiiA by gentic engineering against bacterial soft rot (Qian et al. (2010) Similar attempt has been made by Li et al. (2011) to introduce aiiA gene in *P. putida*. The genetically engineered strain of *P. putida* provided protection potato infected with *Pe. carotovorum*.

17.3.3 Disruption of Quorum Sensing Volatile Compounds of Bacteria

Volatile organic compounds (VOCs) produced by plant associated bacteria are low molecular mass and have received attention by researchres (Insam and Seewald 2010) due to their role in inter kingdom communications (Mendes et al. 2013). Thses compounds exhibits antibiotic action (Wenke et al. 2010), inducses systemic resistance in plants (Farag et al. 2013) and associated with plant growth (Blom et al. 2011; Bailly and Weisskopf 2012). A chemically many of these volatiles compounds are dimethyl disulfide (DMDS), dimethyl sulfide, and dimethyl trisulfide (Schulz and Dickschat 2007).

The growth inhibitory activity of VOCs by *P. fluorescens* and *S. plymuthica* has been demonstrated against *A. tumefaciens* and *A. vitis*. The major volatile compounds identified as dimethyl disulfide (DMDS) produced by *S. plymuthica* IC1270 whereas 1-Undecene, the volatile compound was produced by *P. fluorescens*. A.

tumefaciens could be effectively controlled by the use of *S. plymuthica* and *P. fluorescens* (Dandurishvili et al. 2010a, b).

Production of DMDS has been reported both from microbes like *Burkholderia ambifaria* (Groenhagen et al. 2013), *S. plymuthica* (Müller et al. 2009), and others such as *Allium* and *Brassica* (Kyung and Lee 2001). Interestingly, QQ activity of VOCs has been reported by Chernin (2011). The mechanism of interference of QS by VOCs produced by bacteria was found to inhibition of AHL synthesis reversibly. This has suggested that VOCs and AHL can compete in the same bacterium affecting its ability to induce the QS response. Therefore, interaction between these two signals can influence quorum sensing induction (Grandclément et al. 2015).

17.3.4 Transgenic Plants Development with Genes Encoding AHL- Degrading Enzymes

Studies conducted in the recent past have indicated that transgenic plants with AHL degrading gene expression provided protection to specific phytopathogens (Ouyang and Li 2016). *Eucalyptus europhylla* and potato plants reported expression of aiiA genes showed a significant reduction in pathogenesis of *Pe. carotovorum* espressed on the leaves (tobaco) or tubers (potato). This attenuation of virulence provide enough time to activate and buildup its defence system that could eventually overcome pathogen. Figure 17.2 demonstrate that the possible ways of quorum sensing interference mechanism in plants.

Ban et al. (2009) have attempted a similar strategy to transform *Amorphophallus konjac* using aiiA gene from *Bacillus thuringiensis* and demonstrated enhanced resistance in transgenic plants against pathogenic bacteria. It is interesting to note that QQ enzymes are naturally synthesized by some plants (Amara et al. 2011; Barea et al. 2013). Plant extracts from legumes such as clover, alfalfa, yam beans and lotus exhibited ability to degrade N-hexanoyl-HSL (C6-HSL) but inactive against long acyl chain AHL (Delalande et al. 2005; Götz et al. 2007). The degradation activity of above plant extract was found sensitive to temperature indicating its enzymatic natures. Production of different quorum quenching (QQ) compounds by plants has been documented by various workers as indicated in Table 17.2. However, enzymatic nature of compounds are yet to verified.

17.4 Conclusion

It is now clear that the majority of phytopathogenic bacteria such as *Agrobacterium tumefaciens, Xanthomonas campestris pv. Campestris, Pectobacterium atrosepticum, Pectobacterium carotovorum, Ralstonia solanacearum, Erwinia amylovora*

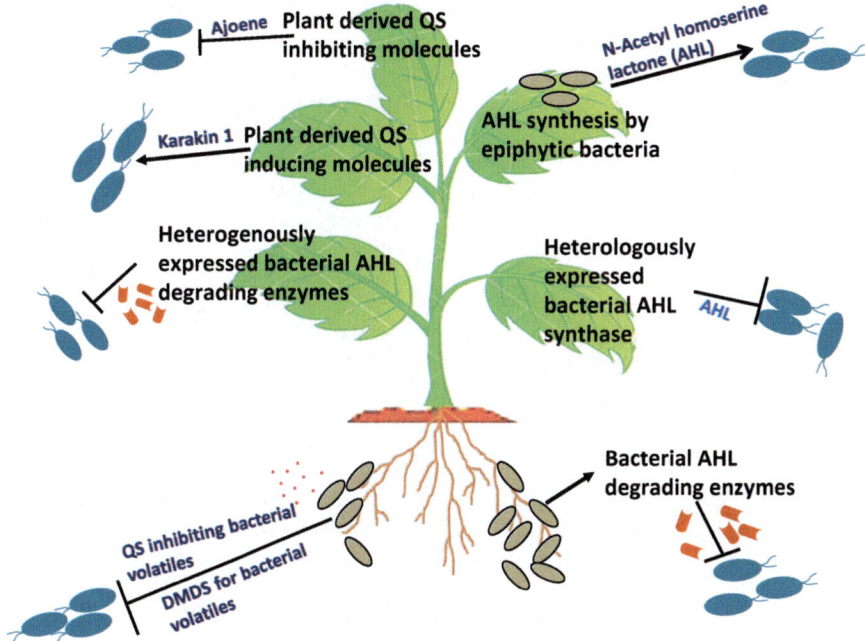

Fig. 17.2 Overview of the different bacterial quorum sensing (QS) based interactions in association with plants

and *Xylella fastidiosa* etc. regulates largely their virulence and pathogenicity through quorum sensing. The role of different regulatory mechanism in production and recognitions of signal molecules are also demonstrated. Considering the significant role of QS in bacterial plant pathogenesis, effective strategy to control plant disease can be devised by the targeting quorum sensing through one or other ways. These strategies can be based on using biocontrol agents naturally or engineered AHL degrading abilities, disruption of quorum sensing by bacterial volatiles and development of suitable transgenic plants with aii genes (Valente and Xavier 2015).

However, for effective implementation on of above approaches in depth molecular understanding on QQ mechanism have to study. Further, effect of environmental factors on the performance of biocontrol agent and transgenic plants need to be investigated. Nonetheless, interference of quorum sensing is novel approach to the altering plant–microbe interaction in favour of plant health. Further, in the development of transgenic, full understanding of gene expression in host plant and successful field trials are needed to evaluate the efficacy. It is interesting to see the development of such approaches which will certainly reduce the dependency on synthetic chemicals for protection of plant and environmental in an eco-friendly way.

Table 17.2 Quorum quenching (QQ) compounds produced by different plants

Plants	Source	QQ compound	References
Allium sativum (garlic)	Bulb extracts	Ajoene (4,5,9-trithiadodeca-1,6,11-triene-9-oxide)	Kyung and Lee (2001), Chernin (2011) and Jakobsen et al. (2012b)
		DMDS (dimethyl disulfide, CH3–S–S–CH3)	
		p-Coumaric acid	
Armoracia rusticana(horseradish)	Root extracts	Isothiocyanate iberin [1-isothiocyanato-3-(methylsulfinyl)propane]	Jakobsen et al. (2012a)
Brassica oleracea (broccoli)	Extracts, syntheticpreparations	Isothiocyanates sulforaphane (4-methyl sulfinyl butyl isothiocyanate) and its precursor erucin (4-methyl thiobutyl isothiocyanate), analogues of iberin	Ganin et al. (2013)
Carex pumila (sand sedge)	Extracts	Resveratrol dimer ε-viniferin (5-[(2R,3R)-6-hydroxy-2-(4-hydroxyphenyl)-4-[(E)-2-(4 hydroxyphenyl) ethenyl]-2,3-dihydro-1-benzofuran-3-yl] benzene-1,3-diol]	Cho et al. (2013)
Citrus spp.	Extracts	O-glycosylated flavonoids naringenin [5,7-dihydroxy-2-(4-hydroxyphenyl)chroman-4-one], neohesperidin, hesperidin	Vikram et al. (2010) and Truchado et al. (2012)
Curcuma longa (turmeric)	Extracts	Curcumin [(1E,6E)-1,7-bis(4-hydroxy-3-methoxyphenyl)-1,6-heptadiene-3,5-dione	Rudrappa and Bais (2008)
Elettaria cardamomun (green cardamom)	Essential oil	Cineol, syn/eucalyptol (1,3,3-trimethyl-2-oxabicyclo [2,2,2]octane)	Jaramillo-Colorado et al. (2012)
Fruits (e.g. apple, pear, peach, banana, pineapple, grape)	Extracts	Patulin (4-hydroxy-4H-furo[3,2-c]pyran-2(6H)-one)	Rasmussen et al. (2005)
Forest plants	Smoke	Karrikins (a family of butenolides related to 3-methyl-2H-furo[2,3-c] pyran-2-one)	Mandabi et al. (2014)
Lippia alba (bushy Lippia)	Essential oil	Limonene-carvone and citral (geranial-neral)	Jaramillo-Colorado et al. (2012)
Medicago sativa (alfalfa)	Seed exudates	An arginine analogue L-canavanine ((2S)-2-amino-4-{[(diaminomethylidene) amino]oxy}butanoic acid)	Keshavan et al. (2005)

(continued)

Table 17.2 (continued)

Plants	Source	QQ compound	References
Minthostachys mollis (muña)	Essential oil	Monoterpene pulegone [(R)-5-methyl-2-(1-methylethylidine) cyclohexanone]	Jaramillo-Colorado et al. (2012)
Myristica cinnamomea (nutmeg)	Nut extracts	Malabaricone C	Chong et al. (2011)
Ocimum basilicum (sweet basil)	Root exudates	Rosmarinic acid (α-o-caffeoyl-3,4-dihydroxyphenyl lactic acid)	Walker et al. (2004)
Ocotea sp.	Essential oil	Terpene α-pinene [(1S,5S)-2,6,6-trimethylbicyclo[3.1.1] hept-2-ene ((−)-α-pinene)]	Jaramillo-Colorado et al. (2012)
Origanum vulgare (oregano)	Essential oil	Carvacrol	Burt et al. (2014)
Psidium guajava (guava)	Extracts	Quercetin, quercetin-3-O-arabinoside	Vasavi et al. (2014)
Swinglea glutinosa (Tabog)	Essential oil	Monoterpene β-pinene (6,6-dimethyl-2-methylenebicyclo[3.1.1] heptane)	Jaramillo-Colorado et al. (2012)
Zingiber officinale (ginger)	Essential oil	Monocyclic sesquiterpene α-zingiberene [2-methyl-5-(6-methylhept-5-en-2 yl) cyclohexa-1,3-diene]	Jaramillo-Colorado et al. (2012)

17.5   Opinion

With the increase number of genetic regulatory mechanisms discovered in synthesis and recognition of different types of signal molecules and complex nature of bacterial pathogenicity, and environmental conditions under which plant is cultivated, poses the major challenge. Therefore, biocontrol agent or antagonist released in the plant soil environment to interfere QS regulated pathogenicity of bacterial pathogens must be stable. Similarly developing transgenic plant with AHL degrading enzymes to control bacterial disease in an excellent approach, but this will requires approval from regulatory agency which may be difficult in many countries. Overall we are in the stage of infancy regarding the exploitation of QS interference to combat bacterial diseases of plant. We expect that with the increase understanding on diversity of regulatory mechanism of QS regulated functions and advances made in molecular and biotechnological approaches, the QS based approach will mature in future for practical application in disease management under Integrated plant disease management.

Acknowledgement One of the author Mr. Firoz Ahmad Ansari is thankful to University Grant Commission, New Delhi for providing Maulana Azad National Fellowship (MAN-JRF) for pursuing PhD at Aligarh Muslim University.

References

Amara N, Krom BP, Kaufmann GF, Meijler MM (2011) Macromolecular inhibition of quorum sensing: enzymes, antibodies, and beyond. Biomed Eng (NY) 111:195–208. https://doi.org/10.1021/cr100101c

An SQ, Allan JH, McCarthy Y, Febrer M, Dow JM, Ryan RP (2014) The PAS domain-containing histidine kinase RpfS is a second sensor for the diffusible signal factor of Xanthomonas campestris. Mol Microbiol 923:586–597. https://doi.org/10.1111/mmi.12577

Andersen AS, Joergensen B, Bjarnsholt T, Johansen H, Karlsmark T, Givskov M, Krogfelt KA (2010) Quorum-sensing-regulated virulence factors in Pseudomonas aeruginosa are toxic to Lucilia sericata maggots. Microbiology 156:400–407. https://doi.org/10.1099/mic.0.032730-0

Bailly A, Weisskopf L (2012) The modulating effect of bacterial volatiles on plant growth: current knowledge and future challenges. Plant Signal Behav 7:79–85. https://doi.org/10.4161/psb.7.1.18418

Ban H, Chai X, Lin Y, Zhou Y, Peng D, Zou Y, Yu Z, Sun M (2009) Transgenic Amorphophallus konjac expressing synthesized acyl-homoserine lactonase (aiiA) gene exhibit enhanced resistance to soft rot disease. Plant Cell Rep 28:1847–1855. https://doi.org/10.1007/s00299-009-0788-x

Barea JM, Pozo MJ, Azcón R, Azcón-Aguilar C (2013) Microbial interactions in the rhizosphere. In: de Bruijn FJ (ed) Molecular microbial ecology of the rhizosphere, vol 2. Wiley, Hoboken, pp 29–44. https://doi.org/10.4067/S0718-95162015005000021

Barnard AML, Salmond GPC (2007) Quorum sensing in Erwinia species. Anal Bioanal Chem 387:415–423. https://doi.org/10.1007/s00216-006-0701-1

Barras F, Vangijsegem F, Chatterjee AK (1994) Extracellular enzymes and pathogenesis of soft-rot erwinia. Annu Rev Phytopathol 32:201–234. https://doi.org/10.1146/annurev.py.32.090194.001221

Bauer WD, Mathesius U (2004) Plant responses to bacterial quorum sensing signals. Curr Opin Plant Biol 7:429–433. https://doi.org/10.1016/j.pbi.2004.05.008

Blom D, Fabbri C, Connor E, Schiestl F, Klauser D, Boller T, Eberl L, Weisskopf L (2011) Production of plant growth modulating volatiles is widespread among rhizosphere bacteria and strongly depends on culture conditions. Environ Microbiol 13:3047–3058. https://doi.org/10.1111/j.1462-2920.2011.02582.x

Bocsanczy AM, Nissinen R, OH CS, Beer SV (2008) HrpN of Erwinia amylovora functions in the translocation of DspA/E into plant cells. Mol Plant Pathol 94:425–434. https://doi.org/10.1111/j.1364-3703.2008.00471.x

Burt SA, Ojo-Fakunle VT, Woertman J, Veldhuizen EJ (2014) The natural antimicrobial carvacrol inhibits quorum sensing in Chromobacterium violaceum and reduces bacterial biofilm formation at sub-lethal concentrations. PLoS One 9(4):e93414

Büttner D, Bonas U (2010) Regulation and secretion of Xanthomonas virulence factors. FEMS Microbiol Rev 342:107–133. https://doi.org/10.1111/j.1574-6976.2009.00192.x

Chen X, Schauder S, Potier N, Van Dorsselaer A, Pelczer I, Bassler BL, Hughson FM (2002) Structural identification of a bacterial quorum sensing signal containing boron. Nature 415:545–549. https://doi.org/10.1038/415545a

Chen CN, Chen CJ, Liao CT, Lee CY (2009) A probable aculeacin A acylase from the Ralstonia solanacearum GMI1000 is N-acyl-homoserine lactone acylase with quorum-quenching activity. BMC Microbiol 9(1):89. https://doi.org/10.1186/1471-2180-9-89

Cheng H et al (2016) The F-box protein Rcy1 is involved in the degradation of histone H3 variant Cse4 and genome maintenance. J Biol Chem 291(19):10372–10377

Chernin L (2011) Quorum-sensing signals as mediators of PGPRs' beneficial traits. In: Maheshwari DK (ed) Bacteria in Agrobiology: Plant Nutrient Management. Springer, Berlin/Heidelberg, pp 209–236. https://doi.org/10.1007/978-3-642-21061-7

Cho HS, Lee JH, Ryu SY, Joo SW, Cho MH, Lee J (2013) Inhibition of Pseudomonas aeruginosa and Escherichia coli O157:H7 biofilm formation by plant metabolite ε-viniferin. J Agric Food Chem 61:7120–7126. https://doi.org/10.1021/jf4009313

Chong YM, Yin WF, Ho CY, Mustafa MR, Hadi AH, Awang K, Narrima P, Koh CL, Appleton DR, Chan KG (2011) Malabaricone C from *myristica cinnamomea* exhibits anti-quorum sensing activity. J Nat Prod 74:2261–2264. https://doi.org/10.1021/np100872k

Cirou A, Mondy S, An S, Charrier A, Sarrazin A, Thoison O, DuBow M, Faure D (2012) Efficient biostimulation of native and introduced quorum-quenching *Rhodococcus erythropolis* populations is revealed by a combination of analytical chemistry, microbiology, and pyrosequencing. Appl Environ Microbiol 78:481–492. https://doi.org/10.1128/AEM.06159-11

Corbett M, Virtue S, Bell K, Birch P, Burr T, Hyman L, Lilley K, Poock S, Toth I, Salmond G (2005) Identification of a new quorum-sensing-controlled virulence factor in *Erwinia carotovora subsp. atroseptica* secreted via the type II targeting pathway. Mol Plant-Microbe Interact 18:334–342. https://doi.org/10.1094/MPMI-18-0334

Costa ED, Chai Y, Winans SC (2012) The quorum-sensing protein TraR of *Agrobacterium tumefaciens* is susceptible to intrinsic and TraM-mediated proteolytic instability. Mol Microbiol 84:807–815. https://doi.org/10.1111/j.1365-2958.2012.08037.x

Crépin A, Beury-Cirou A, Barbey C, Farmer C, Hélias V, Burini JF, Faure D, Latour X (2012) N-acyl homoserine lactones in diverse *Pectobacterium* and *Dickeya* plant pathogens: diversity, abundance, and involvement in virulence. Sensors (Basel) 12:3484–3497. https://doi.org/10.3390/s120303484

Dandurishvili N, Toklikishvili N, Ovadis M, Eliashvili P, Giorgobiani N, Keshelava R, Tediashvili M, Szegedi E, Khmel I, Vainstein A, Chernin L (2010a) Broad-range antagonistic rhizobacteria *Pseudomonas fluorescens* and *Serratia plymuthica* suppress agrobacterium crown-gall tumors on tomato plants. J Appl Microbiol 110:341–352. https://doi.org/10.1111/j.1365-2672.2010.04891.x

Dandurishvili N, Toklikishvili N, Ovadis M, Eliashvili P, Giorgobiani N, Keshelava R, Tediashvili M, Szegedi E, Khmel I, Vainstein A, Chernin L (2010b) Broad-range antagonistic rhizobacteria *Pseudomonas fluorescens* and *Serratia plymuthica* suppress *Agrobacterium* crown-gall tumors on tomato plants. J Appl Microbiol 110:341–352. https://doi.org/10.1111/j.1365-2672.2010.04891.x

Davidsson PR, Kariola T, Niemi O, Palva ET (2013) Pathogenicity of and plant immunity to soft rot pectobacteria. Front Plant Sci 4:191. https://doi.org/10.3389/fpls.2013.00191

Delalande L, Faure D, Raffoux A, Uroz S, D'Angelo-Picard C, Elasri M, Carlier A, Berruyer R, Petit A, Williams P, Dessaux Y (2005) N-Hexanoyl-L-homoserine lactone, a mediator of bacterial quorum-sensing regulation, exhibits a plant-dependent stability in the rhizosphere and may be inactivated by germinating lotus corniculatus seedlings. FEMS Microbiol Ecol 52:13–20. https://doi.org/10.1016/j.femsec.2004.10.005

Evans TJ, Perez-Mendoza D, Monson R, Stickland HG, Salmond GPC (2009) Secretion systems of the enterobacterial phytopathogen, erwinia. In: Wooldridge K (ed) Bacterial secreted proteins. Caister Academic Press, Norfolk, pp 479–503

Farag MA, Zhang H, Choong-Min Ryu CM (2013) Dynamic chemical communication between plants and bacteria through airborne signals: induced resistance by bacterial volatiles. J Chem Ecol 39:1007–1018. https://doi.org/10.1007/s10886-013-0317-9

Fekete A, Kuttler C, Rothballer M, Hense BA, Fischer D, Buddrus-Schiemann K, Lucio M, Müller J, Schmitt-Kopplin P, Hartmann A (2010) Dynamic regulation of N-acyl-homoserine lactone production and degradation in *Pseudomonas putida* IsoF. FEMS Microbiol Ecol 72:22–34. https://doi.org/10.1111/j.1574-6941.2009.00828.x

Flavier AB, Clough SJ, Schell MA, Denny TP (1997) Identification of 3 hydroxypalmitic acid methyl ester as a novel autoregulator controlling virulence in *Ralstonia solanacearum*. Mol Microbiol 26:251–259. https://doi.org/10.1046/j.1365-2958.1997.5661945.x

Fuqua C, Parsek MR, Peter Greenberg E (2001) Regulation of gene expression by cell-to-cell communication: acyl-homoserine lactone quorum sensing. Annu Rev Genet 351:439–468. https://doi.org/10.1146/annurev.genet.35.102401.090913

Ganin H, Rayo J, Amara N, Levy N, Pnina Krief P, Meijler MM (2013) Sulforaphane and erucin, natural isothiocyanates from broccoli, inhibit bacterial quorum sensing. Med Chem Commun 4:175–179. https://doi.org/10.1039/C2MD20196H

Gao R, Krysciak D, Petersen K, Utpatel C, Knapp A, Schmeisser C, Daniel R, Voget S, Jaeger KE, Streit WR (2015) Genome-wide RNA sequencing analysis of quorum sensing-controlled regulons in the plant-associated Burkholderia glumae PG1 strain. Appl Environ Microbiol 81(23):7993–8007

Garge SS, Nerurkar AS (2016) Attenuation of quorum sensing regulated virulence of *Pectobacterium carotovorum subsp. carotovorum* through an AHL lactonase produced by *Lysinibacillus sp.* Gs50. PLoS One 11:167–344. https://doi.org/10.1371/journal.pone.0167344

Gelencser Z, Choudhary KS, Coutinho BG, Hudaiberdiev S, Galbats B, Venturi V (2012) Classifying the topology of AHL-driven quorum sensing circuits in proteobacterial genomes. Sensors 12:5432–5444. https://doi.org/10.3390/s120505432

Genin S, Denny TP (2012) Pathogenomics of the *Ralstonia solanacearum* species complex. Annu Rev Phytopathol 50:67–89. https://doi.org/10.1146/annurev-phyto-081211-173000

Gohlke J, Deeken R (2014) Plant responses to *Agrobacterium tumefaciens* and crown gall development. Front Plant Sci 5:155. https://doi.org/10.3389/fpls.2014.00155

Götz C, Fekete A, Gebefuegi I, Forczek ST, Fuksova K, Li X, Englmann M, Gryndler M, Hartmann A, Matucha M, Schmitt-Kopplin P, Schroder P (2007) Uptake, degradation and chiral discrimination of N-acyl-D/L-homoserine lactones by barley (*Hordeum vulgare*) and yam bean (*Pachyrhizus erosus*) plants. Anal Bioanal Chem 389:1447–1457. Not available

Grandclément C, Tannières M, Moréra S, Dessaux Y, Faure D (2015) Quorum quenching: role in nature and applied developments. FEMS Microbiol Rev 401:86–116. https://doi.org/10.1093/femsre/fuv038

Groenhagen U, Baumgartner R, Baily A, Gardiner A, Eberl L, Schulz S, Weisskopf L (2013) Production of bioactive volatiles by different *Burkholderia ambifaria* strains. J Chem Ecol 39:892–906. https://doi.org/10.1007/s10886-013-0315-y

Ham JH (2013) Intercellular and intracellular signalling systems that globally control the expression of virulence genes in plant pathogenic bacteria. Mol Plant Pathol 14:308–322. https://doi.org/10.1111/mpp.12005

Han Y, Chen F, Li N, Zhu B, Li XZ (2010) *Bacillus marcorestinctum sp* nov., a novel soil acyl-homoserine lactone quorum-sensing signal quenching bacterium. Int J Mol Sci 11:507–520. https://doi.org/10.3390/ijms11020507

Han SW, Sriariyanun M, Lee SW, Sharma M, Bahar O, Bower Z, Ronald PC (2011) Small protein-mediated quorum sensing in a gram-negative bacterium. PLoS One 6(12):e29192. https://doi.org/10.1371/journal.pone.0029192

Hanano A, Harba M, Al-Ali M, Ammouneh H (2014) Silencing of *Erwinia amylovora* sy69 AHL-quorum sensing by a *bacillus* simplex AHL-inducible aiiA gene encoding a zinc-dependent N-acyl-homoserine lactonase. Plant Pathol 634:773–783. https://doi.org/10.1111/ppa.12142

Haudecoeur E, Tannières M, Cirou A, Raffoux A, Dessaux Y, Faure D (2009) Different regulation and roles of lactonases AiiB and AttM in *Agrobacterium tumefaciens* C58. Mol Plant-Microbe Interact 22:529–537. https://doi.org/10.1094/MPMI-22-5-0529

Hayward AC, Fegan N, Fegan M, Stirling GR (2010) *Stenotrophomonas* and *Lysobacter*: ubiquitous plant-associated gamma-proteobacteria of developing significance in applied microbiology. J Appl Microbiol 108:756–770. https://doi.org/10.1111/j.1365-2672.2009.04471.x

He YW, Wu J, Zhou L, Yang F, He YQ, Jiang BL, Bai L, Xu Y, Deng Z, Tang JL, Zhang LH (2011) *Xanthomonas campestris* diffusible factor is 3-hydroxybenzoic acid and is associated with xanthomonadin biosynthesis, cell viability, antioxidant activity, and systemic invasion. Mol Plant-Microbe Interact 24:948–957. https://doi.org/10.1094/MPMI-02-11-0031

Holden MTG, Ram Chhabra S, De Nys R, Stead P, Bainton NJ, Hill PJ, Manefield M, Kumar N, Labatte M, England D, Rice S, Givskov M, Salmond GPC, Stewart GSAB, Bycroft BW, Kjelleberg S, Williams P (1999) Quorum-sensing cross talk: isolation and chemical characterization of cyclic dipeptides from *Pseudomonas aeruginosa* and other gram-negative bacteria. Mol Microbiol 33:1254–1266. https://doi.org/10.1046/j.13652958.1999.01577.x

Hughes DT, Sperandio V (2008) Inter-kingdom signalling: communication between bacteria and their hosts. Nat Rev Microbiol 6:111–120. https://doi.org/10.1038/nrmicro1836

Hussain MBBM, Zhang HB, Xu JL, Liu QG, Jiang Z, Zhang LH (2008) The acyl-homoserine lactone-type quorum-sensing system modulates cell motility and virulence of *Erwinia chrysanthemi pv. zeae*. J Bacteriol 190:1045–1053. https://doi.org/10.1128/JB.01472-07

Insam H, Seewald MSA (2010) Volatile organic compounds (VOCs) in soils. Biol Fertil Soils 46:199–213. Not available

Ionescu M, Zaini PA, Baccari C, Tran S, da Silva AM, Lindow SE (2014) Xylella fastidiosa outer membrane vesicles modulate plant colonization by blocking attachment to surfaces. Proc Natl Acad Sci 111(37):E3910–E3918

Jakobsen TH, Bragason SK, Phipps RK, Christensen LD, van Gennip M, Alhede M, Skindersoe M, Larsen TO, Hoiby N, Bjarnsholt T, Givskov M (2012a) Food as a source for quorum sensing inhibitors: iberin from horseradish revealed as a quorum sensing inhibitor of *Pseudomonas aeruginosa*. Appl Environ Microbiol 78:2410–2421. https://doi.org/10.1128/AEM.05992-11

Jakobsen TH, van Gennip M, Phipps RK, Shanmugham MS, Christensen LD, Alhede M, Skindersoe ME, Rasmussen TB, Friedrich K, Uthe F, Jensen PO, Moser C, Nielsen KF, Eberl L, Larsen TO, Tanner D, Hoiby N, Bjarnsholt T, Givskov M (2012b) Ajoene, a sulfur-rich molecule from garlic, inhibits genes controlled by quorum sensing. Antimicrob Agents Chemother 56:2314–2325. https://doi.org/10.1128/AAC.05919-11

Jaramillo-Colorado B, Olivero-Verbel J, Stashenko EE, Wagner-Döbler I, Kunze B (2012) Anti-quorum sensing activity of essential oils from Colombian plants. Nat Prod Res 26(12):1075–1086. https://doi.org/10.1080/14786419.2011.557376

Kai K, Ohnishi H, Shimatani M, Ishikawa S, Mori Y, Kiba A, Ohnishi K, Tabuchi M, Hikichi Y (2015) Methyl 3-hydroxymyristate, a diffusible signal mediating phc quorum sensing in *Ralstonia solanacearum*. Chem Bio Chem 16:2309–2318. https://doi.org/10.1002/cbic.201500456

Kakkar A, Nizampatnam NR, Kondreddy A, Pradhan BB, Chatterjee S (2015) Xanthomonas campestris cell–cell signalling molecule DSF (diffusible signal factor) elicits innate immunity in plants and is suppressed by the exopolysaccharide xanthan. J Exp Bot 66(21):6697–6714

Kang BR, Lee JH, Ko SJ, Lee YH, Cha JS, Cho BH, Kim YC (2004) Degradation of acyl-homoserine lactone molecules by *Acinetobacter sp.* strain C1010. Can J Microbiol 50:935–941. https://doi.org/10.1139/w04-083

Keshavan ND, Chowdhary PK, Haines DC, Gonzalez JE (2005) L-Canavanine made by Medicago sativa interferes with quorum sensing in *Sinorhizobium meliloti*. J Bacteriol 187:8427–8436. https://doi.org/10.1128/JB.187.24.8427-8436

Kleerebezem M, Quadri LEN, Kuipers OP, De Vos WM (1997) Quorum sensing by peptide pheromones and two-component signal-transduction systems in gram-positive bacteria. Mol Microbiol 24:895–904. https://doi.org/10.1046/j.1365-2958.1997.4251782

Koczan JM, McGrath MJ, Zhao Y, Sundin GW (2009) Contribution of *Erwinia amylovora* exopolysaccharides amylovoran and levan to biofilm formation: implications in pathogenicity. Phytopathology 99:1237–1244. https://doi.org/10.1094/PHYTO-99-11-1237

Koutsoudis MD, Tsaltas D, Minogue TD, von Bodman SB (2006) Quorum-sensing regulation governs bacterial adhesion, biofilm development, and host colonization in Pantoea stewartii subspecies stewartii. Proc Natl Acad Sci 103(15):5983–5988

Kyung KH, Lee YC (2001) Antimicrobial activities of sulfur compounds derived from S-alk (en) yl-L-cysteine sulfoxides in *Allium* and *Brassica*. Food Rev Int 17:183–198. https://doi.org/10.1081/FRI-100000268

LaSarre B, Federle MJ (2013) Exploiting quorum sensing to confuse bacterial pathogens. Microbiol Mol Biol Rev 771:73–111. https://doi.org/10.1128/MMBR.00046-12

Li Q, Ni H, Meng S, He Y, Yu Z, Li L (2011) Suppressing *Erwinia carotovora* pathogenicity by projecting N-acyl homoserine lactonase onto the surface of *Pseudomonas putida* cells. J Microbiol Biotechnol 21:1330–1335. Not available

Lin Y, Xu J, Hu J, Wang L, Ong SL, Leadbetter JR (2003) Acyl homoserine lactone acylase from *Ralstonia* strain XJ12B represents a novel and potent class of quorum quenching enzymes. Mol Microbiol 47:849–860. https://doi.org/10.1046/j.1365-2958.2003.03351.x

Liu H, Coulthurst SJ, Pritchard L, Hedley PE, Ravensdale M, Humphris S, Burr T, Takle G, Brurberg MB, Birch PRJ, Salmond GPC, Toth IK (2008) Quorum sensing coordinates brute force and stealth modes of infection in the plant pathogen *Pectobacterium atrosepticum*. PLoS Pathog 4:e1000093. https://doi.org/10.1371/journal.ppat.1000093

Mandabi A, Ganin H, Krief P, Rayo J, Meijler MM (2014) Karrikins from plant smoke modulate bacterial quorum sensing. Chem Commun 50:5322–5325. https://doi.org/10.1039/C3CC47501H

Mei GY, Yan XX, Turak A, Luo ZQ, Zhang LQ (2010) AidH, an alpha/beta-hydrolase fold family member from an *Ochrobactrum sp.* strain, is a novel N-acylhomoserine lactonase. Appl Environ Microbiol 76:4933–4942. https://doi.org/10.1128/AEM.00477-10

Mendes R, Garbeva P, Raaijmakers JM (2013) The rhizosphere microbiome: significance of plant beneficial, plant pathogenic, and human pathogenic microorganisms. FEMS Microbiol Rev 37:634–663. https://doi.org/10.1111/1574-6976.12028

Molina L, Constantinescu F, Michel L, Reimmann C, Duffy B, Defago G (2003) Degradation of pathogen quorum-sensing molecules by soil bacteria: a preventive and curative biological control mechanism. FEMS Microbiol Ecol 45:71–81. https://doi.org/10.1016/s0168-6496

Mori Y, Ishikawa S, Ohnishi H, Shimatani M, Morikawa Y, Hayashi K, Ohnishi K, Kiba A, Kai K, Hikichi Y (2017) Involvement of ralfuranones in the quorum sensing signalling pathway and virulence of *Ralstonia solanacearum* strain OE1-1. Mol Plant Pathol 19:454. https://doi.org/10.1111/mpp.12537

Mukherjee A, Cui Y, Liu Y, Chatterjee AK (1997) Molecular characterization and expression of the *Erwinia carotovora* hrpNEcc gene, which encodes an elicitor of the hypersensitive reaction. Mol Plant-Microbe Interact 10:462–471. https://doi.org/10.1094/MPMI.1997.10.4.462

Müller H, Westendorf C, Leitner E, Chernin L, Riedel K, Eberl L, Berg G (2009) Quorum sensing effects in the antagonistic rhizosphere bacterium *Serratia plymuthica* HRO-C48. FEMS Microbiol Ecol 67:468–478. https://doi.org/10.1111/j.1574-6941.2008.00635.x

Nasser W, Dorel C, Wawrzyniak J, Van Gijsegem F, Groleau MC, Déziel E, Reverchon S (2013) Vfm a new quorum sensing system controls the virulence of *Dickeya dadantii*. Environ Microbiol 15:865–880. https://doi.org/10.1111/1462-2920.12049

Ouyang LJ, Li LM (2016) Effects of an inducible aiiA. Trans Res 25(4):441–452. https://doi.org/10.1007/s11248-016-9940-x

Papenfort K, Bassler BL (2016) Quorum sensing signal-response systems in gram-negative bacteria. Nat Rev Microbiol 149:576–588. https://doi.org/10.1038/nrmicro.2016.89

Park SY, Lee SJ, Oh TK, Oh JW, Koo BT, Yum DY, Lee JK (2003) AhlD, an N-acylhomoserine lactonase in *Arthrobacter sp.*, and predicted homologues in other bacteria. Microbiology 149:1541–1550. https://doi.org/10.1099/mic.0.26269-0

Park SY, Hwang BJ, Shin MH, Kim JA, Kim HK, Lee JK (2006) N-acylhomoserine lactonase producing *Rhodococcus spp.* with different AHL degrading activities. FEMS Microbiol Lett 261.102–108. https://doi.org/10.1111/j.1574-6968.2006.00336.x

Park CJ, Kazunari N, Ronald PC (2010) Quantitative measurements of *Xanthomonas oryzae* pv. *oryzae* distribution in rice using fluorescent-labelling. J Plant Biol 15:595–599. https://doi.org/10.1007/s12374-011-9164-9

Pereira CS, Thompson JA, Xavier KB (2013) AI-2-mediated signalling in bacteria. FEMS Microbiol Rev 37:156–181. https://doi.org/10.1111/j.1574-6976.2012.00345.x

Pérombelon MCM (2002) Potato diseases caused by soft rot erwinias: an overview of pathogenesis. Plant Pathol 51:1–12. https://doi.org/10.1046/j.0032-0862.2001.Shorttitle.doc.x

Pierce BK, Voegel T, Kirkpatrick BC (2014) The *Xylella fastidiosa* PD1063 protein is secreted in association with outer membrane vesicles. PLoS One 9(11):e113504. https://doi.org/10.1371/journal.pone.0113504

Piqué N, Miñana-Galbis D, Merino S, Tomás JM (2015) Virulence factors of Erwinia amylovora: a review. Int J Mol Sci 16(6):12836–12854

Põllumaa L, Alamäe T, Mäe A (2012) Quorum sensing and expression of virulence in pectobacteria. Sensors 12(3):3327–3349

Qian GL, Fan JQ, Chen DF, Kang YJ, Han B, Hu BS, Liu FQ (2010) Reducing pectobacterium virulence by expression of an N-acyl homoserine lactonase gene P-lpp-aiiA in *Lysobacter enzymogenes* strain OH11. Biol Control 52:17–23. https://doi.org/10.1016/j.biocontrol.2009.05.007

Qin Y, Su S, Farrand SK (2007) Molecular basis of transcriptional antiactivation. TraM disrupts the TraR–DNA complex through stepwise interactions. J Biol Chem 282:19979–19991. https://doi.org/10.1074/jbc.M703332200

Rai R, Ranjan M, Pradhan B, Chatterjee S (2012) Atypical regulation of virulence-associated functions by a diffusible signal factor in *Xanthomonas oryzae* pv. *oryzae*. Mol Plant-Microbe Interact 25:789–801. https://doi.org/10.1094/MPMI-11-11-0285-R

Rasmussen TB, Bjarnsholt T, Skindersoe ME, Hentzer M, Kristoffersen P, Köte M, Nielsen J, Eberl L, Givskov M (2005) Screening for quorum sensing inhibitors (QSI) by use of a novel genetic system, the QSI selector. J Bacteriol 187:1799–1814. https://doi.org/10.1128/JB.187.5.1799-1814.2005

Rudrappa T, Bais HP (2008) Curcumin, a known phenolic from Curcuma longa, attenuates the virulence of *Pseudomonas aeruginosa* PAO1 in whole plant and animal. J Agric Food Chem 56:1955–1962. https://doi.org/10.1021/jf072591j

Ryan RP, Dow JM (2008) Diffusible signals and interspecies communication in bacteria. Microbiology 154:1845–1858. https://doi.org/10.1099/mic.0.2008/017871-0

Schulz S, Dickschat JS (2007) Bacterial volatiles: the smell of small organisms. Nat Prod Rep 24:814–842. https://doi.org/10.1039/B507392H

Seet Q, Zhang LH (2011) Antiactivator QslA defines the quorum sensing threshold and response in *Pseudomonas aeruginosa*. Mol Microbiol 80:951–965. https://doi.org/10.1111/j.1365-2958.2011.07622

Taguchi F, Takeuchi K, Katoh E, Murata K, Suzuki T, Marutani M, Kawasaki T, Eguchi M, Katoh S, Kaku H, Yasuda C (2006) Identification of glycosylation genes and glycosylated amino acids of flagellin in Pseudomonas syringae pv. tabaci. Cell Microbiol 8(6):923–938

Takano E (2006) Gamma-butyrolactones: *Streptomyces* signalling molecules regulating antibiotic production and differentiation. Curr Opin Microbiol 9:287–294. https://doi.org/10.1016/j.mib.2006.04.003

Tannières M, Lang J, Barnier C, Shykoff JA, Faure D (2017) Quorum-quenching limits quorum-sensing exploitation by signal-negative invaders. Sci Rep 7:40126

Teplitski M, Mathesius U, Rumbaugh KP (2011) Perception and degradation of N-acyl homoserine lactone quorum sensing signals by mammalian and plant cells. Chem Rev 111:100–116. https://doi.org/10.1021/cr100045m

Terwagne M, Mirabella A, Lemaire J, Deschamps C, De Bolle X, Letesson JJ (2013) Quorum sensing and self-quorum quenching in the intracellular pathogen *Brucella melitensis*. PLoS One 8(12):e8251482. https://doi.org/10.1371/journal.pone.0082514

Torres M, Uroz S, Rafael S, Laure F, Emilia Q, Inmaculada L (2017) HqiA, a novel quorum-quenching enzyme which expands the AHL lactonase family. Sci Rep 7(1):943. https://doi.org/10.1038/s41598-017-01176-7

Toth IK, Bell K, Holeva MC, Birch PRJ (2003) Soft rot erwiniae: from genes to genomes. Mol Plant Pathol 4:17–30. https://doi.org/10.1046/j.1364-3703.2003.00149.x

Toth IK, Newton JA, Hyman LJ, Lees AK, Daykin M, Williams P, Fray RJ (2004) Potato plants genetically modified to produce N-acylhomoserine lactones increase susceptibility to soft rot erwiniae. Mol Plant-Microbe Interact 17:880–888. https://doi.org/10.1094/MPMI.2004.17.8.880

Truchado P, Gimenez-Bastida JA, Larrosa M, Castro-Ibanez I, Espin JC, Tomas-Barberan FA, Garcia-Conesa MT, Allende A (2012) Inhibition of quorum sensing (QS) in *Yersinia enterocolitica* by an orange extract rich in glycosylated flavanones. J Agric Food Chem 60:8885–8894. https://doi.org/10.1021/jf301365a

Uroz S, D'Angelo-Picard C, Carlier A, Elasri M, Sicot C, Petit A, Oger P, Faure D, Dessaux Y (2003) Novel bacteria degrading N-acylhomoserine lactones and their use as quenchers of quorum-sensing regulated functions of plant-pathogenic bacteria. Microbiology 149:1981–1989. https://doi.org/10.1099/mic.0.26375-0

Vakulskas CA, Potts AH, Babitzke P, Ahmer BM, Romeo T (2015) Regulation of bacterial virulence by Csr (Rsm) systems. Microbiol Mol Biol Rev 792:193–224. https://doi.org/10.1128/MMBR.00052-14

Valente RS, Xavier KB (2015) The Trk potassium transporter is required for RsmB-mediated activation of virulence in the phytopathogen *Pectobacterium wasabiae*. J Bacteriol 198:248–255. https://doi.org/10.1128/JB.00569-15

Valente RS, Nadal-Jimenez P, Carvalho AFP, Vieira FJD, Xavier KB (2017) Signal integration in quorum sensing enables crossspecies induction of virulence in *Pectobacterium wasabiae*. mBio 8:398–417. https://doi.org/10.1128/mBio.00398-17

Vanneste JL (2000) In: Vanneste JL (ed) What is fire blight? Who is *Erwinia amylovora*? how to control it? In: Fire Flight; The Disease and its Causative Agent *Erwinia amylovora*. CAB International London, London, pp 01–08. https://doi.org/10.1079/9780851992945.0000

Vasavi HS, Arun AB, Rekha PD (2014) Anti-quorum sensing activity of Psidium guajava L. flavonoids against Chromobacterium violaceum and Pseudomonas aeruginosa PAO1. Microbiol Immunol 58(5):286–293

Vikram A, Jayaprakasha GK, Jesudhasan PR, Pillai SD, Patil BS (2010) Suppression of bacterial cell–cell signalling, biofilm formation and type III secretion system by citrus flavonoids. J Appl Microbiol 109:515–527. https://doi.org/10.1111/j.1365-2672.2010.04677.x

Walker TS, Bais HP, Déziel E, Schweizer HP, Rahme LG, Fall R, Vivanco JM (2004) Pseudomonas aeruginosaplant root interactions. Pathogenicity, biofilm formation, and root exudation. Plant Physiol 134(1):320–331

Wang WZ, Morohoshi T, Someya N, Ikeda T (2012) Diversity and distribution of N acylhomoserine lactone (AHL)-degrading activity and AHL-lactonase (AiiM) in genus microbacterium. Microbes Environ 27:330–333. https://doi.org/10.1264/jsme2.ME11341

Wenke K, Kai M, Piechulla B (2010) Belowground volatiles facilitate interactions between plant roots and soil organisms. Planta 231:499–506. https://doi.org/10.1007/s00425-009-1076-2

Whitehead NA, Byers JT, Commander P, Corbett MJ, Coulthurst SJ, Everson L, Harris AKP, Pemberton CL, Simpson NJL, Slater H, Smith DS, Welch M, Williamson N, Salmond GPC (2002) The regulation of virulence in phytopathogenic Erwinia species: quorum sensing, antibiotics and ecological considerations. Antonie van Leeuwenhoek 81:223–231. doi: Not available

Wu J, Jiao Z, Guo F, Chen L, Ding Z, Qiu Z (2016) Constitutive and secretory sxpression of the AiiA in *Pichia pastoris* inhibits *Amorphophallus konjac* Soft Rot Disease. Am J Mol Biol:602–679. https://doi.org/10.4236/ajmb.2016.62009

Yu X, Liang X, Liu K, Dong W, Wang J, Zhou MG (2015) The thiG gene is required for full virulence of *Xanthomonas oryzae pv. oryzae* by preventing cell aggregation. PLoS One 10(7):e0134237. https://doi.org/10.1371/journal.pone.0134237

Zheng D, Yao X, Duan M, Luo Y, Liu B, Qi P, Ruan L (2016) Two overlapping two-component systems in *Xanthomonas oryzae pv. oryzae* contribute to full fitness in rice by regulating virulence factors expression. Sci Rep 6:22–768. https://doi.org/10.1038/srep22768

Chapter 18
Scope of Pathogenesis-Related Proteins Produced by Plants in Interrupting Quorum Sensing Signaling

Pratheep Chinnappan, Saisundar Rajan, Shaarath Thondanure, Leena Champalal, and Pachaiappan Raman

Abstract Quorum sensing (QS) is a microbial consortia growth gene regulation process mediated by hormone like signaling molecules known as auto inducers (AI). QS is involved in regulating virulent factors of pathogens that is required for the successful establishment of bacterial infections. Therefore, interrupting the QS signaling process can be exploited as an effective strategy for antimicrobial therapy. Plants produce several compounds like pathogenesis-related proteins as a result of pathogen exposure and stress factors that might affect the quorum sensing signaling process. Plant-pathogen interactions during pathogenic attack leads to the production of secondary metabolites, which in turn can target the bacterial QS system by either inhibiting synthesis of signal transducers, degrading the transducers and/or receptor targeting mechanism. Since plants are safe for human consumption, their usage is considered as a safer alternative to the conventional antibiotic mediated approach to treat infections.

Keywords Plant · Pathogenesis · Quorum sensing · Signal transducers · Virulence

P. Chinnappan · S. Rajan · S. Thondanure · L. Champalal
Department of Biotechnology, School of Bioengineering, SRM Institute of Science and Technology, Kattankulathur, Tamil Nadu, India

P. Raman (✉)
Department of Biotechnology, School of Bioengineering, SRM Institute of Science and Technology, Kattankulathur, Tamil Nadu, India

Visiting Faculty, Metabolomics, Proteomics and Mass Spectrometry Core Facilities, University of Utah, Salt Lake City, UT, USA
e-mail: pachaiappan.ra@ktr.srmuniv.ac.in

© Springer Nature Singapore Pte Ltd. 2018   371
V. C. Kalia (ed.), *Biotechnological Applications of Quorum Sensing Inhibitors*,
https://doi.org/10.1007/978-981-10-9026-4_18

18.1   Introduction

Infectious diseases are generally caused by biofilm forming microbes, through Quorum sensing process (Agarwala et al. 2014; Gui et al. 2014; Hemaiswarya et al. 2008). Bacteria residing in the biofilms resist antibiotic concentration up to thousand times more than those which are sufficient to kill their free-living counterparts (Rasmussen and Givskov 2006; Arasu et al. 2015). Quorum sensing is a population density-dependent intercellular communication mechanism by which bacterial moieties exhibit coordinated behaviour. Quorum signalling has been found responsible for the expression of various virulence traits in pathogenic bacteria.

Bacterial quorum sensing developed amongst microbial consortia to rapidly alter genome expression in response to environmental factor (Bandyopadhyay et al. 2015). Single cell in a microbial consortium react to a signalling transducer called an autoinducer (AI) that acts as an indicator of the population density (Kalia 2013). The autoinducer molecule in the QS system of gram negative bacteria is N-acyl homoserine lactone (AHL). Increased synthesis of AHL is correlated with large population, which results in regulation of gene expression. Emerging research studies have suggested that functions such as secretion of virulence factors, biofilm formation, swarming, and acquiring competency of bacteria play an important role infecting the living systems. (Schauder and Bassler 2001; Vattem et al. 2007).

Plants get exposed to diverse environmental stresses, both biotic and abiotic stresses and hostile conditions. Abiotic factors affecting plants include nutrient deficiency, hypoxia, drought, extreme hydration, salinity, adverse temperature fluctuations, high illumination by sunlight, and anthropogenic factors, such as use of chemical fertilizers, insecticides which are considered as pollutants and irradiation by UV. (Iriti and Faoro 2009; Suzuki et al. 2014). Furthermore, several living (biotic) factors are also induce stress depending upon the interaction with plants. The interaction usually be pathogenic when quorum sensing is involved. This kind of interaction leads to activation of defence pathways in plants. Living organisms, biotic in nature such as bacteria, fungi, nematodes, pests and insecticides releases chemical agents as antigens act as stress-inducers (Rausher 2001). Plants respond to stress induction by signal molecules or elicitors interaction through strengthening of cell wall, protein cleaving enzymes – proteases, biosynthesis of secondary metabolites and synthesis of **Pathogenesis Related Proteins (PR)**.

PR proteins are natural defence products of plants which is activated by induction of a biotic stress especially by pathogens. These act as first line of defence in plants by curtailment of pathogen development in them. The study related to PR proteins has not been fully overviewed, but with their defence role there is a huge scope of PR proteins acting as quorum sensing inhibitors.

Higher order organisms having interactions with high microbial cell density observed to have natural defensive mechanisms against them. Due to their vast defence system plants control biofouling by QS inhibition. (War et al. 2012; Fürstenberg-Hägg et al. 2013) Small molecules from plants have been isolated and characterized for their QS inhibition potential by interfering in signal transducers of

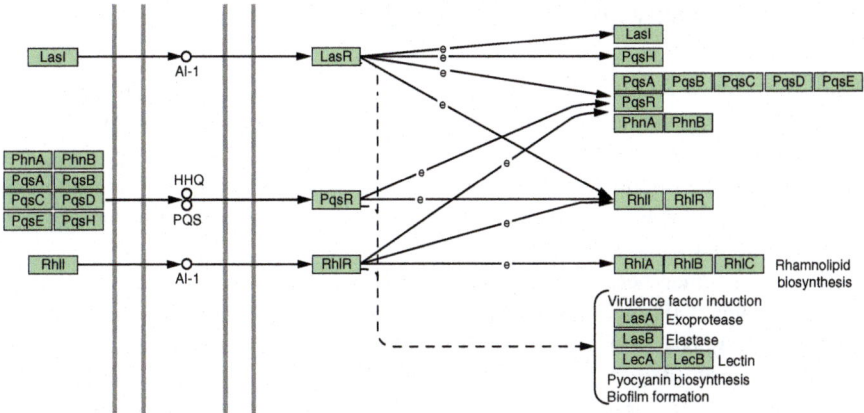

Fig. 18.1 Genes responsible for QS activity in *P. aeruginosa* [KEGG Pathway 2017]

Fig. 18.2 Genes responsible for QS activity in *Chromobacterium violaceum* [KEGG Pathway 2017]

bacteria (Vanetten et al. 1981). AHL dependent transcription factors are inhibited by various small molecules. These small molecules maybe of flavonoids, alkaloids, fatty acids and structure oriented compounds like furanones etc. (González-Lamothe et al. 2009). The natural inhibitors are obtained as antibiotics from plants are highly reactive and sensitive to toxicity. Furanone analogues from fungal species are good example for the natural inhibitors. These metabolites at high concentration sometimes lethal to human cells in an *in vitro* study (Dehghan et al. 2016; Akram et al. 2014). The state of art methodologies are developed to discover and develop novel antimicrobials to inhibit QS having the qualities of non-toxicity and have therapeutic activities, which has the potentiality to be treated for bacterial infections in humans (Dehghan et al. 2016; Balakrishnan et al. 2015).

Secondary metabolites produced by plants - dietary phytochemicals - Ajoene, Iberin, limonoids, and furocoumarins, through their antimicrobial and QSI activities are known to provide health benefits (Kazemian et al. 2015; Sarkar et al. 2015; Brackman et al. 2016). Plants that have been able to affect short chain AHL regulated violacein in *Chromobacterium violaceum*, were tested for inhibition of Rhl regulated swarming motility, pyocyanin and biofilm development in *P. aeruginosa* (Figs. 18.1 and 18.2). Plants carry out a different synergic interaction to defend against virulence of microbes. Thus, plant products are potential candidates which can act as an effective anti-quorum sensing and antimicrobial compound in co-therapies with antibiotics.

18.2 Significance of Plant Based QSI

Plants have been found to have complex interactions with microorganisms which make them as a potential study for compounds interfering QS system (Du Fall and Solomon 2011).

Secondary metabolites produced by different variety of plants play an important role in resistance against microbes and pests but many compounds have large scope of applications such as in food and medicine (Nazzaro et al. 2013).

Solanaceae members in plant family synthesis bioactive proteins which are against virulence through quorum quenching. Bioactive proteins disable, inhibit or disrupt QS signals from microbes to inhibit interaction between the cells, thus inhibiting the virulence factors in *Pseudomonas aeruginosa*. The exploitation of these bioactive active proteins which are quorum quenchers are less susceptible for bacteria to develop resistance against them. The virulence factors and defence proteins are mainly expressed in stress conditions (Gurpreet et al. 2015).

Plants under stress conditions developed by micro-organism, down regulate certain signal molecules and enhance the regulation of defense system by enhancing the plant hormone signal transduction (War et al. 2012). Also, this interaction leads to accumulation of plant signal hormones like salicylic acid. The defence-related genes are over expressed due to this accumulation factor (Pieterse and Van Loon 2004). The salicylic acid accumulation derives a hypothesis in which if the threshold concentration of salicylic acid for quorum sensing inhibition is not attained, they induce the synthesis of PR proteins as a defense against the pathogens (Rivas-San Vicente and Plasencia 2011).

In plant seeds undergoing germination, the protrusion of radical tip occurs at phase 2. The radical tip undergoes thorough production of nutrient requirements such as carbohydrates monomers. Even though the presence of nutrients encourages the growth of microorganism which may lead to antagonistic effect, the growth is inhibited by PR proteins and other defense molecules. These growth inhibitors showing bacteriostatic effect but not cidal effects rather can also act as quorum sensing inhibitors.

Plant compounds which are considered as the products of green chemistry are sustainable source of new molecules as effective biofilm inhibitor (Siddiqui et al. 2015; Karumuri et al. 2015). Phytochemicals possess structural variety along with multi-target mechanism of action, which remarkably differs from the mechanism of conventional antibiotics (Szweda et al. 2015; Begum et al. 2016; Jeyanthi and Velusamy 2016). These unique properties can help in avoiding the multi-drug resistance (MDR) problem. In fact, the occurrence of pathogen resistance to phytochemicals is evidently lesser than conventional antibiotics (Borges et al. 2015). Thus, the exploitation of plant products could be an effective alternative approach for combating infectious disease that is otherwise treated with conventional antibiotics.

18.3   Interations of Plant Based QSI with Microbes

Plant bioactive molecules play an important role for treatment and prevention of infectious diseases (Koehn and Carter 2005). The plant derived compounds interact with the bacterial QS system by either blocking the signal transducers produced by AHL synthase, degrading the AHL molecule and also by targeting the luxR signal receptor (Huma et al. 2011).

Plant molecules acts as a QS inhibitor mimic AHL which interfere in the interaction of AHL to LuxR receptor. Signal reception-interfering molecules competitively bind to the AHL receptor and non-competitively to other non-AHL binding sites of the receptor (Ahiwale et al. 2017; Sharma and Lal 2017; Kalia et al. 2017). The effective competitive binding inhibits the activation of AHL- mediated QS system (Kalia and Kumar 2015a, b; Koul et al. 2016; Sanchart et al. 2017; Koul and Kalia 2017).

Halogenated furanones synthesized as plant products undergo competitive binding with LuxR type proteins (Givskov et al. 1996). Furanones are strong AI-1 and AI-2 mediated inhibitors which enhance the proteolytic degradation during bacteriostatic process (Salini and Pandian 2015). These inhibit QS system (Kalia et al. 2011; Kalia and Purohit 2011). Other natural products such as fatty acids, furocoumarins, carotenoids, and limonoids exhibit antibacterial and antifungal activity. Furan moiety act as a reactive site which has QSI abilities (Lönn-Stensrud et al. 2007). These inhibit biofilm formation as well as are strong antagonists of EHEC (Entero haemorrhagic *E. coli*) virulence in *E. coli* (Vikram et al. 2010).

GABA defence system is suppressed by accumulation of proline residues. The accumulation of proline residues as well as salicylic acid is by abiotic stresses (Chevrot et al. 2006; Zhang et al. 2002). These accumulation induces the activation of defence pathways which inhibits OHC8HSL which is an AHL signal is disrupted by lactonase causing the antagonist mechanism in *Agrobacterium tumefaciens* infection property. (Haudecoeur et al. 2009). *Medicago truncatula* seedlings alter LuxR, CviR and AhyR, reporter mechanisms in various organisms (Gao et al. 2003). Dhurrin extracted from *Sorghum bicolor* down regulates chiC genes responsible for chitinase enzyme which is the primary defence enzyme of *Pseudomonas* spp. Isovitexin extracted from *Vigna mungo* down regulates Las B which is an activator of AI 2 in *Pseudomonas* spp.

One of the most pathogenic activities of microbes is biofilm formation. The metal ions play a vital role as signaling molecules in the pathogenesis and virulence such as biofilm formation. Iron sensing is the first step of biofilm formation (O'Toole and Kolter 1998; Singh et al. 2002). Iron molecules impacting the cell adherence with the matrix or intracellular play vital role in microbial growth (Dobrucka and Długaszewska 2015). Calcium and iron are emphasized in inter link properties among the components of biofilm by maintaining the matrix integrity (Kumar et al. 2015; Chen and Stewart 2002). Another way of inhibiting the biofilm formation is by metal binding agents targeting bacteria. Chelating compounds are also being extricated from plant source. Phenolic acids, polyphenols, and flavonoids with 6,7-dihydroxy iron chelation site are some of the chelators (Kalia 2014; Azman et al. 2017). Lin et al. (2012) observed that the antibiofilm activity of gallotannin class has the ability to chelate iron and this antibiofilm activity was tested for *S. aureus*.

Plants have an extensive defense system and high level of interaction with microorganisms both at proteome and genome level. Following this interaction, plants synthesize defense molecules which can act on various virulence expressions of pathogens. Thus, plants are considered as a vital source for quorum sensing inhibitors.

18.4   Sources of Plant-Based QSI

The sources varies depending on the type of QSI and properties of it. Since most of the QSI from plants have been plant's secondary metabolites such as flavonoids, alkaloids and fatty acids, new studies gives insight on the scope of macro molecules which are primarily PR-proteins as QSI. No proper source yet has been found for these PR-proteins, the mechanism to induce the biosynthesis of these have been found out mainly by stress induction using salicylic acid, jasmonic acid, methyl jasmonate and ethylene.

18.4.1   Micromolecules

Secondary metabolites are not growth regulators and doesn't involve in cellular proliferation, but they are often involved with plant defence (Pooja et al. 2015; Shiva Krishna et al. 2015; Varsha et al. 2016). Plants have evolved an arrangement of chemical defence molecules referred as secondary metabolites. Phytoanticipins (including cyanogenic glycosides, saponins, and glucosinolates) are produced by pre-existing precursors producing infection. (Morant et al. 2008). Vacuolar membrane is found to be the source of phytoanticipins other than the major source which is cell wall. They are released by a hydrolysing enzyme after pathogenic interaction (Kalia 2015; Go et al. 2015; Bose and Chatterjee 2015). Phytoalexins (non-glycosides compounds such as terpenoids, polyphenols etc.) are small molecules. These molecules are synthesized on pattern recognition of elicitors during pathogenic encounter and get accumulated in the cells.

The secondary metabolites in plants are of three large chemical classes: terpenoids, phenolics, and alkaloids. Ajoene from garlic, catechin from *Combretum albiflorum* and iberin from horseradish specifically inhibits the bacterial QS in various reporter strains (Vandeputte et al. 2010; Jakobsen et al. 2012). As an acquired resistance by biotic interactions, several plants synthesize metabolites that modulate

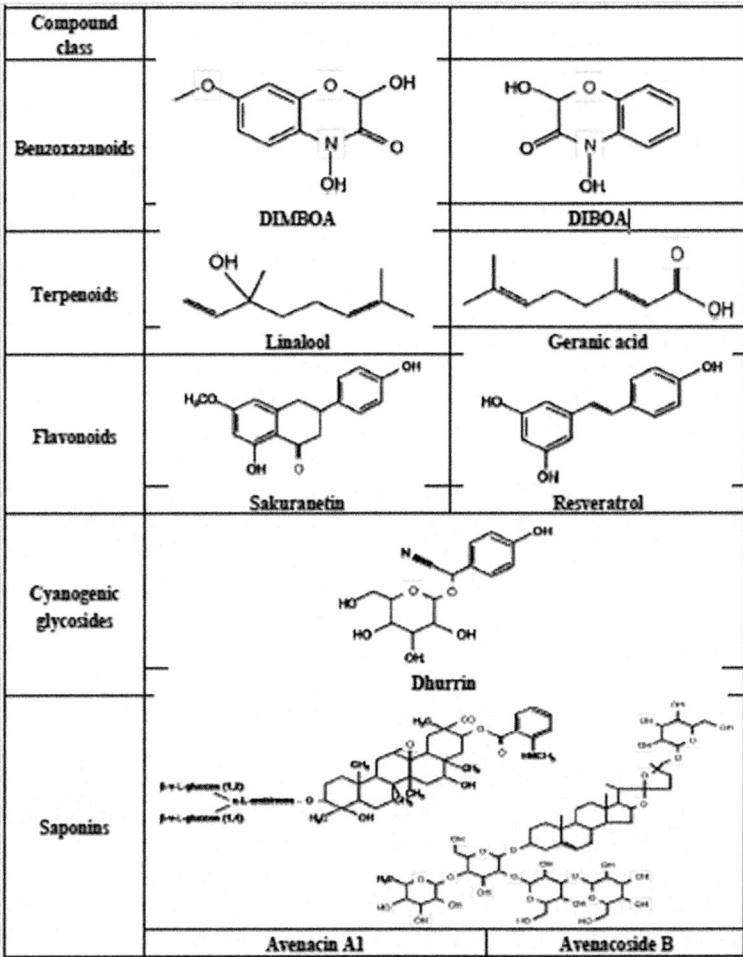

Compound class		
Benzoxazanoids	DIMBOA	DIBOA
Terpenoids	Linalool	Geranic acid
Flavonoids	Sakuranetin	Resveratrol
Cyanogenic glycosides	Dhurrin	
Saponins	Avenacin A1	Avenacoside B

Fig. 18.3 Diagram illustrating the structures of several plant secondary metabolites belonging to the major classes of defence related compounds from cereal crops

growth of microbes and have been conventionally used to treat microbial infections. Epigallocatechin-3-gallate (EGCG), a major catechin in *Camellia sinensis,* green tea leaves are known for their anti-oxidative, anti-cancerous and anti-microbial properties (Kumar et al. 2011).

The advantage of exploiting plant secondary metabolites to identify QS inhibitors is that plants are extensively being exposed to pathogenic attacks (Stout et al. 2006) and are hence hypothesised to have developed sophisticated plant defence mechanisms (Fig. 18.3). Plant-pathogen interactions during pathogenic attack leads to the production of secondary metabolites (Du Fall and Solomon 2011). Plants synthesize several secondary metabolites which phenol and its derivatives, tannins, anthocyanins, phytosterols, avenenathramides, policosanols. These secondary metabolites are not exploited for plant metabolism but they defend plants from pathogenic attacks. These metabolites can be damper compounds, allelopathic compounds and antimicrobial agents, they are either pathogen/stress induced compounds (phytoalexins), or innately produced (phytoanticipans) and these cereal secondary metabolites have been identified to induce pathogenic, insect and bird resistance in plants.

18.4.2 Macromolecules

Plants have different mechanism to counteract during threats like physical, chemical and biological stresses. PR proteins synthesis during a stress induction or pathogenic attack is an important phenomenon of plant's response. PR proteins gets accumulated in the site of infection and neighbouring tissues. Production of PR proteins in the parts of plants which are not attacked by pathogens can act as a preventing agent for spreading of infection, thus acting as a major barrier for infection and play a pivotal role in immune response in plants (Ryalls et al. 1996). Most PR proteins of plant species are low molecular weight, acid-soluble, and proteins that resist protease activity. Recently, PR-proteins till now have been segregated into 17 families according to their properties and function, including β-1,3-glucanases mostly expressed in endosperm cap tissues (Antoniw et al. 1980), chitinases expressed in radical tip tissues (Van Loon 1982; Me'traux et al. 1988; Melchers et al. 1994), thaumatin-like proteins expressed by viroid attack on plants (Van Loon 1982; Vera and Conejero 1988), peroxidases-antioxidant enzymes (Lagrimini et al. 1987), ribosome-inactivating proteins which are protein synthesis inhibitor (Green and Ryan 1972), defensins (Terras et al. 1995), thionins or gamma thionins having high antimicrobial activity (Epple et al. 1995), nonspecific lipid transfer proteins referred also as trypsin- alpha amylase inhibitor (García-Olmedo et al. 1995; Green and Ryan 1972), oxalate oxidase (Zhang et al. 1995) and oxalate-oxidase-like proteins belonging to oxido reductase family (Wei et al. 1998).

Though PR proteins have been found to have exhaustive defence activity against various streams of micro-organisms, studies are yet to define the QSI activity of these proteins. There are various identified PR proteins families (War et al. 2012).

Lectins are glycosidic-binding proteins or macromolecules, pervasive in nature, and exhibits essential defence activity, due to which they are hypothesized to have the ability to act as a QS inhibitor. Jasmonic acid induced the expression of NICTABA lectin in tobacco leaves (Lannoo et al. 2007; Epple et al. 1995). Proteinase inhibitors (PI) are the extensively available defence proteins in plants. Proteinase inhibitors are present along with storage proteins in seeds and tubers because of one its major function to degrade the storage proteins and by inhibiting the enzymes for metabolism. 1–10% of total proteins comprises of PIs. PIs activity can be studied through pathogen interactions in virulence factor inhibition. Chitinase being an effective virulence factor of C. violaceum, can be examined for its inhibition by PR proteins like lipid transfer proteins, thionins and proteinase inhibitors, which in turn can act as important QS inhibitory candidates (García-Olmedo et al. 1995).

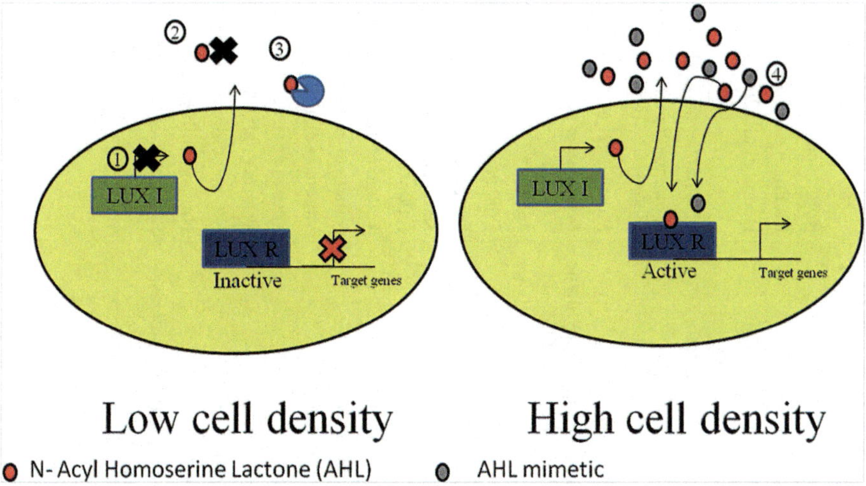
LUX I
LUX R
Inactive
Target genes
LUX I
LUX R
Active
Target genes
Low cell density
High cell density
N-Acyl Homoserine Lactone (AHL)
AHL mimetic

Fig. 18.4 Quorum sensing can be disrupted by: ① Inhibition of AHL synthesis, ② Degradation of AHL, ③ AHL sequestration and ④ AHL mimicking

The general character of bioactive proteins/peptides unexplored leaves a research promotion on PR-proteins to have QS inhibition (Fig. 18.4). Though lack of information on plant bioactive proteins/peptides specific to its activity makes small molecules a potential selection as QSI, the undeniable property of bioactive proteins/peptides to be a therapeutic agent gives high scope of it being a QSI (Table 18.1).

18.4.3 Quorum Sensing Inhibitors Extracted from Plant Source

18.5 Biochemistry of QSI Synthesis in Plants (Fig. 18.5)

18.5.1 Small Molecules or Micro Molecules

Secondary metabolites having anti-bacterial properties are richly populated in the seed coat. Since the first site of interaction of pathogens to utilize the nutrients such as storage proteins and carbohydrates is the testa (seed coat), accumulation of defence molecules can be observed. Phenolic compounds accumulation in exotesta play a static or cidal inhibition of micro-organisms having antagonistic effect.

Table 18.1 QSIs identified from plants and their effect on QS-mediated mechanisms

Source	Nature of compound	Microorganism	Mechanism	References
Allium sativum	Extract	*C. violaceum*	Violaceum production inhibition	Bodini et al. (2009)
	Extract	*P. aeruginosa*	Alginate elastase and biofilm production inhibition	Bodini et al. (2009)
Ananas comosus	Extract	*P. aeruginosa*	Inhibition of biofilmand virulence factor	Musthafa et al. (2010)
Alyssum maritimum (Leaf)	Extract	*C. violaceum* CV0blu	Inhibition	Karamanoli and lindow (2006)
Ananas comosus	Extract	*C. violaceum*	Violacein production inhibition	Musthafa et al. (2010)
		P. aeruginosa	Inhibition of Pyocyanin pigment, staphylolytic protease, elastase production and biofilm formation	Musthafa et al. (2010)
Arabidopsis exudate:	Extract	*A. tumefaciens*	AHL signaling	Chai et al. (2007)
Blueberry extracts	Extract	*C. violeceum*	Violacein production inhibition	Vattem et al. (2007)
Brassica napus (Leaf)	Extract	*C. violaceum* CV0blu	Slight inhibition	Karamanoli and Lindow (2006)
Brassica oleracea	Extract	*C. violeceum*	Violacein production inhibition	Vattem et al. (2007)
Basil	Extract	*C. violeceum*	Violacein production inhibition	Vattem et al. (2007)
Thyme	Extract	*C. violeceum*	Violacein production inhibition	Vattem et al. (2007)
Rosemary	Extract	*C. violeceum*	Violacein production inhibition	Vattem et al. (2007)
Ginger, turmeric	Extract	*C. violeceum*	Violacein production inhibition	Vattem et al. (2007)
Cinnamomum zeylacium	Cinnamaldehyde	*P. aeruginosa*	Inhibition of biofilm formation	Niu and Gilbert (2004)
		V. harveyi	AHL and AI2 mediated QS	Niu and Gilbert (2004)
	Cinnamaldehyde, $4-NO_2-$ cinnamaldehyde	*Vibrio spp.*	AI2 mediated QS- bioluminescense, protease activity, pigment formation	Brackman et al. (2008)

Plant	Type	Target organism	Activity	Reference
Combretam albiflorum	Flavonoid	P. aeruginosa	C4-HSL perception by RhlR	Schaefer et al. (2008)
Grape extract	Extract	C. violeceum	Violacein production inhibition	Vattem et al. (2007)
Grape fruit juice	Furocoumarins	E. coli	Biofilm inhibition	Girennavar et al. (2008)
		P. aeruginosa	Biofilm inhibition	Girennavar et al. (2008)
Lotus corniculatus (seedlings)	Extract	A. tumefaciens NTLR	Beta-galactosidase	Delalande et al. (2005)
Manikara zapota	Extract	C. violeceum CVO26	Violacein production inhibition	Musthafa et al. (2010)
Musa paradiciaca	Extract	C. violeceum	Violacein production inhibition	Musthafa et al. (2010)
		P. aeruginosa PAO1	Inhibition of Pyocyanin pigment, staphylolytic protease, elastase production and biofilm formation	Musthafa et al. (2010)
Medicago sativa	Seed exudate L-canavine	C. violeceum	Violacein production inhibition	Keshavan et al. (2005)
		Sinorhizobium melitoti	EPS II	Keshavan et al. (2005)
Medicago truncatula	Extract	E. coli	LuxR receptor, AhyR receptor	Gao et al. (2003)
		C. violaceum	CviR receptor	Gao et al. (2003)
	Extract	Salmonella enterica	LuxR reporter to AI2 Signals	Gao et al. (2003)
	Extract	S. meliloti 1021	QSI in general	Mathesius et al. (2003)
Ocimum basilicum	Rosamaric acid	P. aeruginosa	Inhibition of Pyocyanin pigment, staphylolytic protease, elastase production and biofilm formation	Walker et al. (2004)

(continued)

Table 18.1 (continued)

Source	Nature of compound	Microorganism	Mechanism	References
Ocimum sanctum	Extract	*C. violaceum*	Violacein production inhibition	Musthafa et al. (2010)
	Extract	*P. aeruginosa*	Inhibition of Pyocyanin pigment, staphylolytic protease, elastase production and biofilm formation	Musthafa et al. (2010)
Passiflora incarnate (leaf)	Extract	*C. violaceum* (CV0blu)	Violecein production inhibition	Karamanoli and lindow (2006)
Pisum sativum (seedlings)	Extract	*C. violaceum* CV026	C4HSL inducible protease N-acetylglucosaminidase, violacein production inhibition	Teplitski et al. (2000)
P. sativum (roots)	Extract	*C. violaceum* CV0blu	Violacein production inhibition	Karamanoli and Lindow (2006)
Prunus armeniaca	Extract	*C. violaceum* CV026 and *P. aeruginosa* PA01	Violacein production and swarming motoility inhibition	Koh and Tham (2011)
Raspberry extract	Extract	*C. violaceum*	Violacein production inhibition	Vattem et al. (2007)
Romneya trichoclyx (leaf)	Extract	*C. violaceum* CV0blu	Violacein production inhibition	Karamanoli and Lindow (2006)
Ruta graveolens	Extract	*C. violaceum* CV0blu	Slight inhibition	Karamanoli and Lindow (2006)
Squash exudate	y-hydroxybutyra e	*A. tumefaciens*	AHL signaling	Chai et al. (2007)
Tomato Seedling	y-hydroxybutyrae	*A. tumefaciens*	AHL signaling	Chai et al. (2007)
Vanilla planifolia	Extract	*C. violaceum* CV026	Violacein production inhibition	Choo et al. (2006)
Ballota nigra	Extract	*MRSA* (NRS385)	Inhibition of haemolysin production through agr gene interference	Quave et al. (2011)

Salvadora persica	Extract	*S. mutanscariogenic* isolates	Inhibition of adhaesion and biofilm	Al-Sohaibani and Murugan (2012)
Emblica officinalis	Extract	*S. mutans* MTCC 497	Suppresses the expression of genes involved in biofilm formation	Hasan et al. (2012)
Hydrastis canadenis	Extract	*S. aureus*	Inhibit toxin production and prevent keratinocyte damage	Cech et al. (2012)
Achyranthes aspera	Extract	*S. mutans*	Interation through OmpR QS regulators	Murugan et al. (2013)
Amphyterygium adstringens	Extract	*P. aeruginosa* PA14	Inhibition Pyocyanin and rhamnolipid production	Castillo-Juárez et al. (2013)
Terminalia chebula	Ellagic acid derivatives	*P. aeruginosa*	Downregulates laslr and rhIIR genes	Sarabhai et al. (2013)
Dalbergia trichocarpa	Extract	*P. aeruginosa*	Biofilm inhibition	Rasamiravaka et al. (2013)
Rosa rugosa	Polyphenols	*E. coli* K-12, *P. aeruginosa* *C. violaceum*	Swarming motility and violacein production inhibition	Zhang et al. (2014)
Wheat bran	Extract	*P. fluorescenes*	Degradation of AHL and biofilm inhibition	González-Ortiz et al. (2014)
Chamemeleum nobile	Extract	*P. aeruginosa* PA01	Inhibit biofilm formation	Kazemian et al. (2015)
Kalanchoe blossfeldiana	Extract	*P. aeruginosa*	Reduces virulence factors secretion	Sarkar et al. (2015)
Amomum tsaoka	Extract	*C. violaceum*	Violacein production, swarming motility inhibition	Rahman et al. (2017)
Anethum graveolens	3-O-methyl ellagic acid	*S. marcescens*	Downregulates QS genes fimC, bsmA and FlhD.	Salini and Pandian 2015
Sygygium aromaticum	Extract	*C. violaceum*	Biofilm formation and proteolytic activity inhibition	Mutungwa et al. (2015)
Castanea sativa	Oleanene derivatives	*S. aureus*	Haemolytic activity and biofilm formation	Quave et al. (2011)

(continued)

Table 18.1 (continued)

Source	Nature of compound	Microorganism	Mechanism	References
Usnea longissimi	Orcinol, arabitol, apigenin and usnic acid	*C. violaceum* CV12472	Violacein production inhibition, reduction of virulence factor secretion.	Singh et al. (2015)
Z. officinale	Extract	*S. mutans*	Biofilm Inhibition	Hasan et al. (2015)
Glycyrrhiza glabra	Licoricone, glyeyrin and glyzarin	*A. boumannii*	Reduce production of QS regulated virulence factors.	Bhargava et al. (2015)
Piper betle	Extract	*P. aeruginosa*	Reduces swarming effect, pyocyanin production inhibition	Datta et al. (2016)
Green tea	Epigallocatechin	*P. aeruginosa*	Synergistic activity with ciprofloxacin in biofilm treatment	Yang et al. (2010)
Cambretum	Catechin and naringenin	*P. aeruginosa*	Pyocyanin production and elastase production inhibition	Vandeputte et al. (2010)
	Saponins, ginseroides and polysaccharides	*P. aeruginosa*	Suppression of production of LasA and LasB	Wu et al. (2011)
Rheum officinale	Chrysophanol, nodakenetin, shikonin	*Stenotrophomonas maltrophilia*	Proteolysis of QS signal receptor TraR	Ding et al. (2011)
Cuminum cymimum	Methyl eugenol	*P. aeruginosa*	Motility and EPS production inhibition	Packiavathy et al. (2012)
	Rosamarinic acid, naringin and mangiferin	*P. aeruginosa*	Biofim formation and virulence factor inhibition	Annapoorani et al. (2012)

Curcuma langa	Curcumin	E. coli	Biofilm inhibition and QS dependent factors	Packiavathy et al. (2013), Packiavathy et al. (2014)
		P. aeruginosa		
		P. mirabilis		
		S. marcescenes		
		Vibrio harveyi		
		Vibrio vulnifus		
Cecropia pachystachya	Chlorogenic acid, isoorientin, orientin, isovitexin, vitexin	C. violaceum	Violacein production inhibition	Brango-Vanegas et al. (2014)
	Hamamelitannin	MRSA Mu50	Increases susceptibility to vancomycin through TraP receptor	Brackman et al. (2016)
Clove essential oil	Extract	P. aeruginosa PA01 A. hydrophilia WAF-28	Inhibition of LasB, Protease, chitinase activity	Husain et al. (2013)
Murraya koenigii	Essential oil	C. violaceum	Violacein production inhibition	Bai and Vittal (2014)
Cinnamon oil	Essential oil	P. aeruginosa	Alginate and EPS production	Kalia et al. (2015a)
	Eugenol	C. violaceum	Violacein production nhibition	Zhou et al. (2013)
	Carvacrol	C. violaceum	Biofilm inhibition and down regulation of CviI genme	Burt et al. (2014)

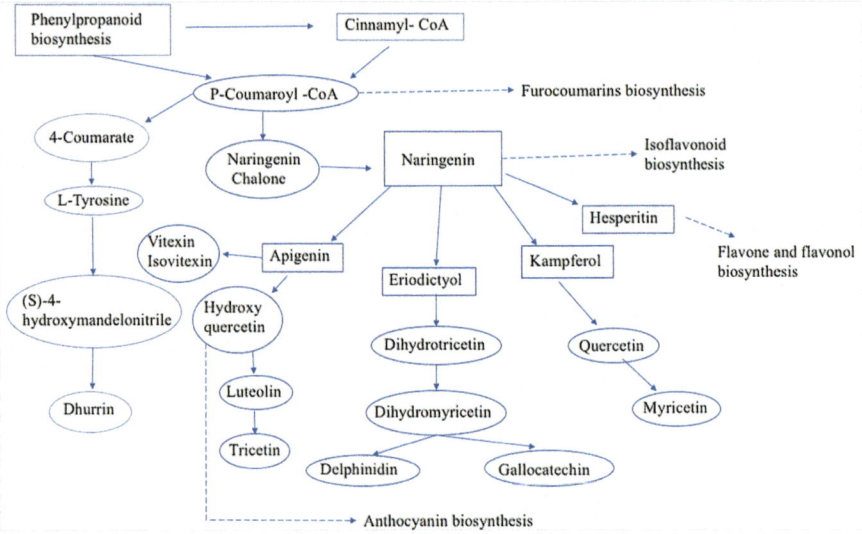

Fig. 18.5 Biosynthesis pathway of secondary metabolites in plants (QSIs)

Isovitexin present in seed coat of *Vigna mungo* has been tested for anti-quorum sensing activity. These types of secondary metabolites are the first line of defense action in seeds. Secondary metabolites thus increase the shelf life of the seeds. In addition to that, some cyanogenic glycosides such as, Dhurrin are synthesized during the process of seed germination. This has been now identified as a quorum sensing inhibitor which was extricated from *Sorghum bicolor*.

18.5.2 Macromolecules (PR Proteins)

Salicylic acid being an important phytohormone is involved in the regulation of plant defence mechanism by inducing the SAR pathway. During the plant pathogen interaction, accumulation of salicylic acid induces Non-Expressor of Pathogenesis related genes 1 (NPR-1), which is activated through REDOX pathway by salicylic acid accumulation and translocated to nucleus. It acts as a transcription factor and a non-binder of DNA, activates PR protein expression (Pieterse and Van Loon 2004). Salicylic acid and jasmonic acid act antagonistically by inhibiting each other's activity (Maffei et al. 2007). Methyl- Salicylic acid is a volatile signal to trigger induced response in plants. It induces the arthropods defensin's mechanism (Maffei et al. 2007; De Boer et al. 2008).

18.6   Conclusion

Several plant-derived inhibitors of *P. aeruginosa* QS systems have been discovered, and *in-vivo* studies of some of these inhibitors (garlic, *C. erectus, P. ginseng*) are demonstrated to be significant in treating infections of *P. aeruginosa*. Further investigation for these plant sources may focus on the evaluation of the synergistic effects of the QSI compounds or crude extracts along with the antibiotics. Alternatively, a combination of two or more QSI compounds on bacterial growth, QS signalling, and its virulence may also be examined. There is a need to understand the PR proteins signal molecules, their identification, mode of action, and signal transduction. An understanding of induced resistance in plants can be utilized for interpreting the ecological interactions between plants and microbes. The future challenge is to exploit the elicitors of induced defence mechanism in plants and identify the genes encoding for PR proteins which is hypothesized to have QS inhibition. This activity takes place by regulating the genes responsible for QS in bacteria during plant microbe interaction. The current trends in QS inhibition is synthetic QSI analogue to plant based. This has high specificity since the interaction is signal molecule based and receptor- ligand based. Synthetic compounds having similar mechanism to that of plant based QSI has an effective role in QSI activity.

18.7   Opinion

Among Quorum sensing inhibitors, plant derived QSI shows promising activity due to the biochemistry of plant's defence mechanism. The abundance of secondary metabolites in plants make it the most significant source to obtain the QSI. The current trend focuses on PR proteins and peptides which are less toxic and have high specificity on the interaction with the virulence gene pool of different QS bacteria. Due to the generalized defence system such as PAMP induction, the QSI from plants can counteract the genetic diversity of pathogens. Recent analysis has revealed the resistance mechanism of QS bacteria against synthetic QSI there is an urgency in finding a new tool to tackle QS. The synthetic QSI analogue to plant-based QSI is now being studied on gene silencing mechanism. Also, now Dietary phytochemicals are now being analysed to study the plant-human-microbiome. This study can reveal information on various applications such as health, food, agriculture, aquaculture etc.

References

Agarwala M, Choudhury B, Yadav RN (2014) Comparative study of antibiofilm activity of copper oxide and iron oxide nanoparticles against multidrug resistant biofilm forming uropathogens. Indian J Microbiol 54:365–368. https://doi.org/10.1007/s12088-014-0462-z

Ahiwale SS, Bankar AV, Tagunde S, Kapadnis BP (2017) A bacteriophage mediated gold nanoparticle synthesis and their anti-biofilm activity. Indian J Microbiol 57:188–194. https://doi.org/10.1007/s12088-017-0640-x

Akram M, Hamid A, Khalil A, Ghaffar A, Tayyaba N, Saeed A, Naveed A (2014) Review on medicinal uses, pharmacological, phytochemistry and immunomodulatory activity of plants. Int J Immunopathol Pharmacol 27:313–319. https://doi.org/10.1177/039463201402700301

Al-Sohaibani S, Murugan K (2012) Anti-biofilm activity of *Salvadora persica* on cariogenic isolates of Streptococcus mutans: in vitro and molecular docking studies. Biofouling 28:29–38. https://doi.org/10.1080/08927014.2011.647308

Annapoorani A, Umamageswaran V, Parameswari R, Pandian SK, Ravi AV (2012) Computational discovery of putative quorum sensing inhibitors against LasR and RhlR receptor proteins of *Pseudomonas aeruginosa*. J Comput Aided Mol Des 26:1067–1077. https://doi.org/10.1007/s10822-012-9599-1

Antoniw JF, Ritter E, Pierpoint WS, Van Loon LC (1980) Comparison of three pathogenesis-related proteins from plants of two cultivars of tobacco infected with TMV. J Gen Virol 47:79–87. https://doi.org/10.1099/0022-1317-47-1-79

Arasu MV, Al-Dhabi NA, Rejiniemon TS, Lee KD, Huxley VAJ, Kim DH, Duraipandiyan V, Karuppiah P, Choi KC (2015) Identification and characterization of *Lactobacillus brevis* P68 with antifungal, antioxidant and probiotic functional properties. Indian J Microbiol 55:19–28. https://doi.org/10.1007/s12088-014-0495-3

Azman CA-S, Othman I, Fang C-M, Chan K-G, Goh B-H, Lee L-H (2017) Antibacterial, anti-cancer and neuroprotective activities of rare actinobacteria from mangrove forest soils. Indian J Microbioldoi 57:177. https://doi.org/10.1007/s12088-016-0627-z

Bai AJ, Vittal RR (2014) Quorum sensing inhibitory and anti-biofilm activity of essential oils and their in vivo efficacy in food systems. Food Biotechnol 28:269–292. https://doi.org/10.1080/08905436.2014.932287

Balakrishnan D, Bibiana AS, Vijayakumar A, Santhosh RS, Dhevendaran K, Nithyanand P (2015) Antioxidant activity of bacteria associated with the marine sponge *Tedania anhelans*. Indian J Microbiol 55:13–18. https://doi.org/10.1007/s12088-014-0490-8

Bandyopadhyay P, Mishra S, Sarkar B, Swain SK, Pal A, Tripathy PP, Ojha SK (2015) Dietary *Saccharomyces cerevisiae* boosts growth and immunity of IMC *Labeo rohita* (Ham.) juveniles. Indian J Microbiol 55:81–87. https://doi.org/10.1007/s12088-014-0500-x

Begum IF, Mohankumar R, Jeevan M, Ramani K (2016) GC–MS analysis of bioactive molecules derived from *Paracoccus pantotrophus* FMR19 and the antimicrobial activity against bacterial pathogens and MDROs. Indian J Microbiol 56:426–432. https://doi.org/10.1007/s12088-016-0609-1

Bhargava N, Singh SP, Sharma A, Sharma P, Capalash N (2015) Attenuation of quorum sensing-mediated virulence of *Acinetobacter baumannii* by *Glycyrrhiza glabra* flavonoids. Future Microbiol 10:1953–1968. https://doi.org/10.2217/fmb.15.107

Bodini SF, Manfredini S, Epp M, Valentini S, Santori F (2009) Quorum sensing inhibition activity of garlic extract and p-coumaric acid. Lett Appl Microbiol 49:551–555. https://doi.org/10.1111/j.1472-765X.2009.02704

Borges A, Abreu AC, Ferreira C, Saavedra MJ, Simões LC, Simões M (2015) Antibacterial activity and mode of action of selected glucosinolate hydrolysis products against bacterial pathogens. J Food Sci Technol 52:4737–4748. https://doi.org/10.1007/s13197-014-1533-1

Bose D, Chatterjee S (2015) Antibacterial activity of green synthesized silver nanoparticles using Vasaka (*Justicia adhatoda* L.) leaf extract. Indian J Microbiol 55:163–167. https://doi.org/10.1007/s12088-015-0512-1

Brackman G, Defoirdt T, Miyamoto C, Bossier P, Van Calenbergh S, Nelis H, Coenye T (2008) Cinnamaldehyde and cinnamaldehyde derivatives reduce virulence in *Vibrio spp.* by decreasing the DNA-binding activity of the quorum sensing response regulator LuxR. BMC Microbiol 8:149. https://doi.org/10.1186/1471-2180-8-149

Brackman G, Breyne K, De Rycke R, Vermote A, Van Nieuwerburgh F, Meyer E, Coenye T (2016) The quorum sensing inhibitor Hamamelitannin increases antibiotic susceptibility of *Staphylococcus aureus* biofilms by affecting peptidoglycan biosynthesis and eDNA release. Sci Rep 6:20321. https://doi.org/10.1038/srep20321

Brango-Vanegas J, Costa GM, Ortmann CF, Schenkel EP, Reginatto FH, Ramos FA, Castellanos L (2014) Glycosylflavonoids from *Cecropia pachystachya Trécul* are quorum sensing inhibitors. Phytomedicine 21:670–675. https://doi.org/10.1016/j.phymed.2014.01.001

Burt SA, Ojo-Fakunle VTA, Woertman J, Veldhuizen EJA (2014) The natural antimicrobial carvacrol inhibits quorum sensing in *Chromobacterium violaceum* and reduces bacterial biofilm formation at sub-lethal concentrations. PLoS ONE 9:1–6. https://doi.org/10.1371/journal.pone.0093414

Castillo-Juárez I, García-Contreras R, Velázquez-Guadarrama N, Soto-Hernández M, Martínez-Vázquez M (2013) *Amphypterygium adstringens* anacardic acid mixture inhibits quorum sensing-controlled virulence factors of *Chromobacterium violaceum* and *Pseudomonas aeruginosa*. Arch Med Res 44:488–494. https://doi.org/10.1016/j.arcmed.2013.10.004

Cech N, Junio H, Ackermann L, Kavanaugh J, Horswill A (2012) Quorum quenching and antimicrobial activity of goldenseal (*Hydrastis canadensis*) against methicillin-resistant *staphylococcus aureus* (MRSA). Planta Med 78:1556–1561. https://doi.org/10.1055/s-0032-1315042

Chai Y, Ching ST, Cho H, Winans SC (2007) Reconstitution of the biochemical activities of the AttJ repressor and the AttK, AttL, and AttM catabolic enzymes of *Agrobacterium tumefaciens*. J Bacteriol 189:3674–3679. https://doi.org/10.1128/JB.01274-06

Chen X, Stewart PS (2002) Role of electrostatic interactions in cohesion of bacterial biofilms. Appl Microbiol Biotechnol 59:718–720. https://doi.org/10.1007/s00253-002-1044-2

Chevrot R, Rosen R, Haudecoeur E, Cirou A, Shelp BJ, Ron E, Faure D (2006) GABA controls the level of quorum-sensing signal in *Agrobacterium tumefaciens*. Proc Natl Acad Sci 103:7460–7464. https://doi.org/10.1073/pnas.0600313103

Choo JH, Rukayadi Y, Hwang JK (2006) Inhibition of bacterial quorum sensing by vanilla extract. Lett Appl Microbiol 42:637–641. https://doi.org/10.1111/j.1472-765X.2006.01928

Datta S, Jana D, Maity TR, Samanta A, Banerjee R (2016) Piper betle leaf extract affects the quorum sensing and hence virulence of *Pseudomonas aeruginosa PAO1*. 3 Biotech 6:1–6. https://doi.org/10.1007/s13205-015-0348-8

De Boer JG, Hordijk CA, Posthumus MA, Dicke M (2008) Prey and non-prey arthropods sharing a host plant: effects on induced volatile emission and predator attraction. J Chem Ecol 34:281–290. https://doi.org/10.1007/s10886-007-9405-z

Dehghan H, Sarrafi Y, Salehi P (2016) Antioxidant and antidiabetic activities of 11 herbal plants from Hyrcania region, Iran. J Food Drug Anal 24:179–188. https://doi.org/10.1016/j.jfda.2015.06.010

Delalande L, Faure D, Raffoux A, Uroz S, D'Angelo-Picard C, Elasri M, Dessaux Y (2005) N-hexanoyl-L-homoserine lactone, a mediator of bacterial quorum-sensing regulation, exhibits plant-dependent stability and may be inactivated by germinating *Lotus corniculatus* seedlings. FEMS Microbiol Ecol 52:13–20. https://doi.org/10.1016/j.femsec.2004.10.005

Ding X, Yin B, Qian L, Zeng Z, Yang Z, Li H, Zhou S (2011) Screening for novel quorum-sensing inhibitors to interfere with the formation of *Pseudomonas aeruginosa* biofilm. J Med Microbiol 60:1827–1834. https://doi.org/10.1099/jmm.0.024166-0

Dobrucka R, Długaszewska J (2015) Antimicrobial activities of silver nanoparticles synthesized by using water extract of *Arnicae anthodium*. Indian J Microbiol 55:168–174. https://doi.org/10.1007/s12088-015-0516-x

Du Fall LA, Solomon PS (2011) Role of cereal secondary metabolites involved in mediating the outcome of plant-pathogen interactions. Metabolites 1:64–78. https://doi.org/10.3390/metabo1010064

Epple P, Apel K, Bohlmann H (1995) An Arabidopsis thaliana thionin gene is inducible via a signal transduction pathway different from that for pathogenesis-related proteins. Plant Physiol 109:813–820. https://doi.org/10.2307/4276871

Fürstenberg-Hägg J, Zagrobelny M, Bak S (2013) Plant defense against insect herbivores. Int J Mol Sci 14:10242–10297. https://doi.org/10.3390/ijms140510242

Gao M, Teplitski M, Robinson JB, Bauer WD (2003) Production of substances by *Medicago truncatula* that affect bacterial quorum sensing. Mol Plant Microbe Interact MPMI 16:827–834. https://doi.org/10.1094/MPMI.2003.16.9.827

García-Olmedo F, Molina A, Segura A, Moreno M (1995) The defensive role of nonspecific lipid-transfer proteins in plants. Trends in Microbiol 3:72–74. https://doi.org/10.1016/s0966-842x(00)88879-4

Girennavar B, Cepeda ML, Soni KA, Vikram A, Jesudhasan P, Jayaprakasha GK, Patil BS (2008) Grapefruit juice and its furocoumarins inhibits autoinducer signaling and biofilm formation in bacteria. Int J Food Microbiol 125:204–208. https://doi.org/10.1016/j.ijfoodmicro.2008.03.028

Givskov M, de Nys R, Manefield M, Gram L, Maximilien R, Eberl L, Kjelleberg S (1996) Eukaryotic interference with homoserine lactone-mediated prokaryotuc signalling. J Bact 178:6618–6622. https://doi.org/10.1128/JB.178.22.6618-6622.1996

Go T-H, Cho K-S, Lee S-M, Lee O-M, Son H-J (2015) Simultaneous production of antifungal and keratinolytic activities by feather-degrading *Bacillus subtilis* S8. Indian J Microbiol 55:66–73. https://doi.org/10.1007/s12088-014-0502-8

González-Lamothe R, Mitchell G, Gattuso M, Diarra MS, Malouin F, Bouarab K (2009) Plant antimicrobial agents and their effects on plant and human pathogens. Int J Mol Sci 10:3400–3419. https://doi.org/10.3390/ijms10083400

González-Ortiz G, Quarles Van Ufford HC, Halkes SBA, Cerdà-Cuéllar M, Beukelman CJ, Pieters RJ, Martín-Orue SM (2014) New properties of wheat bran: anti-biofilm activity and interference with bacteria quorum-sensing systems. Environ Microbiol 16:1346–1353. https://doi.org/10.1111/1462-2920.12441

Green TR, Ryan CA (1972) Wound-induced proteinase inhibitor in plant leaves: a possible defense mechanism against insects. Science 175:776–777. https://doi.org/10.1126/science.175.4023.776

Gui Z, Wang H, Ding T, Zhu W, Zhuang X, Chu W (2014) Azithromycin reduces the production of α-hemolysin and biofilm formation in Staphylococcus aureus. Indian J Microbiol 54:114–117. https://doi.org/10.1007/s12088-013-0438-4

Gurpreet S, Ekant T, Aurovind A, Chellan K, Kanchana K, Pachaiappan R (2015) Bioactive proteins from Solanaceae as quorum sensing inhibitors against virulence in Pseudomonas aeruginosa. Med Hypotheses 84(6):539–542. https://doi.org/10.1016/j.mehy.2015.02.019

Hasan S, Danishuddin M, Adil M, Singh K, Verma PK, Khan AU (2012) Efficacy of E. Officinalis on the cariogenic properties of streptococcus mutans: a novel and alternative approach to suppress quorum-sensing mechanism. PLoS ONE 7:1–12. https://doi.org/10.1371/journal.pone.0040319

Haudecoeur E, Planamente S, Cirou A, Tannières M, Shelp BJ, Moréra S, Faure D (2009) Proline antagonizes GABA-induced quenching of quorum-sensing in *Agrobacterium tumefaciens*. Proc Natl Acad Sci U S A106:14587–14592. https://doi.org/10.1073/pnas.0808005106

Hemaiswarya S, Kruthiventi AK, Doble M (2008) Synergism between natural products and antibiotics against infectious diseases. Phytomedicine 15:639–652. doi: https://doi.org/10.1016/j.phymed.2008.06.008

Huma N, Shankar P, Kushwah J, Bhushan A, Joshi J, Mukherjee T, Raju SC, Purohit HJ, Kalia VC (2011) Diversity and polymorphism in AHL-lactonase gene (*aiiA*) of *Bacillus*. J Microbiol Biotechnol 21:1001–1011. https://doi.org/10.4014/jmb.1105.05056

Husain FM, Ahmad I, Asif M, Tahseen Q (2013) Influence of clove oil on certain quorum-sensing-regulated functions and biofilm of *Pseudomonas aeruginosa* and *Aeromonas hydrophila*. J Biosci 38:835–844. https://doi.org/10.1007/s12038-013-9385-9

Iriti M, Faoro F (2009) Chemical diversity and defence metabolism: how plants cope with pathogens and ozone pollution. Int J Mol Sci 10:3371–3399. https://doi.org/10.3390/ijms10083371

Jakobsen TH, Van Gennip M, Phipps RK, Shanmugham MS, Christensen LD, Alhede M, Givskov M (2012) Ajoene, a sulfur-rich molecule from garlic, inhibits genes controlled by quorum sensing. Antimicrob Agents Chemother 56:2314–2325. https://doi.org/10.1128/AAC.05919-11

Jeyanthi V, Velusamy P (2016) Anti-methicillin resistant *Staphylococcus aureus* compound isolation from halophilic *Bacillus amyloliquefaciens* MHB1 and determination of its mode of action using electron microscope and flow cytometry analysis. Indian J Microbiol 56:148–157. https://doi.org/10.1007/s12088-016-0566-8

Kalia VC (2013) Quorum sensing inhibitors: an overview. Biotechnol Adv 31:224–245. https://doi.org/10.1016/j.biotechadv.2012.10.004

Kalia VC (2014) In search of versatile organisms for quorum-sensing inhibitors: acyl homoserine lactones (AHL)-acylase and AHL-lactonase. FEMS Microbiol Letts 359:143. https://doi.org/10.1111/1574-6968.12585

Kalia VC (2015) Microbes: The most friendly beings? In: Kalia VC (ed) Quorum sensing vs quorum quenching: a battle with no end in sight. Springer India, New Delhi, pp 1–5. https://doi.org/10.1007/978-81-322-1982-8_1

Kalia VC, Kumar P (2015a) Potential applications of quorum sensing inhibitors in diverse fields. In: Kalia VC (ed) Quorum sensing vs quorum quenching: a battle with no end in sight. Springer India, New Delhi, pp 359–370. https://doi.org/10.1007/978-81-322-1982-8_29

Kalia VC, Kumar P (2015b) The Battle: Quorum-sensing inhibitors versus evolution of bacterial resistance. In: Kalia VC (ed) Quorum sensing vs quorum quenching: a battle with no end in sight. Springer India, New Delhi, pp. 385–391. https://doi.org/10.1007/978-81-322-1982-8_31

Kalia VC, Purohit HJ (2011) Quenching the quorum sensing system: potential antibacterial drug targets. Critical Rev Microbiol 37:121–140. https://doi.org/10.3109/1040841X.2010.532479

Kalia VC, Raju SC, Purohit HJ (2011) Genomic analysis reveals versatile organisms for quorum quenching enzymes: acyl-homoserine lactone-acylase and –lactonase. Open Microbiol J 5:1–13. https://doi.org/10.2174/1874285801105010001

Kalia M, Yadav VK, Singh PK, Sharma D, Pandey H, Narvi SS, Agarwal V (2015) Effect of cinnamon oil on quorum sensing-controlled virulence factors and biofilm formation in *Pseudomonas aeruginosa*. PLoS ONE 10:1–18. https://doi.org/10.1371/journal.pone.0135495

Kalia VC, Prakash J, Koul S, Ray S (2017) Simple and rapid method for detecting biofilm forming bacteria. Indian J Microbiol 57(1):109–111. https://doi.org/10.1007/s12088-016-0616-2

Karamanoli K, Lindow SE (2006) Disruption of N-acyl homoserine lactone-mediated cell signaling and iron acquisition in epiphytic bacteria by leaf surface compounds. Appl Environ Microbiol 72:7678–7686. https://doi.org/10.1128/AEM.01260-06

Karumuri S, Singh PK, Shukla P (2015) *In silico* analog design for terbinafine against *Trichophyton rubrum*: a preliminary study. Indian J Microbiol 55:333–340. https://doi.org/10.1007/s12088-015-0524-x

Kazemian H, Ghafourian S, Heidari H, Amiri P, Yamchi JK, Shavalipour A, Sadeghifard N (2015) Antibacterial, anti-swarming and anti-biofilm formation activities of *Chamaemelum nobile* against *Pseudomonas aeruginosa*. Rev Soc Bras Med Trop, 48:432–436. https://doi.org/10.1590/0037-8682-0065-2015

KEGG Pathway (2017) http://www.genome.jp/kegg-bin/show_pathway?map=ko02024&show_description=show

Keshavan ND, Chowdhary PK, Donovan C, González JE (2005) L -Canavanine made by *Medicago sativa* interferes with quorum sensing in *Sinorhizobium meliloti*. J Bacteriol 187:8427–8436. https://doi.org/10.1128/JB.187.24.8427

Koehn FE, Carter GT (2005) The evolving role of natural products in drug discovery. Nat Rev Drug Discov 4:206–220. https://doi.org/10.1038/nrd1657

Koh KH, Tham FY (2011) Screening of traditional Chinese medicinal plants for quorum-sensing inhibitors activity. J Microbiol Immunol Infect 44:144–148. https://doi.org/10.1016/j.jmii.2009.10.001

Koul S, Kalia VC (2017) Multiplicity of quorum quenching enzymes: a potential mechanism to limit quorum sensing bacterial population. Indian J Microbiol 57:100–108. https://doi.org/10.1007/s12088-016-0633-1

Koul S, Prakash J, Mishra A, Kalia VC (2016) Potential emergence of multi-quorum sensing inhibitor resistant (MQSIR) bacteria. Indian J Microbiol 56:1–18. https://doi.org/10.1007/s12088-015-0558-0

Kumar A, Saigal K, Malhotra K, Sinha KM, Taneja B (2011) Structural and functional characterization of Rv2966c protein reveals an RsmD-like methyltransferase from *Mycobacterium tuberculosis* and the role of its N-terminal domain in target recognition. J Biol Chem 286:19652–19661. https://doi.org/10.1074/jbc.M110.200428

Kumar P, Koul S, Patel SKS, Lee JK, Kalia VC (2015) Heterologous expression of quorum sensing inhibitory genes in diverse organisms. In: Kalia VC (ed) Quorum sensing vs quorum quenching: a battle with no end in sight. Springer India, New Delhi, pp 343–356. https://doi.org/10.1007/978-81-322-1982-8_28

Lagrimini LM, Burkhart W, Moyer M, Rothstein S (1987) Molecular cloning of complementary DNA encoding the lignin-forming peroxidase from tobacco: molecular analysis and tissue-specific expression. Proc Natl Acad Sci 84:7542–7546. https://doi.org/10.1073/pnas.84.21.7542

Lannoo N, Vandenborre G, Miersch O, Smagghe G, Wasternack C, Peumans WJ, Van Damme EJM (2007) The jasmonate-induced expression of the *Nicotiana tabacum* leaf lectin. Plant Cell Physiol 48:1207–1218. https://doi.org/10.1093/pcp/pcm090

Lin MH, Shu JC, Huang HY, Cheng YC (2012) Involvement of iron in biofilm formation by *Staphylococcus aureus*. PLoS ONE 7:3–9. https://doi.org/10.1371/journal.pone.0034388

Lönn-Stensrud J, Petersen FC, Benneche T, Scheie AA (2007) Synthetic bromated furanone inhibits autoinducer-2-mediated communication and biofilm formation in oral Streptococci. Oral Microbiol Immunol 22:340–346. https://doi.org/10.1111/j.1399-302X.2007.00367

Maffei ME, Mithofer A, Boland W (2007) Insects feeding on plants: rapid signals and responses preceding the induction of phytochemical release. Phytochemistry 68:2946–2959. https://doi.org/10.1016/j.phytochem.2007.07.016.0

Mathesius U, Mulders S, Gao M, Teplitski M, Caetano-Anolles G, Rolfe BG, Bauer WD (2003) Extensive and specific responses of a eukaryote to bacterial quorum-sensing signals. Proc Natl Acad Sci U S A A100:1444–1449. https://doi.org/10.1073/pnas.262672599

Melchers LS, Groot MAD, Knaap JA, Ponstein AS, Sela-Buurlage MB, Bol JF, Linthorst HJ (1994) A new class of tobacco chitinases homologous to bacterial exo-chitinases displays antifungal activity. Plant J 5:469–480. https://doi.org/10.1046/j.1365-313x.1994.05040469.x

Me'traux JP, Streit L, Staub TH (1988) A pathogenesis-related protein in cucumber is a chitinase. Physiol Mol Plant Path 33:1 9. https://doi.org/10.1016/0885-5765(88)90038-0

Morant AV, Jørgensen K, Jørgensen C, Paquette SM, Sánchez-Pérez R, Møller BL, Bak S (2008) β-Glucosidases as detonators of plant chemical defense. Phytochemistry 69:1795–1813. https://doi.org/10.1016/j.phytochem.2008.03.006

Murugan K, Sekar K, Sangeetha S, Ranjitha S, Sohaibani SA (2013) Antibiofilm and quorum sensing inhibitory activity of *Achyranthes aspera* on cariogenic *Streptococcus* mutans: an in vitro and in silico study. Pharm Biol 51:728–736. https://doi.org/10.3109/13880209.2013.764330

Musthafa KS, Ravi AV, Annapoorani A, Packiavathy IASV, Pandian SK (2010) Evaluation of anti-quorum-sensing activity of edible plants and fruits through inhibition of the n-acyl-homoserine lactone system in *Chromobacterium violaceum* and *Pseudomonas aeruginosa*. Chemotherapy 56:333–339. https://doi.org/10.1159/000320185

Mutungwa W, Alluri N, Majumdar M (2015) Anti-quorum sensing activity of some commonly used traditional indian spices. Int J Pharm Pharm Sci 7:80–83

Nazzaro F, Fratianni F, Coppola R (2013) Quorum sensing and phytochemicals. Int J Mol Sci 14:12607–12619. https://doi.org/10.3390/ijms140612607

Niu C, Gilbert ES (2004) Colorimetric method for identifying plant essential oil components that affect biofilm formation and structure. Society 70:6951–6956. https://doi.org/10.1128/AEM.70.12.6951

O'Toole GA, Kolter R (1998) Initiation of biofilm formation in *Pseudomonas fluorescens WCS365* proceeds via multiple, convergent signalling pathways: a genetic analysis. Mol Microbiol 28:449–461. https://doi.org/10.1046/j.1365-2958.1998.00797.x

Packiavathy IASV, Agilandeswari P, Musthafa KS, Karutha Pandian S, Veera Ravi A (2012) Antibiofilm and quorum sensing inhibitory potential of *Cuminum cyminum* and its secondary metabolite methyl eugenol against Gram negative bacterial pathogens. Food Res Int 45:85–92. https://doi.org/10.1016/j.foodres.2011.10.022

Packiavathy IASV, Sasikumar P, Pandian SK, Veera Ravi A (2013) Prevention of quorum-sensing-mediated biofilm development and virulence factors production in *Vibrio spp.* by curcumin. Appl Microbiol Biotechnol 97:10177–10187. https://doi.org/10.1007/s00253-013-4704-5

Packiavathy IASV, Priya S, Pandian SK, Ravi AV (2014) Inhibition of biofilm development of uropathogens by curcumin – an anti-quorum sensing agent from *Curcuma longa*. Food Chem 148:453–460. https://doi.org/10.1016/j.foodchem.2012.08.002

Pieterse CMJ, Van Loon LC (2004) NPR1: the spider in the web of induced resistance signaling pathways. Curr Opin Plant Biol 7:456–464. https://doi.org/10.1016/j.pbi.2004.05.006

Pooja S, Pushpananthan M, Jayashree S, Gunasekaran P, Rajendhran J (2015) Identification of periplasmic a-amlyase from cow dung metagenome by product induced gene expression profiling (Pigex). Indian J Microbiol 55:57–65. https://doi.org/10.1007/s12088-014-0487-3

Quave CL, Plano LRW, Bennett BC (2011) Quorum sensing inhibitors for *Staphylococcus aureus* from Italian medicinal plants. Planta Med 77:188–195. https://doi.org/10.1055/s-0030-1250145

Rahman MRT, Lou Z, Yu F, Wang P, Wang H (2017) Anti-quorum sensing and anti-biofilm activity of *Amomum tsao-ko* (Amommum tsao-ko Crevost et Lemarie) on foodborne pathogens. Saudi J Biol Sci 24:324–330. https://doi.org/10.1016/j.sjbs.2015.09.034

Rasamiravaka T, Jedrzejowski A, Kiendrebeogo M, Rajaonson S, Randriamampionona D, Rabemanantsoa C, Vandeputte OM (2013) Endemic *Malagasy dalbergia* species inhibit quorum sensing in *Pseudomonas aeruginosa PAO1*. Microbiol (United Kingdom) 159:924–938. https://doi.org/10.1099/mic.0.064378-0

Rasmussen TB, Givskov M (2006) Quorum sensing inhibitors: a bargain of effects. Microbiol 152:895–904. https://doi.org/10.1099/mic.0.28601-0

Rausher MD (2001) Co-evolution and plant resistance to natural enemies. Nature 411:857–864. https://doi.org/10.1038/35081193

Rivas-San Vicente M, Plasencia J (2011) Salicylic acid beyond defence: its role in plant growth and development. J Exp Bot 62:3321–3338. https://doi.org/10.1093/jxb/err031

Ryalls JA, Neuenschwander UH, Willits MG, Molina A, Steiner HY, Hunt MD (1996) Systemic acquired resistance. Plant Cell 8:1809–1819. https://doi.org/10.1105/tpc.8.10.1809

Salini R, Pandian SK (2015) Interference of quorum sensing in urinary pathogen *Serratia marcescens* by *Anethum graveolens*. Pathog Dis 73:1–32. https://doi.org/10.1093/femspd/ftv038

Sanchart C, Rattanaporn O, Haltrich D, Phukpattaranont P, Maneerat S (2017) *Lactobacillus futsaii* CS3, a new GABA-producing strain isolated from Thai fermented shrimp (*Kung–Som*). Indian J Microbiol 57:211–217. https://doi.org/10.1007/s12088-016-0632-2

Sarabhai S, Sharma P, Capalash N (2013) Ellagic acid derivatives from Terminalia chebula Retz. Downregulate the expression of quorum sensing genes to attenuate *Pseudomonas aeruginosa PAO1* virulence. PLoS ONE 8:1–11. https://doi.org/10.1371/journal.pone.0053441

Sarkar R, Mondal C, Bera R, Chakraborty S, Barik R, Roy P, Sen T (2015) Antimicrobial properties of *Kalanchoe blossfeldiana*: a focus on drug resistance with particular reference to quorum sensing-mediated bacterial biofilm formation. J Pharm Pharmacol 67:951–962. https://doi.org/10.1111/jphp.12397

Schaefer AL, Greenberg EP, Oliver CM, Oda Y, Huang JJ, Bittan-Banin G, Harwood CS (2008) A new class of homoserine lactone quorum-sensing signals. Nature 454:595–599. https://doi.org/10.1038/nature07088

Schauder S, Bassler BL (2001) The languages of bacteria. Genes Dev 15:1468–1480. https://doi.org/10.1101/gad.899601

Sharma A, Lal R (2017) Survey of (Meta)genomic approaches for understanding microbial community dynamics. Indian J Microbiol 57:23–38. https://doi.org/10.1007/s12088-016-0629-x

Shiva Krishna P, Sudheer Kumar B, Raju P, Murty MSR, Prabhakar Rao T, Singara Charya MA, Prakasham RS (2015) Fermentative production of pyranone derivate from marine *Vibrio* sp. SKMARSP9: isolation, characterization and bioactivity evaluation. Indian J Microbiol 55:292–301. https://doi.org/10.1007/s12088-015-0521-0

Siddiqui MF, Rzechowicz M, Harvey W, Zularisam AW, Anthony GF (2015) Quorum sensing based membrane biofouling control for water treatment: a review. J Water Process Eng 30:112–122. https://doi.org/10.1016/j.jwpe.2015.06.003

Singh G, Tamboli E, Acharya A, Kumarasamy C, Mala K, Raman P (2015) Bioactive proteins from Solanaceae as quorum sensing inhibitors against virulence in Pseudomonas aeruginosa. Med Hypotheses 84(6):539–542

Singh PK, Parsek MR, Greenberg EP, Welsh MJ (2002) A component of innate immunity prevents bacterial biofilm development. Nature 417:552–555. https://doi.org/10.1038/417552a

Stout MJ, Thaler JS, Thomma BPHJ (2006) Plant-mediated interactions between pathogenic microorganisms and herbivorous anthropods. Annu Rev Entomol 51:663–689. https://doi.org/10.1146/annurev.ento.51.110104.151117

Suzuki N, Rivero RM, Shulaev V, Blumwald E, Mittler R (2014) Abiotic and biotic stress combinations. New Phytol 203:32–43. https://doi.org/10.1111/nph.12797

Szweda P, Gucwa K, Kurzyk E, Romanowska E, Dzierżanowska-Fangrat K, Jurek AZ, Kuś PM, Milewski S (2015) Essential oils, silver nanoparticles and propolis as alternative agents against fluconazole resistant *Candida albicans, Candida glabrata* and *Candida krusei* clinical isolates. Indian J Microbiol 55:175–183. https://doi.org/10.1007/s12088-014-0508-2

Teplitski M, Robinson JB, Bauer WD (2000) Plants secrete substances that mimic bacterial N-acyl homoserine lactone signal activities and affect population density-dependent behaviors in associated bacteria. Mol Pl Microbe Inter 13(6):637–648. https://doi.org/10.1094/MPMI.2000.13.6.637

Terras FR, Eggermont K, Kovaleva V, Raikhel NV, Osborn RW, Kester A, Vanderleyden J (1995) Small cysteine-rich antifungal proteins from radish: their role in host defense. Plant Cell 7:573–588. https://doi.org/10.1105/tpc.7.5.573

Van Loon LC (1982) Regulation of changes in proteins and enzymes associated with active defence against virus infection. In Active defense mechanisms in plants. Springer US, pp 247–273. https://doi.org/10.1007/978-1-4615-8309-7_14

Vandeputte OM, Kiendrebeogo M, Rajaonson S, Diallo B, Mol A, MEl J, Baucher M (2010) Identification of catechin as one of the flavonoids from combretum albiflorum bark extract that reduces the production of quorum-sensing-controlled virulence factors in *Pseudomonas aeruginosa* PAQ1. Appl Environ Microbiol 76:243–253. https://doi.org/10.1128/AEM.01059-09

Vanetten HD, Mansfield JW, Bailey JA, Farmer EE (1981) Letter To the Editor. Phytopathol:106–109. https://doi.org/10.1105/tpc.6.9.1191

Varsha KK, Nishant G, Sneha SM, Shilpa G, Devendra L, Priya S, Nampoothiri KM (2016) Antifungal, anticancer and aminopeptidase inhibitory potential of a phenazine compound produced by Lactococcus BSN307. Indian J Microbiol 56:411–416. https://doi.org/10.1007/s12088-016-097-1

Vattem DA, Mihalik K, Crixell SH, McLean RJC (2007) Dietary phytochemicals as quorum sensing inhibitors. Fitoterapia, 78:302–310. https://doi.org/10.1016/j.fitote.2007.03.009

Vera P, Conejero V (1988) Pathogenesis-related proteins of tomato P-69 as an alkaline endoproteinase. Plant Physiol 87:58–63. https://doi.org/10.1104/pp.87.1.58

Vikram A, Jayaprakasha GK, Jesudhasan PR, Pillai SD, Patil BS (2010) Suppression of bacterial cell-cell signalling, biofilm formation and type III secretion system by citrus flavonoids. J Appl Microbiol 109:515–527. https://doi.org/10.1111/j.1365-2672.2010.04677.x

Walker TS, Walker TS, Bais, HP, Bais HP, De E, De E, Vivanco JM (2004) *Pseudomonas aeruginosa*. Plant Society 134:320–331. https://doi.org/10.1104/pp.103.027888.such

War AR, Paulraj MG, Ahmad T, Buhroo AA, Hussain B, Ignacimuthu S, Sharma HC (2012) Mechanisms of plant defense against insect herbivores. Plant Signal Behav 7:1306–1320. https://doi.org/10.4161/psb.21663

Wei Y, Zhang Z, Andersen CH, Schmelzer E, Gregersen PL, Collinge DB, Thordal-Christensen H (1998) An epidermis/papilla-specific oxalate oxidase-like protein in the defence response of barley attacked by the powdery mildew fungus. Plant Mol Bio 36:101–112. https://doi.org/10.1023/a:1005955119326

Wu H, Lee B, Yang L, Wang H, Givskov M, Molin S, Song Z (2011) Effects of ginseng on *Pseudomonas aeruginosa* motility and biofilm formation. FEMS Immunol Med Microbiol 62:49–56. https://doi.org/10.1111/j.1574-695X.2011.00787.x

Yang L, Liu Y, Sternberg C, Molin S (2010) Evaluation of enoyl-acyl carrier protein reductase inhibitors as *Pseudomonas aeruginosa* quorum-quenching reagents. Molecules 15:780–792. https://doi.org/10.3390/molecules15020780

Zhang Z, Collinge DB, Thordal-Christensen H (1995) Germin-like oxalate oxidase, a H2O2-producing enzyme, accumulates in barley attacked by the powdery mildew fungus. Plant J 8:139–145. https://doi.org/10.1046/j.1365-313x.1995.08010139.x

Zhang R, Pappas KM, Brace JL, Miller PC, Oulmassov T, Molyneaux JM, Joachimiak A (2002) Structure of a bacterial quorum-sensing transcription factor complexed with pheromone and DNA. Nature 417:971–974. https://doi.org/10.1038/nature10294

Zhang J, Rui X, Wang L, Guan Y, Sun X, Dong M (2014) Polyphenolic extract from *Rosa rugosa* tea inhibits bacterial quorum sensing and biofilm formation. Food Control 42:125–131. https://doi.org/10.1016/j.foodcont.2014.02.001

Zhou L, Zheng H, Tang Y, Yu W, Gong Q (2013) Eugenol inhibits quorum sensing at sub-inhibitory concentrations. Biotechnol Lett 35:631–637. https://doi.org/10.1007/s10529-012-1126-x

Chapter 19
Bioactive Phytochemicals Targeting Microbial Activities Mediated by Quorum Sensing

Beatriz Ximena Valencia Quecán, Milagros Liseth Castillo Rivera, and Uelinton Manoel Pinto

Abstract Bacteria engage on a cell-density dependent communication mechanism known as quorum sensing (QS) in order to regulate important phenotypes, including virulence. QS inhibition through the use of small interfering molecules has become attractive as an anti-virulence strategy. There is great potential presented by plant secondary metabolites due to multiple interactions between plants and microbes. The broad classes of phenolic compounds are valuable resources since their interference with QS has been suggested. This chapter will present studies dealing with plant compounds that can interfere with QS and how this interaction happens at the molecular level. We also discuss those compounds or plant extracts in which QS interference has been suggested based upon inhibition of QS controlled phenotypes, even though details about the mechanism of action are still missing. Quorum sensing inhibition has emerged as an antivirulence strategy and phytochemicals are among the main compounds that can be used in biotechnological applications directed to the food and pharmaceutical industries.

Keywords Quorum sensing · Quorum quenching · Antivirulence strategy · Phytochemicals

B. X. V. Quecán · M. L. C. Rivera · U. M. Pinto (✉)
Food Research Center, Department of Food and Experimental Nutrition,
Faculty of Pharmaceutical Sciences, University of Sao Paulo, São Paulo, SP, Brazil
e-mail: uelintonpinto@usp.br

© Springer Nature Singapore Pte Ltd. 2018
V. C. Kalia (ed.), *Biotechnological Applications of Quorum Sensing Inhibitors*,
https://doi.org/10.1007/978-981-10-9026-4_19

Abbreviations

AHLs	Acyl homoserine lactones
AI	Autoinducer
AI-1	Autoinducer-1
AI-2	Autoinducer-2
AIP	Autoinducing peptides
EPS	Exopolysaccharide
HSL	Homoserine lactone
OHHL	N-3-oxohexanoyl-L-homoserinelactone
QQ	Quorum quenching
QS	Quorum sensing
C4HSL	N-butanoyl-L-homoserine lactone
C6HSL	N-hexanoyl-L-HSL
3-oxo-C12HSL	N-(3-oxododecanoyl)-L-HSL
3-oxo-C6HSL	N-(3-oxohexanoyl)-L-HSL

19.1   Introduction

Quorum sensing is a communication mechanism by which bacteria synchronize gene expression in response to small signaling molecules, also known as autoinducers, which accumulate according to cell density (Papenfort and Bassler 2016). There are different signaling molecules making quorum sensing systems quite diverse (Table 19.1). In Gram-negative bacteria, signaling is commonly mediated by acyl-homoserine lactones (AHLs), while small peptides generally facilitate cell-to-cell communication in Gram-positive cells (Platt and Fuqua 2010; Papenfort and Bassler 2016). Another small molecule termed autoinducer-2 (AI-2) is associated with quorum sensing in both types of bacteria due to the widespread presence of its synthase gene *luxS* (Fuqua and Greenberg 2002). Other types of signaling molecules have also been described such as p-coumaryl-homoserine lactone, unsaturated fatty acids, quinolone, among others (Tsai and Winans 2010).

Quorum sensing involves the synthesis of an autoinducer by an appropriate synthase, release and accumulation of the autoinducer according to cell density, and detection by an appropriate autoinducer receptor protein. Therefore, disrupting any of these steps would hinder bacterial communication and all the associated phenotypes. This approach has been extensively investigated and is part of a broader field related to antivirulence strategies (Dickey et al. 2017). Quorum sensing inhibition or as commonly referred to as quorum quenching can be accomplished by inhibiting autoinducer synthesis or its secretion, degradation of these molecules by means of enzymatic or chemical reactions, and by hampering autoinducer recognition by their receptor proteins. The advantage of inhibiting quorum sensing arises from the fact that many virulence factors would also be inhibited by this approach which

Table 19.1 Bacterial quorum sensing systems

QS system	Signaling molecules	Bacterial group
AI-1	Many types of AHLs which vary in acyl chain length and substitution on carbon 3	Gram-negative
AI-2	Furanosyl borate diester or (2R,4S)-2-methyl-2,3,3,4-tetrahydroxytetrahidrofuran	Gram-negative and Gram-positive
AI-3	Unknown structure	*Salmonella*
AIP	Many different types of signaling peptides	Gram-positive
Other	Quinolones, Diketopiperazines, indole Hydroxyketones	Gram-negative or Gram-positive
	Gamma-butyro-lactones	Gram-positive (Actinobacteria)

Source: Bai and Rai (2011), Roy et al. (2011), and Skandamis and Nychas (2012)
AI autoinducer, *AIP* autoinducer peptide, *AHL* acyl homoserine lactone

would have important consequences on bacterial fitness and present lower selective pressure compared to treatment with conventional antibiotics (LaSarre and Federle 2013; Allen et al. 2014).

Many natural compounds have been shown to interfere with quorum sensing controlled phenotypes. These compounds can be either small molecules or enzymes produced by plants, bacteria, algae, fungi and animals (Kalia 2013; Ta and Arnason 2016; Saurav et al. 2017). In fact, in the last few years, many studies have been published with plant extracts or plant derived molecules for their ability to inhibit quorum sensing controlled phenotypes (Borges et al. 2016; Ta and Arnason 2016). These plants include those from medicinal origin as well as foods such as garlic, ginger, berries, basil, thyme, and many others. However, a concern has recently been raised bringing into question the specificity of many tested molecules or extracts towards quorum sensing (Defoirdt et al. 2013). We discuss that in more detail in the next sections.

In this review, we discuss the different types of quorum sensing inhibiting molecules derived from plants and provide insights into their mechanism of action as bona fide quorum sensing inhibitors, meaning that they directly inhibit the quorum sensing circuitry. We also discuss those inhibitors that may indirectly influence quorum sensing by affecting QS controlled phenotypes and not directly affecting the QS communication circuit.

19.2 Quorum Sensing and Quorum Quenching Approaches

There is great need for developing alternatives to combat bacterial antimicrobial resistance. Studies aiming at elucidating the mechanisms of bacterial virulence and cell-to-cell communication have revealed promising strategies in the development of drugs. The interference in virulence mechanisms and signaling pathways, termed antivirulence strategy, has caught the attention of the scientific community since selective pressure posed by this approach should be lower than that from traditional antimicrobials, besides presenting specificity to the target organism (Rasko and Sperandio 2010; Kalia 2013; Allen et al. 2014).

Bacterial cell-to-cell signaling or quorum sensing (QS) is mediated by chemical compounds that accumulate according to population density triggering bacterial community responses. The phenotypes regulated by QS directly contribute to pathogenesis through the synchronized production of virulence factors such as toxins, proteases, motility, biofilm formation among others, having important implications for human and veterinary medicine, agriculture, as well as food safety and quality (Skandamis and Nychas 2012; LaSarre and Federle 2013).

The mechanism of QS communication varies according to the organism and the system presents specificities in each species. It is believed that through QS communication, bacteria are able to differentiate self from others and to maintain a fidelity to their own communication circuits. Therefore, an autoinducer of one species would have low effect on gene expression of another species, agreeing with the notion that some autoinducers promote intraspecies communication (Federle and Bassler 2003; Taga and Bassler 2003; Lazdunski et al. 2004).

However, autoinducers share common links even among distantly related species, with similarities to autocrine and paracrine signaling (Doganer et al. 2016). Typically, a QS circuit comprises a signal synthase, a signaling molecule and a signal responsive protein that regulates the expression of target genes. In Gram-negative bacteria, communication is generally mediated by acyl-homoserine lactone (AHL) molecules, also known as autoinducer-1 (AI-1), and in Gram-positives signaling is accomplished by autoinducer peptides (AIP) (Table 19.1). Another molecule involved in communication between both groups of bacteria is called autoinducer-2 (AI-2), which is believed to mediate communication interspecifically (Rui and Monnet 2015).

For Gram-negative microbes, the signal synthase is encoded by *luxI* homologues and transcription regulation is mediated by LuxR-type proteins by interaction with AHLs. Model systems in this class of bacteria are exemplified by LuxI/LuxR proteins from *Vibrio fischeri*, RhlI/RhlR and LasI/LasR proteins from *Pseudomonas aeruginosa*, TraI/TraR from *Agrobacterium tumefaciens,* and CviI/CviR from *Chromobacterium violaceum* (Pinto and Winans 2009; Schuster et al. 2013). In every case, specific signaling compounds are produced and detected by their partner proteins. Homologous systems are found in other bacterial species controlling a range of phenotypes including biofilm formation, conjugation, pigment production, bioluminescence, as well as many virulence factors (Tsai and Winans 2010; Schuster et al. 2013). In Gram-positives, AIPs are usually detected by transmembrane histidine kinases which phosphorylate a response regulator leading to changes in transcription regulation (Rui and Monnet 2015).

The model system for the AHL mediated quorum sensing mechanism is found in the marine bacterium *V. fischeri* (Turan et al. 2017). In this bacterium, quorum sensing relies on the production of N-3-oxohexanoyl-L-homoserinelactone (OHHL) by a protein called LuxI (Eberhard et al. 1981). OHHL can diffuse through the bacterial cell membrane, accumulating in the light organ of certain species of squid and fish. When its concentration reaches a critical threshold, which coincides with high cell density, the molecule is able to bind to a cytoplasmic receptor protein called LuxR. LuxR bound to OHHL is a transcriptional activator that induces bioluminescence in *V. fischeri* (Waters and Bassler 2005). LuxR bound to OOHL is a dimeric protein and works activating transcription similarly to TraR from *A. tumefaciens* by recruiting RNA polymerase to promoters of QS target genes (Costa et al. 2009; Pinto and Winans 2009). LuxR/LuxI homologous systems have been found to regulate diverse phenotypes in many proteobacteria, coordinating group behavior that ranges from pathogenesis to biofilm formation, pigment production, conjugation, production of secondary metabolites, to symbiosis (Pinto et al. 2012; Skandamis and Nychas 2012; Papenfort and Bassler 2016).

LuxR homologues are usually activated by a cognate AHL, however in a few cases these proteins are actually inactivated by their signal (Winans et al. 2016). Another variation to the theme is the presence of LuxR homologues that are not coupled to a LuxI partner. These proteins, usually referred to as orphan regulators, are thought to either work in a ligand-independent fashion or to recognize endogenous or exogenous AHLs, as well as to detect other types of small molecules (Subramoni and Venturi 2009; Nguyen et al. 2015; Almeida et al. 2016). They can also function as dominant-negative inhibitors as in the case of TlhR, a truncated TraR homologue from *A. tumefaciens* (Chai et al. 2001).

There is great interest in interfering with QS due to the importance of the QS-controlled phenotypes. There are many aspects of the QS circuitry that can be targeted such as the autoinducer synthesis, its inactivation through enzymatic or chemical actions, and by reducing activity of the receptor protein through the use of autoinducer antagonists (Fig. 19.1) (LaSarre and Federle 2013; Allen et al. 2014).

The mechanism of QS inhibition through autoinducer antagonists can be achieved through binding of the inhibitor to the enzyme that synthesizes the autoinducer or through binding of the inhibitor to the receptor protein (LuxR homolog) which may undergo protein destabilization or structural rearrangement ultimately hindering DNA binding and transcriptional control (Chen et al. 2011; Chung et al. 2011).

The strategy usually undertaken to study quorum sensing inhibition has been through evaluation of easily measured quorum sensing controlled phenotypes such as bioluminescence, pigment production, swarming motility, biofilm formation, among others. However, there are limitations of such approach since some phenotypes may be affected by other factors apart from QS, including the metabolic state of the cell. This is particularly true for complex phenotypes such as bioluminescence which depends upon many co-factors in order to be expressed (Miyashiro and Ruby 2012). It is also important to consider the general toxicity of the tested compound, even at low concentrations. QS inhibition studies usually make use of subinhibitory concentrations, but even those may present unexpected and unperceived effects on growth curve experiments. Additionally, some studies have only considered end-point bacterial growth measurements which might mask toxic effects during growth of the organism and those effects may be perceived as inhibition of QS but may actually be side effects (Rasmussen and Givskov 2006; Defoirdt et al. 2013).

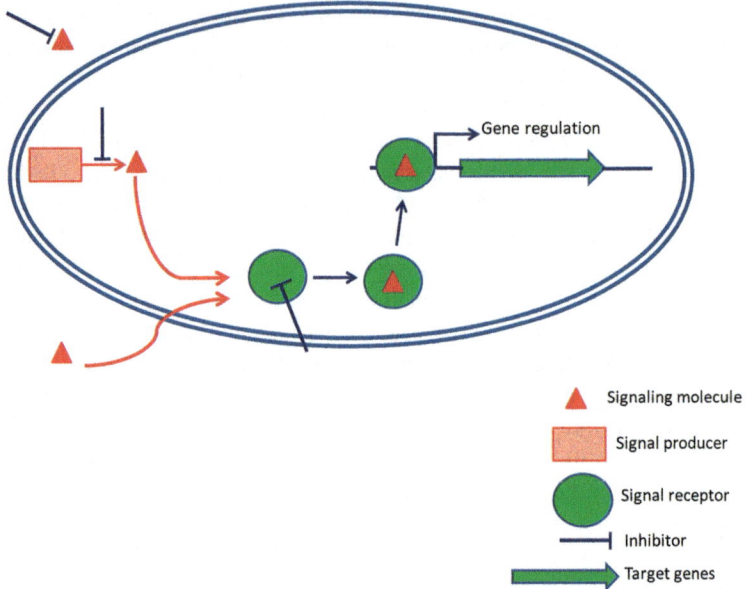

Fig. 19.1 Quorum sensing interference mechanisms in a bacterial cell. The canonical quorum sensing system is exemplified by a signal producer (also known as autoinducer synthase), the signaling molecule, the signal receptor protein and the regulated targeted genes, often referred to as quorum sensing regulon. QS inhibition can occur at the synthesis stage of the signaling molecule through enzymatic inhibition at the **signal producer**, through inactivation or sequestration of the **signaling molecule** by enzymes, antibodies or other molecules, and through signal interference, commonly referred as antagonism, with the receptor protein (**signal receptor**) hindering gene regulation

Thus, a true QS inhibitor should have high specificity either towards the synthase (LuxI homologue), or the signal molecule itself (AHL) or the signal receptor protein (LuxR homologue).

19.3 Plant Derived Compounds with Quorum Sensing Inhibition Properties

Quorum sensing interference by means of signal antagonism has been an intense area of research in the last 10 years. However, this approach dates back to 1986 when Eberhard and collaborators (1986) evaluated the effect of a chemically synthesized AHL analog library over *V. fischeri* bioluminescence. The tested analogs had structures that were similar to the cognate *V. fischeri* autoinducer and in some cases inhibited bioluminescence while in others they actually induced it. The authors concluded that the autoinducer action site was not sterically constrained, which could explain signal interference, even though no structural mechanism was available at that time. In a similar work, Zhu and collaborators (1998) tested 32 compounds with structural similarities to the *A. tumefaciens* autoinducer molecule and found that only the cognate signal was able to induce quorum sensing controlled phenotypes while a few were able to induce but only when at high concentrations. They also found that the majority of the molecules were inactive. However, some analogs were actually potent antagonists in the tested conditions.

In 1996, a pioneering report was published describing secondary metabolites known as furanones produced by the marine alga *Delisea pulchra* with inhibitory action over AHL-controlled phenotypes (Givskov et al. 1996). This work inaugurated the search for natural products with potential for quorum sensing interference. About 20 years later, the quorum quenching field greatly expanded and molecules with this ability have been searched in all domains of life, all throughout the world.

The potential presented by plant secondary metabolites at inhibiting QS is of special notice due to dynamic interactions between plants and microbes. In fact, studies on plant-bacterial interactions in the rhizosphere have shown that plants respond to bacterial autoinducers by secreting phenolic compounds which can act as inhibitors against pathogens and interfere with bacterial signaling (Brencic and Winans 2005; Cho and Winans 2005; Haichar et al. 2014). Phenolic compounds are phytochemicals grouped into diverse classes such as phenolic acids, lignans, stilbenes and flavonoids, based on their structural variations (Cheynier 2005). These compounds are related to anti-oxidative stress, besides presenting important biological functions including antimicrobial activity (Manach et al. 2004).

Many studies have shown the potential of plant extracts rich in phenolic compounds to interfere with QS controlled phenotypes in different bacteria. However, not that many have investigated the mechanism by which QS inhibition takes place. Besides, little is known about the global interference of individual or combined phenolics on bacterial physiology, as well as their specificity over QS circuitries. Even less is known on the effects of these potential interfering molecules over bacterial adaptation and fitness in the environment. A number of studies based on plant extracts are shown on Table 19.2, mainly focusing on the effects of plant extracts on QS regulated phenotypes in model organisms such as *P. aeruginosa* and *C. violaceum*.

Some studies have investigated the potential of some spices like dill, curcuma, oregano, peppermint and vanilla on QS controlled phenotypes. Packiavathy and collaborators (2012) evaluated the methanolic extract of *C. cyminum* using the model bacteria *C. violaceum* ATCC 12472, *C. violaceum* CV026, *P. aeruginosa* PA01, *P. mirabilis* ATCC7002 and *S. marcescens* FJ584421. The study showed a maximum of 90% inhibition in violacein production without affecting bacterial growth. Additionally, the extract showed a reduction in biofilm biomass, EPS production and swarming migration of *P. mirabilis*, *P. aeruginosa* and *S. marcescens*, according to the concentrations used. A molecular docking study indicated that the quorum quenching activity of the *C. cyminum* extract was likely due to methyl eugenol. The same group evaluated if the agent curcumin from *Curcuma longa* (turmeric) inhibited the biofilm formation of uropathogens such as *E. coli*, *P. aeruginosa*, *P. mirabilis* and *S. marcescens*, possibly by interfering with their QS systems. The treatment with curcumin in a concentration of 100 µg/ml attenuated the QS-dependent factors,

Table 19.2 Selected studies showing interference of plant extracts on quorum sensing regulated phenotypes

Plant extract	Isolated compound	Target microorganisms	Inhibited phenotypes and/or mechanism	References
Berries	Phenolic compounds	*C. violaceum* *Escherichia coli* *P. aeruginosa* *Serratia marcescens* *Aeromonas hydrophila*	Violacein production, swarming motility, biofilm formation	Vattem et al. (2007), Oliveira et al. (2016), and Rodrigues et al. (2016a, b)
Broccoli (*Brassica oleracea*)	Not found	*E. coli* O157:H7	It reduces AI-2 synthesis, swimming, and swarming motility	Lee et al. (2011)
Caper berries (*Capparis spinosa*)	Not found	*S. marcescens* *P. aeruginosa* *E. coli* *P. mirabilis*	Biofilm formation, EPS (exopolysaccharide) production Swimming/swarming motility	Abraham et al. (2011)
Senna (*Cassia alata*)	Not found	*C. violaceum* *P. aeruginosa*	Violacein production inhibition of virulence factors and biofilm formation.	Rekha et al. (2017)
Cumin (*Cuminum cyminum*)	Eugenol	*P. aeruginosa* *Proteus mirabilis* *S. marcescens*	Biofilm formation, motility, EPS production	Packiavathy et al. (2012)
Turmeric (*Curcuma longa*)	Curcumin	*C. violaceum* *P. aeruginosa* *P. mirabilis* *S. marcescens*	Inhibition seems to be on AHL synthase	Rudrappa and Bais (2008) and Packiavathy et al. (2014)
Chamomile (*Chamaemelum nobile*)	Not found	*P. aeruginosa*	Biofilm formation and swarming motility	Kazemian et al. (2015)
Dill (*Anethum graveolens*)	Not found	*C. violaceum*	Violacein production	Makhfian et al. (2015)
Common fig (*Ficus carica*) and perilla (*Perilla frutescens*)	Not found	*C. violaceum* *P. aeruginosa*	Violacein production, swimming motility	Sun et al. (2014)
Garlic (*Allium sativum*)	Not found	*E. coli (pSB401/pSB53)* *A. tumefaciens* *C. violaceum* *Pseudomonas putida* *Pseudomonas chlororaphis*	It antagonized the activity of the QS receptors LuxR, AhyR and TraR	Bjarnsholt et al. (2005) and Bodini et al. (2009)

(continued)

Table 19.2 (continued)

Plant extract	Isolated compound	Target microorganisms	Inhibited phenotypes and/or mechanism	References
Ginger (*Zingiber officinale*)	6-Gingerol	*C. violaceum* *P. aeruginosa*	Violacein and pyocyanin production	Kumar et al. (2014) and Kim et al. (2015)
Green tea (*Camellia sinensis*)	Not found	*Shewanella baltica*	Biofilm formation, production of EPS, swimming motility and extracellular protease activity	Yin et al. (2015) and Zhu et al. (2015)
Horseradish (*Armoracia rusticana*)	Iberin	*P. aeruginosa*	It inhibits the expression of *lasB*	Jakobsen et al. (2012a)
Sausage tree (*Kigelia africana*)	Not found	*C. violaceum*	Violacein production, LuxI and LuxR activities affected	Chenia (2013)
Orange (*Citrus reticulata*)	Naringenin	*C. violaceum* *Yersinia enterocolitica* *P. aeruginosa*	Violacein production Biofilm maturation	Truchado et al. (2012a) and Vandeputte et al. (2011)
Oregano (*Origanum vulgare*)	Not found	*C. violaceum*	Violacein production	Alvarez et al. (2014)
Peppermint (*Mentha piperita*)	Menthol	*P. aeruginosa* *A. hydrophila*	Biofilm formation	Husain et al. (2015)
Japanese rose tea (*Rosa rugosa*)	Not found	*C. violaceum* *E. coli* K-12 *P. aeruginosa*	Violacein production, swarming motility, biofilm formation	Zhang et al. (2014)
Rosemary (*Rosmarinus officinalis*)	Rosmarinic acid	*P. aeruginosa*	AHL mimic acting as a QS inducer by binding to RhlR and activating transcription of QS genes.	Corral et al. (2016)
Vanilla (*Vanilla planifolia*)	Vanillin	*C. violaceum*	Violacein production	Choo et al. (2006) and Ponnusamy et al. (2009)

such as exopolysaccharide production, alginate production, swimming and swarming motilities of the uropathogens (Packiavathy et al. 2014). An earlier study also revealed downregulation of 31 QS regulated genes in *P. aeruginosa* in the presence of curcumin (Rudrappa and Bais 2008).

Many other works have focused on fruits and vegetables as possible sources of QS inhibitors, especially their phenolic compounds. Truchado and colleagues (2012a, 2015) tested the inhibitory activity of an orange extract enriched in O-glycosylated flavanones using *C. violaceum* (CECT 494) and *Y. enterocolitica* (CECT 4315) as model organisms. The orange extract showed an inhibition of violacein production in C. *violaceum* and swimming and swarming motilities in

Y. enterocolitica, and it reduced the production of N-acylhomoserine lactones and biofilm formation. Other authors such as Vattem and colleagues (2007) determined if extracts of common fruits have QS inhibition activity at sub-lethal concentrations in model microorganisms like *C. violaceum* CV026, *C. violaceum* ATCC 31532, *P. aeruginosa* PAO1 and *E. coli* O157:H7. Their results indicated that extracts of raspberry, blueberry, grape and strawberry inhibited violacein production in different levels according to the fruit. They also showed that these fruit extracts reduced the swarming motility of *P. aeruginosa*. The authors believed that phytochemicals present in the extracts could have different mechanisms of inhibition, even though additional studies are needed in order to confirm the mode of action (Vattem et al. 2007). Similar results have been found by us working with polyphenolic rich fruit extracts from Brazil (Oliveira et al. 2016 and Rodrigues et al., 2016a, b). However, we have not investigated the mechanism of inhibition and the need for additional studies is evident.

A study with iberin, a compound present in horseradish, found its ability to block the QS-regulated gene expression in *P. aeruginosa* PAO1, namely *lasB*, in a concentration of 100 μM (16 μg/ml) of iberin. Moreover *rhlA* expression was reduced 50% at the same concentration, without growth interference. Rhamnolipid production was attenuated in wild-type *P. aeruginosa*, showing that treatment with iberin could have the potential to decrease the virulence of this microorganism (Jakobsen et al. 2012a).

Some medicinal plants such as *Moringa oleifera, K. africana, C. alata, F. carica* and *P. frutescens* have been studied in Asia and Africa, for their potential inhibitory activity over QS of *C. violaceum* and *P. aeruginosa* (Singh et al. 2009; Chenia 2013; Sun et al. 2014; Rekha et al. 2017). These studies have also showed inhibition of violacein production in *C. violaceum* and biofilm formation in *P. aeruginosa*. It is claimed that these plants are efficient for the treatment of diseases such as psoriasis, gastrointestinal problems, fungal infections and treatment of sexual diseases (Khan et al. 2001; Banno et al. 2004; Bhanushali et al. 2014). However, no QS inhibitory mechanism or the isolated compounds have been found so far.

Many other works have been performed with isolated phytochemicals and their effects over QS regulated phenotypes (Table 19.3). Some of these studies have worked out how the compounds interact with the quorum sensing components of the cell, while many others have not attained such level of depth.

As observed in Table 19.3, a wide variety of phytochemicals have been investigated for their possible QS inhibition properties. Gopu and collaborators (2015a, 2016) found that the anthocyanins malvidin and petunidin at sub-MIC concentrations are potential QSIs in *K. pneumoniae,* acting as competitive inhibitors of LasR. Carvacrol, a monoterpenoid phenol, significantly reduced biofilm formation in *C. violaceum, S. enterica* Typhimurium and *S. aureus*, even though their QS systems are different (Burt et al. 2014). It is possible that inhibition could be via AI-2 type QS or through a mechanism that is not dependent on cell density communication. Considering that carvacrol did not kill bacterial cells, more assays are required to elucidate if this phytochemical interferes in any of these QS circuits and how the interference would take place in different microbes.

Table 19.3 Selected studies showing the interference of plant derived compounds on quorum sensing regulated phenotypes

Compounds	Effects	Proposed mechanism	References
Ajoene (Sulfur-containing compound)	Reduced fluorescence in monitor strains *P. aeruginosa* (*rhlA-gfp, lasB-gfp*) and *E. coli* (*luxI-gfp*). Reduction of rhamnolipid production. Decreased C4HSL synthesis. Enhanced tobramycin effect in *P. aeruginosa* biofilm.	Possibly interfering with small regulatory RNAs in the downstream in the QS hierarchy	Jakobsen et al. (2012b)
Acetosyringone (Phenolic compound)	Growth rate slightly impaired. Induction of virulence genes from the *vir* regulon in *A. tumefaciens*.	Induction of the promoter region of *repABC*, increasing tumour inducing (Ti) plasmid copy number (a QS co-regulated phenotype).	Cho and Winans (2005)
Baicalein (Flavonoid)	Inhibited biofilm formation in *P. aeruginosa*. Proteolysis of TraR in *E. coli* biosensor.	A proteolysis due to destabilization of receptor is suggested.	Zeng et al. (2008)
Carvacrol (Monoterpenoid phenol)	Inhibition of violacein production, chitinolytic activity and biofilm formation in *C. violaceum*. Inhibition of biofilm formation in *S. enterica* subs. Typhimurium DT104 and *S. aureus*. Inhibition of *cviI* expression.	Not determined.	Burt et al. (2014)
Catechin (Flavonoid)	Inhibition of violacein production in *C. violaceum*. Inhibition of pyocyanin, elastase and biofilm formation on *P. aeruginosa*. Reduced transcription of *lasI/R* and *rhlI/R*. Reduced luminescence in *E. coli* pAL101 in C4HSL presence.	Not proven. A possible interference on the detection of C4HSL by RhlR.	Vandeputte et al. (2010)
Cinnamaldehyde Ellagic acid Resveratrol Rutin (Phenolic compounds)	Inhibition of violacein production in *C. violaceum*. Decreased concentration of 3-oxo-C6HSL and C6HSL in cultures of *Y. enterocolitica* and *Erwinia carotovora*.	Degradation and/or inhibition of synthesis of 3-oxo-C6HSL and C6HSL.	Truchado et al. (2012b)

(continued)

Table 19.3 (continued)

Compounds	Effects	Proposed mechanism	References
Diverse flavonoids	Inhibition of bioluminescence and luciferase activity in *E. coli* carrying LasR and *lasB-luxCDABE* and *E. coli* carrying RhlR and *rhlA-luxCDABE*. Inhibition of *rhlA* transcription, pyocyanin production and swarming motility in *P. aeruginosa* PA14.	Allosteric inhibition of LasR/RhlR.	Paczkowski et al. (2017)
Malvidin, Petunidin Anthocyanins (Flavonoids)	No growth kinetics tests. Inhibited violacein production in *C. violaceum*. Inhibition of biofilm formation and EPS production in *Klebsiella pneumoniae*.	Potential competitive inhibitor of LasR receptor (according to molecular docking[a]).	Gopu et al. (2015a, 2016)
Naringenin Flavanone (Flavonoid)	Inhibition of AI-1 and AI-2-induced bioluminescence in *Vibrio harveyi* BB886. Inhibition of biofilm formation in *V. harveyi* BB120 and *E. coli* O157:H7. Inhibition of type III secretion system in *V. harveyi* BB120.	No proven mechanism.	Vikram et al. (2010)
Quercetin (Flavonoid)	No growth kinetics tests. Inhibition of biofilm formation, EPS production, swarming and swimming motility in *K. pneumoniae, P. aeruginosa* and *Y. enterocolitica*. Inhibition of violacein production in *C. violaceum*.	Potential competitive inhibitor of LasR (according to molecular docking)	Gopu et al. (2015b)
Salicylic acid (Phenolic acid)	Inhibition of *vir* gene induction. Inhibition of Ti plasmid *repABC* operon	Inhibition of the expression of *virA/G* two component regulatory system. Induction of a lactonase	Yuan et al. (2007)

(continued)

Table 19.3 (continued)

Compounds	Effects	Proposed mechanism	References
Trans-cinnamaldehyde (Aldehyde)	No detection of C4HSL in supernatants of *E. coli* MG1655 [pBAD-*rhlI*] cultures. Reduction of pyocyanin production, but no effect on swarming motility in *P. aeruginosa*.	Binding to LasI and EsaI at their substrate binding site (according to molecular docking).	Chang et al. (2014)
Zeaxanthin (Carotenoid)	No growth kinetics tests. Inhibition of biofilm formation in *P. aeruginosa*. Inhibition of bioluminescence in *lasB-gfp* and *rhlA-gfp* strains. Downregulation of *lasB* and *rhlA* expression.	Ligation near to the active site of RhlR and LasR as possible mechanism (molecular docking).	Gökalsın et al. (2017)

[a]Molecular docking is an in silico analysis performed with the crystal structure of a protein and its interactions with potential ligands

Flavonoids have been extensively explored as potential QSIs due to their inhibitory effect on QS regulated phenotypes when applied at subinhibitory concentrations, without perceived interference on bacterial growth kinetics in some cases. For example, baicalein, quercetin, naringenin and catechin interfered in biofilm formation and/or the expression of virulence factors in diverse bacteria such as the opportunistic pathogen *P. aeruginosa*, *V. harveyi* and *C. violaceum* (Zeng et al. 2008; Vandeputte et al. 2010, 2011; Vikram et al. 2010; Gopu et al. 2015b). In a recent study by Paczkowski and colleagues (2017) the inhibitory effect on LasR and RhlR of nine flavonoids was validated biochemically. They initially performed a screening of 60,000 molecules from a library by using *E. coli* reporter strain carrying LasR and a *lasB-luxCDABE* construct. It was found that those flavonoids, namely, phloretin, chrysin, naringenin, quercetin, baicalein, apigenin, 7,8-dihydroxyflavone, 3,5,7-trihydroxyflavone and pinocembrin act by binding to the LasR/RhlR in a noncompetitive mechanism, in other words, these molecules bind to a site on LasR/RhlR without accessing the autoinducer binding site. This was the first work to bring to light a mechanism that differs from the literature in the sense that many investigators have proposed that the inhibition by many molecules would be at the LuxR (or its homologues) target site, even though the majority of the tested compounds do not present structural similarity to AHLs. Interestingly, many studies have used molecular docking to indicate such possible interactions. Therefore, it is important that after a molecular docking prediction, a biochemical validation be performed.

As previously discussed, flavonoids with QS inhibition properties, as shown in Table 19.3, might be confirmed as QS inhibitors according to mechanism proposed by Paczkowski et al. (2017). Vandeputte and colleagues (2011) have detailed the QS inhibitory effects of the flavanone naringenin in the pyocyanin, elastase, C4HSL and 3-oxo-C12HSL production and the expression of QS-controlled genes such as *lasI/rhlI* and *lasR/rhlR* in *P. aeruginosa*. Vandeputte et al. (2011) explained these findings as a consequence of the decreased production of AHLs and a more difficult binding between RhlR and C4HSL. Curiously, this was not observed between LasR and 3-oxo-C12HSL, coinciding with the research by Paczkowski et al. (2017).

Interestingly, Paczkowski et al. (2017) assessed flavonoids specificity pointing out that quercetin and naringenin were specific inhibitors of LasR/RhlR and LuxN (except naringenin) but not of CviR, suggesting that the inhibition of violacein production reported by other authors such as Gopu et al. (2015b) and Vandeputte et al. (2011) might be explained by a non-related QS mechanism. Besides naringenin has exhibited inhibition properties over the QS system regulated by AI-2 according to Vikram and collaborators (2010). Nonetheless this flavanone has not shown to be specific for LuxN. Therefore, more assays might be necessary in order to understand how exactly naringenin interferes in this QS circuit. Finally, Paczkowski and colleagues (2017) established that the number of hydroxyl groups and their position in one specific ring of the flavone structure are required to inhibit LasR/RhlR receptors. Taken together, these results suggest that the structure of a flavonoid should be considered before concluding it is a true quorum sensing inhibitor.

19.4 Conclusion

Considering that studies have suggested the role of plant phenolics as bacterial signaling inhibitors, and few have investigated the mechanism of inhibition or the overall effect on bacterial fitness, studies aiming at thoroughly examining the kinds of phenolic compounds that can specifically interfere with QS and how this interference is mediated at the molecular level are still necessary. It is important to take into consideration the diversified class of phenolic compounds available which should be tested systematically over different QS models. Additionally, studies on the concomitant use of multiple and effective inhibition approaches are needed since there is a large body of literature linking many classes of molecules with potential for inhibiting QS. The study on the interaction of these molecules and the possibility for synergism need to be addressed. It is likely that, by using molecules with different mechanisms of action, bacterial resistance will be minimized, but that also needs to be addressed. The influence of QS inhibition on fitness of bacterial pathogens needs to be explored as a means of creating additional hurdles that could be used with already implemented technologies in the food sector, improving food safety throughout the food chain. It is not currently known the extent of cellular damage caused by QS inhibition over bacterial behavior to commonly used processing methods employed by the industry such as survival to thermal treatment and cleaning/sanitizing procedures. Studies involving a holistic approach are needed to stimulate the biotechnological application of the knowledge that has been generated in basic field of research.

19.5 Opinion

There has been a substantial amount of work performed in the QQ field which reflects the great need for developing alternative approaches to control bacterial infections due to antimicrobial resistance. Different research groups throughout the world have been working with many kinds of sources for their potential QS inhibition, as it has been shown here and in many other review articles published in the last few years. Still, many aspects of the QS inhibition mechanism have been overlooked which highlight the need for additional studies that take into account potential toxic effects that could interfere with the QS regulated phenotypes. A true quorum sensing inhibitor should target the signaling circuit, exemplified in Gram negative bacteria by the LuxI and LuxR homologues or directly block the AI-1 signal. It is clear from our evaluation that many of the studies presented here do not arrive at the QS molecular mechanism of inhibition. Even so, it is important to point out that if a particular plant extract or an isolated compound inhibits virulence (in the case of this work, a QS regulated phenotype), additional studies should be pursued for their potential application in controlling virulence. Apart from the mode of action, it is still very important to find virulence inhibitor molecules. We expect that biotechnological applications of quorum sensing inhibitors should emerge with the increase in knowledge about the mechanism of inhibition, as well as with additional studies that take into account practical applications of those inhibitors in the food and pharmaceutical sectors.

Acknowledgements The authors thank São Paulo Research Foundation (FAPESP) for financial support to the Food Research Center - FoRC (2013/07914-8) and a grant from CNPq-Brazil (457794/2014-3). Scholarships to B.X.Q and M.L.R. were provided by CNPq.

References

Abraham S, Palani A, Ramaswamy B, Shunmugiah K, Arumugam V (2011) Antiquorum sensing and antibiofilm potential of *Capparis spinosa*. Arch Med Res 42:658–668. https://doi.org/10.1016/j.arcmed.2011.12.002

Allen RC, Popat R, Diggle SP, Brown SP (2014) Targeting virulence: can we make evolution-proof drugs? Nat Rev Microbiol 12:300–308. https://doi.org/10.1038/nrmicro3232

Almeida FA, Pinto UM, Vanetti MC (2016) Novel insights from molecular docking of SdiA from *Salmonella enteritidis* and *Escherichia coli* with quorum sensing and quorum quenching molecules. Microb Pathog 99:178–190. https://doi.org/10.1016/j.micpath.2016.08.024

Alvarez MV, Ortega-Ramirez LA, Gutierrez-Pacheco MM, Bernal-Mercado AT, Rodriguez-Garcia I, Gonzalez-Aguilar GA, Ponce A, Moreira MR, Roura SI, Ayala-Zavala JF (2014) Oregano essential oil-pectin edible films as anti-quorum sensing and food antimicrobial agents. Front Microbiol 5:699. https://doi.org/10.3389/fmicb.2014.00699

Bai AJ, Rai VR (2011) Bacterial quorum sensing and food industry. Compr Rev Food Sci Food Saf 10:183–193. https://doi.org/10.1111/j.1541-4337.2011.00150.x

Banno N, Akihisa T, Tokuda H, Yasukawa K, Higashihara H, Ukiya M, Nishino H (2004) Triterpene acids from the leaves of *Perilla frutescens* and their anti-inflammatory and antitumor-promoting effects. Biosci Biotechnol Biochem 68:85–90. https://doi.org/10.1271/bbb.68.85

Bhanushali MM, Makhija DT, Joshi YM (2014) Central nervous system activity of an aqueous acetonic extract of *Ficus carica L.* in mice. J Ayurveda Integr Med 5:89–96. https://doi.org/10.4103/0975-9476.131734

Bjarnsholt T, Jensen PØ, Rasmussen TB, Christophersen L, Calum H, Hentzer M, Hougen HP, Rygaard J, Moser C, Eberl L, Høiby N, Givskov M (2005) Garlic blocks quorum sensing and promotes rapid clearing of pulmonary *Pseudomonas aeruginosa* infections. Microbiology 151:3873–3880. https://doi.org/10.1099/mic.0.27955-0

Bodini SF, Manfredini S, Epp M, Valentini S, Santori F (2009) Quorum sensing inhibition activity of garlic extract and p-coumaric acid. Lett Appl Microbiol 49:551–555. https://doi.org/10.1111/j.1472-765X.2009.02704.x

Borges A, Abreu A, Dias C, Saavedra MJ, Borges F, Simões M (2016) New perspectives on the use of phytochemicals as an emergent strategy to control bacterial infections including biofilms. Molecules 21:877. https://doi.org/10.3390/molecules21070877

Brencic A, Winans SC (2005) Detection of and response to signals involved in host-microbe interactions by plant-associated bacteria. Microbiol Mol Biol Rev 69:155–194. https://doi.org/10.1128/MMBR.69.1.155-194.2005

Burt SA, Ojo-Fakunle VT, Woertman J, Veldhuizen EJ (2014) The natural antimicrobial carvacrol inhibits quorum sensing in *C. violaceum* and reduces bacterial biofilm formation at sublethal concentrations. PLoS One 9:e93414. https://doi.org/10.1371/journal.pone.0093414

Chai Y, Zhu J, Winans SC (2001) TrlR, a defective TraR-like protein of *Agrobacterium tumefaciens*, blocks TraR function *in vitro* by forming inactive TrlR:TraR dimers. Mol Microbiol 40:414–421. https://doi.org/10.1046/j.1365-2958.2001.02385.x

Chang CY, Krishnan T, Wang H, Chen Y, Yin WF, Chong YM, Tan LY, Chong TM, Chan KG (2014) Non-antibiotic quorum sensing inhibitors acting against N-acyl homoserine lactone synthase as druggable target. Sci Rep 4:7245. https://doi.org/10.1038/srep07245

Chen G, Swem LR, Swem DL, Stauff DL, O'Loughlin CT, Jeffrey PD, Bassler BL, Hughson FM (2011) A strategy for antagonizing quorum sensing. Mol Cell 42:199–209. https://doi.org/10.1016/j.molcel.2011.04.003

Chenia HY (2013) Anti-quorum sensing potential of crude *Kigelia africana* fruit extracts. Sensors (Basel) 13:2802–2817. https://doi.org/10.3390/s130302802

Cheynier V (2005) Polyphenols in foods are more complex than often thought. Am J Clin Nutr 81:223S–229S. https://doi.org/10.1016/j.molcel.2011.04.003

Cho H, Winans SC (2005) VirA and VirG activate the Ti plasmid *repABC* operon, elevating plasmid copy number in response to wound-released chemical signals. Proc Natl Acad Sci U S A 102:14843–14848. https://doi.org/10.1073/pnas.0503458102

Choo JH, Rukayadi Y, Hwang JK (2006) Inhibition of bacterial quorum sensing by vanilla extract. Lett Appl Microbiol 42:637–641. https://doi.org/10.1111/j.1472-765X.2006.01928.x

Chung J, Goo E, Yu S, Choi O, Lee J, Kim J, Hwang I (2011) Small-molecule inhibitor binding to an N-acyl-homoserine lactone synthase. Proc Natl Acad Sci U S A 108:12089–12094. https://doi.org/10.1073/pnas.1103165108

Corral A, Daddaoua A, Ortega A, Urgel M, Krell T (2016) Rosmarinic acid is a homoserine lactone mimic produced by plants that activates a bacterial quorum-sensing regulator. Sci Signal 9:1–10. https://doi.org/10.1126/scisignal.aaa8271

Costa E, Cho H, Winans S (2009) Identification of amino acid residues of the pheromone-binding domain of the transcription factor TraR that are required for positive control. Mol Microbiol 73:341–351. https://doi.org/10.1111/j.1365-2958.2009.06755.x

Defoirdt T, Brackman G, Coenye T (2013) Quorum sensing inhibitors: how strong is the evidence? Trends Microbiol 21:619–624. https://doi.org/10.1016/j.tim.2013.09.006

Dickey SW, Cheung GY, Otto M (2017) Different drugs for bad bugs: antivirulence strategies in the age of antibiotic resistance. Nat Rev Drug Discov 16:457–471. https://doi.org/10.1038/nrd.2017.23

Doganer BA, Yan L, Youk H (2016) Autocrine signaling and quorum sensing extreme ends of a common spectrum. Trends Cell Biol 26:262–271. https://doi.org/10.1016/j.tcb.2015.11.002

Eberhard A, Burlingame A, Eberhard C, Kenyon G, Nealson K, Oppenheimer N (1981) Structural identification of autoinducer of *Photobacterium fischeri* luciferase. Biochemistry 20:2444–2449. https://doi.org/10.1021/bi00512a013

Eberhard A, Widrig CA, McBath P, Schineller J (1986) Analogs of the autoinducer of bioluminescence in *Vibrio fischeri*. Arch Microbiol 146:35–40. https://doi.org/10.1007/BF00690155

Federle MJ, Bassler BL (2003) Interspecies communication in bacteria. J Clin Invest 112:1291. https://doi.org/10.1172/JCI200320195

Fuqua C, Greenberg E (2002) Listening in on bacteria: acyl-homoserine lactone signalling. Nat Rev Mol Cell Biol 3:685–695. https://doi.org/10.1038/nrm907

Givskov M, de Nys R, Manefield M, Gram L, Maximilien R, Eberl L, Molin S, Steinberg PD, Kjelleberg S (1996) Eukaryotic interference with homoserine lactone-mediated prokaryotic signalling. J Bacteriol 178:6618–6622. https://doi.org/10.1128/jb.178.22.6618-6622.1996

Gökalsın B, Aksoydan B, Erman B, Sesal N (2017) Reducing virulence and biofilm of *Pseudomonas aeruginosa* by potential quorum sensing inhibitor carotenoid: zeaxanthin. Microb Ecol 74:466–473. https://doi.org/10.1007/s00248-017-0949-3

Gopu V, Kothandapani S, Shetty PH (2015a) Quorum quenching activity of *Syzygium cumini* (L.) Skeels and its anthocyanin malvidin against *Klebsiella pneumoniae*. Microb Pathog 79:61–69. https://doi.org/10.1016/j.micpath.2015.01.010

Gopu V, Meena CK, Shetty PH (2015b) Quercetin influences quorum sensing in food borne bacteria: *in-vitro* and *in-silico* evidence. PLoS One 10:e0134684. https://doi.org/10.1371/journal.pone.0134684

Gopu V, Meena CK, Murali A, Shetty PH (2016) Petunidin as a competitive inhibitor of acylated homoserine lactones in *Klebsiella pneumoniae*. RSC Adv 6:2592–2601. https://doi.org/10.1039/C5RA20677D

Haichar FZ, Santaella C, Heulin T, Achouak W (2014) Root exudates mediated interactions belowground. Soil Biol Biochem 77:69–80. https://doi.org/10.1128/mBio.02429-14e02429e14

Husain F, Ahmad I, Khan MS, Ahmad E, Tahseen Q, Khan MS, Alshabib NA (2015) Sub-MICs of *Mentha piperita* essential oil and menthol inhibits AHL mediated quorum sensing and biofilm of Gram-negative bacteria. Front Microbiol 6:420. https://doi.org/10.3389/fmicb.2015.00420

Jakobsen TH, Bragason SK, Phipps RK, Christensen LD, Gennip M, Alhede M, Givskov M (2012a) Food as a source for QS inhibitors: iberin from horseradish revealed as a quorum sensing inhibitor of *Pseudomonas aeruginosa*. Appl Environ Microbiol 78:2410–2421. https://doi.org/10.1128/AEM.05992-11

Jakobsen TH, Gennip M, Phipps RK, Shanmugham MS, Christensen LD, Alhede M, Jensen P (2012b) Ajoene, a sulfur-rich molecule from garlic, inhibits genes controlled by quorum sensing. Antimicrob Agents Chemother 56:2314–2325. https://doi.org/10.1128/AAC.05919-11

Kalia VC (2013) Quorum sensing inhibitors: an overview. Biotechnol Adv 31:224–245. https://doi.org/10.1016/j.biotechadv.2012.10.004

Kazemian H, Ghafourian S, Heidari H, Amiri P, Yamchin JK, Shavalipour A, Sadeghifard N (2015) Antibacterial, anti-swarming and anti-biofilm formation activities of *Chamaemelum nobile* against *Pseudomonas aeruginosa*. Rev Soc Bras Med Trop 48:432–436. https://doi.org/10.1590/0037-8682-0065-2015

Khan MR, Kihara M, Omoloso AD (2001) Antimicrobial activity of *Cassia alata*. Fitoterapia 72:561–564. https://doi.org/10.1016/0378-8741(94)01200-J

Kim HS, Lee SH, Byun Y, Park HD (2015) 6-Gingerol reduces *Pseudomonas aeruginosa* biofilm formation and virulence via quorum sensing inhibition. Sci Rep 5(8656):1–11. https://doi.org/10.1038/srep08656

Kumar NV, Murthy PS, Manjunatha JR, Bettadaiah BK (2014) Synthesis and quorum sensing inhibitory activity of key phenolic compounds of ginger and their derivatives. Food Chem 159:451–457. https://doi.org/10.1016/j.foodchem.2014.03.039

LaSarre B, Federle MJ (2013) Exploiting quorum sensing to confuse bacterial pathogens. Microbiol Mol Biol Rev 77:73–111. https://doi.org/10.1128/MMBR.00046-12

Lazdunski AM, Ventre I, Sturgis JN (2004) Regulatory circuits and communication in gram-negative bacteria. Nat Rev Microbiol 2:581–592. https://doi.org/10.1038/nrmicro924

Lee KM, Lim J, Nam S, Yoon MY, Kwon YK, Jung BY, Yoon SS (2011) Inhibitory effects of broccoli extract on *Escherichia coli* O157: H7 quorum sensing and *in vivo* virulence. FEMS Microbiol Lett 321:67–74. https://doi.org/10.1111/j.1574-6968.2011.02311.x

Makhfian M, Hassanzadeh N, Mahmoudi E, Zandyavari N (2015) Anti-quorum sensing effects of Ethanolic crude extract of *Anethum graveolens*. J Essent Oil Bear Pl 18:687–696. https://doi.org/10.1080/0972060X.2014.998718

Manach C, Scalbert A, Morand C, Rémésy C, Jiménez L (2004) Polyphenols: food sources and bioavailability. Am J Clin Nutr 79:727–747. doi: Not Available

Miyashiro T, Ruby EG (2012) Shedding light on bioluminescence regulation in *Vibrio fischeri*. Mol Microbiol 84:795–806. https://doi.org/10.1111/j.1365-2958.2012.08065.x

Nguyen Y, Nguyen NX, Rogers JL, Liao J, MacMillan JB, Jiang Y, Sperandio V (2015) Structural and mechanistic roles of novel chemical ligands on the SdiA quorum-sensing transcription regulator. MBio 6:e02429-14. https://doi.org/10.1128/mBio.02429-14

Oliveira B, Rodrigues AC, Cardoso B, Ramos A, Bertoldi MC, Taylor J, Cunha LR, Pinto UM (2016) Antioxidant, antimicrobial and anti-quorum sensing activities of *Rubus rosaefolius* phenolic extract. Ind Crop Prod 84:59–66. https://doi.org/10.1016/j.indcrop.2016.01.037

Packiavathy I, Agilandeswari P, Musthafa KS, Pandian SK, Ravi AV (2012) Antibiofilm and quorum sensing inhibitory potential of *Cuminum cyminum* and its secondary metabolite methyl eugenol against gram negative bacterial pathogens. Food Res Int 45:85–92. https://doi.org/10.1590/1678-457X.0089

Packiavathy I, Priya S, Pandian SK, Ravi AV (2014) Inhibition of biofilm development of uropathogens by curcumin – an anti-quorum sensing agent from *Curcuma longa*. Food Chem 148:453–460. https://doi.org/10.1016/j.foodchem.2012.08.002

Paczkowski JE, Mukherjee S, McCready AR, Cong JP, Aquino CJ, Kim H, Henke BR, Smith CD, Bassler BL (2017) Flavonoids suppress *Pseudomonas aeruginosa* virulence through allosteric inhibition of quorum-sensing receptors. J Biol Chem 292:4064–4076. https://doi.org/10.1074/jbc.M116.770552

Papenfort K, Bassler BL (2016) Quorum sensing signal–response systems in Gram-negative bacteria. Nat Rev Microbiol 14:576–588. https://doi.org/10.1038/nrmicro.2016.89

Pinto UM, Winans SC (2009) Dimerization of the quorum-sensing transcription factor TraR enhances resistance to cytoplasmic proteolysis. Mol Microbiol 73:32–42. https://doi.org/10.1111/j.1365-2958.2009.06730.x

Pinto UM, Pappas KM, Winans SC (2012) The ABCs of plasmid replication and segregation. Nat Rev Microbiol 10:755–765. https://doi.org/10.1038/nrmicro2882

Platt TG, Fuqua C (2010) What's in a name? The semantics of quorum sensing. Trends Microbiol 18: 383–387. https://doi.org/10.1016/j.tim.2010.05.003

Ponnusamy K, Paul D, Kweon JH (2009) Inhibition of quorum sensing mechanism and *Aeromonas hydrophila* biofilm formation by vanillin. Environ Eng Sci 26:1359–1363. https://doi.org/10.1089/ees.2008.0415

Rasko DA, Sperandio V (2010) Anti-virulence strategy to combat bacteria-mediated disease. Nat Rev Drug Discov 9:117–128. https://doi.org/10.1038/nrd3013

Rasmussen TB, Givskov M (2006) Quorum sensing inhibitors: a bargain of effects. Microbiology 152:895–904. https://doi.org/10.1099/mic.0.28601-0d

Rekha PD, Vasavi HS, Vipin C, Saptami K, Arun AB (2017) A medicinal herb *Cassia alata* attenuates quorum sensing in *Chromobacterium violaceum* and *Pseudomonas aeruginosa*. Lett Appl Microbiol 64:231–238. https://doi.org/10.1111/lam.12710

Rodrigues AC, Oliveira B, Silva ER, Sacramento N, Bertoldi MC, Pinto UM (2016a) Anti-quorum sensing activity of phenolic extract from *Eugenia brasiliensis* (Brazilian cherry). Food Sci Technol (Campinas) 36:337–343. https://doi.org/10.1590/1678-457X.0089

Rodrigues AC, Zola FG, Oliveira BD'A, Sacramento NTB, Silva ER, Bertoldi MC, Taylo JG, Pinto UM (2016b) Quorum quenching and microbial control through phenolic extract of *Eugenia uniflora* fruits. J. Food Sci 81:M2538-M2544. https://doi.org/10.1111/1750-3841.13431

Roy V, Adams BL, Bentley WE (2011) Developing next generation antimicrobials by intercepting AI-2 mediated quorum sensing. Enzym Microb Technol 49:113–123. https://doi.org/10.1016/j.enzmictec.2011.06.001

Rudrappa T, Bais HP (2008) Curcumin, a known phenolic from *Curcuma longa*, attenuates the virulence of *Pseudomonas aeruginosa* PAO1 in whole plant and animal pathogenicity models. J Agric Food Chem 56:1955–1962. https://doi.org/10.1021/jf072591j

Rui F, Monnet V (2015) How microbes communicate in food: a review of signaling molecules and their impact on food quality. Curr Opin Food Sci 2:100–105. https://doi.org/10.1016/j.cofs.2015.03.003

Saurav K, Costantino V, Venturi V, Steindler L (2017) Quorum sensing inhibitors from the sea discovered using bacterial N-acyl-homoserine lactone-based biosensors. Mar Drugs 15:53. https://doi.org/10.3390/md15030053

Schuster M, Sexton DJ, Diggle SP, Greenberg EP (2013) Acyl-homoserine lactone quorum sensing: from evolution to application. Annu Rev Microbiol 67:43–63. https://doi.org/10.1146/annurev-micro-092412-155635

Singh BN, Singh BR, Singh RL, Prakash D, Dhakarey R, Upadhyay G, Singh HB (2009) Oxidative DNA damage protective activity, antioxidant and anti-quorum sensing potentials of *Moringa oleifera*. Food Chem Toxicol 47:1109–1116. https://doi.org/10.1016/j.fct.2009.01.034

Skandamis PN, Nychas GJ (2012) Quorum sensing in the context of food microbiology. Appl Environ Microbiol 78:5473–5482. https://doi.org/10.1128/AEM.00468-12

Subramoni S, Venturi V (2009) LuxR-family 'solos': bachelor sensors/regulators of signalling molecules. Microbiology 155:1377–1385. https://doi.org/10.1099/mic.0.026849-0

Sun S, Li H, Zhou W, Liu A, Zhu H (2014) Bacterial quorum sensing inhibition activity of the traditional Chinese herbs, *Ficus carica* L. and *Perilla frutescens*. Chemotherapy 60:379–383. https://doi.org/10.1159/000440946

Ta CAK, Arnason JT (2016) Mini review of phytochemicals and plant taxa with activity as microbial biofilm and quorum sensing inhibitors. Molecules 21:29. https://doi.org/10.3390/molecules21010029

Taga ME, Bassler BL (2003) Chemical communication among bacteria. Proc Natl Acad Sci U S A 100:14549–14554. https://doi.org/10.1073/pnas.1934514100

Truchado P, Giménez JA, Larrosa M, Castro I, Espín JC, Tomás FA, Allende A (2012a) Inhibition of quorum sensing (QS) in *Yersinia enterocolitica* by an orange extract rich in glycosylated flavanones. J Agric Food Chem 60:8885–8894. https://doi.org/10.1021/jf301365a

Truchado P, Tomás FA, Larrosa M, Allende A (2012b) Food phytochemicals act as quorum sensing inhibitors reducing production and/or degrading autoinducers of *Yersinia enterocolitica* and *Erwinia carotovora*. Food Control 24:78–85. https://doi.org/10.1016/j.foodcont.2011.09.006

Truchado P, Larrosa M, Castro-Ibáñez I, Allende A (2015) Plant food extracts and phytochemicals: their role as quorum sensing inhibitors. Trends Food Sci Technol 43:189–220. https://doi.org/10.1016/j.tifs.2015.02.009

Tsai CS, Winans SC (2010) LuxR-type quorum sensing regulators that are detached from common scents. Mol Microbiol 77:1072–1082. https://doi.org/10.1111/j.1365-2958.2010.07279.x

Turan NB, Chormey DS, Buyukpinar Ç, Engin GO, Bakirdere S (2017) Quorum sensing: little talks for an effective bacterial coordination. Trends Analyt Chem 91:1–11. https://doi.org/10.1016/j.trac.2017.03.007

Vandeputte OM, Kiendrebeogo M, Rajaonson S, Diallo B, Mol A, El Jaziri M, Baucher M (2010) Identification of catechin as one of the flavonoids from *Combretum albiflorum* bark extract that reduces the production of quorum-sensing-controlled virulence factors in *Pseudomonas aeruginosa* PAO1. Appl Environ Microbiol 76:243–253. https://doi.org/10.1128/AEM.01059-09

Vandeputte OM, Kiendrebeogo M, Rasamiravaka T, Stevigny C, Duez P, Rajaonson S, El Jaziri M (2011) The flavanone naringenin reduces the production of quorum sensing-controlled virulence factors in *Pseudomonas aeruginosa* PAO1. Microbiology 157:2120–2132. https://doi.org/10.1099/mic.0.049338-0

Vattem DA, Mihalik K, Crixell SH, McLean RJC (2007) Dietary phytochemicals as quorum sensing inhibitors. Fitoterapia 78:302–310. https://doi.org/10.1016/j.fitote.2007.03.009

Vikram A, Jayaprakasha GK, Jesudhasan PR, Pillai SD, Patil BS (2010) Suppression of bacterial cell–cell signalling, biofilm formation and type III secretion system by citrus flavonoids. J Appl Microbiol 109:515–527. https://doi.org/10.1111/j.1365-2672.2010.04677.x

Waters CM, Bassler BL (2005) Quorum sensing: cell-to-cell communication in bacteria. Annu Rev Cell Dev Biol 21:319–346. https://doi.org/10.1146/annurev.cellbio.21.012704.131001

Winans SC, Tsai CS, Ryan GT, Flores AL, Costa E, Shih KY, Winans TC, Kim Y, Jedrzejczak R, Chhor G (2016) LuxR-type Quorum-sensing regulators that are antagonized by cognate pheromones. In: de Bruijn FC (ed) Stress and environmental regulation of gene expression and adaptation in bacteria. Wiley, pp 1221–1231. ISBN: 9781119004813 https://doi.org/10.1002/9781119004813.ch118

Yin H, Deng Y, Wang H, Liu W, Zhuang X, Chu W (2015) Tea polyphenols as an antivirulence compound disrupt quorum-sensing regulated pathogenicity of *Pseudomonas aeruginosa*. Sci Rep 5:16158. https://doi.org/10.1038/srep16158

Yuan ZC, Edlind MP, Liu P, Saenkham P, Banta LM, Wise AA, Nester EW (2007) The plant signal salicylic acid shuts down expression of the *vir* regulon and activates quormone-quenching genes in *Agrobacterium*. Proc Natl Acad Sci U S A 104:11790–11795. https://doi.org/10.1073/pnas.0704866104

Zeng Z, Qian L, Cao L, Tan H, Huang Y, Xue X, Zhou S (2008) Virtual screening for novel quorum sensing inhibitors to eradicate biofilm formation of *Pseudomonas aeruginosa*. Appl Microbiol Biotechnol 79:119. https://doi.org/10.1007/s00253-008-1406-5

Zhang J, Rui X, Wang L, Guan Y, Sun X, Dong M (2014) Polyphenolic extract from *Rosa rugosa* tea inhibits bacterial quorum sensing and biofilm formation. Food Control 42:125–131. https://doi.org/10.1016/j.foodcont.2014.02.001

Zhu J, Beaber JW, Moré MI, Fuqua C, Eberhard A, Winans SC (1998) Analogs of the autoinducer 3-oxooctanoyl-homoserine lactone strongly inhibit activity of the TraR protein of *Agrobacterium tumefaciens*. J Bacteriol 180:5398–5405

Zhu J, Huang X, Zhang F, Feng L, Li J (2015) Inhibition of quorum sensing, biofilm, and spoilage potential in *Shewanella baltica* by green tea polyphenols. J Microbiol 53:829. https://doi.org/10.1007/s12275-015-5123-3

Chapter 20
Quorum Sensing Interference by Natural Products from Medicinal Plants: Significance in Combating Bacterial Infection

Mohammad Shavez Khan, Faizan Abul Qais, and Iqbal Ahmad

Abstract Plant derived natural products and phytocompounds are known for their broad spectrum biological activities and are of great therapeutic value in traditional system of medicine. The role of medicinal plants and phytocompounds in the treatment of various diseases including bacterial infection are widely documented. Anti-infective compounds from medicinal plants may provide new drug leads. Bacterial cell to cell communication has been become attractive target for the development of novel anti-infective measures that do not rely on the use of antibiotics. Targeting Quorum sensing has been emerge as promising strategy to combat bacterial infections as it is unlikely to develop multidrug resistance pathogens since it does not impose any selection pressure. In this review, we have surveyed the recent literature available on plant extracts, essential oils and phytocompounds exhibiting anti-quorum sensing properties. Further, significance of phytocompounds to combat bacterial infections caused by MDR bacteria has been discussed.

Keywords Anti-QS activity · Autoinducer · Bacterial cell to cell communication · Infectious disease · *N*-acylhomoserine lactone · Medicinal plants · Phytocompounds

Abbreviations

AHL	N-Acyl homoserine lactone
AI	Auto Inducer
CF	Cystic Fibrosis
EPS	Extracellular Polymeric Substances

M. S. Khan (✉) · F. A. Qais · I. Ahmad
Department of Agricultural Microbiology, Faculty of Agricultural Sciences,
Aligarh Muslim University, Aligarh, UP, India

© Springer Nature Singapore Pte Ltd. 2018
V. C. Kalia (ed.), *Biotechnological Applications of Quorum Sensing Inhibitors*,
https://doi.org/10.1007/978-981-10-9026-4_20

HSL   Homoserine Lactone
MDR   Multi Drug Resistance
MIC   Minimum Inhibitory Concentration
QSI   Quorum Sensing Inhibitors

20.1   Introduction

Infectious diseases are one of the major cause of mortality and morbidity. The development of multi antibiotic resistance among pathogenic bacteria has created immense clinical problem. Development of new antibacterial drugs with novel mechanism of action and use of new treatment strategies is urgently needed to combat the problem of infection caused by MDR-bacteria. Targeting virulence and pathogenicity is one of such strategy (Aqil et al. 2006).

Density dependent, small diffusible signal mediated gene regulation system in bacteria, that controls expression of virulence traits is termed as "Quorum sensing". These small diffusible signal molecules or autoinducers allows bacteria to regulate expression of large sub-set of genes related to pathogenicity with change in bacterial population (Camara et al. 2002). This signalling system has distinct architecture necessary for signal dissemination, revelation and response and has been well characterized in Gram-negative and Gram-positive bacteria (Miller and Bassler 2001; Lazdunski et al. 2004). Dismantling QS networks without having toxicity on target bacteria is considered as one of the important alternative strategy for combating bacterial infections with advantageously decreasing the risk of selection pressure (Bjarnsholt and Givskov 2007; LaSarre and Federle 2013). Thus identification, characterization and development of effective quorum sensing inhibitors (QSI) considerably gain attention from scientific community worldwide as a promising remedy to combat infection caused by multi-resistance bacteria. A potential QSI would have act through three possible mechanisms, (**a**) targeting the signal generation, (**b**) degradation of signal molecule and/or by (**c**) blocking the signal receptors. A model QSI having potential to be transformed into successful anti-infective drug should be attributed with high target specificity and negative toxicity towards the target bacterium as well as the eukaryotic host (Rasmussen and Givskov 2006; Ahmad et al. 2011; Kalia 2013; LaSarre and Federle 2013).

Despite of remarkable development in combinatorial chemistry for providing wide range of lead molecules, naturally occurring plant products are still considered as preferred choice in pharmaceutical industry (Silva et al. 2016). Considering the extensive use of plants for medicinal and dietary purposes in human, efforts have been focussed towards plants and their derived products to explore potential therapeutic principles. Plants and their derived natural products represent a vast, unexplored and comparatively safe reservoirs of diverse bioactive compounds

(Atanasov et al. 2015). These plant secondary metabolites spanning from simple phenols to highly diverse terpenes are involve in various biological process relating to plant defence against pathogens, assisting central metabolic processes, response against external stimulants etc. (Koh et al. 2013; Yarmolinsky et al. 2015). Interestingly, many plants products/phytochemicals offer entirely different mechanism of actions when compare to conventional antibacterial, representing an effective alternative to aging antibiotics (Cegelski et al. 2008; Nazzaro et al. 2013). Besides, plants derived natural products plays indispensable role in traditional medicine system worldwide including Chinese, Indian, and other folk medicine, thus research focussing plants derived products would help in bridging the gap between traditional wisdom and modern therapeutics (Adonizio et al. 2006; Ahmad et al. 2011; Zahin et al. 2010a). Role of medicinal plant products and phytocompounds against infectious agents and their diverse mode of action have been investigated and reviewed by several workers (Cowan 1999; Aqil et al. 2006; Palombo 2011; Husain and Ahmad 2013; Silva et al. 2016).

In this chapter, we present an update and brief review on anti QS activity of plants and its derived products including phytocompounds. We systematically covered plant derived natural products including plant extracts (crude or enriched), essential oils and phytocompounds, highlighting their anti-quorum sensing potential along with underlying their mechanism of actions. In the later sections, while addressing the significance of plant based QSI against infectious disease, an attempt have been made to highlight the development of successful plant based QSI that could possibly be used as anti-infective agent for management bacterial infections.

20.2 Plant Extracts with Anti QS Activity: A Recent Update

Studies in past have reported the ability of the plant extracts used in dietary or therapeutic purposes to interfere in intra and inter species QS communication. Plants are considered as vast reservoirs of bioactive chemicals and enzymes which could be act as natural quorum sensing inhibitors. The co-existence of bacteria and plant date back to millions of year which significantly results in the development and evolution of defensive mechanism against pathogenicity primarily involving the disruption of molecular communication. Among the natural strategies unfolded during past investigations includes enzymatic degradation of QS signals, the inhibition of autoinducer synthesis, receptor antagonism and disruption of signal secretion. However, recent advances in the field of plant based QSI research are more specifically focused on gene expression variation along with targeting global regulatory factors controlling overall QS mechanism. Table 20.1 shows list of common dietary and medicinal plants which have demonstrated QS inhibitory activity.

Table 20.1 Anti-QS activity of dietary and medicinal plants

Plant name	Part used for extraction	Biosensor strain(s) used	Virulence factor(s) inhibited	References
Syzygium jambos	Leaves	*C violaceum* DMST 21761	Pyoverdin	Musthafa et al. (2017)
Syzygium antisepticum		*P. aeruginosa* ATCC 27853		
Pelargonium hortorum	Aerial parts	*P. aeruginosa*	Motility	Elmanama and Al-Reefi (2017)
			Pyocyanin	
Punica granatum				
Artemisia absinthium			LasA protease	
Hibiscus sabdariffa				
Momordica charantia				
Forsythia suspense	Aerial parts	*C. violaceum* ATCC 12472	Pyocyanin, Protease, biofilm and motility	Zhang and Chu (2017)
		P. aeruginosa		
Amomum tsaoko	Fruit pods	*C. violaceum*	Pyocyanin, biofilm and motility	Rahman et al. (2017)
		P. aeruginosa		
Pistacia atlantica	Leaves	*P. aeruginosa* PA01	Pyocyanin	Kordbacheh et al. (2017)
Astilbe rivularis	Leaves	*C. violaceum* MTCC 2656	Pyocyanin and swarming motility	Tiwary et al. (2017)
Fragaria nubicola		*P. aeruginosa* MTCC 2297		
Osbeckia nepalensis				
Piper betle	Leaves	*V. harveyi* MTCC 3438	EPS, swimming motility and biofilm	Srinivasan et al. (2017)
Terminalia bellerica	Leaves	*C. violaceum* ATCC 12472	EPS, pyocyanin and biofilm	Ganesh and Rai (2017)
		P. aeruginosa PA01		
Salvadora persica	Fruit and leaves	*C. violaceum* ATCC 12472	Violecein	Noumi et al. (2017)
		C. violaceum CV026		
Cassia alata	Leaves	*C. violaceum* ATCC 12472	Swarming motility, pyocyanin, LasB elastase, protease and biofilm	Rekha et al. (2017)
		C. violaceum CV026		
		C. violaceum ATCC 31532		
		P. aeruginosa PAO1		
Mangifera indica	Leaves	*C. violaceum* ATCC 12472	Elastase, protease, pyocyanin, EPS, chitinase, swarming and biofilm	Husain et al. (2017)
		P. aeruginosa PAO1		

(continued)

Table 20.1 (continued)

Plant name	Part used for extraction	Biosensor strain(s) used	Virulence factor(s) inhibited	References
Vaccinium macrocarpon	Fruit	*P. aeruginosa* PA14 QS mutants	Elastase (LasA and LasB), alkaline protease	Maisuria et al. (2016)
		C. violaceum ATCC 31532		
Euodia ruticarpa	Fruit	*C. jejuni* NCTC 11168 QS mutants	Luminescence and biofilm formation	Bezek et al. (2016)
		V. harveyi BB 170		
Punica granatum	Peel	*C. violaceum*	Violecein and biofilm	Yang et al. (2016)
Eugenia brasiliensis	Fruit	*C. violaceum* ATCC6357	Violecein and swarming motiliy	Rodrigues et al. (2016a)
Eugenia Uniflora	Fruit	*C. violaceum* ATCC6357	Violecein	Rodrigues et al. (2016b)
Glycyrrhiza glabra	Root	*C. violaceum* ATCC 12472	Violecein	Cosa et al. (2016)
Apium graveolens	Leaves & Stalk			
Capsicum annuum	Fruit			
Syzygium anisatum	Seed			
Anogeissus leiocarpus	Stem bark	*C. violaceum* CV026	Pyocyanin and violecein	Ouedraogo and Kiendrebeogo (2016)
		P. aeruginosa PAO1 QS mutants		
Berberis aristata	Stem bark	*E. coli* (Isolated)	Adhesion and biofilm	Thakur et al. (2016)
Camellia sinensis	Leaves			
Holarrhena antidysenterica	Stem			
Piper betle	Leaves	*S. marcescens* (Isolted)	Protease, lipase, biofilm, swarming motility and EPS	Srinivasan et al. (2016)
Acer monspessulanum	Leaves	*C. violaceum* ATCC 12472	Violecein and swarming motility	Ceylan et al. (2016)
		C. violaceum CV026		
		P. aeruginosa PAO1		
Syzygium cumini	Fruit	*C. violaceum* CV026	Violecein	Gopu et al. (2015b)
Rubus rosaefolius	Fruit	*C. violaceum* ATCC6357	Violecein, biofilm and swarming motility	Oliveira et al. (2016)

(continued)

Table 20.1 (continued)

Plant name	Part used for extraction	Biosensor strain(s) used	Virulence factor(s) inhibited	References
Trigonella foenum-graceum	Seed	*C. violaceum* ATCC 12472	Elastase, protease, pyocyanin, EPS, chitinase, swarming and biofilm	Husain et al. (2015a, b)
		C. violaceum CV026		
		P. aeruginosa PAO1		
Adenanthera pavonina	Leaves	*C. violaceum* ATCC 12472	Elastase, protease, pyocyanin, swarming and biofilm	Vasavi et al. (2015)
		C. violaceum CV026		
		P. aeruginosa PAO1		
Hyptis suaveolens	Leaves	*C. violaceum* ATCC 12472	Protease, heomolysin, swimming and swarming motility	Salini et al. (2015)
Nymphaea tetragona	Leaves & Stalk	*C. violaceum* ATCC 12472	Swarming motility, pyocyanin, biofilm and LasA protease	Hossain et al. (2015)
		C. violaceum CV026		
		P. aeruginosa PAO1		
Psidium guajava	Leaves	C. violaceum MTCC 2656	Violecein and swarming motility	Ghosh et al. (2014)
		P. aeruginosa MTCC 2297		
Ficus carica	Leaves	*C. violaceum* CV026	Violecein and swarming motility	Sun et al. (2014)
Perilla frutescens		*P. aeruginosa* PAO1		
Citrus limon	Peel	*C. jejuni* NCTC 11168	Swarming motility and biofilm	Castillo et al. (2014)
Citrus medica		*V. harveyi* BB170		
Citrus aurantium				
Kigelia africana	Fruit	*C. violaceum* ATCC 12472	Violecein	Chenia (2013)
		A. tumefaciens A136, KYC6		
Fructus gardenia	Whole plant	C. violaceum ATCC 12472	Protease, elastase, pyocyanin, swimming motility and biofilm	Chu et al. (2013)
Andrographis paniculata		*P. aeruginosa* PAO1		
Dalbergia trichocarpa	Bark	*P. aeruginosa* PAO1 QS mutants	Pyocyanin, LasB, protease, biofilm and swarming motility	Rasamiravaka et al. (2013)

20.2.1 Dietary Plants

Recently, extracts of different botanical groups of dietary plant including fruits, vegetables and culinary herb and spices have been demonstrated as strong QSI (Table 20.1). Primarily, fruits received special attention, as they are important part of diet and exert beneficial effects on human health. Fruit extracts, owing to their rich dietary phytochemical content, have been recognized as promising source in the search of novel anti-infective (Truchado et al. 2015).

Aqueous extract of fruits of *Ananas comosus* (pineapple) and *Manilkara zapota* (sapodilla) were found to inhibit the QS system in *Chromobacterium violaceum* CV026 and *P. aeruginosa* PA01, inhibiting QS dependent virulence factors (pyocyanin, staphylolytic protease, elastase and biofilm formation) production (Musthafa et al. 2010). Apple extract when evaluated for QS inhibition against *C. violaceum*, showed a dose-dependent inhibition in pigment production (Fratianni et al. 2011). Koh and Tham (2011) showed that extract of *Prunus armeniaca* (apricot) inhibits AHL production in *C. violaceum* CV026 biosensor strain as well QS dependent swarming motility in *P. aeruginosa* PA01. Quorum sensing regulated production of AHLs and biofilm formation in *Yersinia enterocolitica* was found to be inhibited by flavonones rich orange extract (Truchado et al. 2012).

Berries are considered as rich source of phenolic phytochemicals such as flavonoids including anthocyanins, flavanols and flavonols, tannins, stilbenoids, phenolic acid and lignans (Vattem et al. 2007). It was observed that the production of AHL and AI-2 signal molecule was depleted under the influence of extracts of three berries when tested with *C. violaceum* CV026 and *Vibrio harveyi* BB170 tester strains and the effect was found to be concentration dependent. These phenolic rich extracts of raspberry and cloudberry also caused significant reduction of biofilm formation of *Obesumbacterium proteus* (Priha et al. 2014). Similarly, phenolic extracts of *Rubus rosaefolius*, a berry indigenous to Himalayan region of South Asia, inhibited all the phenotypes typically related to quorum sensing including violacein production, swarming motility and biofilm formation in *C. violaceum*, *Aeromonas hydrophila* and *Serratia marcescens*. Authors further observed that QS disruptive principle were predominantly natural phenols (Oliveira et al. 2016).

Another edible berry, *Syzygium cumini*, derived anthocyanin rich extracts examined for the anti-QS potential against different QS linked phenotypes in *C. violaceum* and *Klebsiella pneumoniae*. Anthocyanin constituent, malvidin, was found to interrupt the QS mechanism via binding with LasR regulatory protein as revealed from docking studies. Malvidin was also shown to inhibit virulence determinants e.g. biofilm formation and EPS (extracellular polymeric substances) production in *K. pneumoniae* (Gopu et al. 2015a). Phenolic rich extracts from Brazilian berries *Eugenia brasiliensis* and *Eugenia uniflora* have been confirmed for their quorum quenching capacities using *C. violaceum* biosensor strain (Rodrigues et al. 2016a, b). Moreover, the phenolic extract of grumixama (*Eugenia brasiliensis*) inhibited the QS regulated swarming motility in *A. hydrophila* and *S. marcesens* at concentration non-inhibitory to bacterial growth. Authors hypothesize that the observed QS inhibitory potential of the extract may have related to its phenolic constituents which have act synergistically or individually, producing the biological effect (Rodrigues et al. 2016a). Maisuria and co-workers (2016), evaluated the proanthocyanidins rich extract of cranberry (*Vaccinium macrocarpon*) against QS-linked virulence traits of *P. aeruginosa* in host *Drosophila melanogaster*. It was observed that application of the extract inhibited the production of virulence determinants and subsequently protected the host organism from fatal infection of *P. aeruginosa*. LC-MS analysis of culture supernatant revealed that levels of autoinducers (AHL) was depleted significantly. QS signalling genes including AHL synthesase LasI/RhlI and transcriptional regulators LasR/RhlR were also inhibited by proanthocyanidins rich extract. Additionally, *in silico* molecular modelling suggested that proanthocyanidins binds to QS transcriptional regulatory proteins, ultimately refraining the ligand molecules (autoinducers) to bind with active sites.

Herbs and spices are also considered as rich source of bioactive phytochemicals and extensively used in various food preparations as well as ethno-pharmaceuticals. Latest reports regarding biological activity prospect of these herbs and spices highlight their anti QS and anti-virulence potential (Table 20.1). Makhfian et al. (2015) screened 31 plant species including numerous spices and herbs for analysing their inhibitory potential against QS related behaviours, violacein pigment production in *C. violaceum* CV026 and plant tissue maceration caused by *Pectobacterium carotovorum*. Among them, the dill (*Anethum graveolens*) extract demonstrate AHL mimicking activity and subsequently inhibit violecein production to significant levels. Moreover, the extract was also effective against *Pectobacterium carotovorum* mediated tissue maceration in potato tubers and calla-lily slices. *Trigonella foenumgraceum* seed's (commonly known as fenugreek) methanol extracts demonstrated inhibition of AHL regulated virulence factors e.g. protease, LasB, elastase, pyocyanin, chitinase, EPS and swarming motility in *P. aeruginosa* PA01 and PAF79 as well as QS regulated swarming motility in aquatic pathogen *A. hydrophila* WAF38. Further, the application of the bioactive extract subsequently downregulate lasB gene as evident from β-galactosidase luminescence in *E. coli* MG4/pKDT17. Additionally, *in vivo* infection model (*Caenorhabditis elegans*), the extract exhibit significant reduction in mortality rate in infected nematode. Caffeine was identified as a major volatile constituent in the extract, also showed significant reduction in QS regulated virulence traits including biofilm formation in pathogenic bacteria (Husain et al. 2015a).

Plant derived beverages including tea and coconut water have been also found to be effective against pathogenic bacteria via modulating their QS related response. Tea polyphenols interfered with autoinducers (AI-2 and diketopiperazines) activities of *Shewanella baltica*, promoted the degradation of AI-2 and was found to be function of epigallocatechin gallate constituent of the extract. Reduction in QS phenotypes such as biofilm development, swarming motility and exopolysaccharide production were suggested to be linked with downregulation of luxS and torA genes as revealed from transcriptional analysis (Zhu et al. 2015). Two-furaldehyde diethyl acetal (2FDA) was identified as potential QSI by screening of identified compounds from water of *Coccus nucifera* via molecular docking studies using transcriptional regulator proteins LasR and RhlR as target. Subsequent transcriptional analysis revealed down regulation of autoinducer genes (lasI/rhlI) and transcriptional regulator genes (lasR/rhlR) that corresponds to observed reduction in QS regulated traits including biofilm formation, aeruginolysin, LasA, LasB, elastase, protease, swarming motility, pigment and haemolysin production in *P. aeruginosa*. Authors also demonstrated the synergistic activity of palmitic acid, another bioactive constituent obtained from coconut water, increases overall QS inhibitory and anti-biofilm potential of 2-FDA (Sethupathy et al. 2015). Celery (*Apium graveolens*) leaf's extract was found to inhibit pigment production in *C. violaceum* at significant levels, however other spices under the examination such as *Syzygium anisatum*, *Glycyrrhiza glabra* and *Capsicum annuum* showed moderate activity. The active constituent 3-n-butyl-4,5-dihydrophthalide (sedanenolide) was isolated from *Apium graveolens* extract and identified using preparative HPLC-MS followed by LC-ToF-MS analysis (Cosa et al. 2016).

20.2.2 Medicinal Plants

Apart from food plants, abundant accounts of quorum sensing inhibitory activity were also reported from medicinal plants. Ethnobotanical use of plants in treatment of various ailment including bacterial infection was one of the important convincing argument which scientific community put forward to direct the search of novel anti-infective or more precisely anti-QS agent from medicinal plants (Adonizio et al. 2006). With special reference to Indian medicinal plants, first attempt was made in 2006, screening the commonly used medicinal plants for their quorum sensing inhibitory properties (Sameena 2006). Subsequently, studies undertaken by several workers (Fatima et al. 2010; Musthafa et al. 2010; Harjai et al. 2010; Zahin et al. 2010b; Husain et al. 2013; Packiavathy et al. 2014; Tiwary et al. 2017), highlighted that Indian medicinal plants as a potential source of quorum sensing inhibitors.

Till date hundreds of medicinal plant extracts have been shown to exhibit anti-QS potential in one or more screening system (Kalia 2013; Yarmolinsky et al. 2015; Silva et al. 2016; Tiwary et al. 2017). The search of potential QSI from medicinal plants become more considerable keeping in view their long history of human use, thus toxicity issues, at least hypothetically, can be ruled out. Furthermore, this search will also boost the concept of evidence based complementary medicine and shed new light on important biological mechanism through which these plant based non-conventional remedy system works (Kothari et al. 2017). Development of medicinal plants into successful anti-pathogenic alternative will also eliminate the present day's shortcomings with conventional antibiotics including development of resistance apart from other side-effects.

As discussed in earlier sections, owing to their rich, diverse and continuously evolving bioactive chemical source, these plants are considered to most effective and most important candidate therapeutics. Table 20.1 presents a brief account of recent reports concerning quorum sensing modulatory effect of various medicinal plant extracts.

20.2.3 Essential Oils

Essential oil derived from plants are widely been recognised as flavouring agent and antimicrobials. Their biological efficacy has been attributed to their constiuent phenolics, terpenes and terpenoid volatiles (Bakkali et al. 2008). These essential oils have also been reported to possess anti-QS properties (Khan et al. 2009; Szabó et al. 2010). Table 20.2 summaries some of the recent reports on QS inhibiting potential of essential oils.

Khan et al. (2009) screened 21 commonly used essential using biosensor strains, *C. violaceum* CV12472 and CVO26. Among them, four essential oils, clove, cinnamon, lavender and peppermint oil showed varying level of QS inhibitory effects in a dose dependent manner. Yap and co-workers (2014) highlight the role of QS inhibitory and membrane disruption action of lavender (*Lavandula angustifolia*) oil as possible mechanism for the observed antibacterial effects. The oil was assayed employing two bioluminescence biosensor strains *Escherichia coli* [pSB1075] and *E. coli* [pSB401] and found effective against the LasR receptor containing strain *E. coli* [pSB1075]. While evaluating different chemotype of essential oil obtained from *Lippia alba*, Olivero-Verbel and co-worker (2014) observed that oil containing high ratio of geranial:neral and limonene:carvone were found to be most effective against QS-controlled violacein pigment production. Moreover, among the two, geranial/neral chemotype was also found effective against *Staphylococcus aureus*. Sub-lethal concentrations of cinnamon oil were found reduce the production of both short and long chain AHL molecules as evident from CV026 and PA01 bioassay. Subsequently, the lower levels of AHL signals were also found in relationship with observed inhibition of pyocyanin, swarming motility, alginate and total exoprotease activity in PA01. Authors suggested that cinnamaldeyde, a major volatile constituent of cinnamon oil could be responsible for observed anti QS activity, nonetheless, other constituent such as eugenol may synergistically or solitary targeted the QS circuit in the tested bacterium (Kalia et al. 2015).

Coriander essential and its major constituent were shown to significantly inhibit the biofilm formation in *Campylobacter jejuni* and *Campylobacter coli*. The oil and its major component linalool also demonstrate inhibitory effects against the production of QS-controlled violecein pigment in *C. violaceum* biosensor strains. Authors concluded that observed anti biofilm activity against the food borne pathogen could possibly the outcome of quorum sensing disruption (Duarte et al. 2016). Luis and co-workers (2016) evaluate anti-QS properties of two eucalypt essential oils namely *E. radiata* and *E. globulus* containing limonene and eucalyptol as major constituent respectively. *E. Radiate* was shown to exhibit greater QS interfering potential.

Table 20.2 Anti-QS activity of essential oils

Source plant	Family	Major component (s)	References
Ellettaria cardamomum	Zingiberaceae	α-terpinyl acetate, 1.8-cineole, Linalool acetae, sabinene	Asghar et al. (2017)
Coriandrum sativum	Apiaceae	Linalool	Duarte et al. (2016)
Thymus vulgare	Lamiaceae	Carvacrol, and thymol	Myszka et al. (2016)
Eucalyptus globulus	Myrtaceae	Eucalyptol	Luis et al. (2016)
Eucalyptus radiata		Limonene	
Citrus reticulata	Rutaceae	Limonene,	Luciardi et al. (2016)
Cryptocaria massoia	Lauraceae	Massoialactone	Pratiwi et al. (2016)
Murraya koenigii	Rutaceae	Caryophyllene, caryophyllene oxide, cinnamaldehyde, α- and β-phellandrene	Ganesh and Rai (2016) and Bai and Vittal (2014)
Ferula asafoetida	Apiaceae		Sepahi et al. (2015)
Dorema aucheri Boiss			
Mentha piperita	Lamiaceae	Menthol	Husain et al. (2015a, b)
Aloysia triphylla	Verbenaceae	Z-citral and E-citral	Cervantes-Ceballos et al. (2015)
Cymbopogon nardus	Poaceae	Citronellal, geraniol, citronelol	
Lippia origanoides	Verbenaceae	trans-β-caryophyllene, p-cymene, Limonene	
Hyptis suaveolens	Lamiaceae	Sabinene, β-caryophyllene, terpinolene, β-pinene	
Swinglea glutinosa	Rutaceae	β-pinene, α-pinene, sabinene	
Eucalyptus globulus	Myrtaceae	1.8-cineole, α-pinene	
Cinnamomum verum	Lauraceae	Cinnamaldehyde	Ganesh and Rai (2015) and Kalia et al. (2015)
Lavandula angustifolia	Lamiaceae	Linalyl anthranilate and linalool	Yap et al. (2014)
Lippia alba	Verbenaceae	Limonene, neral, carvone, geraniol, bicyclosesquitelandrene	Olivero-Verbel et al. (2014)
Minthostachys mollis	Lamiaceae	Pulegone and D-Menthene	Pellegrini et al. (2014)

(continued)

Table 20.2 (continued)

Source plant	Family	Major component (s)	References
Rosa damascene	Rosaceae	Citrenellol, geraniol, nonadekan	Eris and Ulusoy (2013)
Matricaria recutita	Asteraceae	Lillyl aldehyde, geraniol, linalool	
Eugenia caryophyllata	Myrtaceae	Propylene glycol, eugenol	
Pinus sylvestris	Pinaceae	α-pinene, β-pinene, limonene	
Syzygium aromaticum	Myrtaceae	Eugenol	Husain et al. (2013) and Khan et al. (2009)

Citrus reticulate essential oils and limonene, a cyclic monoterpene, have been shown to decrease the autoinducer (AHL) reduction levels by 33% as evident from *P. aeruginosa* qsc 119 bioreporter strain, and subsequently decreasing the elastase enzyme in *P. aeruginosa* HT5 by 75%. However, authors suggested, as revealed by the relationship between elastase production and AHL inhibition, that overall reduction in enzymatic activity was not only due to QS disruption, but also because of direct inhibitory effect of essential oils and its component on the enzyme activity (Luciardi et al. 2016). Pratiwi et al. (2016), while investigating the quorum quenching effect of *Cryptocaria massoia* essential oil observed that the oil effectively impeded the QS-dependent swarming motility of *P. aeruginosa* PA01 and violacein reduction in *C. violaceum* CV026 biosensor. Similarly, *Ferula asafoetida* and *Dorema aucheri* essential oil have also been found to be active against *P. aeruginosa* associated virulence factors including pyoverdine, pyocyanin, elastase and biofilm at significantly low concentration (25 µg/ml). Homoserin lactones (HSL) inhibition and down regulation of QS related genes were considered as apparent mode of action of the oils as observed from *in vitro* assessments. Additionally, keeping in view the diverse of chemical composition of the oil, authors speculated that multiple QS associated molecular target might be affected or global regulator like Vfr or GacA might be impeded (Sepahi et al. 2015). Bai and Vittal (2014) and Ganesh and Rai (2016), independently evaluated the quorum sensing inhibitory activity of essential oil of *Murraya koenigii* and found to be active against AHL mediated cell communication. Biofilm formation, cell attachment, EPS production and biofilm maturation by *P. aeruginosa* were shown to be attenuated at sub-MICs of the oil. Food spoilage by psychotropic *P. psychrophila* PSPF19 was also found to be delayed under the influence of the oil and related to its observed anti-QS activity (Bai and Vittal 2014). Ganesh and Rai (2016) examined the antipathogenic efficacy of *Murraya koenigii* essential oil *in vivo* using nematode *C. elegans–Pseudomonas aeruginosa* infection model. Along with rescuing 60% of *C. elegans*, the oil was also shown to inhibit other virulence factors such as pyocyanin and LasA staphylotic activity significantly. Asghar et al. (2017), demonstrated the activity of *Elletaria cardamomum* (zingiberaceae) essential oil against the pigment production in *C. violaceum* at the concentrations non-inhibitory to bacterial growth. Authors suggested that volatile components of the oil such as α-terpinyl acetate, 1,8-cineole, linalool acetate etc., could be responsible for the observed biological effect.

20.3 Plant Derived Phytocompounds as QSI

Different plant based chemical molecules comprising of diverse chemical groups including flavonoids, fatty acid derivatives, alkaloids, coumarins, lignans, terpenoids etc., have been identified as potent anti-quorum sensing agent (Nazzaro et al. 2013; Silva et al. 2016; Asfour 2017). Recent reports on bioactive photochemicals against QS mechanism and chemical structures has been presented in Table 20.3 and Fig. 20.1. A brief description of some of important chemical groups is presented as follows.

Table 20.3 Anti-QS activity and mode of actions of plant derived phytocompounds

Phytocompound	Source plant(s)	Test organisms	Mode of action	References
Naringenin	*Combretum albiflorum*	*Pseudomomas aeruginosa* PA01	**1.** Reducing the production of signal molecules	Paczkowski et al. (2017) and Vandeputte et al. (2011)
			2. Affecting the proper functioning of signal-receptor complex	
Diarylheptanoids	*Alnus viridis* *Alnus glutinosa*	*Pseudomonas aeruginosa*	Reducing the production of signal molecules	Ilic-Tomic et al. (2017)
Tannic acid	(Pure)	*Aeromonas hydrophila*	Down regulating the expression of QS related genes	Patel et al. (2017)
Zeaxanthin	(Pure)	*Pseudomonas aeruginosa*	Down regulating the expression of QS related genes	Gökalsın et al. (2017)
Carvacrol and thymol	*Thymus vulgare*	*Pseudomonas fluorescens* *Pseudomonas aeruginosa*	Reducing the production of signal molecules	Myszka et al. (2016) and Tapia-Rodriguez et al. (2017)
Eugenol	*Syzygium aromaticum*	*Pseudomonas aeruginosa*	Down regulating the expression of QS related genes	Al-Shabib et al. (2017) and Zhou et al. (2013)
Petunidin	(Pure)	*K. pneumoniae*	Inhibiting receptor protein	Gopu et al. (2016)

(continued)

Table 20.3 (continued)

Phytocompound	Source plant(s)	Test organisms	Mode of action	References
Baicalein	*Scutellaria baicalensis*	*Staphylococcus aureus*	Down regulating the expression of QS related genes	Chen et al. (2016)
Rosmarinic acid	(Pure)	*Aeromonas hydrophila*	Down regulating the expression of QS related genes	Rama Devi et al. (2016)
Phytol	*Piper betle*	*Serratia marcescens*	Down regulating the expression of QS related genes	Srinivasan et al. (2016)
Linalool	*Coriandrum sativum*	*Acinetobacter baumannii*		Alves et al. (2016)
Quercetin	(Pure)	*Pseudomonas aeruginosa*	Mimicking the QS signal	Gopu et al. (2015b) and Paczkowski et al. (2017)
Quercetin	(Pure)	*Candida albicans*	Inhibiting the protein involve in quorum sensing	Singh et al. (2015)
Flavonoid rich fraction	*Glycyrrhiza glabra*	*Acinetobacter baumannii*	Inhibiting the expression of QS related genes	Bhargava et al. (2015)
Menthol	*Mentha piperita*	*Pseudomonas aeruginosa*	Inhibiting receptor protein	Husain et al. (2015b)
Zingerone	*Zingiber officinale*	*Pseudomonas aeruginosa*	Inhibiting receptor protein	Kumar et al. (2015)
Oxyresveratrol	*Smilax china*	*Pseudomonas aeruginosa*	–	Sheng et al. (2015)
Tea polyphenols	*Camellia sinensis*	*Pseudomonas aeruginosa*	–	Yin et al. (2015)
6-Gingerol	*Zingiber officinale*	*Pseudomonas aeruginosa*	Down regulating the expression of QS related genes	Kim et al. (2015)
Methoxy flavones and methoxy chalcones	*Piper delineatum*	*Vibrio harveyi*	Disturbing signal transduction.	Martín-Rodríguez et al. (2015)
2-Furaldehyde diethyl acetal	*Cocos nucifera* water	*Pseudomonas aeruginosa*	Down regulating the expression of QS related genes	Sethupathy et al. (2015)
Coumarin		*Pseudomonas aeruginosa*	Down regulating the expression of QS related genes	Gutiérrez-Barranquero et al. (2015)
		Aliivibrio fischeri		
Quercitin, quercetin-3-O-arabinoside	*Psidium guajava*	*Pseudomomas aeruginosa* PA01	–	Vasavi et al. (2014)
		Chromobacterium violaceum		

(continued)

Table 20.3 (continued)

Phytocompound	Source plant(s)	Test organisms	Mode of action	References
Curcumin	*Curcuma longa*	*Vibrio spp.*	–	Abraham et al. (2013) and Packiavathy et al. (2014)
Glycosylflavonoids	*Cecropia pachystachya*	*Chromobacterium violaceum*	–	Brango-Vanegas et al. (2014)
Punicalagin	*Punica granatum*	*S. typhimurium*	Down regulating the expression of QS related genes	Li et al. (2014)
Ellagic Acid Derivatives	*Terminalia chebula*	*Pseudomonas aeruginosa* PA01	Down regulating the expression of QS related genes	Sarabhai et al. (2013)
(*R*)-Bgugaine	*Arisarum vulgare*	*Pseudomomas aeruginosa*		Majik et al. (2013)
Methyl eugenol	*Cuminum cyminum*	*Pseudomonas aeruginosa* PA01, *P. mirabilis, S. marcesence, V. harveyi*	1. Interfering with AHL dependent cell differentiation	Packiavathy et al. (2012)
			2. Inhibiting receptor protein	
Naringin	Citrus fruits	*Yersinia enterolitica*	Reducing the production of signal molecules	Truchado et al. (2012)
Malabaricone C	*Myristica cinnamomea*	*Pseudomonas aeruginosa* PA01	Interfering with receptor protein	Chong et al. (2011)
Catechin	*Combretum albiflorum* Bark	*Pseudomomas aeruginosa* PA01	Mimicking the QS signal	Vandeputte et al. (2010)

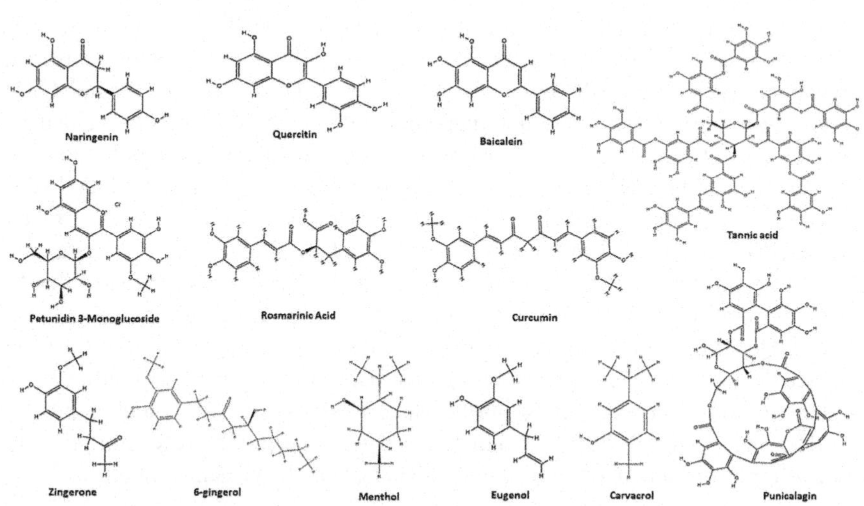

Fig. 20.1 Chemical structures of some of the plant derived QSIs

20.3.1 Flavonoids

Flavonoid, a widely distributed plant secondary metabolites known for their antioxidant and antibacterial efficacy primarily engaged in root elongation process of various plants. These polyphenolic compounds having characteristic benzo-Υ-pyrene ring are synthesised via phenylproponoid pathway (Kumar and Pandey 2013). Among other pharmacological potentials, this class of plant secondary metabolites have also been shown to possess anti-virulence and anti-QS properties (Paczkowski et al. 2017).

O-glycosylated flavonoids isolated from orange peel including naringin, neohesperidin and hesperidin showed anti QS capacity in *C. violaceum* biosensor strain. These flavonoids were also shown to inhibit QS mediated virulence factors in human enteropathogen *Y. enterocolitica* by decreasing the expression of different genes involves in virulence production (Truchado et al. 2012). Brango-Vanegas et al. (2014) identified numerous flavonoids from *Cecropia pachystachya* Trécul, a widely distributed medicinal plant in Latin America. These C-glycosyl flavonoids such as chlorogenic acid, isoorientin, orientin, isovitexin, vitexin and rutin were shown to act as QS antagonist when screened against *C. violaceum* ATCC 31532 and *E. coli* pSB403 biosensor strains. Two structurally related flavonoids first time isolated from *Piper delineatum* were shown to downregulates the QS regulated bioluminescence in *V. harveyi* reporter strain. These flavonoids broadly classified into methoxy flavones and methoxychalcones groups targets downstream LuxO component in *V. harveyi* at significantly low concentration without affecting the bacterial growth (Martín-Rodríguez et al. 2015). Baicalein (5,6,7-trihydroxyflavone) was shown to reduce the levels of enterotoxin A (SEA) and α-hemolysin (hla) in clinical strain of biofilm forming *S. aureus* (Chen et al. 2016). Notably, baicalein treatment significantly downregulated the quorum sensing regulators *agrA*, RNAlll and *sarA* as well as expression of *ica* gene at sub-inhibitory concentrations (32 and 64 μg/ml). Conformational changes in 3D structure of LasR protein of *K. Pneumonia* was evident on binding with petunidin (flavonoid) and demonstrated using simulation studies (Gopu et al. 2016). Root Mean Square Deviation (RMSD) value of thermal dynamism representing the deviation of 3D structure of LasR–OHL and LasR–petunidin complexes revealed that interaction with petunidin results into closing of active sites of the receptor protein and its further interaction which would otherwise remain open when OHL (auto inducer) molecule was present. The structural activity relationship analysis of different flavonoids reveals that the presence of two hydroxyl groups in the flavone A ring are essential for inhibition of QS related self-regulatory proteins in *P. aeruginosa*. Biochemically it was also established that flavonoids prevent LasR/RhlR-DNA binding non-competitively (Paczkowski et al. 2017). Quercitin, a ubiquitous flavonoids, binds to QS receptor protein LasR and reduce their ability to bind promoter region of DNA thus by down regulating overall expression of QS related genes. Further, Paczkowski et al. (2017) and Gopu et al. (2015b) independently demonstrated that binding of quercetin to receptor protein of *P. aeruginosa* results into conformational changes in the protein structure.

20.3.2   Other Phenolics

Other bioactive phenolic secondary metabolites such as phenolic acids blocks the expression of virulence factors in pathogenic bacterium via modulating QS machinery (Truchado et al. 2012). Truchado and co-workers (2012) evaluated the potential of several food phytochemicals to inhibit QS signals in the biosensor strain *C. violaceum*. A preliminary screening using three different concentrations showed that gallic acid and vanilic acid reduced violacein through QS inhibition. Borges and co-workers (2014) also observed the QSI potential of gallic acid, which inhibited pigment production, although all concentrations tested were cytotoxic to mouse lung fibroblasts. Anti QS properties of coumarin, another polyphenolic compound belongs to benzopyrene class, in three biosensor strains (*S. marcescens*, for short chain AHLs, *C. violaceum* for medium chain AHLs and A. tumefaciens for long chain AHLs) was analysed by Gutiérrez-Barranquero et al. (2015). The compound showed varying level of inhibitory activity against different AHLs and downregulated the expression of QS controlled genes *pqsA* and *rhlI*.

The ability of rosmarinic acid, a phenolic acid predominantly found in members of Lamiaceae family was tested in different pathogenic strains of *A. hydrophila* isolated from infected zebra fish. Rosmarinic acid influenced QS regulated virulence factors such as biofilm formation, haemolysin, lipase and elastase production. Gene expression analysis confirmed the down regulation of virulence genes such as ahh1, aerA etc. It was also observed that *in vivo* treatment of rosmarinic acid in zebra fish infection model subsequently enhances the survival rate (Rama Devi et al. 2016). Tannic acid significantly downregulates Ahyl and AhyR transcript in the tested bacterium as revealed from qRT-PCR results. The phenolic compounds were also tested in *Catla catla* when co-stimulated with pathogenic bacterium (*A. hydrophila*) and shown to decline the extent of pathogen induced skin haemorrhage and enhanced the survival rate up to 86.6% (Patel et al. 2017). Ilic-Tomic and co-workers (2017), examined the QS inhibitory effect of diarylheptanoids isolated from barks of *Alnus viridis*, it was demonstrated that among the isolated diarylheptanoids, hirsutenone was able to inhibit the QS dependent pigment production in *C. violaceum* CV026 and *P. aeruginosa* PA01 along with reduction in the levels of the auto inducer molecule (2-alkyl-4-quinolones) in *P. aeruginosa* PA01.

Among the plant derived pigments, zeaxathin was screened for potential QS inhibitory activity using two *lasB*-gfp and *rhlA*-gfp fluorescent monitor strains of *P. aeruginosa*. Gene expression levels of *lasB* and *rhlA* was observed to decrease in concentration dependent manner. Further, *In silico* modelling approach revealed the significantly favourable stabilizing binding energy of zeaxathin with the active sites of lasB and rhlA. Authors thus suggested that the potential QSI (zeaxathin) imparts its effect via inhibition of regulatory proteins involves in QS circuit (Gökalsın et al. 2017).

20.3.3 Essential Oil's Compounds

Essential oil's compounds, primarily mono-terpenes and sesquiterpenes have been primarily screened for their anti-QS potential against *C. violaceum* and *P. aeruginosa* PA01 bioreporter strains (Ahmad et al. 2015). It was revealed that stereochemical variations among the structurally similar components could be responsible for obtained activity. It has been shown (+)-enantiomers of carvone, limonene and borneol increased violacein and pyocyanin production, on the contrary levo (−) analogs such as a-terpineol and cis-3-nonen-1-ol showed more that 90% inhibition in the pigment production. It was also emphasized that mimicking the autoinducer signals structurally could be considered as possible mechanism of action in virulence inhibition (Ahmad et al. 2015). Two important pungent constituents of ginger oil, 6-gingerol and zingerone were also been investigated to evaluate their ability to interfere with QS and regulated traits in *P. aeruginosa* (Kim et al. 2015; Kumar et al. 2015). Kim et al. (2015) observed that application of 6-gingerol successfully reduced QS regulated virulence factors such as exoprotease, rhamnolipid, biofilm formation etc., in the target bacterium. *In-silico* analysis suggested that the test molecule interfere with the active sites of QS receptor protein lasR via multiple weak interactions. Authors also observed the reduction in QS-induced genes expression under the effect of 6-gingerol thus by confirming the role of 6-gingerol at molecular levels interfering the QS system. Kumar and co-workers (2015), virtually examined the interaction of zingerone with different QS receptors (TraR, LasR, RhlR and PqsR) and observed comparative good docking score to respective autoinducer, thus by proposing zingerone as a promising anti-infective drug candidate. Menthol (5-Methyl-2-(propan-2-yl)cyclohexan-1-ol) has been putatively identified as potent QSI among other component of peppermint oil using molecular docking technique. Further *in vitro* assessments reveals that menthol significantly reduces expression of QS-controlled traits at sub-MICs in *C. violaceum* and *P. aeruginosa* PA01 in a concentration dependent manner. The addition of menthol to *E. coli* biosensors, MG4/pKDT17 and pEAL08-2 reduced the β-galactosidase luminescence indicating the direct inhibition of *las* and *pqs* transcription respectively (Husain et al. 2015b). Carvacrol and thymol a monoterpenoid phenol was described to inhibit the QS evidently through inhibiting the production of signal molecules (Myszka et al. 2016; Tapia-Rodriguez et al. 2017). In *Pseudomonas fluorescens* KM121biosensor system, carvacrol and thymol were examined for their ability to interfere the production of quorum sensing autoinducers (AHLs) and flagellar gene (*flgA*) expression. Both the components (carvacrol and thymol) of *Thymus vulgare* essential oil significantly modulates the levels of AHLs along with downregulating the expression of motility (flagellar) genes at sub-MIC concentrations (Myszka et al. 2016). Al-Shabib and co-workers (2016), showed that eugenol significantly inhibited the QS regulated violecin production in *C. violaceum* CV026 along with virulence factors in *P. aeruginosa* PA01. Additionally transcriptional regulation assay in *E. coli* MG4/pKDT17 indicates that observed anti QS activity of eugenol is associated with downregulation of *las* system. Inhibition of QS-controlled production of prodigiosin pigment in *S. marcescens* was observed when treated with sub-inhibitory concentrations of phytol, an acyclic diterpene alcohol isolated from *Piper betle* (Srinivasan et al. 2016). Other related virulence traits including biofilm, hydrophobicity and protease production were also shown to reduce significantly at non-toxic concentration of the volatile component. Alves et al. (2016) screened major components of essential oil of *Coriandrum sativum* including linalool, α-pinene, p-cymene, camphor and geranyl acetate. It was observed that linalool exhibits significant inhibition in the pigment production in *C. violaceum* and QS-dependent bacterial adhesion and biofilm formation in *Acinetobacter baumannii* without affecting the growth of test bacterium.

20.4   Significance of Plant Based QSIs in Combating Bacterial Infection

Development of plant based QSI which can be successfully used as an anti-infective is considered as an ultimate health benefit (Fig. 20.2). Numerous potential plant derived products and synthetic QSIs have been evaluated in animal models (Rumbaugh et al. 1999; Christensen et al. 2007; Nidadavolu et al. 2012). Various animal based *in vivo* models mimicking actual disease conditions like dermonecrosis, lung infection, wound infection, device associated infection and biofilm formation have been developed to test the efficacy of candidate QSI for their possible use in human subjects (Papaioannou et al. 2013). Alternatively, these models can be

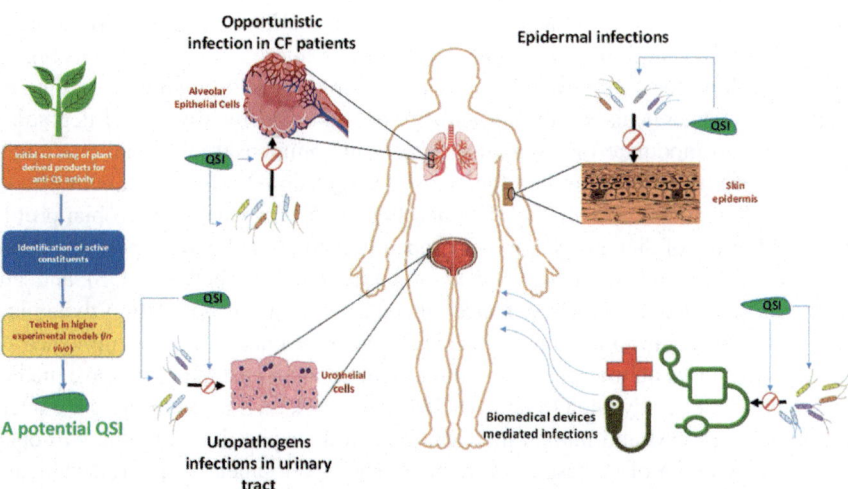

Fig. 20.2 Applications of plant based QSIs in combating bacterial infections

used for validating the use of plants based alternative medicine and substantiate their efficacy with scientific proof. In a study involving mouse infection model and human cell lines, Muhs and co-workers (2017), demonstrate the skin infection treatment ability of *S. terebinthifolia*, an exotic South American weed used locally for treatment of skin infection. The flavone rich active fraction obtained from the plant was shown to possess anti-virulence ability against *S. aureus* via modulating the bacterial *agr* quorum sensing system. The fraction was shown to be well tolerated by human keratinocytes in cell culture and mouse skin *in vivo*. The fraction also inhibited virulent MRSA mediated dermonecrosis in mouse skin model. Baicalin, one of the major flavonoid isolated from roots *Scutellaria baicalensis* have been shown to strengthen the immune response against *P. aeruginosa* infection in mouse model. During preliminary investigation, baicalin was evaluated against different virulence factors including biofilm formation in *P. aeruginosa* and was found be effective in a concentration dependent manner. The authors further observed QSI application results in overall improvement in immune response which was marked by the decrease in interleukin (IL-4) levels, increase in interferon (IFN-Υ) and reversing interlukin/interferon ratio thus by activating Th-1 induced immune response in the infected host (Luo et al. 2017). Ruffin et al. (2016), observed that LasB and LasA protease secreted by *P. aeruginosa* in airways of patients with CF or other lung disease could further hamper the ability of epithelia to repair. Authors demonstrated that, furaneol, a known QSI isolated from *Fragaria ananassa* impeded the LasR transcriptional regulators thereby inhibiting the production of elastase in *P. aeruginosa* cultures. Further, they confirmed the effect of the QSI in patients with CF, a dose dependent wound healing response was observed following the application of the candidate QSI.

Another important aspects of QSI affecting the human health involves the possible effect of QSI on the dynamics of microbial community in human gut (McCarthy and O'Gara 2015). It has been well established microbial community in human body produce signalling molecules including AHLs (Yin et al. 2012; Swearingen et al. 2013). Thus inhibitory effect of QSIs on these microbial community is highly probable. However, this possibility and the outcome yet to be studied in human. Dietary phytochemicals exhibiting anti-QS or more specifically signal degrading activity could modulate the signalling system in commensal microbial communities. As these microbial communities are also responsible for host metabolic condition such as diabetes and obesity (Hur and Lee 2015), the shift in microbial profile under the effect of dietary QSI or as therapeutic is likely to be more possible. Thus understanding and gaining deep view of interconnection between QSI and gut microbial flora would clearly enhance our understanding of population dynamics, and their role in development of specific metabolic condition. The influence of dietary substances on microbial signalling within human host may also have medicinal impact (McCarthy and O'Gara 2015). Future work on the determining the interaction between QSI and human microbiome and their possible outcome will open entirely new arena of disease management. The development of pharmaceuticals, nutraceuticals and functional food with their possible role in modulating over all human microbiome will not only substantiate our battle against bacterial infection but also proved to be effective against other pathophysiological conditions.

20.5   Conclusion

Bacterial pathogenicity is a multifactorial phenomenon involving complex process of host-pathogen interaction. The degree of the pathogenicity is controlled through various cell structures and extracellular products called virulence factors. Expression of various virulence factors are known to be regulated through Quorum sensing. It is expected that attenuating virulence through QS interference in pathogenic bacteria may successfully result in disease control especially where antibiotic is ineffective due development of multi drug resistance. The literature survey in this article revealed that several plant derived products and phytocompounds can be considered as promising candidate for anti-infective drug development. However, further evaluation of these compounds using *in vivo* infection model for their proved therapeutic effectiveness are needed. To exploit anti-QS strategy in the control of bacterial infections, what is more challenging is to obtain active compound or formulation effective against number of QS network present in pathogenic bacteria to attenuate its virulence effectively, so that host defence system can clear the infectious agent.

20.6   Opinion

Considering the current global scenario of emergence and spread of MDR pathogens and slow discovery of new antibiotics with novel mode of action, it is necessary to develop alternative mode of bacterial infection control. The progress made so far on the role of QSI in the treatment of bacterial infection still weak and clumsy. Therefore, more concerted efforts should be directed to understand the linkage of various virulence factors with QS and how to effectively target virulence of pathogen through QS inhibition as well as use of other strategy in combination with QSIs.

References

Abraham I, Packiavathy V, Sasikumar P, Pandian SK, Ravi AV (2013) Prevention of quorum-sensing-mediated biofilm development and virulence factors production in *Vibrio* spp. by curcumin. Appl Microbiol Biotechnol 97(23):10177. https://doi.org/10.1007/s00253-013-4704-5

Adonizio AL, Downum K, Bennett BC, Mathee K (2006) Anti-quorum sensing activity of medicinal plants in southern Florida. J Ethnopharmacol 105(3):427–435. https://doi.org/10.1016/j.jep.2005.11.025

Ahmad I, Khan MSA, Husain FM, Zahin M, Singh M (2011) Bacterial quorum sensing and its interference: methods and significance. In: Ahmad I, Ahmad F, Pichtel J (eds) Microbes and microbial technology. Springer, New York, pp 127–161. https://doi.org/10.1007/978-1-4419-7931-5_6

Ahmad A, Viljoen AM, Chenia HY (2015) The impact of plant volatiles on bacterial quorum sensing. Lett Appl Microbiol 60(1):8–19. https://doi.org/10.1111/lam.12343

Al-Shabib NA, Husain FM, Ahmad I, Baig MH (2016) Eugenol inhibits quorum sensing and biofilm of toxigenic MRSA strains isolated from food handlers employed in Saudi Arabia. Biotechnol Biotechnol Equip 31(2):387–396. https://doi.org/10.1080/13102818.2017.1281761

Alves S, Duarte A, Sousa S, Domingues FC (2016) Study of the major essential oil compounds of *Coriandrum sativum* against *Acinetobacter baumannii* and the effect of linalool on adhesion, biofilms and quorum sensing. Biofouling 32(2):155–165. https://doi.org/10.1080/08927014.2 015.1133810

Aqil F, Ahmad I, Owais M (2006) Ch. 9: targeted screening of bioactive plant extracts and Phytocompounds against problematic groups of multidrug-resistant bacteria. In: Ahmad I, Aqil F, Owais M (eds) Modern phytomedicine: turning medicinal plants into drugs. Wiley-VCH, Weinheim. https://doi.org/10.1002/9783527609987

Asfour HZ (2017) Anti-quorum sensing natural compounds. J Microsc Ultrastruct. https://doi. org/10.1016/j.jmau.2017.02.001

Asghar A, Butt MS, Shahid M, Huang Q (2017) Evaluating the antimicrobial potential of green cardamom essential oil focusing on quorum sensing inhibition of *Chromobacterium violaceum*. Food Sci Technol 54:2306. https://doi.org/10.1007/s1319

Atanasov AG, Waltenberger B, Pferschy-Wenzig EM, Linder T, Wawrosch C, Uhrin P, Temml V, Wang L, Schwaiger S, Heiss EH, Rollinger JM (2015) Discovery and resupply of pharmacologically active plant-derived natural products: a review. Biotechnol Adv 33(8):1582–1614. https://doi.org/10.1016/j.biotechadv.2015.08.001

Bai AJ, Vittal RR (2014) Quorum sensing inhibitory and anti-biofilm activity of essential oils and their in vivo efficacy in food systems. Food Biotechnol 28(3):269–292. https://doi.org/10.108 0/08905436.2014.932287

Bakkali F, Averbeck S, Averbeck D, Idaomar M (2008) Biological effects of essential oils–a review. Food Chem Toxicol 46(2):446–475. https://doi.org/10.1016/j.fct.2007.09.106

Bezek K, Kurinčič M, Knauder E, Klančnik A, Raspor P, Bucar F, Smole Možina S (2016) Attenuation of adhesion, biofilm formation and quorum sensing of *Campylobacter jejuni* by *Euodia ruticarpa*. Phytother Res 30(9):1527–1532. https://doi.org/10.1002/ptr.5658

Bhargava N, Singh SP, Sharma A, Sharma P, Capalash N (2015) Attenuation of quorum sensing-mediated virulence of *Acinetobacter baumannii* by *Glycyrrhiza glabra* flavonoids. Future Microbiol 10(12):1953–1968. https://doi.org/10.2217/fmb.15.107

Bjarnsholt T, Givskov M (2007) Quorum-sensing blockade as a strategy for enhancing host defences against bacterial pathogens. Philos Trans R Soc Lond Ser B Biol Sci 362(1483):1213–1222. https://doi.org/10.1098/rstb.2007.2046

Borges A, Serra S, Cristina Abreu A, Saavedra MJ, Salgado A, Simões M (2014) Evaluation of the effects of selected phytochemicals on quorum sensing inhibition and in vitro cytotoxicity. Biofouling 30(2):183–195. https://doi.org/10.1080/08927014.2013.852542

Brango-Vanegas J, Costa GM, Ortmann CF, Schenkel EP, Reginatto FH, Ramos FA, Arévalo-Ferro C, Castellanos L (2014) Glycosylflavonoids from Cecropia pachystachya Trécul are quorum sensing inhibitors. Phytomedicine 21(5):670–675. https://doi.org/10.1016/j. phymed.2014.01.001

Camara M, Williams P, Hardman A (2002) Controlling infection by tuning in and turning down the volume of bacterial small-talk. Lancet Infect Dis 2(11):667–676

Castillo S, Heredia N, Arechiga-Carvajal E, García S (2014) Citrus extracts as inhibitors of quorum sensing, biofilm formation and motility of *Campylobacter jejuni*. Food Biotechnol 28(2):106–122. https://doi.org/10.1080/08905436.2014.895947

Cegelski L, Marshall GR, Eldridge GR, Hultgren SJ (2008) The biology and future prospects of antivirulence therapies. Nat Rev Microbiol 6(1):17. https://doi.org/10.1038/nrmicro1818

Cervantes-Ceballos L, Caballero-Gallardo K, Olivero-Verbel J (2015) Repellent and anti-quorum sensing activity of six aromatic plants occurring in Colombia. Nat Prod Commun 10(10):1753–1757

Ceylan O, Sahin MD, Akdamar G (2016) Antioxidant and anti-quorum sensing potential of *Acer monspessulanum* subsp. *monspessulanum* extracts. Planta Med 82(15):1335–1340. https://doi. org/10.1055/s-0042-105294

Chen Y, Liu T, Wang K, Hou C, Cai S, Huang Y, Du Z, Huang H, Kong J, Chen Y (2016) Baicalein inhibits *Staphylococcus aureus* biofilm formation and the quorum sensing system in vitro. PLoS One 11(4):e0153468. https://doi.org/10.1371/journal.pone.0153468

Chenia HY (2013) Anti-quorum sensing potential of crude *Kigelia africana* fruit extracts. Sensors (Basel) 13(3):2802–2817. https://doi.org/10.3390/s130302802

Chong YM, Yin WF, Ho CY, Mustafa MR, Hadi AH, Awang K, Narrima P, Koh CL, Appleton DR, Chan KG (2011) Malabaricone C from *Myristica cinnamomea* exhibits anti-quorum sensing activity. J Nat Prod 74(10):2261–2264. https://doi.org/10.1021/np100872k

Christensen LD, Moser C, Jensen PØ, Rasmussen TB, Christophersen L, Kjelleberg S, Kumar N, Høiby N, Givskov M, Bjarnsholt T (2007) Impact of *Pseudomonas aeruginosa* quorum sensing on biofilm persistence in an in vivo intraperitoneal foreign-body infection model. Microbiology 153(Pt 7):2312–2320. https://doi.org/10.1099/mic.0.2007/006122-0

Chu W, Zhou S, Jiang Y, Zhu W, Zhuang X, Fu J (2013) Effect of traditional Chinese herbal medicine with antiquorum sensing activity on *Pseudomonas aeruginosa*. Evid Based Complement Alternat Med 2013:648257. https://doi.org/10.1155/2013/648257

Cosa S, Viljoen AM, Chaudhary SK, Chen W (2016) Hacking in to the social media network of bacteria: the antiquorum sensing properties of herbs and spices. Planta Med 82(S01):S1–S381. https://doi.org/10.1055/s-0036-1596316

Cowan MM (1999) Plant products as antimicrobial agents. Clin Microbiol Rev 12(4):564–582

Duarte A, Luís Â, Oleastro M, Domingues FC (2016) Antioxidant properties of coriander essential oil and linalool and their potential to control *Campylobacter* spp. Food Control 61:115–122. https://doi.org/10.1016/j.foodcont.2015.09.033

Elmanama AA, Al-Reefi MR (2017) Antimicrobial, anti-biofilm, anti-quorum sensing, antifungal and synergistic effects of some medicinal plants extracts. IUG J Nat Stud 25(2):198–207

Eris R, Ulusoy S (2013) Rose, clove, chamomile essential oils and pine turpentine inhibit quorum sensing in *Chromobacterium violaceum* and *Pseudomonas aeruginosa*. J Essent Oil Bear Plants 16(2):126–135. https://doi.org/10.1080/0972060X.2013.794026

Fatima Q, Zahin M, Khan MS, Ahmad I (2010) Modulation of quorum sensing controlled behaviour of bacteria by growing seedling, seed and seedling extracts of leguminous plants. Indian J Microbiol 50(2):238–242. https://doi.org/10.1007/s12088-010-0025-x

Fratianni F, Coppola R, Nazzaro F (2011) Phenolic composition and antimicrobial and antiquorum sensing activity of an ethanolic extract of peels from the apple cultivar Annurca. J Med Food 14(9):957–963. https://doi.org/10.1089/jmf.2010.0170

Ganesh PS, Rai VR (2015) Evaluation of anti-bacterial and anti-quorum sensing potential of essential oils extracted by supercritical CO_2 method against *Pseudomonas aeruginosa*. J Essent Oil Bear Plants 18(2):264–275. https://doi.org/10.1080/0972060X.2015.1025295

Ganesh PS, Rai RV (2016) Inhibition of quorum-sensing-controlled virulence factors of *Pseudomonas aeruginosa* by *Murraya koenigii* essential oil: a study in a *Caenorhabditis elegans* infectious model. J Med Microbiol 65(12):1528–1535. https://doi.org/10.1099/jmm.0.000385

Ganesh PS, Rai VR (2017) Attenuation of quorum-sensing-dependent virulence factors and biofilm formation by medicinal plants against antibiotic resistant *Pseudomonas aeruginosa*. J Tradit Complement Med. https://doi.org/10.1016/j.jtcme.2017.05.008

Ghosh R, Tiwary BK, Kumar A, Chakraborty R (2014) Guava leaf extract inhibits quorum-sensing and *Chromobacterium violaceum* induced lysis of human hepatoma cells: whole transcriptome analysis reveals differential gene expression. PLoS One 9(9):e107703. https://doi.org/10.1371/journal.pone.0107703

Gökalsın B, Aksoydan B, Erman B, Sesal NC (2017) Reducing virulence and biofilm of *Pseudomonas aeruginosa* by potential quorum sensing inhibitor carotenoid: zeaxanthin. Microb Ecol 74(2):466–473. https://doi.org/10.1007/s00248-017-0949-3

Gopu V, Kothandapani S, Shetty PH (2015a) Quorum quenching activity of *Syzygium cumini* (L.) Skeels and its anthocyanin malvidin against *Klebsiella pneumoniae*. Microb Pathog 79:61–69. https://doi.org/10.1016/j.micpath.2015.01.010

Gopu V, Meena CK, Shetty PH (2015b) Quercetin influences quorum sensing in food borne bacteria: in-vitro and in-silico evidence. PLoS One 10(8):e0134684. https://doi.org/10.1371/journal.pone.0134684

Gopu V, Meena CK, Murali A, Shetty PH (2016) Petunidin as a competitive inhibitor of acylated homoserine lactones in *Klebsiella pneumoniae*. RSC Adv 6(4):2592–2601. https://doi.org/10.1039/C5RA20677D

Gutiérrez-Barranquero JA, Reen FJ, McCarthy RR, O'Gara F (2015) Deciphering the role of coumarin as a novel quorum sensing inhibitor suppressing virulence phenotypes in bacterial pathogens. Appl Microbiol Biotechnol 99(7):3303–3316. https://doi.org/10.1007/s00253-015-6465-9

Harjai K, Kumar R, Singh S (2010) Garlic blocks quorum sensing and attenuates the virulence of *Pseudomonas aeruginosa*. FEMS Immunol Med Microbiol 58(2):161–168. https://doi.org/10.1111/j.1574-695X.2009.00614.x

Hossain MA, Lee SJ, Park JY, Reza MA, Kim TH, Lee KJ, Suh JW, Park SC (2015) Modulation of quorum sensing-controlled virulence factors by *Nymphaea tetragona* (water lily) extract. J Ethnopharmacol 174:482–491. https://doi.org/10.1016/j.jep.2015.08.049

Hur KY, Lee MS (2015) Gut microbiota and metabolic disorders. Diabetes Metab J 39(3):198–203. https://doi.org/10.4093/dmj.2015.39.3.198

Husain FM, Ahmad I (2013) Quorum sensing inhibitors from natural products as potential novel anti-infective agents. Drugs Future 38(10):691. https://doi.org/10.1358/dof.2013.038.10.2025393

Husain FM, Ahmad I, Asif M, Tahseen Q (2013) Influence of clove oil on certain quorum-sensing-regulated functions and biofilm of *Pseudomonas aeruginosa* and *Aeromonas hydrophila*. J Biosci 38(5):835–844

Husain FM, Ahmad I, Khan MS, Al-Shabib NA (2015a) *Trigonella foenum-graceum* (seed) extract interferes with quorum sensing regulated traits and biofilm formation in the strains of *Pseudomonas aeruginosa* and *Aeromonas hydrophila*. Evid Based Complement Alternat Med 2015:879540. https://doi.org/10.1155/2015/879540

Husain FM, Ahmad I, Khan MS, Ahmad E, Tahseen Q, Khan MS, Alshabib NA (2015b) Sub-MICs of *Mentha piperita* essential oil and menthol inhibits AHL mediated quorum sensing and biofilm of Gram-negative bacteria. Front Microbiol 6:420. https://doi.org/10.3389/fmicb.2015.00420

Husain FM, Ahmad I, Al-thubiani AS, Abulreesh HH, AlHazza IM, Aqil F (2017) Leaf extracts of *Mangifera indica* l. inhibit quorum sensing–regulated production of virulence factors and biofilm in test bacteria. Front Microbiol 8:727. https://doi.org/10.3389/fmicb.2017.00727

Ilic-Tomic T, Sokovic M, Vojnovic S, Ciric A, Veljic M, Nikodinovic-Runic J, Novakovic M (2017) Diarylheptanoids from *Alnus viridis* ssp. *viridis* and *Alnus glutinosa*: modulation of quorum sensing activity in *Pseudomonas aeruginosa*. Planta Med 83(01/02):117–125. https://doi.org/10.1055/s-0042-107674

Kalia VC (2013) Quorum sensing inhibitors: an overview. Biotechnol Adv 31(2):224–245. https://doi.org/10.1016/j.biotechadv.2012.10.004

Kalia M, Yadav VK, Singh PK, Sharma D, Pandey H, Narvi SS, Agarwal V (2015) Effect of cinnamon oil on quorum sensing-controlled virulence factors and biofilm formation in *Pseudomonas aeruginosa*. PLoS One 10(8):e0135495. https://doi.org/10.1371/journal.pone.0135495

Khan MS, Zahin M, Hasan S, Husain FM, Ahmad I (2009) Inhibition of quorum sensing regulated bacterial functions by plant essential oils with special reference to clove oil. Lett Appl Microbiol 49(3):354–360. https://doi.org/10.1111/j.1472-765X.2009.02666.x

Kim HS, Lee SH, Byun Y, Park HD (2015) 6-Gingerol reduces *Pseudomonas aeruginosa* biofilm formation and virulence via quorum sensing inhibition. Sci Rep 5. https://doi.org/10.1038/srep08656

Koh KH, Tham FY (2011) Screening of traditional Chinese medicinal plants for quorum-sensing inhibitors activity. J Microbiol Immunol Infect 44(2):144–148. https://doi.org/10.1016/j.jmii.2009.10.001

Koh CL, Sam CK, Yin WF, Tan LY, Krishnan T, Chong YM, Chan KG (2013) Plant-derived natural products as sources of anti-quorum sensing compounds. Sensors 13(5):6217–6228. https://doi.org/10.3390/s130506217

Kordbacheh H, Eftekhar F, Ebrahimi SN (2017) Anti-quorum sensing activity of *Pistacia atlantica* against *Pseudomonas aeruginosa* PAO1 and identification of its bioactive compounds. Microb Pathog 110:390–398. https://doi.org/10.1016/j.micpath.2017.07.018

Kothari V, Patel P, Joshi C (2017) Bioactive natural products: an overview, with particular emphasis on those possessing potential to inhibit microbial quorum sensing. In: Kalia VC (ed) Microbial applications, vol 2. Springer, Cham, pp 185–202. https://doi.org/10.1007/978-3-319-52669-0_10

Kumar S, Pandey AK (2013) Chemistry and biological activities of flavonoids: an overview. Scientific World J 2013:162750. https://doi.org/10.1155/2013/162750

Kumar L, Chhibber S, Kumar R, Kumar M, Harjai K (2015) Zingerone silences quorum sensing and attenuates virulence of *Pseudomonas aeruginosa*. Fitoterapia 102:84–95. https://doi.org/10.1016/j.fitote.2015.02.002

LaSarre B, Federle MJ (2013) Exploiting quorum sensing to confuse bacterial pathogens. Microbiol Mol Biol Rev 77(1):73–111. https://doi.org/10.1128/MMBR.00046-12

Lazdunski AM, Ventre I, Sturgis JN (2004) Regulatory circuits and communication in Gram-negative bacteria. Nature Rev Microbiol 2:581–592. https://doi.org/10.1038/nrmicro924

Li G, Yan C, Xu Y, Feng Y, Wu Q, Lv X, Yang B, Wang X, Xia X (2014) Punicalagin inhibits *Salmonella* virulence factors and has anti-quorum-sensing potential. Appl Environ Microbiol 80(19):6204–6211. https://doi.org/10.1128/AEM.01458-14

Luciardi MC, Blázquez MA, Cartagena E, Bardón A, Arena ME (2016) Mandarin essential oils inhibit quorum sensing and virulence factors of *Pseudomonas aeruginosa*. LWT Food Sci Technol 68:373–380. https://doi.org/10.1016/j.lwt.2015.12.056

Luis Â, Duarte A, Gominho J, Domingues F, Duarte AP (2016) Chemical composition, antioxidant, antibacterial and anti-quorum sensing activities of *Eucalyptus globulus* and *Eucalyptus radiata* essential oils. Ind Crop Prod 79:274–282. https://doi.org/10.1016/j.indcrop.2015.10.055

Luo J, Dong B, Wang K, Cai S, Liu T, Cheng X, Lei D, Chen Y, Li Y, Kong J, Chen Y (2017) Baicalin inhibits biofilm formation, attenuates the quorum sensing-controlled virulence and enhances *Pseudomonas aeruginosa* clearance in a mouse peritoneal implant infection model. PLoS One 12(4):e0176883. https://doi.org/10.1371/journal.pone.0176883

Maisuria VB, Lopez-de Los Santos Y, Tufenkji N, Déziel E (2016) Cranberry-derived proanthocyanidins impair virulence and inhibit quorum sensing of *Pseudomonas aeruginosa*. Sci Rep 6:30169. https://doi.org/10.1038/srep30169

Majik MS, Naik D, Bhat C, Tilve S, Tilvi S, D'Souza L (2013) Synthesis of (R)-norbgugaine and its potential as quorum sensing inhibitor against *Pseudomonas aeruginosa*. Bioorg Med Chem Lett 23(8):2353–2356. https://doi.org/10.1016/j.bmcl.2013.02.051

Makhfian M, Hassanzadeh N, Mahmoudi E, Zandyavari N (2015) Anti-quorum sensing effetcs of ethanolic crude extract of *Anethum graveolens* L. J Essent Oil Bear Plants 18(3):687–696. https://doi.org/10.1080/0972060X.2014.998718

Martín-Rodríguez AJ, Ticona JC, Jiménez IA, Flores N, Fernández JJ, Bazzocchi IL (2015) Flavonoids from *Piper delineatum* modulate quorum-sensing-regulated phenotypes in *Vibrio harveyi*. Phytochemistry 117:98–106. https://doi.org/10.1016/j.phytochem.2015.06.006

McCarthy RR, O'Gara F (2015) The impact of phytochemicals present in the diet on microbial signalling in the human gut. J Funct Foods 14:684–691. https://doi.org/10.1016/j.jff.2015.02.032

Miller MB, Bassler BL (2001) Quorum sensing in bacteria. Annu Rev Microbiol 55(1):165–199. https://doi.org/10.1146/annurev.micro.55.1.165

Muhs A, Lyles JT, Parlet CP, Nelson K, Kavanaugh JS, Horswill AR, Quave CL (2017) Virulence inhibitors from brazilian peppertree block quorum sensing and abate dermonecrosis in skin infection models. Sci Rep 7:42275. https://doi.org/10.1038/srep42275

Musthafa KS, Ravi AV, Annapoorani A, Packiavathy IS, Pandian SK (2010) Evaluation of anti-quorum-sensing activity of edible plants and fruits through inhibition of the N-acyl-homoserine lactone system in *Chromobacterium violaceum* and *Pseudomonas aeruginosa*. Chemotherapy 56(4):333–339. https://doi.org/10.1159/000320185

Musthafa KS, Sianglum W, Saising J, Lethongkam S, Voravuthikunchai SP (2017) Evaluation of phytochemicals from medicinal plants of Myrtaceae family on virulence factor production by *Pseudomonas aeruginosa*. APMIS 125(5):482–490. https://doi.org/10.1111/apm.12672

Myszka K, Schmidt MT, Majcher M, Juzwa W, Olkowicz M, Czaczyk K (2016) Inhibition of quorum sensing-related biofilm of *Pseudomonas fluorescens* KM121 by *Thymus vulgare* essential oil and its major bioactive compounds. Int Biodeter Biodegr 114:252–259. https://doi.org/10.1016/j.ibiod.2016.07.006

Nazzaro F, Fratianni F, Coppola R (2013) Quorum sensing and phytochemicals. Int J Mol Sci 14(6):12607–12619. https://doi.org/10.3390/ijms140612607

Nidadavolu P, Amor W, Tran PL, Dertien J, Colmer-Hamood JA, Hamood AN (2012) Garlic ointment inhibits biofilm formation by bacterial pathogens from burn wounds. J Med Microbiol 61(Pt 5):662–671. https://doi.org/10.1099/jmm.0.038638-0

Noumi E, Snoussi M, Merghni A, Nazzaro F, Quindós G, Akdamar G, Mastouri M, Al-Sieni A, Ceylan O (2017) Phytochemical composition, anti-biofilm and anti-quorum sensing potential of fruit, stem and leaves of *Salvadora persica* L. methanolic extracts. Microb Pathog 109:169–176. https://doi.org/10.1016/j.micpath.2017.05.036

Oliveira BD, Rodrigues AC, Cardoso BM, Ramos AL, Bertoldi MC, Taylor JG, da Cunha LR, Pinto UM (2016) Antioxidant, antimicrobial and anti-quorum sensing activities of *Rubus rosaefolius* phenolic extract. Ind Crop Prod 84:59–66. https://doi.org/10.1016/j.indcrop.2016.01.037

Olivero-Verbel J, Barreto-Maya A, Bertel-Sevilla A, Stashenko EE (2014) Composition, anti-quorum sensing and antimicrobial activity of essential oils from *Lippia alba*. Braz J Microbiol 45(3):59–767

Ouedraogo V, Kiendrebeogo M (2016) Methanol extract from *Anogeissus leiocarpus* (DC) Guill. Et Perr.(Combretaceae) stem bark quenches the quorum sensing of *Pseudomonas aeruginosa* PAO1. Medicines (Basel) 3:26. https://doi.org/10.3390/medicines3040026

Packiavathy IA, Agilandeswari P, Musthafa KS, Pandian SK, Ravi AV (2012) Antibiofilm and quorum sensing inhibitory potential of *Cuminum cyminum* and its secondary metabolite methyl eugenol against Gram negative bacterial pathogens. Food Res Int 45(1):85–92. https://doi.org/10.1016/j.foodres.2011.10.022

Packiavathy IA, Priya S, Pandian SK, Ravi AV (2014) Inhibition of biofilm development of uropathogens by curcumin–an anti-quorum sensing agent from *Curcuma longa*. Food Chem 148:453–460. https://doi.org/10.1016/j.foodchem.2012.08.002

Paczkowski JE, Mukherjee S, McCready AR, Cong JP, Aquino CJ, Kim H, Henke BR, Smith CD, Bassler BL (2017) Flavonoids suppress *Pseudomonas aeruginosa* virulence through allosteric inhibition of quorum-sensing receptors. J Biol Chem 292(10):4064–4076. https://doi.org/10.1074/jbc.M116.770552

Palombo EA (2011) Traditional medicinal plant extracts and natural products with activity against oral bacteria: potential application in the prevention and treatment of oral diseases. Evid Based Complement Alternat Med 2011:680354. https://doi.org/10.1093/ecam/nep067

Papaioannou E, Utari PD, Quax WJ (2013) Choosing an appropriate infection model to study quorum sensing inhibition in Pseudomonas infections. Int J Mol Sci 14(9):19309–19340. https://doi.org/10.3390/ijms140919309

Patel B, Kumari S, Banerjee R, Samanta M, Das S (2017) Disruption of the quorum sensing regulated pathogenic traits of the biofilm-forming fish pathogen *Aeromonas hydrophila* by tannic acid, a potent quorum quencher. Biofouling 33(7):580–590. https://doi.org/10.1080/08927014.2017.1336619

Pellegrini MC, Alvarez MV, Ponce AG, Cugnata NM, De Piano FG, Fuselli SR (2014) Anti-quorum sensing and antimicrobial activity of aromatic species from South America. J Essent Oil Res 26(6):458–465. https://doi.org/10.1080/10412905.2014.947387

Pratiwi SU, Hertiani T, Idroes R, Lagendijk EL, de Weert S, Hondel CV (2016) Quorum quenching and biofilm-degrading activity of massoia oil against *Candida albicans* and *Pseudomonas aeruginosa*. Planta Med 81(S01):S1–S381. https://doi.org/10.1055/s-0036-1596813

Priha O, Virkajärvi V, Juvonen R, Puupponen-Pimiä R, Nohynek L, Alakurtti S, Pirttimaa M, Storgards E (2014) Quorum sensing signalling and biofilm formation of brewery-derived bacteria, and inhibition of signalling by natural compounds. Curr Microbiol 69(5):617–627. https://doi.org/10.1007/s00284-014-0627-3

Rahman MR, Lou Z, Yu F, Wang P, Wang H (2017) Anti-quorum sensing and anti-biofilm activity of Amomum tsaoko (Amommum tsao-ko Crevost et Lemarie) on foodborne pathogens. Saudi J Biol Sci 24(2):324–330. https://doi.org/10.1016/j.sjbs.2015.09.034

Rama Devi K, Srinivasan R, Kannappan A, Santhakumari S, Bhuvaneswari M, Rajasekar P, Prabhu NM, Veera Ravi A (2016) In vitro and in vivo efficacy of rosmarinic acid on quorum sensing mediated biofilm formation and virulence factor production in *Aeromonas hydrophila*. Biofouling 32(10):1171–1183. https://doi.org/10.1080/08927014.2016.1237220

Rasamiravaka T, Jedrzejowski A, Kiendrebeogo M, Rajaonson S, Randriamampionona D, Rabemanantsoa C, Andriantsimahavandy A, Rasamindrakotroka A, Duez P, El Jaziri M, Vandeputte OM (2013) Endemic Malagasy Dalbergia species inhibit quorum sensing in *Pseudomonas aeruginosa* PAO1. Microbiology 159(Pt 5):924–938. https://doi.org/10.1099/mic.0.064378-0

Rasmussen TB, Givskov M (2006) Quorum sensing inhibitors: a bargain of effects. Microbiology 152(4):895–904. https://doi.org/10.1099/mic.0.28601-0

Rekha PD, Vasavi HS, Vipin C, Saptami K, Arun AB (2017) A medicinal herb Cassia alata attenuates quorum sensing in *Chromobacterium violaceum* and *Pseudomonas aeruginosa*. Lett Appl Microbiol 64(3):231–238. https://doi.org/10.1111/lam.12710

Rodrigues AC, Oliveira BD, Silva ER, Sacramento NT, Bertoldi MC, Pinto UM (2016a) Anti-quorum sensing activity of phenolic extract from *Eugenia brasiliensis* (Brazilian cherry). Food Sci Technol (Campinas) 36(2):337–343. https://doi.org/10.1590/1678-457X.0089

Rodrigues AC, Zola FG, Ávila Oliveira BD, Sacramento NT, da Silva ER, Bertoldi MC, Taylor JG, Pinto UM (2016b) Quorum quenching and microbial control through phenolic extract of *Eugenia uniflora* fruits. J Food Sci 81(10):M2538–M2544. https://doi.org/10.1111/1750-3841.13431

Ruffin M, Bilodeau C, Maillé É, LaFayette SL, McKay GA, Trinh NT, Beaudoin T, Desrosiers MY, Rousseau S, Nguyen D, Brochiero E (2016) Quorum-sensing inhibition abrogates the deleterious impact of *Pseudomonas aeruginosa* on airway epithelial repair. FASEB J 30(9):3011–3025. https://doi.org/10.1096/fj.201500166R

Rumbaugh KP, Griswold JA, Iglewski BH, Hamood AN, Barbara H (1999) Contribution of quorum sensing to the virulence of *Pseudomonas aeruginosa* in burn wound infections. Infect Immun 67(11):5854–5862

Salini R, Sindhulakshmi M, Poongothai T, Pandian SK (2015) Inhibition of quorum sensing mediated biofilm development and virulence in uropathogens by *Hyptis suaveolens*. Antonie Van Leeuwenhoek 107(4):1095–1106. https://doi.org/10.1007/s10482-015-0402-x

Sameena H (2006) Quorum sensing inhibition and antimicrobial properties of certain medicinal plants and natural products. MSc Dissertation (Ag. Microbiology). Submitted to Aligarh Muslim University, Aligarh, India

Sarabhai S, Sharma P, Capalash N (2013) Ellagic acid derivatives from *Terminalia chebula* Retz. Downregulate the expression of quorum sensing genes to attenuate *Pseudomonas aeruginosa* PAO1 virulence. PLoS One 8(1):53441. https://doi.org/10.1371/journal.pone.0053441

Sepahi E, Tarighi S, Ahmadi FS, Bagheri A (2015) Inhibition of quorum sensing in *Pseudomonas aeruginosa* by two herbal essential oils from Apiaceae family. J Microbiol 53(2):176. https://doi.org/10.1007/s12275-015-4203-8

Sethupathy S, Nithya C, Pandian SK (2015) 2-Furaldehyde diethyl acetal from tender coconut water (*Cocos nucifera*) attenuates biofilm formation and quorum sensing-mediated virulence of *Chromobacterium violaceum* and *Pseudomonas aeruginosa*. Biofouling 31(9–10):721–733. https://doi.org/10.1080/08927014.2015.1102897

Sheng JY, Chen TT, Tan XJ, Chen T, Jia AQ (2015) The quorum-sensing inhibiting effects of stilbenoids and their potential structure–activity relationship. Bioorg Med Chem Lett 25(22):5217–5220. https://doi.org/10.1016/j.bmcl.2015.09.064

Silva LN, Zimmer KR, Macedo AJ, Trentin DS (2016) Plant natural products targeting bacterial virulence factors. Chem Rev 116(16):9162–9236. https://doi.org/10.1021/acs.chemrev.6b00184

Singh BN, Upreti DK, Singh BR, Pandey G, Verma S, Roy S, Naqvi AH, Rawat AK (2015) Quercetin sensitizes fluconazole-resistant *Candida albicans* to induce apoptotic cell death by modulating quorum sensing. Antimicrob Agents Chemother 59(4):2153–2168. https://doi.org/10.1128/AAC.03599-14

Srinivasan R, Devi KR, Kannappan A, Pandian SK, Ravi AV (2016) *Piper betle* and its bioactive metabolite phytol mitigates quorum sensing mediated virulence factors and biofilm of nosocomial pathogen *Serratia marcescens* in vitro. J Ethnopharmacol 193:592–603. https://doi.org/10.1016/j.jep.2016.10.017

Srinivasan R, Santhakumari S, Ravi AV (2017) In vitro antibiofilm efficacy of *Piper betle* against quorum sensing mediated biofilm formation of luminescent *Vibrio harveyi*. Microb Pathog 110:232–239. https://doi.org/10.1016/j.micpath.2017.07.001

Sun S, Li H, Zhou W, Liu A, Zhu H (2014) Bacterial quorum sensing inhibition activity of the traditional Chinese herbs, *Ficus carica* L. and *Perilla frutescens*. Chemotherapy 60(5–6):379–383. https://doi.org/10.1159/000440946

Swearingen MC, Sabag-Daigle A, Ahmer BM (2013) Are there acyl-homoserine lactones within mammalian intestines? J Bacteriol 195(2):173–179. https://doi.org/10.1128/JB.01341-12

Szabó MÁ, Varga GZ, Hohmann J, Schelz Z, Szegedi E, Amaral L, Molnár J (2010) Inhibition of quorum-sensing signals by essential oils. Phytother Res 24(5):782–786. https://doi.org/10.1002/ptr.3010

Tapia-Rodriguez MR, Hernandez-Mendoza A, Gonzalez-Aguilar GA, Martinez-Tellez MA, Martins CM, Ayala-Zavala JF (2017) Carvacrol as potential quorum sensing inhibitor of *Pseudomonas aeruginosa* and biofilm production on stainless steel surfaces. Food Control 75:255–261. https://doi.org/10.1016/j.foodcont.2016.12.014

Thakur P, Chawla R, Tanwar A, Chakotiya AS, Narula A, Goel R, Arora R, Sharma RK (2016) Attenuation of adhesion, quorum sensing and biofilm mediated virulence of carbapenem resistant *Escherichia coli* by selected natural plant products. Microb Pathog 92:76–85. https://doi.org/10.1016/j.micpath.2016.01.001

Tiwary BK, Ghosh R, Moktan S, Ranjan VK, Dey P, Choudhury D, Dutta S, Deb D, Das AP, Chakraborty R (2017) Prospective bacterial quorum sensing inhibitors from Indian medicinal plant extracts. Lett Appl Microbiol 65(1):2–10. https://doi.org/10.1111/lam.12748

Truchado P, Giménez-Bastida JA, Larrosa M, Castro-Ibáñez I, Espín JC, Tomás-Barberán FA, Garcìa-Conesa MT, Allende A (2012) Inhibition of quorum sensing (QS) in *Yersinia enterocolitica* by an orange extract rich in glycosylated flavanones. J Agric Food Chem 60(36):8885–8894. https://doi.org/10.1021/jf301365a

Truchado P, Larrosa M, Castro-Ibáñez I, Allende A (2015) Plant food extracts and phytochemicals: their role as quorum sensing inhibitors. Trends Food Sci Technol 43(2):189–204. https://doi.org/10.1016/j.tifs.2015.02.009

Vandeputte OM, Kiendrebeogo M, Rajaonson S, Diallo B, Mol A, El Jaziri M, Baucher M (2010) Identification of catechin as one of the flavonoids from *Combretum albiflorum* bark extract that reduces the production of quorum-sensing-controlled virulence factors in *Pseudomonas aeruginosa* PAO1. Appl Environ Microbiol 76(1):243–253. https://doi.org/10.1128/AEM.01059-09

Vandeputte OM, Kiendrebeogo M, Rasamiravaka T, Stevigny C, Duez P, Rajaonson S, Diallo B, Mol A, Baucher M, El Jaziri M (2011) The flavanone naringenin reduces the production of quorum sensing-controlled virulence factors in *Pseudomonas aeruginosa* PAO1. Microbiology 157(7):2120–2132. https://doi.org/10.1099/mic.0.049338-0

Vasavi HS, Arun AB, Rekha PD (2014) Anti-quorum sensing activity of *Psidium guajava* L. flavonoids against *Chromobacterium violaceum* and *Pseudomonas aeruginosa* PAO1. Microbiol Immunol 58(5):286–293. https://doi.org/10.1111/1348-0421.12150

Vasavi HS, Arun AB, Rekha PD (2015) Anti-quorum sensing potential of *Adenanthera pavonina*. Pharm Res 7(1):105–109. https://doi.org/10.4103/0974-8490.147220

Vattem DA, Mihalik K, Crixell SH, McLean RJ (2007) Dietary phytochemicals as quorum sensing inhibitors. Fitoterapia 78(4):302–310. https://doi.org/10.1016/j.fitote.2007.03.009

Yang Q, Wang L, Gao J, Liu X, Feng Y, Wu Q, Baloch AB, Cui L, Xia X (2016) Tannin-rich fraction from pomegranate rind inhibits quorum sensing in *Chromobacterium violaceum* and biofilm formation in *Escherichia coli*. Foodborne Pathog Dis 13(1):28–35. https://doi.org/10.1089/fpd.2015.2027

Yap PS, Krishnan T, Yiap BC, Hu CP, Chan KG, Lim SH (2014) Membrane disruption and anti-quorum sensing effects of synergistic interaction between Lavandula angustifolia (lavender oil) in combination with antibiotic against plasmid-conferred multi-drug-resistant Escherichia coli. J Appl Microbiol 116(5):1119–1128. https://doi.org/10.1111/jam.12444

Yarmolinsky L, Bronstein M, Gorelick J (2015) Inhibition of bacterial quorum sensing by plant extracts. Isr J Plant Sci 62(4):294–297. https://doi.org/10.1080/07929978.2015.1067076

Yin WF, Purmal K, Chin S, Chan XY, Chan KG (2012) Long chain N-acyl homoserine lactone production by Enterobacter sp. isolated from human tongue surfaces. Sensors 12(11):14307–14314. https://doi.org/10.3390/s121114307

Yin H, Deng Y, Wang H, Liu W, Zhuang X, Chu W (2015) Tea polyphenols as an antivirulence compound disrupt quorum-sensing regulated pathogenicity of *Pseudomonas aeruginosa*. Sci Rep 6:17987. https://doi.org/10.1038/srep17987

Zahin M, Aqil F, Khan MS, Ahmad I (2010a) Ethnomedicinal plants derived antibacterials and their prospects. In: Chattopadhyay D (ed) Ethnomedicine: a source of complementary therapeutics. Research Signpost, Trivandrum, pp 149–178

Zahin M, Hasan S, Aqil F, Khan M, Ahmad S, Husain FM, Ahmad I (2010b) Screening of certain medicinal plants from India for their anti-quorum sensing activity. Indian J Exp Biol 48(12):1219–1224

Zhang A, Chu WH (2017) Anti-quorum sensing activity of Forsythia suspense on *Chromobacterium violaceum* and *Pseudomonas aeruginosa*. Pharmacogn Mag 13(50):321–325. https://doi.org/10.4103/0973-1296.204547

Zhou L, Zheng H, Tang Y, Yu W, Gong Q (2013) Eugenol inhibits quorum sensing at sub-inhibitory concentrations. Biotechnol Lett 35(4). https://doi.org/10.1007/s10529-012-1126-x

Zhu J, Huang X, Zhang F, Feng L, Li J (2015) Inhibition of quorum sensing, biofilm, and spoilage potential in *Shewanella baltica* by green tea polyphenols. J Microbiol 53(12):829–836. https://doi.org/10.1007/s12275-015-5123-3

Chapter 21
Enzymatic Quorum Quenching for Virulence Attenuation of Phytopathogenic Bacteria

Ashtaad Vesuna and Anuradha S. Nerurkar

Abstract Quorum sensing is a process of cell-cell communication by which the bacteria vary their gene expression mediated by a self-produced signalling molecule in a population density dependent manner to coordinate their behaviour which is virulence in phytopathogens. Quorum quenching or quorum sensing inhibition is a broad term used to include many diverse mechanisms that interfere with the quorum sensing signalling where signal degrading enzymes comprise one of the important mechanism. The signalling molecule, AHL (N-acyl homoserine lactone) is central to the process of quorum sensing in many Gram negative phytopathogenic bacteria. With constant rise in the number of resistant bacterial pathogens, enzymatic quorum quenching is an effective alternative way to attenuate their virulence. Quorum quenching does not kill the pathogen but stops its quorum sensing activity and thus does not introduce selective pressure which leads to development of resistance. Quorum sensing signal degrading strategy can thus be an effective means of biocontrol of phytopathogens. This chapter briefly discusses the virulence processes mediated by quorum sensing in Gram negative plant pathogens particularly in *Pectobacterium carotovorum* subsp. *carotovorum.* and their virulence attenuation by AHL degrading enzymes such as lactonases, acylases and oxido-reductases produced by various bacteria including Actinobacteria. The main focus is on enzymes produced by the quorum quenching bacteria, the genes that control the enzymes and the mechanism by which they attenuate quorum sensing mediated virulence in Gram negative phytopathogenic bacteria.

Keywords Quorum sensing · Quorum quenching · Phytopathogens · N-acyl Homoserine lactone

A. Vesuna · A. S. Nerurkar (✉)
Department of Microbiology and Biotechnology Centre, Faculty of Science,
The Maharaja Sayajirao University of Baroda, Vadodara, India

© Springer Nature Singapore Pte Ltd. 2018
V. C. Kalia (ed.), *Biotechnological Applications of Quorum Sensing Inhibitors*,
https://doi.org/10.1007/978-981-10-9026-4_21

Abbreviations

AHL	N-acylhomoserine lactone
C-4-HSL	N-butanoyl-L-homoserine lactone,
C-6-HSL	N-hexanoyl-L-HSL,
C-10-HSL	N-decanoyl-L-HSL,
C-12-HSL	N-dodecanoyl-L-HSL,
3-O-C6-HSL	N-(3-hydroxyhexanoyl)-L-HSL
3-O-C6-HSL	N-(3-oxohexanoyl)-L-HSL,
3-O-C8-HSL	N-(3-oxooctanoyl)-L-HSL,
3-O-C12-HSL	N-(3-oxododecanoyl)-L-HSL
C-12-HSL	N-dodecanoyl-L-HSL
HSL	Homoserine lactone,
Pcc	*Pectobacterium carotovorum* subsp. *carotovorum,*
PCWDE	Plant Cell Wall Degrading Enzyme
QS	Quorum Sensing,
QQ	Quorum Quenching,
QSI	Quorum sensing Inhibitor

21.1   Introduction

Quorum sensing (QS) is a cell population density based mechanism which controls the gene expression in bacteria related to varied processes like bioluminescence, biofilm formation, virulence, competence development etc. in response to a threshold level concentration of a diffusible signaling molecule produced by the bacteria on buildup of sizable population. Quorum sensing occurs in both gram positive and gram negative bacteria but the QS mechanisms in both gram negative and gram positive bacteria differ to a great extent. (Miller and Bassler 2001; Rutherford and Bassler 2012; Helman and Chernin 2015). Gram negative bacteria regulate their quorum sensing pathways by a small autoinducer molecule called N-acyl Homoserine Lactone (AHL or HHL) signaling molecule while gram positive bacteria use autoinducing oligopeptides(AIPs) signaling molecules to regulate their quorum sensing mechanism (Bassler 2002; Waters and Bassler 2005; Rutherford and Bassler 2012; Zhou et al. 2017; Utari et al. 2017). Gram negative bacteria and their quorum sensing systems have been widely studied in the last couple of decades. *Vibrio fischeri, Vibrio harveyi, Pectobacterium carotovorum* (previously *Erwinia carotovora*), *Pseudomonas* sp., *Agrobacterium tumefaciens, Burkholderia* sp., *Chromobacterium violaceum, Ralstonia solanacearum, Rhizobium leguminosarum, Salmonella typhimurium, Serratia liquefaciens, Yersinia pestis* etc. are few gram negative bacteria which use quorum sensing as a mechanism for their various functions(Miller and Bassler 2001; Boyer and Wisniewski-Dyé 2009) out of which *Pectobacerium* sp., *Pseudomonas* sp., *Ralstonia solanacearum, Agrobacterium tumefaciens, Pantoea stewartii, Xanthomonas campestris* pv. *campestris* are well reported phytopathogens which use the QS mechanism to cause virulence against plants (von Bodman et al. 2003). Such bacteria produce Plant Cell Wall Degrading Enzymes (PCWDE) such as pectate lyase, pectin lyase, protease and cellulase that are QS regulated, causing diseases in plants (Cui et al. 2005; Põllumaa et al. 2012; Moleleki et al. 2017)

Any mechanism that disrupts the quorum sensing signaling, preventing the gene expression leading to loss of its function is termed as Quorum Quenching (QQ) (Grandclément et al. 2015). On a broader term it is also known as Quorum Sensing Inhibition (QSI) as there are various methods by which the process of quorum sensing in bacteria can be inhibited (Kalia 2013). Quorum quenching strategies and mechanisms against AIPs produced by gram positive bacteria has been studied extensively by Singh et al. (2016). QQ in gram negative bacteria can be brought about by chemicals (Borges et al. 2014), external changes (pH or temperature) (Papenfort and Bassler 2016) or enzymes (Kalia 2013; Grandclément et al. 2015). The enzymatic quorum quenching activity against gram negative bacteria has been reported by various genera including those belonging to Actinobacteria. The quorum quenching enzymes are of three major types (i) AHL lactonases (ii) acylases and (iii) oxido-reductases. They act on the signal molecule AHL, produced by quorum sensing gram negative phytopathogenic bacteria and inactivate it, which in turn inhibits the production of Plant Cell Wall Degrading Enzymes (PCWDE) which are the main virulence factors of the pathogen that cause the damage to the plants resulting in soft rot (Uroz et al. 2009; Kalia and Purohit 2011; Grandclément et al. 2015; Polkade et al. 2016; Garge and Nerurkar 2016). Extensive reviews on the topic of quorum quenching mechanism are available (Zhang and Dong 2004; Dong and Zhang 2005; Uroz et al. 2009; Kalia and Purohit 2011). In this chapter we have focused on the enzymatic quorum quenching mechanism with special reference to virulence attenuation of plant pathogens in the last 5 years.

21.2 Quorum Sensing in Phytopathogens

Quorum sensing was first reported in bacteria *Vibrio fischeri* and *Vibrio harveyi* around 45 years ago (Nealson et al. 1970), though the exact words were not used at that time. Later on, transcriptional regulator LuxR and the AHL synthase LuxI were found out to be responsible for bioluminescence in *V. fischeri* and *V. harveyi* in a cell density dependent manner. The signaling molecule AHL is synthesized by the *lux* I gene product, autoinducer synthase. *lux* R and *lux*I mediate the cell density dependent control of transcription of the genes in *lux* operon (Fuqua et al. 1994). The basic model involves the stimulation of gene expression by LuxR protein when it binds to their promoter after it is complexed with AHL (Loh et al. 2002).

AHL molecules produced by different bacteria are of different lengths depending on number of carbons in them and the presence and absence of oxygen atom in the structure (Ng and Bassler 2009). Gram negative phytopathogenic bacteria mostly

Table 21.1 The virulence factors produced by the quorum sensing gram negative phytopathogenic bacteria, their quorum sensing system and the cognate signal molecules

Organism	Regulatory proteins	Major signal molecule	Virulence factors	References
Agrobacterium tumefaciens	TraI/TraR	3-Oxo-C8-HSL	Ti plasmid conjugation	Lang and Faure (2014)
Burkholderia glumae	EanI/EanR	C8-HSL	Production of EPS and infection of onion leaves	Jang et al. (2014), Chen et al. (2015), Gao et al. (2015)
Dickeya dadantii	VfmE/VfmI	3-Oxo-C6-HSL	Plant cell wall degradation	Nasser et al. (2013)
Pantoea stewartii	EsaI/EsaR	3-Oxo-C6-HSL	Production of EPS and Pectate lyases	Burke et al. (2015)
Pectobacterium carotovorum subsp. *carotovorum*	ExpI/ExpR CarI/CarR	3-Oxo-C6-HSL	Virulent exoenzymes	Joshi et al. (2016), Ham (2013)
Pectobacterium carotovorum subsp. *atrosepticum*	ExpI/VirR	3-Oxo-C6-HSL	Production of virulent extracellular enzymes	Monson et al. (2013)
Pseudomonas aeruginosa	LasI/LasR; RhlI/RhlR	3-Oxo-C12-HSL C4-HSL	Extracellular enzymes, RhlR, RpoS, Xep, rhamnolipid, biofilm formation, secondary metabolites	Lee and Zhang (2015), Bondí et al. (2017)
Pseudomonas syringae	AhlI/AhlR	3-Oxo-C6-HSL	Production of EPS	Yu et al. (2014), Scott and Lindow (2016)
Ralstonia solanacearum	SolI/SolR PhcB/RalA	C8-HSL	Action of ralfuranone, resulting in plant virulence	Mori et al. (2017)
Xanthomonas campestris	RpfC/RpfG	DSF	Plant virulence	Cai et al. (2017)

have AHL based quorum sensing systems which regulate the virulence in plants with the exception of *Xanthomonas campestris* which has a DSF based quorum sensing system (Cai et al. 2017). The common gram negative phytopathogenic bacteria that possess quorum sensing regulated virulence mechanism are listed in Table 21.1. On the other hand the gram positive bacterial pathogens such as *Staphylococcus aureus, Clostridium botulinum, Enterococccus faecalis, Bacillus subtilis, Lactobacillus plantarum, Streptococcus* sp. use AIPs as signaling molecules (Rizzello et al. 2014; Teixeira et al. 2013; O'Rourke et al. 2014; Ihekwaba et al. 2015; Wolf et al. 2015; Ishii et al. 2017).

21.2.1 *Pectobacterium Species as Quorum Sensing Mediated Phytopathogens*

Pectobacterium sp. is a plant pathogen having a diverse host range, comprising agriculturally important crops like carrot, potato, cucumber, eggplant (brinjal), tomato, leafy greens etc. *P. carotovorum* and *P. atrosepticum* have been listed as one of the top ten pathogens (Mansfield et al. 2012). *Pectobacterium carotovorum* subsp. *carotovorum* and other *Pectobacterium* spp. such *as P. wasabiae, P. carotovorum* subsp. *atrosepticum* cause soft rot disease as a result of damage to the succulent parts of the plant like fruits,stem, tuber, bulbs etc. leading to necrosis. They rely upon production of PCWDE, like cellulases, hemicellulases, proteases and pectinases which hydrolyze the cellulose and pectin that are components of plant cell wall that keep the plant together. The PCWDEs disrupt the plant cell integrity that eventually destroys the plant structure and lead to destruction of the tissue. Soft rot starts as fluid soaked lesions on plant parts that increase in size rapidly. The plant tissue softens and disintegrates causing fluid to ooze out or sometimes the skin remains intact. The decaying tissue is creamish, blackish or brownish and emanates foul odor. Depending upon the susceptibility of the host and potency of the bacterial strain, different levels of rotting is observed in the host though factors such as temperature and humidity also play an important role in the process of rotting (Põllumaa et al. 2012; Ham 2013; Monson et al. 2013).

The soft rot causing *Pectobacteria* has been divided into 2 classes on the basis of AHL it produces: Class I strains produce N-3-oxooctanoyl-L-homoserine lactone (3-O-C6-HSL) and low quantity of 3-oxohexanoyl-L-homoserine lactone (3-O-C6-HSL). On the other hand, Class II strains mainly produce 3-O-C6-HSL, whereas little or none of 3-O-C8-HSL (Rajesh and Rai 2016). *Pectobacteria* has a complicated regulation of its quorum sensing system. Quorum sensing in *Pectobacteria* is responsible for regulation of plant cell wall degrading enzymes (PCWDE), Type three secretion system (T3SS) and Carbapenem antibiotic production. ExpI, is the LuxI synthase homolog in *Pectobacteria* and can synthesize 3-O-C6-HSL or 3-O-C8-HSL CarR, ExpR1 and ExpR2 are LuxR homologs in *Pectobacteria*. ExpI interacts with CarR, ExpR1 and ExpR2 once the AHLs accumulate and achieve a threshold concentration limit (Joshi et al. 2016).The coordinated production of the virulence factors of *Pectobacteria* is through two main pathways (i) the AHL system and (ii) GacS/Rsm system. As shown in Fig. 21.1, two LuxR homologs, ExpR1 and ExpR2, upregulate RsmA at low levels of AHLs and directly inhibit PCWDE production and the AHL synthase ExpI. rsmA is the global repressor gene that controls extracellular enzymes and pathogenicity in soft rotting *Pectobacteria* (Cui et al. 1995). RsmA belongs to a post transcriptional Rsm system and is responsible for the destabilization of mRNA transcripts that encode PCWDEs. GacA/GacS system is active at high cell density and promotes the transcription of rsmB, a non-

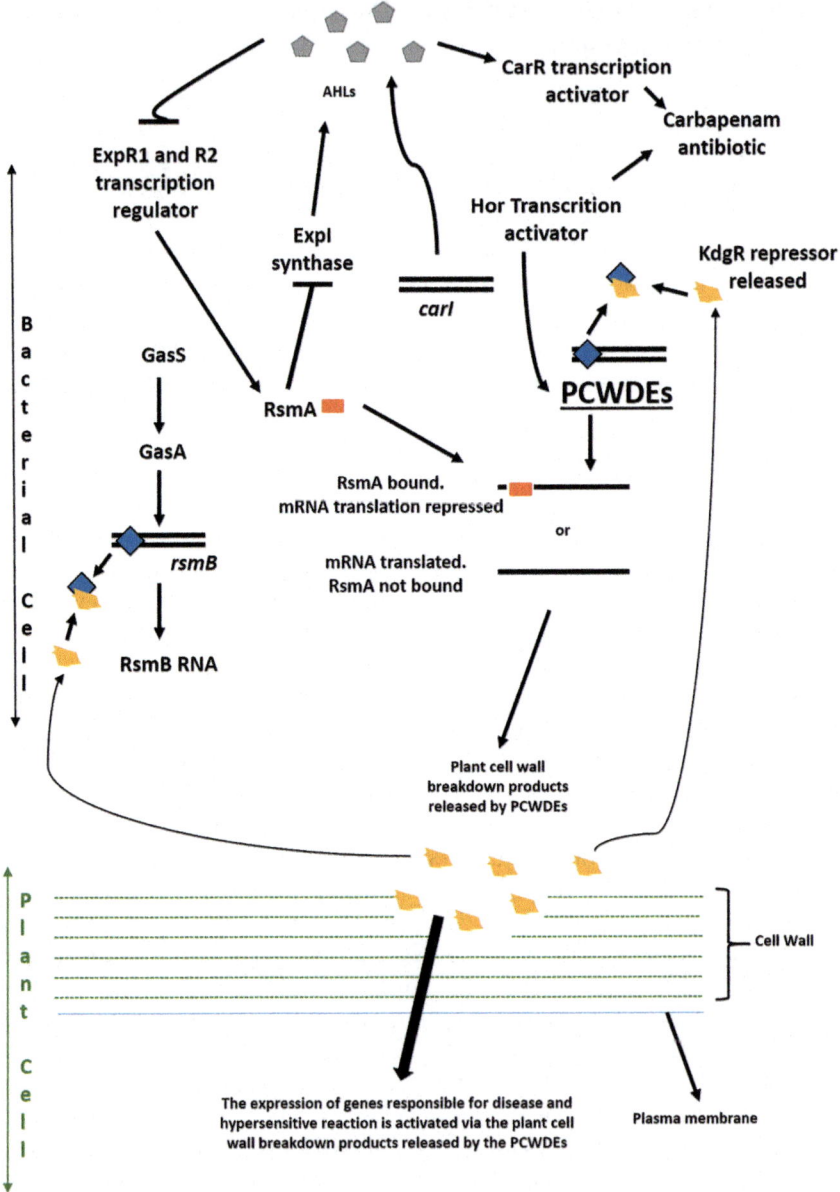

Fig. 21.1 Common regulatory systems controlling virulence in *Pectobacterium carotovorum* subsp. *carotovorum*

coding RNA which is also part of Rsm system and blocks the RsmA activity by sequestration, allowing translation of rsmA-targeted mRNAs. At adequate levels, 3-O-C8-HSL bind to ExpR1, whereas 3-O-C6-HSL binds to ExpR2. When AHLs bind to ExpR1 and ExpR2, they inhibit the expression of rsmA, as a result, the mRNA transcripts that encode PCWDEs become free and the enzymes are expressed. A two component system of ExpS (sensor kinase) and ExpA (response regulator) regulates the transcription of rsmB (Valente et al. 2017). ExpS and ExpA are homologs of GacS and GacA of various other gram negative bacteria. In *Pectobacterium* sp. quorum sensing synchronizes the virulence for prolific infection through synergistic negative regulation of ExpR1 and ExpR2. On the other hand, the communication of AHL with CarR regulator is much simpler. CarR binds to the 3-O-C6-HSL and then binds the carA promotor, which controls the Car operon that encodes the carbapenem antibiotic. Transcription regulator Hor controls the Car operon; the mechanism is not completely understood at the moment (Mole et al. 2007; Tichy et al. 2014).

The quorum sensing system gets more intricate due to the expression of many PCWDEs which are regulated positively by the breakdown of products of the pant cell wall produced by the activity of the bacterial pectinases on the tissues of plants (Fig. 21.1). These products include 5-keto-4-deoxouronate, 2,5-diketo-3-deoxygluconate and 2-keto-3-deoxygluconate (KDG). In the presence of these metabolites, regulation occurs due to the release of the transcriptional repressor KdgR. Operators of many of the PCWDEs and rsmB genes contain the binding sites for KdgR repressor. As a result, the initial production of pectinases causes further induction of virulence genes both transcriptionally as well as post-transcriptionally (Valente and Xavier 2015). Breakdown products of the substrates released by the action of the bacterial PCWDEs also perform the role of signaling molecules for the plant, indicating the presence of a pathogen that prompts the hypersensitive disease response. Quorum sensing places pathogenicity associated genes under density dependent control which avoid the activation of host plant's defense systems (Newton and Fray 2004). A relatively high inoculum of *Pectobacterium* sp. is required for successful infection, and the advancement of the disease is then a competition between the development of plant resistance and bacterial multiplication. Thus, the production of PCWDEs prematurely when the cell densities are low would give rise to an unsuccessful infection and on the contrary would induce local and systemic plant defense response, which in turn would resist subsequent infections. Ingeniously the *Pectobacterium* sp. uses AHLs to gauge its cell density and starts a pathogenic attack only when its population density is above a critical level, which makes sure that there is a high probability of thwarting host resistance (Põllumaa et al. 2012). Thus it offers the pathogen a fined tuned temporally controlled approach to overwhelm the host by evading the host response as well as copiously producing the virulence factors.

In experiments with *Pectobacterium carotovorum* mutant, it was observed that inactivation of ExpI resulted in less production of PCWDEs, inability to produce AHL and overall decrease in virulence in potato tubers and stems (Moleleki et al. 2017). Gene expression analysis proved that flagella are also a part of the QS regulon and is positively regulated while fimbriae and pili are negatively regulated by QS. Various pathogenicity related factors (genes) which work within the quorum sensing system are required for the virulence of *Pcc*, these factors were found out and grouped into (i) production of plant cell-wall-degrading enzymes (PCWDEs) (expI, expR), (ii) nutrient utilization (pyrD, purH, purD, leuA and serB), (iii) biofilm formation (expI, expR and qseC), (iv) susceptibility to antibacterial plant chemicals (tolC), and (v) motility (flgA, fliA and flhB) (Lee et al. 2013). Bearing in mind the crucial role QS plays in *Pcc* virulence, we initiated a diversity study in which strains were isolated from soil and characterized by species specific PCR and 16SrRNA sequencing, pathogenicity on different hosts, virulence enzyme production and biochemical characteristics; out of these a highly virulent *Pcc* BR1 strain showing broad host specificity was obtained (Maisuria and Nerurkar 2013). This strain was later used for QQ studies. Virulence determinant enzymes polygalacturonase (Maisuria et al. 2010) and pectate lyase (Maisuria and Nerurkar 2012) of *Pcc* BR1 were purified and biochemical characterization as well as thermodynamic characterization was carried out.

In *Pectobacterium atrosepticum*, VirR, is the LuxR-type repressor which controls virulence of the bacteria. Here, VirR was found to be auto-repressing and in turn activating the transcription of rsmA in the absence of AHL. VirR was also found out be regulating the production of siderophores and swimming motility. This was one of the initial findings where VirR, a LuxR-type protein can act in vivo as both repressor and activator of transcription in the absence of its signaling molecule AHL (Monson et al. 2013). Again in *P. atrosepticum*, it was found that by moderating RsmA levels, (p)ppGpp exerted regulation through the moderation of the RsmA antagonist, RsmB and thereby have a QS controlled virulence in the bacteria. It was also observed that the ratio of RsmA protein to its RNA rival, rsmB, was controlled independently by QS and (p)ppGpp (Bowden et al. 2013). Surface swarming motility of *P. atrosepticum* is also controlled by QS. It required motility and O antigen biosynthesis for the process (Bowden et al. 2013).

21.2.2   Pseudomonas Species as Quorum Sensing Mediated Phytopathogens

Bacteria of the *Pseudomonas* species have been isolated from diverse habitats including soil, plants, animals and water (Arumugam et al. 2017; Sun et al. 2014; Qi et al. 2014). It includes both plant as well as animal pathogens. In humans, it causes infections in cystic fibrosis patients and also leads to nosocomial infections (De Simone et al. 2014; Stoltz et al. 2015). In many plants, *Pseudomonas* has been reported to cause soft rot and other diseases (McCann et al. 2013) *P. aeruginosa* has been reported to be a potent opportunistic pathogen and can infect diverse plant host via various virulence factors. These virulence factors include exoproteases, siderophores, lipases and exotoxins which are regulated by quorum sensing. It causes virulence in *Arabidopsis*, lettuce etc. (Hilker et al. 2015) *Pseudomonas aeruginosa* uses a compact network of quorum sensing receptors and regulators. The main quorum sensing pathways in *P. aeruginosa* are two LuxI and LuxR-type systems called LasI & LasR and RhlI & RhlR. LasI/LasR systems detects 3-O-C12-HSL and RhlI/RhlR systems detects C4-HSL. In Rhl system, RhlR guides both RhlI dependent and independent regulons. RhlR reins the expression of genes required for biofilm formation and genes encoding virulence factors in absence of RhlI. Bacteria like *P. aeruginosa* use multiple QS signals to control gene expression of various physical properties of their environment (Cornforth et al. 2014; Lee et al. 2013; Mukherjee et al. 2017). In *P. aeruginosa,* LasR activates expression of rsaL, which in turn code for RsaL protein which is a transcriptional repressor of lasI. It is reported that a 3-O-C12-HSL controlling type 1 incoherent feed forward loop (IFFL-1) created by LasR and RsaL splits the QS regulon into two different sub-regulons which increases the phenotypic plasticity of *P. aeruginosa* helping it to adapt in changing environment (Bondí et al. 2017). In an exciting recent report, it was stated that *P. aeruginosa* used QS system to activate cas gene in turn stimulating the prokaryotic adaptive immune system, the CRISPR-Cas system (Høyland-Kroghsbo et al. 2016).

Pseudomonas syringae is a pathogen on leaves of beans (*Phaseolus vulgaris*) and causes spots on them.Its quorum sensing circuit is regulated by AhlR/AefR system, which affect genes *in planta* (Yu et al. 2014). Water availability and its loss via diffusion greatly affects the QS system in *P. syringae*. Along with the presence of water, nutrient availability on the leaf surface also effects the pathogenicity via its QS circuit (Pérez-Velázquez et al. 2015.

21.2.3   Other Bacteria as Quorum Sensing Mediated Phytopathogens

Agrobacterium tumefaciens, a phytopathogenic bacteria is the main cause of crown gall disease in plants. Its pathogenicity is due to horizontal conjugative transfer and vegetative replication of oncogenic Ti plasmids which are under the control of a quorum sensing system. The quorum sensing system in *A. tumefaciens* is simple, it has LuxI/LuxR homologs which are TraI which acts in the synthesis of AHL molecules and activate transcriptional regulator TraR (Lang and Faure 2014). *A. tumefaciens* also has a TraM gene which works against TraR, interrupting the quorum sensing circuit. To counter this, the bacteria also has a secondary quorum sensing circuit TraI2 and TraR2 which can synthesize the 3-O-C8-HSL used by the Tra QS system and continue the QS circuit, this second circuit may play a role in replication and conjugation of Ti plasmid in *A. tumefaciens* (Wang et al. 2014). A motile phytopathogen, *Burkholderia glumae* is the main causal agent of rice panicle blight. It has multiple polar flagella and they are controlled by TofR/TofI quorum sensing system. It was observed that the quorum sensing and temperature play a role in flagellar morphogenesis which is important for the virulence of *B. glumae* (Jang et al. 2014).

Other process controlled by QS-regulated genes in *B. glumae* were linked to metabolic activities. QS regulation was noted in type VI secretion system rhamnolipid, genes linked to CRISPER-cas gene cluster, and Flp pilus biosynthesis (Gao et al. 2015). It was later found that regulation of rhamnolipids is important for control of swarming motility in *B. glumae* via quorum sensing (Nickzad et al. 2015). *Dickeya dadantii*, previously called *Erwinia chrysanthemi* is another well-known plant pathogen which has been known to have the similar QS system like the *Pectobacterium* sp. but recently another system has been identified in *D. dadantii* which has been known to carry out virulence in plants via QS. The newly found system is a VfmE/VfmI two component system which is responsible for transcription of PCWDE genes via expression of vfmE genes (Nasser et al. 2013). *Pantoea stewartii* produces stewertan, an exopolysaccharide (EPS) and colonizes the xylem of maize and causes Stewart's wilt disease in maize. The regulation of production of EPS and consequent virulence is done by a QS system which has a AHL synthase, EsaI that forms 3-O-C6-HSL and transcription regulation EsaR homologous to the usual LuxI/LuxR (Ramachandran and Stevens 2013). At low bacterial cell densities, in the absence of its (AHL) ligand, EsaR suppresses or activates expression of a number of genes. EsaR binds an AHL signal which is synthesized constitutively during growth at high bacterial cell densities and all of its regulation is released. For the causation of disease in the plant QS-dependent gene expression is critical (Ramachandran et al. 2014; Burke et al. 2015). *Ralstonia solanacearum* is a soil borne phytopathogenic organism that cause bacterial wilt in plants. It has PhcA, a virulence regulator which acts through the QS system and positively controls ralA, which encodes for furanone synthase which produces ralfuranones, an aryl-furanone secondary metabolite. Ralfuranones play a mojor role in the virulence of *R. solanacearum* (Mori et al. 2017). *Xanthomonas campestris* is one of the unique gram negative phytopathogenic bacteria which controls the virulence via QS system but does not use N-acyl Homoserine lactone (AHL) as a signaling molecule but rather uses medium chain fatty acid diffusible signal factor (DSF) instead. A medium chain FA DSF binds to the sensor region of a receptor histidine kinase (HK) RpfC of a two component system and activates it by causing an allosteric change (Cai et al. 2017).

21.3   Quorum Quenching of Phytopathogenic Bacteria

Having discussed the strategic position of the quorum sensing systems in regulation of virulence, quorum sensing blockage can be an efficient approach to attenuate the virulence of these phytopathogens. This has been validated in transgenic plants that were found to show resistance to infection by *P. carotovorum* subsp. *carotovorum* when the AHL degrading enzyme gene was cloned (Dong et al. 2001; Helman and Chernin 2015). Any process that inhibits or blocks the process of quorum sensing in any manner is known as Quorum Quenching. Quorum quenching is an effective mechanism to keep in check pathogens (usually phytopathogens) that control their virulence by quorum sensing. In the current scenario of evolving resistance of pathogens to antibiotics and drugs has forced us to look into alternate strategies and

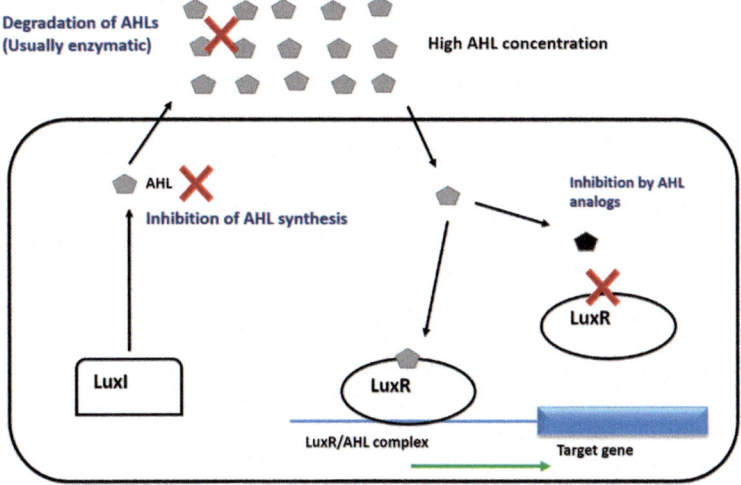

Fig. 21.2 Interruption of quorum sensing circuit at different targets to inhibit the virulence gene expression in phytopathogenic bacteria

by quorum quenching based biocontrol approach the pathogenicity of the bacteria can be controlled without killing the bacteria or the bacteria developing any resistance. By not affecting the growth of the pathogen but by inhibiting the quorum sensing regulated virulence of the pathogen, this strategy applies very limited selection pressure on the pathogen. Cui and Harling (2005) have discussed the targets in quorum sensing circuit that can be interrupted to inhibit the pathogenesis of bacterial plant pathogens which as depicted in Fig. 21.2 are categorized as (i) QS signal (AHL) inactivation by degrading enzymes (Chen et al. 2013). (ii) AHL-LuxR binding inhibition with AHL analogs as antagonists (Koch et al. 2005). (iii) Interruption of the AHL biosynthesis pathway either inhibiting the specific AHL synthase or other enzymes in the pathway of AHL biosynthesis precursors (Uroz et al. 2009). Thus quorum quenching against signal molecule induced quorum sensing phytopathogens have been derived from bacteria as well as eukaryotes (Kalia 2013). Based on their mechanisms they fall into two major categories as follows.

1. Chemical compound which inhibits the function of the quorum sensing system. This type of quorum quenching is usually termed as chemical interference based quorum sensing inhibition and such chemical compounds are termed as Quorum Sensing Inhibitors (QSI). These can be inhibitors of the targets mentioned in ii) and iii) above.
2. The more frequently studied and explored type of quorum quenching is enzyme based which depends directly on disrupting the quorum sensing signal molecule AHL by different bacterial enzymes. This chapter focusses on enzymatic quorum quenching, its mechanisms and the work done in this field in detail in relation to phytopathogenic bacteria.

21.3.1   Chemical Interference Based Quorum Quenching

Chemical compounds from natural products have been known to act as QSIs. Defoirdt et al. (2013) have done a detailed review on all different chemical compound based QSIs. Some examples of such compound which inhibit the quorum sensing induced virulence of phytopathogenic bacteria are discussed here. A study revealed that extracts from the plant *C. asiatica* were effective QSIs against *P. aeruginosa*. It was found out that the *C. asiatica* extracts had abundant amount of flavonoids in them and they had shown specificity towards the las and rhl QS systems in *P. aeruginosa* (Vasavi et al. 2016). In another study, various plants extracts were screened and identified for their QSI activity against *P. aeruginosa* and it was found out that three compounds trans-cinnamaldehyde, tannic acid and salicylic acid were found to be potential QS inhibiting compounds- Trans- cinnamaldehyde was found out to be most effective of the three with further studies (Chang et al. 2014). ZnO nanoparticles also have been reported to decrease elastase, pyocyanin and biofilm formation, molecules and process regulated by QS in *Pseudomonas aeruginosa* strains, in turn decreasing the QS based virulence in the bacteria and was proposed as an alternative treatment against *P. aeruginosa* (García-Lara et al. 2015).

Some Volatile organic compounds (VOCs) produced by bacteria have also been known to been used as QSI. Helman and Chernin (2015) in their review stated and reported a group of volatile suphur compounds like dimethyl sulphide, dimitheyl disulphide(DMDS) and dimethyl trisuphide, produced by bacteria suppressing growth of *Agrobacterium* strains. Dimitheyl disulphide also reduced the amount of AHLs produced by various *Pseudomonas* sp. lowering their virulence and pathogenicity.

Most reports of QSIs produced either by plants or bacteria have been reported against phytopathogenic AHL signal molecule based quorum sensing system of different *Pseudomonas* sp. Supporting this claim, a study revealed that the use of cranberry extract rich of proanthocyanidins (cerPAC) was a very efficient anti-virulence agent against *P. aeruginosa* as it interfered with its QS system. The AHL synthases LasI/RhlI and LasR/RhlR, the QS transcriptional regulators were repressed and antagonized by the cerPAC (Maisuria et al. 2016). The phytopathogenic bacteria *Pseudomonas savastanoi*'s QS system was inhibited by polyphenolic extracts from plants such as *Olea europaea, Cynara scolymus and Vitis vinifera* without hampering the viability of the bacteria (Biancalani et al. 2016).

Xanthomonas oryzae, a causative agent of bacterial rice leaf blight, uses DSF signal molecule based quorum sensing system for its virulence. By inhibiting the QS pathway and the histidine pathway which is involved in QS circuit in *X. oryzae*, bismerthiazol, a thiadiazole molecule reduced the rice leaf blight disease (Liang et al. 2016).

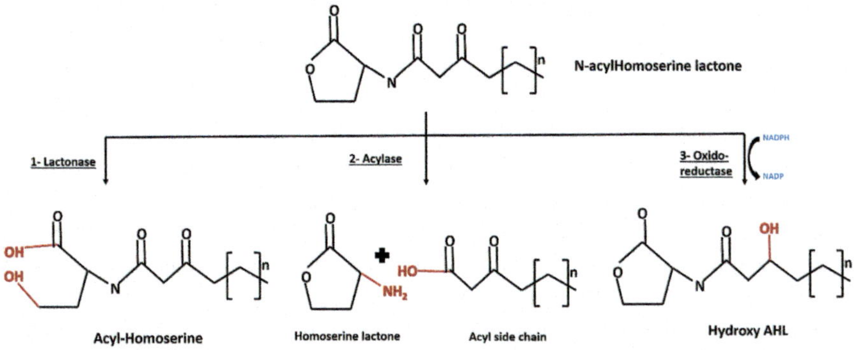

Fig. 21.3 Major classes of AHL inactivating enzymes produced by bacteria

21.3.2 Enzymatic Quorum Quenching for Virulence Attenuation

We have discussed earlier the process of quorum sensing and how the mechanism is used by different gram negative phytopathogenic bacteria for virulence function. Quorum quenching Bacteria produce enzymes that interfere with the AHL based quorum sensing in gram negative phytopathogenic bacteria as mentioned earlier. A wide diversity is observed amongst quorum quenching bacteria even though nearly 50% are covered by *Bacillus* spp. An extensive account of the diversity of QQ bacteria is presented by Chen et al. (2013) where they have given the phylogenetic relationship of the QQ enzymes and have noted that quorum quenching enzymes are present in both the QS and non-QS bacteria. *Bacillus* species are usually not harmful when added as biocontrol agents to the soil. *Bacillus* sp. has been given the designation of GRAS (Generally Regarded As Safe) by the FDA which affords an added advantage of using them to control QS mediated virulence and hence their QQ ability has been studied in great detail (Koul and Kalia 2017).

Enzymatic quorum quenching depends upon bacteria producing enzymes which attack the signaling molecule N-acyl Homoserine lactone (AHL), cleaving or cutting it or changing its orientation in such a way that it can no longer work as a signal molecule for the process of quorum quenching. As of yet as shown in Fig. 21.3, three types of enzymes have been reported to act on AHLs and can be termed as quorum quenching enzymes (Utari et al. 2017).

1. <u>AHL lactonase</u> – AHL lactonase cleaves the lactone ring of the signal molecule. As the structure of the AHL molecule is changed, it is unable to act as a signal molecule since it cannot bind to its target transcriptional regulators, thus inhibiting quorum sensing. AHL lactonases enzymes found in QQ bacteria belong to various different

protein families such as Metallo-β-lactamase-like lactones, Phosphotriesterase-like lactones (PLLs) and Paraoxonases (PONs). Metallo-β-lactamase-like lactones super-family consists of a Zn^{2+}-binding HXHXDH motif and is present in AiiA, AttM/AiiB, AhlD, AhlS, AidC, QlcA etc. PLLs belong to the amidohydrolase superfamily, containing a binuclear metal center within a $(β/α)_8$-barrel structural scaffold. They work against a broad spectrum of AHLs but are known to prefer hydrophobic lactones. QsdA, GKL, GsP etc. belong to this super-family. Usually mammalian lactonases belong to the Paraoxonases family. They have a structural and catalytic Ca^{2+} ion and a six-bladed β propeller fold. (Fetzner 2015). AdeH, an AHL lactonase cloned from *Lysinibacillus* sp. in *E.coli* was found to be a unique one as it shows similarity to lactonases belonging to metallo-β-lactamase superfamily as well as the amidohydrolase superfamily. It is also different in the sense that most Bacillus isolates show the presence of lactonase AiiA, while AdeH shows only 30% similarity to AiiA (Garge and Nerurkar 2016).

2. <u>AHL acylase</u> – AHL acylases also called amidase or aminohydrolase catalyzes and cleaves N-acyl homoserine lactone into homoserine lactone and an acyl side chain. The cleavage of the AHL molecule, inactivates its function as signal molecule and inhibits quorum sensing in the bacteria. Unlike AHL lactonases, AHL acylases belong to a single protein superfamily consisting of a N-terminal nucleophile hydrolase fold with an exception of AiiO from *Ochrobactrum* sp. which has a α/β hydrolase fold (Czajkowski et al. 2011). AHL acylases have been reported to have a preference to long-chain AHLs (Fetzner 2015; Utari et al. 2017). AHL acylase has been purified from *Delftia* sp. VM4in our lab (Maisuria and Nerurkar 2015).

3. <u>AHL oxido-reductase</u> – AHL oxidoreductase do not degrade the AHL molecule like lactonases or acylases but they simply substitute the oxo-group at C3 carbon position with a hydroxyl group, which is degraded easily by amidohydrolases (Chen et al. 2013).

Very few organisms have been reported to possess both AHL lactonase and AHL acylase enzymes. With the help of genomic and metabolic studies three organism *Deinococcus radiodurans, Hyphomonas neptunium* and *Photorhabdus luminescens* subsp. *laumondii* have been reported to have AHL lactonase and acylase in addition to actionobacteria *R. erythropolis* (Kalia et al. 2011). Several bacteria have been reported to be producing the quorum quenching enzymes. Table 21.2 enlists the bacteria that produce the quorum quenching enzymes and have been reported recently to be attenuating the virulence of phytopathogens which use quorum sensing for their pathogenicity. Fetzner (2015) have provided in their studies a comprehensive list from earlier studies of QQ enzymes and the protein family to which they belong. Table 21.2 also includes this information about the QQ information reported in last 5 years. Koul and Kalia (2017) have performed a comparative genomic analysis and expressed that there exists multiplicity of genes for AHL lactonases and acylases which emphasizes that it is a potential mechanism that limits the target bacterial population.

Table 21.2 Various classes of quorum quenching enzyme producing representative bacteria

Enzyme	Host	Substrate	References
AHL lactonase			
AiiA	*Bacillus cereus* *Bacillus thuringiensis*	C4-, C6-, C8-, C10-HSL; 3OC4-, 3OC6-, 3OC8-, 3OC12-HSL; 3-OH-C4-HSL	Safari et al. (2014)
AdeH	*Lysinibacillus* sp.		Garge and Nerurkar (2016)
AttM AttJ	*Agrobacterium tumefaciens*	C4-, C6-, C7-, C8-, C10-HSL; 3OC6-, 3OC8-HSL	Wang et al. (2015)
AiiS	*Agrobacterium radiobacter*	Broad	Uroz et al. (2009)
AhlD	*Arthrobacter* sp.	C6-, C8-, C10-HSL, 3OC6-, 3OC12-HSL	Park et al. (2003)
QlcA	*Acidobacteria*	C6-HSL	Riaz et al. (2008)
AiiM	*Microbacterium testaceum*	C6-, C8-, C10, C12-HSL; 3OC6-, 3OC8-, 3OC10, 3OC12-HSL	Wang et al. (2010)
QsdA	*Rhodococcus erythropolis*	C4-HSL, C6- to C14-HSLs, with or without substitution at C3	Latour et al. (2013)
AidH	*Ochrobactrum* sp.	C4-, C6-, C10-HSL; 3OCC6-, 3OC8-HSL; 3-OH-C6-HSL	Zhang et al. (2016)
DlhR, QsdR1	*Rhizobium* sp.	3OC8-HSL	Krysciak et al. (2011)
AhlS	*Solibacillus silvestris*	C10-HSL	Morohoshi et al. (2012)
SsoPox	*Sulfolobus solfataricus*	C4-, C6-, 8-, C12-HSL; 3OC6-, 3OC8-, 3OC10-, 3OC12-HSL	Jacquet et al. (2016)
GKL	*Geobacillus kaustophilus*	C6-12-HSL	Xue et al. (2013)
AHL acylase			
AiiD	*Ralstonia eutropha*	3OC6-, 3OC8-, 3OC10-, 3OC12-HSL	Du et al. (2014)
AiiC	*Anabaena* sp.	C4-, C6-, C8-, C10-, C12-, C14-HSL; 3OC4-, 3OC6-, 3OC8-, 3OC10-, 3OC12-, 3OC14-HSL and corresponding 3OH-Cx-HSLs	Rolland et al. (2016)
AhlM	*Streptomyces* sp.	Chain length more than C8	Park et al. (2005)
Aac	*Ralstonia solanacearum*	C7-, C8-, C10-HSL; 3OC8-HSL	Chen et al. (2009)
HacB	*Pseudomonas syringae*	C4-, C6-, C8-, C10-, C12-HSL; 3OC6-, 3OC8-HSL	Shepherd and Lindow (2009)
HacA	*Pseudomonas syringae*	C8-, C10-, C12-HSL; 3OC8-HSL	
AiiO	*Ochrobactrum* sp.	C4- to C14- with or without oxo group	Czajkowski et al. (2011)

(continued)

Table 21.2 (continued)

Enzyme	Host	Substrate	References
–	*Comomonas* sp.	C6-, C12-, C16-HSL; 3OC6-, 3OC8-, 3OC10-, 3OC12-, 3OC14-HSL; 3OH-C12-HSLto C16 AHLs, with or without substitution at C3	Chen et al. (2013)
–	*Rhodococcus erythropolis*	3OC6-, 3OC10-HSL; 3-OH-C10HSL	Uroz et al. (2005)
–	*Delftia* sp.	C6-, 3OC6-, 3OC8-HSL	Maisuria and Nerurkar (2015)
AHL oxidoreductases			
–	*Rhodococcus erythropolis*	3OC8-, 3OC10-, 3OC12-, 3OC14-HSL (reduction to 3OH-HSL)	Uroz et al. (2005)
–	*Burkholderia* sp.	3OC4-, 3OC6-, 3OC8-HSL (reduction to 3OH-HSL)	Chan et al. (2011)

21.4 Biocontrol Studies

Pectobacterium carotovorum subsp. *carotovorum* (*Pcc*) has been the bacterium of choice as a model QS regulated pathogen to conduct the quorum quenching studies as their pathosystem yields promising results. Most reported literature therefore involve studies on this bacterium and its QS mechanism as target for biocontrol agents like QQ enzyme producing bacteria. QQ bacteria can be used as biocontrol agents since the QS enzymes are located intracellularly in majority of these bacteria. Biocontrol studies reported are generally *in planta* and in very few cases pot studies. Barring the field applications of quorum quenching *Rhodococcus* as biocontrol agent (Des Essarts et al. 2016) translational studies are sparse. An interesting bio stimulation approach of biocontrol is suggested (Cirou et al. 2007, 2012) where they have demonstrated that the soil AHL degrading *Rhodococcus erythropolis* population can be enriched to exercise *is situ* biocontrol in potato rhizosphere against *Pcc*. Faure and coworkers also have shown investigations involving large scale hydroponics where they could stimulate *in situ* the growth of *R. erythropolis* (Cirou et al. 2012).

21.4.1 Quorum Quenching Bacillus *sp. as Biocontrol against Phytopathogens*

Members of the *Bacillus* species have been widely reported as a major genera to have quorum quenching activity, inhibiting virulence of pathogenic bacteria including many phytopathogens. Studies have been reported regarding quorum quenching enzymes and their types in the *Bacillus* sp. Recently, in our laboratory various AHL degrading *Bacillus* sp. were isolated and checked for their biocontrol activity against soft rot causing phytopathogen *Pcc*. The *bacillus* sp. bacteria were identified to be

B. subtilis, B. firmus and *B. thuringiensis* and were effective in stopping maceration in potato, carrot and tomato caused by *Pectobacterium carotovorum* subsp. *carotovorum* (Garge and Nerurkar 2017).

In another study from our laboratory, *Lysinibacillus* sp. isolated from garden soil was reported to disable the AHL, via hydrolysis of the lactone ring and was found out to have a AHL lactonase enzyme AdeH. This AHL lactonase gene AdeH from *Lysinibacillus* sp. could degrade AHL produced by *Pectobacterium carotovorum* subsp. *carotovorum*, interfered with the QS system of *Pcc* and inhibited the PCWDE but did not hamper the growth of the plant pathogen and was reported as an effective QQ based biocontrol agent in both preventive as well as curative manner against phytopathogen *Pectobacterium carotovorum* subsp. *carotovorum*. (Garge and Nerurkar 2016).

Bacillus thuringiensis' AHL lactonase AiiA, a metallo-gamma-lactonase was reported to have broad substrate specificity. The authors reported that AiiA from *B. thuringiensis* to be one of the most effective QQ enzyme (Liu et al. 2013). *B. cereus, B. thuringiensis* and *Brevibacillus brevis* which had the AiiA AHL lactonase gene were able to reduce the pathogenicity of *Dickeya dadantii* (Khoiri et al. 2017).

Geobacillus kaustophilus has a phosphotriesterase-like lactonase called GKL and has been reported to control QS pathways of disease causing bacteria (Xue et al. 2013). *Solibacillus silvestris*, isolated from potato leaf has been reported as a QQ biocontrol agent against soft rot causing *Pcc*. The enzyme responsible for the QQ activity was found out to be a AHL lactonase which showed faint similarity to AiiA-like AHL lactonase from *B. cereus* group. The expression of AHL lactonase gene from *S. silvestris*, AhlS was observed in plant pathogen *Pcc* and concomitantly found to reduce potato slice maceration indicating an efficient biocontrol (Morohoshi et al. 2012).

A vast range of *Bacillus* isolates were screened and the biodiversity of the aiiA gene which codes for AHL lactonases in *Bacillus* sp. was explored in detail. Among all isolates screened Bacillus sp. strain MBG11 showed high stability in its lactonase which could be used for heterologous expression and mass production (Huma et al. 2011).

21.4.2 Quorum Quenching Actinobacteria as Biocontrol Against Phytopathogens

Actinobacteria is an interesting phylum with regard to its quorum sensing and quorum quenching potential. A well explored review on the quorum sensing ability of Actinobacteria and its quorum quenching ability has been reported recently (Polkade et al. 2016).

Arthrobacter sp. grew on 3-oxo-C6-HSL and degraded several other AHLs with diverse lengths. *Arthrobacter* sp. reduced the virulence by reducing the AHL and pectate lyase when grown together in a co-culture with soft rot causing *Pcc*. It has an AHL lactonase gene, AhlD, which has 25%, 26% and 21% similar identity to other known AHL-degrading enzymes, *Bacillus* sp. 240B1 AiiA, a *Bacillus thuringiensis* subsp. *kyushuensis* AiiA homologue and *Agrobacterium tumefaciens* AttM (Park et al. 2003).

Microbacterium testaceum, isolated from leaf surface of potato showed presence of AHL-degradation activity and the gene responsible for the AHL-degradation here was AiiM, which was a part of α/β hydrolase fold family of *actionobacteria*. AiiM was found out to be an AHL lactonase and it was used against phytopathogen *Pcc* where it reduced the pectinase activity and the soft rot symptoms in potato *in vitro* (Wang et al. 2010).

Rhodococcus erythropolis is another widely used bacteria belonging to phylum *actinobacteria* as a quorum quenchers against QS phytopathogens. *R. erythropolis* is an exceptional bacteria in which all three AHL degrading enzymes AHL lactonase, acylase and oxido-reductase been reported. It encodes a unique class of quorum quenching lactonases which do not show similarity with lactonases which have been previously reported like AiiA, AhlD etc. The AHL lactonase gene QsdA (**Q**uorum-**s**ensing **s**ignal **d**egradation) degrades a wide range of AHLs and this gene is related to phosphotriesterases (Uroz et al. 2008; Cirou et al. 2012). Apart from the AHL lactonase activity by qsdA gene in *R. erythropolis* specific genes for AHL acylase and oxido-reductase haven't been identified yet, however vast range of AHL signal molecules have been reported to be degraded by these two types of enzymes in *R. erythropolis* (Uroz et al. 2005). Bio-stimulated by γ-caprolactone and isolated from potato plants cultivated in hydroponic conditions, AHL degrading *R. erythropolis* was obtained and its efficiency as a quorum quencher against *Pectobacterium* was found to be good (Cirou et al. 2012). Further studies on their quorum quenching activity against gram negative soft-rot causing plant pathogens, by degrading their signal molecules AHL have provided an idea on the γ-lactone catabolic pathway. γ-lactone catabolic pathway is accountable for cleaving of the lactone ring in the AHL molecule which is linked to an alkyl or acyl chain. This pathway is controlled by the availability of γ-lactone, therefore, stimulating it with food flavoring like γ-caprolactone increased the biocontrol potential of *R. erythropolis* and this promoted plant protection *in vivo* (Latour et al. 2013).

Streptomyces sp. has also been reported to have an AHL degrading enzyme, an acylase which plays a role in keeping virulence factors of pathogens in check. *Streptomyces* sp. which is found in abundance in soil can indeed be a useful biocontrol agent against phytopathogens. It shows the presence of AhlM gene which is an AHL acylase and shows minor similarities to AHL acylases of *Ralstonia* (AiiD) and *Pseudomonas aeruginosa* (PvdQ). The AhlM has been able to reduce the QS based virulence factors like elastase, protease and the gene LasA in *P. aeruginosa*, decreasing its pathogenicity and making it an important biocontrol agent to mediate AHL-based pathogenicity (Park et al. 2005).

Thus, *Actinobacteria* has been an important phylum which has some very effective quorum quenching bacteria which work against gram negative based plant pathogens which use quorum sensing for their pathogenicity.

21.4.3   Other Quorum Quenching Bacteria as Biocontrol Against Phytopathogens

The bacteria belonging to the genus *Pseudomonas* are very unique since they possess their own quorum sensing system but also have quorum quenching abilities via the presence of AHL lactonase as well as acylase enzymes. Out of the species of *Pseudomonas*, *Pseudomonas syringae* has been known to produce two different AHL acylases, HacA and HacB. HacA is an acylase responding to long chain AHLs while HacB degrades most reported AHLs. Being a plant associated bacteria, *P. syringae* can be used as quorum quenching against virulent QS dependent pathogens in the soil (Shepherd and Lindow 2009). *Agrobacterium tumefaciens* is known for its QS ability but it also has a quorum quenching system in place. It consist of two components, one is the AHL lactonase which is AttM and a transcriptional factor AttJ, whose function impacts the expression of AttM (Wang et al. 2015). The gene, qlcA, an AHL-lactonase was identified from *Acidobacteria* and its activity was checked against pathogenic *P. carotovorum*. It was noted that it quenched the virulence of *P. carotovorum* (Riaz et al. 2008). *Rhizobium* sp. is an interesting bacteria, it has been reported to have five distinct AHL- degrading loci found out by performing various screenings. It has five genes showing AHL degradation activity viz. dhlR, qsdR1, qsdR2, aldR and hydR-hitR. All have been reported to degrade 3-O-C8-HSL and have been known to affect QS mediated functions and pathogenicity of *P. aeruginosa*. DlhR and QsdR1 are lactonases from *Rhizobium* sp. and they were also reported to have quorum quenching activity against QS mediated virulence of *A. tumefaciens*. *Rhizobium* sp. might also play a role in colonization of cowpea plants and bacterial fitness by degrading AHLs which may lead to harmful QS circuits (Krysciak et al. 2011). *Ochrobactrum* sp. has a AHL-lactonase, AidH, which hydrolyses the lactone ring and degrades AHL. AidH uses C6-HSL as its substrate. Further studies are required for this AHL lactonase for its effect directly on plant pathogens QS system but it can certainly be considered to be an option to create novel treatment against infections relying upon AHL signaling (Zhang et al. 2016). *Ralstonia solanacearum*, a bacteria belonging to phylum proteobacteria, produces an AHL acylase Aac, which is coded by aac gene. It hydrolyses the amide bond of AHL, releasing homoserine lactone and fatty acids, in turn degrading the AHL molecule, making it incompetent to be used in QS circuits of pathogenic bacteria. AHL acylase Aac showed more specificity towards long chain AHLs than short chain AHLs (Chen et al. 2009). Another *Ralstonia* strain, *Ralstonia eutropha* is reported for its AHL acylase enzyme activity, its enzyme AiiD was one of the first acylases to be reported (Du et al. 2014). Soil bacterial isolate *Delftia* sp. has been reported from our laboratory to attenuate virulence of soft rot pathogen *Pcc* via degradation of AHL and quorum quenching that uses an AHL-acylase enzyme for its QQ activity and showed optimum activity at 20–40 °C and pH 6.2 (Maisuria and Nerurkar 2015). Earlier it is mentioned that *Ochrobactrum* sp. produces an AHL lactonase, AidH. It has also been reported to produce a novel AHL acylase, AiiO belonging to the α/β hydrolase superfamily. It breaks down acyl chains for a broad range of AHLs. AHL acylase from *Ochrobactrum* degrades quorum sensing signal molecules of pathogen and has been proved as an effective quorum quencher of AHL dependent virulence of *Pcc in planta* (Czajkowski et al. 2011). QQ has also been seen in nitrogen-fixing cyanobacterium *Anabaena*. It shows the presence of an AHL acylase, AiiC which has homology to *Pseudomonas aeruginosa* acylase QuiP. Its overall effect on a quorum sensing system is yet to be observed (Romero et al. 2008; Rolland et al. 2016). *Burkholderia* sp. isolated from ginger rhizosphere reduced 3-oxo-AHLs to 3-hydroxy compounds, showing an AHL oxidoreductase activity. It was one of the few cases, the other being *R. erythropolis* that showed oxidoreductase activity which helped in inhibiting QS based virulence in phytopathogens (Chan et al. 2011).

21.5   Conclusion

QS is a well-studied phenomenon of regulation of gene expression in many gram negative phytopathogens. *Pcc* has been studied in detail as regards its virulence mechanism among the various phytopathogens and it is the model bacterium used in most of the virulence attenuation studies. Disruption in QS is by either small inhibitors of enzymes in QS circuit or by AHL signal degrading enzymes. QQ enzymes are produced by diverse group of bacteria where the GRAS bacterium belonging to *Bacillus* are predominant genera. QQ enzymes have been reported from bacteria and metagenome. Bacterial enzymes have been structurally studied and classified into protein families. Majority of studies with QQ enzymes are restricted to laboratory and very few field studies are available except for large scale studies by Faure and coworkers.

21.6   Opinion

Variously the QQ enzymes have been reported, cloned and purified from different bacteria still the AHL degrading bacteria has yet to mature into a biocontrol agent. Although quorum quenching has been known as a mechanism of virulence attenuation since long, actual biocontrol studies are restricted to the laboratory rather than taken to the field. Use of QQ bacteria as biocontrol agent either alone or as a consortium need to be tested at the field level to understand its efficacy for sustainable agriculture. The bio-stimulation approach where enrichment of the AHL degrading bacteria *in situ* affords biocontrol also holds promise. QQ can have non-target effects too where they may affect the good bacteria and this can be assessed only in the field. An integrated chemical and biological approach can also be envisaged for better effect.

References

Arumugam K, Ramalingam P, Appu M (2017) Isolation of *Trichoderma viride* and *Pseudomonas fluorescens* organism from soil and their treatment against rice pathogens. J Microbiol 3:77–81

Bassler BL (2002) Small talk: cell-to-cell communication in bacteria. Cell 109:421–424. https://doi.org/10.1016/S0092-8674(02)00749-3

Biancalani C, Cerboneschi M, Tadini-Buoninsegni F, Campo M, Scardigli A, Romani A, Tegli S (2016) Global analysis of type three secretion system and quorum sensing inhibition of *Pseudomonas savastanoi* by polyphenols extracts from vegetable residues. PLoS One 11:e0163357. https://doi.org/10.1371/journal.pone.0163357

Bondí R, Longo F, Messina M, D'Angelo F, Visca P, Leoni L, Rampioni G (2017) The multi-output incoherent feed forward loop constituted by the transcriptional regulators LasR and RsaL confers robustness to a subset of quorum sensing genes in *Pseudomonas aeruginosa*. Mol BioSyst 13:1080–1089. https://doi.org/10.1039/c7mb00040e

Borges A, Serra S, Cristina Abreu A, Saavedra MJ, Salgado A, Simões M (2014) Evaluation of the effects of selected phytochemicals on quorum sensing inhibition and in vitro cytotoxicity. Biofouling 30:183–195. https://doi.org/10.1080/08927014.2013.852542

Bowden SD, Hale N, Chung JC, Hodgkinson JT, Spring DR, Welch M (2013) Surface swarming motility by *Pectobacterium atrosepticum* is a latent phenotype that requires O antigen and is regulated by quorum sensing. Microbiology 159:2375–2385. https://doi.org/10.1099/mic.0.070748-0

Boyer M, Wisniewski-Dyé F (2009) Cell–cell signaling in bacteria: not simply a matter of quorum. ISME J 70:1–19. https://doi.org/10.1111/j.1574-6941.2009.00745

Burke AK, Duong DA, Jensen RV, Stevens AM (2015) Analyzing the transcriptomes of two quorum-sensing controlled transcription factors, RcsA and LrhA, important for *Pantoea stewartii* virulence. PLoS One 10:e0145358. https://doi.org/10.1371/journal.pone.0145358

Cai Z, Yuan ZH, Zhang H, Pan Y, Wu Y, Tian XQ, Wang FF, Wang L, Qian W (2017) Fatty acid DSF binds and allosterically activates histidine kinase RpfC of phytopathogenic bacterium *Xanthomonas campestris* pv. *campestris* to regulate quorum-sensing and virulence. PLoS Pathog 13:e1006304. https://doi.org/10.1371/journal.ppat.1006304

Chan KG, Atkinson S, Mathee K, Sam CK, Chhabra SR, Cámara M et al (2011) Characterization of N-acylhomoserine lactone-degrading bacteria associated with the *Zingiber officinale* (ginger) rhizosphere: co-existence of quorum quenching and quorum sensing in *Acinetobacter* and *Burkholderia*. BMC Microbiol 11:51. https://doi.org/10.1186/1471-2180-11-51

Chang CY, Krishnan T, Wang H, Chen Y, Yin WF, Chong YM, Tan LY, Chong TM, Chan KG (2014) Non-antibiotic quorum sensing inhibitors acting against N-acyl homoserine lactone synthase as druggable target. Sci Rep 4:7245. https://doi.org/10.1038/srep07245

Chen CN, Chen CJ, Liao CT, Lee CY (2009) A probable aculeacin A acylase from the *Ralstonia solanacearum* GMI1000 is N-acyl-homoserine lactone acylase with quorum-quenching activity. BMC Microbiol 9:89. https://doi.org/10.1186/1471-2180-9-89

Chen F, Gao Y, Chen X, Yu Z, Li X (2013) Quorum quenching enzymes and their application in degrading signal molecules to block quorum sensing-dependent infection. Int J Mol Sci 14:17477–17500. https://doi.org/10.3390/ijms140917477

Chen R, Barphagha IK, Ham JH (2015) Identification of potential genetic components involved in the deviant quorum-sensing signaling pathways of *Burkholderia glumae* through a functional genomics approach. Front Cell Infect Microbiol 5:22. https://doi.org/10.3389/fcimb.2015.00022

Cirou A, Diallo S, Kurt C, Latour X, Faure D (2007) Growth promotion of quorum-quenching bacteria in the rhizosphere of *Solanum tuberosum*. Environ Microbiol 9:1511–1522. https://doi.org/10.1111/j.1462-2920.2007.01270.x

Cirou A, Mondy S, An S, Charrier A, Sarrazin A, Thoison O, Faure D (2012) Efficient biostimulation of native and introduced quorum-quenching *Rhodococcus erythropolis* populations is revealed by a combination of analytical chemistry, microbiology, and pyrosequencing. Appl Environ Microbiol 78:481–492. https://doi.org/10.1128/AEM.06159-11

Cornforth DM, Popat R, McNally L, Gurney J, Scott-Phillips TC, Ivens A, Diggle SP, Brown SP (2014) Combinatorial quorum sensing allows bacteria to resolve their social and physical environment. Proc Natl Acad Sci USA 111:4280–4284. https://doi.org/10.1073/pnas.1319175111

Cui X, Harling R (2005) N-acyl-homoserine lactone-mediated quorum sensing blockage, a novel strategy for attenuating pathogenicity of Gram-negative bacterial plant pathogens. Eur J Plant Pathol 111:327–339. https://doi.org/10.1007/s10658-004-4891-0

Cui Y, Chatterjee A, Liu Y, Dumenyo CK, Chatterjee AK (1995) Identification of a global repressor gene, rsmA, of *Erwinia carotovora* subsp. *carotovora* that controls extracellular enzymes, N-(3-oxohexanoyl)-L-homoserine lactone, and pathogenicity in soft-rotting *Erwinia* spp. J Bacteriol 177:5108–5115. https://doi.org/10.1128/jb.177.17

Cui Y, Chatterjee A, Hasegawa H, Dixit V, Leigh N, Chatterjee AK (2005) ExpR, a LuxR homolog of *Erwinia carotovora* subsp. *carotovora*, activates transcription of rsmA, which specifies a global regulatory RNA-binding protein. J Bacteriol 187:4792–4803. https://doi.org/10.1128/JB.187.14.4792–4803.2005

Czajkowski R, Krzyżanowska D, Karczewska J, Atkinson S, Przysowa J, Lojkowska E, Jafra S (2011) Inactivation of AHLs by *Ochrobactrum* sp. A44 depends on the activity of a novel class of AHL acylase. Environ Microbiol Rep 3:59–68. https://doi.org/10.1111/j.1758-2229.2010.00188.x

De Simone M, Spagnuolo L, Lorè NI, Rossi G, Cigana C, De Fino I, Iraqi FA, Bragonzi A (2014) Host genetic background influences the response to the opportunistic *Pseudomonas aeruginosa* infection altering cell-mediated immunity and bacterial replication. PLoS One 9:e106873. https://doi.org/10.1371/journal.pone.0106873

Defoirdt T, Brackman G, Coenye T (2013) Quorum sensing inhibitors: how strong is the evidence? Trends Microbiol 21:619–624. https://doi.org/10.1016/j.tim.2013.09.006

Des Essarts YR, Cigna J, Quêtu-Laurent A, Caron A, Munier E, Beury-Cirou A et al (2016) Biocontrol of the potato blackleg and soft rot diseases caused by *Dickeya dianthicola*. Appl Environ Microbiol 82:268–278. https://doi.org/10.1128/AEM.02525-15

Dong YH, Zhang LH (2005) Quorum sensing and quorum-quenching enzymes. J Microbiol 43:101–109

Dong YH, Wang LH, Xu JL, Zhang HB (2001) Quenching quorum-sensing-dependent bacterial infection by an N-acyl homoserine lactonase. Nature 411:813. https://doi.org/10.1038/35081101

Du Y, Li T, Wan Y, Liao P (2014) Signal molecule-dependent quorum-sensing and quorum-quenching enzymes in bacteria. Crit Rev Eukaryot Gene Expr 24:117–132. https://doi.org/10.1615/CritRevEukaryotGeneExpr.2014008034

Fetzner S (2015) Quorum quenching enzymes. J Biotechnol 201:2–14. https://doi.org/10.1016/j.jbiotec.2014.09.001

Fuqua WC, Winans SC, Greenberg EP (1994) Quorum sensing in bacteria: the LuxR-LuxI family of cell density-responsive transcriptional regulators. J Bacteriol 176:269–275

Gao R, Krysciak D, Petersen K, Utpatel C, Knapp A, Schmeisser C, Daniel R, Voget S, Jaeger KE, Streit WR (2015) Genome-wide RNA sequencing analysis of quorum sensing-controlled regulons in the plant-associated *Burkholderia glumae* PG1 strain. Appl Environ Microbiol 81:7993–8007. https://doi.org/10.1128/AEM.01043-15

García-Lara B, Saucedo-Mora MÁ, Roldán-Sánchez JA, Pérez-Eretza B, Ramasamy M, Lee J, Coria-Jimenez R, Tapia M, Varela-Guerrero V, García-Contreras R (2015) Inhibition of quorum-sensing-dependent virulence factors and biofilm formation of clinical and environmental *Pseudomonas aeruginosa* strains by ZnO nanoparticles. Lett Appl Microbiol 61:299–305. https://doi.org/10.1111/lam.12456

Garge SS, Nerurkar AS (2016) Attenuation of quorum sensing regulated virulence of *Pectobacterium carotovorum* subsp. *carotovorum* through an AHL lactonase produced by *Lysinibacillus* sp. Gs50. PLoS One 11:e0167344. https://doi.org/10.1371/journal.pone.0167344

Garge SS, Nerurkar AS (2017) Evaluation of quorum quenching *Bacillus* spp. for their biocontrol traits against *Pectobacterium carotovorum* subsp. *carotovorum* causing soft rot. Biocatal Agric Biotechnol 9:48–57. https://doi.org/10.1016/j.bcab.2016.11.004

Grandclément C, Tannières M, Moréra S, Dessaux Y, Faure D (2015) Quorum quenching: role in nature and applied developments. FEMS Microbiol Lett 40:86–116. https://doi.org/10.1093/femsre/fuv038

Ham JH (2013) Intercellular and intracellular signalling systems that globally control the expression of virulence genes in plant pathogenic bacteria. Mol Plant Pathol 14:308–322. https://doi.org/10.1111/mpp.12005

Helman Y, Chernin L (2015) Silencing the mob: disrupting quorum sensing as a means to fight plant disease. Mol Plant Pathol 16:316–329. https://doi.org/10.1111/mpp.12180

Hilker R, Munder A, Klockgether J, Losada PM, Chouvarine P, Cramer N, Davenport CF, Dethlefsen S, Fischer S, Peng H, Schönfelder T (2015) Interclonal gradient of virulence in the *Pseudomonas aeruginosa* pangenome from disease and environment. Environ Microbiol 17:29–46. https://doi.org/10.1111/1462-2920.12606

Høyland-Kroghsbo NM, Paczkowski J, Mukherjee S, Broniewski J, Westra E, Bondy-Denomy J, Bassler BL (2016) Quorum sensing controls the *Pseudomonas aeruginosa* CRISPR-Cas adaptive immune system. Proc Natl Acad Sci USA 114:131–135. https://doi.org/10.1073/pnas.1617415113

Huma N, Shankar P, Kushwah J, Bhushan A, Joshi J, Mukherjee T, Raju SC, Purohit HJ, Kalia VC (2011) Diversity and polymorphism in AHL-lactonase gene (aiiA) of Bacillus. J Microbiol Biotechnol 21:1001–1011. https://doi.org/10.4014/jmb.1105.05056

Ihekwaba AE, Mura I, Peck MW, Barker GC (2015) The pattern of growth observed for *Clostridium botulinum* type A1 strain ATCC 19397 is influenced by nutritional status and quorum sensing: a modelling perspective. FEMS Pathogens Dis 73:ftv084. https://doi.org/10.1093/femspd/ftv084

Ishii S, Fukui K, Yokoshima S, Kumagai K, Beniyama Y, Kodama T, Fukuyama T, Okabe T, Nagano T, Kojima H, Yano T (2017) High throughput screening of small molecule inhibitors of the *Streptococcus* quorum-sensing signal pathway. Sci Rep 7:4029. https://doi.org/10.1038/s41598-017-03567-2

Jacquet P, Daudé D, Bzdrenga J, Masson P, Elias M, Chabrière E (2016) Current and emerging strategies for organophosphate decontamination: special focus on hyperstable enzymes. Environ Sci Pollut Res 23:8200–8218. https://doi.org/10.1007/s11356-016-6143-1

Jang MS, Goo E, An JH, Kim J, Hwang I (2014) Quorum sensing controls flagellar morphogenesis in *Burkholderia glumae*. PLoS One 9:e84831. https://doi.org/10.1371/journal.pone.0084831

Joshi JR, Khazanov N, Senderowitz H, Burdman S, Lipsky A, Yedidia I (2016) Plant phenolic volatiles inhibit quorum sensing in pectobacteria and reduce their virulence by potential binding to ExpI and ExpR proteins. Sci Rep 6:38126. https://doi.org/10.1038/srep38126

Kalia VC (2013) Quorum sensing inhibitors: an overview. Biotechnol Adv 31:224–245. https://doi.org/10.1016/j.biotechadv.2012.10.004

Kalia VC, Purohit HJ (2011) Quenching the quorum sensing system: potential antibacterial drug targets. Crit Rev Microbiol 37:121–140. https://doi.org/10.3109/1040841X.2010.532479

Kalia VC, Raju SC, Purohit HJ (2011) Genomic analysis reveals versatile organisms for quorum quenching enzymes: Acyl-homoserine lactone-acylase and –lactonase. Open Microbiol J 5:1–13. https://doi.org/10.2174/1874285801105010001

Khoiri S, Damayanti TA, Giyanto G (2017) Identification of quorum quenching bacteria and its biocontrol potential against soft rot disease bacteria, *Dickeya dadantii*. Agrivita 39:45. https://doi.org/10.17503/agrivita.v39i1.633

Koch B, Liljefors T, Persson T, Nielsen J, Kjelleberg S, Givskov M (2005) The LuxR receptor: the sites of interaction with quorum-sensing signals and inhibitors. Microbiol 151:3589–3602. https://doi.org/10.1099/mic.0.27954-0

Koul S, Kalia VC (2017) Multiplicity of quorum quenching enzymes: a potential mechanism to limit quorum sensing bacterial population. Indian J Microbiol 57:100–108. https://doi.org/10.1007/s12088-016-0633-1

Krysciak D, Schmeisser C, Preuss S, Riethausen J, Quitschau M, Grond S, Streit WR (2011) Involvement of multiple loci in quorum quenching of autoinducer I molecules in the nitrogen-fixing symbiont *Rhizobium (Sinorhizobium)* sp. strain NGR234. Appl Environ Microbiol 77:5089–5099. https://doi.org/10.1128/AEM.00112-11

Lang J, Faure D (2014) Functions and regulation of quorum-sensing in *Agrobacterium tumefaciens*. Front Plant Sci 5:14. https://doi.org/10.3389/fpls.2014.00014

Latour X, Barbey C, Chane A, Groboillot A, Burini JF (2013) *Rhodococcus erythropolis* and its γ-lactone catabolic pathway: an unusual biocontrol system that disrupts pathogen quorum sensing communication. Agronomy 3:816–838. https://doi.org/10.3390/agronomy3040816

Lee J, Zhang L (2015) The hierarchy quorum sensing network in *Pseudomonas aeruginosa*. Protein Cell 6:26–41. https://doi.org/10.1007/s13238-014-0100

Lee DH, Lim JA, Lee J, Roh E, Jung K, Choi M, Oh C, Ryu S, Yun J, Heu S (2013) Characterization of genes required for the pathogenicity of *Pectobacterium carotovorum* subsp. *carotovorum* Pcc21 in Chinese cabbage. Microbiology 159:1487–1496. https://doi.org/10.1099/mic.0.067280-0

Liang X, Yu X, Pan X, Wu J, Duan Y, Wang J, Zhou M (2016) A thiadiazole reduces the virulence of *Xanthomonas oryzae* pv. *oryzae* by inhibiting the histidine utilization pathway and quorum sensing. Mol Plant Pathol. https://doi.org/10.1111/mpp.12503

Liu CF, Liu D, Momb J, Thomas PW, Lajoie A, Petsko GA, Fast W, Ringe D (2013) A phenylalanine clamp controls substrate specificity in the quorum-quenching metallo-γ-lactonase from *Bacillus thuringiensis*. Biochemistry 52:1603–1610. https://doi.org/10.1021/bi400050j

Loh J, Pierson EA, Pierson LS, Stacey G, Chatterjee A (2002) Quorum sensing in plant-associated bacteria. Curr Opin Plant Biol 5:285–290. 12179960

Maisuria VB, Nerurkar AS (2012) Biochemical properties and thermal behaviour of pectate lyase produced by *Pectobacterium carotovorum* subsp. *carotovorum* BR1 with industrial potentials. Biochem Eng J 63:22–30. https://doi.org/10.1016/j.bej.2012.01.007

Maisuria VB, Nerurkar AS (2013) Characterization and differentiation of soft rot causing *Pectobacterium carotovorum* of Indian origin. Eur J Plant Pathol 136:87–102. https://doi.org/10.1007/s10658-012-0140-0

Maisuria VB, Nerurkar AS (2015) Interference of quorum sensing by *Delftia* sp. VM4 depends on the activity of a novel N-Acylhomoserine lactone-acylase. PLoS One 10:e0138034. https://doi.org/10.1371/journal.pone.0138034

Maisuria VB, Patel VA, Nerurkar AS (2010) Biochemical and thermal stabilization parameters of polygalacturonase from *Erwinia carotovora* subsp. *carotovora* BR1. J Microbiol Biotechnol 20:1077–1085. 20668400

Maisuria VB, Lopez-de Los Santos Y, Tufenkji N, Déziel E (2016) Cranberry-derived proanthocyanidins impair virulence and inhibit quorum sensing of *Pseudomonas aeruginosa*. Sci Rep 6:30169. https://doi.org/10.1038/srep30169

Mansfield J, Genin S, Magori S, Citovsky V, Sriariyanum M, Ronald P et al (2012) Top 10 plant pathogenic bacteria in molecular plant pathology. Mol Plant Pathol 13:614–629. https://doi.org/10.1111/j.1364-3703.2012.00804.x

McCann HC, Rikkerink EH, Bertels F, Fiers M, Lu A, Rees-George J, Andersen MT, Gleave AP, Haubold B, Wohlers MW, Guttman DS (2013) Genomic analysis of the kiwifruit pathogen *Pseudomonas syringae* pv. *actinidiae* provides insight into the origins of an emergent plant disease. PLoS Pathog 9:e1003503. https://doi.org/10.1371/journal.ppat.1003503

Miller MB, Bassler BL (2001) Quorum sensing in bacteria. Annu Rev Microbiol 55:165–199. https://doi.org/10.1146/annurev.micro.55.1.165

Mole BM, Baltrus DA, Dangl JL, Grant SR (2007) Global virulence regulation networks in phyto-pathogenic bacteria. Trends Microbiol 15:363–371. https://doi.org/10.1016/j.tim.2007.06.005

Moleleki LN, Pretorius RG, Tanui CK, Mosina G, Theron J (2017) A quorum sensing-defective mutant of *Pectobacterium carotovorum* ssp. *brasiliense* 1692 is attenuated in virulence and unable to occlude xylem tissue of susceptible potato plant stems. Mol Plant Pathol 18:32–44. https://doi.org/10.1111/mpp.12372

Monson R, Burr T, Carlton T, Liu H, Hedley P, Toth I, Salmond GP (2013) Identification of genes in the VirR regulon of *Pectobacterium atrosepticum* and characterization of their roles in quorum sensing-dependent virulence. Environ Microbiol Rep 15:687–701. https://doi.org/10.1111/j.1462-2920.2012.02822

Mori Y, Ishikawa S, Ohnishi H, Shimatani M, Morikawa Y, Hayashi K, Hikichi Y (2017) Involvement of ralfuranones in the quorum sensing signalling pathway and virulence of *Ralstonia solanacearum* strain OE1-1. Mol Plant Pathol. https://doi.org/10.1111/mpp.12537

Morohoshi T, Tominaga Y, Someya N, Ikeda T (2012) Complete genome sequence and characterization of the N-acylhomoserine lactonedegrading gene of the potato leaf-associated Solibacillus silvestris. J Biosci Bioeng 113(1):20–25. https://doi.org/10.1016/j.jbiosc.2011.09.006

Mukherjee S, Moustafa D, Smith CD, Goldberg JB, Bassler BL (2017) The RhlR quorum-sensing receptor controls *Pseudomonas aeruginosa* pathogenesis and biofilm development independently of its canonical homoserine lactone autoinducer. PLoS Pathog 13:e1006504. https://doi.org/10.1371/journal.ppat.1006504

Nasser W, Dorel C, Wawrzyniak J, Van Gijsegem F, Groleau MC, Déziel E, Reverchon S (2013) Vfm a new quorum sensing system controls the virulence of *Dickeya dadantii*. Environ Microbiol 15:865–880. https://doi.org/10.1111/1462-2920.12049

Nealson KH, Platt T, Hastings JW (1970) Cellular control of the synthesis and activity of the bacterial luminescent system. J Bacteriol 104:313–322

Newton JA, Fray RG (2004) Integration of environmental and host-derived signals with quorum sensing during plant–microbe interactions. Cell Microbiol 6:213–224. https://doi.org/10.1111/j.1462-5822.2004.00362.x

Ng WL, Bassler BL (2009) Bacterial quorum-sensing network architectures. Annu Rev Genet 43:197–222. https://doi.org/10.1146/annurev-genet-102108-134304

Nickzad A, Lépine F, Déziel E (2015) Quorum sensing controls swarming motility of *Burkholderia glumae* through regulation of rhamnolipids. PLoS One 10:e0128509. https://doi.org/10.1371/journal.pone.0128509

O'Rourke JP, Daly SM, Triplett KD, Peabody D, Chackerian B, Hall PR (2014) Development of a mimotope vaccine targeting the *Staphylococcus aureus* quorum sensing pathway. PLoS One 9:e111198. https://doi.org/10.1371/journal.pone.0111198

Papenfort K, Bassler BL (2016) Quorum sensing signal-response systems in gram-negative bacteria. Nat Rev Microbiol 14:576–588. https://doi.org/10.1038/nrmicro.2016.89

Park SY, Lee SJ, Oh TK, Oh JW, Koo BT, Yum DY, Lee JK (2003) AhlD, an N-acylhomoserine lactonase in *Arthrobacter* sp., and predicted homologues in other bacteria. Microbiology 149:1541–1550. https://doi.org/10.1099/mic.0.26269-0

Park SY, Kang HO, Jang HS, Lee JK, Koo BT, Yum DY (2005) Identification of extracellular N-acylhomoserine lactone acylase from a *Streptomyces* sp. and its application to quorum quenching. Appl Environ Microbiol 71:2632–2641. https://doi.org/10.1128/AEM.71.5.2632-2641.2005

Pérez-Velázquez J, Quiñones B, Hense BA, Kuttler C (2015) A mathematical model to investigate quorum sensing regulation and its heterogeneity in *Pseudomonas syringae* on leaves. Ecol Complex 21:128–141. https://doi.org/10.1016/j.ecocom.2014.12.003

Polkade AV, Mantri SS, Patwekar UJ, Jangid K (2016) Quorum sensing: an under-explored phenomenon in the phylum Actinobacteria. Front Microbiol 7:131. https://doi.org/10.3389/fmicb.2016.00131

Põllumaa L, Alamäe T, Mäe A (2012) Quorum sensing and expression of virulence in *Pectobacteria*. Sensors 12:3327–3349. https://doi.org/10.3390/s120303327

Qi J, Li L, Du Y, Wang S, Wang J, Luo Y, Che J, Lu J, Liu H, Hu G, Li J (2014) The identification, typing, and antimicrobial susceptibility of *Pseudomonas aeruginosa* isolated from mink with hemorrhagic pneumonia. Vet Microbiol 170:456–461. https://doi.org/10.1016/j.vetmic.2014.02.025

Rajesh PS, Rai VR (2016) Inhibition of QS-regulated virulence factors in *Pseudomonas aeruginosa* PAO1 and *Pectobacterium carotovorum* by AHL-lactonase of endophytic bacterium *Bacillus cereus* VT96. Biocatal Agric Biotechnol 7:154–163. https://doi.org/10.1016/j.bcab.2016.06.003

Ramachandran R, Stevens AM (2013) Proteomic analysis of the quorum-sensing regulon in *Pantoea stewartii* and identification of direct targets of EsaR. Appl Environ Microbiol 79:6244–6252. https://doi.org/10.1128/AEM.01744-13

Ramachandran R, Burke AK, Cormier G, Jensen RV, Stevens AM (2014) Transcriptome-based analysis of the *Pantoea stewartii* quorum-sensing regulon and identification of EsaR direct targets. Appl Environ Microbiol 80:5790–5800. https://doi.org/10.1128/AEM.01489-14

Riaz K, Elmerich C, Moreira D, Raffoux A, Dessaux Y, Faure D (2008) A metagenomic analysis of soil bacteria extends the diversity of quorum quenching lactonases. Environ Microbiol 10:560–570. https://doi.org/10.1111/j.1462-2920.2007.01475

Rizzello CG, Filannino P, Di Cagno R, Calasso M, Gobbetti M (2014) Quorum-sensing regulation of constitutive plantaricin by *Lactobacillus plantarum* strains under a model system for vegetables and fruits. Appl Environ Microbiol 80:777–787. https://doi.org/10.1128/AEM.03224-13

Rolland J, Stien D, Sanchez-Ferandin S, Lami R (2016) Quorum sensing and quorum quenching in the phycosphere of phytoplankton: a case of chemical interactions in ecology. J Chem Ecol 42:1201–1211. https://doi.org/10.1007/s10886-016-0791-y

Romero M, Diggle SP, Heeb S, Camara M, Otero A (2008) Quorum quenching activity in *Anabaena* sp. PCC 7120: identification of AiiC, a novel AHL-acylase. FEMS Microbiol Lett 280:73–80. https://doi.org/10.1111/j.1574-6968.2007.01046.x

Rutherford ST, Bassler BL (2012) Bacterial quorum sensing: its role in virulence and possibilities for its control. Cold Spring Harb Perspect Med 2:a012427. https://doi.org/10.1101/cshperspect.a012427

Safari M, Amache R, Esmaeilishirazifard E, Keshavarz T (2014) Microbial metabolism of quorum-sensing molecules acyl-homoserine lactones, γ-heptalactone and other lactones. Appl Microbiol Biotechnol 98:3401–3412. https://doi.org/10.1007/s00253-014-5518-9

Scott RA, Lindow SE (2016) Transcriptional control of quorum sensing and associated metabolic interactions in *Pseudomonas syringae* strain B728a. Mol Microbiol 99:1080–1098. https://doi.org/10.1111/mmi.13289

Shepherd RW, Lindow SE (2009) Two dissimilar N-acyl-homoserine lactone acylases of *Pseudomonas syringae* influence colony and biofilm morphology. Appl Environ Microbiol 75:45–53. https://doi.org/10.1128/AEM.01723-08

Singh RP, Desouky SE, Nakayama J (2016) Quorum quenching strategy targeting gram-positive pathogenic bacteria. Adv Microbiol Infect Dis Pub Health 2:109–130. https://doi.org/10.1007/5584_2016_1

Stoltz DA, Meyerholz DK, Welsh MJ (2015) Origins of cystic fibrosis lung disease. N Engl J Med 372:351–362. https://doi.org/10.1056/NEJMra1300109

Sun K, Liu J, Gao Y, Jin L, Gu Y, Wang W (2014) Isolation, plant colonization potential, and phenanthrene degradation performance of the endophytic bacterium *Pseudomonas* sp. Ph6-gfp. Sci Rep 4:5462. https://doi.org/10.1038/srep05462

Teixeira N, Varahan S, Gorman MJ, Palmer KL, Zaidman-Remy A, Yokohata R, Nakayama J, Hancock LE, Jacinto A, Gilmore MS, Lopes MDFS (2013) Drosophila host model reveals new *Enterococcus faecalis* quorum-sensing associated virulence factors. PLoS One 8:e64740. https://doi.org/10.1371/journal.pone.0064740

Tichy EM, Luisi BF, Salmond GPC (2014) Crystal structure of the carbapenem intrinsic resistance protein CarG. J Mol Biol 426:1958–1970. https://doi.org/10.1016/j.jmb.2014.02.016

Uroz S, Chhabra SR, Camara M, Williams P, Oger P, Dessaux Y (2005) N-Acylhomoserine lactone quorum-sensing molecules are modified and degraded by *Rhodococcus erythropolis* W2 by both amidolytic and novel oxidoreductase activities. Microbiology 151:3313–3322. https://doi.org/10.1099/mic.0.27961-0

Uroz S, Oger PM, Chapelle E, Adeline MT, Faure D, Dessaux Y (2008) A *Rhodococcus* qsdA-encoded enzyme defines a novel class of large-spectrum quorum-quenching lactonases. Appl Environ Microbiol 74:1357–1366. https://doi.org/10.1128/AEM.02014-07

Uroz S, Dessaux Y, Oger P (2009) Quorum sensing and quorum quenching: the yin and yang of bacterial communication. Chembiochem 10:205–216. https://doi.org/10.1002/cbic.200800521

Utari PD, Vogel J, Quax WJ (2017) Deciphering physiological functions of AHL quorum quenching acylases. Front Microbiol 8:1123. https://doi.org/10.3389/fmicb.2017.01123

Valente R, Xavier K (2015) The Trk potassium transporter is required for RsmB-mediated activation of virulence in the phytopathogen pectobacterium wasabiae. J Bacteriol 198(2):248–255. http://dx.doi.org/10.1128/JB.00569-15

Valente RS, Nadal-Jimenez P, Carvalho AF, Vieira FJ, Xavier KB (2017) Signal integration in quorum sensing enables cross-species induction of virulence in *Pectobacterium wasabiae*. MBio 8:e00398-17. https://doi.org/10.1128/mBio.00398-17

Vasavi HS, Arun AB, Rekha PD (2016) Anti-quorum sensing activity of flavonoid-rich fraction from *Centella asiatica* L. against *Pseudomonas aeruginosa* PAO1. J Microbiol Immunol Infect 49:8–15. https://doi.org/10.1016/j.jmii.2014.03.012

von Bodman SB, Baue WD, Coplin DL (2003) Quorum sensing in plant-pathogenic bacteria. Annu Rev Phytopathol 41:455–482. https://doi.org/10.1146/annurev.phyto.41.052002.095652

Wang WZ, Morohoshi T, Ikenoya M, Someya N, Ikeda T (2010) AiiM, a novel class of N-acylhomoserine lactonase from the leaf-associated bacterium *Microbacterium testaceum*. Appl Environ Microbiol 76:2524–2530. https://doi.org/10.1128/AEM.02738-09

Wang C, Yan C, Fuqua C, Zhang LH (2014) Identification and characterization of a second quorum-sensing system in *Agrobacterium tumefaciens* A6. J Bacteriol 196:1403–1411. https://doi.org/10.1128/JB.01351-13

Wang C, Yan C, Gao YG, Zhang LH (2015) D101 is critical for the function of AttJ, a repressor of quorum quenching system in *Agrobacterium tumefaciens*. J Microbiol 53:623–632. https://doi.org/10.1007/s12275-015-5100-x

Waters CM, Bassler BL (2005) Quorum sensing: cell-to-cell communication in bacteria. Annu Rev Cell Dev Biol 21:319–346. https://doi.org/10.1146/annurev.cellbio.21.012704.131001

Wolf D, Rippa V, Mobarec JC, Sauer P, Adlung L, Kolb P, Bischofs IB (2015) The quorum-sensing regulator ComA from *Bacillus subtilis* activates transcription using topologically distinct DNA motifs. Nucleic Acids Res 44:2160–2172. https://doi.org/10.1093/nar/gkv1242

Xue B, Chow JY, Baldansuren A, Yap LL, Gan YH, Dikanov SA, Robinson RC, Yew WS (2013) Structural evidence of a productive active site architecture for an evolved quorum-quenching GKL lactonase. Biochemistry 52:2359–2370. https://doi.org/10.1021/bi4000904

Yu X, Lund SP, Greenwald JW, Records AH, Scott RA, Nettleton D, Lindow SE, Gross DC, Beattie GA (2014) Transcriptional analysis of the global regulatory networks active in *Pseudomonas syringae* during leaf colonization. MBio 5:e01683-14. https://doi.org/10.1128/mBio.01683-14

Zhang LH, Dong YH (2004) Quorum sensing and signal interference: diverse implications. Mol Microbiol 53:1563–1571. https://doi.org/10.1111/j.1365-2958.2004.04234

Zhang S, Su H, Ma G, Liu Y (2016) Quantum mechanics and molecular mechanics study of the reaction mechanism of quorum quenching enzyme: N-acyl homoserine lactonase with C6-HSL. RSC Adv 6:23396–23402. https://doi.org/10.1039/c6ra00328a

Zhou L, Zhang LH, Cámara M, He YW (2017) The DSF family of quorum sensing signals: diversity, biosynthesis, and turnover. Trends Microbiol 25:293–303. https://doi.org/10.1016/j.tim.2016.11.013

Printed by Printforce, the Netherlands